SCIENCE and TECHNOLOGY of ZIRCONIA V

EDITED BY

S. P. S. Badwal
M. J. Bannister
R. H. J. Hannink

CRC Press
Taylor & Francis Group
Boca Raton London New York

CRC Press is an imprint of the
Taylor & Francis Group, an **informa** business

Sponsored by:

The Australasian Ceramic Society

Co-Sponsors:

ANSTO Advanced Materials Program
BHP Refractories Limited
Ceramic Fuel Cells Ltd.
Chichibu Cement Company
Commercial Minerals Limited
CRA Limited
CSIRO Division of Building Construction and Engineering
CSIRO Division of Materials Science and Technology
Department of Industry, Technology and Commerce,
    Commonwealth Government of Australia
Department of Manufacturing and Industry Development,
    State Government of Victoria
ESK Engineered Ceramics Pty Ltd
General Furnace Construction Pty Ltd
ICI Advanced Ceramics
Matsearch and the Archaeology Research Unit, Deakin University
The Japanese Solid Oxide Fuel Cell Society
The Manufacturing and Materials Technology Committee, DITAC
The National Advanced Materials Analytical Centre

International Advisory Committee:

Australia
S. Badwal, R. Hannink, M. Swain

Germany
N. Claussen, H. Schubert

Japan
K. Funatani, S. Sōmiya, H. Yanagida

U.K.
R. Brook, B. Steele

U.S.A.
A. Heuer, D. Marshall, R. Ritchie

Organising Committee:

J. Cullen (Chairman)	J. Sellar	R. Richards	R. Hughan
J. Bell	S. Badwal	J. Van der Hayden	M. Scully
R. Bowman	B. Ben-Nissan	J. Bannister	V. White
S. Pratt	R. Hannink	K. Bettles	

Program Sub-Committee and ZIRCONIA V Editors:

S. Badwal   J. Bannister   R. Hannink

*To the memory
of Ronald Charles Garvie
1930–1991*

Table of Contents

Foreword

The Fifth International Conference on the Science and Technology of Zirconia (ZIRCONIA V) was held on 16–21 August, 1992 in Melbourne, Australia, in conjunction with AUSTCERAM 92, the biennial conference of the Australasian Ceramic Society. Previous Zirconia conferences have been held in Cleveland, Ohio (1980), Stuttgart, Germany (1983), Tokyo, Japan (1986) and Anaheim, California (1989), the latter as part of the First International Ceramic Science and Technology Congress.

Australia had been represented on the Organising Committee of the first conference, and on the International Advisory Committee of each subsequent conference, by Ron Garvie of the CSIRO Division of Materials Science and Technology. Ron Garvie, with his colleagues Hannink and Pascoe, had been the first to report the phenomenon of transformation toughening in a zirconia-based alloy system, on which much of today's use of zirconia as an engineering ceramic depends. It was in recognition of that early work, plus his active role in all previous Zirconia conferences together with his persuasive advocacy on behalf of Melbourne, that the decision was taken to hold ZIRCONIA V in Australia. Ron was active in the initial planning for ZIRCONIA V and would have been its Chairman, but he died one year before it began; thus the Conference itself, and these Proceedings, are dedicated to him.

ZIRCONIA V drew 122 contributions from 19 countries. The papers presented, either orally or in poster sessions, served to demonstrate that the field of zirconia ceramics remains one of scientific challenge and technical attraction. The papers, most of which appear in these Proceedings, provide an up-to-date picture of zirconia research and development around the world. As might have been expected, there is still considerable interest in the theory and practice of transformation toughening together with the application of zirconia toughening to increasingly more complex composite systems. These Proceedings also reflect a prominent development of recent years, namely the resurgence of international interest in the zirconia-based Solid Oxide Fuel Cell. It is clear from the many papers given at ZIRCONIA V on various aspects of SOFC technology that this could become the most significant zirconia application of all.

As editors of the Proceedings we wish to thank all the authors and those who helped to review the manuscripts. All contributions were subjected to peer review, in most cases two independent reviews being obtained. This is the first Zirconia Proceedings which has been produced in camera-ready form, and we are grateful to the authors for their care in following the detailed instructions to permit, so far as possible, a uniformity of style and presentation throughout this volume. Finally, we extend our thanks to all the conference participants, particularly those who chaired the various technical sessions, the International Advisory Committee, the conference sponsors and the AUSTCERAM 92/ZIRCONIA V Organising Committee of the Australasian Ceramic Society.

S. P. S. BADWAL
M. J. BANNISTER
R. H. J. HANNINK

PHASE TRANSFORMATIONS

Isothermal Martensitic Transformation in ZrO$_2$ Ceramics

G. BEHRENS, J. MARTINEZ-FERNANDEZ, G. W. DRANSMANN
and A. H. HEUER

ABSTRACT

The stress-assisted martensitic transformation in ZrO$_2$ ceramics contains both an athermal and an isothermal component. The thermodynamics and kinetics of the isothermal (thermally activated) transformation has been investigated in one MgO- and two Y$_2$O$_3$-ZrO$_2$ alloys. In MgO-partially-stabilized ZrO$_2$ with an appropriate thermal history, isothermal $t \Rightarrow m$ and $t \Rightarrow o$ transformations take place at room temperature over many months. In Y$_2$O$_3$-stabilized alloys, relatively rapid $t \Rightarrow m$ transformation occurs at 300 to 800K within the residual stress fields of Vickers indentations. The transformation rate is controlled by the thermally activated, stress-assisted nucleation of the martensitic product.

INTRODUCTION

Martensitic transformations have traditionally been considered to display athermal kinetics, *i.e.* the transformation starts during cooling at a well-defined M$_S$ temperature (even upon rapid quenching), and transformation continues only upon further decreases in temperature. However, a number of metallic alloys exhibit martensitic transformations that display isothermal kinetics [1,2], where the extent of transformation varies with time at a given temperature and the transformation can be suppressed by rapid quenching. This isothermal component of martensitic transformations exists in addition to the athermal component, and is often obscured by the latter. The transformation kinetics are usually controlled by the thermally activated nucleation of the martensitic product; it is possible to view the athermal character usually found in a cooling transformation as being due to rapid isothermal transformation of the most easily nucleated regions at a succession of temperatures. In other words, a range of activation energies must exist for such martensitic transformations.

G. Behrens and A. H. Heuer, Department of Materials Science and Engineering, Case Western
    Reserve University, Cleveland, OH 44106, USA.
J. Martinez-Fernandez, Dpto. de Fisica de la Materia Condensada, Universidad de Sevilla, Apdo.
    1065, 41080 Sevilla, Spain.
G. W. Dransmann, Institut für Reaktorwerkstoffe, Forschungszentrum Jülich GmbH, D-5170
    Jülich, Germany.

The activation energy for nucleation of a martensitic transformation in a given material depends on the temperature and stress state and can be written [3, 4]:

$$Q = A - B \, \Delta G_{ch} - \sigma V^*$$ (1)

where A and B are constants, ΔG_{ch} is the molar chemical driving force for the transformation, σ the stresses that interact with the transformation strains (σ may be internal, arising from thermal expansion mismatch, for example, or the appropriate component of the applied stress), and V* the activation volume for nucleation. The nucleation rate is then:

$$N = N_0 \exp\left(- \frac{Q}{RT} \right) = N_0 \exp\left(- \frac{A - B \, \Delta G_{ch} - \sigma V^*}{RT} \right)$$ (2)

where N_0 is a prefactor and RT has its usual meaning.

MARTENSITIC TRANSFORMATION IN ZrO_2

ZrO_2 undergoes a martensitic transformation between the elevated-temperature tetragonal (t) form and the monoclinic (m) structure stable at room temperature. In the pure material, this transformation is athermal and displays a large thermal hysteresis between 1200 and 1500K. The transformation temperature is lowered by the addition of stabilizing oxides, such as Y_2O_3 or MgO. The $t{\Rightarrow}m$ transformation involves a dilatational strain of $\approx$4% and a (unit cell) shear strain of $\approx$ 16%. In some cases, t-ZrO_2 can also transform to an orthorhombic (o) polymorph; this transformation, too, involves a volume increase, which is smaller ($\approx$1%) than that of the $t{\Rightarrow}m$ transformation.

Many ZrO_2 alloys at room temperature contain small particles or precipitates of t-ZrO_2; although this structure is metastable at room temperature (M_s is below room temperature), it does not transform to the stable monoclinic polymorph because the activation energy for nucleation of this transformation is insurmountable without stress assistance. The $t{\Rightarrow}m$ transformation in these partially-stabilized ZrO_2 (PSZ) alloys is usually thought to occur either athermally, during cooling from high temperatures, or instantaneously, under an applied stress (*transformation toughening*). However, isothermal kinetics have been reported in Ca-PSZ [5] and Mg-PSZ [6,7] at room temperature, and in Y_2O_3-stabilized t-ZrO_2 polycrystals (Y-TZP) at 400-600K [8-10]. In this paper, we summarize our recent research on the thermally activated transformation of t-ZrO_2 in three different ZrO_2 ceramics, namely, a 3^m/o-Y_2O_3 (3Y)-TZP, a 3.4Y-PSZ single crystal, and a 9.7^m/o Mg-PSZ.

DETECTING THE TRANSFORMATION

The extent of $t{\Rightarrow}m$ (and $t{\Rightarrow}o$) transformation in all the materials studied was determined using optical microscopy and Raman spectroscopy. Because the martensitic transformation occurs by the migration of an invariant plane strains (IPS) interface, it produces surface uplifts on previously flat surfaces. Such uplifts were used to detect the occurrence of the transformation and were imaged using Nomarski interference contrast and quantified with Tolansky interferometry.

The Raman spectrometer (Dilor XY, Dilor Instruments, Lille, France) utilized a 1-W krypton laser (λ=568nm) and an entrance slit width of 100 μm. Traditional (macroprobe) spectroscopy was employed to determine the phase assemblage in bulk samples. The extent of transformation around Vickers indentations was determined with a Raman microprobe by acquiring spectra at 5-10 μm intervals across the region of interest; the sampling volume is less than 5 μm in diameter [11].

Both m-, o- and t-ZrO$_2$ are strong Raman scatterers and exhibit well-defined and well-separated Raman peaks in the 100-200 cm^{-1} wavenumber range (figure 1) [12,13]. The stabilized cubic phase produces only a broad, low-intensity band above 400 cm^{-1} [14], and no information about the cubic content in the samples was obtained. We arbitrarily chose the tetragonal Raman peak at 147 cm^{-1}, the monoclinic doublet at 178 and 190 cm^{-1}, and the orthorhombic peak at 120 cm^{-1} to represent the relative amounts of t-ZrO$_2$, m-ZrO$_2$ and o-ZrO$_2$, respectively. After subtracting a linear background for each spectrum as shown in figure 1, the absolute intensities of the relevant peaks, I_{147}, I_{178}, I_{190} and I_{120}, were determined. Finally, the monoclinic and orthorhombic intensity fractions were defined as:

$$R_m = \frac{I_{178}+I_{190}}{I_{120}+I_{147}+I_{178}+I_{190}} \quad \text{and} \quad R_o = \frac{I_{120}}{I_{120}+I_{147}+I_{178}+I_{190}} \qquad (3)$$

Because of a lack of standards, we will assume for convenience in the remainder of this paper that R_m and R_o are equal to the actual fraction of m- and o-ZrO$_2$, respectively, in the sample, excluding the cubic phase.

MATERIAL #1: Y-TZP

Thermally activated and isothermal $t \Rightarrow m$ transformation in Y-TZP was revealed by studying the transformation zone induced by room-temperature Vickers indents. A fuller description is published elsewhere [15].

The two-phase microstructure of the 3Y-TZP samples (provided by Coors Ceramics Co., Golden, CO.) consists of ≈80% of fine (<1 μm) t-ZrO$_2$ grains and ≈20% of coarser (≈5 μm) c-ZrO$_2$ grains [11]. 500N Vickers indentations were placed on polished surfaces at room temperature with an indenter displacement rate of 50μm/min and a 15s hold, followed by rapid unloading (using a Model 1125 Instron testing machine and a specially-fabricated Vickers indenter (made by Gilmore Diamond Tools, Needham Hts., MA)). With this load, the indent diameter

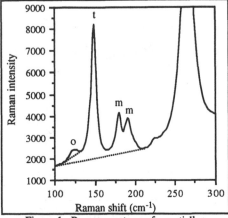

Figure 1. Raman spectrum of a partially-transformed Mg-PSZ. The Raman peaks used for phase content determinations are marked. The subtracted background is indicated as dashed lines.

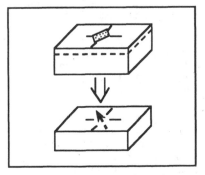

Figure 2. Schematic of sample preparation of polished away indents in 3Y-TZP; the indent is polished away, leaving behind only the traces of the four radial cracks. The regions along which Raman and surface displacement measurements were obtained are indicated as a dashed arrow.

is ≈270 µm and the edge of the indent is ≈190 µm in length. The indented surfaces were carefully ground away to a few microns below the bottom of the indent and polished with diamond paste, finishing with a 1µm grit (figure 2). Care was taken to avoid heating the samples during polishing.

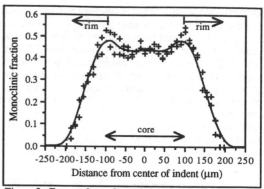

Figure 3. Extent of transformation under a polished away indent in 3Y-TZP. Crosses (+) are raw data, and the line represents the fit after averaging two scans and smoothing (factor=5). The standard deviation of the data is 0.02.

Indented and polished samples were subsequently annealed for various times at 400 to 1300K; each sample was heated to temperature in less than 5min, annealed for 5min, and quenched by removal from the furnace. The samples were subsequently re-annealed at the same temperature for another 10 min, and finally for another 60 min in the same manner, for a total of 75 min at each temperature. After each anneal, the surface displacements resulting from the transformation were studied using optical microscopy. After the 60-min anneal, the extent of transformation in the indented region was mapped with Raman microprobe spectroscopy. The data were collected along a line at 45° to the indentation cracks (the dashed arrow in figure 2); two perpendicular Raman scans were obtained for each indent.

RESULTS

During indentation at room temperature, the indent-induced stresses cause significant $t \Rightarrow m$ transformation around and below the indent. The distribution of m-ZrO_2 on a surface a few microns below the base of the indent, as measured with Raman spectroscopy, is shown in figure 3. The zone of stress-assisted transformation is ≈400 µm in diameter, twice as large as the original indent edge. Directly underneath the indent, in the *core*, the extent of transformation is ≈43%. A maximum in monoclinic fraction of ≈48% occurs at a distance of about 100 µm from the center of the indent, which is approximately the position of the edge of the indent at the original free surface. The monoclinic fraction gradually decreases to zero (the bulk value) within a zone, the *rim*, extending ≈100 µm from the indent edge.

After such a sample is stored at room temperature for several days, a slight surface uplift develops in the indented region on the polished-away surface.

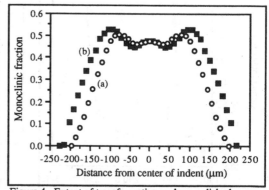

Figure 4. Extent of transformation under a polished away indent in 3Y-TZP (a) immediately after indentation and polishing, and (b) 3 weeks later. The points represent curves fitted to the Raman data using the method described for figure 3.

The surface uplift is evidence for slow isothermal $t \Rightarrow m$ transformation in the presence of the residual stress field of the indent. The increase in monoclinic content measured after 3 weeks of room-temperature aging is shown in figure 4. In order to avoid complications resulting from this room-temperature aging, samples were polished and annealed within 1-2 days after indentation.

Annealing for 75 min at 400 to 1300K causes considerable changes in the monoclinic fraction in the transformation zone under the indent. Anneals up to 500K increase the size of the transformation zone, as well as the amount of m-ZrO_2 within the zone (figure 5a). The 600K anneal causes further transformation in the rim, while the monoclinic content in the core decreases slightly. Higher-temperature anneals, significantly, cause gradual *decreases* in the monoclinic content of both core and rim, *i.e.* a general shrinkage of the transformation zone (figure 5b). After the 900K anneal, the extent of transformation has been reduced to smaller values than in the as-indented transformation zone. Finally, the 1300K anneal decreases the monoclinic content in the sample to zero.

As already noted, the anneals also produce surface displacements in the indented region of the polished surface (figure 6), whose dimensions are intimately related to the extent of transformation just discussed. Two regions of uplift can again be distinguished -- the core and the surrounding rim. The core is approximately the same size as the original indent, and is bounded by the four radial cracks. The rim region forms an almost circular zone around the core and extends approximately one-third along the length of the cracks. (We initially attempted to correlate the two zones of uplift with the extent of indent-induced cracks. However, very similar, albeit smaller, core-and-rim structures developed with lower-load (40N) indents, which did not induce any indentation cracks.) The surface displacements (*i.e.* volume changes) in the transformation zone, studied with optical microscopy, are directly related to the changes in monoclinic content caused by the anneals [15], but the exact nature of this transformation plasticity, *i.e.* the spatial distribution of plastic deformation, is not yet clear.

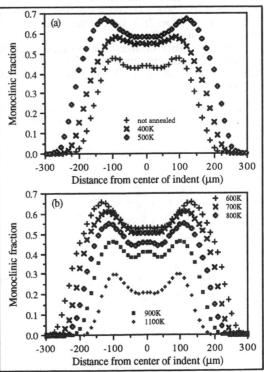

DISCUSSION

The thermodynamic and kinetic aspects of the thermally activated $t \Rightarrow m$ transformation can be analyzed with reference to equations 1 and 2. Several separate stages of

Figure 5. Extent of transformation (fitted curves) under polished away indents in 3Y-TZP annealed for 75min at various temperatures (a) up to 500K, and (b) above 500K. The points represent curves fitted to the Raman data using the method described for figure 3.

transformation take place during the indentation and annealing experiments. Firstly, extensive stress-assisted $t \Rightarrow m$ transformation is induced during indentation at room temperature (figure 3); the indentation stresses reduce the activation energy for nucleation of m-ZrO_2 by the factor σV^* and allow some transformation to take place. Therefore, M_d (the martensite start temperature for *deformed* materials) must be above room temperature.

Further $t \Rightarrow m$ and $m \Rightarrow t$ transformations take place during annealing of the indented and polished samples. Annealing up to 800K results in an increase in the monoclinic fraction relative to the unannealed sample (figure 7). This additional $t \Rightarrow m$ transformation in the presence of the residual stress field of the indent occurs in spite of a decrease in the chemical driving force (ΔG_{ch}) with increasing temperature, and is therefore clearly due to thermal activation (*cf.* equation 2). The maximum amount of additional transformation occurs at 500-600K. This coincides with the temperature range at which TZP materials are known to undergo extensive isothermal $t \Rightarrow m$ transformation on their surfaces in the presence of moisture [8,10].

Anneals at higher temperatures (900K and above) cause the monoclinic content to decrease to below its original as-indented value (*cf.* figures 3 and 5). We interpret this as indicating that some or all of the m-ZrO_2 undergoes the reverse transformation to tetragonal symmetry upon heating to temperatures above 800 to 900K. The temperature at which t-ZrO_2 starts to form from the martensite phase m-ZrO_2 during heating is conventionally known as A_s (austenite start temperature); in our case, it is $\approx$850K. The austenite finish temperature (A_f) must be between 1100 and 1300K, as all m-ZrO_2 has re-transformed to t-ZrO_2, *i.e.* the $m \Rightarrow t$ transformation is complete, at 1300K.

The possibility of some transformation occurring during cooling from the annealing temperature must also be considered. However, no m-ZrO_2 exists in samples quenched from 1300K, where t-ZrO_2 is stable. This suggests that the martensite that had formed during indentation and undergone the reverse $m \Rightarrow t$ transformation upon heating above A_s did not re-transform to m-ZrO_2 during quenching from the annealing temperature. In other words, no athermal transformation occurs during quenching, and the extent of transformation measured after quenching to room temperature (figure 5) is equal to that at the annealing temperature.

On the other hand, furnace cooling from 1300K, or quenching from

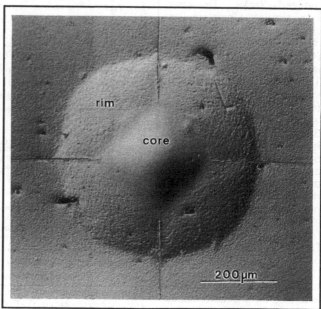

Figure 6. Nomarski interference micrograph of a polished away indent in 3Y-TZP annealed for 75 min at 500K, revealing surface uplift caused by the thermally activated martensitic transformation.

1300K followed by reannealing at 500K both result in a considerable amount of *m*-ZrO$_2$ in the indented region (data not shown here). This increase in monoclinic content must be a result of the thermally activated *t*⇒*m* transformation taking place while the samples were cooled or re-annealed below M$_d$.

By definition, the isothermal transformation is time-dependent. Extremely slow isothermal transformation occurs in indented and polished samples over several days even at room temperature (figure 4). The rate of the thermally activated transformation increases with increasing annealing temperature (figure 7). At 400K, for example, the volume increase resulting from a 5-min anneal is small compared to that produced by the second 10-min anneal, which in turn is much smaller than that caused by the final 60-min anneal. At increasingly higher temperatures, the second and third anneals have less and less effect, indicating that most of the *t*⇒*m* transformation occurs during the first 5 min at temperature, and saturation is achieved rather quickly. For anneals at 700K and above, the transformation does not show any time-dependence on the time scale employed (5 to 75 min).

In summary, we have observed the following sequence of phase transformations:
1. Stress-assisted *t*⇒*m* transformation during room-temperature indentation.
2. Thermally activated (isothermal) *t*⇒*m* transformation at temperatures up to ≈800K.
3. Reverse *m*⇒*t* transformation above ≈850K (A$_s$) up to ≈1200K (A$_f$).
4. Thermally-activated *t*⇒*m* transformation during slow cooling or re-annealing below M$_d$.

MATERIAL #2: Y-PSZ

A similar study of the thermally activated *t*⇒*m* transformation was performed on 3.4Y-PSZ single crystals by analyzing the transformation zone around high-temperature Vickers indentations; some of these results have been previously published [16].

Cubic ZrO$_2$ single crystals containing 3.4 ^{m}/$_o$ Y$_2$O$_3$ were grown by skull melting (by Ceres Corp., N. Billerica, MA, USA). The crystals were oriented using Laue X-ray back-reflection, cut, and polished to a 1μm finish on their {100}$_c$ faces. Samples were aged for 150 h at 1873K in air to allow precipitation and growth of *t*-ZrO$_2$ precipitates in the fully-stabilized *c*-ZrO$_2$ matrix. The *t*-ZrO$_2$ particles form elongated "colony" precipitates up to 2 μm long with {101}$_c$ habit planes [17]. Each colony precipitate consists of alternating twin lamellae whose *c*$_t$-axes are at ≈90° to each other; the twin plane is {101}$_t$.

10N Vickers indentations were placed on {100}$_c$ surfaces using a high-temperature microhardness tester. In a typical experiment, a sample was heated to 1273K, indented, cooled to 1173K, indented, and so on at 100K intervals during cooling to room temperature. The indent diagonals were positioned along <110>$_c$ directions, the hold time was 15s, and the furnace vacuum was

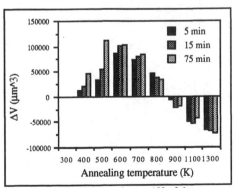

Figure 7. Volume change ΔV of the transformation zone (proportional to the monoclinic content) under polished away indents in 3Y-TZP after the 5-min, the 10-min, and the 60-min anneal.

1kPa. The Vickers hardness in this orientation decreases from ≈15 GPa at room temperature to ≈3 MPa at 1273K. No indentation-induced cracking occurred, no doubt because of the low loads and the transformation toughening resulting from the stress-induced transformation around indents.

RESULTS

After indentation at temperatures above 473K, significant surface uplifts (up to 300 nm high), evidence for the martensitic transformation, are formed around indents (figure 8). The zones of indent-induced surface uplifts are not uniform around the indents but are confined to four prominent "wings" along directions near to $<210>_c$, which extend up to 400 μm from the indent; this is about 10 times as large as the plastic zone around the indent. Serial sectioning of the indents (as in figure 2) revealed that the indent-induced $t{\Rightarrow}m$ transformation is a bulk and not a surface phenomenon; the zones extend up to 150-200 μm below the surface. The maximum extent of the uplift (the length of the "wings") is plotted as a function of indentation temperature in figure 9. (When the experiment was repeated by indenting at 100K intervals during *heating* to 1273K, instead of during cooling, similar surface uplifts resulted, but the size of the uplift zones was reduced.)

In situ observation of the indents during cooling revealed that the uplifted zone around the indents did not form immediately upon high-temperature indentation, but only began to be visible during cooling to below ≈720K, and stopped growing upon further cooling to below ≈580K. In a related experiment, one sample, which had been indented during cooling, was annealed at 1273K for a few minutes. During heating, the surface uplifts around the indents disappeared in a narrow temperature range at ≈870K, and during subsequent cooling, the surface uplifts reappeared when the temperature decreased to below ≈720K, but the extent of the uplift was reduced by ≈20%. A second anneal of this sample reduced the uplift by another 10%.

The extent of $t{\Rightarrow}m$ transformation in the uplifted regions around indents was determined by acquiring Raman spectra at 5 μm intervals across one of the wings surrounding a 773K indent. The results of two such scans in figure 10 reveal that, although extensive transformation has occurred inside the wing, the transformation is by no means exhausted.

Figure 8. Nomarski interference micrograph of two 10N Vickers indents in 3.4Y-PSZ formed at 773K, displaying surface uplift ("wings") along the $<210>_c$ directions (the micrograph was taken after cooling to room temperature).

DISCUSSION

The surface uplift in the wings surrounding Vickers indentations is clearly due to the indentation-induced $t \Rightarrow m$ transformation. This transformation behavior is similar to, albeit more complex than, that in 3Y-TZP; different explanations are necessary to understand this phenomenon at low and high temperatures.

As the indentation temperature is raised to ≈773K, the extent of transformation around indents increases (figure 9) because the transformation is thermally-activated: as in the 3Y-TZP, the increase in thermal energy (the RT term in equation 2) dominates over the decrease in the chemical driving force (ΔG_{ch}) as the temperature is raised, and consequently allows more transformation to take place at higher temperatures. From *in situ* observations of the transformation zones, we deduce that the A_s temperature, at which the uplift starts to disappear during heating, is ≈870K; A_f is only slightly higher. Similarly, the M_d and M_f temperatures are ≈720K and ≈580K, respectively. The A_s temperature is surprisingly close to A_s in the 3Y-TZP (≈850K), but the temperature of maximum transformation in the PSZ (≈770K) is much higher than that in the TZP (≈550K).

The high-temperature (≥773K) indentation data requires another explanation. We believe that the size of the transformation zone formed during cooling beyond M_d is directly related to the dislocation density produced by the indentation process. As the indentation temperature is increased, the dislocation zone around indents becomes less dense and localized, and consequently less effective in inducing the stress-assisted $t \Rightarrow m$ transformation during subsequent cooling. Higher-temperature indents therefore produce smaller transformation zones (as measured after cooling to room temperature), resulting in the observed decrease in wing size with increasing indentation temperature above 773K (figure 9).

The results of our other experiments can now be explained as well. Annealing of a previously indented sample at 1273K allows sufficient dislocation recovery that a smaller transformation zone is formed during subsequent cooling to room temperature. For the same reason, the transformation zones are smaller around indents that are formed during heating than those formed during cooling.

To summarize our results, high-temperature indentation tests have revealed the influence of stress-assistance on the $t \Rightarrow m$ transformation, whereas lower-temperature indents provided insight into the thermal activation of the process. We are, however, still impressed by the large size and peculiar shape of the indent-induced transformation zones in this material. The wings are believed to be caused

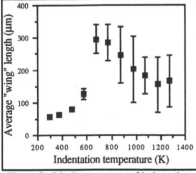

Figure 9. Maximum extent of indentation-induced surface uplift (measured after cooling to room temperature) as a function of indentation temperature in 3.4Y-PSZ crystals. Error bars indicate the standard deviation.

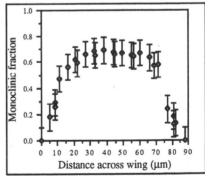

Figure 10. Extent of transformation, as measured with Raman microprobe spectroscopy, across a "wing" around a 773K indent in a 3.4Y-PSZ single crystal. Error bars indicate the standard deviation.

by autocatalytic transformation in only specific crystallographic directions [16], but further studies of the crystallography of the transformation zones are necessary to understand this phenomenon.

MATERIAL #3: Mg-PSZ

High-toughness Mg-PSZ undergoes an isothermal martensitic transformation to both m- and o-ZrO_2 at room temperature. The 9.7 $^{m}/o$ Mg-PSZ used for this study (TS grade, Nilcra Ceramics Pty Ltd., Northcote, Victoria 3070, Australia) consists of fully-stabilized c-ZrO_2 grains about 60 μm in diameter, containing oblate precipitates up to $\approx$250 μm in maximum diameter. Most of these precipitates have the tetragonal structure; some may also have monoclinic or orthorhombic symmetries. The grains also contain smaller precipitates of $Mg_2Zr_5O_{12}$, a metastable decomposition product [18]. In addition, larger particles of m-ZrO_2 exist along the grain boundaries; they are a product of the eutectoid decomposition reaction (c-$ZrO_2 \Rightarrow t$-ZrO_2 + MgO) during processing and the subsequent rapid martensitic transformation (t-$ZrO_2 \Rightarrow m$-ZrO_2) during cooling [19]. The exact phase assemblage in this material has not been determined; measurements of the fraction of c-ZrO_2 + $Mg_2Zr_5O_{12}$ in similar Mg-PSZ's using neutron and X-ray diffraction (XRD) range from $\approx$30 [20] to 53 wt.% [21]. In this study, we are only concerned with the relative amounts of the remaining three phases, t-, m-, and o-ZrO_2.

Unfortunately, transmission electron microscopy is not suitable for phase characterization of this material because the t-ZrO_2 precipitates are so transformable that they readily transform during thin foil preparation. The exact distribution of these three phases at the grain boundaries and within the grains is thus not known. Similarly, XRD is less than ideal because the X-rays sample only the top 5 μm or so of the PSZ, and this layer may have undergone substantial transformation during grinding and polishing. (The difficulties with utilizing XRD to characterize this material will be further discussed in a future publication).

Polished samples were annealed at 1273K for 20 min, followed by a furnace cool ($\approx$200 K/h). This is a standard anneal commonly used before testing this material in order to reverse any $t \Rightarrow m$ transformation that may have been induced during sample preparation. The changes in the phase assemblage caused by the anneal and subsequent aging at room temperature were monitored with bulk Raman spectroscopy on four samples.

RESULTS

The relative Raman intensity fractions of m-ZrO_2 (R_m) and o-ZrO_2 (R_o) in the as-received sample are 0.35 and 0.03, respectively. The anneal reduces the monoclinic content to 0.25 and the orthorhombic content to zero.

After the annealed sample is stored at room temperature for several months, surface rumpling develops on the previously polished flat surface [8], indicating that some martensitic transformation has occurred. This isothermal room-temperature transformation was monitored with Raman spectroscopy in bulk samples, with the results shown in figure 11. Both the monoclinic and orthorhombic contents increase slowly with time after the anneal and level off near their respective original values before the anneal.

DISCUSSION

The presence of o-ZrO_2 at room temperature in bulk Nilcra TS samples has been previously reported [20,22]. A detailed explanation of the phase assemblage is not yet in hand, as it is not yet clear whether o-ZrO$_2$ is a stable phase or a metastable reaction product.

The anneal at 1273K causes a previously unreported change in the phase assemblage, which can be explained as follows: During heating to the annealing temperature, all the monoclinic and orthorhombic regions in the sample transform to the high-temperature tetragonal structure. During subsequent cooling from the annealing temperature, some of the t-ZrO$_2$ particles become unstable and rapidly (*i.e.* athermally) transform to m-ZrO$_2$; no o-ZrO$_2$ is formed during cooling. It appears reasonable to assume that all of the coarser grain-boundary particles are transformed to m-ZrO$_2$ but that most of the intergranular precipitates retain their tetragonal structure during cooling.

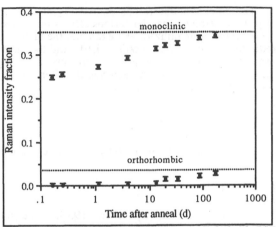

Figure 11. Relative fractions of m-ZrO$_2$ and o-ZrO$_2$ in Mg-PSZ during room-temperature aging after the 1000°C anneal. The error bars represent the standard deviation of the data; the dotted lines correspond to the phase assemblage before the anneal.

During subsequent room-temperature aging, these t-ZrO$_2$ precipitates transform isothermally to m- and o-ZrO$_2$. As before, the transformation rate is believed to be controlled by the thermally activated, stress-assisted nucleation step. The t-ZrO$_2$ precipitates display a range of activation energies for nucleation because of the different particle sizes and local stress states (due to thermal expansion mismatch or coherency strains) of the individual precipitates; both of these factors control their transformability. Immediately after the anneal, a large number of precipitates are sufficiently unstable to readily undergo the martensitic transformation. As the transformation proceeds, the population of easily transformable precipitates is depleted and the reaction rate decreases. After many months, the point is reached when the thermal energy and local stresses available to nucleate the transformation in the remaining t-ZrO$_2$ precipitates are too low to overcome the activation barrier, and the transformation effectively ceases.

In view of these results, it is apparent that the standard 1273K anneal reverses not only any transformation caused during sample preparation but also the isothermal bulk transformation occurring over long periods of time at room temperature.

SUMMARY

Isothermal (thermally activated) martensitic transformation occurs in all three ZrO$_2$ ceramics investigated here. The t-ZrO$_2$ particles in the Y$_2$O$_3$ alloys are more stable and exhibit different transformation characteristics than the highly transformable t-ZrO$_2$ precipitates in Mg-PSZ; in the former samples, M$_s$ is below room temperature and substantial stress-assistance and thermal energy are required for the transformation to proceed. In Mg-PSZ, there is a range of M$_s$ temperatures because of the variety of particle sizes, and some transformation takes place even at and above room temperature.

The isothermal transformation in ZrO_2 alloys is unfortunate because the change in phase assemblage has deleterious effects on the mechanical properties for which these ceramics are utilized. For example, the high strength of Y-TZP's is drastically reduced by aging at 500 to 600K [9,10], and the superior toughness of Mg-PSZ is a sensitive function of the anneal and subsequent aging [6,23]. It is therefore imperative to carefully characterize the thermal history of ZrO_2 ceramics when acquiring mechanical property data.

ACKNOWLEDGMENTS

The research was sponsored by NSF under Grant No. DMR-90-19383.

REFERNCES

1.   Kaufman, L. and M.Cohen, 1958. "Thermodynamics and Kinetics of Martensitic Transformations," in Progress in Metals Physics, Vol. 7, B.Chalmers and R.King, eds., New York, NY, USA: Pergamon Press, pp. 165-246.

2.   Thadhani, N.N. and M.A.Meyers,1986. "Kinetics of Isothermal Martensitic Transformations." Progress in Mater. Sci., 30:1-37.

3.   Olsen, G.B. and M.Cohen, 1976. "A General Mechanism of Martensitic Nucleation: Part III. Kinetics of Martensitic Nucleation." Metall. Trans. A, 7(12): 1915-23.

4.   Chen, I-W. and P.E.Reyes-Morel, 1987. "Transformation Plasticity and Transformation Toughening in Mg-PSZ and Ce-TZP." Mater. Res. Soc. Symp. Proc., 78:75-88.

5.   Dickerson, R.M., M.V.Swain and A.H.Heuer, 1987. "Microstructural Evolution in Ca-PSZ and the Room-Temperature Instability of Tetragonal ZrO_2." J. Am. Ceram. Soc., 70(4):214-20.

6.   Heuer, A.H., M.J.Readey and R.W.Steinbrech, 1988. "Resistance Curve Behavior of Supertough MgO-partially-stabilized ZrO_2." Mater. Sci. Engg. A, 105/106:83-89.

7.   Shaw, M.C., D.B.Marshall, A.H.Heuer and E.Inghels, 1992. "Recovery of Crack-Tip Transformation Zones in Zirconia After High-Temperature Annealing." J. Am. Ceram. Soc., 75(2):474-76.

8.   Matsui, M., T.Soma and I.Oda, 1984. "Effect of Microstructure on the Strength of Y-TZP Components," in Advances in Ceramics, Vol. 12, Science and Technology of Zirconia II, N.Claussen, M.Rühle and A.H.Heuer, eds., Columbus, OH, USA: American Ceramic Society, Inc., pp. 371-81.

9.   Tsukuma, K., Y.Kubota and T.Tsukidate, 1984. "Thermal and Mechanical Properties of Y_2O_3-Stabilized Tetragonal Zirconia Polycrystals," ibid, pp. 382-90.

10. Watanabe, M., S.Iio and I.Fukuura, 1984. "Aging Behavior of Y-TZP," ibid, pp. 391-98.

11. Kaliszewski, M.S., G.Behrens, A.H.Heuer, M.Shaw, D.B.Marshall, G.W.Dransmann, R.W.Steinbrech, A.Pajares, F.Guiberteau, F.L.Cumbrera and A.Dominguez-Rodriguez, 1992. "Indentation Studies on Y_2O_3-Stabilized ZrO_2: I, Development of Indentation-Induced Cracks." Submitted to J. Am. Ceram. Soc.

12. Ishigame, M. and T.Sakurai, 1977. "Temperature Dependence of the Raman Spectra of ZrO_2." J. Am. Ceram. Soc., 60(7-8):367-69.

13. Marshall, D.B., M.R.James and J.R.Porter, 1989. "Structural and Mechanical Property Changes in Toughened Magnesia-Partially-Stabilized Zirconia at Low Temperatures." J. Am. Ceram. Soc., 72(2):218-27.

14. Feinberg, A. and C.H.Perry, 1981. "Structural Disorder and Phase Transitions in ZrO_2-Y_2O_3 System." J. Phys. Chem. Solids, 42(6):513-18.

15. Behrens, G., G.W.Dransmann and A.H.Heuer, 1993. "On the Isothermal Martensitic Transformation in 3Y-TZP." J. Am. Ceram. Soc., in press.

16. Martinez-Fernandez, J., M.Jimenez-Melendo, A.Dominguez-Rodriguez and A.H.Heuer, 1991. "Microindentation-Induced Transformation in 3.5-mol%-Yttria-Partially Stabilized Zirconia Single Crystals." J. Am. Ceram. Soc., 74(5):1071-81.

17. Heuer, A.H., V.Lanteri and A.Dominguez-Rodriguez, 1989. "High-Temperature Precipitation Hardening of Y_2O_3-Partially-Stabilized Zirconia (Y-PSZ) Single Crystals." Acta Metall., 37(2):559-67.

18. Farmer, S.C., L.H.Schoenlein and A.H.Heuer, 1983. "Precipitation of $Mg_2Zr_5O_{12}$ in MgO-Partially Stabilized ZrO_2." J. Am. Ceram. Soc., 62(7):C107-C109.

19. Hannink, R.H.J., 1983. "Microstructural Development of Sub-Eutectoid Aged MgO-ZrO_2 Alloys." J. Mater. Sci., 18:457-70.

20. Howard, C.J., E.H.Kisi, R.B.Roberts and R.J.Hill, 1990. "Neutron Diffraction Studies of Phase Transformations between Tetragonal and Orthorhombic Zirconia in Magnesia-Partially-Stabilized Zirconia." J. Am. Ceram. Soc., 73(10): 2828-33.

21. Marshall, D.B., M.C.Shaw, R.H.Dauskardt, R.O.Ritchie, M.J.Readey and A.H.Heuer, 1990. "Crack-Tip Transformation Zones in Toughened Zirconia." J. Am. Ceram. Soc., 73(9):2659-66.

22. Lim, C.S., T.R.Finlayson, F.Ninio and J.R.Griffiths, 1992. "In-Situ Measurement of the Stress-Induced Phase Transformations in Magnesia-Partially-Stabilized Zirconia Using Raman Spectroscopy." J. Am. Ceram. Soc., 75(6):1570-73.

23. Readey, M.J., A.H.Heuer and R.W.Steinbrech, 1988. "Annealing of Test Specimens of High-Toughness Magnesia-Partially-Stabilized Zirconia." J. Am. Ceram. Soc., 71(1):C2-C6.

Nucleation in Stress-Induced Tetragonal-Monoclinic Transformation of Constrained Zirconia

S.-K. CHAN

ABSTRACT

A theory for stress-induced tetragonal→monoclinic transformation of constrained zirconia is presented based on the assumption that when forcibly strained to a regime of absolute instability where the free energy density of the tetragonal phase has a negative curvature, the constrained tetragonal zirconia becomes unstable with respect to the development of a modulated strain pattern that will evolve into a band of twinned monoclinic domains. The temperature range for such an instability, the critical size of the inclusion, the corresponding critical strain, and the periodicity of the modulation are derived in terms of parameters that can be related to the elastic stiffness coefficients of various orders of the inclusion and the shear modulus of the host matrix. An entirely different mechanism is suggested for the reverse monoclinic→tetragonal transformation because the monoclinic phase is metastable when the extrinsic stress is removed. Estimates for the parameters are inferred from a variety of experimental data for pure zirconia and the numerical values for the predicted physical quantities are obtained.

INTRODUCTION

The stress-induced tetragonal (t) → monoclinic (m) transformation of constrained zirconia has been of intense interest since the heyday of the discovery of the phenomenon of transformation toughening [1]. Various theories have been attempted based on different nucleation perspectives [2,3] and on end-point thermodynamics [4]. Although a generally

S.-K. Chan, Materials Science Division, Argonne National Laboratory, 9700 S. Cass Avenue, Argonne, IL, 60439-4838, U.S.A.

16

satisfactory quantitative description of the transformation is still an elusive goal, these attempts explored the consequences of useful ideas, leading to information for later research to draw on. Recently, experimental and theoretical studies have been carried out for unconstrained zirconia to obtain a basic perspective of the lattice properties [5,6,7], vibrational modes [8,9], the interplay between symmetry and collective atomic movements [10], nonclassical nucleation from partially soft lattice modes [11,12], and stability limits for the t and m phases [13]. Based on the understanding obtained in these efforts for the unconstrained situation, some ideas will be explored and results presented in this paper for stress-induced t→m transformation of constrained zirconia. The objectives of this exposition are to determine (1) the factors that control the critical strain, (2) the temperature ranges for the t→m transformation and for the switching of t domains, (3) the critical size of a transformable inclusion, and (4) the difference between the plausible mechanisms of the t→m and the reverse m→t transformations.

THERMODYNAMIC INSTABILITIES AND MODES OF TRANSFORMATION

A thermodynamic state of a system is stable if its corresponding free energy F is at a minimum $\delta^2 F > 0$, i.e., the second order variation with respect to an appropriate parameter is positive definite. When the minimum is a global one, i.e., it has the lowest F compared to all other minima, the state is absolutely stable against all manners of probing by perturbation. However, when the minimum is a local one, the state is only metastable in the sense that the system is stable against perturbations of small amplitudes but unstable against perturbations of larger amplitudes that can take the system from the local minimum over the intervening barrier into the global minimum. When a state ceases to correspond to a minimum in free energy because of changing parameters and becomes associated with an inflexion or a saddle point, i.e., $\delta^2 F = 0$ for variation in at least one degree of freedom, the system is absolutely unstable against even small amplitude perturbations. Based on this difference in response to large and small-amplitude perturbations, Gibbs [14,15] introduced a classification scheme that divides phase transformations into two types: (I) those that start with a perturbation that is large in the degree of transformation but small in spatial extent, and (II) those that start with a perturbation that is small in the degree of transformation but large in spatial extent. The former type corresponds to first-order transformations that begin with the nucleation of critical nuclei (or embryos) of the new phase and are followed by their subsequent growth (figure 1, upper). The latter type corresponds to all homogeneous transformations including spinodal decomposition and other modes involving long-wavelength modulations

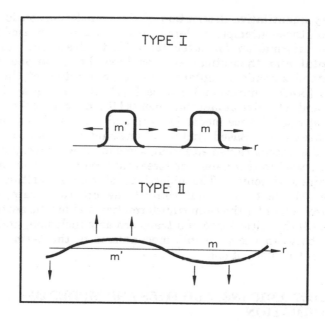

Figure 1. Schematic illustration of Gibbs' classification. Upper (type I): Transformation proceeds by nucleation via localized fluctuation of large amplitude followed by growth of nuclei. Lower (type II): Transformation proceeds by the development of long wavelength modulations of infinitesimal amplitude followed by increase of these amplitudes.

(figure 1, lower) that proceed by the amplification of the initially small amplitudes. The term nucleation was originally used exclusively for type I transformations only. However, in recent times, it is used frequently in a more generalized sense to cover the beginning of any transformation, including type II ones that involve no critical nuclei at all. The intention of this paper is to examine the onset of modulated strain instability (type II) when a constrained inclusion of t zirconia is forcibly taken from a stable thermodynamic state ($\delta^2 F > 0$) into an unstable regime ($\delta^2 F \leq 0$) by an extrinsic stress field and the possible reversal when the stress is removed. The term nucleation should be understood here in the generalized sense.

LATTICE MODES AND DRIVING PARAMETER FOR THE T→M TRANSFORMATION

In general, displacive and martensitic transformations involve minute but cooperative movements of atoms from their original equilibrium positions and are describable in terms of the symmetry-allowed, lattice-vibrational modes of the parent structure. For the t→m transformation of zirconia, the m structure is related to the parent t structure by a small number of collective motions. Three of these quantify the macroscopic size and shape changes in terms of nonlinear Lagrange strains [10,11]

$$\eta_{ij} = \frac{1}{2}\left(\frac{\partial u_i}{\partial r_j} + \frac{\partial u_j}{\partial r_i} + \sum_k \frac{\partial u_k}{\partial r_i}\frac{\partial u_k}{\partial r_j}\right) \tag{1}$$

where r_i (i = 1,2,3) are the Cartesian coordinates attached to the axes of the unit cell of the t structure and u_i (i = 1,2,3) are components of a displacement vector. They are:

(1) uniform change of the basal square, $\eta_{xx} + \eta_{yy}$,

(2) uniform change of the 4-fold axis, η_{zz}, and

(3) homogeneous shears of the xz and yz planes, the doublet (η_{xz}, η_{yz}).

The energies associated with these motions are embodied in the elastic constants of the t phase $c_{11} + c_{12}$, c_{33}, and c_{44}, respectively [10,11]. They are associated with long-wavelength, acoustic phonons (in-phase vibrations) at the center of the Brillouin zone of the reciprocal lattice of the t structure. There are five other motions that together quantify the sublattice movements (shuffles) within the unit cell of the t structure and produce the correct periodicity of the m structure [8]. They are optical phonons (out-of-phase vibrations) $M_1, M_2, \ldots, M_5$ at the M point (110) of the Brillouin zone boundary. These are associated with five different frequencies.

In principle, the free energy density f_0 relevant to the t→m transformation is a function of all these lattice modes

$$f_0 = f_0\,[\eta_{xx} + \eta_{yy};\ \eta_{zz};\ (\eta_{xz}, \eta_{yz});\ M_1, \ldots, M_5] \tag{2}$$

However, these normal mode motions are not independent of one another but are connected via anharmonic (i.e., higher than second order) coupling coefficients. As such, not all of them are needed explicitly to formulate a quantitative theory. It has been suggested [8] that since the five optical phonon modes are triggered off simultaneously at the t→m transformation, they are passive motions dragged along by the driving shear motion (η_{xz}, η_{yz}). Furthermore, the changes in the size represented by $\eta_{xx} + \eta_{yy}$ and η_{zz} can also be absorbed as an implicit parameter [12] in the coefficients of the Taylor expansion of f_0 in terms of η_{xz} and η_{yz}. The free energy density can therefore be expressed in terms of the explicit parameters of the shears η_{xz}, η_{yz} as, up to the sixth order,

$$f_0 = f_0 (\eta_{xz}, \eta_{yz}) = A\eta^2 + B_1\eta^4 + B_2\eta^4 \cos 4\phi + C_1\eta^6 + C_2\eta^6 \cos 4\phi \quad (3)$$

where $\eta_{xz} = \eta \sin \phi$, $\eta_{yz} = \eta \cos \phi$, and $\eta^2 = \eta_{xz}^2 + \eta_{yz}^2$. The coefficients A, B's, and C's involve, respectively, elastic constants of the second, fourth, and sixth orders, the explicit forms of which may be found in reference 11.

Depending on how these coefficients are related to one another as the external parameters (such as temperature, hydrostatic pressure, composition, etc.) are varied, equation 3 can depict (figure 2) four degenerate minima at $\eta \neq 0$ representing the four variants of the m phase that are strained differently with respect to the t phase at $\eta = 0$.

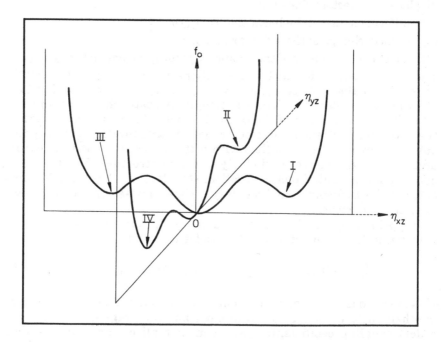

Figure 2. Two sections of the surface of the free energy density f_0 for a homogeneous material of tetragonal symmetry. The two pairs of homogeneous strains $\pm\eta_{yz}$, $\pm\eta_{xz}$ can lead to four variants I, II, III, and IV of monoclinic symmetry. The four degenerate local minima can become competitive with the minimum at $\eta = 0$ corresponding to the tetragonal symmetry.

The behavior can best be visualized by focusing on one representative planar section $\phi = 0$ (figure 3) of the three-dimensional plot. We have [11], in this plane,

$$f_0 = A\eta^2 - B\eta^4 + C\eta^6 \qquad (4)$$

where $B = -(B_1 + B_2)$ and $C = C_1 + C_2$. Above an upper critical temperature T^*, the t phase is absolutely stable and the minima corresponding to the m phase have yet to develop from the inflexion points of the free energy for T^*. Just below T^*, four minima with $\eta \neq 0$ (two being in the plane $\phi = 0$) representing the m phase start to evolve until at T_0 when they are at the same free energy as the minimum at $\eta = 0$ for the t phase. This temperature T_0 is the coexistence temperature between the t and m phases. Below T_0 the m phase becomes more stable than the t phase. Finally, at the lower critical temperature T_c, the original minimum at $\eta = 0$ becomes an inflexion point. The temperatures T^* and T_c represent, respectively, the upper stability limit of the m phase and the lower stability limit of the t phase [11,13].

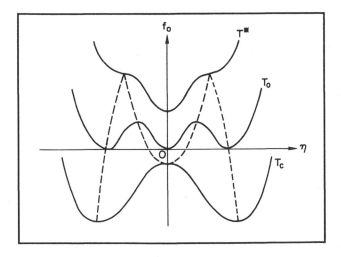

Figure 3. One particular section of the surface of the free energy density f_0 showing the relative changes of the three extrema as the temperature varies. T^* is the upper critical temperature above which the monoclinic phase becomes absolutely unstable. T_0 is the coexistence temperature at which the tetragonal and monoclinic phases can coexist in equilibrium. T_c is the lower critical temperature below which the tetragonal phase becomes absolutely unstable. The dotted curves indicate how the inflexion points evolve into minima and maxima as the temperature varies between T^* and T_c.

THE ONSET OF STRESS-INDUCED TRANSFORMATION IN A CONSTRAINED INCLUSION

Suppose an inclusion is homogeneously sheared and held at a strain η_0 that corresponds to a negative curvature of the free energy density (figure 4), the interface with the similarly strained surrounding material being coherent. We want to determine the criterion under which the inclusion, together with its immediate environment, would become unstable with respect to the development of a modulation of the shear strain. For simplicity, we assume that the original inclusion is a sphere of radius R and the strain is in the planar section $\phi = 0$. After being sheared, a modulation of amplitude a and wavelength λ along the direction of the original polar axis z is allowed to develop:

$$\eta(z) = \eta_0 + a.\sin(2\pi z/\lambda) \tag{5}$$

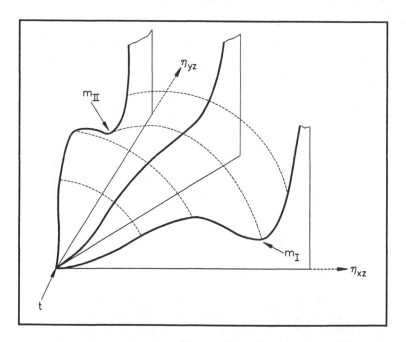

Figure 4. Schematic illustration of the surface of the free energy density f_0 in one quadrant. If an extrinsic stress brings about a homogeneous strain η_0 of the material that corresponds to a negative curvature f_0'' of the surface such as on or between the dotted curves, a modulated strain pattern will develop which eventually leads to a band of alternating m_I and m_{II} domains.

The local free energy density inside the inclusion depends on both the local strain η and its spatial gradient $\nabla\eta$ [11,12]. We have, to second order,

$$
\begin{aligned}
f_0 &= f_0(\eta, \nabla\eta) \\
&= f_0(\eta_0) + f_0'(\eta_0)(\eta - \eta_0) + \frac{1}{2}f_0''(\eta_0)(\eta - \eta_0)^2 + \kappa(d\eta/dz)^2
\end{aligned}
\tag{6}
$$

where $f_0' = df_0/d\eta$ and $f_0'' = d^2f_0/d\eta^2$ are the first and second derivatives, respectively, of the homogeneous part of the free energy density $f_0(\eta_0)$, given by equation (3), and κ is a coefficient that can be related to the interfacial free energy. The change in the total free energy of the inclusion becomes, to second order in the amplitude a,

$$
\Delta F_I = \int dV \left[\frac{1}{2}f_0''(\eta_0)(\eta - \eta_0)^2 + \kappa(d\eta/dz)^2 \right]
\tag{7}
$$

The first-order term involving $(\eta - \eta_0)$ drops out of the integral because η_0 is the average strain of the inclusion. Substituting equation 5 into equation 7, we have

$$
\Delta F_I = \int dV \left[\frac{1}{2}a^2 f_0''(\eta_0)\sin^2(2\pi z/\lambda) + \kappa(2\pi a/\lambda)^2 \cos^2(2\pi z/\lambda) \right]
\tag{8}
$$

Since the volume integration averages over many wavelengths of the modulation, each sinusoidal square term contributes to the same factor $\frac{1}{2}$ and we have

$$
\Delta F_I = \frac{1}{3}\pi R^3 a^2 \left[f_0''(\eta_0) + 8\pi^2\kappa/\lambda^2 \right]
\tag{9}
$$

There will be a concomitant layer of modulated strain developed in the host matrix around the inclusion (figure 5). This modulated strain field will decay away from the surface of the inclusion because of destructive interference from the ups and downs of the modulation, in accordance with the St. Venant principle in elasticity, and vanish over a distance $\sim\lambda$ [16,17]. Within this layer, the host matrix will have the additional strain $a.\sin(2\pi z/\lambda)$ leading to an additional average strain energy density $\mu a^2/4$, where μ is the shear modulus of the host matrix which is assumed to be an isotropic material. The total change in the strain energy stored in the layer is therefore

$$
\Delta F_M = \pi R^2 \mu a^2 \lambda
\tag{10}
$$

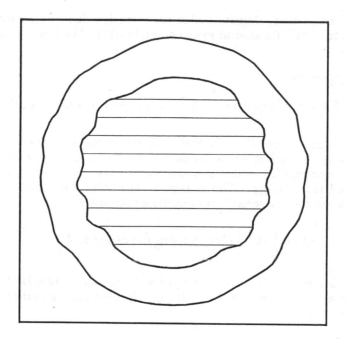

Figure 5. Schematic illustration of a spherical inclusion (inside) with a sinusoidal strain pattern superposed on a stress induced homogeneous strain η_0. The modulation leads to destructive interference of the strain field outside the inclusion which decays completely beyond the shell (outside the inclusions) of thickness of a wavelength of the modulation.

Adding this to equation 9, the total change in free energy as a result of the modulation is, to second order in the amplitude a,

$$\Delta F = \Delta F_I + \Delta F_M = \frac{1}{3}\pi R^3 a^2 \left[f_0''(\eta_0) + 8\pi^2 \kappa / \lambda^2 + 3\mu\lambda / R \right] \qquad (11)$$

For any amplitude a, the most energetically favorable wavelength λ corresponds to $\partial \Delta F / \partial \lambda = 0$, i.e.,

$$\lambda_m^3 = \frac{16\pi^2 \kappa}{3\mu} R \qquad (12)$$

For this particular wavelength, the total change in free energy is

$$\Delta F = \frac{1}{3}\pi R^3 a^2 \left[f_0''(\eta_0) + \beta / R^{\frac{2}{3}} \right] \tag{13}$$

where $\beta = \left(486\pi^2\mu^2\kappa\right)^{\frac{1}{3}}$. Provided that the curvature f_0'' at η_0 is sufficiently negative, the criterion for instability with respect to strain modulation is

$$R > R_c = \sqrt{486\kappa}\ \pi\mu/(-f_0'')^{\frac{3}{2}} \tag{14}$$

The precise geometry of the free energy surface represented by equation 3 and depicted in figure 4 changes with external parameters (e.g., temperature and hydrostatic pressure) in a rather complex manner. To make the task of obtaining a numerical estimate of the critical radius R_c simple, we shall assume that reasonable parametric dependence of the curvature f_0'' can be obtained from consideration of the planar section $\phi = 0$ (figure 3) alone. The largest possible negative curvature of $f_0(\eta)$ corresponds to one of the two maxima of equation 4 which, by setting to zero the first derivative $f_0'(\eta)$, can be obtained as

$$\eta = \frac{B}{3C}\left(1 - \sqrt{x}\right) \tag{15a}$$

where

$$x = 1 - 3AC/B^2 \tag{15b}$$

Substituting equations 15a and 15b into the second derivative $f_0''(\eta)$ of equation 4, we find that the largest possible negative curvature is

$$-f_0'' = \frac{8B^2}{3C}\left(\sqrt{x} - x\right) \tag{15c}$$

If we assume that only the second order coefficient A in equations 3 and 4 carry sensitive temperature dependence which can be linearized about the lower critical temperature T_c as

$$A(T) = A'(T - T_c) \tag{16a}$$

where

$$A' = \left(\frac{dA}{dT}\right)_{T_c}$$

then by differentiation of equation 4, we find at the upper critical temperature T^*,

$$A(T^*) = \frac{B^2}{3C} \qquad\qquad (16b)$$

and at the coexistence temperature T_0,

$$A(T_0) = \frac{B^2}{4C} \qquad\qquad (16c)$$

Substituting equations 16a, 16b and 16c into equation 15, we find that the negative curvature $-f_0''$ is zero at T^* and, with decreasing T, rises sharply first and then much more gradually reaching a maximum value of $2B^2/(3C)$ at T_0.

The above treatment addresses only the onset of modulated instability superposed on a homogeneously strained inclusion. To predict the final form of the product, one would have to carry out a numerical computation of the kinetics of evolution based on irreversible statistical thermodynamics. However, even without such an investigation, one can draw the following conclusions from the incipient instability:

(1) Since T^* is the temperature above which the curvature f_0'' ceases to be negative, for a given t inclusion, however large, no macroscopic strain η_0 can induce it to transform into a modulated pattern of m domains. Above T^*, the extrinsic stress can only switch some of the original t domains in the inclusion into ones favored by the biasing stress.

(2) Somewhat below T^*, when the curvature f_0'' becomes sufficiently negative at η_0, inclusions exceeding the critical size of radius R_c will transform into a modulated pattern while those below will not.

(3) Since the curvature f_0'' is temperature dependent, being zero at T^* and attaining the largest negative value at T_0, the critical radius R_c varies with temperature, being infinite at T^* and, with decreasing T, drops sharply first and then much more gradually, reaching the smallest value at T_0.

(4) If the incipient instability leads finally to a twin band of the m phase inside the transformed inclusion, then the periodicity λ_m of the band varies with the radius R of the inclusion as $\lambda_m^3 \sim R$.

THE REVERSE M→T TRANSFORMATION AFTER STRESS IS RELEASED

In the temperature regime for which the t phase is stable, an inclusion with a twin band of m domains previously formed under the influence of an extrinsic stress is thermodynamically metastable after the stress is released. Each m domain, the free energy density of which is only a local minimum, will eventually transform back to the t phase, the free energy density of which is a global minimum. However, these m domains are stable against small amplitude perturbation of large spatial extent (i.e., type II of Gibbs) but unstable only against much larger amplitude perturbation of small spatial extent (i.e., type I of Gibbs). Since the precursors to the former perturbation, the harmonic lattice waves, are always present while those for the latter, the anharmonic solitary waves, are rare except in the vicinity of extended defects where the harmonic lattice waves can pile up to form solitary waves, the reverse m→t transformation which proceeds via the growth of critical nuclei may have a rather long incubation time. The reverse process is possible provided that the host matrix surrounding the inclusion has not undergone any irreversible change that can retain some residual stress when the inclusion originally transformed from the t phase into the m with the accompanying volume expansion.

DISCUSSION

It is appropriate to discuss how the theoretical results may be related to an actual material under hypothetical service condition. The free energy function equation 4 with the gradient term in equation 6 is isomorphic to the Landau-Ginzburg-Devonshire generic type [18] of even power free energy expansion with a negative fourth order coefficient which can lead to a first order transformation. For such a function, it is straightforward to derive the following measurable quantities in terms of the parameters A′, B, C, and κ:

(1) latent heat at $T_0 = A' BT_0/(2C)$,

(2) surface energy at $T_0 = \dfrac{1}{8}B^2\sqrt{\kappa/C^3}$,

(3) discontinuous jump of self-strain $\Delta\eta = 2B/(3C)$, and

(4) temperature gradient of the square of the self-strain
    $d\eta^2/dT = A'/(2B)$.

Data on these quantities for stabilized zirconias are not available. We shall instead illustrate the results with parameters for pure zirconia. The latent heat (1420 cal mole^{-1}) and the surface energy (360 erg cm^{-2}) can be obtained from reference 19. The temperature dependence of the self-strain η can be deduced from the x-ray data of reference 20 leading to $A'/(2B) \cong 5 \times 10^{-6}/°C$ and $2B/(3C) \cong 2.25 \times 10^{-2}$. From these we obtain the estimates $\kappa = 1.5 \times 10^{-2}$ erg cm^{-1}, $B = 10^{13}$ erg cm^{-3}, and $C = 2.5 \times 10^{14}$ erg cm^{-3}. Assuming that spherical inclusions of pure zirconia are dispersed in a hypothetical matrix with a shear modulus of 10^{12} erg cm^{-3}, we have for the critical radius of a transformable inclusion $R_c \cong 0.6$ μm at the largest negative curvature

$-f_0'' = 2B^2/(3C)$ with a corresponding stress induced strain $\eta_0 = \sqrt{B/(6C)} \cong 0.08$. The wavelength of the modulation is $\lambda_m = 0.05$ µm for an inclusion of radius $R = 1$ µm. These numerical results for constrained pure zirconia in a hypothetical matrix are rather reasonable compared to experimental measurements for stabilized zirconia in Al_2O_3 matrix [21]. A more stringent test would require measurements of the higher order elastic stiffness coefficients for the stabilized zirconias.

ACKNOWLEDGMENT

This work was supported by the U.S. Department of Energy, BES-Materials Sciences, under contract #W-31-109-ENG-38.

REFERENCES

1.   Green, D.J., R.H.J.Hannink and M.V. Swain, 1989. Transformation Toughening of Ceramics, Boca Raton, Florida, USA: CRC Press.

2.   Heuer, A.H. and M.Rühle, 1985. "On the Nucleation of the Martensitic Transformation in Zirconia." Acta Metallurgica, 33:2101-2112.

3.   Chen, I.-W., Y.-H.Chiao and K.Tsuzaki, 1985. "Statistics of Martensitic Nucleation." Acta Metallurgica, 33:1847-1859.

4.   Garvie, R.C., 1985. "Thermodynamic Analysis of the Tetragonal to Monoclinic Transformation in a Constrained Zirconia Microcrystal." Journal of Materials Science, 20:3479-3486.

5.   Nevitt, M.V., S.-K.Chan, J.Z.Liu, M.H.Grimsditch and Y.Fang, 1988. "The Elastic Properties of Monoclinic ZrO_2." Physica B, 150:230-233.

6.   Chan, S.-K., Y.Fang, M.Grimsditch, Z.Li, M.V.Nevitt, W.M.Robinson and E.S.Zouboulis, 1991. "Temperature Dependence of the Elastic Moduli of Monoclinic Zirconia." Journal of the American Ceramic Society, 74:1742-1744.

7.   Nevitt, M.V., Y.Fang and S.-K.Chan, 1990. "Heat Capacity of Monoclinic Zirconia Between 2.75 and 350 K." Journal of the American Ceramic Society, 73:2502-2504.

8.   Chen, C.H. and S.-K.Chan. "Lattice Dynamics and Elastic Properties of Tetragonal Zirconia I. Dynamics of Harmonic Vibrations," to be published.

9. Chan, S.-K. and C.H. Chen. "Lattice Dynamics and Elastic Properties of Tetragonal Zirconia II. Elastic Properties from Long Wave Limit," to be published.

10. Chan, S.-K., 1988. "The Polymorphic Transformation of Zirconia." Physica B, 150:212-222.

11. Chan, S.-K., 1988. "Theory of the Energetics and Nonclassical Nucleation for the Tetragonal-Monoclinic Transformation of Zirconia," in Advances in Ceramics, Vol. 24, S.Sōmiya, N.Yamamoto and H.Yanagida, eds., Westerville, OH, USA: The American Ceramic Society, pp. 983-995.

12. Chan, S.-K., 1990. "Theory of the Energetics and Nonclassical Nucleation for Martensitic Transformations from Partially Soft Lattice Modes." Materials Science Forum, 56-58:101-106.

13. Garvie, R.C. and S.-K.Chan, 1988. "Stability Limits in the Monoclinic-Tetragonal Transformations of Zirconia." Physica B, 150:203-211.

14. Gibbs, J.W., 1961, in Scientific Papers of J. W. Gibbs, New York, NY, USA: Dover, pp. 105 and 252.

15. de Fontaine, D., 1979. "Configurational Thermodynamics of Solid Solutions," in Solid State Physics, Vol. 34, H.Ehrenreich, F.Seitz and D.Turnbull, eds., New York, NY, USA: Academic Press, pp. 73-274.

16. Roitburd, A.L., 1978. "Martensitic Transformation as a Typical Phase Transformation in Solids," in Solid State Physics, Vol. 33, H.Ehrenreich, F.Seitz and D.Turnbull, eds., New York, NY, USA: Academic Press, pp. 317-390.

17. Khachaturyan, A.G., 1983. Theory of Structural Transformations in Solids, New York, NY, USA: Wiley.

18. Chan, S.-K., 1990. "Some Theoretical Aspects of Displacive Transformations in Ceramic Oxides," in Science of Advanced Materials, H.Wiedersich and M.Meshii, eds., Materials Park, Ohio, USA: ASM International, pp. 387-405.

19. Garvie, R.C., 1965. "The Occurrence of Metastable Tetragonal Zirconia as a Crystallite Size Effect." Journal of Physical Chemistry, 69:1238-1243.

20. Patil, R.N. and E.C.Subbarao, 1969. "Axial Thermal Expansion of ZrO_2 and HfO_2 in the Range Room Temperature to 1400°C." Journal of Applied Crystallography, 3:281-288.

21. Heuer, A.H., N.Claussen, W.M.Kriven and M.Rühle, 1982. "Stability of Tetragonal ZrO_2 Particles in Ceramic Matrices." Journal of the American Ceramic Society, 65:642-650.

Defect Chemistry, Phase Stability and Properties of Zirconia Polycrystals

PARIS KOUNTOUROS and GÜNTER PETZOW

ABSTRACT

Recent investigations indicate that oxygen ion vacancies play an important role in the phase stability, phase transformation, and properties of zirconia polycrystals. The effective double positive charged oxygen vacancies are considered as quasi-chemical species which contribute in the same manner to the stability of zirconia as low-valent cations. The oxygen vacancy concentration is primarily responsible for the metastability of the t- and c-ZrO_2 phases at room temperature. Variation of the oxygen vacancy concentration in the oxygen sublattice of zirconia can be used to control the stabilization of any single phase at room temperature. Decreasing the oxygen vacancy concentration leads, for example, to a breakdown of the c-phase which transforms to the t- or m-phase. The oxygen vacancies appear to be closely involved with the grain growth. The results indicate that the single zirconia phases are stabilized in part by a crystal-chemical mechanism and in part by a physical mechanism. Phase stability mechanisms and the correlation between oxygen vacancies, phase stability, phase transformation, and properties will be discussed.

INTRODUCTION

Zirconia polycrystals are one of the most interesting and at the same time demanding ceramic materials. The excellent mechanical properties at low temperature and the excellent electrical properties at high temperature of the metastable tetragonal phase provide promising applications for this material. Research efforts have mainly concentrated for years on improvement of these properties and on understanding the nature and specific manipulation of this complex material. However, one of the questions that is not yet well understood, is the stabilization of the high-temperature structures, tetragonal (t) and cubic (c), at low temperatures. Normally, the monoclinic (m) structure is the thermodynamically stable phase at room temperature.

The metastability of the high-temperature phases at low temperatures either in powder state or in compact form were proposed by different theories, hypotheses, and explanations. Starting from the thermal decomposition of zirconium salts and of amorphous hydrated zirconia, Davis [1] and Srinivasan [2,3] attribute the

Paris Kountouros and Günter Petzow, Max-Planck-Institut für Metallforschung

Institut für Werkstoffwissenschaft, Pulvermetallurgisches Laboratorium

W-7000 Stuttgart 80, Federal Republic of Germany

metastability of t-ZrO_2 to the source of zirconium salt, the pH-value during precipitation and additionally to the rate of precipitation [3]. Livage *et al.* [4] and others [5,6] considered the metastability of the t-phase to be related to the structural similarity between the amorphous and crystalline state. Ruff and Ebert [7] and others [5, 8-12] proposed that anion impurities such as $(Cl)^-$, $(SO_4)^{2-}$, and $(NO_3)^{2-}$, depending on the anion-containing precursor, the OH^- anions from the amorphous hydrated zirconia, or cation impurities from the base, such as Na^+, cause the metastability of the t- or c-phase. Weber and Schwartz [13] mentioned that the particle size has a considerable influence on the t→m transformation temperature, while Krauth and Meyer [14] and Garvie [15-17] developed a concept of t-ZrO_2 metastability that was based upon surface free energy arguments. They estimated that the critical crystallite size for the stabilization of the t-phase is about 170 Å [14] and 300 Å [15-17]. Crystallites larger than the critical value would transform to the stable m-phase. The metastability of the c-phase requires such a small crystallite size that only an amorphous state can be observed [15]. The addition of certain amounts of various divalent (e.g., Ca^{2+}, Mg^{2+}), trivalent (e.g., Y^{3+}, Sc^{3+}, Ln^{3+}) or tetravalent (e.g., Ce^{4+}) oxides of cubic symmetry allow the stabilization of t- and c-ZrO_2 at low temperatures. The heterovalent cations replace a part of the Zr^{4+} ions and create vacancies in the oxygen sublattice in order to maintain charge neutrality. However, Subbarao [18] suggested that size, charge and concentration of dopant cations may influence the stabilization. Sheu [19] believed that the metastability of the t-phase is related to the anisotropic thermal expansion behavior and not to the tetragonality suggested by Kim and Tien [20, 21]. Mitsuhashi *et al.* [22] and others [17, 23, 24] argued that the metastability of the t-phase depends upon the strain energy whereas Heuer *et al.* [23] and Yoshimura [25] postulate a dependence on kinetic factors at low temperatures. Morinaga *et al.* [26] pointed out that the displacement of the anions from the ideal positions induces a change in the electronic structure, i. e., covalency changes, which influence the stability. Livage *et al.* [4] and others [18, 27-31] suggested that the oxygen vacancies play an important role on the stabilization of the high temperature phases, in both doped and nonstoichiometric ZrO_2. However, some authors believe that the metastability is a simultaneous effect of several mechanisms. For example, Hannink *et al.* [32] maintain the metastability of the t-phase to be a balance between chemical, strain, and interfacial energies. Unfortunately, the controlling mechanisms responsible for the metastability of t- and c-ZrO_2 are still unknown.

The purpose of the present study was to investigate the importance of the oxygen anion vacancies on the stabilization and to clarify the factors influencing the stability of the high-temperature phases at low temperatures, comparing some of the proposed mechanisms with the presence of the anion vacancies. Furthermore, an attempt was undertaken to correlate the presence of the anion vacancies to natural effects and properties of doped ZrO_2.

EXPERIMENTAL PROCEDURE

The investigated powders were prepared by a coprecipitation technique as described in detail elsewhere [33]. Some compositions were prepared by conventional oxide mixing. The density of the sintered specimens was determined by Archimedes method. The low temperature corrosion experiments were carried out in an autoclave with distilled water at 200°C and 20 bar. The fracture toughness was evaluated by the indentation technique using the equation proposed by Anstis *et al.* [34]. For the crystal phase determination, the scans were taken in the 2Θ range 28° to 36° and 71° to 77° using CuKα radiation in steps of 0.02°.

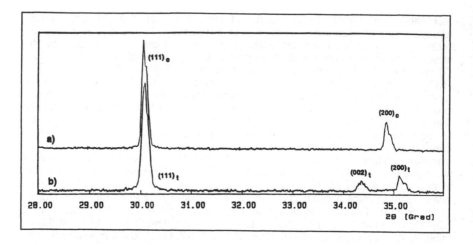

Figure 1. X-ray diffraction patterns of (a) 8 mol% Y_2O_3-ZrO_2 and (b) after doping with 6 mol% Nb_2O_5.

RESULTS AND DISCUSSION

STABILIZATION AND DESTABILIZATION OF t- AND c-ZrO_2

The influence of the extrinsic oxygen vacancies on the metastability of the high temperature t- and c-ZrO_2 was investigated by doping with aliovalent cations. Oxygen vacancies are introduced into the ZrO_2 lattice in order to maintain charge neutrality. For each mol% Y_2O_3 in ZrO_2, "one mol%" of vacancies is created. The vacancy concentration in the crystal lattice was controlled by specific doping with lower-valent (Y^{3+}) and higher-valent (Nb^{5+}) cations than the host Zr^{4+} cations. Using the nomenclature by Kröger and Vink [35] the substitutions can be formulated as follows:

$$Y_2O_3 \xrightarrow{ZrO_2} 2Y_{Zr}' + V_O^{\cdot\cdot} + 3O_O^x \tag{1}$$

$$Y_2O_3 + Nb_2O_5 \xrightarrow{ZrO_2} (2Y_{Zr}' + 2Nb_{Zr}^{\cdot})^x + 8O_O^x \tag{2}$$

The basic idea was that if the vacancy concentration actually plays a decisive role in the metastability, it would then be possible to transform the c-phase of ZrO_2 to the t- or m-phase due to the reduction of the vacancy concentration in the lattice. The opposite transformation from the m-phase to the t- and c-phase which appears with increasing Y_2O_3 concentration is well known. In fact, the c-phase of 8 mol% Y_2O_3-doped ZrO_2 was transformed to the t-phase after doping with 6 mol% Nb_2O_5 (Figure 1). It is possible to obtain the t-phase directly from coprecipitated powders, thus avoiding having to obtain the t-phase from the already existing c-phase by oxide mixing. In order to confirm this result, several Nb_2O_5-Y_2O_3-ZrO_2 compositions were prepared. The dopant concentration of Nb_2O_5 was kept 2 to 3 mol% less than the Y_2O_3 content. This difference was chosen to introduce the same amount of oxygen vacancies in ZrO_2 as with 2 to 3 mol% Y_2O_3 which are essential for the existence of the metastable t-phase. The

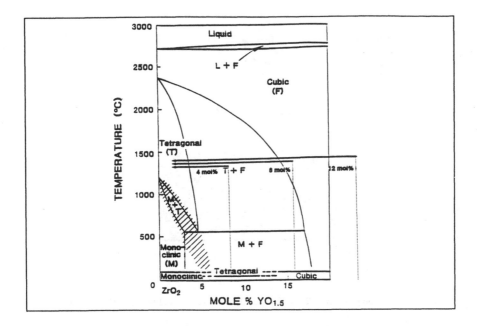

Figure 2. Phase diagram of the zirconia-rich region in the system Y_2O_3-ZrO_2 with the back shifted phases (arrows) after certain additions of Nb_2O_5. See text for explanation.

investigated Y_2O_3-ZrO_2 compositions extended from the t-phase region over the t+c- to the c-phase region. Starting from the Y_2O_3-doped t-ZrO_2, an amount of 1 mol% Nb_2O_5 in 2Y-TZP was enough to fracture the sample during cooling due to the t→m transformation. Doping with 0.5 mol% Nb_2O_5 transformed only 17% of the t-phase at 1350°C and 40% at 1450°C to the m-phase. As expected, the doping of 1 mol% Nb_2O_5 into 3Y-TZP does not transform the t-phase while 3 mol% Nb_2O_5 allows complete transformation to the m-phase. The appearance of the m-phase is attributed to the vacancy annihilation due to the charge compensation Y^{3+}/Nb^{5+} related to the host Zr^{4+} cations. In the case of doping with 1 mol% and 2 mol% Nb_2O_5 in 2Y-TZP the t-phase is "shifted back" to the thermodynamically stable m-phase region. The annihilation of "1 mol%" of vacancies by doping with 1 mol% Nb_2O_5 in 3Y-TZP shifted back the t-phase near to the t/t+m phase boundary with a vacancy concentration of "2 mol%" which corresponds to that of 2Y-TZP. Also, the doping of certain amounts of Nb_2O_5, depending on the Y_2O_3 content in ZrO_2, allows a shifting from t+c- and c-phase regions to the t- or m-phase regions (Figure 2). At approximately 16 mol% addition of Nb_2O_5 in 18 mol% Y_2O_3-doped ZrO_2, there is a small amount of niobate detected next to the dominant t-phase. This is a sign that the solubility of niobia in the zirconia lattice is reached. Kim [20] and Kim and Tien [21] found also that with increasing amounts of Nb_2O_5 and Ta_2O_5 in 2Y-TZP and 8Y-CSZ, the stability of the t- and c-phase decreases, respectively. These results were attributed to increasing tetragonality (c/a-ratio). However, the vacancy concentration also plays a decisive role in the metastability as well as the destabilization of t- and c-ZrO_2.

The reduction of vacancies in the oxygen sublattice and the simultaneous change of the structure lead to the conclusion that a certain number of oxygen

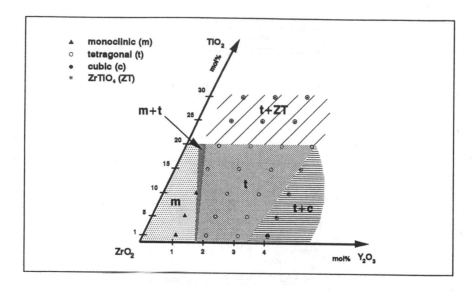

Figure 3. Phase composition of TiO_2-Y_2O_3-ZrO_2 samples sintered between 1300 and 1450°C up to 4 h.

vacancies are responsible for the metastability of t- and c-phase. In general, there is a *critical minimum and a critical maximum vacancy concentration* for each ZrO_2 phase. For example, the single t-phase can exist between the critical minimum (about "2 mol%") and maximum vacancy concentration (about "3 mol%"). Under the critical minimum, the thermodynamically stable m-phase occurs. The critical values for each single phase are expected to be as high as about "1.5 mol%" oxygen vacancies for the m-phase (critical maximum) and about "1.7 mol%" for the critical minimum of the t-phase. The critical maximum of the latter phase is approximately "3.3 mol%" and about "8 mol%" for the critical minimum of the c-phase. Assuming that the oxygen vacancies are homogeneously distributed, the sharp critical values between m- and t-phase as well as between t- and c-phase would lie in the two-phase regions m+t and t+c, and likely at the equilibrium temperature T_0^{m-t} and T_0^{t-c}, respectively. The equilibrium temperature means that both phases have the same free energy. However, for the sake of simplicity the critical maximum values for m- and t-phase and the critical minimum of the c-phase are chosen to be "2, 3, and 8 mol%", respectively.

The phase stability region of t-ZrO_2 in the system TiO_2-Y_2O_3-ZrO_2 for compacted samples sintered between 1300°C and 1450°C up to 4 h was determined as shown in figure 3. The solubility of TiO_2 in t-ZrO_2 was found to be approximately 20 mol% at 1450°C in contrast to about 15 mol% at 1400°C [36] and 1600°C [37] and 28 mol% at 1300°C [38]. Above the solubility limit, the orthorhombic $ZrTiO_4$ structure appears. However, the strongest peak (111) of the new phase is overlapped by the (111) peak of the t-phase. X-ray measurements with stringent monochromatized $CuK\alpha$ radiation for lattice constant determination is in progress to determine the solubility limit of TiO_2 in Y_2O_3-ZrO_2.

It has been found that an increased TiO_2 content in Y_2O_3-ZrO_2 has several effects. For example, the sintering temperature decreases, the grain size

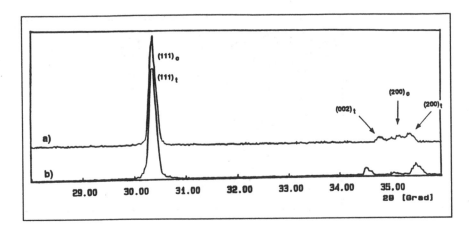

Figure 4. X-ray diffraction patterns of (a) 4 mol% Y_2O_3-ZrO_2 (t+c) and (b) after doping with 20 mol% TiO_2 (t).

increases, and the c-phase destabilizes (see phase boundary between t- and t+c-phase region in figure 3). These results confirm those found in [36] and [37]. Figure 4 shows the splitting of the cubic $(200)_c$ peak to tetragonal $(002)_t$ and $(200)_t$ peaks which shift with increasing TiO_2 content to lower and higher 2Θ values, respectively. In the case of the c-phase destabilization, the c-axis of the tetragonal structure increases while the a-axis decreases. This is quite interesting because the ionic radius of Ti^{4+} is smaller than that of Zr^{4+} (r_{Ti4+}=0.74 Å and r_{Zr4+}=0.84 Å for coordination number=8) [39]. The destabilization of the c-phase is attributed to the reduced coordination number [37]. Besides this explanation, an additional reason for the expansion of the c-axis and contraction of the a-axis with increasing TiO_2 content is believed to be due to the preferred repulsion between Ti^{4+} cations and the effective double positive charged oxygen vacancies in the [001] direction (it is known that the smaller the ion and the higher its charge, the higher the repulsion between the same charged component parts). The length change in both directions [100] and [001] suggests that the Ti^{4+} cations are not incorporated randomly in the lattice, but in preferred sites which cause a repulsion between the dopant cations and the vacancies. Consequently, the c-axis expands when the a-axis contracts. It seems that the dopant cations would be incorporated in energetically favourable sites. This explanation is supported by X-ray absorption nearedge spectroscopy (XANES) studies [40] on Ti-Y-TZP powders, where it was found that the Ti^{4+} cations doped in the ZrO_2 crystal lattice are not statistically distributed. For a better understanding, the immediate vicinity of an oxygen vacancy is shown schematically in figure 5. The host and/or dopant cations as the nearest neighbors to the previous oxygen (effective positive charged vacancy) move outwards because of the electrostatic repulsion between the same charged species. This repulsion can cause the c-axis of the lattice to increase. Conversely, the vacancy allows the surrounding oxygen ions to draw inward because of Coulomb attraction and simultaneous repulsion from the next nearest oxygens. The local bonding is disrupted and the immediate vicinity of the oxygen vacancy relaxes. The local lattice relaxation around the vacancy perturbs several lattice sites.

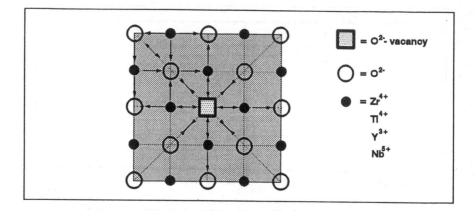

Figure 5. Lattice relaxation around anion vacancy in doped ZrO_2. See text for explanation.

STABILIZATION OF THE "LOW-TEMPERATURE" t- AND c-ZrO_2

It has been shown that the metastability of the t- and c-phase can occur after thermal decomposition of zirconium salts or amorphous hydrated zirconia in low temperature treatments. The metastable phases transform at about 800°C or higher to the stable m-phase. For example, the transformation of amorphous zirconia to the stable m-phase either via the metastable t-phase [1, 15] or via the metastable c- and t-phase [41] was observed also in Y_2O_3- and CaO-doped ZrO_2 [42, 43]. This was suggested to be a crystal size effect. However, to understand this, it is necessary to follow the process occurring upon heating of the amorphous hydrated zirconia. X-ray patterns of amorphous zirconia show a very broad peak around $2\Theta \approx 30°$ which is attributed to the extremely fine crystallites, which are most likely smaller than 40 Å (compare figure 2 in [15]). That the crystallites could be in a quasi-crystalline state is supported by the fact that the interionic Zr-Zr and the Zr-O distances in the amorphous phase are similar to the crystalline phase [4].

On heating, the amorphous hydrated powder undergoes several stages. At lower temperatures, between 200° and 300°C, the powder starts to slowly dehydrate. After this external dehydration, the powder is still amorphous. At about 400°C, the "crystal water" begins to be driven off from the amorphous powder and the powder gradually transforms to the crystalline state. Although this dehydration is a complex reaction process, it can be assumed that it proceeds in two steps. In the first step, in which the crystallization takes place, water molecules form by proton transfer between neighbouring hydroxyl groups ($OH^- + OH^- \rightarrow H_2O + O^{2-}$); the second step is the removal of these water molecules from the lattice (Figure 6). It is more likely that the dehydration process starts at the surface. The water molecules diffuse outwards from the quasi-lattice via the already-formed anionic vacancies on the surface, and leave the quasi-crystal in a gaseous state. The water molecules leave O^{2-} ions behind in the lattice and the origin of the OH^- sites as anion vacancies. It is assumed that these vacancies could be preserved because of the low temperature. This intermediate state of the "quasi-crystal" with a certain number of anion vacancies would stabilize the low-temperature c- and t-phase. Increasing the temperature and/or time, the process of the water removal, the reorientation of the already existing ions (Zr^{4+} and O^{2-}) and the disappearance of the new added component (vacancies) will be complete

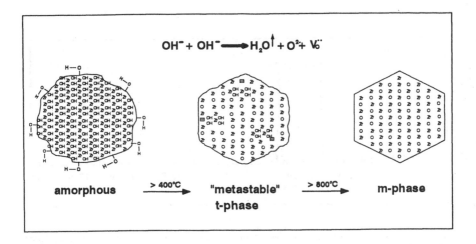

Figure 6. Low-temperature metastability of t-phase during crystallization from the amorphous phase. See text for explanation.

and the expected m-phase occurs. In the case of doped ZrO_2, the existence of anion vacancies is necessary in order to neutralize the charge differences.

The proton transfer (water formation) and diffusion to the surface could be delayed by a water skin on the powder surfaces which desorbs slower than water from the quasi-crystal interior supplies. Therefore, the influence of some parameters, such as precursors, pH value, precipitation rate etc., could delay or accelerate the amorphous → metastable phase → m-phase transformation. The calcination temperature and time, as well as the atmosphere (nonstoichiometry) would affect the crystalline phase-state. Consequently, the resultant structure is attributed to the gradual formation (c- and/or t-phase) and complete annihilation (m-phase) of the created anion vacancies in the crystal lattice, and not on the mechanisms proposed earlier [1-12].

The concept advanced by Krauth and Meyer [14] and Garvie [15-17] that the crystallites of t-ZrO_2 have a lower surface free energy than the m-phase is not universally accepted. For example, tetragonal microcrystals were found to be much larger, 2000 Å [22] or 900 Å [30], or monoclinic microcrystals smaller [2, 3, 44, 45] than the critical crystallite size (300 Å) for the t-phase given by Garvie. It is believed that the parameters just mentioned above also play a very important role in the microcrystal growth, and thus determine the crystal size. Therefore, the critical crystallite size does not have any influence on the metastability of the low-temperature t- and c-phases.

ORTHORHOMBIC STABILIZED ZrO_2 BY CATION VACANCIES?

It has been found that apart from the temperature-dependent phases, m, t, t' (nontransformable t-phase), and c, there is a pressure-dependent phase with orthorhombic (o) symmetry. The o-phase was not observed as an original phase in the bulk material, but only as dispersed particles in thin ZrO_2-films. Early work on the ZrO_2-rich end of the system Nb_2O_5-ZrO_2 [46] has shown that an anion-excess orthorhombic phase exists.

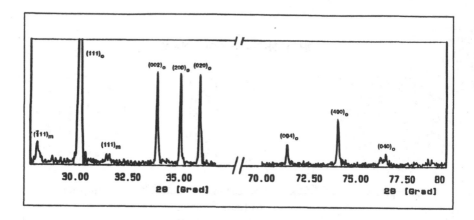

Figure 7. X-ray diffraction pattern of the "cation vacancies"-stabilized orthorhombic ZrO_2 doped with 10 mol% Nb_2O_5.

However, due to the doping of ZrO_2 with 10 mol% Nb_2O_5, it is possible to stabilize the high-pressure o-phase to room temperature and atmospheric pressure (Figure 7). In this case, each Nb^{5+} ion in the ZrO_2 crystal lattice substitutes for a Zr^{4+} cation and increases simultaneously the amount of positive charge. An interstitial position in the lattice as suggested by Kim et al. [47] in Nb_2O_5-Y_2O_3-ZrO_2 is unlikely because of the relative high ionic radius ($r_{Nb5+} = 0.74$ Å) [39], high charge valency, and the relatively close-packed structure. Furthermore, it is believed that the formation of interstitial Nb^{5+} is energetically unfavorable. For the charge balance, there are two possibilities: 1) creation of cation vacancies in the Zr^{4+} sublattice or 2) partial reduction of Nb^{5+} cations. EPR spectra do not give any sign of reduced Nb^{5+} in air-sintered and HIP'ed Nb_2O_5-Y_2O_3-ZrO_2 tetragonal compositions. Therefore, it is believed that the heterovalent cations substitute for the cation host positions and create cation vacancies in the cation sublattice in order to maintain charge neutrality. The substitution can be formulated as follows:

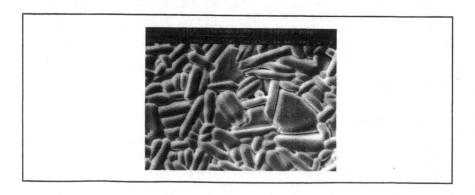

Figure 8. Scanning electron micrograph of the orthorhombic stabilized 10 mol% Nb_2O_5-ZrO_2.

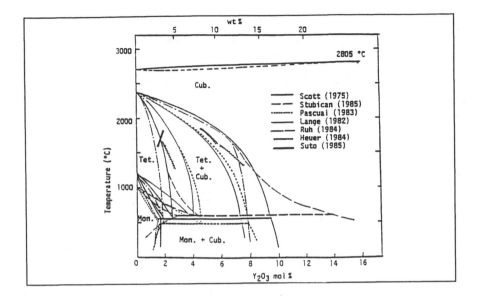

Figure 9. Phase diagrams of the zirconia-rich region in the system ZrO_2-Y_2O_3 (summarized by Yoshimura [25]).

$$2Nb_2O_5 \xrightarrow{ZrO_2} 4Nb_{Zr}{}^{\cdot} + V_{Zr}{}^{''''} + 10O_O{}^{x} \qquad (3)$$

The doped Nb^{5+} and the created cation vacancies are effective charged, $+1$ and -4, respectively, relative to the lattice. Figure 8 shows the microstructure of the 10 mol% Nb_2O_5-doped ZrO_2 sintered at 1650°C for 6 h.

PHASE EQUILIBRIA IN THE SYSTEM Y_2O_3 - ZrO_2

The homogeneous distribution of the dopant cations as well as the oxygen vacancies play a very important role in the nature of stabilized ZrO_2. For example, regions with low Y^{3+} content or low vacancy concentration favour the m-phase or promote the t→m transformation, whereas regions of high Y^{3+} content or high vacancy concentration tend to form the c-phase with large grains. Consequently, disagreements which exist in the precise position of the phase boundaries of the adjacent phases between m, m+t, t, t+c, and c in the Y_2O_3-ZrO_2 system (Figure 9) [25] are due to the "non-ideal" distribution of the oxygen vacancies as well as to the non-ideal dopant concentration. Also, discrepancies on the exact location of the equilibrium temperature T_0 exist due to the factors mentioned above.

CeO_2-STABILIZED ZrO_2

From the above conclusions, it is clear that the oxygen vacancies play a decisive role in the stabilization of the high temperature phases of ZrO_2. Normally, substitution of Zr^{4+} ions by Ce^{4+} ions would not create any oxygen

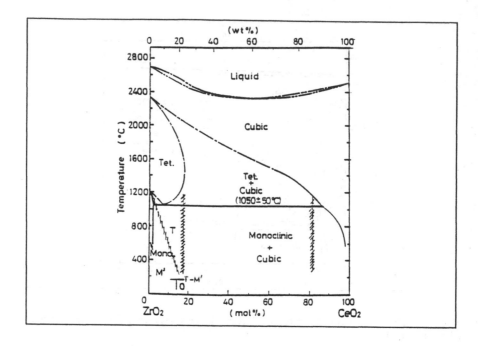

Figure 10. Phase diagram of the system ZrO_2-CeO_2 [48].

vacancies. Therefore, it appears that the stability of the t- and c-phase, by tetravalent Ce^{4+} dopant cations, are not affected by oxygen vacancies. However, the t-phase in the system ZrO_2-CeO_2 (Figure 10) [48] is extended up to 20 mol% CeO_2. The metastability of t-ZrO_2 at room temperature is possible by a minimum CeO_2 content of about 9 mol%, whereas the c-ZrO_2 occurs above about 20 mol% CeO_2. It has been found that CeO_2 either in pure or in solid solution with ZrO_2 can be reduced at high temperatures in reducing and in air atmosphere [49-51]. Theunissen et al. [50] and Schmid et al. [51] found that in CeO_2-stabilized TZP after sintering at 1400°C in air, Ce^{4+} ions partially reduced to Ce^{3+}.

Whatever the reason, the autoreduction of Ce^{4+} to Ce^{3+} is associated with the creation of oxygen vacancies for charge neutrality. Although the reduced amount of Ce^{4+} to Ce^{3+} in the ZrO_2 lattice is unknown, it can be assumed that a certain fraction for each mol% doped Ce^{4+} in ZrO_2 lattice is reduced to Ce^{3+}. The reduced Ce^{3+} cations increased with increasing CeO_2 content. The reason that t-ZrO_2 could not be stabilized down to room temperature with lower content than approximately 9 mol% CeO_2 and certain sintering temperature is, that the total amount of reduced Ce^{3+} cations is not enough to introduce the minimum number of oxygen vacancies (critical minimum), which is essential for the metastability. For example, the amount of reduced Ce^{3+} cations and resulting oxygen vacancies in 9 mol% CeO_2-doped t-ZrO_2 is believed to be approximately the same as the vacancy concentration resulting from 2 mol% Y_2O_3. Consequently, the number of oxygen vacancies increases with increasing CeO_2 content and number of reduced cations, and after reaching the content of about "8 mol%" oxygen vacancies, i. e. about 20 mol% CeO_2-doped ZrO_2, the c-phase occurs. Vacancy annihilation experiments, as shown above, confirm these facts. The

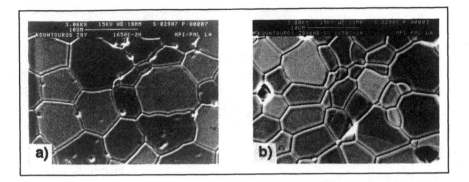

Figure 11. Scanning electron micrographs of 8 mol% Y_2O_3-ZrO_2 (a) before and (b) after doping with 6 mol% Nb_2O_5.

doping of 1 mol% Nb_2O_5 in 12Ce-TZP does not affect the t-phase while 3 mol% Nb_2O_5 transformed the t-phase.

The argument that the m-phase occurs after partial reduction of Ce^{4+} in t-ZrO_2 is favoured by the decrease of CeO_2 content [52] in 12Ce-TZP which was confirmed by Theunissen [50] and Schmid [51]. They found that the reduced cations segregate from the lattice into the grain boundaries.

OXYGEN VACANCIES, GRAIN GROWTH, AND c → t → m TRANSFORMATIONS

On the condition that Y_2O_3-doped ZrO_2 is chemically homogeneous, there is a variation of the grain size depending on the Y_2O_3 content. The grain size and growth rate increases with increasing Y_2O_3 content in the ZrO_2 lattice. The increasing amount of Y_2O_3 results in an increase of the vacancies in the oxygen sublattice and a decrease in the density. The oxygen anions due to this nonstoichiometry are more mobile at the sintering temperatures than the host and dopant cations. Figures 11a and 11b show the microstructure of 8 mol% Y_2O_3-stabilized c-ZrO_2 before and after doping with 6 mol% Nb_2O_5 sintered at 1650°C for 2 h. Although the sintering conditions were the same, it is obvious that the c-ZrO_2 grains (high vacancy concentration) are larger than the t-phase grains with lower vacancy concentration ("2 mol%"). The same observation was found in samples with lower and higher Y_2O_3 content after doping with Nb_2O_5. In the latter case, it seems that the grain size decreases more strongly than at lower contents. These results indicate that besides the cation type, amount, and sintering conditions, the oxygen vacancies as well as highly mobile oxygen ions play a decisive role in the grain size and growth of metastable t- and c-ZrO_2.

The simultaneous reducing vacancy concentration and grain size lead to the following conclusions. The critical vacancy concentration between the minimum and maximum is responsible for the metastability of the grains (t and c) and for their size. For example, an increase of the vacancy concentration from "2 mol%" to "8 mol%" leads to a breakdown of the t-phase and to a strong grain growth. The opposite effect could be reached if 8Y-ZrO_2 is doped with 6 mol% Nb_2O_5. Here, the c-phase completely destabilizes and the t-phase with smaller grains occurs (see also figure 1).

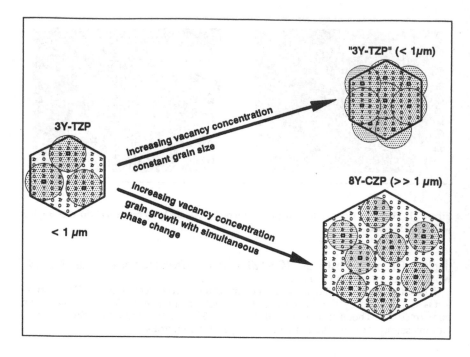

Figure 12. The influence of the increased oxygen vacancy concentration on a 3 mol% Y_2O_3-doped ZrO_2 grain. See text for explanation.

It is believed that each oxygen vacancy has an area of responsibility or affected area in the crystal lattice. The sum of these, dependent on their concentration, determine the phase stability (critical vacancy concentration) and the grain size of ZrO_2. Figure 12 shows a t-ZrO_2 grain with 3 mol% Y_2O_3 as well as "3 mol%" created oxygen vacancies (one square represents "1 mol%" vacancies). The affected/responsible area of each vacancy is represented by a shaded circle around the vacancy. Continuous increase of the vacancy concentration with constant grain size could be possible up to a certain concentration (maximum "vacancy solubility"). After exceeding the maximum concentration responsible for the t-phase, the phase as well as the grain size has to change because the small grain can not afford such high vacancy concentration. Therefore, after exceeding the maximum vacancy concentration, the t-phase transforms to the c-phase (continuous decrease of the Helmholtz free energy through the entropy contribution) and simultaneously the grain size increases (high "vacancy solubility"). In contrast to this explanation, the nontransformable t'-phase could be a tetragonal grain with a high oxygen vacancy concentration which normally occurs in the c-phase.

The opposite effect, i.e. a decrease of vacancy concentration of 8Y-ZrO_2 with constant grain size, leads to a breakdown of the c-phase because the affected/responsible area of the vacancies can not cover up the whole area. This suggests that after falling below the critical minimum vacancy concentration ("8 mol%"), the c-phase must transform to the t-phase with a simultaneous grain size decrease.

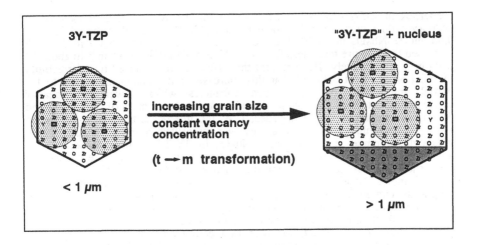

Figure 13. The influence of the increased grain size with constant vacancy concentration. See text for explanation.

On the other hand, increasing the grain size with a constant vacancy concentration ("3 mol%"), the grain could not remain tetragonal (Figure 13) because the affected/responsible area of the vacancies can not cover up the extended area and a nucleus is formed (dark shaded region). The nucleus could be an area of the crystal lattice in which the vacancy concentration is not high enough (lower than the critical minimum) in order to keep the phase metastable. Therefore, the nucleation (free energy) barrier ΔF^* is surmounted (see Figure 6 in [53]) and the grain transforms into the m-phase. A similar effect could be found in grains with inhomogeneous vacancy distributions where the higher the grain size, the higher the possibility of inhomogeneous regions. Therefore, the critical nucleus size is exceeded (ΔF^* is surmounted) and the grain transforms to the m-phase. This concept can be also applied to the c→t phase transformation.

The start of the martensitic transformation (M_s) is lowered down to room temperature if the critical minimum vacancy concentration exists (i. e., about "2 mol%"). The martensitic temperature increases with decreasing vacancy concentration. At the M_s of pure ZrO_2 (1150°C), there are no extrinsic vacancies. This transformation temperature can be suppressed below the temperature of liquid nitrogen if the vacancy concentration reaches the critical maximum of the t-phase.

In general, under the critical minimum vacancy concentrations, a phase transformation, c→t or t→m, occurs. Exceeding the critical maximum vacancy concentration, a change from m→t or t→c occurs. Therefore, the expression "critical grain size" for the t→m transformation must be treated as equivalent to the critical minimum vacancy concentration or declared invalid.

PROPERTIES

Empirical knowledge allows the conclusion that the mechanical properties depend strongly on microstructure, without it being clear how this improvement affects the properties. However, transformation toughening through the martensitic t→m transformation is an effective way to improve the structural properties in ZrO_2 ceramics. The fracture toughness of Y-TZP is known to decrease with increasing Y_2O_3 content. Kim [20, 47] found that the addition of 0.45 mol% Nb_2O_5 in 2Y-TZP and of 1.5 mol% Ta_2O_5 or Nb_2O_5 in 3Y-TZP increased the fracture toughness from 6 to 13.3 MPa $m^{1/2}$ and from 5 to about 14 MPa $m^{1/2}$, respectively. He attributed this effect to the enhanced transformability of the t-phase. It is believed that this increase in the toughness is due to the partial annihilation of the oxygen vacancies which lower the vacancy concentration and shift the t-phase around the critical minimum of the t-phase and the critical maximum of the m-phase, i. e. the m/t phase boundary. As shown above, the annihilation of the vacancies over the critical minimum of the t-phase yields the m-phase (e. g., 2 mol% Nb_2O_5 in 3Y-TZP). It is found that the amount of the solute cations affects, in a disadvantageous way, the mechanical properties. For example, recovering the t-phase from the pure c-phase by doping with 6 mol% Nb_2O_5 in 8 mol% Y_2O_3-ZrO_2 shows a very low fracture toughness, namely 2.1 MPa $m^{1/2}$, despite the approximately critical minimum of oxygen vacancies ("2 mol%"). Decreasing the amount of cations (Y^{3+} and Nb^{5+}) the fracture toughness continuously increases.

In contrast to this effect, the increased cation amount in the ZrO_2 lattice increases the resistivity against water attack at about 200°C and the t-phase is overstabilized. Degradation experiments have shown that increasing the TiO_2 content (>10 mol%) in 2 to 4 mol% Y_2O_3-ZrO_2 as well as Nb_2O_5 content (e. g., Z8Y6Nb) opposes the low temperature degradation. However, at low cation contents (2Y- or 3Y-TZP) the t-phase transforms to the m-phase. This is attributed to the partial annihilation of the oxygen vacancies due to their occupation, where the concentration falls short of the critical minimum of the t-phase. The occupation of the vacancies could take place as follow:

$$H_2O(g) + V_o^{\cdot\cdot} \rightarrow 2H_i^{\cdot} + O_o^{x} \tag{4}$$

$$\text{or} \quad H_2O(g) + V_o^{\cdot\cdot} + O_o^{x} \rightarrow 2(OH)_o^{\cdot} \tag{5}$$

In both reactions, the oxygen vacancy plays a decisive role. In the first reaction, the oxygen of the water gas occupies the vacancy and the hydrogen is taken interstitially (i) into the lattice. In the second reaction, it is possible that the water gas reacts with an oxygen vacancy and an oxygen of the lattice to form two hydroxyl ions which occupy both the vacancy and the oxygen site. In both reactions, the vacancy-occupying forces diffuse into the interior of the lattice. These results are supported by the findings that annealing degraded samples above the crystallization temperature (>400°C) allows continuous recovery of the t-phase with increasing temperature and time. In the case of higher cation content in t-ZrO_2 (e.g., increased TiO_2, >10 mol% or Z8Y6Nb), both reactions could be impeded due to the change of the bonding conditions in the lattice arising from the radii of the dopant cations.

The change of the bonding conditions could be one of the factors influencing the ionic conductivity of this overstabilized t-ZrO_2. It has been found that the ionic conductivity decreases with increasing cation concentration. In the case of Ti-Y-TZP materials, the activation energy of the total conductivity (bulk+grain boundary) increases with increasing TiO_2 content, namely from about 0.97 eV to about 1.16 eV (20 mol%) [54]. The activation energy of Z8Y6Nb lies in the same order of magnitude (1.16 eV). This result indicates that the ionic conductivity depends not only on the vacancy concentration, but also on cation type. The smaller radii and the higher charge of the dopant cations affect the interactions in the lattice and in the same manner the bonding conditions where the conductivity of the oxygen as well of the vacancies decreases.

The physical and chemical properties of stabilized ZrO_2 are determined not only from the microstructure, but also the dopant cations (type, amount, charge), oxygen vacancy concentration, and bond conditions, etc. The anisotropy of the crystal lattice as well as the local anisotropy caused by oxygen vacancies and inhomogeneities could play an important role in properties which are mostly characterized from the atomistic structure, although the compact polycrystalline body shows isotropic behavior. Therefore, an understanding of the properties of the stabilized ZrO_2 is closely related to an understanding of the atomistic structure. These actually control the grain size and grain shape (microstructure) and hence, the mechanical properties, the mobility of the oxygen ions (conductivity), and the attack, e.g., by water, etc., in stabilized ZrO_2 ceramics. The knowledge of these relations would allow the control of the structure of doped ZrO_2 through specific changes in the nonstoichiometry of the oxygen sublattice with certain aliovalent cations.

CONCLUSIONS

It has been shown that the extrinsic oxygen vacancies may be interpreted as quasi-chemical species whose contribution to the crystal lattice is essential for the metastability of the high temperature phases at low temperatures. They also have a decisive influence on the physical and chemical properties of ZrO_2.

ACKNOWLEGMENTS

The authors gratefully acknowledge Mrs. U. Täffner for specimen preparation and toughness measurements, and the Federal Ministry for Science and Technology (BMFT) for financial support (contract No. 03 M 2045).

REFERENCES

1. Davis, B. H., 1984. "Effect of pH on Crystal Phase of ZrO_2 Precipitated from Solution and Calcined at 600°C." J. Am. Ceram. Soc., 67(8):C-168.

2. Srinivasan, R., R. J. De Angelis and B. H. Davis, 1986. "Factors Influencing the Stability of the Tetragonal Form of Zirconia." J. Mater. Res., 1(4):583-588.

3. Srinivasan, R., R. J. De Angelis, M. B. Harris, S. F. Simpson and B. H. Davis, 1988. "Zirconium Oxide Crystal Phase: The Role of the pH and Time to Attain the Final pH for Precipitation of the Hydrous Oxide." J. Mater. Res., 3(4):787-797.

4. Livage, J., K. Doi and C. Mazieres, 1968. "Nature and Thermal Evolution of Amorphous Hydrated Zirconium Oxide." J. Am. Ceram. Soc., 51(6):349-353.

5. Nishizawa, H., N. Yamasaki and K. Matsuoka, 1982. "Crystallization and Transformation of Zirconia under Hydrothermal Conditions." J. Am. Ceram. Soc., 65(7):343-346.

6. Tani, E., M. Yoshimura and S. Somiya, 1983. "Formation of Ultrafine Tetragonal ZrO_2 Powder under Hydrothermal Conditions." J. Am. Ceram. Soc., 66(1):11-14.

7. Ruff, O. and F. Ebert, 1929. "Refractory Ceramics: I. The Forms of Zirconium Dioxide." Z. Anorg. Allg. Chem., 180: 19-41 (in German).

8. Clark, G. L. and D. H. Reynolds, 1937. "Chemistry of Zirconium Dioxide-X-Ray Diffraction Studies." Ind. Eng. Chem., 29(6):711-715.

9. Clabaugh, W. S. and R. Gilchrist, 1952. "Methods for Freeing Zirconium of Common Impurities and for Preparing Zirconium Sulfate and Oxide." J. Am. Chem. Soc., 74:2104.

10. Cyprès, R., R. Wollast and J. Raucq, 1963. "Contribution on the Polymorphic Conversion of Pure Zirconia." Ber. Dtsch. Keram. Ges., 40(9):527-532 (in German).

11. Clearfield, A., 1964. "Crystalline Hydrous Zirconia." Inorg. Chem., 3(1):146-148.

12. Whitney, E. D., 1965. "Kinetics and Mechanism of the Transition of Metastable Tetragonal to Monoclinic Zirconia." Trans. Faraday Soc., 61(9):1991-2000.

13. Weber, B. C. and M. A. Schwartz, 1957. "The Crystal Polymorphism of Zirconium Oxide and Their Suitability for High Temperatures." Ber. Dtsch. Keram. Ges., 34(12):391-396 (in German).

14. Krauth, A. and H. Meyer, 1965. "Modifications Produced by Quenching and Their Crystal Growth in the Systems Containing Zirconium Dioxide." Ber. Dtsch. Keram. Ges., 42(3):61-72 (in German).

15. Garvie, R. C., 1965. "Occurrence of Metastable Tetragonal Zirconia as a Crystallite Size Effect." J. Phys. Chem., 69(4):1238-1243.

16. Garvie, R. C., 1970. "Zirconium Dioxide and Some of Its Binary Systems," in Refractory Materials, Vol. 5-II, High Temperature Oxides, A. M. Alper, ed., New York, NY, USA: Academic Press, pp. 117-166.

17. Garvie, R. C., 1978. "Stabilization of the Tetragonal Structure in Zirconia Microcrystals." J. Phys. Chem., 82(2):218-224.

18. Subbarao, E. C., 1981. "Zirconia-An Overview," in Advances in Ceramics, Vol. 3, Science and Technology of Zirconia, A. H. Heuer and L. W. Hobbs, eds., Columbus, OH, USA: The American Ceramic Society, pp. 1-24.

19. Sheu, T.-S.. 1993. "Thermal Expansion Properties of Tetragonal Zirconia Polycrystals." Submitted to J. Am. Ceram. Soc.

20. Kim, D.-J., 1990. "Effect of Ta_2O_5, Nb_2O_5, and HfO_2 Alloying on the Transformability of Y_2O_3-Stabilized Tetragonal ZrO_2." J. Am. Ceram. Soc., 73(1):115-120 ; Kim, D.-J., 1988. "The Effect of Alloying on the Transformability of Y_2O_3 Stabilized Tetragonal ZrO_2." Ph.D. Dissertation. University of Michigan, Ann Arbor, MI.

21. Kim, D.-J. and T.-Y. Tien, 1991. "Phase Stability and Physical Properties of Cubic and Tetragonal ZrO_2 in the System ZrO_2-Y_2O_3-Ta_2O_5." J. Am. Ceram. Soc., 74(12):3061-3065.

22. Mitsuhashi, T., M. Ichihara and U. Tatsuke, 1974. "Characterization and Stabilization of Metastable Tetragonal ZrO_2." J. Am. Ceram. Soc., 57(2) 97-101.

23. Heuer, A. H., N. Claussen, W. M. Kriven and M. Rühle, 1982. "Stability of Tetragonal ZrO_2 Particles in Ceramic Matrices." J. Am. Ceram. Soc., 65(12):642-650.

24. Schubert, H. and G. Petzow, 1988. "Microstructural Investigations on the Stability of Yttria-Stabilized Tetragonal Zirconia," in Advances in Ceramics, Vol. 24A, S. Somiya, N. Yamamoto and H. Yanagida, eds., Columbus, OH, USA: The American Ceramic Society, pp. 21-28.

25. Yoshimura, M., 1988. "Phase Stability of Zirconia." Bull. Amer. Ceram. Soc., 67(12):1950-1955.

26. Morinaga, M., H. Adachi and M. Tsukada, 1983. "Electronic Structure and Phase Stability of ZrO_2." J. Phys. Chem. Solids, 44(4) 301-306.

27. Torralvo, M. J., J. Soria and M. A. Alario, 1982. "Non-Stoichiometry and the Glow Phenomenon in Zirconia Gels, " in the Proceedings of the 9th International Symposium on the Reactivity of Solids, Vol. 2, K. Dyrek, J. Haber and J. Nowotny, eds., Amsterdam, Holland: Elsevier, pp. 512-516.

28. Osendi, M. I., J. S. Moya, C. J. Serna and J. Soria, 1985. "Metastability of Tetragonal Zirconia Powders." J. Am. Ceram. Soc., 68(3):135-139.

29. Narita, N., S. Leng, T. Inada and K.Higashida, 1988. "Environmental Effect on Phase Stability in Y-TZP Ceramics," in Sintering '87, S. Somiya, M. Shimada, M. Yoshimura and R. Watanabe, eds., London, U.K.: Elsevier, Applied Science, pp. 1130-1135.

30. Badwal, S. P. S. and N. Nardella, 1989. "Formation of Monoclinic Zirconia at the Anodic Face of Tetragonal Zirconia Polycrystalline Solid Electrolytes." Appl. Phys., A49: 13-24.

31. Hillert, M., 1991. "Thermodynamic Model of Cubic→Tetragonal Transition in Nonstoichiometric Zirconia." J. Am. Ceram. Soc., 74(8):2005-2006.

32. Hannink, R. H. J., K. A. Johnston, R. T. Pascoe and R. C. Garvie, 1981. "Microstructural Changes During Isothermal Ageing of a Calcia Partially Stabilized Zirconia Alloy," in [18], pp. 116-136.

33. Kountouros, P. N. and H. Schubert, 1991. "Bloating Effect During Sintering of TZP," in Proceedings of the Second Conference of the European Ceramic Society, Augsburg, Germany. In press.

34. Anstis, G. R., P. Chantikul, B. R. Lawn and D. B. Marshall, 1981. "A Critical Evaluation of Indentation Techniques for Measuring Fracture Toughness: I, Direct Crack Measurements." J. Am. Ceram. Soc., 64(9):533-538.

35. Kröger, F. A. and H. J. Vink, 1956. "Relation Between Concentration of Imperfections in Crystallite Solids," in Solid State Physics, Vol. 3, F. Seitz and D. Turnbull, eds., New York, USA: Academic Press, pp. 307-435.

36. Bannister, M. J. and J. M. Barnes, 1986. "Solubility of TiO_2 in ZrO_2." J. Am. Ceram. Soc., 69(11):C-269-271.

37. Lin, C. L., D. Gan and P. Shen, 1990. "The Effects of TiO_2 Addition on the Microstructure and Transformation of ZrO_2 with 3 and 6 mol% Y_2O_3." Mat. Sci. Eng., A129: 147-155.

38. Pyda, W., K. Haberko, M. M. Bucko and M. Faryna, 1991. "The Study of Tetragonal Polycrystals in the System TiO_2-Y_2O_3-ZrO_2," in [33].

39. Shannon, R. D., 1976. "Revised Effective Ionic Radii and Systematic Studies of Interatomic Distances in Halides and Chalcogenides." Acta Cryst., A32(5):751-767.

40. Zschech, E., P. Kountouros, G. Petzow, P. Behrens, A. Lessmann and R. Frahm, 1993. "Synchrotron Radiation Ti-K XANES Study of TiO_2-Y_2O_3-Stabilized Tetragonal Zirconia Polycrystals." J. Am. Ceram. Soc., 76(1): In press.

41. Mazdiyasni, K. S., C. T. Lynch and J. S. Smith,1966. "Metastable Transitions of Zirconium Oxide Obtained from Decomposition of Alkoxides." J. Am. Ceram. Soc., 49(5):286-287.

42. Haberko, K., A. Giesla and A. Pron, 1975. "Sintering Behavior of Yttria-Stabilized Zirconia Powders Prepared from Gels." Ceramurgia Intern., 1(3):111-116.

43. Pyun, S.-I., H.-J. Jung and G.-D. Kim, 1987. "The Stability of Low-Temperature Metastable Cubic and Tetragonal Phases of Zirconia," in High Tech Ceramics, PART A, Materials Science Monographs, 38A, P. Vincenzini, ed., Amsterdam, Holland: Elsevier, pp. 271-280.

44. Porter, D. L., A. G. Evans and A. H. Heuer, 1979. "Transformation-Toughening in Partially-Stabilized Zirconia (PSZ)." Acta Metall., 27(10):1649-1654.

45. Morgan, P. E. D., 1984. "Synthesis of 6-nm Ultrafine Monoclinic Zirconia." J. Am. Ceram. Soc., 67(10):C-204-205.

46. Phase Diagrams for Ceramists, 1975, E. M. Levin and H. F. McMurdie, eds., Columbus, OH, USA: The American Ceramic Society, Figure 4457.

47. Kim, D.-J., P. F. Becher and C. R. Hubbard,1993. "Effect of Nb_2O_5 on Anisotropic Thermal Expansion of 2 mol% Y_2O_3 Stabilized Tetragonal ZrO_2." Submitted to J. Am. Ceram. Soc.

48. Tani, E., M. Yoshimura and S. Somiya, 1983. "Revised Phase Diagram of the System ZrO_2-CeO_2 Below 1400°C." J. Am. Ceram. Soc., 66(7):506-510.

49. Sorensen, O. T., 1981. Nonstoichiometric Oxides, O. T. Sorensen, ed., New York, USA: Academic Press, pp 1-59.

50. Theunissen, G. S. A. M., A. J. A. Winnubst and A. J. Burggraaf, 1992. "Effect of Dopants on the Sintering Behaviour and Stability of Tetragonal Zirconia Ceramics." J. Eur. Ceram. Soc., 9(4): 251-263.

51. Schmid, H. K., R. Pennefather, S. Meriani and C. Schmid, 1992. "Redistribution of Ce and La During Processing of Ce(La)-TZP/Al_2O_3 Composites." J. Eur. Ceram. Soc., 10(5):381-392.

52. Heussner, K.-H. and N. Claussen, 1989. "Strengthening of Ceria-Tetragonal Zirconia Polycrystals by Reduction-Induced Phase Transformation." J. Am. Ceram. Soc., 72(6):1044-1046.

53. Rühle, M. and A. H. Heuer, 1984. "Phase Transformations in ZrO_2-Containing Ceramics:II, The Martensitic Reaction in t-ZrO_2," in Advances in Ceramics, Vol. 12, N. Claussen, M. Rühle and A. H. Heuer, eds., Columbus, OH, USA: The American Ceramic Society, pp. 14-32.

54. Kountouros, P., H. Schubert, A. Kopp, H. Näfe and W. Weppner, 1993. "Ionic and Electronic Conductivity of TiO_2-Y_2O_3-Stabilized Tetragonal Zirconia Polycrystals." This conference.

The Tetragonal-Monoclinic Transformations of Zirconia Studied by Small Angle Neutron Scattering and Differential Thermal Analysis

Z. LI, J. E. EPPERSON, Y. FANG and S.-K. CHAN

ABSTRACT

The tetragonal-monoclinic transformations of zirconia have been studied on pristine single crystals and on their cycled crystallites. Two complementary techniques have been used. Small angle neutron scattering experiments were carried out to monitor the degree of completion of a transformation under equilibrium conditions for collections of 20–30 large crystals using the total internal and external surface area as an indicator. Differential thermal analysis experiments were carried out on smaller single-domain crystals of different sizes individually during heating and cooling to measure the rates of latent heat absorption and emission. The investigation establishes the upper limit of stability of the monoclinic phase, the lower limit of stability of the tetragonal phase, and the coexistence temperature between the two phases. The characteristics of the transformations are also inferred from these experiments.

INTRODUCTION

The tetragonal (t)–monoclinic (m) transformation of zirconia has been of intense scientific and technological interests since the seminal paper of Garvie, Hannink, and Pascoe [1] showed that fine metastable tetragonal zirconia particles acted as toughening agents in a stable cubic zirconia matrix. In spite of the amount of work done on the material and the understanding attained [2,3], a number of issues remain unresolved [2] in part because most experiments have been performed on polycrystalline samples and some of the results might have been masked by extrinsic factors arising from grain boundaries, voids

Z. Li, J. E. Epperson, Y. Fang, and S.-K. Chan, Materials Science Division, Argonne National Laboratory, 9700 S. Cass Avenue, Argonne, IL, 60439-4838, U.S.A.

and impurities. The availability of sufficiently large single crystals grown by the flux method [4] opens up new opportunities of carrying out experiments that can be free of the influence of extrinsic factors. In this paper, we report results of investigation of (1) the temperature fields of the t→m and the m→t transformations, (2) the characteristics of the nucleation events (i.e., whether random or sample-size dependent), and (3) the nature of the martensitic transformations (i.e., whether isothermal or athermal) based on the use of unconstrained single crystals. Two techniques have been used to complement each other in the pursuit of the aforementioned goals. One involves the use of small angle neutron scattering (SANS) under equilibrium conditions to monitor the amount of internal and external surfaces at selected constant temperatures. The manner in which the total surfaces change at equilibrium temperature steps provides information on the temperature fields of the forward and reverse transformations. The other involves differential thermal analysis (DTA) that measures heat emission or absorption during heating or cooling through the first order transformation. The degree of abruptness of the emission or absorption and the temperature at which it occurs provide clues on the nature of the transformations.

GROWTH OF SINGLE CRYSTALS

Single crystals of m ZrO_2 were grown from a PbF_2/KF flux solution. The composition of the batch material (in mole %) was 10% ZrO_2, 80% PbF_2 and 10% KF. The mixture was placed in a platinum crucible and heated to 1050°C. After being kept at this temperature for 300 hours, the melt was slowly cooled to 850°C at the rate of 1° per hour. The single crystals so grown were generally in the form of thick plates with the largest face in the (100) crystallographic plane. Most crystals are of dimensions of about $3 \times 3 \times 1$ mm^3. Most of these are of single domain. Much larger multidomain single crystals ($10 \times 5 \times 3$ mm^3) can also be obtained.

SANS EXPERIMENTS

Unlike x-ray and electron probes which sample the outer part of a crystal, neutrons penetrate and sample the entire bulk. Furthermore, being scattered by the nuclei rather than by the electrons (as would be the case for x-rays), neutrons are as sensitive to oxygen as zirconium. For scattering angle 2θ very near the forward direction, the scattered intensity S(Q) as a function of wave vector $Q = 4\pi \sin \theta/\lambda$, λ being the wavelength of neutrons, is related to the successive moments of the distribution of scattering centers [5]. It peaks in the forward direction θ = 0 and, for randomly oriented crystals of arbitrary shapes, drops with increasing Q, approaching the "asymptotic slope" according to the Porod law [5,6]

$$S(Q) = 2\pi A(\Delta\rho)^2/Q^4 \qquad (1)$$

Here $\Delta\rho$ is the difference in the scattering length density between the crystal and the outside background (or between two parts of the crystal with a contrast in density separated by an interface) and A is the total external and internal surface areas that separate regions with the contrast $\Delta\rho$. A does not include domain walls because the contrast in density is zero on the opposite sides of such walls. The sum of the observed intensity $\int dQ\, S(Q)$ over the range of Q for which the Porod law holds provides a measure of the product $A(\Delta\rho)^2$. During the t–m transformations of ZrO_2, the volume changes by ~3% so that $(\Delta\rho)^2$ changes correspondingly by ~6%. If a single crystal transforms intact as one piece isothermally by having an interface moving very rapidly across the crystal, then the summed intensity before and after the transformation will be different by about 6%. On the other hand, if a single crystal transforms intact athermally (i.e., with the t and m phases coexisting over a range of temperature and the interface moving in step with change of temperature), then the summed intensity will increase gradually to a maximum when approximately half of the crystal has transformed and then decrease until the transformation is complete. Any large and monotonic increase in the summed intensity must come from a drastic increase in the total surface A (internal and external) resulting from the breaking up of the crystals and/or the opening of cracks because of volume mismatch during the martensitic transformation. Therefore, SANS can provide a means of monitoring the manner and the degree of completion of a martensitic transformation at each equilibrium temperature step using the change in scattering intensity corresponding to the change in total surface areas as an indicator.

The *in situ* measurements of the total surface areas generated during the t–m transformations from the increase in scattering intensities were carried out on three separate samples of ZrO_2 each weighing 16–18 g and containing 20–30 multidomain single crystals with linear dimension 5–12 mm. Samples 1 and 2 were packed loosely into soft fibrous Al_2O_3 holders without windows at the open ends. Sample 3 was packed loosely into a rigid boron nitride holder with amorphous quartz windows to contain the sample. The former configuration has the advantage that it corresponds nearer to the ideal situation of unconstrained crystals while the latter has the advantage that the crystals cannot move accidentally during the experiments. Each experiment was carried out at constant temperature maintained by heat transfer with helium gas in a resistance furnace. To obtain good statistics, scattering of neutrons generated from the Intense Pulsed Neutron Source at the Argonne National Laboratory, in pulses of 60–80 nsec durations at the rate of

30 pulses per sec and a flux of 4×10^4 neutrons cm^{-2} sec^{-1}, were counted for about 90 minutes in each constant temperature experiment. These experiments were therefore not meant to measure the kinetics of transformation but rather the degree of completion of transformation in a collection of 20–30 single crystals at a number of constant temperatures.

Figure 1 is a typical double logarithmic plot of S(Q) vs. Q (obtained from sample 1) showing the Porod law behavior of equation 1 with a constant slope of approximately –4 and an intercept that changes with temperature as the transformation occurs and reaches a saturated value when all crystals have been transformed. The sequence of events is as follows. The pristine m ZrO$_2$ single crystals were untransformed at 816°C and remained so at 1155°C. As the temperature was raised to 1176°C, m→t transformation occurred resulting in a drastic increase of surface area. The m→t transformation became completed before the temperature reached 1298°C. The reverse t→m transformation did not take place when the sample was cooled down to 1144°C. When it occurred at lower temperatures, further large increase in surface area was detected until it became complete when cooled to about 1004°C.

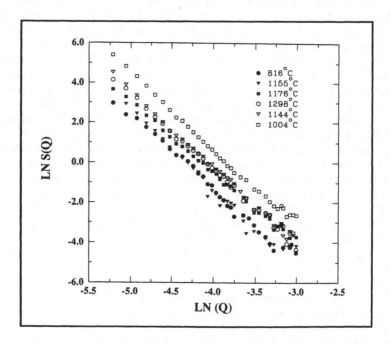

Figure 1. The scattering intensity S(Q) vs Q on logarithmic scales showing the Porod law behavior for various temperatures. Each temperature was maintained for 90 minutes to obtain reasonable statistics.

One notices from figure 1 that (1) there is a small departure from the Porod law for the low temperature (816°C) run manifested in the small bending at low Q and (2) the signal to noise ratio was generally poor at high Q.

The summed intensities are plotted in figure 2 which summarizes the results for all three samples. Sample 1 was heated up from room temperature through the regime of m→t transformation at a number of equilibrium temperatures. One notices (1) the flat initial trend indicating constant surface areas for the original crystals before transformation, (2) a regime of increasing surface areas for which some crystals (or parts of some crystals) have transformed into the t phase leading to cracking or breaking up that produced more surfaces, and (3) a flat final trend indicating constant but larger surface areas when all

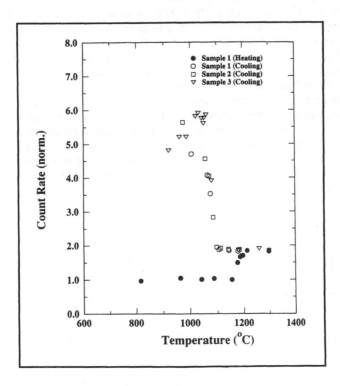

Figure 2. The summed intensity (normalized to the same time duration for each experiment) as a function of temperature for sample 1 during heating up and cooling down in equilibrium steps and for samples 2 and 3 during cooling down in equilibrium steps. Each point represents 90 minutes counting at an equilibrium temperature.

the crystals have transformed and no further increase in surfaces was possible. During cooling, the summed intensity traces the same upper plateau of the heating results but stays flat until a lower temperature is reached when the t→m transformation began. Thereafter the total surface areas increased again because of new cracking or breaking up until the t→m transformation was complete. Sample 2 was heated from room temperature in one step to 1300°C so that all the crystals were transformed into the t phase. It was then cooled down in equilibrium temperature steps and the neutron scattering counted at each constant temperature indicated. Sample 3, which was mounted in a different configuration, was also heated up in one temperature step and then cooled down in successive small steps similar to that for sample 2. Comparing the three sets of results, one notices that they are internally consistent showing a temperature regime from 1170 to 1205°C for the m→t transformation and one from 1100 to 1000°C for the t→m transformation with a possible error of a consistent shift of ±10°C (rather than a random one). The increase in intensities associated with the increase in total surface areas in these two temperature ranges indicates that the forward or reverse transformation was incomplete for the entire collection of crystals but does not answer the question of whether each individual crystal in a sample was fully or only partly transformed at each temperature. The amorphous quartz windows used to contain sample 3 were found after the experiments to have degraded somewhat leading to minor reduction in the transmission coefficient for neutrons. Most likely the degradation occurred towards the end of the series of experiments. The effect is reflected in the anomalous drop of the summed intensity for sample 3 at the three temperatures between 990 and 910°C.

DTA EXPERIMENTS

To determine whether or not each crystal was fully transformed at a given temperature within the two temperature ranges (i.e., whether the transformation was isothermal or athermal), thirteen single-domain single crystals of ZrO_2 of masses ranging from 3 to 120 mg were selected for their very high degree of perfection as indicated by their optical properties. DTA experiments were carried out on each crystal by heating at specified rates up to and beyond the temperature of transformation, then cooling at the same rate past the transformation and finally repeating the heating and cooling cycle again. The results are summarized in Table I for the first cycle and in Table II for the second. Figure 3 is an illustration of the DTA curves during the first and the second heating-cooling cycles for a typical crystal. For all the crystals investigated, the m→t transformation during the heating part of the first cycle took place isothermally within a time duration controlled by heat transport from the crystal to the furnace. The actual temperatures

TABLE I - DTA RESULTS OF PRISTINE SINGLE CRYSTALS

Sample	Weight (mg)	Rate (°C/min)	m→t			t→m		
			Peaks	Temperature (°C)	Time (sec)	Peaks	Temperature (°C)	Time (sec)
1	30.0	1	S	1181.2	10	M	1093–1059	10–25
2	16.8	1	S	1200.6	10	M	1108–1069	10–40
3	39.5	1	S	1191.4	10	M	1110–1067	15–60
4	95.3	1	S	1180.6	12	M	1098–1058	15–40
5	67.7	1	S	1187.1	12	M	1086–1060	10–30
6	43.7	1	S	1165.5	12	M	1088–1060	15–35
7	27.8	1	S	1187.1	10	M	1091–1058	15–40
8	9.0	1	S	1193.9	10	M	1106–1086	20–50
9	3.1	1	S	1189.2	7	M	1089–1066	10–15
10	3.3	0.2	S	1198.3	7	M	1113–1089	15–40
11	120.2	0.2	S	1189.9	15	M	1111–1071	15–40
12	64.3	20	S	1179.4	12	M	1095–1056	15–30
13	22.0	20	S	1183.8	10	M	1094–1059	15–35

S: Single Peak   M: Multiple Peaks   Time: from onset to top of a single peak

TABLE II - DTA RESULTS CYCLED CRYSTALS

Sample	Weight (mg)	Rate (°C/min)	m→t			t→m		
			Peaks	Temperature (°C)	Time (sec)	Peaks	Temperature (°C)	Time (sec)
1	30.0	1	M	1175–1188	10–25	M	1091–1074	15–40
2	16.8	1	M	1184–1191	15–25	M	1099–1094	15–40
3	39.5	1	M	1184–1188	10–25	M	1098–1061	15–50

S: Single Peak   M: Multiple Peaks   Time: from onset to top of a single peak

of transformation for the thirteen different crystals varied from 1181°C to 1201°C and were independent of sample masses. After the m→t transformation, the original single crystal apparently became a polycrystal with several grains of crystallites sticking together. During the cooling part of the first cycle, these crystallites transformed back into the m phase individually at different temperatures within the range 1115°C to 1060°C. The durations of transformation were somewhat longer for the t→m transformation than for the m→t. In the second cycle, the m→t transformation of each sample took place with several sharp absorption peaks, indicating that the crystallites that had transformed once before transformed again but at different temperatures. The number of m→t endothermic peaks in the second cycle appeared to be less than the number of t→m exothermic peaks in

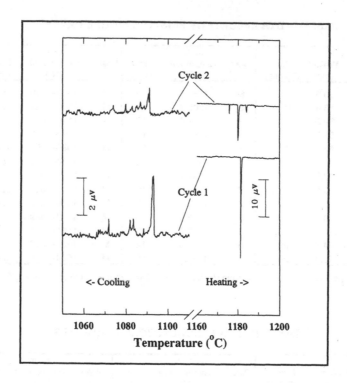

Figure 3. Right: Endothermic peaks at the m→t transformation during heating. Left: Exothermic peaks at the t→m transformation during cooling. The scales (in arbitrary units) are different for the heating (right) and the cooling (left) curves.

the first cycle. The reason is not clear. However, some of the small t→m exothermic peaks in the first cycle could have been noise. There was no correlation of the temperature of m→t transformations between the first cycle and the second for the same sample. Similarly, there was also no correlation of the temperature of the t→m transformations between the two cycles for the same sample.

CONCLUSIONS

The results of the investigations with SANS and DTA techniques point to the following conclusions:

(a) Each pristine single crystal of m ZrO_2 and each cycled crystallite of t ZrO_2 or m ZrO_2 transformed spontaneously at a temperature within the range 1170–1205°C for the m→t transformation and the range 1115–1000°C for the t→m transformation. The upper limit of stability of unconstrained m ZrO_2 is $T^* \sim 1205°C$ and the lower limit of stability of unconstrained t ZrO_2 is $T_c \sim 1000°C$. The mid-point between the starting temperatures 1170°C and 1115°C of the forward and the reverse transformations leads to a hypothetical coexistence temperature between the t and m phases of $T_0 = 1143°C$. These characteristic temperatures T^*, T_0 and T_c are, within the limits of experimental error (±10°C), consistent with what was previously suggested [7] of very asymmetric (1:3) temperature fields for superheating and supercooling based on theoretical considerations and collation of other experimental data.

(b) The DTA results indicate that a transformation, once started, would proceed to completion very rapidly, in agreement with an early investigation based on smaller single crystals [8]. There is no evidence of coexistence of the t and m phases within a pristine single crystal or a grain of a once-transformed polycrystal through a stationary or slowly moving interface.

(c) At least for the pristine single-domain m ZrO_2 crystals, there is no correlation between the mass of a crystal and its temperature of spontaneous transformation, insofar as all the crystals are of mm size. The question of what physical factor controls the nucleation and growth at a given temperature remains intriguing. Our conjecture is that each crystal has a somewhat different geometry with somewhat different stress concentrated regions. These buildups in stress concentrations, which depend on geometry and possibly also extended imperfections but not the mass of the crystal, provide the prerequisite for critical nuclei to grow.

(d) The SANS experiments indicate that some crystals or grains of polycrystals remained untransformed within the aforementioned temperature ranges for the entire duration of the neutron experiments (~ a day) and became transformed only when the temperature was increased to approach T^* or decreased to approach T_c.

ACKNOWLEDGMENT

This work, including the use of the Intense Pulsed Neutron Source, was supported by the U.S. Department of Energy, BES-Materials Sciences, under contract #W-31-109-ENG-38.

REFERENCES

1.   Garvie, R.C., R.H.J.Hannink and R.T.Pascoe, 1975. "Ceramic Steel." <u>Nature</u> (London), 258:703-705.

2.   Green, D.J., R.H.J.Hannink and M.V.Swain, 1989. Transformation Toughening of Ceramics, Boca Raton, Florida, USA: CRC Press.

3.   Heuer, A.H. and M.Rühle, 1985. "On the Nucleation of the Martensitic Transformation in Zirconia." Acta Metallurgica, 33:2101-2112.

4.   Nevitt, M.V., S.-K.Chan, J.Z.Liu, M.H.Grimsditch and Y.Fang, 1988. "The Elastic Properties of Monoclinic ZrO_2." Physica B, 150:230-233.

5.   Glatter, O. and O. Kratky, 1982. Small Angle X-ray Scattering, New York, NY, USA: Academic Press.

6.   Guinier, A., 1963. X-ray Diffraction in Crystals. Imperfect Crystals and Amorphous Bodies, London, UK: Freeman.

7.   Garvie, R.C. and S.-K.Chan, 1988. "Stability Limits in the Monoclinic-Tetragonal Transformations of Zirconia." Physica B, 150:203-211.

8.   Mitsuhashi, T. and Y.Fujiki, 1973. "Phase Transformation of Monoclinic ZrO_2 Single Crystals." Journal of American Ceramic Society, 56:493.

Phase Diagram Calculations for ZrO_2 Based Ceramics: Thermodynamic Regularities in Zirconate Formation and Solubilities of Transition Metal Oxides

HARUMI YOKOKAWA, NATSUKO SAKAI, TATSUYA KAWADA and MASAYUKI DOKIYA

ABSTRACT

Phase diagram calculations have been made using evaluated or estimated thermodynamic properties of stoichiometric compounds and solid solutions in the ZrO_2 – MO (M = Mg, Ca, Sr), the ZrO_2 – $M'O_{1.5}$ (M' = Sc, Y, La, Rare earth) and the ZrO_2 – $M''O_2$ (M" = Ti, Hf, Ce) systems. An emphasis has been placed on finding thermodynamic regularities in the stability of perovskites, pyrochlore–type compounds, and fluorite–type solid solutions. Interaction parameters for the fluorite solid solutions have been found to be well correlated with ionic radii. This has been utilized to calculate the solubilities of the transition metal oxides into Y_2O_3–stabilized ZrO_2 (YSZ). An associated solution model has been adopted to derive the respective contributions from the various valence states of the transition metal ions as functions of temperature and oxygen potential.

INTRODUCTION

Zirconia–based oxides have attracted much attention for technological applications such as a high temperature solid oxide fuel cell; Y_2O_3 stabilized ZrO_2 (YSZ) is used as electrolyte together with Ni/YSZ cermet and $(La,Sr)MnO_3$ electrodes. The chemical stability among the cell components is quite important. In fact, experimental investigations have revealed that some perovskite electrodes can react with YSZ to form zirconates such as $La_2Zr_2O_7$, $SrZrO_3$ or $CaZrO_3$ [1], and that the transition metal oxides, especially manganese oxide can dissolve in YSZ.

We have made a series of chemical thermodynamic analyses of materials problems associated with solid oxide fuel cells, particularly interactions between YSZ and perovskite oxide electrodes [2,3]. In the present study, we report results of phase diagram calculations on the ZrO_2 containing systems. A preliminary set of thermodynamic parameters which were used in our previous calculations [2] have been refined by taking into account correlations based on ionic radii [4]. After determination of the thermodynamic properties in the pseudobinary systems, those in the higher–order systems have then been estimated. In particular, phase relations in the ZrO_2–$YO_{1.5}$–MO_x (M = transition metal) systems have been calculated to

Harumi Yokokawa, Natsuko Sakai, Tatsuya Kawada and Masayuki Dokiya, National Institute of Materials and Chemical Research, Tsukuba Research Center, Tsukuba, Ibaraki 305, JAPAN

derive the solubilities of transition metal oxides into YSZ for the respective valence states. This makes it possible to rationalize the change in solubility among transition metal series in terms of the valence number and the ionic size.

CHEMICAL THERMODYNAMIC REPRESENTATION

Although some thermodynamic properties of zirconia–based ceramics have been investigated experimentally, there have not been accumulated enough data to provide Gibbs–energy derived phase relations. In earlier phase diagram calculations [5–7], many thermodynamic properties have been estimated so as to reproduce experimental phase relations. As suggested by Yoshimura [8], there remain some uncertainties in the ZrO_2–related phase diagrams below 1000°C; slow kinetic rates of diffusion and reaction may preclude the achievement of equilibrium. This implies that it will not be fruitful to make efforts in reproducing precisely low–temperature experimental phase relations by using many adjustable parameters in sophisticated solution models. We adopted the following approach of calculating phase diagrams:

(1) First, we adopted the thermodynamic properties of ZrO_2 given in IVTAN tables [9] and we also relied on experimentally determined thermodynamic properties for $CaZrO_3$, $SrZrO_3$, $La_2Zr_2O_7$, and ZrO_2–Y_2O_3 cubic solutions [10–12].

(2) A simple sub–regular solution model was adopted for representing mixing properties as follows:

$$\Delta_{mix}G = x_1 x_2 [A^0 + A^1 (x_1 - x_2)] + RT \{x_1 \ln x_1 + x_2 \ln x_2\} \tag{1}$$

where A^0 and A^1 are the first and the second coefficients of the interaction parameters. We adopted only the ideal mixing entropy and neglected the temperature dependence of A^0 and A^1. The lattice stability of constituent oxides MO_x in the particular phase, p, can be given as

$$G°(MO_x,p) - G°(MO_x,\text{ref. st.}) = a + b T \tag{2}$$

where the reference state was defined as the most stable state at 298.15 K for each oxide. The thermodynamic properties of a solution in a ZrO_2–MO_x system can be represented by four adjustable parameters, A^0, A^1, a and b.

(3) In order to maintain the consistency throughout the series of MO, $M'O_{1.5}$ and $M''O_2$, the thermodynamic parameters of solutions are assumed to change smoothly with ionic radius of constituent oxides; in this determination, the critical dopant sizes of Kim [4] for respective valence numbers were taken into account. Since selection of parameters a and b affects the choice of A^0 and A^1, we adopted a similar smooth change of a and b. For simplicity, we assumed $b = 0$ when possible.

(4) The transition metal ions may change their valence state also in the fluorite lattice. This effect can be taken into account as follows: First, the thermodynamic parameters were determined for the respective oxides, $MO_{n/2}$ (n = 2,3,4), separately on the basis of the correlation, and then these were used in the multicomponent chemical equilibria calculations using the CTC/SOLGASMIX program [13].

(5) The above approach seems to be quite appropriate for utilization in the complicated chemical equilibria calculations. The calculated equilibrium properties can be related to the valence state and the size of ions, and this makes it easy to extract the chemical meaning of reactions involved.

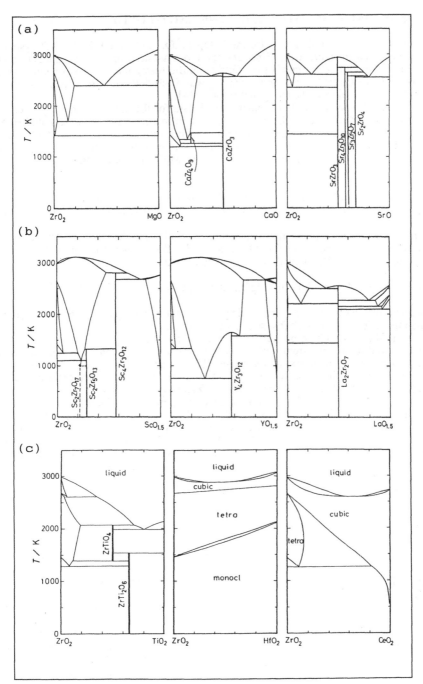

Figure 1. Calculated phase diagrams: (a) ZrO_2–MgO, ZrO_2–CaO and ZrO_2–SrO systems;(b) ZrO_2–$ScO_{1.5}$, ZrO_2–$YO_{1.5}$ and ZrO_2–$LaO_{1.5}$ systems;(c) ZrO_2–TiO_2, ZrO_2–HfO_2, and ZrO_2–CeO_2 systems.

PHASE DIAGRAMS WITH ALKALINE EARTH OXIDES

Calculated phase diagrams for the ZrO_2–MgO, ZrO_2–CaO, ZrO_2–SrO systems are presented in figure 1(a). Since the thermodynamic properties of $CaZrO_3$ have been well established, those of the cubic and the tetragonal phases were adjusted relatively to $CaZrO_3$, and then $CaZr_4O_9$ was included so as to appear between $CaZrO_3$, the cubic and the tetragonal phases [14].

In the absence of available experimental data for the thermodynamic properties in the ZrO_2–MgO system, the thermodynamic properties have been estimated so as to reproduce the eutectoid temperature and composition; that is, T_{eut} = 1679 K and x_{eut}(MgO) = 0.13. Since the cubic solution is in equilibrium with MgO and the tetragonal ZrO_2 containing one mole percent of MgO, the following thermodynamic relation can be written;

$$\mu°(MgO,cub) + RT \ln x + (1-x)^2 [A^0 + A^1(1-4x)] = \mu°(MgO,NaCl-type) \quad (3)$$
$$\mu°(ZrO_2,c) + RT \ln (1-x) + x^2 [A^0 + A^1(3-4x)] = \mu°(ZrO_2,t) + RT \ln(0.99) \quad (4)$$

Unknown values are interaction parameters, A^0 and A^1, in the cubic solution and the lattice stability of MgO in cubic phase, a and b. The former depends on the selection of the latter. We adopted the following estimates;

$$\mu°(MgO,cubic) - \mu°(MgO,NaCl-type) = a + bT = 75,000 \text{ J/mol} \quad (5)$$

where b was assumed to be zero. This provided interaction parameters less negative than those for the ZrO_2–CaO system (see figure 3).

The thermodynamic properties of $SrZrO_3$ and $Sr_{n+1}Zr_nO_{3n+1}$(N = 1,2,3) have been found to be consistent with other Ruddlesden Popper phases in the alkaline earth–transition metal–oxygen systems [10]. Since solubility data have not yet been clarified well in this system, the interaction parameters were determined by comparison with those in the analogous CaO and MgO systems.

The solubility depends both on the mixing properties of the cubic solution and on the stability of zirconates. The stability of the perovskite phases depends strongly on tolerance factors, in other words, on the radii of the alkaline earth ions [10]: $MgZrO_3$ is completely unstable, $CaZrO_3$ exists stably and $SrZrO_3$ is more stable. On the other hand, the interaction parameters show a moderate change with a maximum in the vicinity of the calcium ion (see figure 3). This explains why the most stable compositional region of the cubic phase appears in the CaO–ZrO_2 system.

PHASE DIAGRAMS WITH $MO_{1.5}$

Calculated phase diagrams for the ZrO_2–$ScO_{1.5}$, ZrO_2–$YO_{1.5}$, ZrO_2–$LaO_{1.5}$ systems are shown in figure 1(b). Interaction parameters of the cubic solid solution in the ZrO_2–$YO_{1.5}$ system were determined using EMF results [11]. Subsequently, the interaction parameters and the lattice stability of the tetragonal phase were determined to reproduce a temperature–composition line [8] in which the Gibbs energies of the cubic and the tetragonal phases are in accordance with each other. The present calculated stability field of the tetragonal phase does not extend down to 900 K in contradiction to experimental work. In addition, the calculated composition width for the cubic–tetragonal two–phase region is wider than experimental ones; however, no further adjustment was made. The thermodynamic properties of $Y_4Zr_3O_{12}$ were determined so as to reproduce its decomposition temperature.

The enthalpy change for formation of $La_2Zr_2O_7$ was measured by Korneev *et al.* [12]. In the ZrO_2–$LaO_{1.5}$ system, the cubic phase appears also in the $LaO_{1.5}$–rich

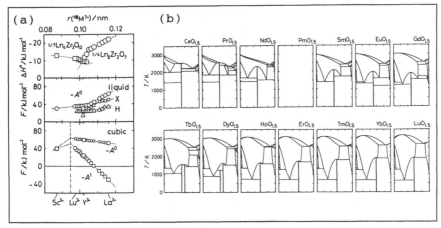

Figure 2. ZrO_2–$LnO_{1.5}$ (Ln = rare earth) systems: (a)Thermodynamic properties of double oxides and solid and liquid solutions in the ZrO_2–$LnO_{1.5}$ systems. (i) The stabilization energies of $Ln_2Zr_2O_7$ and $Ln_4Zr_3O_{12}$, (ii) interaction parameters for liquid, X and H phases; (iii) interaction parameters of cubic solution, A^0 and A^1,(b) Calculated phase diagrams.

region. Although significantly asymmetric interaction parameters are needed to reproduce this, moderately asymmetric values are adopted here.

There is no experimental work on the thermodynamic properties of the ZrO_2–$ScO_{1.5}$ system. The phase diagram was calculated using interaction parameters less negative than those for the ZrO_2–$YO_{1.5}$ system; the calculated stable field of the cubic phase is smaller than the experimentally determined one.

A correlation between interaction parameters and ionic radii in the ZrO_2–$M'O_{1.5}$ systems was used in calculations on the ZrO_2–$LnO_{1.5}$ (Ln = Rare earth) systems; estimation was also made of enthalpy changes for the high temperature phase transition of the rare earth oxides. Figure 2(b) shows calculated phase diagrams, while the thermodynamic data used are plotted in figure 2(a) as a function of ionic radius with 8 fold–coordination. With decreasing radius of rare earth ions, the pyrochlore phase stability decreases markedly, whereas the stability region of the cubic phase is widened in composition and in temperature. In the heavy rare earth systems, the stability of $Ln_4Zr_3O_{12}$ increases. The calculated phase diagrams essentially agree with experimental results.

PHASE DIAGRAMS WITH MO₂

Figure 1(c) shows the calculated phase diagrams for the ZrO_2–TiO_2, ZrO_2–HfO_2 and ZrO_2–CeO_2 systems. For the ZrO_2–TiO_2 system, the interaction parameter of the tetragonal phase was first assumed as $\omega = 10,000$ J/mol. Secondly, the thermodynamic properties of $ZrTiO_4$ and $ZrTi_2O_6$ relative to the cubic TiO_2 phase were determined so as to be consistent with experimental phase relations [35,36]. The lattice stability of cubic TiO_2 was determined as 16,000 J/mol; this can reproduce the decomposition temperature of $ZrTi_2O_6$ to ZrO_2 and rutile around 1200 °C. Here, we neglected the solubility of ZrO_2 in TiO_2. After above adjustment, the thermodynamic properties of the double oxides were converted to those relative to monoclinic ZrO_2 and rutile. The interaction parameter of the cubic solution was estimated as 10000 J/mol; this value is the same as the tetragonal phase. The lattice

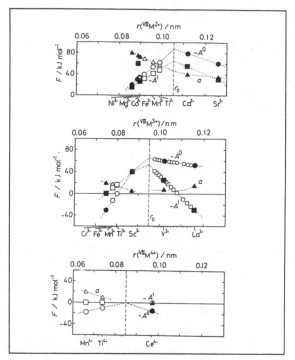

Figure 3. Thermodynamic regularity in interaction parameters of the cubic phase in ZrO$_2$–MO, ZrO$_2$–MO$_{1.5}$, and ZrO$_2$–MO$_2$ systems as a function of radius of M^{n+} with 8–fold coordination. Solid symbols represent those parameters which were derived by fitting to the experimental phase relations and/or thermodynamic properties, open symbols being estimates.

stability of TiO$_2$ in the cubic phase is 10000 J/mol. The interaction parameter of the liquid phase was determined using the above data for the tetragonal phase so as to reproduce the peritectic melting of ZrTiO$_4$ at 1820°C and the eutectic temperature, 1760°C, and composition, x(TiO$_2$) = 0.8.

The phase diagram of the ZrO$_2$–CeO$_2$ system has been investigated experimentally and thermodynamically by Tani *et al.* [15]. We adopted a similar set of parameters. For the ZrO$_2$–HfO$_2$ system, simpler estimates were adopted.

ZrO$_2$–YO$_{1.5}$–MO$_x$(M = TRANSITION METAL) SYSTEMS

Since the correlation between the thermodynamic properties and the ionic radii is satisfactorily good, this leads to an expectation that similar correlations can be applied to other systems containing transition metal ions. Since reported solubilities are quite different among transition metal series and there seems no good explanation for the change in solubilities across the transition metal series, an attempt has been made to explain these solubilities by taking into account the valence number of the transition metal ions.

The present adopted interaction parameters for the ZrO$_2$–MO and ZrO$_2$–MO$_{1.5}$ and ZrO$_2$–MO$_2$ systems are presented in figure 3.

The valence state of chromium ions in YSZ appears to be constant at 3+ at high

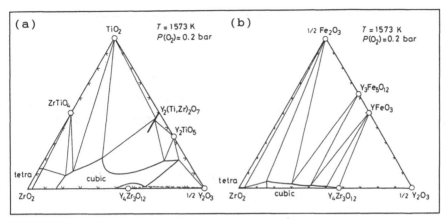

Figure 4. Calculated phase relations in the ZrO_2–$YO_{1.5}$–MO_x systems at 1573 K. (a) $MO_x = TiO_x$, (b) FeO_x.

temperatures. Since solubilities and other phase relations in the ZrO_2–$YO_{1.5}$–$CrO_{1.5}$ system have been experimentally investigated, the interaction parameters for the ZrO_2–$CrO_{1.5}$ system was first determined. The calculated ternary phase diagram using the thermodynamic properties of $YCrO_3$ [16] shows essentially the same Y_2O_3 composition for the $YCrO_3$–Cr_2O_3–cubic three phase combination. Interaction parameters for other transition metal oxides, $FeO_{1.5}$ and $TiO_{1.5}$, were estimated by linear interpolation. In the divalent ions, the solubility of NiO is considered to be about 1 percent at high temperatures. Since the ionic radius of Ni^{2+} is close to that of Mg^{2+}, the interaction parameters of the cubic phase in the ZrO_2–NiO system were determined so as to reproduce the low solubility using similar values to those of the MgO system. As shown in figure 3, we adopted a less negative interaction parameter, A^1, than expected from the corresponding value for MgO. The interaction parameters for Co^{2+}, Fe^{2+}, Mn^{2+}, and Ti^{2+} were estimated similarly as given in figure 3.

SOLUBILITIES OF TRANSITION METAL OXIDES IN YSZ

The calculated phase relations for the pseudoternary systems at 1573 K are presented in figure 4; for $CrO_{1.5}$ and MnO_x solubilities, see reference 17.

In the $YO_{1.5}$–TiO_2 system, $Y_2Ti_2O_7$ and Y_2TiO_5 were reported as stable double oxides and the cubic phase appeared at high temperatures. The thermodynamic properties of $Y_2Ti_2O_7$ were estimated from the other pyrochlores. The present thermodynamic parameters, $-54000 + 16000\,(x_1 - x_2)$ J/mol, were adjusted by taking into account the ionic size mismatch so that the fluorite solution appears above 1663 K in the $YO_{1.5}$ rich region. The solubility of TiO_2 in YSZ was calculated with interaction parameters for the $YO_{1.5}$–TiO_2 system. Figure 4 shows the calculated phase relations. The maximum solubility of TiO_2 in $(ZrO_2)_{0.85}(YO_{1.5})_{0.15}$ was calculated to be about 15 mole% at 1573 K; this is in good agreement with the experimental value obtained at 1873 K [18]. Although there is no pyrochlore phase in the ZrO_2–$YO_{1.5}$ system, the $Y_2Ti_2O_7$ phase is expected to form solid solutions with $Y_2Zr_2O_7$, its thermodynamic properties were estimated using the correlation given in figure 2. The calculated phase diagram shows that there is a range of composition where the fluorite solid solution and the pyrochlore solution compete with each other.

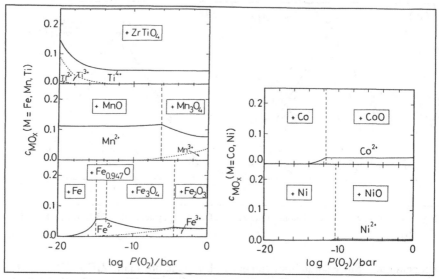

Figure 5. Calculated solubilities of transition metal oxides as a function of oxygen potential at 1273 K.

In the $YO_{1.5}-FeO_{1.5}$ system, there are two double oxides, $YFeO_3$ and $Y_3Fe_5O_{12}$; their thermodynamic properties have been extensively investigated [19]. The solubility of FeO_x was calculated using the interaction parameters for Fe^{2+} and Fe^{3+} simultaneously. Solubility of iron oxides into YSZ have not been conclusively determined experimentally; the solubility of Fe_2O_3 ranges from 2–3 to 5 %. The present calculated value is close to the lower experimental value. The calculated percentage of Fe^{3+} is about 87 % at 1573 K.

VALENCE STABILITY IN YSZ AS A FUNCTION OF OXYGEN POTENTIAL

Figure 5 shows the solubility limit of transition metal oxides in YSZ as a function of oxygen potential. The compounds given in figure 5 are those precipitates which should be in equilibrium with the fluorite solution. In the present calculations, the contributions of respective transition metal ions having different valence number can be separately determined. They depend on the following properties: (1) the valence stability of the particular valence state in the binary systems; (2) the stabilization energy of the respective valence ion in the fluorite solution which can be correlated with ionic size. The difference in the oxygen–potential dependence of the solubilities is due to both effects. For example, Ti^{3+} and Ti^{2+} are expected to have large stabilization energies in the fluorite solution, while their valence stability in an oxidizing atmosphere is not significant. Thus, their contributions become significant in extremely reducing atmospheres [20]. Comparison between iron and manganese may be of interest in view of the similarity in their valence stability for the divalent and trivalent ions. Figure 5 shows that the contribution from the trivalent ions is about the same between the two systems, whereas the behavior of the divalent ions is different. Note that the solubility of manganese divalent ions is relatively large. This is due to the large ionic size of divalent manganese ions and their high valence stability.

SUMMARY AND CONCLUSIONS

The phase diagram behavior of ZrO_2–based binary ceramics can be well represented by the simple subregular solution model with a small number of parameters which change reasonably from the viewpoint of crystal chemistry based on the valence and size of ions. This good correlation can make it possible to estimate the corresponding parameters for systems containing transition metal oxides. The association solution model can provide a good basis for describing the thermodynamic properties of systems containing the transition metal ions whose valence state can change with temperature and oxygen potential. To the first order of magnitude approximation, the solubility of the transition metal ions in YSZ can be explained in terms of the size and valence number.

REFERENCES

1. Tedmon, C. S. Jr., H. S. Spacil and S. P. Mitoff, 1969. "Cathode Materials and Performance in High Temperature Zirconia Electrolyte Fuel Cell." J. Electrochem. Soc., 116:1170–1175.

2. Yokokawa,H., N. Sakai, T. Kawada and M. Dokiya, 1990. "Thermodynamic Analysis on Interface between Perovskite and YSZ Electrolyte." Solid State Ionics, 40/41:398–401.

3. Yokokawa, H., N. Sakai, T. Kawada and M. Dokiya, 1991. "Thermodynamic Analysis on Reaction Profiles between $LaMO_3$(M = Ni,Co,Mn) and ZrO_2." J. Electrochem. Soc., 138:2719–2727.

4. Kim, D.-J., 1989. "Lattice Parameters, Ionic Conductivities, and Solubility Limits in Fluorite–Structure MO_2 Oxide (M = Hf^{4+}, Zr^{4+}, Ce^{4+}, Th^{4+}, U^{4+}) Solid Solutions." J. Am. Ceram. Soc., 72(8):1415–1421.

5. Kaufman, L., 1987. "Calculation of Quasibinary and Quasiternary Ceramic Systems," in User Applications of Phase Diagram Calculations, Materials Mark, OH, USA: ASM International, pp. 145–176.

6. Degtyarev, S. A. and G. F. Voronin, 1988. "Solution of Ill–Posed Problems in Thermodynamics of Phase Equilibria. The ZrO_2–Y_2O_3 System." CALPHAD, 12(1):73–82.

7. Du, Y, Z. Jin and P. Huang, 1991. "Thermodynamic Calculation of the ZrO_2–$YO_{1.5}$–MgO System." J. Am. Ceram. Soc., 74(9):2107–2112.

8. Yoshimura, M., 1988. "Phase Stability of Zirconia." Bull. Am. Ceram. Soc., 67(12):1950–1955.

9. Glushkov, V. P., L. V. Gurvich, G. A. Bergman, I. V. Veitz, V. A. Medvedev, G. A. Khachkuruzov and V. A. Yungman, 1982. "Thermodynamic Data for Individual Substances," Vol 4., Moscow: High temperature Institute, State Institute of Applied Chemistry, National Academy of Sciences of the U. S. S. R.

10. Yokokawa, H., N. Sakai, T. Kawada and M. Dokiya, 1991. "Thermodynamic Stability of Perovskites and Related Compounds in Some Alkaline Earth – Transition Metal – Oxygen System." J. Solid State Chem., 94:106–120.

11. Vintonyak,V. M., Yu. Ya. Skolis, V. A. Leviskii and Ya. I. Gerasimov, 1984. "Thermodynamic Activity of Y_2O_3 in $Y_2O_3-ZrO_2$ Solid Solutions." Russ. J. Phys. Chem., 58(10):1577.

12. Korneev, V. R. and V. B. Keler, 1971. "Heats of Formation of Rare Earth Zirconates." Rus. Inorganic Materials, 7(5):781-782.

13 Yokokawa, H., M. Fujishige, S. Ujiie and M. Dokiya, 1988. "CTC : Chemical Thermodynamic Computation System(in Japanese)." J. Natl. Chem. Lab. Ind., 83(special issue):1-26.

14. Stubican, V. S., 1986. "Phase Equilibria and Metastabilities in the System ZrO_2-MgO, ZrO_2-CaO, and $ZrO_2-Y_2O_3$." in Science and Technology of Zirconia III, S. Somiya, N. Yamamoto and H. Yanagida, eds, Westerville, Ohio, USA: The American Ceramic Society, pp. 71-82.

15. Tani, E. M.,Yoshimura and S. Somiya, 1983. "Revised Phase Diagram of the System ZrO_2-CeO_2 Below 1400 °C." J. Am. Ceram. Soc., 66(7):506-510.

16. Kovba, M. L., Yu. Ya. Skolis, V. M. Vintonyak and V. A. Levitskii, 1984. "Determination of Thermodynamic Properties of $YCrO_3$ by EMF Method with Solid CaF_2 Electrolyte (in Russian)." Dokl. Akad. Nauk SSSR, 277(3):622-625.

17. Yokokawa, H., N. Sakai, T. Kawada and M. Dokiya, 1991. "Chemical Thermodynamic Compatibility of Solid Oxide Fuel Cell Materials." in Proc. Second Internatl. Symp. Solid Oxide Fuel Cells, Luxembourg:Commission of the European Communities, pp. 663-670.

18. Liou, S. S. and W. L. Worrell, 1989. "Electrical Properties of Novel Mixed-Conducting Oxides." Applied Physics A, 49:25-31.

19. Piekarezyk, W., W. Weppner and A. Rabenau, 1979. "Solid State Electrochemical Study of Phase Equilibria and Thermodynamics of the Ternary System Y-Fe-O at Elevated Temperature." Z. Naturforsch., 34a:430-436.

20. Lin, C., D. Gan and P. Shen, 1988. "Stabilization of Zirconia Sintered with Titanium." J. Am. Ceram. Soc., 71(8):624-629.

Further Observations of an Orthorhombic Zirconia in Magnesia-Partially Stabilised Zirconia

DIMITRI N. ARGYRIOU, CHRISTOPHER J. HOWARD
and SRINIVASARAO LATHABAI

ABSTRACT

A neutron powder diffraction investigation has revealed the occurrence of substantial quantities of an orthorhombic (o) zirconia in Mg-PSZ which has been aged at the eutectoid temperature (1440°C) then cooled to 77 K. Though there have been reports of TEM observations of orthorhombic zirconia in very similar materials, this is the first observation of bulk quantities of the o phase in eutectoid aged Mg-PSZ. The neutron diffraction pattern is simpler than those showing orthorhombic in materials aged at sub-eutectoid temperature (1100°C), since there are fewer distinct phases in the material aged at the higher temperature. Refinement from the diffraction pattern confirms the structure (space group $Pbc2_1$) previously reported.

INTRODUCTION

Pure zirconia is known to have three ambient pressure polymorphs, cubic (c) at high temperatures, tetragonal (t) at intermediate temperatures and monoclinic (m) at room temperature. The higher temperature phases can be stabilised at room temperature by the fabrication of alloys with various oxides (MgO, CaO, Y_2O_3, CeO_2 etc) to form cubic stabilised zirconias (CSZ), and zirconia toughened

Dimitri N. Argyriou & Christopher J. Howard. Australian Nuclear Science and Technology Organisation. Private Mail Bag 1, Menai, N.S.W. 2234, Australia.
Srinivasarao Lathabai, CSIRO Division of Materials Science and Technology Locked Bag 33, Clayton, Vic. 3168, Australia.

ceramics (ZTCs) such as partially stabilised zirconia (PSZ), which is a mixture of c and t, or tetragonal zirconia polycrystal (TZP). In Mg-PSZ a rhombohedral phase (δ), related to the cubic phase, has been observed. In the technology of ZTCs, the aim is to produce ceramics containing zirconia in a metastable tetragonal form, which will readily transform to monoclinic zirconia under mechanical stress.

It is also known that two orthorhombic polymorphs of ZrO_2 can occur [1]. The first evidence for these polymorphs was reported by Heuer et al. [2,3] in transmission electron microscopy (TEM) of thin foils. Heuer et al. observed two orthorhombic zirconias (o and o'), where the a parameter of the second was twice that of the first. Orthorhombic phases were observed in Mg-PSZ, Y-PSZ, Ca-PSZ, ternary (Mg, Ca)-PSZ and (Mg, Y)-PSZ as well as zirconia toughened alumina (ZTA), but in every case their occurrence was thought to be an artefact of the TEM sample preparation [3]. Confirmation that o can be produced in bulk quantities came from X-ray and more especially neutron diffraction studies of high toughness Mg-PSZ cooled to liquid nitrogen temperatures [4,5]. The o observed under these conditions was different from the orthorhombic phase obtained by quenching from high pressure - high temperature by Ohtaka et al. [6]; this latter phase corresponds to the o' phase mentioned by Heuer et al. [3]. To date, the only neutron diffraction observations of (bulk quantities of) o in ZTCs have been in sub-eutectoid (1100°C) aged Mg-PSZ, under three conditions: a) after cooling to liquid nitrogen temperatures [4,5], b) in overaged samples [7], and c) after subjecting this high toughness Mg-PSZ to creep testing [8]. The crystal structure of o was determined using neutron diffraction by Kisi et al. [5], from sub-eutectoid aged Mg-PSZ which had been cooled to liquid nitrogen temperature. It was found that o made up 45% of the bulk, the balance consisting of c, t, m and the δ-phase.

The only observation until now of o in eutectoid (1440°C) aged Mg-PSZ was that by Heuer and Schoenlein [2] in TEM thin foils. In that work Mg-PSZ samples, aged at 1440°C for 3 and 6 hours, were quenched from 200, 400 and 1000°C to room temperature, producing o within 10 to 20 µm of the surface.

In the present work we report the observation, using neutron powder diffraction, of bulk quantities of o zirconia in eutectoid aged Mg-PSZ which had been cooled to 77 K. In this material, o is found in a mixture with just c and t, resulting in a relatively simple diffraction pattern. This contrasts with the material analysed by Kisi et al. [5], in which o was mixed with four other phases. In this work we confirm the structure determination of Kisi et al., and provide a new refinement of the o phase.

EXPERIMENTAL PROCEDURE AND RESULTS

The samples were 9.5 mol% MgO-PSZ, sintered at 1700°C, and cooled at 500°C per hour down to 1000°C, followed by a slower cooling to room temperature. They were then aged at the eutectoid temperature, 1440°C, for three hours. The neutron diffraction sample comprised a stack of five disks each 23 mm in diameter and 3 mm thick.

Neutron diffraction patterns were recorded using the high-resolution powder diffractometer (HRPD) [9], at the Australian Nuclear Science and Technology Organisation's HIFAR nuclear reactor. Diffraction patterns were collected at room temperature before and after the sample was cooled to liquid nitrogen temperature. These patterns were recorded at a wavelength of 1.4925 Å, over the 2θ range from 0° to 155°, with a step size of 0.05° (see figure 1).

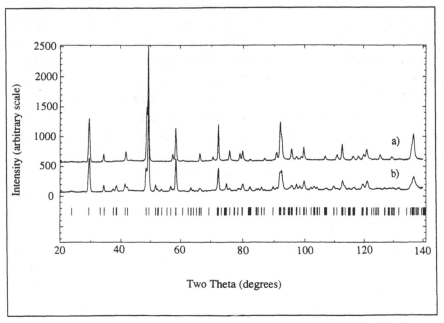

Figure 1. Neutron diffraction patterns recorded at room temperature, using a wavelength of 1.4925 Å, from a) eutectoid aged Mg-PSZ prior to quenching. b) eutectoid aged Mg-PSZ after quenching in liquid nitrogen. The markers underneath indicate the positions of the peaks from o-ZrO_2.

From the diffraction pattern taken prior to cooling it is clear that the sample consisted of c and t. The tetragonal peak at $2\theta=41°$ is clearly evident. We also observe diffuse intensity, at $2\theta \sim 43.5°$ which corresponds approximately to the (312) δ phase reflection. The δ phase is closely related to the cubic phase and is produced by vacancy ordering [10]. It appears that in this sample only short range ordering has taken place, so the observed intensity remains diffuse. Diffuse intensity has also been observed in TEM of thin foils of this sample in selected area diffraction of c grains, a behaviour which is similar to that already reported by Hannink *et al.* in as fired Mg-PSZ [7]. The diffraction pattern recorded after cooling indicates the presence of c and t and also shows new diffraction peaks, for example at $2\theta \sim 38°$ and $53°$. This confirms the presence of o in the material. There is no evidence for the presence of o' which, if present, would be apparent through its (102) reflection at $2\theta = 35°$.

The diffraction patterns were analysed using the Rietveld method [11]. In this method, a least-squares fit is made of a diffraction pattern calculated from the crystal structure models to the entire observed pattern obtained (in this application) from the multiphase mixture. The program LHPM [12] was used for Rietveld refinements. The pattern from the sample prior to cooling was analysed as a two phase mixture of c and t, while the pattern collected after cooling was analysed as a three phase mixture of c, t and o. Scale factors, lattice parameters, atomic coordinates and isotropic thermal parameters were refined together with peak width and shape parameters (Gaussian width, Lorentzian particle size) for each phase (except for the pattern after cooling where thermal parameters and atomic coordinates were not refined for t). The instrumental peak asymmetry was also refined. Unfortunately our polynomial background function was not able to model the diffuse intensity, so it was decided to use an interpolated background (which included the diffuse intensity) based on visual estimates at 30 points, and subtract it from the pattern prior to refinement. The composition of the ceramic, before and after cooling, was determined from the Rietveld scale factors using the relation given by Hill and Howard [13]. As stated earlier, diffuse peaks due to short range ordering appear around positions where δ-phase is expected. These diffuse peaks, attributed to embryonic δ-phase, can be very poorly fitted by peaks from crystalline δ-phase, and the inclusion of this phase in the refinement leads to a gross over-estimate of its amount (~ 10 wt%).

The structure of the o phase was checked by analysing the diffraction pattern obtained after quenching to liquid nitrogen. The starting point for o was the structure in space group $Pbc2_1$ determined by Kisi *et al.* [5]. The final fit to the pattern is shown in figure 2. From the difference plot (bottom of figure 2) it is confirmed that the calculated pattern is in good agreement with the observed data.

The refined values of the structural parameters for the *o* phase are shown in Table I; these are in good agreement with the results of Kisi *et al.* [5] (Table II). Unfortunately our interpolated background has not fully accounted for the diffuse intensity of the δ phase. This manifests itself in ambiguous thermal parameters. The composition of the sample prior to cooling was estimated as 50.2(4) wt% *c* and 49.8(2) wt% *t*, whereas after cooling it was 16.6(4) wt% *t* and 35.2(4) wt% *o* with *c* being relatively unchanged. These figures show that 71 wt% of the *t* polymorph transformed to *o* on cooling to 77 K.

TABLE I. CRYSTAL STRUCTURE PARAMETERS FOR *o*-ZrO_2 OBSERVED IN EUTECTOID (1440°C) AGED Mg-PSZ AFTER COOLING TO 77 K.
LATTICE PARAMETERS: a=5.071(1) Å, b=5.2637(4) Å, c=5.0803(5) Å.
THE NUMBER IN PARENTHESES REPRESENTS THE ERROR IN THE LAST DIGIT.

Atom	Site	x	y	z	B ($Å^2$)
Zr	4a	0.267(1)	0.030(1)	0.249(2)	0.08(5)
O1	4a	0.074(1)	0.366(1)	0.107(2)	1.0(1)
O2	4a	0.535(1)	0.227(1)	0.0	0.14(6)

TABLE II. CRYSTAL STRUCTURE PARAMETERS FOR *o*-ZrO_2 OBSERVED IN SUB-EUTECTOID (1100°C) AGED Mg-PSZ AFTER COOLING (FROM KISI *et al.*[5])
LATTICE PARAMETERS: a=5.068(1) Å, b=5.260(1) Å, c=5.077(1) Å.

Atom	Site	x	y	z	B ($Å^2$)
Zr	4a	0.267(1)	0.030(1)	0.250(1)	0.4(1)
O1	4a	0.068(1)	0.361(1)	0.106(1)	0.8(1)
O2	4a	0.537(1)	0.229(1)	0.0	0.2(1)

DISCUSSION

Kisi *et al.* [5] determined the structure of *o* by analysis of a neutron diffraction pattern from a sample containing four other phases. In this work we have found *o* in a mixture with just two other phases. Accordingly, the neutron diffraction pattern is much simpler than previous patterns involving the *o* polymorph. The pattern has been analysed by the Rietveld method, and the structure proposed by Kisi *et al.* adequately confirmed.

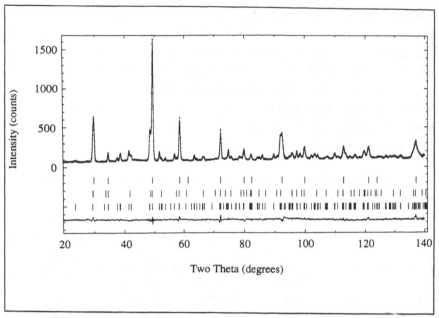

Figure 2. The result from Rietveld method analysis of the diffraction pattern collected after quenching the sample in liquid nitrogen. The observed data are indicated by crosses, while the fitted pattern is shown as a solid line. The peak positions of the *c, t* and *o* polymorphs (in that order) are indicated by the markers underneath the patterns, and the difference between the observed and calculated patterns is also shown. The weighted profile agreement index is R_{wp}=5.9%, and the goodness of fit is 3.2.

Although *o* zirconia has been reported previously in similar material in TEM thin foils [2], this is the first observation of bulk quantities of *o* in eutectoid aged Mg-PSZ. This observation, together with the previously mentioned observations of *o* in sub-eutectoid aged Mg-PSZ, represent the only instances in which bulk quantities of the *o* polymorph have been observed in ZTCs.

In sub-eutectoid aged Mg-PSZ the environment of *t* precipitates is different from that in the eutectoid aged case, the main difference being the presence of large amounts of δ phase in the sub-eutectoid aged material. It was found [5] when that material was cooled to 77 K that 75% of *t* transformed to *o*, which is similar to what was observed in this work. The *t* to *o* transformation occurs in both eutectoid and sub-eutectoid aged Mg-PSZ, in different microstructural environments and conditions, so that the transformation is not rare in this system. So far, however, large scale *t* to *o* transformation has been observed only in Mg-PSZ. In studies on other systems [14,15], we found that cooling Ca-PSZ and Ce-TZP to 77 K induced the *t* to *m* transformation, whereas in Y-TZP and ZTA no transformation occurred.

Currently the factors determining the *t* to *o* transformation are not known. The observations of bulk quantities of *o* are limited to Mg-PSZ, under the conditions of cooling, overaging and creep. One of the objectives of the present research is to understand what factors induce the *t* to *o* transformation as opposed to the martensitic *t* to *m* transformation on which the toughness depends.

REFERENCES

1. Howard, C.J., E.H. Kisi and O. Ohtaka, 1991. "Crystal Structures of Two Orthorhombic Zirconias." Journal of the American Ceramic Society , 74: 2321-2323.

2. Heuer, A.H., L.H. Schoenlein, 1985. "Thermal Shock Resistance of Mg-PSZ." Journal of Materials Science , 20: 3421-3427.

3. Heuer, A.H., V. Lanteri, S.C. Farmer, R. Chaim, R.R. Lee, B.W. Kibbel, R.M. Dickerson, 1989. "On the Orthorhombic Phase in ZrO_2-based Alloys." Journal of Materials Science, 24: 124-132.

4. Marshall, D.B., M.R. James and J.R. Porter, 1989. "Structural and Mechanical Property Changes in Toughened Mg-PSZ at Low Temperatures." Journal of the American Ceramic Society , 72: 218-227.

5. Kisi, E.H., C.J. Howard and R.J. Hill, 1989. "The Crystal Structure of Orthorhombic Zirconia in Partially Stabilized Zirconia." Journal of the American Ceramic Society, 72: 1757-1760.

6. Ohtaka, O., T. Yamanaka, S. Kume, N. Hara, H. Asano and F. Izumi, 1990. "Structural Analysis of Orthorhombic ZrO_2 by High Resolution Neutron Powder Diffraction." Proceedings of the Japanese Academy of Science , 66: 193-196.

This work is supported by the Australian Goverment, Industry Research and Development Board, Generic Technology Grant No. 15042.

7. Hannink, R.H.J., C.J. Howard, E.H. Kisi and M.V. Swain, "Relationship between Fracture Toughness and Phase Assemblage in Mg-PSZ." To be published in the Journal of the American Ceramic Society.

8. Kisi, E.H., J.R. Griffiths, A.K. Gross and T.R. Finlayson (1992). "Stress Induced T to O Transformation in Mg-PSZ." This conference.

9. Howard, C.J., C.J. Ball, R.L. Davis and M.M. Elcombe (1983). "The Australian High Resolution Neutron Powder Diffractometer." Australian Journal of Physics, 36: 507-518.

10. Rossell, H.J. and R.H.J. Hannink, 1984. "The Phase $Mg_2Zr_5O_{12}$ in MgO Partially Stabilized Zirconia." in Advances in Ceramics , Science and Technology of Zirconia II, N. Claussen, M Ruhle, and A.H. Heuer, eds., Columbus, OH, USA: American Ceramic Society, Vol. 12, pp. 139-151.

11. Rietveld, H.M. 1969. "A Profile Refinement Method for Nuclear and Magnetic Structures." Journal of Applied Crystallography, 2: 65-71.

12. Hill, R.J. and C.J. Howard 1986. "A Computer Program for Rietveld Analysis of Fixed-Wavelength X-ray and Neutron Powder Diffraction Patterns." Report No. AAEC/M112, Australian Atomic Energy Commission, Sydney, Australia.

13. Hill, R.J., and C.J. Howard 1987. "Quantitative Phase Analysis from Neutron Powder Diffraction Data Using the Rietveld Method." Journal of Applied Crystallography, 20: 467-474.

14. Kisi, E.H., and C.J. Howard 1992. "Neutron Diffraction Studies of Zirconia-Toughened Engineering Ceramics." Neutron News, 3(1): 24-28.

15. Howard, C.J., E.H. Kisi, D.N. Argyriou and R.J. Hill. Unpublished data.

Stress-Induced Tetragonal→Orthorhombic Transformation in Mg-PSZ

E. H. KISI, J. R. GRIFFITHS, A. K. GROSS
and T. R. FINLAYSON

ABSTRACT

During a study of room temperature creep in high toughness commercial Mg-PSZ, neutron diffraction was used to quantify the phase transitions which had occurred. As expected, substantial transformation of tetragonal to monoclinic zirconia was found but, unexpectedly, the tetragonal to orthorhombic transformation was also discovered. The ortho-rhombic phase was found to represent up to 10% of the strained volume of the cylindrical tensile creep specimens and is crystallographically identical to that formed in Mg-PSZ when it is cooled to below ~200K. The ratio of the orthorhombic phase content to the new monoclinic phase content (that is, additional to that in the starting material) is similar in this series of specimens. The stress-induced orthorhombic phase has only been observed in samples with very high creep strains and hence its presence is related to the transformability of the ceramic. These results add weight to the proposition that the tetragonal to orthorhombic transformation is martensitic.

INTRODUCTION

Magnesia-Partially Stabilised Zirconia (Mg-PSZ) is known for its very high fracture toughness [1]. Whilst there is some debate over the occurrence and effectiveness of the various toughening mechanisms proposed, it is clear that the transformation of metastable tetragonal zirconia to the stable monoclinic form and the associated 4.3% volume increase plays a very important role. During fracture toughness tests,

E.H. Kisi, School of Science, Griffith University, Nathan Qld 4111, Australia.
J.R. Griffiths, CSIRO Division of Manufacturing Technology, P.O. Box 883 Kenmore 4069, Australia.
A.K. Gross and T.R. Finlayson, Dept of Physics, Monash University, Clayton 3168, Australia.

plastic strains are localised adjacent to and immediately ahead of propagating cracks. However, when subjected to tensile creep loading at room temperature, some Mg-PSZs undergo large plastic strains that must be considered to permeate the bulk of the material [2]. The phase changes accompanying such behaviour are of considerable interest.

Mg-PSZ consists of a mixture of cubic (c), tetragonal (t) and monoclinic (m) zirconia and $Mg_2Zr_5O_{12}$ (δ). Bulk quantities of an ortho-rhombic phase (o) have also been reported after cooling below 200K [3,4,5] or overageing [6]. In addition, a stress-induced t→o transformation at surfaces after grinding has been reported from X-ray diffraction evidence [7] and also in TEM thin foils, both adjacent to fractured or ground surfaces and *in situ* due to beam heating stresses [8,9]. These latter observations relate to thin surface layers or thin-foil specimens.

Neutron powder diffraction, on the other hand, has proven to be an invaluable tool in the study of phase transformations and proportions in bulk ceramics [4,5,6]. It has also been applied to the study of creep in Mg-PSZ in a demonstration experiment [2]. In that experiment, the strain calculated from the amount of t→m conversion in the specimen was in excellent agreement with the measured (macroscopic) creep strain. Subsequent experiments on different commercial batches of Mg-PSZ have shown different degrees of t→m transformation. In the course of these further studies, it was noticed that at very high creep strains, the o phase (as in cooled Mg-PSZs) was present in addition to the expected c, t, m and δ phases. We therefore report here the first observations, by neutron diffraction, of substantial stress-induced t→o transformation in Mg-PSZ.

EXPERIMENTAL PROCEDURE

Samples from two differently heat treated batches of commercial 9.7 mol% Mg-PSZ (thermal shock grade - TS[1]) were supplied as cylinders 150mm long and 12.6mm in diameter. These were ground with a 150μm diamond grit tool to produce tensile specimens with a gauge section 35mm long and 6mm in diameter. The gauge section was polished to a 14μm finish. Creep testing was conducted on an Instron machine with self-aligning "Super-Grips" which ensured that bending strains were less than 5% of the tensile strains. Strains were measured by longitudinal and transverse strain gauges giving both the linear and volume creep strains. We focus here on two samples which showed the greatest creep strains and also the largest t→o (and t→m) conversion. Sample A was loaded to 255 MPa and developed a tensile creep strain of 3.4×10^{-3} in 352 hours. Sample B, loaded to 190-200 MPa, produced a tensile creep strain of 5.3×10^{-3} after 197 hours.

After creep testing, neutron diffraction data were collected at $\lambda = 1.500\text{Å}$ on the High Resolution Powder Diffractometer (HRPD) at the Australian Nuclear Science and Technology Organisation's HIFAR reactor, Lucas Heights, Australia [10]. Powder patterns were recorded at $10° \leq 2\theta \leq 150°$ with a 0.1° step size using eight detectors, data from which were corrected

---

[1] ICI Advanced Ceramics, Redwood Drive, Clayton 3168, Victoria, Australia.

for counter offsets and relative efficiencies, and averaged. Data are collected in transmission and the very low absorption of neutrons by zirconia ensures that the pattern is representative of the whole of the irradiated portion of the sample and not just the surface. For each sample, a pattern was recorded both from the entire highly strained gauge region and from one of the unstrained end regions of the test-piece (to provide a reference). The patterns were first subjected to visual analysis with reference to certain diagnostic diffraction peaks (see also [6]). These peaks are depicted in Figure 1; the low-angle region of a calculated neutron diffraction pattern from a mixture of c, t, m, o and δ phases.

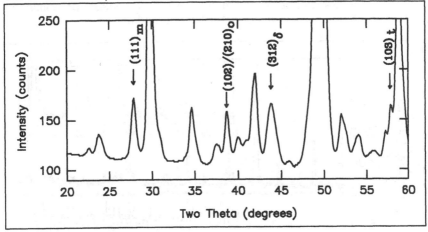

Figure 1. A calculated 1.500 Å neutron diffraction pattern for the low-angle region of a c,t,m,δ and o mixture. Reflections which are useful indicators for each phase are marked.

The m phase is readily identified from its (11-1) reflection, the t phase from its (103) reflection and the δ phase from the broad (312) reflection which overlaps but nonetheless dominates some very weak m peaks. The o phase is recognised from the (102)/(210) reflections which unfortunately overlay some weak m peaks. When the intensity in this region exceeds the intensity of the m peaks immediately to the left and right, the presence of o can be inferred. The c phase has all its reflections heavily overlapped by the other phases although use may sometimes be made of the (220) reflection when large amounts of c are present.

Following this procedure, the patterns were fitted by multi-phase Rietveld analysis, in which a calculated diffraction pattern is matched to all the observed data using a least-squares refinement technique [11]. In the analysis, the intensity of the calculated pattern for each phase is adjusted via a scale factor. These refined scale factors can then be used to determine quantitatively the phase proportions in the sample [12]. In addition, lattice parameters were refined for each phase and diffraction line broadening, which can provide estimates for particle size and residual micro-strains, was refined for the major phases. The crystal structures used in these analyses were summarised in a recent publication [5] and further details of the method have appeared elsewhere [2,4,6].

RESULTS

The neutron diffraction patterns from sample B, that with the higher creep strain, are shown in Figure 2 along with the results of Rietveld fitting. An examination of Figure 2a (particularly the inset enlargement) with reference to the diagnostic peaks of Figure 1 shows that the reference pattern from the sample end piece is typical of a TS grade Mg-PSZ (the most transformable commercial grade). It consists predominantly of the t and δ phases with minor amounts of c and m.

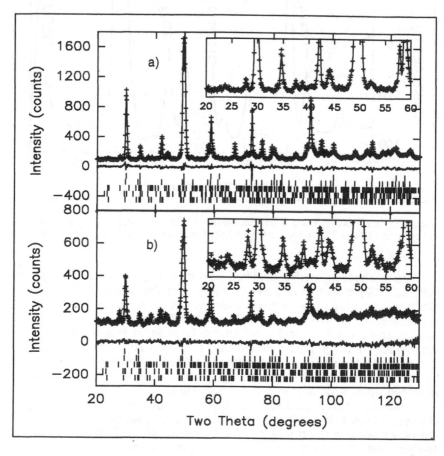

Figure 2.   Rietveld refinement results for the reference (upper pattern) and strained region (lower pattern) of sample B. In each the data are shown as + signs and the calculated pattern appears as the continuous line through them. A difference profile (calculated - observed) on the same scale is provided for each. The vertical bars below the patterns show the reflection positions for the c, t, m, δ and o phases respectively. Inset into each pattern is an expanded view of the low angle region to allow more critical examination.

There is little visual evidence for the o phase and the Rietveld refinement is equally good with and without o included. This puts an upper limit on o in the starting material of ~2.5% and a value of zero was assumed. The difference plot (observed pattern minus calculated pattern) shows that the observed data are well accounted for. Quantitative phase analysis results for this refinement are shown in Table I.

TABLE I - PHASE CONTENTS (vol%) OF Mg-PSZ CREEP SPECIMENS

Phase	Sample A		Sample B	
	Reference	Gauge	Reference	Gauge
c	14.7	11.9	13.5	7.1
t	51.7	38.4	51.0	27.0
m	4.8	14.9	6.3	26.4
δ	28.8	27.9	29.1	29.8
o	~0	6.9	~0	9.7
$\varepsilon_v^{n\ 2}$		5.3×10^{-3}		9.9×10^{-3}
$\varepsilon_v^{\ 3}$		4.9×10^{-3}		10.1×10^{-3}
$R_{wp}^{\ 4}$ (%)	6.23		6.11	4.49
GOF[5]	4.36		4.97	1.95

Considerable differences to the reference pattern are evident in Figure 2b, the pattern from the strained region of the sample, particularly in the inset enlargement. The amount of m has increased and t has decreased. In addition, there is moderately strong intensity in several new places, among them the diagnostic o peaks at $2\theta = 38.6°$. These observations are confirmed by Rietveld analysis with c, t, m, o and δ included. The quantitative phase analysis results appear in Table I and show that nearly 10vol% of the entire gauge region has undergone t→o transformation in addition to the expected t→m transformation. Similar results were observed in sample A (with lower creep strain) and are recorded in Table I. The estimated standard deviations (esd's) from the refined scale factors suggest that the maximum probable errors (ie 3esd) for the phase proportions recorded in Table I are 1.5-1.8%. For the o phase, there is the additional uncertainty in the initial quantity of o (~0-2.5%) to also consider.

The volume strain caused by phase transformations was calculated using equation 1,

$$\varepsilon_v^n = (0.014\Delta o + 0.0425\Delta m) \qquad (1)$$

where Δo and Δm are the changes in o and m contents respectively.
These results are included in Table I for comparison with the macroscopic volume change in the specimen measured by the longitudinal and

---

2 Volume strain calculated from phase transitions.
3 Volume strain measured with strain gauges.
4 Weighted profile R-factor [11].
5 Goodness of fit, the square of the ratio of R_{wp} to its expected value

transverse strain gauges (ε_v^n). In these samples the volume increase during creep can be satisfactorily accounted for by the t→m and t→o transformations. A more detailed set of creep results and discussion of mechanisms will appear elsewhere [13]. It is interesting to note that in these specimens the ratio Δo/Δm is similar for both specimens.

DISCUSSION

Previous reports of stress-induced t→o transformation have relied on evidence from X-ray diffraction [7], TEM [8,9] or Raman Spectroscopy [14]. In the former, the scattering power of zirconium relative to the two oxygens is approximately 12:1. Orthorhombic zirconia is therefore not easily distinguished from the other zirconia polymorphs (except m), when all are present, because the principal differences among them are in the atomic positions of the oxygen atoms [4]. This causes the diagnostic peaks (ref Fig 1) for o-ZrO_2 in the X-ray case to be very weak. Phase identification is much easier in the TEM case using SAED. However it is unclear whether the o phase observed in TEM was transformed during the mechanical testing and grinding thought to produce it, or whether the transformation occurred during thin foil preparation or even during TEM observation as is sometimes observed [9]. Whilst there exists a well-resolved diagnostic peak for o-ZrO_2 in the Raman spectrum, this technique only probes the near-surface region. The observations reported in the present paper are unambiguous confirmation that a stress-induced t to o transformation does occur in some Mg-PSZs and that substantial quantities of the entire stressed volume can undergo this transformation.

Samples A and B have different amounts of o and m but this can be understood because these amounts are approximately in proportion to the creep strain. More puzzling is the behaviour of samples from other commercial batches of Mg-PSZ with different levels of creep strain. Why was discernable transformation to o not observed in earlier neutron diffraction creep studies? One possibility is that, if we scale the observed levels of o production in samples A and B to the lower levels of creep strain in samples from other batches (see for example [2]), we would expect, at most, 3% of o phase to be produced. Since in these other samples there were generally lower levels of transformation, per unit of creep strain, we can see how the o phase might easily have been close to the limit of detection.

Transformation to the o phase involves a volume increase of 1.4% and hence has a potential transformation toughening effect of its own. It makes a small but significant contribution to the observed volume strain (equation 1), but is obviously less effective than the m phase. TEM observations of near-fracture zones [9] suggest that, in Mg-PSZ at least, t→o occurs in the lower stress regions near the boundary between the transformation zone and the untransformed material. The distribution of o (and m) in our creep samples is not known but from the large quantities of transformation it would appear that the m at least (and probably the o) is widely distributed throughout the sample.

The nature of the t→o transformation requires some comment. After the initial neutron diffraction studies, it was suggested that t→o may be a martensitic transformation (like t→m) [15]. The present results, confirming that the transformation can be driven by an applied stress, add extra weight to this suggestion. The t→o transformation is now known to behave similarly to t→m in all of the relevent respects; that is, it is diffusionless [9], athermal and reversible with considerable temperature hysteresis [5], involves both a size and a shape change [4] and may be induced to occur at temperatures above M_s by an applied stress [this work,14].

In an earlier neutron diffraction study of o-ZrO_2 produced by cooling Mg-PSZ [5], the possibility was canvassed that the t→o transformation on cooling is triggered by thermal expansion mismatch between the t particles and the largely c+δ matrix. The thermal expansion of the t phase is anisotropic with that of the c axis being 1.45 times as great as that of the a axis and 1.27 times as great as that of the c+δ matrix in the low temperature region where data are available [5]. During the latter parts of the fabrication cooling ramp, where no stress relaxation is expected to occur, part or all of the mismatch will accumulate as residual stresses provided the interface remains adherent (i.e. no voids are formed in the t/matrix interface). The broad lenticular faces of t precipitates (approximately {001} planes) will therefore be in tension whilst the edges of the precipitates ({hk0} planes) will be in compression. Data from [5] suggest that the residual tensile stress may be hundreds of MPa. This is comparable to the critical stress for the t→m transformation and presumeably also the t→o transformation. The transformation pathway followed would then appear to depend upon the nature of the stress field experienced by each t precipitate. If the stress is primarily dilational, t→o would be favoured. In other cases, where there is a significant shear component to the microstresses experienced by the particles, the t→m transformation occurs instead, aided by the, quite distinct, pre-stressing effect of δ-phase particles [6].

Finally it may be useful to consider the practical implications of studying the t→o transformation. Firstly, in the Mg-PSZ system the o phase is undesirable because ceramics where many t precipitates have already undergone t→o have lower fracture toughness than those where only t→m occurs [3]. It is therefore desirable to avoid its formation in this system. Secondly, in a system like Ce-TZP where the t→m transformation is so vigorous as to sometimes fracture the sample (for example on cooling to 77K), if we can determine conditions for inducing t→o instead, this might provide a gentler transformation toughening mechanism and increase the usefulness of the TZPs.

ACKNOWLEDGEMENTS

The authors thank AINSE for financial assistance, Dr R.L. Davis for collecting some of the neutron diffraction data, and also the National Research Fellowship (EHK) and Australian Postgraduate Research Scholarship (AKG) schemes. The support of ICI, Advanced Ceramics in providing specimens is also gratefully acknowledged.

REFERENCES

1. Green D.G., Hannink R.H.J. and Swain M.V., 1989. Transformation Toughening of Ceramics, Boca Raton, Florida, USA: CRC Press Inc.

2. Kisi E.H., Finlayson T.R. and Griffiths J.R., 1990. "Phase Determination in Partially Stabilised Zirconia Creep Specimens." Materials Science Forum, 56-58:351-356.

3. Marshall D.B., James M.R. and Porter J.R., 1989. "Structural and Mechanical Property Changes in Mg-PSZ at Low Temperatures." J. Am. Ceram. Soc, 72[2]:218-227.

4. Kisi E.H., Howard C.J. and Hill R.J. 1989. "Crystal Structure of Orthorhombic Zirconia in Partially Stabilised Zirconia" J. Am. Ceram. Soc., 72(9):1757-1760, and an abstract in Mater. Sci. Forum, 56-58:391-392.

5. Howard C.J., Kisi E.H., Roberts R.B. and Hill R.J., 1990. "Neutron Diffraction Studies of Phase Transformations Between Tetragonal and Orthorhombic Zirconia in Magnesia-Partially-Stabilised-Zirconia." J. Am. Ceram. Soc., 73(10):2828-2833.

6. Hannink R.H.J., Howard C.J., Kisi E.H. and Swain M.V., 1992. "Relationship Between Phase Assemblage and Fracture Toughness in Mg-PSZ." J. Am. Ceram. Soc,, (in press).

7. Hill R.J. and Reichert B.E., 1990. "Measurement of Phase Abundance in Magnesia-Partially-Stabilised Zirconia by Rietveld Analysis of X-ray Diffraction Data." J. Am. Ceram. Soc., 73(10):2822-2827.

8. Lenz L.H. and Heuer A.H., 1982. "Stress-Induced Transformation During Subcritical Crack Growth in Partially Stabilized Zirconia." J. Am. Ceram. Soc., 65(11):C192-C194.

9. Heuer A.H., Lanteri V., Farmer S.C., Chaim R., Lee R.R., Kithel B.W. and Dickerson R.M., 1989. "On The Orthorhombic Phase in ZrO_2-based Alloys." J. Mater. Sci., 24:124-132.

10. Howard C.J., Ball C.J., Davis R.L. and Elcombe M.M., 1983. "The Australian High Resolution Neutron Powder Diffractometer." Aust J. Phys., 36:507-518.

11. Rietveld H.M, 1969. "A Profile Refinement Method for Nuclear and Magnetic Structures." J. Appl. Crystallogr., 2:65-71.

12. Hill R.J. and Howard C.J., 1987. "Quantitative Phase Analysis from Neutron Powder Diffraction Data Using the Rietveld Method." J. Appl. Crystallogr, 20:467-474.

13. Finlayson T.R, Griffiths J.R., Gross A.K. and Kisi E.H., 1993. "Creep of Magnesia-Partially-Stabilised Zirconia at Room Temperature." J. Am. Ceram. Soc, (submitted).

14. Lim C.S., Finlayson T.R., Ninio F., and Griffiths J.R., 1992. "In-Situ Measurement of the Stress-Induced Phase Transformations in Magnesia Partially Stabilised Zirconia Using Raman Spectroscopy." J. Am. Ceram. Soc, 75:1570-1573.

The Cubic-to-Tetragonal Transformation in Zirconia Alloys

T. SAKUMA

ABSTRACT

The nature of the cubic–to–tetragonal(c–t) transformation is briefly reviewed. Assuming that this transformation is a second–order phase transition, the c–t equilibrium is satisfactorily described with a firm thermodynamic basis. The microstructure evolution caused by this transformation is also reasonably explained on the same assumption. In particular, the modulated structure formed in some zirconia alloys is attributable to a spinodal decomposition. The generation of a fully tetragonal structure in some fine–grained, toughened zirconia is related to the nature of this transformation.

INTRODUCTION

There are two different interpretations of the cubic–to–tetragonal(c–t) transformation in zirconia alloys. The first one assumes that the cubic(c–ZrO_2) and tetragonal(t–ZrO_2) phases are described by two different free energy curves, and the c–t transformation is a first–order phase transition. Most previous articles have discussed the experimental results according to this assumption. Cohen and Schaner found acicular or banded microstructures formed by the diffusionless c–t transformation in ZrO_2–UO_2[1], which may be evidence that this transformation is martensitic[2]. Andersson et al. demonstrated that the Ms temperature can be defined in the binary ZrO_2–Y_2O_3 system, and plate–shaped t–ZrO_2 martensite is formed in the c–ZrO_2 matrix by rapid cooling from high temperatures[3]. It has later been clarified that t–ZrO_2 formed by rapid cooling from high–temperature c–ZrO_2 is characterized by domains separated by antiphase boundaries with curvilinear features and that the plate or banded microstructures are twins induced for mechanical accommodation[4]. The domain structure is always formed throughout the samples, and this transformation has a nature which is very easily completed. Heuer and his coworkers have called it a displacive c–t transformation[4,5].

Precipitation from supersaturated cubic solid solutions has also been discussed from the same assumption. Hannink made the first careful microstructural examinations of aged zirconia alloys, and found that the precipitate morphology depends on the system and is explained in terms of the misfit between

T. Sakuma

University of Tokyo, Department of Materials Science, Bunkyo–ku Hongo7–3–1, Tokyo, 113 Japan.

tetragonal and cubic phases[6]. Among various precipitate morphologies, the generation of tweed contrast has been held to show the nature of this transformation[6–9]. Tweed contrast has been interpreted as the strain field contrast of small coherent $t-ZrO_2$ precipitates in a $c-ZrO_2$ matrix; the habit plane is likely to be {101}[7–9].

The second analysis assumes that the c–t transformation is a second–order phase transition. This assumption was first introduced by the present author's group to explain the modulated structure formed in an early stage of aging in a region inside the cubic–tetragonal two–phase field[10,11]. The modulated structure is developed by aging of metastable, fully–tetragonal ZrO_2 containing about 3–6mol%Y_2O_3[12], and has a habit plane close to {111}[10]. The microstructure evolution in $t-ZrO_2$ generated by rapid cooling from $c-ZrO_2$ has also been explained using this assumption[13,14]. More recently, it has been demonstrated through thermodynamic analysis that the two–phase field may be regarded as a miscibility gap with a sharp maximum and a spinodal[15,16].

The present paper aims to discuss the difference in the two assumptions and to explain the microstructure evolution associated with this transformation.

THE FIRST AND SECOND ORDER PHASE TRANSITION

In most phase transformations, the first derivative of free energy, e. g. enthalpy, entropy, volume etc., is discontinuous at the transition temperature. These are the first–order phase transitions. In some order–disorder transformations or magnetic transformations, the first derivative of free energy is continuous at the transition temperature, but the second derivative is discontinuous. The discontinuity appears in the data of heat capacity, compressibility and so on. This type of phase transformation is termed a second–order phase transition.

The c–t transformation in zirconia alloys is associated with a displacement in the [001] direction of oxygen ions in a cubic fluorite structure of $c-ZrO_2$[17]. The free energy of zirconia at a constant temperature can be described as a function of the displacement. The free energy–displacement relationship is different between the first– and the second–order phase transitions as schematically shown in figure 1, where the state of zero displacement corresponds to the cubic state. In the first–order phase transition depicted in figure 1(a), two minima having the same energy as the cubic state exist at the transition temperature T_c. The situation means that the cubic state and tetragonal state with a displacement η_0 are in equilibrium at T_c. In this case, the jump of oxygen ions from zero to η_0 must take place during the transition. The jump of oxygen ions is accompanied by energy consumption to overcome the potential barrier. This is characteristic of a first–order phase transition.

In the second–order phase transition, the minimum energy at T_c is given for

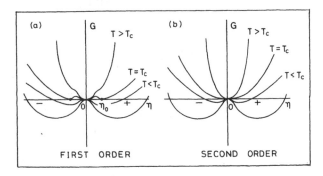

Figure 1. Free energy-displacement relationship for the first- and second-order phase transitions.

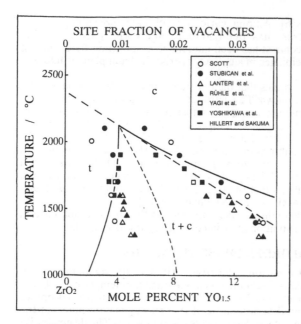

Figure 2. The tetragonal-cubic equilibrium in the system ZrO_2-Y_2O_3. The phase boundaries depicted as solid curves are the ones calculated on the assumption that the c-t transformation is of second-order type [15].

zero displacement as in figure 1(b). The two states, cubic and tetragonal, are identical at T_c. Below T_c, two energy minima, which have lower energy than the cubic state, appear and the tetragonality gradually increases during cooling. It is possible that the crystal symmetry changes abruptly at T_c with a very small displacement of constituent ions in the second–order phase transition[18].

The first– and second–order phase transitions may not necessarily be distinguished clearly in actual transformations. If the energy hump at the transformation temperature in figure 1(a) is sufficiently small, the transformation can be treated as a second–order phase transition. The purpose of this paper is not to identify the energy profile of figure 1(a) or (b) in the strict sense but to discuss whether or not the c–t transformation in zirconia alloys may be approximated as a second–order phase transition.

THE CUBIC–TETRAGONAL EQUILIBRIUM

The phase equilibrium between cubic and tetragonal phases has been studied experimentally by many workers, in particular for the system ZrO_2–Y_2O_3. The experimental data are compiled in figure 2. They agree well except above 1800°C, but the phase boundaries have sometimes been drawn considerably differently[19,20]. The difference is generally attributable to the lack of a firm thermodynamic understanding of the c–t equilibrium.

The first attempt of a thermodynamic analysis of the c–t equilibrium is probably the one based on the assumption that the c–t transformation is a second–order phase transition[15]. In this analysis, stabilization of the cubic state by cations with a lower valency than Zr^{4+} is assumed to be caused by extrinsic vacancies introduced to keep electrical neutrality. The Gibbs free energy for the cubic state is formulated through the compound energy model. The difference in free energy between the tetragonal and cubic states is obtained from the heat capacity data for pure ZrO_2 near T_c[21]. For details on the free energy of mixing in the cubic state and the difference in free energies between the tetragonal and cubic states, refer to the original paper[15]. The calculated phase boundaries for the enthalpy difference

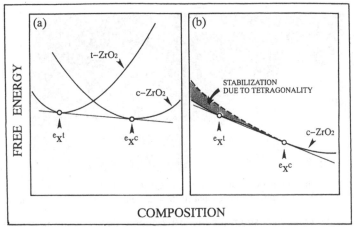

Figure 3. The schematic free energy-composition diagrams; (a) the commonly-used one [9] and (b) the one based on the second-order phase transition [15]. The notations $^ex^t$ and $^ex^c$ represent the equilibrium alloy contents in t-ZrO$_2$ and c-ZrO$_2$, respectively.

ΔH^0_{max} of $-17,000$kJ/mol are shown in figure 2. The value of ΔH^0_{max} was chosen to fit the experimental data, while a value of $-50,000$ kJ/mol is expected if all the abnormal increase in heat capacity near T$_c$ in pure ZrO$_2$ is attributed to a gradual loss of tetragonality. This fact may indicate that the loss of tetragonality is only partly responsible for the abnormal heat capacity increase, and the rest of it is caused by a factor in c-ZrO$_2$[15].

The calculated phase boundaries have a sharp maximum with a spinodal. The phase boundary construction is essentially different from the previous ones, which were drawn on the assumption that t-ZrO$_2$ and c-ZrO$_2$ have different Gibbs free energies as in figure 3(a). Figure 3(a) has generally been assumed to explain the microstructure evolution associated with the c-t transformation in zirconia alloys[9]. In contrast, figure 3(b) is the one used to calculate the phase boundaries of figure 2. It is assumed in the thermodynamic analysis that the introduction of tetragonality slightly stabilizes the crystal and lowers the free energy below T$_c$[15]. The analysis predicts a spinodal in a region inside the miscibility gap as depicted in figure 2.

Another interesting point is that the thermodynamic analysis seems to be successful in describing the c-t two-phase field in nonstoichiometric ZrO$_2$[16], i.e. the experimental phase boundaries in the ZrO$_2$-Y$_2$O$_3$ and ZrO$_2$-Zr$_2$O$_3$ systems agree well with each other[16]. This result suggests that vacancies play the principal role in stabilization of the cubic state, and the alloying effect on the c-t transformation temperature depends primarily on the number of oxygen vacancies[16]. The present author believes that this interpretation is valid for the c-t equilibrium, at least, in the first approximation.

MICROSTRUCTURE OF DIFFUSIONLESSLY TRANSFORMED t-ZrO$_2$

The thermodynamic model discussed above predicts that if high-temperature c-ZrO$_2$ is cooled below the critical temperature T$_c$, then t-ZrO$_2$ with a certain tetragonality spontaneously develops but soon undergoes by spinodal decomposition, and the initial t-ZrO$_2$ state prior to decomposition may not be detected by conventional techniques. However, microstructures different from those formed by a diffusional process are often found in t-ZrO$_2$ induced by rapid cooling from the high-temperature single phase c-ZrO$_2$ field. This fact may

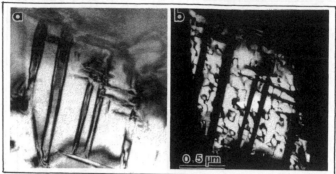

Figure 4. Microstructure of rapidly-solidified ZrO_2-4mol%Y_2O_3 [14]. The two micrographs were taken in the same area, (a) is a bright field image and (b) a dark field image using a 112 reflection.

indicate that the coherency strain may retard spinodal decomposition, enabling the t–ZrO_2 state to be retained at room temperature.

Acicular or banded microstructures which resemble the martensite in some metallic materials were first found in ZrO_2–UO_2 water-quenched from the c–ZrO_2 region[1]. Other variations such as plate–like or lenticular microstructures or herringbone structures have later been found in several zirconia alloys[3,22–25]. These microstructures have been interpreted on the basis of phenomenological crystallographic theory by introducing the lattice invariant deformation of the {011}<011> twinning system. In particular, the analysis seems to be successful in explaining the morphology and habit plane of the herringbone structure[24,25]. Phenomenological analysis is commonly applied to martensite in metallic materials. However, in the present author's view, its success does not directly support the contention that the diffusionless c–t transformation is martensitic. There is disagreement as to whether the twins formed by this transformation are transformation twins or those required for mechanical accommodation.

Another argument to support the martensitic nature of this transformation depends on the assumption that an Ms temperature can be defined below T_0 in binary ZrO_2–base systems[3]. Unfortunately, T_0 has no thermodynamic basis in this analysis. The present author's group has already pointed out that this interpretation is not valid because the diffusionless c–t transformation in the system ZrO_2–Y_2O_3 takes place even above the Ms temperature defined by their method[14]. It may be claimed that this is caused by the uncertainty of the c–t equilibrium in zirconia alloys[26], or that T_0 is not in the middle of the c–t two–phase field as has originally been assumed[3] but shifted towards the cubic side, so that it is possible to define an Ms temperature not inconsistent with the experimental data. In the latter case, T_0 has to be drawn close to the t+c/c boundary. It is, however, quite strange for T_0 to be close to the cubic side because c–ZrO_2 has a wide solid solution range. The interpretation must not be rationalized by assuming suitable thermodynamic parameters.

As first clarified by Heuer's group, it seems reasonable that none of the microstructures discussed above is the dominant one for this transformation[4,5,9]. The transformation is characterized by the domain structure with curved boundaries; the acicular or banded microstructures should be regarded as twins induced for mechanical accommodation. The domain structure is not seen in conventional bright field electron micrographs, but appears in a dark field image taken using a reflection of the type, odd, odd, even, which is forbidden for the cubic fluorite structure.

Figure 4 is an example of microstructures in ZrO_2–4mol%Y_2O_3 rapidly solidified after arc–melting[14]. In the bright field image of figure 4(a), the plate–

like microstructure is seen, comprising twins in t-ZrO$_2$. The domain structure with curvilinear boundaries appears in figure 4(b), which is characteristic of t-ZrO$_2$ formed by this transformation. The domain boundaries are regarded as antiphase boundaries with a displacement vector a/2[101] or a/2[011][4,5], and appear as π-boundaries in electron microscopy. The order of the displacement of the oxygen ion rows in t-ZrO$_2$ is reversed across the boundary. An interesting fact is that the domain structure always appears throughout the samples, both inside and outside the twins[14]. The domain boundaries are not continuous across twin boundaries but are often continuous in a single t-ZrO$_2$ variant across twins[13]. It has, therefore, been postulated that the domain structure was formed first and twins later[4,5,9,13,14]. The nature of the diffusionless c-t transformation is such that it is very easily completed by generating the domain structure[4,5,9,13,14].

The generation of microstructures during this transformation will now be discussed in more detail. Figure 5 shows micrographs of as-cast ZrO$_2$-Er$_2$O$_3$[27]. The microstructure depends markedly on Er$_2$O$_3$ content, found also in Gd$_2$O$_3$-, Y$_2$O$_3$- or Yb$_2$O$_3$-containing alloys[27]. Figure 5(a) is the herringbone structure in ZrO$_2$-2.5mol%Er$_2$O$_3$, in which a faint tweed-like contrast is seen. The contrast must be associated with the microstructure caused by a diffusional process. ZrO$_2$-4 and 6mol%Er$_2$O$_3$ have microstructures consisting of the domain structure and plate-like twins, as seen in figure 5(b) and (c) which resemble the microstructure of figure 4(b). ZrO$_2$-8mol%Er$_2$O$_3$ exhibits fine, dotted contrast as in figure 5(d), which was also taken using a 112 reflection. Three 112 reflections from three t-ZrO$_2$ variants always appeared in this alloy, and each reflection revealed fine, dotted contrast similar to figure 5(d). It seems, therefore, reasonable that this alloy is regarded as t-ZrO$_2$ at room temperature, and is fully occupied with uniformly-distributed microdomains of three t-ZrO$_2$ variants with a size of 10-20 nm, which appear as fine dots in electron microscopy. This result is not consistent with that of

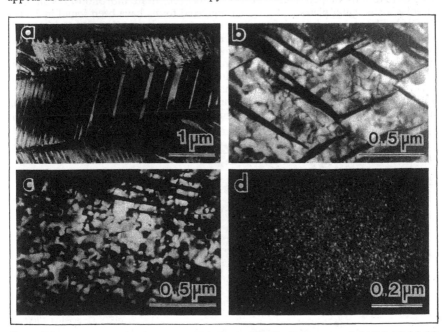

Figure 5. Microstructures of rapidly-cooled ZrO$_2$-Er$_2$O$_3$ [27]. The Er$_2$O$_3$ content is (a) 2.5, (b) 4.0, (c) 6.0 and (d) 8.0mol%, respectively.

conventional X-ray diffraction analysis, in which the peak splitting between 400 and 004 reflections cannot be detected. There are two possibilities for the difference between X-ray and electron diffraction analyses. One is that the tetragonality in t-ZrO_2 is too small in this alloy to be detected by X-ray diffraction, and the other is that the cations maintain cubic symmetry but the oxygen ions are slightly displaced in the [001] direction.

It may be reasonable to assume that the microdomains of figure 5(d) are formed prior to the domain structure in the single t-ZrO_2 variant in figure 5(b) or (c), because the c-t transformation temperature decreases with an increase of cubic-stabilizing oxide. If this is the case, it can be expected that the microdomain structure is not stable probably due to its extremely large interfacial area per unit volume, and thus some microdomains grow rapidly at the expense of others. As a consequence, a fairly large area of single t-ZrO_2 variant is soon developed, which includes the curved domain boundaries and twins subsequently induced for mechanical accommodation. In the present author's view, the initial development of uniform and very fine microdomains throughout samples is related to the fact that there is no, or at least, a negligibly small barrier for t-ZrO_2 nucleation in c-ZrO_2.

MICROSTRUCTURE EVOLUTION DURING AGING BY DIFFUSIONAL PROCESS

It has been reported that precipitation during sintering or aging in the c-t two-phase field results in a change in room-temperature mechanical properties. These properties are influenced by the microstructure and the stability of t-ZrO_2 precipitates. The morphology of tetragonal precipitates in zirconia solid solutions depends on the type of cubic-stabilizing oxide and on aging time. In ZrO_2-MgO, small spherical or disc-shaped precipitates are formed initially[28], which grow into ellipsoidal-shaped ones[6,28]. Various precipitate morphologies such as cuboids, rectangular plates and irregular equiaxed forms have been found in ZrO_2-CaO[6]. Plate-shaped precipitates are often observed in ZrO_2-Y_2O_3[6,10] an example of which is shown in figure 6. The growth morphology of precipitates has been described in terms of the lattice parameter mismatch between t-ZrO_2 and c-ZrO_2[6]. The importance of coherency strain or loss of coherency on the precipitate morphology has been discussed[29]. Most of these precipitates are formed at a fairly late stage of aging, and probably have close to the equilibrium composition. In this paper, the microstructure evolution during an early stage of aging or annealing in the c-t two-phase field is mainly discussed, because analysis of the

Figure 6. Plate-shaped precipitates in ZrO_2-5.2mol%Y_2O_3 aged at 1700°C for 1h [10].

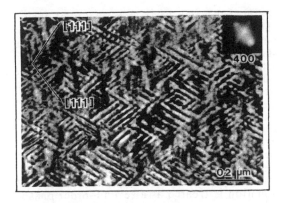

Figure 7. The modulated structure in ZrO_2-4mol%Y_2O_3 aged at 1700°C for 10min [11]. The beam direction is close to [011].

microstructures, which differ in morphology from the above precipitates, is particularly important in understanding the nature of the transformation.

We first reported that the modulated structure is developed during isothermal aging in the c–t two–phase field[10,11]. Figure 6 is a typical microstructure obtained in ZrO_2–4mol%Y_2O_3[11]. It is noted that this microstructure is different from the ordinary precipitation structure such as figure 7, and has the following features:

(1) the microstructure is generated in an early stage of aging, and is sometimes found even in rapidly–cooled samples,

(2) the black and white lamellae have a habit plane close to {111}, which is normal to the elastically–soft direction of c–ZrO_2,

(3) the lamellar spacing decreases with a decrease in aging temperature as predicted from the theory of spinodal decomposition,

(4) the modulated structure is formed in an area within the c–t two–phase field.
We have assessed that the microstructure is formed by spinodal decomposition[10,11]. Our interpretation, however, has been opposed by Heuer's group[9]. They have several reasons but the important points may be described as follows:

(1) the microstructure consisting of black and white lamellae is not the modulated structure but the tweed structure associated with the strain field of coherent t–ZrO_2 precipitates in the c–ZrO_2 matrix,

(2) the habit plane is not {111} but {011},

(3) spinodal decomposition never happens inside the c–t two–phase field, because the two phases have different crystal symmetry.

It is not possible to reach a firm conclusion on the first point solely from microstructural evidence through electron microscopy. The habit plane of the microstructure has been reexamined in ZrO_2–3mol%Y_2O_3 by Hayakawa et al[30]. They have reported that the trace analysis yielded a result consistent with the (223) habit plane, which is compatible both with platelet precipitates formed by a nucleation and growth process, and with a modulated structure induced by spinodal decomposition. They did not discuss whether spinodal decomposition in zirconia alloys is rational or not. The present author speculates that the microstructure they termed the tweed contrast must be a modulated structure associated with spinodal decomposition. The habit plane is about 10° different from {111} but is far from {011}. They have insisted that the difference in habit plane between (223) and {111} is caused by the accuracy of trace analysis. It may, however, be possible to expect that the difference arises not simply from the accuracy of trace analysis but results from the difference in alloys used. Their 3mol%Y_2O_3 alloy is almost the lowest composition exhibiting spinodal decomposition, and has a large tetragonality of about 1.013. The habit plane of {111} was determined in our original work for ZrO_2-5.2mol%Y_2O_3[10] which has a tetragonality of about 1.007. The habit plane must be related to the tetragonality, which determines the elastically–soft direction in crystals. It seems rational that the habit plane varies from (223) to {111} with decreasing tetragonality and, therefore, with increasing Y_2O_3 content. The tweed contrast examined by Heuer's group must have a different microstructural origin from the modulated structure.

The third point is an important one. As has been described in the original paper[31], spinodal decomposition in solids occurs between phases which are crystallographically quite similar to the original phase. The case where the two phases have the same crystal structure and differ only in composition is the simplest case. The two phases may also have slightly different crystal structures, which must be the case for the tetragonal and cubic phases in zirconia alloys. As mentioned previously, spinodal decomposition can be expected inside the c–t miscibility gap. The thermodynamic analysis predicts the sharp maximum as a tricritical point, if this transformation is treated as a second–order phase transition[15,16]. In addition, it is necessary to point out that the modulated structure is developed from t–ZrO_2 with uniform composition, i.e. the compositional fluctuation must be produced from a single tetragonal state. The modulated structure is formed within the domains of t–ZrO_2 as in figure 8[32]. The sample of figure 8 was originally rapidly cooled after arc–melting and subsequently aged at 1200°C. Both curved domain boundaries and the modulated structure are seen in this micrograph. The domains grow slightly during aging, and the modulated structure is developed within them. This fact seems to suggest that the modulated structure is associated with compositional fluctuation in the t–ZrO_2. The two–phase microstructure consisting of t–ZrO_2 and c–ZrO_2 with nearly equilibrium compositions is developed at a later stage of aging.

GRAIN SIZE STABILITY

Tetragonal zirconia polycrystal ceramics have stable, fine grain size. It is quite unique for a single–phase material to have such grain size stability. In this section, the origin of grain size stability is briefly discussed.

It has been clarified that the sluggish grain growth of zirconia containing 3–4mol%Y_2O_3 is caused by annealing in the two–phase field, and by the partitioning

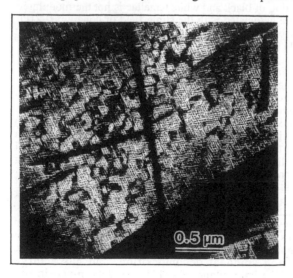

Figure 8. Microstructure of ZrO_2-6mol%Y_2O_3 aged at 1200°C for 120h [32]. The micrograph was taken using a 112 reflection.

of cations between grains[33–36]. According to a recent report[36], the grain growth takes place in two steps; the initial very sluggish growth during the cation partitioning and the latter a little faster growth after the equilibrium partitioning is attained. Even in the latter stage, grain growth is still sluggish and the fine grain size is stable during annealing in the two–phase field. Figure 9 is an example showing the yttria partitioning between grains in fine-grained ZrO_2–4mol%Y_2O_3 annealed at 1700°C for 10h[34]. The numerical figures in this micrograph represent the average yttria content of each grain estimated by TEM–EDS analysis. The yttria content is separated into two groups; one has around 2mol%Y_2O_3 and the other 6mol%Y_2O_3. These yttria contents are close to the equilibrium contents in the tetragonal and cubic phases at 1700°C. This fact indicates that the equilibrium partitioning of yttria is almost completed by the annealing. The microstructure must be the mixed structure of tetragonal grains and cubic grains at 1700°C. This structure is the same as the, so–called, dual–phase or duplex structure in metallic materials. The annealing in the two–phase field may, therefore, be termed intercritical annealing[37]. It is well–known that the generation of a dual–phase structure is very effective in keeping the grain size fine. The grain size stability in fine–grained, toughened zirconia is related to its annealing in the c–t two–phase field, and must be due to cation partitioning.

It is noted that the low–yttria content grains in figure 9 are almost featureless except for bend contours, but the high–yttria content grains include plate–like features. The latter are twins induced by mechanical accommodation caused by the c–t transformation during cooling after the high–temperature anneal. In other words, the c–t dual–phase structure is not retained at room temperature. The room–temperature X–ray diffraction profiles in this alloy show two t–ZrO_2 phases with different tetragonality; one corresponds to t–ZrO_2 containing about 2mol%Y_2O_3 and the other about 6mol%Y_2O_3[34,35]. The high–temperature c–ZrO_2 with about 6mol%Y_2O_3 must be inevitably and completely transformed into t–ZrO_2 during cooling so that a fully tetragonal structure is obtained at room temperature. The generation of a fully–tetragonal structure in fine–grained zirconia after annealing in the two–phase field is also related to the nature of the c–t transformation.

CONCLUDING REMARKS

The c–t transformation may be treated satisfactorily as a second–order phase transition. On this assumption, it is possible to calculate the c–t phase

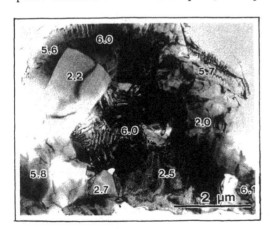

Figure 9. Electron micrograph of fine-grained ZrO_2–4mol%Y_2O_3 annealed at 1700°C for 10h [34].

boundaries in some binary zirconia alloys from a thermodynamic basis. The calculated two-phase field is consistent with experimental data for ZrO_2-Y_2O_3 and nonstoichiometric ZrO_2. This fact seems to suggest that anion vacancies play a principal role in the stability of c-ZrO_2 and in the nature of the c-t transformation, at least, to a first approximation. The two-phase field is regarded as a miscibility gap with a spinodal. The modulated structure with {111} habit formed in an early stage of aging in some ZrO_2 alloys must be associated with the spinodal decomposition. Inside the spinodal, the precipitates with nearly equilibrium composition, t-ZrO_2 precipitates in the c-ZrO_2 matrix, must be developed in a later stage of aging. The diffusionless c-t transformation seems to induce a microdomain microstructure, the domain structure accompanying curved boundaries in a single t-ZrO_2 variant and twins. The grain size stability in fine-grained, toughened zirconia is associated with the partitioning of cations between grains during annealing in the two-phase field. After the annealing, a fully tetragonal structure is developed at room temperature. This phenomenon is also related to the nature of this transformation.

ACKNOWLEDGMENTS

This work is financially supported by the Grant-in-Aid for General Scientific Research B 04453055 for Fundamental Research from the Ministry of Education, Science and Culture, Japan.

REFERENCES

1. Cohen, I. and B.E.Schaner, 1963. "A Metallographic and X-Ray Study of the UO_2-ZrO_2 System." J. Nucl. Mater., 9:18-52.

2. Subbarao, E.C., H.S.Maiti and K.K.Srivastava, 1974. "Martensitic Transformation in Zirconia." Phys. Stat. Sol., (a)21:9-40.

3. Andersson, C.A., J.Greggi Jr. and T.K.Gupta, 1984. "Diffusionless Transformations in Zirconia Alloys," in Advances in Ceramics, Vol. 12, Science and Technology of Zirconia II, N.Claussen, M.Ruhle and A.H.Heuer, eds., Columbus, OH, USA: The Am. Ceram. Soc., pp. 78-85.

4. Lanteri, V., A.Chaim and A.H.Heuer, 1986. "On the Microstructures Resulting from the Diffusionless Cubic-Tetragonal Transformation in ZrO_2-Y_2O_3 Alloys." J. Am. Ceram. Soc., 69: C258-261.

5. Chaim, R., V.Lanteri and A.H.Heuer, 1987. "The Displacive Cubic-Tetragonal Transformation in ZrO_2 Alloys." Acta Metall., 35:661-666.

6. Hannink, R.H.J., 1978. "Growth Morphology of the Tetragonal Phase in Partially-Stabilized Zirconia." J. Mater. Sci., 13:2487-2496.

7. Ruhle, M., N.Claussen and A.H.Heuer, 1984. "Microstructural Studies of Y_2O_3--Containing Tetragonal ZrO_2 Polycrystals(Y-TZP)," in Advances in Ceramics, Vol.12, Science and Technology of Zirconia II, N.Claussen, M.Ruhle and A.H.Heuer, eds., Columbus, OH, USA: The Am. Ceram. Soc., pp.352-370.

8. Chaim, R., M.Ruhle and A.H.Heuer, 1985. "Microstructural Evolution in a 12wt%Y_2O_3-ZrO_2 Ceramic." J. Am. Ceram. Soc., 68: 427-431.

9. Heuer, A.H., R.Chaim and V.Lanteri, 1988. "Review: Phase Transformations and Microstructural Characterization of Alloys in the System Y_2O_3-ZrO_2," in Advances in Ceramics, Vol. 24A, Science and Technology of Zirconia III, S.Somiya,

N.Yamamoto and H.Yanagida, eds., Westerville, OH, USA: The Am. Ceram. Soc., pp. 3-29.

10. Sakuma, T., Y.Yoshizawa and H.Suto, 1985. "The Modulated Structure Formed by Isothermal Ageing in ZrO_2-5.2mol%Y_2O_3 Alloy." J. Mater. Sci., 20:1085-1092.

11. Sakuma, T, Y.Yoshizawa and H.Suto, 1986. "Metastable Two-phase Region in the System ZrO_2-Y_2O_3." J. Mater. Sci., 21:1436-1440.

12. Sakuma, T., 1988. "Phase Transformation and Microstructure of Partially-Stabilized Zirconia." Trans. Jpn. Inst. Met., 29:879-893.

13. Sakuma, T., 1987. "Development of the Domain Structure Associated with the Diffusionless Cubic-to-Tetragonal Transition in ZrO_2-Y_2O_3 Alloys," J. Mater. Sci., 22:4470-4475.

14. Sakuma, T. and H.Hata, 1989. "The Domain Structure of Tetragonal Zirconia in ZrO_2-Y_2O_3 Alloys," in Zirconia 88, Advances in Zirconia Science and Technology, S.Meriani and C.Palmonari, eds., London, UK: Elsevier, pp. 283-292.

15. Hillert, M. and T.Sakuma, 1991. "Thermodynamic Modeling of the c-t Transformation in ZrO_2 Alloys." Acta Metall. Mater., 39:1111-1115.

16. Hillert, M., 1991. "Thermodynamic Model of the Cubic-Tetragonal Transition in Nonstoichiometric Zirconia." J. Am. Ceram. Soc., 74:2005-2006.

17. Teufer, G., 1962. "The Crystal Structure of Tetragonal ZrO_2." Acta Cryst., 15:1187-1188.

18. Landau, L.D. and E.M.Lifshitz, 1958. Statistical Physics, Oxford, UK: Pergamon Press.

19. Scott, H.G., 1975. "Phase Relationships in the Zirconia-Yttria System." J. Mater. Sci., 10:1527-1535.

20. Yoshikawa, N., E. Eda and H. Suto, 1986. "On the Cubic/Tetragonal Phase Equilibrium of the ZrO_2-Y_2O_3 System." J.Jpn.Inst.Met., 50:113-118.

21. Ackermann, R.J. and E. G. Rauh, 1975. "The Thermodynamic Properties of ZrO_2(g)." High Temp. Sci., 7:304-316.

22. Sakuma, T., Y.Yoshizawa and H.Suto, 1985. "The Microstructure and Mechanical Property of Yttria-Stabilized Zirconia Prepared by Arc-Melting." J. Mater. Sci., 20:2399-2407.

23. Ingel, R.P., D.Lewis and B.A.Bender, 1988. "Properties and Microstructures of Rapidly Solidified Zirconia-Based Ceramic Alloys", in Advances in Ceramics, Vol. 24A, Science and Technology of Zirconia III, S.Somiya, N.Yamamoto and H.Yanagida, eds., Westerville, OH, USA: The Am. Ceram. Soc., pp. 385-396.

24. Shibata-Yanagisawa, M., M.Kato, H.Seto, N.Ishizawa, N.Mizutani and M.Kato, 1987. "Crystallographic Analysis of the Cubic-to-Tetragonal Phase Transformation in the ZrO_2-Y_2O_3 System." J. Am. Ceram. Soc., 70:503-509.

25. Hayakawa, M. and M.Oka, 1990. "Strain Analysis of the Herringbone Structures Observed in ZrO_2 Alloys." Mater. Sci. Forum, 56-58:383-388.

26. Yoshimura, M., 1988. "Phase Stability of Zirconia." Ceramic Bulletin, 67:1950-1955.

27. Sakuma, T., T.Seki and T.Yamamoto, 1992. "The Diffusionless Cubic-to-Tetragonal Transition in ZrO_2-R_2O_3 Systems (R:Rare Earths)," a paper presented at ICOMAT-92 Conference, Monterey, USA.

28. Hannink, R.H.J., 1983. "Microstructural Development of Sub-Eutectoid Aged MgO-ZrO_2 Alloys." J.Mater.Sci., 18:457-470.

29. Porter, D.L. and A.H.Heuer, 1979. "Microstructural Development in MgO-Partially Stabilized Zirconia(Mg-PSZ)." J.Am.Ceram.Soc., 62:298-305.

30. Hayakawa, M., K.Adachi and M.Oka, 1990. "Tweed Contrast with (223) Habit in Arc-Melted Zirconia-Yttria Alloys." Acta Metall. Mater., 38:1761-1767.

31. Cahn, J.W., 1968. "Spinodal Decomposition." Trans. Met. Soc. AIME, 242,166-179.

32. Sakuma, T. and H.Hata, 1989. "Diffusionless Cubic-to-Tetragonal Transformation and Microstructure in ZrO_2-Y_2O_3." J. Jpn. Inst. Met., 53:972-979.

33. Lange, F.F., D.B.Marshall and J.R.Porter, 1988. "Controlling Microstructures through Phase Partitioning from Metastable Precursors," in Ultrastructure Processing of Advanced Ceramics, J.D.Mackenzie and D.R.Ulrich, eds., New York, NY, USA: John Wiley & Sons, pp. 519-532.

34. Yoshizawa, Y. and T.Sakuma, 1989. "Evolution of Microstructure and Grain Growth in ZrO_2-Y_2O_3 Alloys." ISIJ Int., 29:746-752.

35. Sakuma, T. and Y. Yoshizawa, 1992. "The Grain Growth of Zirconia during Annealing in the Cubic/Tetragonal Two-Phase Region." Mater. Sci. Forum, 94-- 96:865-870.

36. Stoto, T., M.Nauer and C.Carry, 1991. "Influence of Residual Impurities on Phase Partitioning and Grain Growth Processes of Y-TZP Materials." J. Am. Ceram. Soc., 74:2615-2621.

37. Sakuma, T., Y.Yoshizawa and Y.Zhou, 1989. "Intercritical Annealing in the System ZrO_2-Y_2O_3," in New Materials and Processes for Future, N.Igata, I.Kimpara, T.Kishi, E.Nakata, A.Okura and T.Uryu, eds., Tokyo, Japan: The Nikkan Kogyo Shinbun, pp. 815-820.

On the Stability of Tetragonal Precipitates in Y_2O_3-Partially Stabilized ZrO_2 Single Crystals

J. MARTINEZ-FERNANDEZ, M. JIMENEZ-MELENDO,
A. DOMINGUEZ-RODRIGUEZ, G. BEHRENS and A. H. HEUER

ABSTRACT

The stability of the tetragonal precipitates against the tetragonal-to-monoclinic transformation in Y_2O_3-partially-stabilized ZrO_2 (Y-PSZ) single crystals with two different Y_2O_3 compositions (3.4 and 4.7 ^{m}/o) has been investigated using Raman spectroscopy and transmission electron microscopy. Aging at 1600°C causes precipitation and coarsening of tetragonal particles in the cubic matrix and a concomitant decrease in their stability. The particle stability appears to be unrelated to their size and shape; rather, particle-particle interactions seem to be important in determining the M_S temperature. The internal microstructure of the tetragonal and monoclinic phases are discussed.

INTRODUCTION

Since the discovery of the martensitic tetragonal (t) to monoclinic (m) transformation in ZrO_2 as a toughening mechanism in partially stabilized ZrO_2 (PSZ) alloys [1,2], much effort has been devoted to the study of the transformation. The room temperature toughness of PSZ alloys increases with decreasing stability of the metastable t-ZrO_2 particles, because such particles readily undergo the stress-induced $t \Rightarrow m$ crack tip transformation. However, when the particles are so transformable that the $t \Rightarrow m$ transformation occurs during cooling from high temperatures (*i.e.* the martensite start temperature M_S is above room temperature), the mechanical properties can degrade. Therefore, in order to optimize the mechanical properties, the tetragonal phase must be in an appropriate metastable state at room temperature.

The stability of t-ZrO_2 particles depends on a number of factors, such as particle size and alloy composition [3]. The t-ZrO_2 precipitates in Y_2O_3-containing alloys are particularly stable compared to alloys containing other solute stabilizers [4,5]. We have investigated the stability of such precipitates in two different Y_2O_3-containing alloys as a function of precipitate size, which was varied by high-temperature aging.

J. Martinez-Fernandez, M. Jimenez-Melendo, and A. Dominguez-Rodriguez, Dpto. de Fisica de la Materia Condensada, Universidad de Sevilla, Apdo. 1065, 41080 Sevilla, Spain.
G. Behrens and A. H. Heuer, Department of Materials Science and Engineering, Case Western Reserve University, Cleveland, OH 44106, USA.

EXPERIMENTAL

Cubic (*c*) ZrO$_2$ single crystals containing 3.4 and 4.7 ^{m}/$_o$ Y$_2$O$_3$ and grown by skull-melting (supplied by Ceres Corp., Billerica, MA, USA) were used in this work. Crystals were oriented by the Laue back-reflection X-ray technique so that thin plates with low index surfaces could be cut and polished, finishing with a 1µm grade diamond grit. The samples were aged at 1600°C in air for up to 1000h, followed by quenching to room temperature.

The Y$_2$O$_3$ content of the crystals was determined with energy dispersive X-ray spectroscopy (EDS) attached to electron microscopes. EDS spectra were obtained from bulk samples in a scanning electron microscope (SEM) operating at 75 keV, and from thin foils in a scanning transmission electron microscope (STEM) operating at 200 keV. In all cases, sample areas large compared to the precipitate size were used to measure average concentration. Both methods gave similar results, with the standard deviations being less than 5%.

Raman spectroscopy was applied to bulk crystals to determine the fraction of *t*-ZrO$_2$ that had transformed to monoclinic symmetry on cooling from the aging temperature. The characteristic Raman spectra of *m*- and *t*-ZrO$_2$ are well documented and easily distinguished [6,7].

Thin foil specimens for transmission electron microscopy (TEM) studies were prepared by mechanical thinning and subsequent ion-thinning to electron transparency. TEM was employed to investigate the metastability of *t*-ZrO$_2$ in thin foils, as well as to analyze the morphology of the precipitates and the crystallographic aspects of the transformation.

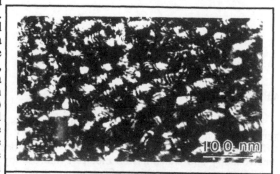

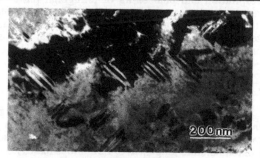

Figure 1. TEM micrograph ((100)$_c$ foil) of the 3.4 ^{m}/$_o$ crystal (a) in the as-received condition, containing small *t*-ZrO$_2$ precipitates; (b) after 2h of aging, illustrating the twinned *t*-ZrO$_2$ colony precipitates; and (c) after 200h of aging, showing transformed precipitates.

RESULTS

The compositions of the as-received crystals are inside the two-phase $c+t$ field of the Y_2O_3-ZrO_2 system at 1600°C [8]. During the heat treatments at 1600°C, the crystals undergo diffusional decomposition to form solute-lean t-ZrO_2 precipitates in a fully-stabilized, solute-rich c-ZrO_2 matrix. According to the phase diagram, the equilibrium compositions of t- and c-ZrO_2 at 1600°C are ~1 and ~7 $^m/o$ Y_2O_3, respectively, and the corresponding equilibrium volume fraction of t-ZrO_2 is then ~0.6 in the 3.4 $^m/o$ crystal, and ~0.4 in the 4.7 $^m/o$ crystal. The actual values may be somewhat different, as the precipitation reaction is believed to take place under *coherent* equilibrium [9].

The as-received crystals contain very small, homogeneously-nucleated t-ZrO_2 precipitates formed during slow cooling through the two-phase $c+t$ field (figure 1a). In the early stages of the heat treatment at 1600°C, the t-ZrO_2 precipitates coarsen to form elongated and internally-twinned "colony" precipitates [10], as shown in figure 1b. Each colony precipitate consists of alternating lamellae whose c_t-axes are at ~90° to each other; the twin plane is $\{101\}_t$. The precipitation reaction is complete after 24h of aging; further aging causes coarsening of the elongated precipitates which often interconnect, forming a zig-zag fibrous structure (figure 1c). The precipitation and coarsening kinetics are essentially equivalent for the two alloy compositions.

Raman spectra were obtained from bulk samples after different heat treatments. In the 3.4 $^m/o$ crystals, the spectra reveal no evidence of $t \Rightarrow m$ transformation in samples aged for up to 150h, but samples annealed for longer times ($\geq$200h) display strong monoclinic Raman peaks. In comparison, Raman spectroscopy on 4.7 $^m/o$ samples shows no evidence of transformation in bulk samples, even after aging for 1000h.

TEM foils of 3.4 $^m/o$ samples aged for 50h contained only monoclinic, *i.e.* transformed, precipitates. Accordingly, a sample aged for 150h was employed for a series of experiments to analyze the precipitate stability in thin foils. The sample was thinned to ~40μm in the center and annealed for 1h at 1300°C, in order to reverse any stress-induced $t \Rightarrow m$ transformation that may have occurred during sample preparation up to this point. After subsequent ion thinning, TEM observations of this foil showed only monoclinic precipitates. Next, this same foil was re-annealed at 1300°C for 1h, and TEM observations revealed that the precipitates were tetragonal again. The $m \Rightarrow t$ (reverse) transformation must have occurred on heating but no forward $(t \Rightarrow m)$ transformation occurred during cooling; this has been confirmed by monitoring the transformation using an *in situ* TEM heating stage. After this anneal, the precipitates were unstable during TEM examination and readily transformed to m-ZrO_2 in the thin foil, due to thermoelastic stresses associated with electron beam irradiation.

TEM observations of 4.7 $^m/o$ foils revealed no $t \Rightarrow m$ transformation during foil preparation, even after 1000h

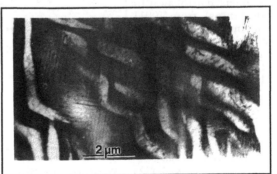

Figure 2. TEM micrograph ($(100)_c$ foil) of precipitates in the 4.7 $^m/o$ crystal aged for 1000h. The particles are oriented with their $\{101\}_t$ twin planes at 45° (trace along $<100>_c$) and 90° (trace along $<110>_c$, the so-called edge-on particles) to the plane of the foil.

of aging. Figure 2 illustrates the microstructure of such a sample; the precipitates have grown to several microns in length due to coarsening and interconnection of individual colony precipitates, forming fibers along $<100>_c$. Despite their large size, the precipitate/matrix interface apparently remains coherent. These precipitates in thin foils do, however, transform under the electron beam. The *in situ* transformation to m-ZrO_2 can be readily observed in figures 3a and b, the two TEM micrographs of the same region of the foil having been taken 5 min apart.

These *in situ* observations allow for detailed studies of the crystallography and morphology of the $t \Rightarrow m$ transformation product in thin foils. The m-ZrO_2 product forms as thin laths that cross the colony precipitates (figure 3), oriented at 30° to the edge-on t-ZrO_2 twin planes. Trace analysis (figure 3c) reveals that the m-ZrO_2/t-ZrO_2 habit plane of the *in situ* transformation is near $\{311\}_c$. This analysis was accomplished as follows. Two foil orientations were used, corresponding to $[010]_c$ and $[\bar{1}30]_c$ zone axes; the twin plane orientation is $(101)_t$. In each orientation, the habit plane of the laths yielded a trace; the plane normals of the true habit planes lay perpendicular to this trace. In the $[010]_c$ foil, the traces of the habit planes were along two well-defined directions, $\pm \sim 15°$ to $[001]$, while in the $[\bar{1}30]_c$ foil, they were oriented 8-15° away (in both senses) from $[001]$. Assuming that the trace of the habit plane occurs to the "left" of $[001]$ in both foils, the habit planes must be $[\bar{3}11]_t$ and $(311)_t$ with regard to one tetragonal variant and $[11\bar{3}]_t$ and $(113)_t$ with regard to the other. The t/m orientation relationship could not be determined because of the small size of the laths.

While the m-ZrO_2 particles grow during TEM observation, they adopt a lenticular shape in which two of the habit planes are near

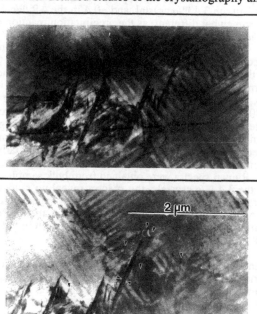

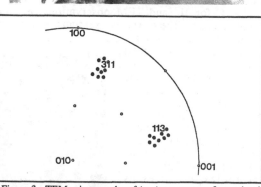

Figure 3. TEM micrographs of *in situ* $t \Rightarrow m$ transformation in the 4.7 $^m/o$ crystal aged for 1000h, induced by electron beam heating; the micrographs in (a) and (b) were taken 5 min apart. The growing m-ZrO_2 laths are arrowed. (c) shows a stereographic trace analysis of the m-ZrO_2/t-ZrO_2 habit planes observed during this *in situ* $t \Rightarrow m$ transformation.

$\{311\}_t$ and the other two are close to $\{110\}_t$ (figure 4a). Occasionally, adjacent particles are twinned on $\{100\}_m$ (figure 4b). For these larger m-ZrO_2 particles, the orientation relationship is $a_m//a_t$ and $b_m//c_t$ for one t-ZrO_2 twin variant, while that for the other variant is generated by a 90° rotation about b_t. The original tetragonal twin planes remain visible inside the m-ZrO_2 particles (figure 5), probably because they remain as disordered planes through the transformation process.

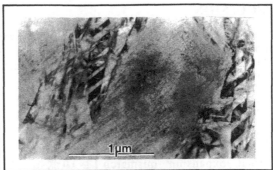

Several TEM micrographs in the $(100)_c$ orientation were employed for a stereological analysis of the development of the precipitate morphology during aging. The particle length (figure 6a) was measured as the longest dimension of the colony

Figure 4. TEM micrographs of the 4.7 $^m/_o$ crystal aged for 1000h; (a) lenticular and (b) twinned monoclinic particles.

precipitates; the interconnection of the precipitates into fibers after the longer aging times was ignored. For a given aging time, the t-ZrO_2 particles in the 3.4 $^m/_o$ crystal are consistently larger than those in the 4.7 $^m/_o$ material. The width of the t-ZrO_2 twin planes (figure 6b) was measured utilizing only edge-on particles in the micrographs. Even after the particles are transformed, the twin spacing can still be determined since the twin planes remain visible in the transformation product (see figure 5). Again, the values are higher in the 3.4 $^m/_o$ material than in the 4.7 $^m/_o$ material. The volume fraction of precipitates (figure 6c) was determined by superimposing a grid onto micrographs and counting the intersections of grid points with precipitates. In both alloys, the precipitate fraction remains approximately constant after 24h of aging but is ~5% higher in the 3.4 $^m/_o$ alloy. The precipitate volume fraction in the 3.4 $^m/_o$ samples aged for >50h may be an underestimate, as the strain contrast due to the $t \Rightarrow m$ transformation may have obscured some m-ZrO_2 particles.

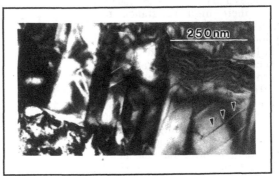

Figure 5. TEM micrographs of the 4.7 $^m/_o$ crystal aged for 1000h showing m-ZrO_2 particles which reveal contrast due to the twin planes of the parent phase.

DISCUSSION

It is now generally believed that the $t \Rightarrow m$ transformation in ZrO_2 ceramics is controlled by the stress-assisted nucleation of the martensitic product [11,12]; local stresses can lower the activation energy for nucleation by interacting with the transformation strains [13]. These stresses may be external, such as crack tip stresses, or internal, resulting from coherency strains or thermal expansion mismatch between the t-ZrO_2 precipitates and the surrounding c-ZrO_2 matrix.

Raman spectroscopy indicates that in the 4.7 $^m/o$ material, M_S remains below room temperature even after 1000h of aging, but that in 3.4 $^m/o$ crystals aged for longer than 150h, M_S is above room temperature. In these slightly over-aged low-Y_2O_3 crystals, the colony precipitates are ~0.3 μm wide and ~2 μm long; the precipitate fibers, formed by colony precipitates growing together, can reach lengths of up to 8 μm. These "critical" precipitate dimensions when M_S increases to above room temperature are considerably larger than those in other ZrO_2 alloys; in MgO- and CaO-partially-stabilized ZrO_2's, for example, the critical particle sizes are ~200 nm [14] and ~60 nm [15], respectively.

The stability of the t-ZrO_2 particles in the Y_2O_3-ZrO_2 system, $i.e.$ their high resistance to transformation, is related to their microstructure and morphology [16]. Each elongated colony precipitate lies on a $\{101\}_c$ habit plane and contains a set of two $\{101\}_t$ twin variants, resulting in a total of 12 possible colony orientations. The lamellar twinning lowers the strain energy (due to the ~0.02% dilatation associated with the $c \Rightarrow t$ transformation); the lattice distortion of one variant compensates that of the twin-related variant, allowing the $\{101\}_t$ twin plane to be essentially free of

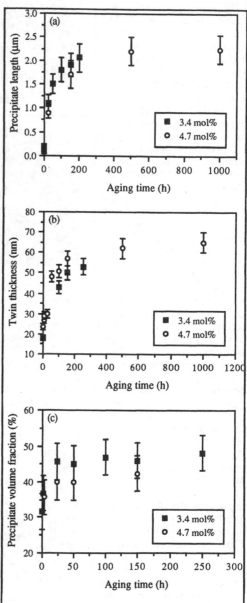

Figure 6. Comparison of stereological data in the two alloys after various aging times: (a) precipitate size, (b) twin spacing, and (c) volume fraction of precipitates.

coherency stresses [17]. This reduced strain energy in the system allows the precipitates to reach large dimensions without loss of coherency, and therefore provides insufficient nucleation sites for the stress-induced martensitic transformation. In MgO- and CaO-alloys, the critical precipitate size for "spontaneous" $t \Rightarrow m$ transformation on cooling to room temperature is smaller

Measurement (approximate values)	3.4 m/o Aged for 24h	4.7 m/o Aged for 1000h
Volume fraction	47 %	41 %
Particle length	1.1 μm	2.3 μm
Particle width	0.2 μm	0.4 μm
Twin spacing	29 nm	65 nm
Aspect ratio	6	6
Particle density	13 /μm³	1.4 /μm³
Particle/matrix interface	9.4 μm²/μm³	4.2 μm²/μm³
No. lamella / vol	50 /μm³	6 /μm³
Twin plane/matrix interface	40 μm/μm³	10 μm/μm³

TABLE I - STEREOLOGICAL DATA ON PEAK-AGED PRECIPITATES

because the much larger lattice distortions of the $c \Rightarrow t$ transformation are not accommodated by twinning of the t-ZrO$_2$ precipitates.

TEM observations of the 3.4 m/o crystals revealed that the precipitates in thin foils are transformed after only 50h of aging. Annealing of such foils at 1300°C caused re-transformation of the precipitates to tetragonal symmetry, but after cooling, these precipitates were generally unstable and readily underwent stress-induced $t \Rightarrow m$ transformation during ion milling and under the electron beam. These crystals are thus peak-aged, with their M_s temperature slightly below room temperature. We conclude that the transformation observed in 3.4 m/o TEM foils aged for 50 to 150h is an artifact, in that particles in *bulk* crystals are stable.

To analyze the particle stability, we chose to compare the microstructures of the two alloy compositions in the same state of transformability, *i.e.* when the $t \Rightarrow m$ transformation can be induced in TEM foils with the electron beam, but not by TEM sample preparation. This occurs after ~24h of aging in the 3.4 m/o alloy, and after 1000h of aging in the more stable 4.7 m/o alloy. The microstructures and morphologies of these peak-aged crystals are compared in Table I. Whereas the volume fraction of precipitates is slightly higher in the 3.4 m/o material, the size and twin spacing in those precipitates is considerably smaller than in the high-solute alloy. This must be because during cooling, the 3.4 m/o alloy enters the two phase $c \Rightarrow t$ phase field at a much higher temperature (~2200K) than the 4.7 m/o alloy (~2000K), which must have resulted in more copious nucleation. The resulting microstructure clearly reflects the pre-aging distribution of nuclei.

A number of possibly relevant stereological factors were calculated, such as the particle concentration, particle/matrix interfacial area, and twin plane density; however similarities between the precipitates in these two alloys that might explain their similar transformabilities were undetectable. This is interpreted as indicating that the particle transformability is relatively independent of the morphology of the individual precipitates; instead, it must be controlled by particle-particle interactions. This idea is supported by the fact that considerable autocatalytic transformation* has been observed in these materials [18].

Trace analyses have revealed that during early stages of transformation, the m-ZrO$_2$/t-ZrO$_2$ habit plane lies near {311}$_t$. Unfortunately, we do not know the appropriate t/m lattice correspondence, so we cannot compare our results with those reported in ref. 19 for Y$_2$O$_3$-stabilized ZrO$_2$ polycrystals. Upon growth of the m-ZrO$_2$ particles to a lenticular shape, additional habit planes lying close to {110}$_t$ are

---

* Autocatalytic transformation occurs when the $t \Rightarrow m$ transformation in one particle produces a sufficiently large strain field around that particle to allow nucleation in neighboring particles.

formed; this plane is near the $(671)_m$ and $(761)_m$ habit planes observed in single crystals of pure ZrO_2 [20].

CONCLUSIONS

The stability of t-ZrO_2 precipitates in Y_2O_3-ZrO_2 single crystals has been analyzed with the following results:
- Single crystals containing 3.4 and 4.7 $^m/o$ Y_2O_3 can be peak-aged by annealing at 1600°C for ~24h and ~1000h, respectively.
- The critical particle size for room-temperature transformation in the 3.4 $^m/o$ alloy is ~2 μm.
- The stability of t-ZrO_2 precipitates appears to be related to particle-particle interactions.
- The $t \Rightarrow m$ transformation observed using TEM in some crystals occurs due to stresses resulting from ion thinning and electron beam heating.

ACKNOWLEDGEMENTS

The research in Seville was sponsored by CICYT under Contract No. MAT-0181-C02, and that in Cleveland by NSF under Grant No. DMR-90-19383. Further support for this collaboration was provided by NATO under Grant No. 0734/88.

REFERENCES

1.   Garvie, R.C., R.H.J.Hannink and R.T.Pascoe, 1975. "Ceramic Steel?" Nature (London), 258: 703-704.

2.   Porter, D.L. and A.H.Heuer, 1977. "Mechanism of Toughening Partially Stabilized Zirconia (PSZ)." J. Amer. Ceram. Soc., 60(3,4): 183-84.

3.   Heuer, A.H., N.Claussen, W.Kriven and M.Rühle, 1982. "The Stability of Tetragonal ZrO2 Particles in Ceramic Matrices". J. Amer. Ceram. Soc., 65(12): 642-50.

4.   Rühle, M., N.Claussen and A.H.Heuer, 1985. "Microstructural Studies of Y_2O_3-Containing Tetragonal ZrO_2 Polycrystals (Y-TZP)," in Advances in Ceramics, Vol. 12. Science and Technology of Zirconia II, N.Claussen, M.Rühle and A.H.Heuer, eds., Columbus, OH, USA: American Ceramic Society, Inc., pp. 352-70.

5   Lanteri, V., A.H.Heuer and T.E.Mitchell, 1984. "Tetragonal Phase in the System ZrO_2-Y_2O_3," ibid., pp. 118-30.

6.   Clarke, D.R. and F.Adar, 1982. "Measurement of the Crystallographically Transformed Zone Produced by Fracture in Ceramics Containing Tetragonal Zirconia." J. Am. Ceram. Soc., 65(6): 284-88.

7.   Marshall, D.B., M.C.Shaw, R.H.Dauskardt, R.O.Ritchie, M.J.Readey and A.H.Heuer, 1990. "Crack-Tip Transformation Zones in Toughened Zirconia." J. Am. Ceram. Soc., 73(9): 2659-66.

8.   Stubican, V.S., 1988. "Phase Equilibria and Metastabilities in the Systems ZrO_2-MgO, ZrO_2-CaO, and ZrO_2-Y_2O_3," in Advances in Ceramics, Vol. 24, Science and Technology of Zirconia III, S.Somiya, N.Yamamoto and H.Yanagida, eds., Westerville, OH, USA: American Ceramic Society, Inc., pp. 71-82.

9.   Heuer, A.H., R.Chaim and V.Lanteri, 1988. "Review: Phase Transformations and Microstructural Characterization of Alloys in the System Y_2O_3-ZrO_2," ibid., pp.3-20.

10. Heuer, A.H., V.Lanteri and A.Dominguez-Rodriguez, 1989. "High-Temperature Precipitation Hardening in Y_2O_3 Partially-Stabilized ZrO_2 (Y-PSZ) Single Crystals," Acta Metall. 37(2): 559-67.

11. Heuer, A.H. and M.Rühle, 1985. "On the Nucleation of the Martensitic Transformation in Zirconia (ZrO_2)," Acta Metall., 33(12): 2101-12.

12. Chen, I-W. and Y.-H.Chiao, 1985. "Theory and Experiment of Martensitic Nucleation in ZrO_2-Containing Ceramics and Ferrous Alloys," Acta Metall., 33(10): 1827-45.

13. Patel, J.R. and M.Cohen, 1953. "Criterion for the Action of Applied Stress in the Martensitic Transformation," Acta Metall., 1(9): 531-38.

14. Porter, D.L. and A.H.Heuer, 1979. "Microstructural Development in MgO-Partially Stabilized Zirconia (Mg-PSZ)," J. Am. Ceram. Soc., 62(5-6): 298-305.

15. Dickerson, R.M., M.V.Swain and A.H.Heuer, 1987. "Microstructural Evolution in Ca-PSZ and the Room-Temperature Instability of ZrO_2," J. Am. Ceram. Soc., 70(4): 214-20.

16. Lanteri, V, T.E.Mitchel and A.H.Heuer, 1986. "Morphology of Tetragonal Precipitates in Partially Stabilized ZrO_2," J. Am. Ceram. Soc., 69(7): 564-69.

17. Heuer, A.H., S.P.Krause-Lanteri, P.A.Labun, V.Lanteri and T.E.Mitchell, 1985. "HREM Studies of Coherent and Incoherent Interfaces in Zirconia Containing Ceramics: A Preliminary Account," Ultramicroscopy, 18(1-4): 335-48.

18. Martinez-Fernandez, J., M.Jimenez-Melendo, A.Dominguez Rodriguez and A.H.Heuer, 1991. "Microindentation-Induced Transformation in 3.5-mol%-Yttria-Partially-Stabilized Zirconia Single Crystals," J. Am. Ceram. Soc., 74(5): 1071-81.

19. Hayakawa, M., N.Kuntani and M.Oka, 1989. "Structural Study on the Tetragonal to Monoclinic Transformation in Arc-Melted ZrO_2-2mol.%Y_2O_3 - I. Experimental Observations," Acta. Metall. 37(8): 2223-35.

20. Bansal, G.K. and A.H.Heuer, 1974. "On a Martensitic Phase Transformation in Zirconia (ZrO_2) - II. Crystallographic Aspects," Acta Metall., 22(4): 409-17.

Cubic-Tetragonal Phase Transformations in the ZrO$_2$-CeO$_2$ System

MASATOMO YASHIMA, KENJI MORIMOTO, NOBUO ISHIZAWA and MASAHIRO YOSHIMURA

ABSTRACT

Single phase tetragonal $Zr_{1-x}Ce_xO_2$ solid solution was prepared by annealing at 627°C the c'- ZrO$_2$, which was obtained by quenching after sintering at 1660° or 1760°C, where the c'- ZrO$_2$ was defined as cubic or tetragonal phases whose axial ratio c/a (tetragonality) equals 1. The lattice parameter and cell volume increased but the tetragonality decreased with CeO$_2$ content. The $c' \rightarrow t'$ transformation was a thermally activated process with an activation energy of 113 kJ/mol, which was estimated by measuring the time dependence of the t'- fraction by X-ray diffraction.

This value indicates that the $c' \rightarrow t'$ transformation is controlled by oxygen diffusion. In fact, it did not occur in vacuum. The $c' \rightarrow t'$ cation-diffusionless transformation was clearly distinguished from the cation-diffusional phase separation $t' \rightarrow (t + c)$ which occurred above 900°C in the TTT diagram.

Reversible $c' \leftrightarrow t'$ transformation was observed in 65mol%CeO$_2$-ZrO$_2$. The $t' \rightarrow c'$ transformation temperatures were located in the vicinity of the cubic solubility limit within the $(c+t)$ two-phase region. The $t' \rightarrow c'$ transformation occurred without oxygen-concentration change in 65mol%CeO$_2$-ZrO$_2$.

INTRODUCTION

The tetragonal phase is formed through either the diffusionless cubic-to-tetragonal (t') phase transition or diffusional decomposition within the tetragonal + cubic two-phase compositional region [1]. The diffusionless c-t' phase transition in the ZrO$_2$-CeO$_2$ system is quite different from that in the ZrO$_2$-RO$_{1.5}$ ones (R: rare earths) [1-3]. In this paper, we discuss the nature of the c-t' transition in the ZrO$_2$-CeO$_2$ system on the basis of x-ray powder diffraction results. Details will be published elsewhere [4,5].

Masatomo YASHIMA, Kenji MORIMOTO, Nobuo ISHIZAWA, and Masahiro YOSHIMURA.: Research Laboratory of Engineering Materials, Tokyo Institute of Technology, 4259, Nagatsuta, Midori, Yokohama, 227, Japan.

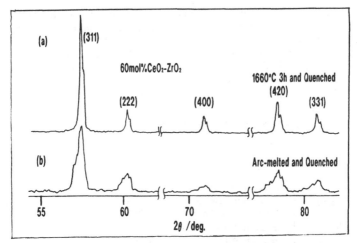

Figure 1.   X-ray powder diffraction pattern 60 mol% CeO$_2$-ZrO$_2$ quenched after firing at 1660°C for 3 h in air (a), and after arc melting (b).

The ZrO$_2$-solid solution has three forms in one tetragonal structure with space group P4$_2$/nmc: t-, t'- and t''-ZrO$_2$[2,3,6], which are conveniently distinguished from each other to describe the phase changes in ZrO$_2$-solid solutions. t'-ZrO$_2$ is defined as a tetragonal form with a tetragonality (axial ratio, c/a) larger than unity, which is formed by the diffusionless phase transition from the high-temperature cubic (c-) phase. t''-ZrO$_2$ [2,3] is an intermediate form between t'- and c-ZrO$_2$, which has no tetragonality but has oxygen ion displacements from its ideal sites in the fluorite-type structure. Since it is difficult for X-ray diffraction to distinguish t''-ZrO$_2$ from c-ZrO$_2$ due to the small atomic scattering factor of oxygen, we refer to both t''-ZrO$_2$ and c-ZrO$_2$ as c'-ZrO$_2$ in this paper.

Homogeneous samples must be used to investigate the diffusionless c'-t' phase transition. The compositional dependence of the lattice parameter has not been known clearly probably due to the difficulty in preparing such homogeneous samples. Therefore, we carefully prepared homogeneous ZrO$_2$-CeO$_2$ solid solutions and investigated the diffusionless phase transition between t'- and c'-ZrO$_2$.

PREPARATION OF HOMOGENEOUS c'-(Zr,Ce)O$_{2-\delta}$

High purity ZrO$_2$ and CeO$_2$ were mixed as dry powders and methanol slurries for 3 hours in an agate mortar. These mixed powders (ZrO$_2$-Xmol% CeO$_2$, X=30-90) were pressed into pellets at 200 MPa, and fired at 1400℃ for 2 h in air, and then crushed and ground thoroughly in an alumina mortar. The obtained powder was again pressed into pellets and fired at 1660℃ for 3 h or at 1760℃ for 2.5 h and then quenched by being dropped into water. The black color of the quenched sample indicates partial reduction of Ce^{4+} into Ce^{3+} at high temperatures [7]. As shown in figure 1 (a), the x-ray powder diffraction profile showed a single phase of c'- ZrO$_2$ with very sharp reflection peaks, which indicated the formation of a compositionally homogeneous sample, where each single peak had doublets due to

$K\alpha_1$ and $K\alpha_2$. The lattice parameter increased continuously with CeO_2 content, also showing the formation of a compositionally homogeneous solid solution between ZrO_2 and $CeO_{2-\delta}$.

Rapid quenching of the melt is a very effective technique to prepare homogeneous samples in the ZrO_2-$RO_{1.5}$ systems (R=rare earths) [8]. In the ZrO_2-CeO_2 system, the arc-melted sample, however, had inhomogeneities as shown in figure 1 (b). This inhomogeneity may be ascribed to the evaporation and precipitation of Ce and valence change from Ce^{4+} to Ce^{3+} at high temperatures. Muroi *et al.* [9] prepared similarly inhomogeneous ZrO_2-$CeO_{2-\delta}$ samples by arc-melting.

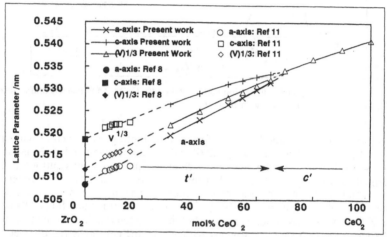

Figure 2. Relationship between CeO_2 concentration and lattice parameters of *"non-defective"* t'- and c'- $(Zr,Ce)O_2$.

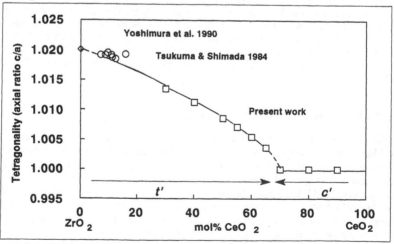

Figure 3. Relationship between CeO_2 concentration and the axial ratio c/a (tetragonality) of *"non-defective"* t'- and c'- $(Zr,Ce)O_2$.

FORMATION OF HOMOGENEOUS *t'*- (Zr,Ce)O$_2$

The *c'*- ZrO$_2$ powder obtained by crushing and grinding the sintered materials at 1660° or 1760°C was annealed at 627℃ for 10 min in air to recover the partial reduction. The sample color changed into white from black. The lattice volume of each specimen decreased during the annealing. This volume decrease and the color change were ascribed to the valence change from Ce^{3+} into the smaller Ce^{4+} ion. The *"defective"* *c'*- ZrO$_2$ containing 30-65 mol% CeO$_2$ transformed into *"non-defective"* *t'*- ZrO$_2$ during this annealing. The lattice parameters, *c, a* and the cube root of the unit cell volume increased continuously with CeO$_2$ content as shown in figure 2. The tetragonality (axial ratio *c/a*) decreased with CeO$_2$ content and became unity in the samples with compositions larger than 65 mol% CeO$_2$ (Figure 3). These compositional dependencies of lattice parameters are similar to those in ZrO$_2$-RO$_{1.5}$ systems [8,10]. Tsukuma and Shimada [11] reported the lattice parameter of *t*- ZrO$_2$, which was not metastable *t'*- ZrO$_2$ but the stable form at high temperatures; however its formation region was limited by the tetragonal solubility limit as shown in Figs. 2 and 3. Meriani [12] prepared *t'*-ZrO$_2$, although the formation region was limited by the cubic solubility limit. We successfully prepared *t'*- ZrO$_2$ utilizing the diffusionless *c'*-*t'* transformation in a wide compositional region.

TEMPERATURE- TIME- TRANSFORMATION DIAGRAM OF OXYGEN-DIFFUSION CONTROLLED *C'→T'* TRANSFORMATION IN Zr$_{0.5}$Ce$_{0.5}$O$_2$ SOLID SOLUTION

c'- ZrO$_2$ samples containing 50 mol% CeO$_2$ were annealed in air at various temperatures and then quenched by dropping into water (cooling rate > 100 K/sec) to investigate the kinetics of the cation-diffusionless *c'*- to-*t'* transformation and cation-diffusional phase separation into *t+c* phases. The {400} XRD profile could be decomposed into those of *t'*- and *c'*- ZrO$_2$ using a program (PRO-FIT) written by Toraya [13]. The *c'*- ZrO$_2$ partially transformed into the *t'*- ZrO$_2$ during annealing. The volume fraction of *t'*- ZrO$_2$ was calculated from the integrated intensity ratio,

$$V_{t'} = \frac{I_{t'}(004) + I_{t'}(400)}{I_{t'}(004) + I_{t'}(400) + I_{c'}(400)} \qquad (1)$$

The $V_{t'}$ increased with annealing time and temperature as shown in figure 4. Time dependence of $V_{t'}$ could be expressed by a first order rate equation, $V_{t'} = 1 - \exp(-k \cdot t)$, where k is the rate constant and t is annealing time. The rate constant k increased with temperature following the Arrhenius equation, $k = A \cdot \exp(-\Delta G * / RT)$, where A is a constant independent of temperature. By the Arrhenius plot and a least squares calculation, the activation energy $\Delta G *$ was estimated to be 113±5 kJ/mol, which was the same order of magnitude as that of oxygen diffusion in stabilized ZrO$_2$ [14-16]. Therefore, this transformation must be controlled by oxygen diffusion in the ZrO$_2$-CeO$_2$ solid solution. This conclusion was also confirmed by annealing *c'*- ZrO$_2$ in a sealed silica-glass capsule (P$_{O_2}$ < 10^{-3} torr): The *c'*- ZrO$_2$ containing 40 mol% CeO$_2$ did not transform into *t'*- ZrO$_2$ during annealing at 400℃ for 30 min. Both the weight and the color (black)

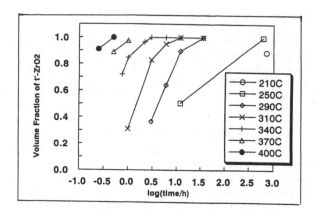

Figure 4. Kinetics of the c'-to-t' phase transformation in $(Zr_{0.5},Ce_{0.5})O_2$.

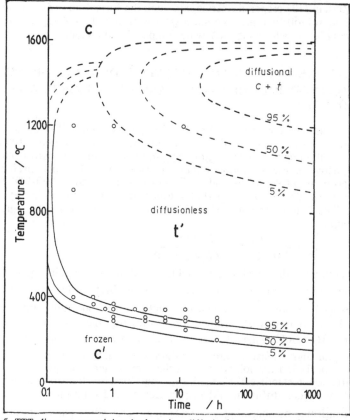

Figure 5. *TTT* diagram containing both cation-diffusionless c'-t' transformation and cation-diffusional $c+t$ phase separation in $(Zr_{0.5},Ce_{0.5})O_2$.

of the c'- ZrO_2 remained unaltered during the anneal, which indicates that the c'-ZrO_2 remained *"defective"*. This c'-t' transition may occur as follows: (1) Oxygen diffuses into the grains of the ZrO_2-CeO_2 solid solution from their surface. (2) The *"defective"* ZrO_2-CeO_2-δ changes into *"non-defective"* ZrO_2-CeO_2 by diffusion of oxygen. (3) Since t'- ZrO_2 is more stable in *"non-defective"* ZrO_2-CeO_2 than c'-ZrO_2, which is more stable in *"defective"* ZrO_2-CeO_2-δ, the c'-ZrO_2 transforms into t'- ZrO_2 from the grain surface.

At 1200℃, c'- ZrO_2 rapidly transformed cation-diffusionlessly into t'-ZrO_2 and then decomposed slowly into $t + c$ phases. These various phase changes are summarized in the *TTT* diagram shown in figure 5, which enables us to clearly distinguish the cation-diffusionless c'-t' phase transformation from the cation-diffusional $c + t$ phase separation. This diagram makes possible to resolve misunderstandings and contradictions in the literature, which are described elsewhere[4].

T'- TO-C' TRANSFORMATION TEMPERATURE

t'-ZrO_2 was annealed in air at various temperatures for 3-5 min to determine the t'-to-c' transformation temperature. As shown in figure 6, the finish point of t'-to-c'

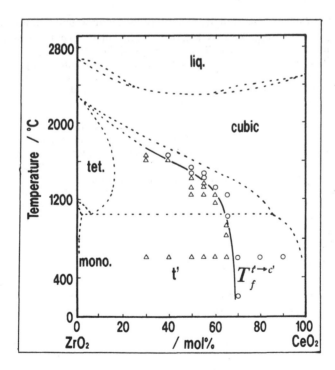

Figure 6. Metastable -stable phase diagram in ZrO_2-CeO_2 system.
○ and △ indicate c'- (100%) and t'-$(+c')$ -ZrO_2, respectively.

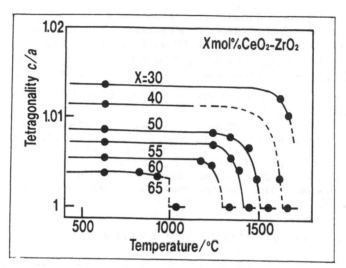

Figure 7.   Temperature dependence of the axial ratio c/a (tetragonality).

transformation $T_f^{t' \to c'}$ is located in the vicinity of the cubic solubility limit within the two-phase ($c + t$) region. This location could not be explained by a simple regular solution model, in which the entropy difference between cubic and tetragonal phases was assumed to be nonzero at the c'-t' transition temperature [17]. However, a phenomenological expression containing an order parameter can express both the location of the c-t transition temperature and the stable diagram (the equilibrium phase boundaries), where the transition was assumed to be second order [17].

The tetragonality was almost independent of temperature when the sample was annealed at temperatures much lower than the transition temperature; however it decreased with temperature in the vicinity of the transition temperature as shown in Figure 7. In 50 and 55 mol% CeO$_2$ samples, the tetragonality seemed to decrease continuously with temperature in the vicinity of transition temperature, although it seemed to decrease discontinuously in 65 mol% samples. We [17] can approximately regard the transition as a second order one due to the continuities in temperature (Figure 7) and composition (Figure 3) dependencies of the tetragonality.

REVERSIBILITY OF C'-T' TRANSFORMATION

The *"non-defective"* $t' \leftrightarrow c'$ transformation, which is accompanied by no oxygen concentration changes, can be observed in 65 mol% CeO$_2$. Both c'- ZrO$_2$ (V_t =0) and t'- ZrO$_2$ (V_t =1) containing 65 mol% CeO$_2$ were annealed in air in the vicinity of the c'-t' phase transformation temperature to investigate the reversibility. The {400} peak profiles broadened during annealing at 860° and 900°C, which is interpreted that c' and t' transformed partially into t' and c', respectively. This broadening seemed to increase with the annealing time. Furthermore, the *"non-defective"* c' could be quenched from 1031°C, although t' is more stable below

1000°C. These isothermal behaviors may be associated mainly with lattice relaxation, which is necessary for the tetragonality increase, rather than with the compositional change by oxygen diffusion. In fact, the color and cell volume of the sample remained unaltered during the *"non-defective"* $t' \rightarrow c'$ transformation, which should be distinguished from the oxygen-controlled $c' \rightarrow t'$ transformation at low temperatures described before.

CONCLUDING REMARKS

It has been demonstrated that the homogeneous t'-ZrO₂ solid solution containing 30 to 65 mol% CeO₂ can be obtained by annealing and quenching. It has been proposed that the c'-t' transformation is cation-diffusionless but needs some thermally activated process and the activation energy has been estimated to be 113±5 kJ/mol, suggesting that the transformation is controlled by oxygen diffusion. In fact, the c'-ZrO₂ did not transform without oxygen. The *TTT* diagram of the 50 mol% CeO₂-ZrO₂ solid solution, outlining the c'-t' and $[t'$-$(t+c)]$ transformation lines, has been determined. The $T_f^{t' \rightarrow c'}$ temperatures have been investigated and found to exist in the vicinity of the cubic solubility limit within the $(t+c)$ two-phase region in the temperature-composition phase diagram. This location would give important information for the thermodynamical model in the ZrO₂-CeO₂ system [17]. Finally, the reversible transformation between c'- and t'-ZrO₂ has been discussed and it appears that the $t' \leftrightarrow c'$ transformation in the vicinity of T_0 line is not accompanied by change in the oxygen concentration.

The relationship between t'-c' and cubic-tetragonal transformations must be clarified in the future. Oxygen displacement is important in the cubic-tetragonal transformation, therefore, we are now in the progress of studying it using neutron diffraction and Raman scattering.

ACKNOWLEDGMENT

We would like to express our thanks to Prof. Ikuma for useful discussions and to Prof. Toraya for supplying his useful program (PRO-FIT).

REFERENCES

1. Heuer,A.H., R.Chaim and V.Lanteri, 1987. "Displacive Cubic→Tetragonal Transformation in ZrO₂ Alloy." <u>Acta Metall.</u>, 35 (3):661-66.

2. Yashima,M., N.Ishizawa and M.Yoshimura, 1993. "High Temperature X-ray Diffraction Study on Cubic-Tetragonal Phase Transition in the ZrO₂-RO₁.₅ Systems (R: Rare Earths)." in this volume.

3. Yashima,M., N.Ishizawa and M.Yoshimura, 1993. "High Temperature X-ray Study of the Cubic-Tetragonal Diffusionless Phase Transition in the ZrO₂-ErO₁.₅ System, I, Phase Change between Two Forms of a Tetragonal Phase, t'-ZrO₂ and t''-ZrO₂ in the Compositionally Homogeneous 14 mol% ErO₁.₅-ZrO2." and "II, Temperature Dependencies of Oxygen Ion Displacement and Lattice Parameter of a Compositionally Homogeneous 12 mol% ErO₁.₅-ZrO₂, "<u>J. Am. Ceram. Soc.</u>, Accepted Sep. 17, 1992 and in press.

4. Yashima,M., K.Morimoto, N.Ishizawa and M.Yoshimura, "Zirconia-Ceria Solid Solutions Synthesis and Temperature- Time- Transformation Diagram of the 1:1 Composition." J. Am. Ceram. Soc., Accepted Nov. 18, 1992 and in press.

5. Yashima,M., N.Ishizawa and M.Yoshimura, "Diffusionless Tetragonal-Cubic Transformation Temperature in the ZrO_2-Cerium Oxide Solid Solutions." submitted to J. Am. Ceram. Soc.

6. Sugiyama,M. and H.Kubo, 1986. "Tetragonal Phase in ZrO_2-Y_2O_3 Ceramic System." Yogyo-Kyokai-Shi (J. Ceram. Soc. Jpn.) 94 (8):726-731.

7. Negas, T., R.S.Roth, C.L.McDaniel, H.S.Parker and C.D.Olson, 1976. "Influence of K_2O on the Cerium Oxide-ZrO_2 System, " Proc. Rare Earth Res. Conf., 12th Vail, Colorado, July 18-22, 1976, ed. Lundin, Charles E., University of Denver, Denver Research Institute, Denver, Colorado, pp.607-614.

8. Yoshimura,M., M.Yashima, T.Noma and S.Somiya, 1990. "Formation of Diffusionlessly Transformed Tetragonal Phases by Rapid Quenching of Melts in ZrO_2-$RO_{1.5}$ Systems (R=Rare Earths)." J. Mater. Sci., 25 (4):2011-16.

9. Muroi,T., J.Echigoya and H.Suto, 1988. "Structure and Phase Diagram of ZrO_2-CeO_2 Ceramics." *Trans Japan Inst. Metals*, 29(8):634-41.

10. Scott,H.G., 1975. "Phase Relationships in the Zirconia-Yttria System." J. Mater. Sci., 10 (9):1527-1535.

11. Tsukuma,K.. and M.Shimada, 1984. "Strength, Fracture Toughness and Vickers Hardness of CeO_2-Stabilized Tetragonal ZrO_2 Polycrystals (Ce-TZP)." J. Mater. Sci., 20 (4):1178-84.

12. Meriani,S., 1985. "A New Single Phase Tetragonal CeO_2-ZrO_2 Solid Solutions", Materials Science and Engineering, 71:369-73.

13. Toraya,H., 1986, "Whole-Powder-Pattern Fitting without Reference to a Structural Model: Application to X-ray Powder Diffractometer Data." J. Appl. Cryst., 19 (6):440-47.

14. Sakka,Y., Y.Oishi, and K.Ando, 1982. "Zr-Hf Interdiffusion in Polycrystalline Y_2O_3-(Zr+Hf)O_2." J. Mater. Sci., 17 [11]:3101-05.

15. Simpson,L.A. and R.E.Carter, 1966. "Oxygen Exchange and Diffusion in Calcia-Stabilized Zirconia, "J. Am. Ceram. Soc., 49(5):139-44.

16. Oishi,Y., and K.Ando, 1985. "Oxygen Self-Diffusion in Cubic ZrO_2 Solid Solutions, " Transport in Nonstoichiometry Compound, ed. by Smikovich and Stubican, Plenum Publishing, pp.189-201.

17. Yashima,M. and M.Yoshimura, 1992. "Thermodynamical Models for Phase Changes between Tetragonal and Cubic Phases in ZrO_2-CeO_2 Solid Solution." Jpn. J. Appl. Phys., 31 [11B]:L1614-17.

Martensitic Transformation in Deoxidized Ceria-Zirconia Ceramics

TOSHIHIKO SHIGEMATSU, NOBUAKI SHIOKAWA,
NOBUYA MACHIDA and NORIHIKO NAKANISHI

ABSTRACT

Ce(IV) ion in ceria doped zirconia ceramics was completely reduced to Ce(III) ion after deoxidation in hydrogen atmosphere above 1073 K. This deoxidation resulted in a weight loss due to the formation of oxygen vacancies. Tetragonal $xCeO_{1.5}$-$(1-x)ZrO_2$ ceramics of $0.06 \leq x \leq 0.16$ were obtained after deoxidation above 1073 K. The martensitic transformation of ceria-zirconia ceramics was affected by this deoxidation. The starting temperature of martensitic transformation (Ms) was lowered by this deoxidation, e.g., Ms=1173 K and 240 K for $0.06CeO_2$-$0.94ZrO_2$ and $0.06CeO_{1.5}$-$0.94ZrO_2$ respectively. These deoxidized ceramics were also stable under mechanical stress and no monoclinic phase appeared after grinding in an agate mortar. The formation of oxygen vacancies may play an important role in the martensitic transformation in zirconia ceramics.

INTRODUCTION

It is well known that the tetragonal to monoclinic (martensitic) transformation is responsible for the high strength and toughness of tetragonal zirconia polycrystals (TZP). For the practical use of TZP, the transformation temperature is lowered by alloying with several oxides, such as calcia, yttria and ceria. In ceria-doped TZP (Ce-TZP) the martensitic transformation occurs athermally or even as a burst type during the cooling of samples [1]. On the other hand, in yttria doped TZP (Y-TZP) the transformation occurs not only athermally but also isothermally at around 473 K and water vapor in air is necessary for this transformation [2]. The

Toshihiko Shigematsu, Nobuaki Shiokawa, Nobuya Machida and Norihiko Nakanishi, Department of Chemistry, Faculty of Science, Konan University, Okamoto 8-9-1, Higashinada, Kobe 658, Japan

formation of M-O-H (M=Zr or Y) and the diffusion of OH ions into the Y-TZP lattice may have an important role for the nucleation of monoclinic phase.

This difference in transformation characteristics can be explained as follows [3]. The addition of yttria to zirconia generates oxygen vacancies in the tetragonal lattice to satisfy charge neutrality in the crystals. These vacancies make the diffusion of oxygen ions much easier, arrest the athermal type of transformation and change the character of this transformation to a bainitic transformation, because the transformation can be controlled by OH ion diffusion. On the other hand, in the case of Ce-TZP, the so-called chemical vacancies are not necessary, only Ce(IV) ions exist, and the athermal behavior of the martensitic transformation can be expected.

Ce(IV) ions in ceria doped zirconia ceramics are easily reduced to Ce(III) ions in a reducing atmosphere at elevated temperatures. Oxygen vacancies may be introduced after the deoxidation of ceria doped zirconia ceramics to satisfy charge neutrality in the crystal lattice. If we compare the transformation characteristics of $xCeO_2$-$(1-x)ZrO_2$ and $xCeO_{1.5}$-$(1-x)ZrO_2$, we can understand the role of oxygen vacancies in zirconia ceramics on the martensitic transformation.

EXPERIMENTAL

The ceria doped zirconia powders with $0.06 \leq x \leq 0.16$ were prepared by a coprecipitation method. These powders were pressed at 100 MPa and calcined at 1673 K for 216 ks in air. The same conditions were used for the powder samples. The mean particle size of the calcined samples of ~1.5 μm was determined by SEM measurements. XRD, DTA and DSC measurements were used to check the martensitic transformation characteristics.

The deoxidation of ceria doped zirconia ceramics was carried out in a

TABLE I. THE ESTIMATED MEAN IONIC CHARGE ON THE Ce IONS IN $xCeO_{2-y}$-$(1-x)ZrO_2$

x	deoxidation temperature / K	weight gain / %	y	mean ionic charge on the Ce ions
0.06	1173	0.39	0.51	+3.0
0.10	1173	0.60	0.48	+3.0
0.12	873	0.22	0.15	+3.7
0.12	973	0.40	0.27	+3.5
0.12	1173	0.70	0.47	+3.1
0.16	1173	0.97	0.49	+3.0

hydrogen atmosphere at temperatures of 573-1273 K for 3.6 ks. After the deoxidation, the samples were quenched to room temperature. Insufficient quenching resulted in the partial re-oxidation of the samples. The deoxidized samples were characterized by XRD and XPS measurements. The re-oxidation of the deoxidized samples was measured by TG-DTA. XRD and DSC measurements were used to check the martensitic transformation upon cooling to liquid nitrogen temperature. The stress assisted martensitic transformation of the as-calcined and deoxidized Ce-TZP was studied by XRD measurements after grinding in an agate mortar or in a planetary type ball mill (zirconia pot and zirconia balls). The amount of monoclinic phases were estimated from the relative intensities of the monoclinic peaks $(111)_m + (11\bar{1})_m$ and the tetragonal plus monoclinic peaks $(111)_t + (111)_m + (11\bar{1})_m$.

RESULTS AND DISCUSSION

DEOXIDATION AND RE-OXIDATION OF CERIA DOPED ZIRCONIA CERAMICS

The as-calcined samples were monoclinic+tetragonal for $x=0.06$ and tetragonal single phase for $x=0.10, 0.12$ and 0.16.

The deoxidation of ceria doped zirconia ceramics in a hydrogen atmosphere occurred above 800 K and was accompanied by a weight loss. The weight loss increased with the increase of deoxidation temperature and it became constant after deoxidation above 1073 K. The deoxidation rate was fast and no depth dependence of lattice parameter was found in the deoxidized samples after 3.6 ks. Those deoxidized samples were easily re-oxidized in air above 400 K with an accompanying weight gain. This re-oxidation was completed at around 700 K. The ceria content showed no change after the deoxidation and re-oxidation process. From the weight gain after the re-oxidation, we can estimate the mean ionic charge on the Ce ions. For this estimate we assumed that the weight gain came from the oxidation of $xCe(IV-2y)O_{2-y}-(1-x)ZrO_2$ to $xCe(IV)O_2-(1-x)ZrO_2$. In other words, the deoxidation resulted in a weight loss due to the reduction of Ce(IV) ions to Ce(III) ions and the formation of oxygen vacancies. The estimated mean ionic charge on the Ce ions is shown in Table I. Ce(IV) ions were completely reduced to Ce(III) ions after deoxidation above 1073 K. The Ce ion in ceria-doped zirconia ceramics seems to reduce more easily than that in CeO_2 [4]. The XPS measurements supported this conclusion. Zr 3d XP spectra were nearly the same for as-calcined and deoxidized samples. On the other hand, Ce 3d peaks shifted about 5.0 eV to higher binding energy and the satellite peaks of Ce 3d spectra disappeared after the deoxidation above 1073 K.

Figure 1 shows the variation in lattice parameter, a and c, unit cell volume and tetragonality c/a of a deoxidized sample with $x=0.12$ as a function of the deoxidation temperature. These values started to change at around 800 K and became constant above 1073 K. The a value increased and the c value

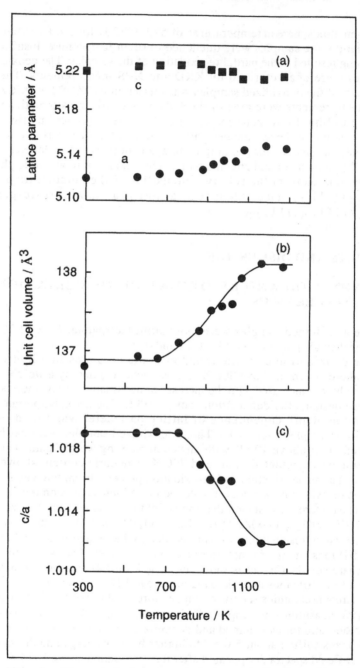

Figure 1.   Change in lattice parameters (a), unit cell volume (b) and tetragonality (c) of $0.12CeO_{2-y}$-$0.88ZrO_2$ after the deoxidation in a hydrogen atmosphere.

decreased with increase in deoxidation temperature, the unit cell volume increased, and the c/a ratio decreased. Similar changes after the deoxidation at elevated temperatures were observed for x=0.10 and 0.16.

The reduction of Ce(IV) ions commenced at around 800 K and Ce(IV) ions were completely reduced to Ce(III) ions above 1073 K. The ionic radius of Ce increased from 111 pm to 128 pm when Ce(IV) was reduced to Ce(III). Oxygen vacancies were introduced with the reduction of Ce(IV) ions to Ce(III) ions to satisfy charge neutrality. The formation of oxygen vacancies and the increase of ionic radius as the Ce ion is reduced can explain the increase of unit cell volume and the decrease of c/a.

THERMALLY ASSISTED MARTENSITIC TRANSFORMATION IN DEOXIDIZED CERIA DOPED ZIRCONIA CERAMICS

Figure 2 shows the DSC curve for as-calcined $(0.12CeO_2-0.88ZrO_2)$ and deoxidized at 1173 K $(0.12CeO_{1.5}-0.88ZrO_2)$ samples. The as-calcined sample transformed to the monoclinic phase at 205 K within a narrow temperature range of <2.0 K. On subsequent heating the monoclinic phase transformed to the tetragonal phase at ~490 K. On the other hand, no martensitic transformation was observed upon cooling to 130 K in the

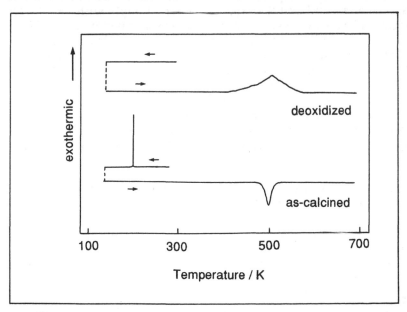

Figure 2. DSC curves for as-calcined $(0.12CeO_2-0.88ZrO_2)$ and deoxidized at 1173 K $(0.12CeO_{1.5}-0.88ZrO_2)$ samples. The cooling and heating rates were 5 K/min.

deoxidized samples. On subsequent heating a broad exothermic peak due to the re-oxidation was observed at around 500 K. About 60% monoclinic phase appeared in the as-calcined sample. The amount of monoclinic phase decreased with increasing deoxidation temperature. The amount of monoclinic phase became 0% after deoxidation above 873 K. The mean ionic charge of the Ce ion was +3.8 for the sample deoxidized at 873 K. A partial reduction of Ce(IV) ion and the formation of oxygen vacancies suppressed the martensitic transformation. Similar stabilization of tetragonal phase by deoxidation was observed in the sample with x=0.10.

The Ms was 1173 K for as-calcined samples and its reverse transformation temperature was above 1300 K. The amount of monoclinic phase that formed in the sample with x=0.06 after deoxidation in a hydrogen atmosphere is shown in figure 3. About 96% of monoclinic phase was observed at room temperature. The amount of monoclinic phase started to decrease after deoxidation at around 800 K and became almost constant above 1200 K. The decrease in the amount of monoclinic phase with the increase of deoxidation temperature corresponds to the change in mean ionic charge of the Ce ions. The stabilization of the tetragonal phase initiated when the partial reduction of Ce(IV) to Ce(III) started. Above 1073 K, Ce(IV) was completely reduced to Ce(III) and the amount of monoclinic phase dropped to about 15%.

A small exothermic peak due to the martensitic transformation was observed at around 240 K in the deoxidized sample with x=0.06. On subsequent heating no reverse transformation was observed up to 300 K. Measurements at higher temperatures were impossible because of re-oxidation of the sample at around 400 K. About 25% of monoclinic phase were observed in the sample after cooling to 77 K. Only one tenth of $0.06CeO_{1.5}$-$0.94ZrO_2$ transformed to the monoclinic phase at around 240 K. This may result from the distribution of particle sizes and/or the concentration fluctuation between particles. The reduction of Ce(IV) to Ce(III) and the formation of oxygen vacancies lowered the Ms from 1173 K to 240 K for some of the particles and below 77 K for the majority of them.

STRESS ASSISTED MARTENSITIC TRANSFORMATION IN DEOXIDIZED CERIA DOPED ZIRCONIA CERAMICS

Ce-TZP easily transforms to the monoclinic phase under mechanical stress. About 85% and 83% of monoclinic phase were formed in the as-calcined sample with x=0.10 and 0.12, respectively, after grinding in an agate mortar for 0.3 ks. Figure 4 shows the amount of monoclinic phase in $0.12CeO_2$-y-$0.88ZrO_2$ after grinding in an agate mortar. The amount of monoclinic phase decreased with increase in the deoxidation temperature. If a small amount of Ce(IV) ion was reduced to Ce(III) ion, such as the sample after deoxidation at 873 K, no monoclinic phase was formed under mechanical stress. Moreover, no monoclinic phase appeared even after the grinding in a planetary type ball mill for 7.2 ks. Similar stabilization of

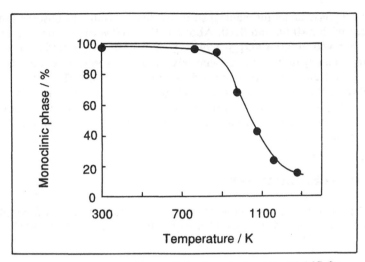

Figure 3.   The amount of monoclinic phase in deoxidized $0.06CeO_2$-y-$0.94ZrO_2$ at room temperature.

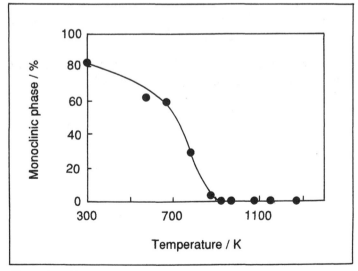

Figure 4.   The amount of monoclinic phase in deoxidized $0.12CeO_2$-y-$0.88ZrO_2$ after grinding in an agate mortar.

tetragonal phase under mechanical stress by deoxidation was observed in the samples with x=0.06 and 0.10. About 15% and 0% of monoclinic phase were observed in $0.06CeO_{1.5}-0.94ZrO_2$ and $0.10CeO_{1.5}-0.90ZrO_2$ after grinding in an agate mortar, respectively. In other words, no stress assisted martensitic transformation occurred in deoxidized Ce-TZP. Not only the lowering of the free energy of the tetragonal phase relative to that of the monoclinic phase, but also an increase of the activation barrier for the tetragonal to monoclinic transformation, results from the deoxidation of ceria doped zirconia ceramics.

ACKNOWLEDGEMENTS

The present work was supported in part by a Grant-in-Aid for Scientific Research No. 0465715 from the Ministry of Education, Science and Culture.

REFERENCES

1. Reyes-Morel, P.E. and I.W. Chen, 1988. "Transformation Plasticity of Ceria-Stabilized Tetragonal Zirconia Polycrystals: 1. Stress Assistance and Autocatalysis." J. Am. Ceram. Soc., 7 (5): 343-353.

2. Shigematsu, T. and N. Nakanishi, 1989. "Tetragonal to Monoclinic Phase Transition in $Y_2O_3-ZrO_2$." Japan. J. Appl. Phys., Series 2: 159-165.

3. Nakanishi, N. and T. Shigematsu, 1991. "Bainite-like Transformation in Zirconia Ceramics." Mater. Trans., JIM, 32 (8): 778-784.

4. Sata, T. and M. Yoshimura, 1968. "Some Material Properties of Cerium Sesquioxide." J. Ceram. Assoc. Japan, 76 (4): 116-122.

High Temperature X-Ray Diffraction Study on Cubic-Tetragonal Phase Transition in the ZrO_2-$RO_{1.5}$ Systems (R:Rare Earths)

MASATOMO YASHIMA, NOBUO ISHIZAWA
and MASAHIRO YOSHIMURA

ABSTRACT

The cubic-tetragonal phase transition in ZrO_2-X mol% $RO_{1.5}$ ($X=12$, 14 for R=Er and Y) prepared by rapid quenching of its melt was investigated mainly by high temperature x-ray diffraction. (1) Three forms existed in the tetragonal phase of ZrO_2 solid solutions, t-ZrO_2, t'-ZrO_2 and t''-ZrO_2. The t''-ZrO_2 was defined as a tetragonal form which had no tetragonality (axial ratio, $c/a=1$) but had an oxygen ion displacement from the ideal fluorite site. (2) The t''-to-t' change was isothermal, because there existed an energy barrier between t''- and t'-ZrO_2. (3) An isothermal feature was also found in the increase of the tetragonality. (4) The t'-form disappeared at about 1200°C in 14mol%$ErO_{1.5}$-ZrO_2. The t'-to-t'' change and/or t''-to-c transition temperatures were found to be about 1425°C and 1400°C in 12mol%$ErO_{1.5}$- and 12mol%$YO_{1.5}$- samples, respectively. (5) The t''-t' and/or t''-c transition temperature existed in the vicinity of the cubic solubility limit within the ($c+t$) two-phase region in the temperature-composition phase diagram. (6) The tetragonality and the oxygen ion displacement seemed to decrease continuously with temperature in ZrO_2-12mol%$RO_{1.5}$.

INTRODUCTION

Tetragonal zirconia is formed through either diffusionless cubic (c)-to-tetragonal (t') phase transition or diffusional phase decomposition within the {tetragonal (t)+cubic} two-phase compositional region. In this paper, we briefly discuss the nature of the diffusionless c-t' transition in the ZrO_2-$RO_{1.5}$ systems mainly on the basis of high temperature x-ray powder diffraction study. Details will be published elsewhere [1]. In the ZrO_2-$YO_{1.5}$ solid solution, two forms, t-ZrO_2, and t'-ZrO_2 [1-4] are known in one tetragonal structure with space group $P4_2/nmc$; which are conventionally distinguished from each other to describe the phase changes in specimens. t-ZrO_2 is defined as a tetragonal form prepared by diffusional phase separation within the ($t + c$) two-phase region. The t'- form

Masatomo YASHIMA, Nobuo ISHIZAWA and Masahiro YOSHIMURA: Research Laboratory of Engineering Materials, Tokyo Institute of Technology, 4259, Nagatsuta, Midori-ku, Yokohama, 227, Japan.

($P4_2/nmc$) is prepared through diffusionless phase transition from the high-temperature c-phase ($Fm3m$). Here we propose one more tetragonal form: t''-ZrO_2 ($P4_2/nmc$, the same space group as that of the t'-ZrO_2), which has no tetragonality (axial ratio, $c/a = 1$) but has oxygen displacement from its ideal site in the fluorite structure ($Fm3m$) [1]. The phase stability between t'- and t''- ZrO_2 is studied. In the 14mol% $RO_{1.5}$ samples, the t'-t'' change behaves as a first order transition with isothermal changes. On the other hand, the transition in 12 mol% $RO_{1.5}$ samples behaves as a higher-order or a first order/near second-order one.

EXPERIMENTAL PROCEDURE

ZrO_2-X mol% $RO_{1.5}$ (X =12 and 14 for R=Er and Y) were prepared by rapid quenching of melts [2,5,6]. High purity ZrO_2 and $RO_{1.5}$ powders were mixed in an agate mortar and pressed into pellets. The pellet was melted in air using an arc-imaging furnace and then quenched rapidly with a hammer-anvil apparatus made of copper. The 15 mm diameter and 20 μm thick films thus obtained were crushed and ground in a B_4C mortar. Phase changes of rapidly quenched powder at high temperatures in air were studied by a θ-θ type Seemann-Bohlin diffractometer with a $LaCrO_3$ furnace [1,7]. The heating rate was 5°C/min and each measurement was conducted at a constant temperature after a 30min keeping. Temperature and Bragg angle accuracies and temperature stability were within 8°C, 0.01°deg, and 1°C, respectively [1]. The microstructure was observed through TEM (Hitachi, H-800, 200kV) for rapidly quenched ZrO_2-14mol% $ErO_{1.5}$. ICP (Inductively Coupled Plasma) emission spectroscopy for melted specimens showed a negligible contamination of silica (0.02 ≤ weight %) during grinding in agate mortar and no compositional changes during melting and quenching [6].

PHASE CHANGE BETWEEN T'' - AND T'- ZRO_2 IN ZRO_2- 14MOL% $RO_{1.5}$

The ZrO_2- 14mol% $RO_{1.5}$ showed a characteristic isothermal phase change between the t'- and t''- ZrO_2, which behaved as a first order transition. The colors of rapidly quenched samples were pink (R=Er) and white (R=Y), which indicates that they had no non-equilibrium vacancies, which might occur during the melting.

In rapidly quenched 14 mol% $ErO_{1.5}$ there was a single peak for {400} reflections with no splitting between (004) and (400) (axial ratio c/a=1, the error value of c/a, 2×10^{-5} estimated by a profile fitting program: PRO-FIT [8] was less than the value 0.001 estimated from the full width of half maximum of NBS Si) as shown in figure 1 (A) [1]. However, the electron diffraction pattern was the same as that of the t'-form ($P4_2/nmc$), which has oxygen displacements along the c-axis. The (112) reflection, which is forbidden for the fluorite-type structure ($Fm3m$), was observed everywhere in the sample by electron diffraction. The dark field image of the (112) reflection showed a domain structure as shown in figure 1(B). Therefore the existing phase is a tetragonal phase without tetragonality ($c/a = 1$), which has been observed in previous work [9,10]. We call this tetragonal form t''- ZrO_2 for convenience. Although the crystal systems of the t''- and t'- ZrO_2 belong to the same space group $P4_2/nmc$, the t''- ZrO_2 (c/a=1) should be distinguished from the t'- ZrO_2 (c/a >1), because there exists an energy barrier between them as described below.

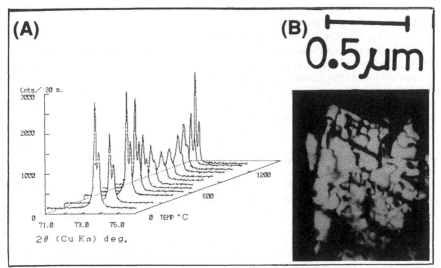

Figure 1. (A): XRD patterns of {400} peaks on heating the rapidly quenched ZrO_2 -14 mol% $ErO_{1.5}$ (B): Dark field image taken by a (112) reflection in rapidly quenched ZrO_2 -14 mol% $ErO_{1.5}$.

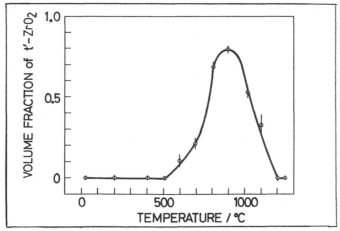

Figure 2. Relationship between temperature and fraction of t'-ZrO_2 on heating the rapidly quenched ZrO_2-14 mol% $ErO_{1.5}$.

When the rapidly quenched ZrO_2-14mol% $ErO_{1.5}$ was reheated, the {400} peak profile broadened and the shoulder at the low-angle side of the peak profile increased with temperature and annealing time {Figure 1(A)}, which indicates that the t''- ZrO_2 changed isothermally and partially into t'- ZrO_2 without any cation diffusion. The {400} peak profile could be decomposed into $(004)_{t'}$, $(400)_{t''}$ and $(400)_{t'}$ peaks using a computer program, PRO-FIT [8], where the integrated intensity ratio $I_{t'}(400)/I_{t'}(004)$ was assumed to be 2.33, which was calculated using a computer program, powd10 [11]. The volume fraction of t'- ZrO_2, $V_{t'}$ was calculated by using an equation:

$$V_{t'} = \{I_{t'}(004) + I_{t'}(400)\} / \{I_{t'}(004) + I_t(400) + I_{t''}(400)\} \tag{1}$$

The $V_{t'}$ increased with temperature above 500°C, had a maximum at about 800-900°C, and decreased during further heating as shown in figure 2. The t'-form disappeared at about 1200°C. The equilibrium temperature between t''- and t'-ZrO_2 may exist at about 800-900°C. As shown in figure 3 (A), $V_{t'}$ increased during isothermal treatment at 700°C. These phase changes indicate that (1): t'- ZrO_2 is more stable than t''- ZrO_2 and that (2): t''- ZrO_2 changes isothermally into t'-ZrO_2 due to the existence of an energy barrier between them as shown in a schematic free energy-order parameter diagram {Figure 3 (B)} where the order parameter is defined as a linear combination of the oxygen displacement from its ideal fluorite site $c(0.25-z)$ and the tetragonality c/a-1 [1]. The magnitude of the activation energy for the t''- t' change must have a distribution, because $V_{t'}$ saturated a certain value after a long anneal at 700°C {Figure 3 (A)} [1]. This behavior in the $t'' \rightarrow t'$ change is similar with that in the stress-induced $t \rightarrow m$ transformation [12].

Similar results for the t'- t'' change were observed also in rapidly quenched ZrO_2-14mol% $YO_{1.5}$. The rapidly quenched 14 mol% $YO_{1.5}$ sample had no tetragonality (c/a=1). This t''- ZrO_2 partially changed into t'-ZrO_2 in the quenched sample after firing at 800°C for 24 h. The t'-ZrO_2 in this fired sample changed back into t''-ZrO_2 in the quenched sample after further firing for 10 min at 1525°C, well above the t'-t'' change temperature. Therefore, the t'-t'' change is not affected by non-equilibrium oxygen deficiency, which might occur during melting. This was confirmed also by annealing of the rapidly

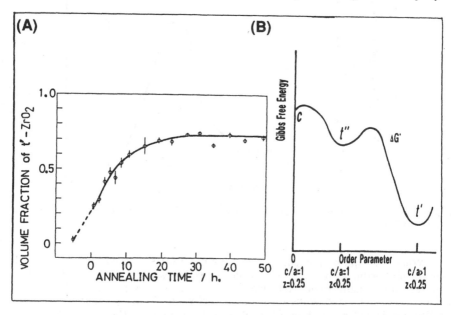

Figure 3 (A): Relationship between annealing time and the t'-fraction when the rapidly quenched ZrO_2-14 mol% $ErO_{1.5}$ was annealed at 700°C, (B): Schematic Gibbs free energy-order parameter diagram to explain the phase changes.

quenched t''- ZrO_2 at 700°C for 50 h in a sealed silica glass tube ($P_{O_2} < 10^{-3}$ torr), because the t''- ZrO_2 changed into t'- ZrO_2 by such annealing without oxygen supply. In the present study, the coexistence of t''- and t'- ZrO_2 was observed over a wide temperature range. Such coexistence was observed in rapidly quenched ZrO_2-$YO_{1.5}$ at room temperature in the literature [2,5], where the t''- ZrO_2 has been regarded as a cubic phase.

TEMPERATURE DEPENDENCE OF OXYGEN DISPLACEMENT AND TETRAGONALITY IN ZRO₂ - 12 MOL% RO₁.₅ (R=ER AND Y)

High temperature x-ray diffraction measurements were conducted for the rapidly quenched ZrO_2 - 12 mol% $RO_{1.5}$ (R=Er and Y) samples. In these samples, only t'- ZrO_2 was observed below 1400°C. As shown in figure 4, the tetragonality increased during heating the rapidly quenched samples, but remained constant during cooling slowly from 700-1000°C to room temperature, which indicates that the smaller tetragonality at high temperatures could be frozen to some extent by rapid quenching due to its thermally activated nature. The tetragonality decreased continuously with temperature above 1000°C and became unity ($c/a = 1$) at about 1425°C and 1400°C for 12 mol% $RO_{1.5}$ for R= Er and Y, respectively, which were identified to be t''- and/or c- ZrO_2, although we could not distinguish them from each other because of the very weak (112) reflection detected by XRD. This continuous decrease with temperature indicates that the $t' \rightarrow t''$ (or c) change in ZrO_2 - 12 mol% $RO_{1.5}$ (R=Er and Y) is a higher-order transition or a first/near second-order one.

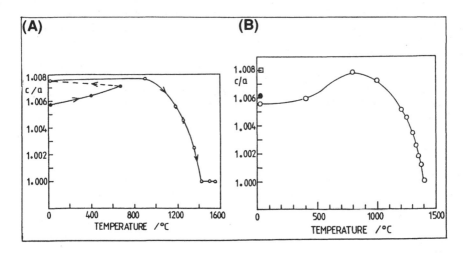

Figure 4. Relationship between temperature and the axial ratio c/a when the ZrO_2-12 mol% $RO_{1.5}$ was heated. (A) R=Er and (B) R=Y.

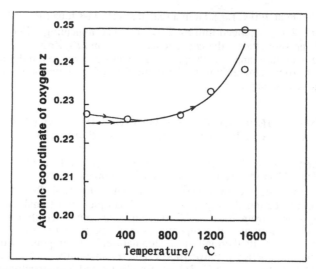

Figure 5 . Relationship between temperature and the atomic coordinate of oxygen z, when the ZrO_2-12 mol% $ErO_{1.5}$ was heated.

The z parameter of the atomic coordinate of oxygen ions is important because the oxygen displacement from its ideal site ($z=0.25$) in the fluorite-type structure can be expressed by $c \cdot (0.25 - z)$ using z where c is the length of the c- axis. The $(0.25 - z)$ should be a suitable order parameter of the cubic-tetragonal phase transition. z was calculated from the integrated intensity ratio $I(112)/I(111)$, where the full width of half maximum of the (112) reflection was determined using a computer program PFLS for Rietveld analysis of total patterns [13]. As shown in figure 5, large z (small oxygen displacement) seems to be frozen by the rapid quenching. z became smaller during heating up to about 400°C and increased above 1000°C during heating after cooling to room temperature. Therefore the oxygen displacement also has an isothermal nature to some extent similar to that of the tetragonality. z seems to increase continuously with temperature above 1000°C.

DISCUSSION

Figures 6 and 7 show the stable (the equilibrium phase boundaries) and metastable (the equilibrium temperature between two phases with the same composition) diagrams of the ZrO_2-rich part in the ZrO_2-$RO_{1.5}$ systems (R= Er [14] and Y [2, 15-22]. We could first show quantitatively the diffusionless cubic-tetragonal transition temperature (metastable phase boundary), which has been drawn schematically in previous works [3,4,23,24]. The t'-t'' and/or t'-c change temperatures exist in the vicinity of the cubic solubility limit within the $(t+c)$ two-phase region in both systems. This location could not be explained using the previous expressions of Gibbs free energies for tetragonal and cubic phases [25,26], because the T_0^{c-t} calculated by us [27] using these expressions was located near the center of the $(t+c)$ two-phase compositional region. The calculated T_0^{c-t} using the sublattice model proposed by Hillert and Sakuma [24] roughly agrees with the present T_0^{c-t} determined experimentally, but their model used a number of drastic assumptions and did not take into direct

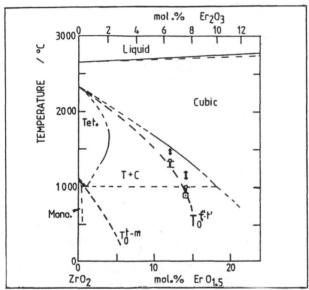

Figure 6 . Stable [14] and metastable diagram in ZrO$_2$ -ErO$_{1.5}$ system.

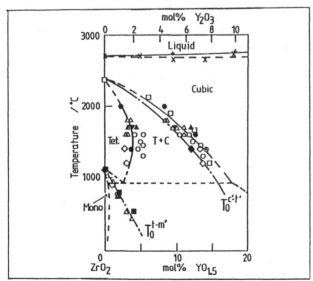

Figure 7 . Stable and metastable diagram in the ZrO$_2$ -YO$_{1.5}$ system: ● [2], □[15], ○[16], ◇[17], ▼ [18], △[19,20], ▲[21], × [21], ◆$T_0^{t'-t''}$ or $T_0^{t'-c}$ obtained in the present study.

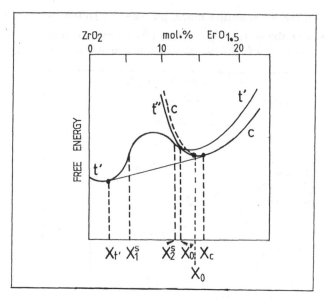

Figure 8. A schematic Gibbs free energy-composition diagram in the ZrO_2-$RO_{1.5}$ system.

consideration the order parameter in the mathematical expression. More thermodynamical data are required for progress to be made.

We propose a schematic Gibbs free energy-composition (G-X) diagram (figure 8 [1]), which is generally useful to discuss and predict the phase formation. This diagram can explain the following results, which have been observed experimentally [1]: (1) The phase separation occurs within the compositional region between X_t and X_C, [2,14-21]. (2): The t'-c equilibrium composition (metastable phase boundary T_0^{c-t}) exists in the vicinity of the cubic solubility limit within the ($c+t$) two-phase region [present work]. (3): The t'-t'' change has an isothermal nature, which indicates

$$\frac{dG_{t'}}{dX} \neq \frac{dG_{t''}}{dX} \qquad (2)$$

at the equilibrium composition X_0 between t'- and t''-ZrO_2 [present work]. (4) Spinodal phase separation occurs within the compositional region between X_1^s and X_2^s [29]. (5): The modulated structure due to the spinodal decomposition is always accompanied by (112) reflections that are forbidden for the fluorite-type structure and that are characteristic of tetragonal symmetry [29]. The present G-X diagram explains this decomposition as $t'(X) \rightarrow t'(X_1) + t'(X_2)$, where $X_1^s < X_1 < X < X_2 < X_2^s$. The results of (2), (3), (4) and (5) can not be explained by the G-X diagram after Andersson [3,4,23,24]. The G-X diagram after Sakuma et al.[29] can not explain the present t'-t'' change of (3). The present G-X diagram is the most probable to explain these results. More detail results and discussions will be published elsewhere [1].

ACKNOWLEDGMENT

We would like to express our thanks to Dr. M.Sugiyama and Dr. H.Kubo (Nippon Steel Co.) and to Mr. Y.Tabira (RICOH, Co.) for the TEM observations. We would like to express our thanks to Prof. Toraya for our use of his programs (PRO-FIT and PFLS). Many thanks to Prof. S.Sasaki (Tokyo Institute of Tech.) Dr. H.G.Scott (CSIRO), Dr. E.Kisi (Griffith Univ.) and Dr. T.Noma (Tokyo Univ. Agriculture and Tech.) for useful comments and discussions

REFERENCES

1. Yashima,M. N.Ishizawa and M.Yoshimura, 1993. "High Temperature XRD Study of Cubic-Tetragonal Diffusionless Phase Transition in the ZrO_2-$ErO_{1.5}$ System, I, Phase Change between Two Forms of a Tetragonal Phase, t'-ZrO_2 and t''-ZrO_2 in the Compositionally Homogeneous 14 mol% $ErO_{1.5}$-ZrO_2." and "II, Temperature Dependencies of Oxygen Ion Displacement and Lattice Parameter of a Compositionally Homogeneous 12 mol% $ErO_{1.5}$-ZrO_2." J. Am. Ceram. Soc. 76 to be published.

2. Scott,H.G., 1975. "Phase Relationships in the Zirconia-Yttria System." J. Mater. Sci., 10 (9):1527-1535.

3. Andersson,C.A. and T.K.Gupta, 1981. "Phase Stability and Transformation Toughening in Zirconia." Advances in Ceramics, Vol.3, Science and Technology of Zirconia, Ed. A.H.Heuer and L.W.Hobbs, Columbus, Ohio, USA: The American Ceramic Society Inc., pp.184-201.

4. Lanteri,V., A.H.Heuer and T.E.Mitchell, 1984. "Tetragonal Phase in the System ZrO_2-Y_2O_3." Advances in Ceramics, 12, Science and Technology of Zirconia II, N.Claussen, M.Rhüle and A.H.Heuer, eds., Columbus, Ohio, USA, The American Ceramic Society Inc., pp.118-130.

5. Noma,T., M.Yoshimura, S.Sōmiya, M.Kato, M.Shibata, and H.Seto, 1989. "Stability of Diffusionlessly Transformed Tetragonal Phases in Rapidly Quenched ZrO_2-Y_2O_3." Advances in Ceramics, 24A, Science and Technology of Zirconia III, S.Sōmiya, N.Yamamoto, and H.Yanagida, eds., Columbus, Ohio, USA, The American Ceramic Society, pp.377-384.

6. Yoshimura,M., M.Yashima, T.Noma and S.Sōmiya, 1990. "Formation of Diffusionlessly Transformed Tetragonal Phases by Rapid Quenching of Melts in ZrO_2-$RO_{1.5}$ Systems (R=Rare Earths)." J. Mater. Sci. 25(4):2011-2016.

7. Marumo,F., H.Morikawa, Y.Shimizugawa, M.Tokonami, M.Miyake, K.Ohsumi and S.Sasaki, 1989. Rev. Scientific Instruments, 60(7):2421-2424.

8. Toraya,H., 1986. "Whole-Powder-Pattern Fitting without Reference to a Structural Model: Application to X-ray Powder Diffractometer Data." J. Appl. Cryst., 19(6):440-447.

9. Sugiyama,M. and H.Kubo, 1986. "Tetragonal Phase in ZrO_2-Y_2O_3 Ceramics System." J. Ceram. Soc. Jpn., 94 (8):726-81.

10. Zhou,Y.,T.C.Lei, and T.Sakuma, 1991. "Diffusionless Cubic-to-Tetragonal Phase Transition and Microstructural Evolution in Sintered Zirconia-Yttria Ceramics." J. Am. Ceram. Soc., 74 (3):633-640.

11. Smith,D.K., M.C.Nichols, and M.E.Zolensky, 1983. "powd10, A FORTRAN IV Program for Calculating X-ray Powder Diffraction Patterns-Version 10." The Pennsylvania State Univ., University Park, Pennsylvania.

12. Yashima,M., T.Noma, N.Ishizawa and M.Yoshimura, 1991. "Effects of Noncompositional Inhomogeneity on $t \to m$ Phase Transformation During Grinding of Various Rare-Earths-Doped Zirconias." J. Am. Ceram. Soc., 74(12):3011-3016.

13. Toraya,H. and F.Marumo, 1980. "Application of Total Pattern-Fitting to X-Ray Powder Diffraction Data." Report of the Research Laboratory of Engineering Materials, Tokyo Institute of Technology, 5:55-64.

14. Yashima,M., N.Ishizawa, T.Noma and M.Yoshimura, 1991. "Stable and Metastable Phase Relationships in the System ZrO_2-$ErO_{1.5}$." J. Am. Ceram. Soc., 74 (3):510-513.

15. Stubican,V.S., R.C.Hink, and S.P.Ray, 1978. "Phase Equilibria and Ordering in the System ZrO_2-Y_2O_3." J. Am. Ceram. Soc., 61(1-2):17-21.

16. Heuer,A.H. and M.Rühle, 1984. "Phase Transformations in ZrO_2-Containing Ceramics: I, The Instability of c-ZrO_2 and the Resulting Diffusion-Controlled Reactions", pp.1-13, in Ref.3

17. Ruh,R., K.S.Mazdiyasni, P.G.Valentine, and H.O.Bielstein, 1984. "Phase Relations in the System ZrO_2-Y_2O_3 at Low Y_2O_3 Contents." J. Am. Ceram. Soc., 67(9):C190-C192.

18. Yagi,T., A.Saiki, N.Ishizawa, N.Mizutani and M.Kato, 1986. "Analytical Electron Microscopy of Yttria-Partially Stabilized Zirconia Crystal." J. Am. Ceram. Soc., 69 (1):C3-C4.

19. Yoshikawa,N., N.Eda, and H.Suto, 1986. "On the Cubic/Tetragonal Phase Equilibrium of the ZrO_2-Y_2O_3 System." J. Jpn. Inst. Metals, 50 (1):113-118.

20. Yoshikawa,N. and H.Suto, 1986. "Transformation Behavior of Y_2O_3-PSZ Investigated by Thermal Dilatometry." J. Jpn. Inst. Metals, 50 (1):108-113.

21. Noma,T., M.Yoshimura, M.Yanagisawa, H.Seto, M.Kato and S.Sōmiya, 1988. "Phase Separation of Rapidly Quenched Y-TZP by High Temperature Annealing." Extended Abstracts of 26th Symposium of Basic Science of Ceramics Jan. 20-22, Nagoya, Japan, The Ceramic Soc. Jpn., p.96.

22. Noguchi,T., M.Mizuno, and T.Yamada, 1970. "The Liquidus Curve of the ZrO_2-Y_2O_3 System as Measured by a Solar Furnace." Bull. Chem. Soc. Jpn., 43(8):2614-2616.

23. Sheu,T.S., T.Y.Tien and I.W.Chen, 1992. "Cubic-to-Tetragonal (t') Transformation in Zirconia-Containing Systems." J. Am. Ceram. Soc., 75(5):1108-1116.

24. Yoshimura,M., 1988. "Phase Stability of Zirconia." Ceram. Bull., 67 (12):1950-1955.

25. Degtyarev,S.A. and G.G.Voronin, 1988. "Solution of Ill-posed Problems in Thermodynamics of Phase Equilibria. The ZrO_2-Y_2O_3 System." CALPHAD, 12 (1);73-82

26. Du,Y., J.Zhanpeng and P.Huang, 1991. "Thermodynamic Assessment of the ZrO_2-Y_2O_3 System." J. Am. Ceram. Soc.. 74 (7):1569-1577.

27. Yashima,M. and M.Yoshimura, 1992. "Thermodynamical Models of Phase Changes between Tetragonal and Cubic Phases in ZrO_2-CeO_2 Solid Solution." Jpn. J. Appl. Phys., 31(11B): L1614-1617.

28. Hillert,M. and T.Sakuma, 1991. "Themodynamic Modelling of the $c \rightarrow t$ Transformation in ZrO_2 Alloys." Acta metall. mater.. 39 (6):1111-1115.

29. Sakuma, T. Y.Yoshizawa, and H.Suto, 1986. "The Metastable Two-Phase Region in the Zirconia-Rich Part of the ZrO_2-Y_2O_3." J. Mater. Sci.. 21 (4):1436-1440.

TZPs in the TiO_2-Y_2O_3-ZrO_2 System

W. PYDA, K. HABERKO, M. M. BUĆKO and M. FARYNA

ABSTRACT

Hydrothermal technique was used to obtain the TZP powders in the TiO_2 - Y_2O_3 - ZrO_2 system.

Within the studied composition range (Y_2O_3 from 0.5 to 3 mole % and TiO_2 from 3 to 28 mole %) the field of fully or nearly fully tetragonal polycrystals was found. Zirconium titanate ($ZrTiO_4$) crystallizing in the form of elongated grains in the samples of the highest TiO_2 concentration inhibits the growth of the tetragonal zirconia s.s. grains. The critical grain size of the tetragonal phase vs. TiO_2 concentration was measured. These measurements confirm that TiO_2 plays the role of the tetragonal phase stabilizer.

INTRODUCTION

It has been demonstrated that several binary and ternary oxide systems can be a potential source of tetragonal zirconia polycrystals (TZP) of good mechanical properties. Relatively little attention has been paid to the systems containing TiO_2. Although TiO_2 dissolves in zirconia, the attempts to use it as a stabilizer of the tetragonal phase at room temperature have failed [1-2]. However, positive results, as for the retention of the tetragonal phase down to room temperature, have been gained in the TiO_2 - Y_2O_3 - ZrO_2 system [2-6].

In the present work further investigations of the system are presented.

W.Pyda, K.Haberko, M.M.Bućko
University of Mining and Metallurgy, Department of Special Ceramics, Cracow, al.Mickiewicza 30, Poland,
M.Faryna
Jagiellonian University,
Regional Laboratory for Physicochemical Analyses and Structural Research, Cracow, ul.Reymonta 25,

TABLE I. MONOCLINIC PHASE CONTENT (vol%) IN SAMPLES SINTERED AT 1400°C

$Y_2O_3 \rightarrow$ $\downarrow TiO_2$	0.5	1.0	1.5	2.0	3.0
3	100	100	95.5 ±0.4	81.1 ±0.4	37.9 ±0.5
8	100	100	95.1 ±0.5	–	18.9 ±0.4
13	100	96.2 ±0.4	1.0 ±0.9	0	0
18	91.7 ±0.5	0	0	–	–
23	–	0 (ZT)	0 (ZT)	–	–
28	–	0 (ZT)	0 (ZT)	–	–

± – denotes confidence interval at the confidence level of 0.95
(ZT) stands for ZrTiO4

EXPERIMENTAL PROCEDURE

The studied TiO₂ and Y₂O₃ concentrations ranged from 3 to 28 mole 5 and from 0.5 to 3.0 mole %, respectively.
The powder preparation route is described in detail elsewhere [7]. The coprecipitated Ti-Y-Zr hydroxide gels were subjected to the hydrothermal treatment at 250°C for 4 h, in order to prepare crystalline powders. The powders were then dried at 120°C and deagglomerated using a small vibratory mill.

Cylindrical samples of 12 mm diameter and ~3 mm thickness were uniaxially compacted under 50 MPa and then subjected to cold isostatic pressing under 350 MPa. Sintering was performed in air atmosphere in an electrically heated furnace at temperatures ranging from 1350°C to 1450°C. The rate of temperature increase was 6°C/min and the soaking time at the peak temperature was 2 h. The samples were furnace cooled.

The Porter and Heuer X-ray diffraction method [8] (CuKα) was used to determine the phase composition of the sintered samples. The Pt shaded carbon replicas, taken from the as received surface, were used to obtain the "true" grain size distribution by the Saltykov method [9].

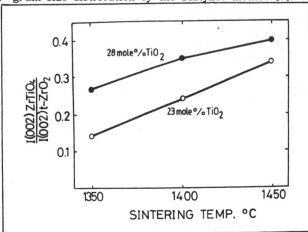

Figure 1. ZrTiO₄ and t-ZrO₂ X-ray reflection intensity ratio vs. sintering temperature. TiO₂ concentration indicated. Yttria content of 1.5 mole %.

The critical grain size of the tetragonal phase was assessed on the basis of the grain size distribution curves and the X-ray phase analyses of the sintered samples, assuming that the monoclinic grains were bigger than the tetragonal ones. The method is described in [10]. The K_{Ic} values were determined by the Vickers indentation under the load of 196 N using the Palmqvist crack model [11].

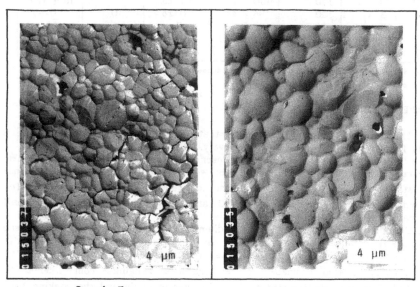

a - 8 mole %   b - 18 mole %

c - 23 mole %

Figure 2. TEM micrographs of the surface replicas. Samples sintered at 1400°C. Y_2O_3 concentration of 1.5 mole %. TiO_2 concentrations: a - 8 mole %, b - 18 mole %, c - 23 mole %.

RESULTS AND DISCUSSION

Table I shows monoclinic phase content found on the as received surface of the samples sintered at 1400°C. The balance is the tetragonal phase, and in the compositions of the highest TiO_2 concentrations – zirconium titanate ($ZrTiO_4$). The data indicate that fully or nearly fully tetragonal bodies (TiY-TZP) could be obtained within the studied chemical compositions range. The same conclusion comes from the data for the other sintering temperatures (not shown here). It is evident that the tetragonal phase is stabilized by the both alloying components. Zirconium titanate appears in the compositions of 23 and 28 mole % TiO_2 concentration.Its fraction increases with TiO_2 concentration and with sintering temperature. This is substantiated by the data of figure 1.

They show the intensity of the (002) $ZrTiO_4$ reflection in relation to the intensity of the (002) reflection of the major phase of the system, i.e. of the tetragonal zirconia solid solution. The TiO_2 concentration dependence of these relationships is obvious. However, the temperature dependence seems to indicate that the phase equilibrium is not reached, at least at the lower sintering temperatures. If it were not the case the lower fractions of $ZrTiO_4$ should be observed in the samples sintered at higher temperatures, because the increased solubility of TiO_2 in the tetragonal zirconia lattice with increasing temperature should be expected.

Typical microstructures of the sintered samples are shown in figure 2.

The tetragonal to monoclinic phase transformation is responsible for the cracks occurring in the sample of 8 mole % TiO_2 content (1.5 mole % Y_2O_3) – see figure 2a. This is substantiated by the 96 % content of the monoclinic phase found on the as received surface of this sample. No cracks are visible in the material of TiO_2 concentration of 18 mole % (figure 2b).

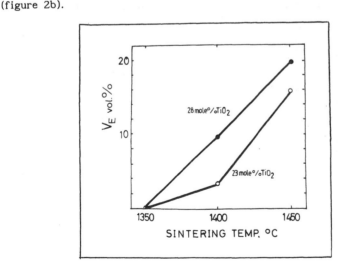

Figure 3. Volume fraction of the elongated grains (V_E) vs. sintering temperature. TiO_2 concentration indicated. Yttria content of 1.5 mole %.

In this case only small fraction of the monoclinic phase (7.2 %) has been found on the as received surface of the sample. Elongated grains occur in the sample of still higher TiO2 concentration (23 mole % – figure 2c). Using micrographs like these in Fig.2c, the volume fraction of the elongated grains has been measured.

Results in figure 3 indicate qualitatively the same sintering temperature and composition dependence of the elongated grains volume fraction as it is observed in figure 1 for the relative intensity of the (002) ZrTiO4 reflection. So, it seems plausible to assume that zirconium titanate is the phase crystallizing in the form of the elongated grains.

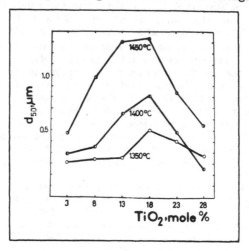

Figure 4. Median grain size (d_{50}) vs. TiO$_2$ concentration. Y$_2$O$_3$ content - 1.5 mole %. Sintering temperature indicated.

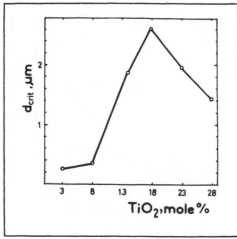

Figure 5. Critical grain size (d_{crit}) of the tetragonal phase vs. TiO$_2$ concentration in the solid solution. Y$_2$O$_3$ content in each case - 1.5 mole %. The plot is based on the data for a sintering temperature of 1450°C.

Using the grain size distribution curves (not shown here) the median grain sizes (d₅₀) were determined. They are demonstrated in figure 4. vs. TiO₂ concentration for the all applied sintering temperatures. We notice that TiO₂ promotes grain growth of the systems but only up to its concentration of 18 mole %. At higher concentrations grains of the system become smaller. This can plausibly be explained if we realize that at these TiO₂ concentrations zirconium titanate crystallizes at the tetragonal grain boundaries (cf. Fig.2c). This is possibly the effect by which the growth rate of the zirconia solid solution is inhibited.

In figure 5 the critical grain sizes (d$_{crit}$) vs. TiO₂ concentration are shown. The plot indicates that d$_{crit}$ is the greater the higher is the TiO₂ concentration. So, titania plays the role of the tetragonal phase stabilizer. The maximum d$_{crit}$ value occurs in the system of 18 mole % TiO₂ concentration and at its higher concentrations it decreases. This is an unusual behavior if we assume that at this composition the border line between the tetragonal field and the two phase field (i.e. the tetragonal ZrO₂ s.s + ZrTiO₄) is reached. In the analogous situation of the Ca-TZP materials the compositions situated within the two phase field (tetragonal + cubic) show the constant d$_{crit}$ value [10]. It happens so because of the redistribution of the alloying component between the two phases. Under such circumstances the concentration of the alloying component remains constant at the constant sintering temperature. It results in the constant d$_{crit}$ value. So, the observed behavior suggests that under the applied sintering conditions phase equilibrium is not reached in the system studied, as it has already been pointed out.

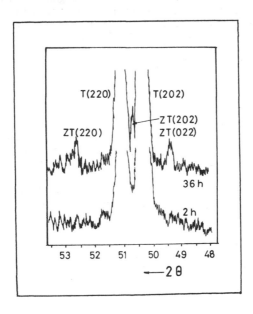

Figure 6. X-ray diffraction pattern of the sample of 18 mole % TiO₂ and 1.5 mole % Y₂O₃ content heated at 1350°C for 2 h and 36 h. T - stands for tetragonal ZrO₂ s.s., ZT - for ZrTiO₄.

TABLE II. FRACTURE TOUGHNESS, K_{Ic}, OF 1.5 mole % Y_2O_3 SAMPLES SINTERED AT INDICATED TEMPERATURES, [MPa.m$^{1/2}$].

TiO$_2$ mole %	1350°C	1400°C	1450°C
3	5.34 ± 0.42	6.13 ± 0.88	–
8	4.34 ± 0.28	5.74 ± 0.50	–
13	5.94 ± 0.72	6.33 ± 0.47	6.01 ± 0.34
18	9.85 ± 0.71	9.46 ± 1.84	6.63 ± 0.80
23	4.37 ± 0.16	4.05 ± 0.11	4.18 ± 0.14
28	3.72 ± 0.14	3.89 ± 0.20	3.90 ± 0.23

If so, the sample of 18 mole % TiO2 + 1.5 mole % Y2O3 should be recognized to be oversaturated with TiO2. To verify this supposition the sample under discussion has been subjected to the prolonged heat treatment at 1350°C. As it should be expected the X-ray diffraction pattern (figure 6) shows clearly that the ZrTiO4 peaks appear in the system heated for 36 h. This phase could not be detected after 2 h soaking at the sintering temperature. The most interesting fracture toughness values have been obtained in the systems of Y2O3 concentration of 1.5 mole %. The data presented in Table II demonstrate that the influence of the chemical composition on KIc is a complicated function. On the one hand, chemical composition affects both the powder characteristics [6] and hence its behavior during sintering and grain growth. On the other hand, fracture toughness depends on the transformability of the tetragonal grains. This transformability depends both on the chemical composition of a body and on the distance of the actual grain size from its critical size [12]. This complex situation makes the precise analysis of the results presented in Table II difficult.

CONCLUSIONS

X-ray diffraction study of the sintered bodies indicated the composition range in which the tetragonal phase dominated. Simultaneously this is the range of the highest KIc values. Critical tetragonal grain size measurements vs. their chemical composition substantiate the statement that TiO2 plays the role of the tetragonal phase stabilizer. Titania promotes grain growth of the tetragonal phase. However in the compositions of the high TiO2 concentration (23 and 28 mole %) it is inhibited. It is suggested that the elongated grains of zirconium titanate crystallizing at the tetragonal grain boundaries are responsible for this phenomenon. The phase equilibrium is not reached under the applied conditions of the heat treatment, but these conditions are sufficient to prepare dense materials of good mechanical properties.

REFERENCES

1.   Bannister, M.J. and J.M. Barnes, 1986. "Solubility of TiO$_2$ in ZrO$_2$." J. Am. Ceram. Soc., 69(11):C269-C271.

2.   Haberko, K., W. Pyda and M. Baćko, 1991. "Zirconia Polycrystals in the System ZrO_2-TiO_2-Y_2O_3," in Materials Science Monographs 66C, Ceramics Today - Tomorrow's Ceramics, P. Vincenzi, ed., Amsterdam - Oxford - New York - Tokyo: Elsevier, pp. 1563-1572.

3.   Hofmann, H., B. Michel and L.J. Gauckler, 1989. "Zirconia Powder for TZP-Ceramics Ti-Y-TZP," in Advances in Zirconia Science and Technology - Zirconia '88, S. Meriani and C. Palmonari, eds., Elsevier Applied Science, pp. 119-129.

4.   Lin, L.C., D. Gan and P. Shen, 1990. "The Effects of TiO_2 Additions on the Microstructure and Transformation of ZrO_2 with 3 and 6 mole % Y_2O_3." Mat. Sci. and Engineering, A129:147-155.

5.   Haberko, K., W. Pyda, M.M. Baćko and M. Faryna, 1991. "Tough Materials in the System TiO_2-Y_2O_3-ZrO_2," in Proc. European Ceramic Society Second Conference, Augsburg, FRG, September 11-14, in print.

6.   Pyda, W., K. Haberko, M.M. Baćko and M. Faryna, 1992. "A Study on Preparation of Tetragonal Polycrystals (TZP) in the TiO_2-Y_2O_3-ZrO_2 System." Ceramics International, 18(5):321-326.

7.   Pyda, W., K. Haberko and M.M. Baćko, 1991. "Hydrothermal Crystallization of Zirconia and Zirconia Solid Solutions." J. Am. Ceram. Soc., 74(10):2622-2629.

8.   Porter, D.L. and A.H. Huer, 1979. "Microstructural Development in MgO-Partially Stabilized Zirconia." J. Am. Ceram. Soc., 62:298-305.

9.   Saltykov, S.A., 1970. Stereometric Metallography, 3rd Edition, Moscow, Russia: Metallurgy (in Russian).

10.   Pyda, W. and K. Haberko, 1987. "CaO-Containing Tetragonal Zirconia Polycrystals (Ca-TZP)." Ceramics International, 13:113-118.

11.   Niihara, K.A., 1983. "A Fracture Mechanics Analysis of Indentation-Induced Palmqvist Crack in Ceramics." J. Mater. Sci. Let., 2:221.

12.   Haberko, K. and R. Pampuch, 1983. "Influence of Yttria Content on Phase Composition and Mechanical Properties of Y-PSZ." Ceramics International, 9(1):8-12.

The Effect of Low Temperature Aqueous Environments on Magnesia-Partially Stabilised Zirconia (Mg-PSZ)

J. DRENNAN, A. J. HARTSHORN and S. W. THOMPSON

ABSTRACT

Mg-PSZ may suffer some degradation of properties when subjected to high pressure aqueous environments for prolonged periods of time. The effect is far less pronounced and more difficult to study than the degradation of Y-TZP ceramics under similar conditions. In order to quantify this phenomenon and to go some way to predicting lifetimes of parts used in industrial applications, a series of experiments using autoclaves has been designed. The extent of corrosion was found to be very dependent on the aging condition of the Mg-PSZ with an inverse correlation existing between the stability of tetragonal particles and the susceptibility to attack. Temperatures >250°C are required before significant corrosion takes place in the most stable materials and grain boundaries play a pivotal role in the corrosion process.

INTRODUCTION

In 1981 Kobayashi et al [1] reported that the ceramic yttria tetragonal zirconia Polycrystals (Y-TZP) suffered severe degradation of physical properties after being exposed to relatively low temperatures. Since this first report our knowledge of this problem with Y-TZP has expanded to a point where the phenomenon is now well characterised and a number of theories have been developed to explain the mechanisms taking place. For more comprehensive reviews of the literature and the various theories pertaining to this phenomenon the reader is directed to the publications of Lilley [2] and Swab [3].

Briefly summarising the effects observed for Y-TZP, there is a temperature range over which the ceramic is most susceptible to degradation, ie. 150-400°C. The degradation process is accelerated in moist environments and is manifested by an increase in the monoclinic phase content of the surface of the ceramic which is a consequence of the triggering of the tetragonal to monoclinic phase transformation. The transformation initiates at the surface and proceeds into the bulk with continuing exposure to moisture and as a result there is a significant reduction in flexural strength and fracture toughness of the ceramic body. More disturbingly, Y-TZP

J Drennan, CSIRO, Division of Materials Science and Technology, Locked Bag 33, Clayton, 3168, Australia.
A J Hartshorn, S W Thompson, ICI Advanced Ceramics, 8 Redwood Drive, Monash Business Park, Clayton, 3168, Australia.

ceramic pieces are seen to become severely cracked and even reduced to powder after exposure to these relatively mild environments.

In the case of magnesia partially stabilised zirconia (Mg-PSZ), the problem is nowhere near so severe. However there are reports of detrimental effects on the material after prolonged exposure to moist environments. Swain [4] examined the effect of 4 hour exposures at 400°C and 6 atm. of steam on various grades of Mg-PSZ. It should be noted that no examination of any control samples was undertaken to study the effect of 400°C on these materials under dry conditions. He observed an increase in the monoclinic content of the surface of both grades of Mg-PSZ but the increase was far greater for the material which had been aged for longer periods of time (TS grade). In addition he reported an accompanying reduction in strength of the materials which had been subjected to these environments.

Sato et al [5] in a more comprehensive study examined the effects of a range of solutions including water over a range of pressures and temperatures for once again two varieties of Mg-PSZ. They observed that the tetragonal to monoclinic phase transformation in both materials occurred in water above 200°C. The bend strength of the materials degraded under most of the conditions used and Mg+ ions were detected in the residual water and 1M HCl used in the studies.

This paper describes a systematic examination of the effect of low temperature high pressure moist environments on the stability of the tetragonal phase present in the Mg-PSZ ceramic. In addition by combining both stress and a moist environment we have shown that a stress induced corrosion process might be taking place, and detailed microstructural evidence recorded from the effected regions has suggested a mechanism by which the degradation proceeds. These observations also provide an insight into how the effect of moist environments on Mg-PSZ can be minimised.

EXPERIMENTAL

Various grades of Mg-PSZ used in this study were supplied by ICI Advanced Ceramics. Samples were prepared by following the procedure outlined in [6], which enables specimens to be prepared in such a way that the degree of metastability of the tetragonal precipitates can be varied in a controlled manner. The degree of metastability was determined by measuring the monoclinic content of carefully ground surfaces by x-ray diffraction. Both bars and pellets were prepared.

In the case of the static stress tests, a bar (2.2 x 8 x 43 mm) of MS-grade Mg-PSZ was placed in an especially designed four point bending rig in which a specific load could be applied and maintained. The strain levels were measured by attaching strain gauges to the tensile surface of the specimen. Once loaded the rig was placed in an autoclave and sealed for each run. Temperatures were maintained at 175°C for various times and the starting stress was 250MPa. On removal of the specimen after each run, x-ray diffraction was performed on both the compressive and tensile surfaces to determine the amount of transformation which had taken place by monitoring the monoclinic content of each surface. In addition the Nomarski contrast facility of an optical microscope was used to examine the extent of surface rumpling. Once completed the samples were reloaded, new strain gauges attached and the experiment repeated. In each case two experiments were run in parallel: one in the autoclave, and a further identical rig set up in an evacuated glass tube which was maintained at the same temperature as the autoclave specimen.

For the low temperature studies in which the static fatigue rig was not able to be used, variously aged specimens with highly polished sides were subjected to 300°C for a period of days in an autoclave which was maintained at a pressure of 7.7 MPa. The monoclinic content before and after exposure to the moist environment was monitored as above. In addition an identical set of control samples was maintained at

300°C in an evacuated glass tube for the same period of time as the autoclaved samples.

Analytical electron microscopy was performed on ion beam thinned specimens using the NAMAC [7] Philips CM30 electron microscope operated at an accelerating voltage of 300keV. Cross sectioned specimens were prepared by following the method outlined in [8] in such a way that regions directly affected by the moist environments could be examined.

RESULTS

Figure 1 shows the results obtained using the static fatigue rig under both autoclave and dry conditions. The measured monoclinic values were recorded from the tensile surface of the specimens in both cases. Clearly the combined effect of stress and high pressure aqueous environment has significantly increased the amount of transformation taking place on the surface of the specimens. The amount of monoclinic phase detected on the compressive surface was also measured and found to be significantly lower than the tensile surface as might be expected.

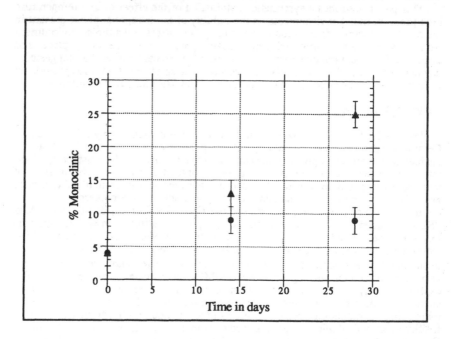

Figure 1. Plot of the effect of the combination of steam and stress on MS-grade Mg-PSZ. Autoclave conditions were 175°C and 0.8 MPa pressure, and the applied static stress was 250 MPa. ▲ Autoclaved sample, ● Dry sample.

A further observation which was of interest was that, when the sample, which had been under stress for approximately 30 days in the autoclave, was removed from the rig it was noticeably deformed as can clearly be seen in figure 2. This deformation was found to be reversible by annealing the specimen at 1000°C for 2h . It should be noted that the specimen which had been stressed under dry conditions showed no such deformation.

This shape memory effect has been previously reported for Y-TZP specimens [9] and Swain [10] produced similar deformations of Mg-PSZ with 200 MPa at 800°C in air. Using the same formula as Swain, the permanent deformation strains recorded for the autoclaved samples examined in this study were of the order of 0.51%, larger than those reported by Swain (0.42%) and at a significantly lower temperature.

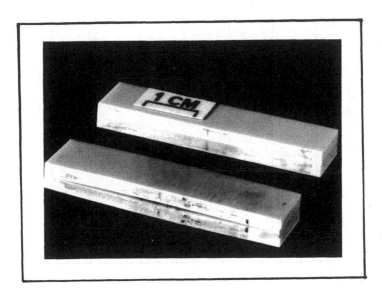

Figure 2. Bars of Mg-PSZ which had been subjected to the same static stress for 30 days. The bent bar was subjected to this stress under autoclave conditions.

To obtain significant changes in the amount of surface transformation for Mg-PSZ specimens in the absence of a tensile stress it was necessary to raise the autoclave temperature to 300°C and even then periods of up to 3 days were required before changes were observed. In this set of experiments only highly polished specimens were used. The previously described static fatigue rig was not able to withstand the corrosive nature of these higher temperatures and pressures. Figure 3 plots the polished surface monoclinic content recorded after autoclave exposure versus the ground surface monoclinic recorded before the autoclave runs. The ground surface monoclinic is a measure of the ease of transformation of the tetragonal precipitates present in the specimen and is directly related to the aging conditions to which the specimen has been subjected. This effect can be seen more graphically in figure 4. Here the polished surfaces of the specimens used to obtain figure 3 are presented after being subjected to autoclave treatment. The surfaces show an increasing degree of rumpling as a function of their original aging time .

Transmission electron microscopy of regions which had been in direct contact with the steam atmosphere revealed features which give some insight into the processes which take place when high pressure moisture comes in contact with Mg-PSZ. Regions close to the grain boundary (approximately 1μm) show tetragonal precipitates outlined by what appears to be a loss of material or coherency at the interface with the matrix. Figure 5 shows this effect clearly at various magnifications.

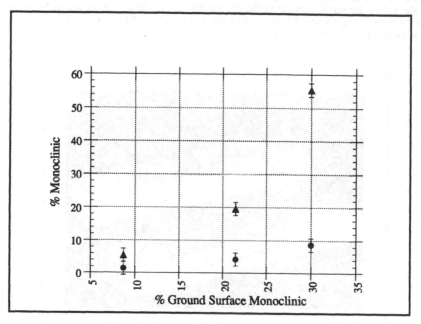

Figure 3. Plot showing the effect of the aging characteristics of Mg-PSZ. Autoclave conditions were 300°C at 7.7 MPa steam for 3 days. ▲ Autoclaved sample, ● Dry sample.

DISCUSSION

The results presented above clearly show that the transformation of tetragonal to monoclinic structure of the precipitates in the exposed surface grains of Mg-PSZ is promoted by the presence of high pressure low temperature moist environments. This effect may be accelerated by the presence of a static or applied stress and is dependant on the relative stability of the precipitates (aging characteristics).

Previous studies on zirconia systems tend to suggest that an intrinsic property of these types of materials which contain a transformable tetragonal phase is that the action of water vapour gives rise to a reduction in the stability of the tetragonal phase at room temperature. A full discussion on the exact mechanism as to why this occurs is outside the scope of this paper. However, if this is an intrinsic property of the metastable tetragonal phase, then the rumpling observed in the micrographs recorded from affected surfaces is simply caused by transformation of those surface grains which are the most unstable. Consequently, for materials with smaller more stable precipitates, the effect will be less pronounced. It should be pointed out that a transformation layer on the surface of ceramics parts is often a desirable situation because the resultant compressive layer acts to prevent crack growth. In the case of the TZP materials this of course is catastrophic since transformation of a whole grain of the material will immediately give rise to a local loss of coherency and the subsequent formation of a critical flaw.

An additional and possibly related process also appears to be taking place as evidenced by the micrographs of figure 5. It appears that the grain boundaries provide a pathway for the penetrating moisture which in turn acts on the precipitates immediately surrounding the boundary. What remains are transformed precipitates outlined by what appears to be material denuded zones. The amount of material loss is not detectable by energy dispersive studies but previous work [5] has shown that Mg can be detected in the aqueous residues from autoclave studies suggesting some dissolution is taking place. If dissolution takes place in these regions then the constraining matrix is removed allowing the metastable tetragonal precipitates to transform. The accompanying microcracking may then provide a continuous pathway for the penetrating moisture on to the adjoining precipitates.

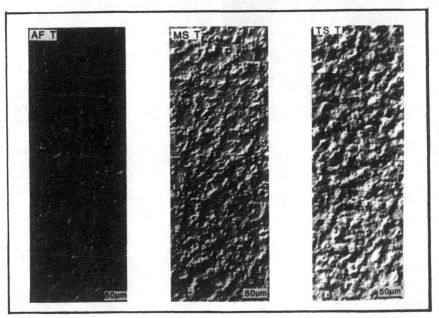

Figure 4. Nomarski contrast micrographs showing the development of rumpling as a function of the aging time to which the Mg-PSZ sample had been subjected. AF stands for as-fired, MS maximum strength and TS, thermal shock. TS is aged for a longer time than MS-grade PSZ.

The region around the growing precipitates in Mg-PSZ has been well characterised for a range of heat treatments. Rossell and Hannink [11] and Hannink [12] have shown that the Mg content immediately adjacent to a growing precipitate can reach such a high level that the metastable compound $Mg_2Zr_5O_{12}$ is formed. The slow cation diffusion processes at the aging temperature prevent the equilibrium redistribution of the Mg into the surrounding matrix as it is expelled from the growing precipitate. It is this Mg-rich region that we suggest is the site of attack by low temperature moisture. In addition the non uniform distribution of Mg immediately surrounding a precipitate will be much more significant for materials containing larger precipitates and hence the materials which have been subjected to lengthy aging conditions are more susceptible.

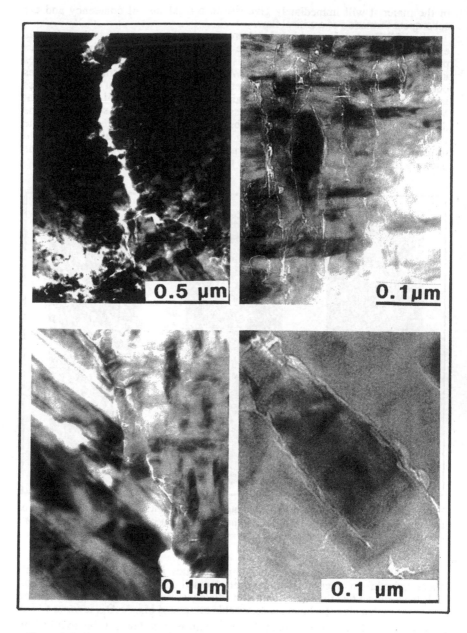

Figure 5. A series of transmission electron micrographs showing the effect of moisture attack near the grain boundaries in a Mg-PSZ sample.

CONCLUSION

Mg-PSZ suffers degradation under the severe conditions of low temperature, high pressure aqueous environments. However the level of degradation is very dependant on the stability of the tetragonal precipitates present. Attack appears to take place at the Mg-rich regions around well developed precipitates. The ability to alter the microstructure of Mg-PSZ by judicious aging and heat treatments provides the possibility to tailor the material to minimise the effects of low temperature high pressure moist environments, a situation that is not possible with the TZP class of compounds.

REFERENCES

1. Kobayashi, K., H. Kuwajima, and T. Masaki, 1981. "Phase Change and Mechanical Properties of ZrO_2-Y_2O_3 Solid Electrolyte After Aging." Solid State Ionics, 3(4): 489-495.

2. Lilley, E., 1990. "Review of Low-Temperature Degradation of Y-TZPs." Ceram. Trans, 10 : 387-407.

3. Swab, J. J.,1990. "Low Temperature Degradation of Y-TZP Materials," U. S. Department of Energy Report, No. MTL TR 90-4.

4. Swain, M. V., 1985. "Stability of Mg-PSZ in High Temperature Steam Environments," J. Mat. Sci. Letts, 4 : 848-850.

5. Sato, T., T. Endo, M. Shimada, T. Mitsudome and N. Otabe, 1991. " Hydrothermal Corrosion of Magnesia-Partially-Stabilised Zirconia." J. Mat. Sci., 26 : 1346-1350.

6. Garvie, R. C., R. H. J. Hannink and N. A. McKinnon, 1981. "Partially Stabilised Zirconia Ceramics," U. S. Patent No. 4,279,655.

7. National Advanced Materials Analytical Centre, Locked Bag 33, Clayton,Victoria, Australia.

8. Alani, R., J. Jones and P. Swann, 1990. "Chemically Assisted Ion Beam Etching (CAIBE)- A New Technique for TEM Specimen Preparation of Materials." Mat. Res. Soc. Symp. Proc., 199 : 85-101.

9. Soma, T., and M. Matsui, 1988. "Ceramic Shape Memory Element," U. S. Patent No. 4,767,730.

10. Swain ,M. V., 1986. "Shape Memory Behaviour in Partially Stabilised Zirconia Ceramics," Nature, 322 : 234-236.

11. Rossell, H. J., and R. H. J. Hannink, 1984. " The Phase $Mg_2Zr_5O_{12}$ in MgO Partially Stabilised Zirconia," in Advances in Ceramics Vol. 12. Science And Technology of Zirconia, N Clausen, M Ruhle and A H Heuer, eds., Columbus, Ohio, USA : The American Ceramic Society, Inc., pp 139-152.

12. Hannink, R. H. J.,1983. "Microstructural Development of Sub-eutectoid Aged MgO-ZrO_2 Alloy," J. Mat. Sci., 18: 457-470.

Degradation of Y-TZP in Moist Environments

A. E. HUGHES, F. T. CIACCHI and S. P. S. BADWAL

ABSTRACT

The moisture sensitive sensitive t → m transformation in Y-TZP has been examined by X-ray diffraction with X-ray photoelectron spectroscopy (XPS). Pre-sintered (both as-fired and after polishing) discs of 3 mol% Y_2O_3-ZrO_2 from two sources, containing different levels of impurities, were annealed in both dry and moist air in the 200-300 °C range. On annealing in moist air a considerable amount of m-ZrO_2 is formed accompanied by increases in the Y/Zr ratio by a factor of 2-3 and changes in the O 1s regions and Zr 3d. These changes suggest large increases in the amount of $O^=$ and Zr^{3+}.

INTRODUCTION

Over the years there have been many reports of the moisture sensitivity of pure ZrO_2 [1] and yttria-tetragonal zirconia polycrystal (Y-TZP). For example, Kobayashi et al. [2] observed that sintered and fully dense Y-TZP undergoes an unusual surface degradation in moist environments at low temperatures (150-400°C). The degradation manifests itself as microcracking at the surface [3-5] and a diminution of mechanical properties [6-9] and has been shown to be due to transformation of the surface tetragonal (t) grains to monoclinic (m) phase. Specifically, the amount of m-ZrO_2 phase, as monitored using X-ray diffraction (XRD), increases in the first few microns of the surface at temperatures between 100 to 400°C and with treatment time in moist environments [4,5,7,10-13]. The transformation has been shown to be dependent on the grain size, and the yttria content of the Y-TZP [8,14,15].

A.E. Hughes, F.T. Ciacchi and S.P.S. Badwal, CSIRO Division of Materials Science and Technology, Locked Bag 33, Clayton, Victoria, Australia, 3168

Indeed there appears to be a critical grain size below which the transformation does not proceed. Watanabe et al. [8], for example, found that the critical grain size increased from 0.2 to 0.6μm as the yttria content of the Y-TZP increased from 2 to 5 mol% Y_2O_3.

Sato and Shimada [13], and Chen and Lu [14,15] have shown that the production of m-ZrO_2 is, however, completely reversible and it can be removed by heating above the martensitic start temperature, M_s, which defines the t→m transformation temperature in the constrained matrix and which varies with grain size and yttria content.

Results published from different laboratories vary significantly due to the limited control over ceramic and powder processing variables. The extent of transformation depends on the grain size, the yttria content and distribution between and within grains. Consequently, specimen preparation techniques, which control the ceramic microstructure, become important variables. In addition, the microstructure will be influenced by the type and distribution of impurities in the starting powders. Impurity phases in zirconia-based ceramics have a variety of compositions [16-20], it is these compositional differences which will lead to different rates of yttrium redistribution [16] and may well modify the kinetics of the t → m transformation. Different impurity phase compositions will mean that even after firing at the same sintering temperatures, times, atmospheres and cooling rates, there is no guarantee that the impurity phase distribution will be the same. These variables affect the stability of the tetragonal phase by setting up inhomogenous distribution of Y leading to different grain size distributions in different ceramics.

This paper aims to identify how various parameters influence the development of the monoclinic phase in Y-TZP in moist environments. XRD has been used to monitor the generation of m-ZrO_2, while X-ray photoelectron spectroscopy (XPS) has been used to study chemical changes at the surface.

EXPERIMENTAL

Discs were prepared from two starting powders of tetragonal yttria-zirconia with different impurity profiles [16]. TS3Y, a high purity powder, was obtained from TOSOH Corporation, Japan and HS3Y, a low purity powder, from Daiichi Kigenso Kagaku Co. Ltd, Japan. Discs were heated to 1500°C at a rate of 300°Ch^{-1}, sintered for 4 hours then quenched in air. The average grain sizes for TS3Y and HS3Y were 0.8 and 0.9 microns respectively.

Several series of experiments were performed and the monoclinic phase content was determined by XRD comparing the m(111), m(11$\bar{1}$) to the t(111) reflections (Cu Kα radiation). In the first series of experiments the %m-ZrO_2 was determined for discs as-sintered, polished and polished and heated to 600°C. In the second series, polished discs were exposed to air saturated with water at ambient temperature (22°C), hereafter called moist air, for 48hrs at 200, 250 and 300°C after which the %m-ZrO_2 was determined. In the third series of experiments, polished discs were first heated to 600°C for 5hrs to ensure that the surface was entirely free of m-ZrO_2, cooled to 200°C, then exposed to moist air for various times. The %m-ZrO_2 was determined after each exposure.

The experimental setup and quantification of results for XPS has been described previously [16]. XPS data were collected on duplicate specimens from the above series at $\theta = 0$ and 70°. Measurement at $\theta = 70°$ probes only a third of the depth (~2.4nm) at $\theta = 0°$ (~7.0nm) thus giving higher surface sensitivity. The data reduction has been treated in detail elsewhere [21].

XPS specimens were exposed in-situ to moist air at 200°C for various times thus avoiding contamination from the external laboratory atmosphere. However, to determine the %m-ZrO_2 an initial set of specimens were treated in-situ, removed and examined by XRD. Subsequent series were then treated for the same period of time but examined directly by XPS assuming the %m-ZrO_2 from the previous set of experiments.

TABLE I - %m-ZrO_2 GENERATED UNDER VARIOUS TREATMENT CONDITIONS

TREATMENT CONDITIONS	TS3Y	HS3Y
As-sintered	0	0
Polished	0	0
Polished + 600°C (5h Dry Air)		
600°C (1h Dry Air) + 200°C (48h Dry Air)	-	11
Polished + 600°C (5h dry air) + 200°C (48h dry air)	62,61	44,43
250°C (48h moist air)	11,0	58,57
300°C (48h moist air)	0	55
120°C (70h autoclave)	73	73
170°C (70.5h autoclave)	80	77

RESULTS

As can be seen in Table I, no m-ZrO_2 was detected on the surface of the discs of either TS3Y or HS3Y after they were either sintered, polished or polished and heated to 600°C. However, 11%m-ZrO_2 was detected for HS3Y after exposure to dry air for 48hrs at 200°C which may be due to some contaminant water vapour in the gas stream.

The %m-ZrO_2 generated after treatment in moist air for 48 hours at 200, 250 and 300°C was quite different for each ceramic. TS3Y displayed the most transformation at 200°C and little transformation at 250°C or 300°C whereas

HS3Y displayed considerable transformation at all temperatures with a maximum at 250°C. Hence the martensitic start temperature for TS3Y is lower than HS3Y which is consistent with its smaller grain size.

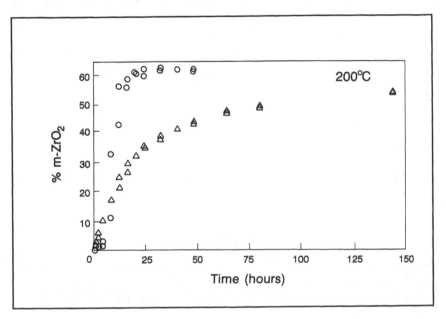

Figure 1. %m-ZrO$_2$ generated as a function of aging time at 200°C in a moist air stream. (o) - TS3Y and (▲) - HS3Y.

The %m-ZrO$_2$ vs exposure time to moist air at 200°C for TS3Y is compared to HS3Y in figure 1. It is obvious that at 200°C the kinetics of transformation for TS3Y are more favourable than for HS3Y. At times less than 10 minutes the %m-ZrO$_2$ phase was higher in HS3Y than TS3Y. After 12 minutes, however, there was a dramatic increase in the %m-ZrO$_2$ in TS3Y whereby at ~ 25 minutes the %m-ZrO$_2$ phase content of the surface had saturated at around 60%. For HS3Y, there was a steady increase in the %m-ZrO$_2$ phase with increasing treatment time.

The difference in kinetics can be eliminated by examining changes as a function of %m-ZrO$_2$ rather than time, thereby identifying some of the common features of the moisture induced t → m transformation. Hence in figure 2 the XPS Y/Zr atomic ratios at θ=0° and 70° are plotted against the %m-ZrO$_2$ in the surface. At θ=0° the Y/Zr ratios displayed complex behaviour between 0 and 15%m-ZrO$_2$. Between 0 and ~6%m-ZrO$_2$ the Y/Zr ratios increased to a similar value of ~0.074 for both TS3Y and HS3Y. The rate of increase was larger for TS3Y than HS3Y. Between 6 and 12%m-ZrO$_2$ there was an apparent discontinuity in the Y/Zr ratio, but above 15%m-ZrO$_2$ both ceramics displayed similar values. For θ=70° the Y/Zr ratios decreased

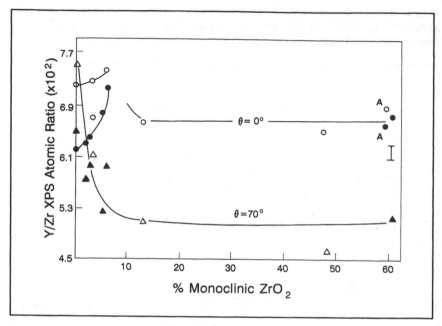

Figure 2. XPS Y/Zr ratio vs %m-ZrO$_2$ generated in a moist air stream at 200°C. The rate of generation is given in figure 1. ($\bullet$,$\blacktriangle$) - TS3Y and ($\circ$,$\triangle$) - HS3Y at θ = 0° ($\circ$,$\bullet$) and 70° ($\triangle$,$\blacktriangle$).

from 0.075 and 0.064 for HS3Y and TS3Y respectively to an equilibrium value of ~0.052 at 15%m-ZrO$_2$ indicating a surface depletion of Y.

O 1s spectra were fitted with four components at 529.8 (O$_I$) due to O$^=$, 530.5 (O$_{II}$) due to oxygen anions adjacent to vacancies [22], 531.5 (O$_{III}$) due to O$^=$ in the triangular configuration in the monoclinic phase [22] and 532.2 (O$_{IV}$)eV due to physisorbed water [23]. O$_{III}$ may however, have a significant contribution from bound hydroxyl groups.

The dependence O$_I$/Zr and O$_{III}$/Zr on %m-ZrO$_2$ for TS3Y and HS3Y after treatment at 200°C is displayed in figure 3. There was a significant increase in O$_I$ for both ceramics at low %m-ZrO$_2$ content which was accompanied by a loss of O$_{II}$. O$_{III}$ (531.5eV) gradually increased with exposure time confirming a contribution from surface hydroxyl groups since if it arose entirely from the m-ZrO$_2$ phase then it would saturate at much shorter exposure times. This is further confirmed after treatment of TS3Y at 300°C where both O$_{II}$ as well as O$_{III}$ were evident but no m-ZrO$_2$ was present on the surface.

Treatment of either ceramic at 200°C for various times, resulted in a loss of resolution in the normally well resolved Zr 3d5/2-3d3/2 spin-orbit doublet suggesting the presence of additional components. Analysis of the Zr 3d spectra (θ=0°), for either TS3Y or HS3Y as a function of treatment time, revealed the presence of Zr^{3+} which reflects the change in the vacancy

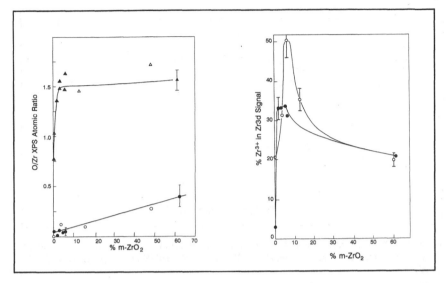

Figure 3. Percentage of O_I (▲,△) and O_{II} (●,○) O 1s components generated in a moist air stream at 200°C vs %m-ZrO$_2$. The rate of generation is given in figure 1. (●,▲) - TS3Y and (○,▲) - HS3Y.

Figure 4. % Zr^{3+} vs %m-ZrO$_2$ generated in a moist air stream at 200°C at θ=0°. The rate of generation is given in figure 1. (●) - TS3Y and (○) - HS3Y.

concentration in the near surface region. The subsurface enrichment of Y (figure 2) will be accompanied by an enrichment of anion vacancies:

$$Y_2O_3(ZrO_2) \rightarrow 2Y_{Zr} + 3O_O + V_O^{\cdot\cdot} \qquad (1)$$

These vacancies, which are generally adjacent to Zr^{4+} [24], are efficient electron traps for any low energy electrons, including photoelectrons generated in the XPS experiment [25]. Hence:

$$Zr^{4+} + nV_O^{\cdot\cdot} + e^- \rightarrow (n-1)V_O^{\cdot\cdot} + [Zr^{4+} + V_O^{\cdot\cdot}]'. \qquad (2)$$

where $[Zr^{4+} + nV_O^{\cdot\cdot}]'$ is a trapped electron on a Zr cation-oxygen vacancy pair. The percentage of Zr^{3+} versus the %m-ZrO$_2$ content of the surface (Figure 4), peaked at around 35% for TS3Y and 50% for HS3Y at 10%m-ZrO$_2$. For longer exposures the percentage Zr^{3+} decreased to around 20%.

The dependence of the Si/Zr ratio on the %m-ZrO$_2$ for HS3Y and TS3Y at θ=0 and 70° showed no distinguishable difference between the two ceramics despite their different impurity profiles (Figure 5). There was an increase in the silicon levels up to 6%m-ZrO$_2$ beyond which the Si/Zr ratio levelled out. The ratios at θ=70° are roughly three times larger than at θ=0° indicating that the silicon was enriched on the external surface.

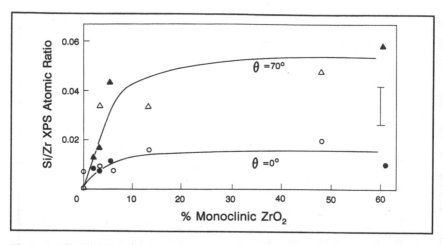

Figure 5. XPS Si/Zr ratio vs %m-ZrO_2 generated in a moist air stream at 200°C. The rate of generation is given in figure 1. (●,▲) - TS3Y and (o,△) - HS3Y at θ = 0° (o,●) and 70° (△,▲).

DISCUSSION

The rate of the t→m phase transformation is dependent on the kinetics of chemical processes and the thermodynamic stability of various phases. As a function of temperature, HS3Y has a higher rate of conversion at higher temperatures compared to TS3Y (Table I). The temperature dependence of conversion has been examined thoroughly by Sato and Shimada [13] and Chen and Lu [14,15] as a function of grain size. They found that as the grain size decreased the maximum temperature at which the t→m transition occurred also decreased which was attributed to a lowering of the martensitic start temperature (M_s) in the constrained matrix of a sintered material. Hence the negligible conversion of TS3Y to m-ZrO_2 at higher temperatures suggests that M_s is lower than HS3Y, consistent with the smaller grain size of TS3Y.

Clearly, there are significant differences between the kinetics of conversion for the two ceramics. Contributing factors may include the Y_2O_3 content and distribution within and between grains, the presence of intergranular glassy phases, the presence of large cubic grains in HS3Y, differences in the defect distributions or a combination of all these factors. Lu and Chen [16] and Watanabe et al. [8] have found that the rate of transformation depends on the grain size as well as the Y_2O_3 content. The larger rate of transformation for TS3Y at lower temperatures (200°C compared to 250° or 300°C for HS3Y) must be due solely to a combination of lower initial surface yttria content and the impurity phase distribution since TS3Y has the smaller grain size which would help to retard the transformation.

The difference in kinetics can be eliminated by examining changes as a function of %m-ZrO$_2$ rather than time, thereby identifying some of the common features of the moisture induced t → m transformation. Although the kinetics are different for both ceramics, once the transformation has occurred, i.e., at the same surface m-ZrO$_2$ phase content, then there are some extraordinary similarities between the two ceramics. Hence the route to transformation is the same in both cases, only the rate is different.

The most notable features of the XPS data versus %m-ZrO$_2$ phase content are peaks in the Y/Zr ratio and the %Zr^{3+} at around 8%m-ZrO$_2$ which are heralded by a substantial increase in the O$_1$/Zr ratio (oxygen anions). These changes represent a significant redistribution of species within the surface. It should also be noted that 8%m-ZrO$_2$ as determined by XRD, almost certainly represents a considerably higher m-ZrO$_2$ content for the surface. Furthermore, in the early stages of formation of m-ZrO$_2$ there is considerable preferred orientation of the monoclinic phase in the surface [10]. Hence, given that the XPS only probes the *surface* of the surface grains, then the XPS analysis will be weighted towards the m-ZrO$_2$ grains of preferred orientation.

The complexities of disentangling the processes which contribute to the t→m transformation are obvious. However, the following explanation is offered for the observed changes in the XPS atomic ratios with increases in the m-ZrO$_2$ phase content. At low %m-ZrO$_2$ the production of oxygen anions must be due to the dissociation of H$_2$O. Hence:

$$O-Zr-V_O^{\cdot\cdot}-Zr-O + H_2O \rightarrow 2Zr-OH + O^- \qquad (3)$$

Consequently, the surface quickly becomes saturated with O$^=$ which sets up a strain field within the surface of the grains inducing the t→m transformation. A similar mechanism to the one proposed here was invoked by Badwal and Nardella to explain the presence of a considerable amount of m-ZrO$_2$ at the anodic side of Y-TZP during current flow [26]. It is assumed that this attack proceeds not only at the external surface but more importantly, in the grain boundary network near the surface. In the grain boundary however, the presence of an impurity phase will modify the kinetics possibly through significantly decreasing the population of V$_o^{\cdot\cdot}$ via coordination of oxygen in the glass to these sites. In the case of TS3Y, the likelihood of the transformation is improved by the low impurity content resulting in less impurity phase in the grain boundary network.

As will be demonstrated in a separate paper hydroxides of Zr only develop after long exposure to the moist air stream at 200°C or autoclave environment [21]. Hydroxides of yttrium were only observed after severe autoclave treatment. Hence, they should be viewed as symptomatic of the degradation in moist environments rather than the initiators of the t→m transformation. From the viewpoint of surface energy, the presence of some form of silicon on the external surface of grains could modify the surface energy of the tetragonal phase. Holmes et al. [27] determined from the heats of immersion that the surface energy was much lower for the tetragonal phase (500→600 erg/cm^2) than the monoclinic phase (~1100 erg/cm^2). Earlier work of Harkins and Boyd [28] gave the heats of immersion of SiO$_2$ and ZrSiO$_4$ as 600 and 850 ergs/cm^2 respectively. These results suggest an oxidized form of silicon on the external surface of the ceramic may increase the surface energy of the tetragonal phase thereby reducing the free energy barrier to transformation. Increases in the Si/Zr ratios on the external surface could result from squeezing out the silicate impurity phase after transformation.

Apart from Si there are a number of other impurities in HS3Y and TS3Y which form glassy phases with SiO_2 and are soluble in the lattice. Hughes and Badwal [16,17] found that significant impurity and yttrium concentration gradients develop within the surface of grains. These concentration gradients would significantly alter defect concentrations which are reflected in the $\%Zr^{3+}$. It should be noted that the Y/Zr ratios for both TS3Y and HS3Y are similar but HS3Y had a higher $\%Zr^{3+}$, which may be due to different levels of soluble impurities present in the ZrO_2.

CONCLUSIONS

XPS results indicate that the mechanism of t→m transformation is related to strain built up in the lattice due to the presence of oxygen anions. Concomitant with the buildup of lattice oxygen is a redistribution of yttrium cations as well as significant changes in defect distributions. The role of defect structures in stabilizing t-ZrO_2 is unclear.

ACKNOWLEDGMENT

The authors would like to thank Ms. K. Crane for the preparation of some specimens, Dr. D. Hay for some assistance with the XRD and Dr. J. Drennan for discussion and reviewing this manuscript.

REFERENCES

1.   Argon, P.A., E.L. Fuller, Jr and H.F. Holmes, 1975. "IR Studies of Water Sorption on ZrO_2 Polymorphs. 1." Journal of Colloid and Interface Science, 52(3):553-561.

2.   Kobayashi, K., H. Kuwajima and T. Masaki, 1981. "Phase Change and Mechanical Properties of ZrO_2-Y_2O_3 Solid Electrolyte After Aging." Solid State Ionics, 3/4:489-493.

3.   Wang, J. and R. Stevens, 1989. "Surface Transformation and Toughening of TZP Ceramics by Low Temperature Ageing." British Ceramic Proceedings, 42:167-178.

4.   Sato, T. and M. Shimada, 1984. "Crystalline Phase Change in Yttria-Partially-Stabilized Zirconia by Low-Temperature Annealing." Journal of the American Ceramic Society, 67(10):C212-213.

5. Hernandez, M.T., J.R. Jurado, P. Duran and J.L.G. Fierro, 1991. "Subeutectoid Degradation of Yttria-Stabilized Tetragonal Zirconia Polycrystal and Ceria-Doped Yttria-Stabilized Tetragonal Zirconia Polycrystal Ceramics." Journal of the American Ceramic Society, 74(6):1254-1258.

6. Whalen, P.J., F. Reidinger and R.F. Antrim, 1989. "Prevention of Low-Temperature Surface Transformation by Surface Recrystallization in Yttria-Doped Tetragonal Zirconia." Journal of the American Ceramic Society, 72(2):319-321.

7. Tsukuma, K., Y. Kubota and T.Tsukidate, 1984. "Thermal and Mechanical Properties of Y$_2$O$_3$-Stabilized Tetragonal Zirconia Polycrystals," in Adv. Ceram., 12, Science and Technology of Zirconia II, N. Claussen, M. Rühle and A.H. Heuer, eds., Columbus, OH, USA:The American Ceramic Society, Inc., pp.382-390.

8. Watanabe, M., S.Iio and I. Fukuura, 1984. "Aging Behaviour of Y-TZP," in Adv. Ceram. 12, Science and Technology of Zirconia II, N. Claussen, M. Rühle and A.H. Heuer, eds., Columbus, OH, USA:The American Ceramic Society, Inc., pp.391-398.

9. Matsui, M., T. Soma and I. Oda, 1984. "Effect of Microstructure on the Strength of Y-TZP Components," in Adv. Ceram. 12, Science and Technology of Zirconia II, N. Claussen, M. Rühle and A.H. Heuer, eds., Columbus, OH, USA:The American Ceramic Society, Inc., pp.371-381.

10. Wang, J. and R. Stevens, 1989. "Preferred ZrO$_2$(t) $\rightarrow$ ZrO$_2$ (m) Transformation on the Aged Surface of TZP Ceramics." Journal of Materials Science Letters, 8(10): 1195-98.

11. Lepistö, T.T. and T.A. Mäntylä, 1989. "A Model for Structural Degradation of Y-TZP Ceramics in Humid Atmosphere." Ceramic Engineering and Science Proceedings, 10(7-8):658-67.

12. Lange, F.F., G.L. Dunlop and B.I. Davis, 1986. "Degradation During Aging of Transformation-Toughened ZrO$_2$-Y$_2$O$_3$ Materials at 250°C." Journal of the American Ceramic Society, 69(3):237-40.

13. Sato, T. and M. Shimada, 1985. "Control of the Tetragonal-to-Monoclinic Phase Transformation of Yttria Partially Stabilized Zirconia in Hot Water." Journal of Materials Science, 20(11):3988-92.

14. Chen, S-Y. and H-Y Lu, 1989. "Low Temperature Aging Map for 3 mol% Y$_2$O$_3$-ZrO$_2$." Journal of Materials Science, 24(2):453-56.

15. Lu, H-Y. and S-Y. Chen, 1987. "Low Temperature Aging of t-ZrO$_2$ Polycrystals with 3 mol% Y$_2$O$_3$." Journal of the American Ceramic Society, 70(8):537-41.

16. Hughes, A.E. and S.P.S. Badwal, 1991. "Impurity and Yttrium Segregation in Yttria-Tetragonal Zirconia." Solid State Ionics, 46(3,4):265-74.

17. Hughes, A.E. and S.P.S. Badwal, 1991. "An XPS Investigation of Impurity Glass in Y-TZP." Materials Forum, 15(3):261-67.

18. Nieh, T.G., D.L. Yaney and J. Wadsworth, 1989. "Analysis of Grain Boundaries in Fine-Grained, Super-plastic, Yttria-Containing Tetragonal Zirconia." Scripta Metallurgica, 23(12):2007-012.

19. Theunissen, G.S.A.M., A.J.A. Winnubst and A.J. Burggraaf, 1989. "Segregation Aspects in the ZrO_2-Y_2O_3 Ceramic System." Journal of Materials Science Letters, 8(1):55-57.

20. Ingo, G.M., G. Mattogno, N. Zacchetti, P. Scardi and R. Dal Maschio, 1991. "X-ray Photoelectron Spectroscopic Investigation of Impurity Phase Segregation in Ceria-Yttria-Zirconia." Journal of Materials Science Letters, 10(6):320-22.

21. Hughes, A.E., F.T. Ciacchi and S.P.S. Badwal, to be submitted.

22. Leonov, A.I., Yu. P. Kostikov, I.K. Ivanov, N.S. Andreeva and E.M. Trusova, 1980. "The Nature of Electrical Conductivity of Solid Solutions in the ZrO_2-Y_2O_3 System." Inorganic Materials, 16(9):1076-1079 (English translation of Iz. Akad. Nauk SSSR, Neorganicheskie Materialy).

23. Miller, M.L. and R.W. Linton, 1985. "X-ray Photoelectron Spectroscopy of Thermally Treated SiO_2 Surfaces." Analytical Chemistry, 57(12):2314-19.

24. Catlow, C.R.A., A.V. Chadwick, G.N. Greaves and L.M. Moroney, 1986. "EXAFS Study of Yttria-Stabilized Zirconia." Journal of the American Ceramic Society, 69(3):272-77.

25. Orera, V.M., R.I. Merino, Y. Chen, R. Cases and P.J. Alonso, 1990. "Intrinsic Electron and Hole Defects in Stabilized Zirconia Single Crystals." Physical Review B, 42(16-A):9782-89.

26. Badwal, S.P.S. and N. Nardella, 1989. "Formation of Monoclinic Zirconia at the Anodic Face of Tetragonal Zirconia Polycrystalline Solid Electrolytes." Applied Physics A, 49(1):13-24.

27. Holmes, H.F., E.L. Fuller, Jr. and R.B. Gammage, 1972. "Heats of Immersion in the Zirconium Oxide-Water System." Journal of Physical Chemistry, 76(10):1497-1502.

Characterization of H_2O-Aged TZP by Elastic Recoil Detection Analysis (ERDA)

O. KRUSE, H. D. CARSTANJEN, P. W. KOUNTOUROS,
H. SCHUBERT and G. PETZOW

ABSTRACT

The influence of the microstructural parameters on the corrosion stability of Y–TZP (Yttria–Stabilized Tetragonal Zirconia Polycrstals) has been clarified and can be understood on the basis of martensitic transformations. Nevertheless, the role of water for the degradation mechanism is still unclear. One possible explanation for the enhancement of the degradation in the presence of water is to assume that water radicals (such as hydroxile groups or similar configurations) penetrate into the lattice. In order to test this hypothesis 3Y–TZP (known to be very sensitive to degradation) and 15Ti–3Y–TZP (very corrosion resistant) samples were aged in an autoclave in a D_2O rich water atmosphere. The presence of hydrogen and deuterium in the lattice was then monitored by ERDA (Elastic Recoil Detection Analysis), by applying a beam of 1.9 MeV ^{4}He ions under an incident angle of 20^0. While sintered and polished, but otherwise untreated 3Y–TZP samples show hydrogen mainly at the very surface with only a very small bulk concentration, they exhibit a significantly increased hydrogen concentration after the aging treatment: concentrations of $9 \cdot 10^{20}$ atoms/cm^3 for H and $5 \cdot 10^{20}$ atoms/cm^3 for D were recorded. Within the experimental accuracy the total H and D concentration is of the same order of magnitude as the concentration of stoichiometric vacant oxygen sites. The concentration of hydrogen in the more corrosion resistant Ti–containing material, however, was found to be only slightly higher than in an untreated sample.

INTRODUCTION

Y–TZP has been intensively investigated in the past because of its high toughness and strength. It has been meanwhile well established that both toughness and strength are related to the transformability of grains from the tetragonal (t) to the monoclinic (m) structure [1,2]. As a necessary consequence of this transformability the material is metastable at room

O. Kruse, H.D. Carstanjen, Max–Planck–Institut für Metallforschung, Institut für Physik, Heisenbergstr. 1, D(W) 7000 Stuttgart 80
P.W. Kountouros, H. Schubert, G. Petzow, Max–Planck–Institut für Metallforschung, Institut für Werkstoffwissenschaft, Pulvermetallurgisches Laboratorium, Heisenbergstr. 5, D(W) 7000 Stuttgart 80

temperature. After an aging treatment in the temperature range around 250^0C in a humid atmosphere a degradation of the material is observed [3–6]. This degradation is accompanied by the t–m transformation of the grains. The degraded layer grows in thickness with further aging time, i.e. is not passivating. As an explanation, Matsui [7] suggested that the transformation of large surface grains causes stresses in the adjacent grain boundaries and, thus, microcracking occurs. The water can then proceed deeper into the ceramic via these microcracks.

Different aspects of the degradation have been studied experimentally and the influence of the microstructural parameters on the degradation has been analyzed: Generally, the degradation is enhanced by a larger grain size and smaller stabilizer content. The same large grain size would, however, be very advantageous for high toughness. This means unfortunately a contradiction between the requirements of stability and toughness of the material. Furthermore, influences of the chemical homogeneity [8,9] (homogeneous samples were found to show the better stability) and of second phase additions were observed [10]. For the latter case a hindrance of the autocatalytic reaction (the first grain transforms and imposes stresses on the next grain which again causes a transformation) was discussed.

All the microstructural effects can be traced back to changes in the stress situation in the region of the nucleus for the t–m transformation [11–13]. When these stresses are incorporated in the free energy balance between t and m phase, the difference in free energy between both phases reads [14]:

$$\Delta F = -\Delta F_{chem} - (\sigma_{ij} + 1/2 \ \sigma_{ij}{}^T) \ \epsilon_{ij}{}^T$$
$$+ \ \Delta F_{twin} + \Delta F_{surf}. \tag{1}$$

with ΔF_{chem} = difference in chemical free energy
     σ_{ij} = residual stresses after sintering
     $\sigma_{ij}{}^T$ = transformation stresses
     $\epsilon_{ij}{}^T$ = transformation strain tensor
     ΔF_{twin} = contribution of twin boundaries
     ΔF_{surf} = difference in surface energies.

According to eq. (1) changes in microstructure lead to changes in the difference of free energy between t and m phase.

In spite of this well established understanding of the microstructural influences on the basis of martensitic transformation the role of water in the degradation remained unclear, although several approaches have been undertaken in the past, see e.g. [15,16].

It is the aim of this paper to test the hypothesis of a penetration of water radicals into the lattice. In a first experiment we studied the penetration of hydrogen. For the detection of H and D atoms ERDA (Elastic Recoil Detection Analysis) has been applied as the most appropriate technique.

EXPERIMENTAL

Co–precipitated powders were used as starting materials (3 mole% Y_2O_3–TZP = 3Y–TZP; Tosoh, Japan; 15 mole% TiO_2– 3 mole% Y_2O_3–TZP = 15Ti–3Y–TZP, precipitated from oxychlorides [17]). The powders were isostatically compacted (200 MPa) and sintered in air in the temperature range from 1350⁰C to 1450⁰C. The sintered bodies were cut, lapped and polished (Struers Abramin). The final samples consisted of cylinders 10 mm in length and 10 mm in diameter. The degradation aging was carried out in a PTFE lined autoclave of 700 ml volume (Berghof HR 700, Ehingen, Germany) at 200⁰C in D-enriched water vapour pressure of 15 bar. The surface m-phase content of the samples was detected by x-ray diffractometry (Seifert MZ IV).

The content of H and D in the samples was determined by use of ERDA. This ion beam technique is particularly suitable for the depth resolved analysis of light elements like H and D in solids. A description of its priciples can be found in e.g. Refs. [18–20]; here we shall give a short outline.

The perhaps most widely used ion beam technique is RBS (Rutherford Backscattering Spectroscopy, see e.g. [18]). It relies on the analysis of the energy of MeV ions (usually He) elastically backscattered by a solid, and serves for the analysis of medium and high mass elements. For low mass elements such as H and D this technique fails because of reasons of the scattering kinematics. Instead one uses heavy ions (such as He or N) at grazing incidence to eject these atoms, and analyzes the energies of these recoil particles.

The energy T transfered to a recoil particle in such a collision is given by:

$$T = 4 \cdot \frac{m_1 \cdot m_2}{(m_1 + m_2)^2} \, E \cos^2 \tau \qquad (2)$$

where m_1 is the mass and E the energy of the incident ion, m_2 the mass and τ the emission angle of the recoil particle. As is seen from eq. (2), the recoil energy T depends on the mass m_2 of the recoil particle and is, hence, element specific. In the following this fact will allow us to distinguish between H and D atoms. Information about the depth where the recoil took place is provided by the fact that the incident ion as well as the recoil particle are slowed down on the way through the sample (primarily due to interaction with the target electrons). Hence recoil particles from inside the sample exhibit lower energies than particles which originate from the sample surface. This energy loss can be converted into a depth scale.

For the actual analysis of H and D in the ZrO_2 samples a 1.9 MeV He beam (diameter 1 mm) at grazing incidence (20⁰) was used; an angle of 40⁰ was chosen for the emission angle τ. The H and D recoil particles were counted with a standard silicon surface barrier detector which was covered by a thin aluminium foil (7 μm thick). The thickness of this foil was chosen such as to stop the He ions elastically scattered by the sample completely (they would cause serious background problems), but to allow the recoil H or D particles to penetrate. Together with a slit-type aperture in front of the detector of 1 mm in width this set-up provided a depth resolution of ca 33 nm for H and 25 nm for D. The maximum depth to be analyzed in this experiment was 180 nm for H and 270 nm for D, respectively.

RESULTS AND DISCUSSION

The final sinter temperatures of 3Y–TZP and 15Ti–3Y–TZP materials were found to be quite different. 3Y–TZP was densified at 1450⁰C while 15Ti–3Y–TZP was densified at 1350⁰C resulting in ≥ 99% final density in both cases. The microstructures of both materials are shown in figure 1 a and b. As expected, the grain size of the Ti–Y–TZP was larger than that of Y–TZP. The susceptibility of both materials to degradation was quite different. After a 2h aging the 3Y–TZP material was almost completely transformed whereas the 15Ti–3Y–TZP did not show any indication of the m–phase.

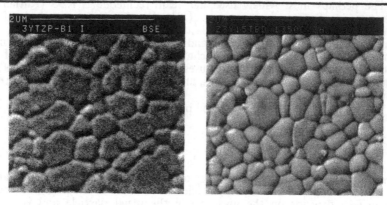

Figure 1 a and b: SEM micrographs of a) 3Y–TZP and b) 3Y–15Ti–TZP.

Figures 2 to 4 illustrate the results of the ERDA analysis of the H and D content of the samples. In all cases the same dose was used; so the spectra in the figures can be compared directly. Figure 2 shows the ERDA spectrum, i.e. the energy spectrum of the recoil H and D atoms from the untreated 3Y–TZP sample. The spectrum consists of two parts: (i) a high peak at about 300 keV which arises from H contamination at the surface (ca $1.2 \cdot 10^{16}$ H atoms/cm^2, most probably due to hydrolysis during surface grinding) and (ii) a more continuous part which extends from 300 to 0 keV (it corresponds to a surveyed depth of 180 nm) and which is due to hydrogen in the bulk (ca $7 \cdot 10^{19}$ H atoms/cm^3). Upon charging the sample with D$_2$O rich water (autoclave test) a new, more continuous part shows up in the spectrum which extends from ca 520 to 0 keV (Fig. 3). This part arises from deuterium (ca $5 \cdot 10^{20}$ D atoms/cm^3) which was absorbed by the sample during the aging procedure. But also the H concentration in the bulk (ca $9 \cdot 10^{20}$ H atoms/cm^3, after subtraction of the D–background yield) and the H contamination at the surface (ca. $1.4 \cdot 10^{16}$ H atoms/cm^3) have increased. For comparison an ERDA spectrum from a 15Ti–3Y–TZP sample is shown in Fig. 4. This sample had experienced the same treatment as the sample of Fig. 3. As can be seen from the comparison of these spectra, the Ti–doped sample has taken up by far less deuterium (only $5.5 \cdot 10^{19}$ D atoms/cm^3) than the undoped sample which clearly shows the positive influence of adding titanium as quoted in the introduction. But also the H content in the bulk (ca $6.2 \cdot 10^{19}$ H atoms/cm^3) and the H contamination at the surface are now lower.

The H and D concentrations found in the ERDA experiment will be discussed in the following with respect to the degradation phenomenon.

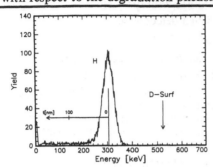

Figure 2: ERDA spectrum of a virgin 3Y–TZP sample

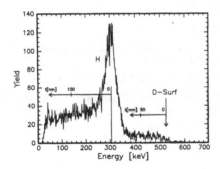

Figure 3: ERDA spectrum of an aged (2h / 200°C / 16 bar) 3Y–TZP sample.

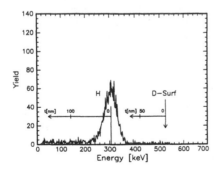

Figure 4: ERDA spectrum of an aged 3Y–15Ti–TZP sample.

The volume concentration of ZrO_2 or "TZP" formula units amounts to

$$C_{ZrO_2} \quad = 2.88 \cdot 10^{22} \text{ formula units / cm}^3 \tag{3}$$

Assuming the well established defect model

$$Y_2O_3 \rightarrow 2Y_{Zr}' + V_O^{\cdot\cdot} + 3O_O^{\ X} \tag{4}$$

the number of vacancies (i.e. vacant oxygen sites) reads

$$C_{vac} \quad = 4.33 \cdot 10^{20} \text{ vacancies / cm}^3 \tag{5}$$

for a given Y_2O_3 content of 3 mole% The same vacancy concentration is assumed for the Ti containing material, because Ti remains predominantly tetravalent as could be shown by XANES studies of Zschech et al. [21]. The bulk concentration of H in the lattice of an as sintered and subsequently polished (i.e. untreated) material,

$$C_{H(new)} \quad = 7.0 \cdot 10^{19} \text{ atoms / cm}^3 \tag{6}$$
is fairly small.

After aging the bulk concentrations were found to be

$$C_{H(aged)} \quad = 9.0 \cdot 10^{20} \text{ atoms / cm}^3 \tag{7}$$
and
$$C_{D(aged)} \quad = 5.0 \cdot 10^{20} \text{ atoms / cm}^3, \tag{8}$$

respectively. The total hydrogen concentration is given by the sum of the D and H concentrations and amounts to

$$C_{hydrogen} \quad = 1.4 \cdot 10^{21} \text{ atoms / cm}^3 \tag{9}$$

In contrast, the concentrations in the Ti containing material are of the same order or only slightly higher than in the as sintered material:

$$C_{H(aged)} \quad = 6.2 \cdot 10^{19} \text{ atoms / cm}^3 \tag{10}$$
$$C_{D(aged)} \quad = 5.5 \cdot 10^{19} \text{ atoms / cm}^3 \tag{11}$$

The comparision of (9) and (5) shows that the hydrogen concentration in the near surface region of the aged sample is very high. Even if there are errors in the measured D and H concentrations because of uncertainties in the scattering cross section, one can say that – after aging – the hydrogen concentration is in the same range as the number of vacancies, i.e. in fact the measured hydrogen concentration is higher (approximately a factor of 3). For most water–vacancy interaction models [22] the number of hydrogen atoms is expected to be higher (factor of 2) than the number of vacancies, provided that all of the vacancies are filled. Thus, the fact that we find a higher number of hydrogen atoms than vacancies implies that there is enough to fill the vacancies and to destabilize the material [22].

OUTLOOK

For the near future we plan to study the questions which of the possible water radicals is likely to penetrate into the material and which position in the lattice is occupied by it. It is not very probable that the hydrogens are located right in the vacancy, because both hydrogen (ions) and the vacancy have the same sign of charge and, hence, coulomb repulsion will occur. More likely, hydroxile groups might penetrate with oxygen occupying the vacancies. For differentiation between these two possibilities tracer experiments with ^{18}O appear to be a suitable technique. Further studies, concerning the position of the radicals in the lattice are in preparation. They, however, require single crystals; hence, these experiments will be performed on cubic stabilized $Y-ZrO_2$.

ACKNOWLEDGEMENT

This work was subsidized by the Deutsche Forschungsgemeinschaft (DFG) under contract Ca 122/1-4, Schu 679/1-3 and by the Federal Minister of Science and Technology (BMFT) under contract 03 M 2045.

REFERENCES

1.   Rühle, M.M. and A.H. Heuer, 1984. "Phase Transformations in ZrO$_2$-Containing Ceramics: II, The Martensitic Reaction in t-ZrO$_2$," in <u>Science and Technology of Zirconia II</u>, N. Claussen, M. Rühle and A.H. Heuer, eds., Columbus, OH, USA: The American Ceramic Society, Inc., pp.14-32.

2.   Evans, A.G. and R. Cannon, 1986. "Toughening of Brittle Solids by Martensitic Transformations." <u>Acta Met</u>, 34(5):761-800.

3.   Kobayashi, K., H. Kuwajima and T. Masaki, 1981. "Phase Change and Mechanical Properties of ZrO$_2$-Y$_2$O$_3$ Solid Electrolyte After Aging." <u>Solid State Ionics</u>, 3/4:489-493.

4.   Tsukuma, K., Y. Kubota and T. Tsukidate, 1984. "Thermal and Mechanical Properties of Y$_2$O$_3$-Stabilized Tetragonal Zirconia Polycrystals," in <u>Science and Technology of Zirconia II</u>, N. Claussen, M. Rühle and A.H. Heuer, eds., Columbus, OH, USA: The American Ceramic Society, Inc., pp.382-390.

5.   Matsui, M., T. Soma and I. Oda, 1984. "Effect of Microstructure on the Strength of Y-TZP Components," in <u>Science and Technology of Zirconia II</u>, N. Claussen, M. Rühle and A.H. Heuer, eds., Columbus, OH, USA: The American Ceramic Society, Inc., pp.371-381.

6.   Watanabe, M., S. Iio and I. Fukuura, 1984. "Aging Behavior of Y-TZP," in <u>Science and Technology of Zirconia II</u>, N. Claussen, M. Rühle and A.H. Heuer, eds., Columbus, OH, USA: The American Ceramic Society, Inc., pp.391-398.

7.   Matsui, M., T. Soma and I. Oda, 1986. "Stress Induced Transformation and Plastic Deformation for Y-TZP." <u>J. Am. Ceram. Soc.</u>, 69(3):198-202.

8.   Matsui, M., M. Masuda, T. Soma and I. Oda, 1988. "Thermal Stability and Microstructure of Y-TZP," in <u>Science and Technology of Zirconia III</u>, S. Sōmiya, N. Yamamoto and H. Yanagida, eds., Westerville, OH, USA: The American Ceramic Society, Inc., pp.607-614.

9.   Schubert, H., 1987. PhD Thesis, University of Stuttgart, Stuttgart, Germany.

10. Kimura, N., H. Okamura and J. Morishita, 1988. "Preparation of Low-Y$_2$O$_3$-TZP by Low-Temperature Sintering," in <u>Science and Technology of Zirconia III</u>, S. Sōmiya, N. Yamamoto and H. Yanagida, eds., Westerville, OH, USA: The American Ceramic Society, Inc., pp.183-191.

11. Schubert, H. and G. Petzow, 1988. "Microstructural Investigation on the Stability of Y-TZP," in <u>Science and Technology of Zirconia III</u>, S. Sōmiya, N. Yamamoto and H. Yanagida, eds., Westerville, OH, USA: The American Ceramic Society, Inc., pp.21-28.

12. Schmauder, S. and H. Schubert, 1986. "Significance of Internal Stresses for the

Martensitic Transformation in Yttria-Stabilized Tetragonal Zirconia Polycrystals During Degradation." J. Am. Ceram. Soc., 69(7):534-540.

13. Schmauder, S., 1987. "The Solution of Selected Problems in Materials Science Through Stress Calculations Using the Finite Element Method." PhD Thesis, University of Stuttgart, Stuttgart, Germany.

14. Christian, J.W., 1975. The Theory of Phase Transformations in Metals and Alloys, Vol. 1, 2nd edition, Elmsford, NY, USA: Pergamon Press.

15. Lange, F.F., G.L. Dunlop and B.I. Davis, 1986. "Degradation During Aging of Transformation-Toughened ZrO_2-Y_2O_3 Materials at 250°C." J. Am. Ceram. Soc., 68(3):237-240.

16. Schubert, H., 1986. "Investigations on the Stability of Yttria Stabilized Tetragonal Zirconia (Y-TZP)." Zirconia Ceramics, 7:65-81.

17. Kountouros, P., H. Schubert and G. Petzow, "Bloating Effect During Sintering of TZP," to be published.

18. Chu, W.-K., J.W. Mayer and M.-A. Nicolet, 1978. Back-Scattering Spectroscopy, London, UK: Academic Press.

19. Nölscher, C., W. Schmidt, K. Brenner, V. Brückner, M. Lehmann, P. Müller and G. Saemann-Ischenko, 1989, in Nuclear Physics Methods in Material Research, K. Bethge, H. Baumann, H. Jex and F. Rauch, eds., Braunschweig, Germany: Viehweg & Sohn, pp.349-352.

20. Nölscher, C., K. Brenner, R. Knauf and W. Schmidt, 1983. "Elastic Recoil Detection Analysis of Light Particles (^{1}H-^{16}O) Using 30 MeV Sulphur Ions." Nucl. Instrum. & Methods Phys. Res., 218(1-3):116-119.

21. Zschech, E., P. Kountouros, G. Petzow, P. Behrens, A. Lessmann and R. Frohn, "Synchrotron Radiation Ti-K XANES Study of TiO_2-Y_2O_3-Stabilized Zirconia Polycrystals," submitted to J. Am. Ceram. Soc.

22. Kountouros, P. and G. Petzow, "Defect Chemistry, Phase Stability, and Properties of Zirconia Polycrystals," this volume.

A Novel Characterisation of Phases in Zirconia-Based Ceramics — ^{91}Zr NMR

T. J. BASTOW, M. E. SMITH and S. N. STUART

ABSTRACT

It is demonstrated that various phases of ZrO_2 can be characterised using ^{91}Zr NMR spectroscopy by their distinctive nuclear quadrupolar powder patterns. We display the ^{91}Zr NMR line shapes of the cubic, monoclinic and tetragonal phases of ZrO_2 and show how zirconium spectra can describe these and the other phases exhibited by magnesia-partially-stabilised zirconia (MgPSZ) during the course of heat treatment subsequent to the preparation of the as-fired material. Specifically, we have observed the growth and ordering of the tetragonal precipitates on isothermal aging at 1400°C, the partial transformation of the cubic to the δ-phase ($Mg_2Zr_5O_{12}$) on annealing at 1100°C, and the transformation of the tetragonal phase into the orthorhombic phase after cooling to 77 K, followed by the reverse transformation on warming to 600°C.

INTRODUCTION

. The various partially stabilised zirconias (ZrO_2) can be a mixture of at least five separate crystallographic phases, viz cubic solid solution (c), tetragonal (t), monoclinic (m), orthorhombic (o), and δ-phase ($Mg_2Zr_5O_{12}$). Detailed understanding of the phase distribution is essential for the optimisation of the thermomechanical properties of these materials.

A quantitative estimate of these phases in a particular sample generally requires a neutron diffraction analysis [1]. However neutron diffraction is not widely available as a characterisation technique for everyday use. X-ray diffraction, which is more generally available, has well documented disadvantages for bulk ceramic samples [2]. Another standard characterisation technique, electron diffraction, samples only a small selection of microcrystals which may be unrepresentative of the bulk. A new method for characterising bulk specimens of these materials

T.J.Bastow, M.E.Smith and S.N.Stuart
CSIRO Division of Materials Science and Technology, Locked Bag 33, Clayton, Vic 3168, Australia

would obviously be valuable. NMR has recently become a powerful tool for examining ceramic phases [3,4]. A useful summary of modern solid state NMR with applications is ref. 5. A basic outline of the pertinent aspects of the technique is given below.

In an isolated atom a nucleus with spin $I=1/2$ has its magnetic sublevels, $m=1/2$ and $m=-1/2$, split by an applied magnetic field B_0 due to the Zeeman interaction. Transitions between these levels can be induced by a radiofrequency field of frequency $f = h^{-1}(E_{-1/2}-E_{1/2})$ which is proportional to B_0. If the atom is in a solid (or a molecule) the nucleus sees a small additional field due to the circulating currents of the bonding electrons, which alters the spectroscopic transition frequency by an amount known as the chemical shift. This effect has been used extensively for molecular and solid state characterisation.

An atomic nucleus with spin $I>1/2$ possesses a quadrupole moment (eQ) which interacts with the electric field gradient (eq) which is present at a site with less than cubic point symmetry. This quadrupole interaction is proportional to the product of the moment and the gradient at the nucleus, both of which tend to increase with atomic number. For a relatively heavy nucleus such as ^{91}Zr ($I=5/2$) in the magnetic field of a modern NMR spectrometer (ca. 10 T), the magnitude of the quadrupole interaction begins to be a significant fraction of its Zeeman interaction and provides the principal NMR signature by which the compound under investigation may be characterised. Since the electric field gradient is a second-order traceless tensor, the quadrupole interaction is completely described by two quantities, viz the coupling constant e^2qQ/h and asymmetry parameter η; note that for axial symmetry $\eta = 0$. Although in principle all the transitions (m,m-1) may now be observed (m taking any integral-spaced values between I and 1-I), in general, for strong quadrupole coupling, only the (1/2,-1/2) transition is observed. The frequency of this transition depends on the orientation of the crystal with respect to the magnetic field, so that in a polycrystalline material a powder pattern is observed. The width of the powder pattern is directly proportional to the coupling constant and inversely proportional to the magnetic field. By comparison with patterns simulated theoretically, the values of e^2qQ/h and η can be determined. A typical theoretical powder pattern is shown in figure 1 for η in the range $0<\eta<1/3$. Note that each crystallographically distinct site will have, in principle, a different quadrupole interaction and the observed pattern will be a superposition of all of them.

For most NMR accessible nuclei (except the heavy nuclei with $I>1/2$, eg. ^{91}Zr) a valuable line narrowing technique exists known as magic angle spinning (MAS), in which the sample is spun at frequencies of order 5 - 15 kHz at an angle of $\cos^{-1}(3^{-1/2}) = 54.7°$ with respect to the magnetic field [5]. This yields high resolution spectra in which many of the line broadening mechanisms occurring in solids are eliminated, including nuclear quadrupole and dipole-dipole interactions and chemical shift anisotropy.

For zirconia two nuclei, ^{17}O and ^{91}Zr, are available for NMR investigation. (Note that for yttria-stabilised zirconia, ^{89}Y, $I=1/2$, is also available.) The oxygen nucleus ^{17}O is an ideal probe, with its relatively large chemical shift range and small nuclear quadrupole moment, and ^{17}O MAS NMR spectra have been shown by their chemical shifts to distinguish clearly between the m, t and c-phases of ZrO_2 (Figure 2) [6]. Note that the two ^{17}O resonances observed for m-ZrO_2 correspond to the two

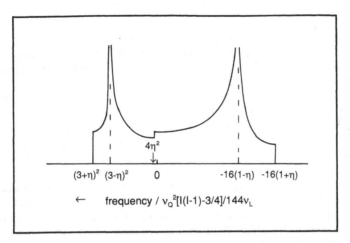

Figure 1. The theoretical powder lineshape for the second-order quadrupole shifted (1/2,-1/2) transition in the range $0 \leq \eta < 1/3$. Frequency reference is a nominal Lamor frequency $\nu_L = 37.2$ MHz; $\nu_Q = 3(e^2qQ/h)/2I(2I-1)$

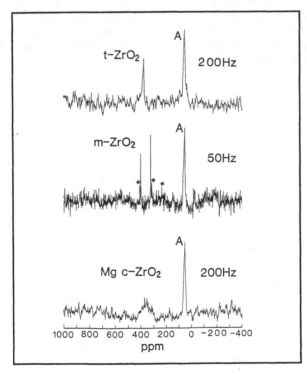

Figure 2. ^{17}O MAS NMR spectra of t, m and magnesia-stabilised c-phase ZrO_2 with ^{17}O in natural abundance. The peak A is from oxygen in the alumina spinner. The line-broadening used for spectrum smoothing is given in hertz. Spinning side-bands are labelled with asterisks. Frequency reference H_2O.

inequivalent oxygen sites in the unit cell. The very broad line shown for the c-phase indicates the disordered nature of the oxygen environments. However without isotopic enrichment, the sensitivity of ^{17}O NMR is limited, for general applications, by its low natural abundance of 0.037%.

This paper examines the suitability of zirconium NMR for investigating the phase content of magnesia-stabilised zirconia in various preparative stages. The only NMR active zirconium isotope is ^{91}Zr with spin I=5/2 and natural abundance 11.2%. Recently it has been shown that the t and m-phases of ZrO_2, and also the t-phase of YTZP (ZrO_2; 3 mole% Y_2O_3), can be characterised by ^{91}Zr NMR, using their distinctive quadrupolar powder patterns [7]. The frequency width of the ^{91}Zr (1/2,-1/2) line profile in these polycrystalline materials in a field of 9.4 T is about 0.4 - 0.5 MHz and is at the upper limit of what can sensibly be determined by direct transform pulse techniques with available digitisation rates. It is certainly too great for line narrowing by MAS. A more efficient method for determining the line profile is found to be a frequency-stepped echo technique [7]. The method uses a static sample, does not rely on high power pulses, and can use dense ceramic specimens to good advantage for improving signal-to-noise ratio. The isotropic chemical shift difference is far too small relative to the linewidth for the phases to be distinguished on this basis. This can be expected since in all the phases Zr has only nearest neighbour oxygens. However it is often the case that a quadrupolar interaction is much more sensitive to local structural differences than is the chemical shift and so can distinguish different sites and phases.

EXPERIMENTAL DETAILS

The MgPSZ was manufactured by Nilcra (Australia) in both as-fired form (AF; slow and fast cooled) and their proprietary MS grade (the preparation of material described as very similar to MS is described in ref [2]). The MgO content of this and the other MgPSZ specimens examined here was ca 9.3 mole%. The AF material was quenched from the cubic phase field at 1720°C to 1000°C at a cooling rate of typically 300°C/hr. The fully stabilised cubic phase at the eutectoid composition (13.5 mole% MgO) was prepared by mixing a commercial monoclinic ZrO_2 powder (Magnesium Elektron) with 13.5 mole% MgO, followed by milling, drying, pressing and sintering at 1700°C. The samples were in the form of solid cylinders (of near theoretical density) of typical dimensions 35 mm length and 18 mm diameter.

The NMR spectrometer was a Bruker MSL 400, operating around a frequency of 37.210 MHz at a field of 9.4 T. The spectra were obtained by stepping the spectrometer frequency through the spectrum and measuring the Fourier-transformed spin echo amplitude as a function of frequency [7]. The reference compound for the zero of the frequency scale is $BaZrO_3$, which yields a sharp ^{91}Zr NMR line. A more extensive discussion is given elsewhere [8].

DISCUSSION

We first identify the ^{91}Zr NMR spectra of the t, m and c-phases corresponding to the ^{17}O spectra in figure 2. The crystal structure of each of these three phases has been re-determined recently by high-resolution powder neutron powder diffraction [9].

The t-phase ^{91}Zr spectrum (Figure 3a) has been identified in MgPSZ samples as the dominant spectrum in samples which have been given secondary annealing at 1400°C for 2 hrs to develop the tetragonal precipitates [10]. The annealing process expels most of the magnesium from the precipitates leaving relatively pure t-phase. The sharply defined $\eta = 0$ pattern indicates the presence of one distinct zirconium atom in the unit cell in a site of axial point symmetry, consistent with the crystal structure. The coupling constant is 19.1 MHz.

The m-phase is the stable phase of pure ZrO_2 at room temperature. The unit cell has one Zr site in a general position. The observed ^{91}Zr spectrum (Figure 3b) has finite asymmetry ($\eta = 0.1$) and a wider frequency spread than the t-phase, indicating a larger coupling constant, viz 23.1 MHz.

Magnesia-fully-stabilised cubic zirconia is a solid solution with an average fluorite structure but with oxygen vacancies to maintain charge balance [9,11]. The Zr and Mg atoms are distributed randomly over the cation sublattice. The anion sublattice has approximately 7% random vacancies, and the oxygens are probably displaced from their ideal fluorite positions by 0.025 Å in the [111] direction [9]. From the zirconium viewpoint it is then reasonable to expect a smooth distribution of quadrupolar coupling constants due principally to the distribution of Zr-O bond lengths that the vacancies cause. The spectrum shown in figure 3c, which is peaked at approximately -25 kHz and decreases monotonically (but asymmetrically) on either side, is consistent with this picture.

The as-fired (AF) PSZ yields spectra characteristic of a mixture of cubic and tetragonal zirconia. The sharpness of the tetragonal component of the spectrum appears to correlate with the rapidity of quenching from the cubic phase field. A spectrum from a rapidly quenched specimen is shown in figure 4a and indicates rather poorly crystallised t-phase The peak at approximately 0 kHz shift is presumably c-phase; the peaks at ca. 120 and 170 kHz possibly represent minor transitory phases. Slower quenching induced secondary recrystallisation of the tetragonal precipitates with better crystallinity and lower magnesium content. The spectrum then approaches that of figure 3a with a peak at -25 kHz from the residual cubic phase.

A δ-phase ($Mg_2Zr_5O_{12}$) component has been identified by neutron diffraction in suitably annealed PSZ [11] - optimally 8 hours at 1100°C [12]. It forms in microdomains through an ordering of vacancies and cations in the basic fluorite structure of the solid solution c-phase which has been enriched in the magnesium expelled from the t-phase platelets during the annealing process. The only ^{91}Zr NMR manifestation of δ-phase formation in the same specimen as used in the neutron diffraction study [12] is the disappearance of the residual c-phase peak at -25 kHz. It appears likely that the random distribution of the Zr and Mg atoms over the cation sites produces a relatively featureless spectrum which underlies the tetragonal pattern.

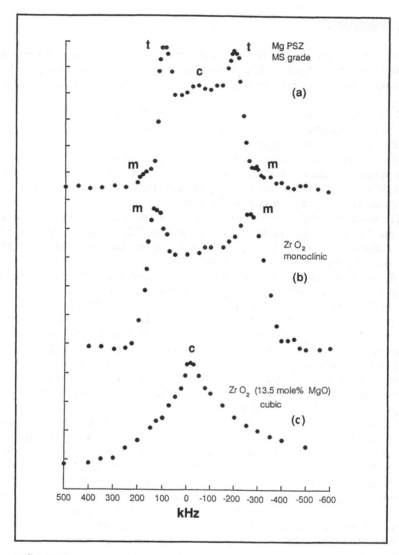

Figure 3. ^{91}Zr NMR spectrum of (a) t-phase ZrO$_2$ in MS grade MgPSZ (9.3 mole% MgO); the minor peaks from the residual c and m-phase are marked c and m: (b) pure m-phase of ZrO$_2$; (c) magnesia-fully-stabilised cubic ZrO$_2$ (13.5% MgO).

A transformation to an orthorhombic phase can be induced in certain MgPSZ preparations, at the expense of the t-phase, by cooling to -200°C and below [1,13]. The unit cell contains one Zr site. The o-phase remains intact on warming to room temperature, but a reverse transformation to the t-phase can be made by further warming to 330°C [1]. Figure 3a shows the spectrum from one half of a sample of MS grade MgPSZ. The other half was immersed in liquid nitrogen for 20 minutes,

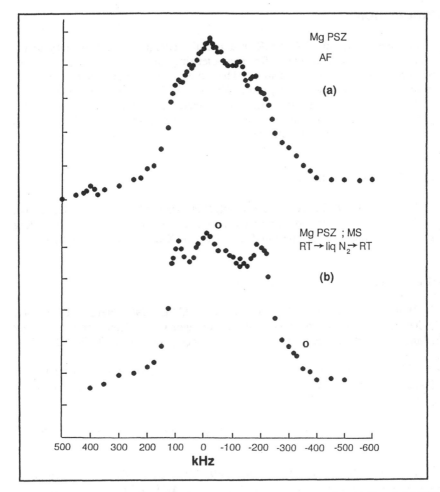

Figure 4. ^{91}Zr NMR spectrum of (a) AF fast cooled MgPSZ (9.3 mole% MgO); (b) MS grade MgPSZ (9.3 mole% MgO) cooled in liquid nitrogen for 20 minutes and then warmed to room temperature; features of the o-phase spectrum are marked o.

then warmed up to room temperature and the spectrum, shown in figure 4b, recorded. The spectrum of the t-phase, clearly defined in figure 3a, is now strongly attenuated, but a central peak near zero shift and a bulge at low frequency have developed and represent features of the spectrum assigned to the o-phase. The total spectrum may be plausibly fitted by the superposition of that for the t-phase (e^2qQ/h = 19.1 MHz, $\eta = 0$), together with one for the o-phase with proposed parameters of e^2qQ/h = 17 MHz and $\eta = 0.8$, in the proportions 2:3 respectively, agreeing approximately with the proportions found by neutron diffraction for similar material given a similar thermal cycle [1]. This specimen was now warmed to 600°C and then cooled to room temperature whereupon the spectrum of figure 3a was regained, indicating the restoration of the transformed t-phase component.

CONCLUSION

We have shown that ^{91}Zr NMR is a useful technique, complementary to the various diffraction techniques, for characterising the phase content of these transformation-toughened zirconias. The cubic, monoclinic, tetragonal and orthorhombic zirconias are clearly distinguished by their quadrupole parameters. The technique does not yet have the quantitative precision of neutron diffraction, but it is relatively fast, the equipment is more commonly available and (like neutron diffraction) it represents the bulk sample.

ACKNOWLEDGEMENTS

We are grateful to G.Mayer, H.J.Rossell, R.H.J.Hannink, J.Drennan, S.Lathabai, R.J.Hill and C.J.Howard for useful conversations and for generously providing accurately characterised samples and to D.Hay for X-ray powder diffraction.

REFERENCES

1.   Howard, C.J., E.H.Kisi, R.B.Roberts and R.J.Hill, 1990. "Neutron Diffraction Studies of Phase Transformations between Tetragonal and Orthorhombic Zirconia in Magnesia-Partially Stabilised Zirconia." J.Amer.Ceram.Soc., 73(10):2828-2833.

2.   Hill, R.J. and B.E.Reichert, 1990. "Measurement of Phase Abundance in Magnesia-Stabilised-Zirconia by Rietveld Analysis of X-ray Diffraction Data." J.Amer.Ceram.Soc., 73(10):2822-2827.

3.   Dupree, R. and D.Holland, 1989, in Glasses and Glass-ceramics, M.H.Lewis ed. London, UK:Chapman and Hall, pp. 1-40.

4.   Hatfield, G.R. and K.R.Carduner, 1989. "Solid State NMR: Applications in High Performance Ceramics." J.Mater.Sci., 24(12):4209-4219.

5.   Engelhardt, G. and D.Michel, 1987. High Resolution Solid State NMR of Silicates and Zeolites, Chichester, UK:John Wiley, pp. 50-67.

6.   Bastow, T.J. and S.N.Stuart, 1990. "^{17}O NMR in Simple Oxides." Chemical Physics, 143:459-467.

7.   Bastow T.J., M.E.Smith and S.N.Stuart, 1992. "Observation of ^{91}Zr NMR in Zirconium-based Metals and Oxides." Chem.Phys.Lett, 191:125-129.

8.   Bastow T.J. and M.E.Smith, 1992. "^{91}Zr NMR Characterisation of Phases in Transformation-Toughened Zirconia." Solid State Nuclear Magnetic Resonance, 1:165-174.

9.   Howard, C.J., R.J.Hill and B.E.Reichert, 1988. "Structure of ZrO_2 Polymorphs at Room Temperature by High-Resolution Neutron Diffraction." <u>Acta Cryst.</u>, B44:116-120.

10. Hannink, R.H.J., 1978. "Growth Morphology of the Tetragonal Phase in Partially-Stabilised Zirconia." <u>J.Mater.Sci.</u>, 13:2487-2496.

11. Rossell, H.J. and R.H.J.Hannink, 1984. "The Phase $Mg_2Zr_5O_{12}$ in MgO-Partially Stabilised Zirconia," in <u>Advances in Ceramics, Vol 12, Science and Technology of Zirconia II</u>, Columbus, Ohio, USA: American Ceramic Society, p.139-151.

12. Hannink, R.J.H., C.J.Howard, E.H.Kisi and M.V.Swain, 1993. "Relation Between Fracture Toughness and Phase Assemblage in Mg-PSZ." <u>J.Amer.Ceram.Soc.</u> (in press).

13. Kisi E.H., C.J.Howard and R.J.Hill, 1989. "Crystal Structure of Orthorhombic Zirconia in Partially Stabilised Zirconia." <u>J.Amer.Ceram.Soc.</u>, 72(9): 1757-1760.

Kinetics and Crystallography of the Monoclinic (B) to Cubic (C) Transformation in Dysprosia (Dy_2O_3)

O. SUDRE, K. R. VENKATACHARI and W. M. KRIVEN

ABSTRACT

The large volume change ($\approx 8\%$) and the fast kinetics of the monoclinic (B) to cubic (C) transformation at 1950°C in dysprosia (Dy_2O_3) causes shattering of ceramic bodies on cooling. This behavior is analogous to that of zirconia with its tetragonal (t) to monoclinic (m) transformation. Dysprosia was thus considered as a potential transformation toughener for high temperature applications.

A laser-melting/roller-quenching technique was used to stabilize the high temperature monoclinic phase down to room temperature. The kinetics of the transformation back to the cubic phase through annealing heat treatments was followed by TGA, X-ray diffraction and electron microscopy techniques. A crystallographic model involving a shear mechanism was proposed.

INTRODUCTION

The first example of high-toughness ceramics relied upon the tetragonal (t) to monoclinic (m) phase transformation occurring in zirconia. The increase in fracture resistance was associated with the crystal volume expansion and shear deformation occurring in the vicinity of the crack tip. However, this toughening effect appears limited to low temperature applications, Therefore, other potential transformation tougheners have been proposed for high temperature applications [1]. The lanthanide sesquioxides (Ln_2O_3) appeared promising for further investigation.

Of particular interest, the monoclinic (B) to cubic (C) phase transformation of dysprosia (Dy_2O_3) occurs with a large positive volume expansion causing shattering upon cooling. The transformation temperature has been reported at around 1950°C which, in one hand, offers a greater potential for high temperature toughening, but on the other hand, makes it a challenging material to study by conventional techniques. Studies on the different polymorphs of the rare-earth

O. Sudre,[¥][#] K. R. Venkatachari[#] and W. M. Kriven
Department of Materials Science and Engineering, University of Illinois at Urbana-Champaign, Urbana, Illinois, 61801, USA.
[¥] Now at: ONERA Direction des Matériaux B.P.72, 92322 Châtillon Cedex, France.
[#] Now at Institute of Materials Processing, Michigan Technical University, Houghton, MI 49931.

sesquioxides has been extensive but results on their transformations were often inconsistent and led to many interpretations [2]. The polymorphs are as follows [3], where H is a hexagonal phase:

$$\text{Melt} \xrightarrow{\quad 2360°C \quad} H \xrightarrow{\quad 2175°C \quad} B \underset{1950°C}{\overset{1860°C}{\underset{\longleftarrow}{\longrightarrow}}} C \qquad [1]$$

Many different factors appeared to affect the stabilization of the monoclinic (B) phase and the B to C transformation, such as, pressure, stress state produced by grinding, some chemical impurities and/or additions, or even oxygen deficiency and cooling rate [2]. Possible surface anisotropy and particle size effects could also stabilize the high temperature monoclinic (B) phase. These various factors also influenced the transformation back to the cubic stable phase. In addition, the crystallographic mechanism for the transformation has not been worked out, but it has been postulated to involve some type of diffusion mechanism in the case of gadolinia [4].

In order to assess the potential use of dysprosia as a transformation toughener, further characterizations of this transformation are presented. This study was primarily focussed on the kinetics of the monoclinic to cubic transformation and on a crystallographic model compatible with experimental observations.

EXPERIMENTAL

The retention of the monoclinic phase was performed using a laser-melting/roller quenching technique. Bars of pure dysprosia powder were uniaxially compacted and the tip of the bars was melted using a laser beam ($T_m = 2400°C$). Once the droplet of liquid reached a critical size, it fell between two counter-spinning titanium rollers that were kept in contact. The liquid was therefore quenched and laminated as flakes. The quenching rate of this technique is thought to be of the order of $10^7°C/s$. The resulting material was polycrystalline, monoclinic dysprosia. A similar technique applied to zirconia and hafnia was not able to retain the high temperature tetragonal phase.

The flakes were subsequently characterized as a function of annealing temperature. Techniques used included differential thermal analysis and thermogravimetry, quantitative X-ray diffraction, and electron microscopy. The quantitative analysis of the phase transformation from the monoclinic to the cubic phase was performed using integrated peaks (two for the cubic phase and six for the monoclinic phase) in the 25°-35° range and a calibration curve. The latter was produced using crushed monoclinic flakes and as-received cubic powder or flakes fully transformed to the cubic phase by heat treatment.

The flakes were up-quenched to different annealing temperatures and quenched back in air to room temperature. The quantity of monoclinic phase remaining was then determined by X-ray diffraction. The results were analysed using an Avrami-type equation [5,6]:

$$f(t) = k \exp(-kt^n) \qquad [2]$$

where $f(t)$ is the fraction of the monoclinic phase left untransformed after a time t, n is the kinetic law constant, also called the order of the transformation. This constant is related to the nucleation and growth mechanism of a cubic nucleus in the monoclinic matrix. The other constant, k, carries the thermal activation of the transformation through the relation:

$$k = k_0 \exp(-E_a/RT) \qquad [3]$$

where E_a is the activation energy and k_0 is a constant.

RESULTS

The laser-melting/roller-quenching technique produced yellowish flakes, 1 to 3 cm^2 in size and 40 to 100 mm thick. The technique was difficult to control which resulted in a variable batch to batch flake size. Chemical analysis showed that they were 99.9% dysprosia. On the other hand, zirconia, hafnia or gadolinia produced in the same conditions resulted in white flakes.

The fast quenching rate effectively retained the high-temperature monoclinic phase of dysprosia down to room temperature. However, the strongest peak of the cubic phase corresponding to the $(222)_C$ plane could be detected by X-ray diffraction which may indicate the presence of a few percent of that phase. It was nevertheless neglected in the analysis. A preferred orientation of the $(20\overline{1})_B$ plane of the monoclinic phase relative to the flake surface was also observed and, therefore, the flakes were slightly crushed before heat treatments. The microstructural observations of the as-quenched flakes revealed a lot of surface features reminiscent of the solidification history of the material. In a number of locations, the grains were revealed and the grain size was estimated by computerized trace analysis to be about 2.4 μm. Observations by transmission electron microscopy showed different types of defects within the grains such as dislocations, low angle grain boundaries and twins. Some of the dislocations were mobile under the electron beam.

The transformation of the flakes from the monoclinic to cubic phase was solely obtained through annealing heat treatments. Grinding or undercooling in liquid nitrogen did not appear to trigger the transformation. Annealing experiments performed in air, nitrogen, in a vacuum furnace or in the hot-stage of a transmission electron microscope always resulted in the transformation of the monoclinic phase to the cubic phase. The thermogravimetric analysis under argon indicated a slight weight gain of about 0.325% as shown in figure 1. The weight gain occurred in two steps: below 600°C, the weight increased slowly, whereas it was more rapid above 600°C. This gain was accompanied by a change of coloration of the flakes from yellow-brown to grey. This result could originate from a partial recovering of the stoichiometry of dysprosia. With this hypothesis, the initial composition of the flakes would be Dy_2O_{3-x}, x=0.075. At the end of this thermal analysis to 1000°C, the monoclinic phase had fully transformed to the cubic phase. T.E.M. observations indicated that most microstructural defects were annealed out at 400-500°C. However, around 600°C, the large volume expansion of the transformation caused cracking and shattering of the flakes. This was observed by S.E.M (figure 2a) and on the hot stage of a high voltage electron microscope (figure 2b). However, no characterisitic peak could be detected by differential thermal analysis.

The kinetic study further characterized the transformation. The quantitative X-ray analysis using the experimental monoclinic/cubic calibration curve provided data that could be fitted to the Avrami equation. Linear regression for different annealing temperatures provided a value of n, the order of the transformation, as a function of the temperature (figure 3a). In addition, two activation energies could be deduced from the data (figure 3b). Below 600°C, the transformation appears sluggish with a high order, n, and a high activation energy of 396 kJ/mole. However, above 600°C, the order of the transformation dropped towards zero and the activation energy was reduced to 30 kJ/mole.

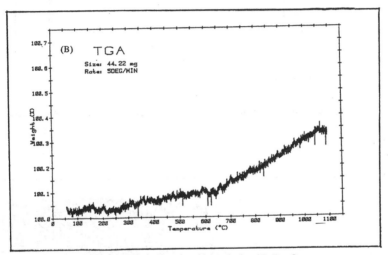

Figure 1. Thermogravimetric analysis of B-Dy$_2$O$_3$

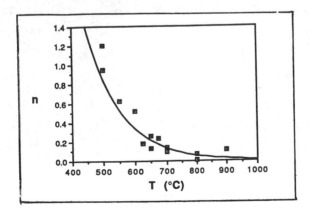

Figure 2(a). Kinetic study yielding the order of the transformation

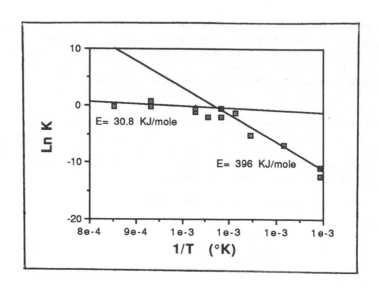

Figure 2(b). Kinetic study yielding activation energy as a function of temperature.

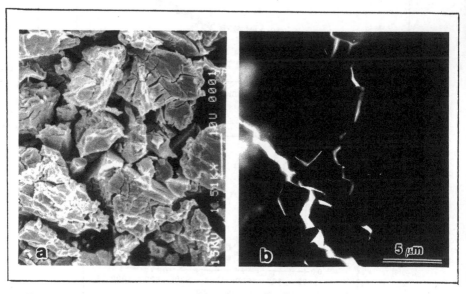

Figure 3(a). SEM micrograph of grains produced after annealing of a monoclinic flake and (b) H.V.E.M. micrograph of cracks in a flake at 600°C.

CRYSTALLOGRAPHIC MODEL

Parallel to the experimental work, a theoretical approach was adopted as it could provide additional insights into the experimental work. Most descriptions of the polymorphs of the rare-earth oxides are based on a (OLn_4) tetrahedral unit [7]. Even though the individual phases are well described by various packings of this structural unit, the transformation itself is more controversial. In the present work, a simple representation of the correspondence between the monoclinic (B) and the cubic (C) phases is proposed.

A complete description of the two crystal phases has been presented elsewhere [7]. The cubic structure is based on an oxygen-deficient fluorite structure, whereas the monoclinic structure is often described as a layered structure. In these two structures, the $(111)_C$ and $(20\bar{1})_B$ planes are often found in the literature in epitaxial relations [8,9]. In addition, the cation sublattice is almost identical in these two planes. A second relationship is found by noticing that the motif in the $(010)_B$ plane of the monoclinic phase (figure 4 a) can be derived from three motifs of each of the four types of $(110)_C$ planes of the cubic phase (figure 4 b). From these two observations, it was possible to establish a lattice correspondence shown in figure 5 which can be formalized mathematically by the following relation, where (x,y,z) are the coordinates of a direction in the respective structures.

$$\begin{pmatrix} xC \\ yC \\ zC \end{pmatrix} = \begin{pmatrix} 3/4 & -1/4 & 0 \\ 3/4 & 1/4 & 0 \\ -1/4 & 0 & 1 \end{pmatrix} \begin{pmatrix} xB \\ yB \\ zB \end{pmatrix} \qquad [4]$$

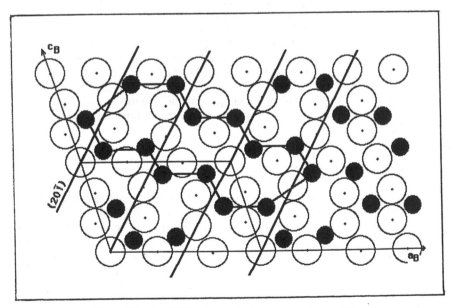

Figure 4(a). The $(010)_B$ plane with the corresponding motif.

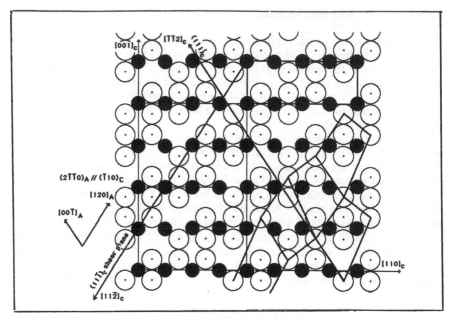

Figure 4(b). One of the four $(110)_C$ planes with major directions and three outlined motifs.

Futhermore, a phase transformation mechanism could also be proposed. The transformation is believed to occur via a shear deformation of the lattice in the $(20\bar{1})_B$ layer plane along the $1/3[\bar{1}02]_B$ direction. During this large shear deformation, a one to one relation exists between the atoms of the $(010)_B$ and the $(110)_C$ planes. The cation sublattice is simply sheared whereas some shuffling of the oxygen atoms is required within the motifs. Additional calculations using an approach developed for martensitic transformations [10], indicated that the change in volume ($\Delta V/V = 8.5\%$) and strain energy ($= 0.3687$) involved during the B to C transformation at room temperature are very large (2 to 3 times larger than those of zirconia).

DISCUSSION

The laser-melting/roller-quenching technique was effective in retaining the high temperature phase of Dy_2O_3 unlike that of ZrO_2 and HfO_2. Several factors may have produced this difference. Assuming that the crystallization of the different materials was done in their respective high-temperature phase fields, the stabilization of monoclinc (B) dysprosia implied that additional stabilizing effects are present in dysprosia compared to the two others. These effects are of several types. First, the monoclinic phase appears to be non-stochiometric, based upon the weight gain on annealing and the change in color from yellowish to white. The nonstoichiometry may partly reduce the free-energy of the phase such as twins or

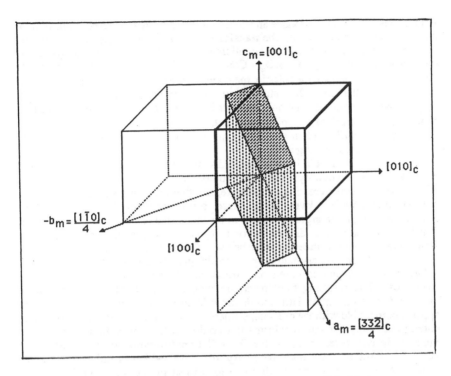

Figure 5. Lattice correspondence between cubic (C) and monoclinic (B) phases.

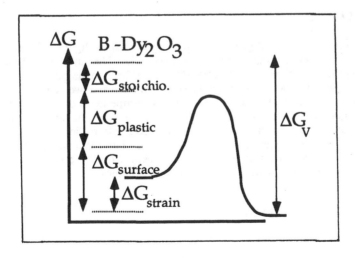

Figure 6. Schematic of the free-energy contributions to the stabilization of the monoclinic phase.

dislocations which could produce an additional free-energy reduction. Annealing of the phase was correlated with the transformation back to the stable cubic phase. Second, the mechanical action of the rollers may have been sufficient to trigger the transformation of ZrO_2 and HfO_2. Conversely, the monoclinic structure of the rare-earth oxides being rather plastic, the material may have relieved the mechanical strain of the roller action by deformation via twinning or dislocation activation. Finally, the preferred orientation of the surface of the flakes relative to the $(20\bar{1})_B$ plane may also produce a stabilizing effect. All these different factors may well have produced sufficient free energy gain to stabilize the monoclinic phase as depicted schematically in figure 6.

The characterization was completed with the kinetic study. The analysis of the data using an Avrami-type equation showed that the order of the transformation falls off to zero with increasing temperature, and that the activation energy is largely reduced above 600°C. Even though these results are difficult to interpret since nucleation and growth arguments have to be accounted for, a comparison with the transformation kinetic studies on zirconia [6] shows both a similar fall-off of the kinetic law exponent with temperature below 0.5 and similar activation energies.

In addition, the crystallographic model proposed a lattice correspondence between the two phases based on similar motifs between planes of the two structures. The resulting transformation mechanism involving a large shear deformation of the cation lattice with some local shuffling of the oxygen atoms appears compatible with the other experimental observations. First, the large strain energy necessary for the transformation could explain the high stability of the monoclinic structure. Second, the B to C transformation occurs from a low symmetry structure to a high symmetry structure, which is rather uncommon. Therefore, a cooperative nucleation and growth at the atomic scale in different planes of the monoclinic structure is necessary to form a critical nucleus.

CONCLUSION

The present investigation attempted to shed some light on the monoclinic (B) to cubic (C) phase transformation of dysprosia. The study was successful in retaining the monoclinic phase and transforming it back to the cubic phase through annealing heat treatments. Both a two-step weight gain and two-step transformation kinetics were observed with a transformation temperature of 600°C. However, these observations are insufficient to fully characterize the transformation. A crystallographic model which involved a large shear deformation of the lattice and a local shuffling of oxygens, was also proposed. Further investigations on this transformation are required to settle the question on the possible use of this transformation for toughening of bulk structural ceramics.

ACKNOWLEDGEMENTS

This work was supported by the U.S. Air Force Office of Scientific Research under Grant number AFOSR 85-0242. Use of the facilities in the Center for Microanalysis of Materials at UIUC is acknowledged.

REFERENCES

1. Kriven, W.M., 1988. "On Possible Transformation Tougheners Alternative to Zirconia: Crystallographic Aspects." J. Amer. Ceram. Soc., 71 [12]: 1021-1030.

2. Sudre, O., 1988. "Investigations of the Monoclinic (B) to Cubic (C) Transformation of Dysprosium Sesquioxide." M.S. Thesis, University of Illinois at Urbana-Champaign, Illinois, USA.

3. Foex, M. et Traverse, J.P., 1966. "Remarques sur les Cristallines Présentées à Hautes Températures par les Sesquioxides de Terres Rares." Rev. Int. Hautes Temp. et Réfract., 3: 429-453.

4. Stecura, S., 1965. "Crystallographic Modifications of Phase Transformation Rates of Five Rare-earth Sesquioxides." U.S. Bureau of Mines Report 6616.

5. Rao, C.N.R. and Rao, K.J., 1967. "Phase Transformations in Solids", in Progress in Solid State Chemistry 4, H. Reiss, ed., Thousand Oaks, California, USA: Pergamon Press, pp 131-185.

6. Whitney, E.D., 1965. "Kinetics and Mechanism of the Transition of Metastable Tetragonal to Monoclinic Zirconia." Trans. Farad. Soc., 61 [9]:1991-2000.

7. Caro, P.E., 1968. "OM_4 Tetrahedra Linkages and the Cationic Group $(MO)nn+$ in the Rare-Earth Oxides and Oxysalts." J. Less Common Metals, 16:367-377.

8. Caro, P.E., Schiffmacher, G., Boulesteix, C., Loir, Ch. and Portier, R.., 1974. "Defects and Impurities Influences on the Phase Transformations in the Rare-Earth Oxides," in Defects and Transport in Oxides, M.S. Seltzer and R.I. Jaffee, eds., Columbus and Salt Fork, Ohio, USA: Plenum Press, pp 519-535.

9. Michel, D., Rouaux, Y. and Perez Y Jorba, 1980. "Ceramic Eutectics in the ZrO_2-Ln_2O_3 (Ln: Lanthanide): Unidirectional Solidification, Microstructural and Crystallographic Characterization." J. Mater. Sci., 15:61-66.

10. Bowles, J.S. and MacKenzie, J.K., 1954. "The Crystallography of Martensite Transformation I." Acta Met., 2 [1]:129-137.

High Temperature Transformation Toughening of Magnesia by Terbia

P. D. JERO and W. M. KRIVEN

ABSTRACT

The monoclinic (B) to cubic (C) transformation in terbia (Tb_2O_3) was examined for possible use as a high temperature transformation toughener. It involves an ~8% volume expansion on cooling and a unit cell shape change of ~10°, and it occurs at ~1650°C. Composites of 20 vol% B-Tb_2O_3 – 80 vol% MgO were prepared and the mechanical properties were evaluated at temperatures up to 1400°C. The high temperature B phase of Tb_2O_3 could only be retained under ambient conditions by rapid quenching or cooling under pressure in a HIP. The B to C transformation could not be stress induced at room temperature, and elevated temperatures were required to overcome the nucleation barrier for transformation.

Single edge notched beam (SENB) specimens were fabricated and broken at ambient and elevated temperatures and the fracture surfaces were examined by X-ray diffraction (XRD). From room temperature to 1000°C, no transformation was observed. From 1075° to 1150°C, the B to C transformation was observed only on fracture surfaces. At 1300°C transformation was also observed on free surfaces and at 1400°C, spontaneous transformation was observed throughout the material. The toughness of specimens increased with increasing temperature above 600°C, with a significant increase in toughness at 1400°C.

INTRODUCTION

The lanthanide sesquioxides (of general formula Ln_2O_3) have been identified as potential transformation tougheners alternative to zirconia [1]. The temperature stability ranges as a function of atomic number (Z) were

P.D.Jero* and W.M.Kriven, Department of Materials Science and Engineering, University of Illinois at Urbana-Champaign, Urbana, Illinois 61801, USA.
* Now at: US Air Force Materials Laboratory, WL/MLLM, Wright Patterson Air Force Base, Ohio 45433-6523, USA.

elucidated by several workers, notably Foex and Traverse [2,3]. As seen in Fig. 1, the compounds exhibit cubic (X), primitive hexagonal (H), hexagonal (A), monoclinic (B) and cubic (C) symmetry. On cooling, the B to C transformation is accompanied by an ~8% volume expansion and 10° angular cell shape change. The monoclinic (B) unit cell has space group symmetry C2/m with Z = 6 and seven-fold coordination. The structure can be derived from fluorite (CaF_2) by removal of one fourth of the oxygens along non-intersecting <111> directions, such that each cation has six-fold rather than eight-fold coordination. Caro has proposed a widely accepted model for viewing the polymorphs in terms of oxygen coordinations [4].

The aim of this work was to investigate the lanthanide sesquioxides as high temperature transformation tougheners of a chemically unreactive matrix, as an analogue to zirconia toughened alumina (ZTA). Terbia was chosen as a representative sesquioxide because of its accessible transformation temperature of 1650°C and rapid B to C kinetics. Unfortunately, terbium is one of several of the lanthanides which exhibit variable valence states (+3 and +4). Because of this, all high temperature processing must be done in inert or reducing atmospheres or in vacuum. In the presence of oxygen, Tb_2O_3 converts to Tb_4O_7 starting at ≈280°C. Although Tb_4O_7 is not stable even in oxygen above ≈1100°C, on cooling the conversion back to Tb_4O_7 occurs quite rapidly. Although no Tb_2O_3 -MgO phase diagram exists in the literature, diagrams of the neighboring lanthanides, Gd_2O_3 and Dy_2O_3 with MgO do exist [5]. They indicate that these systems exhibit no compound formation or significant solid solution.

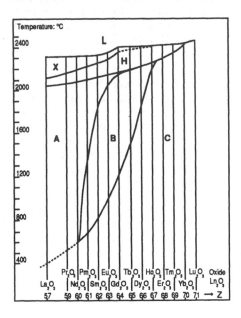

Figure 1. Temperature stability ranges of the lanthanide sesquioxides after Foex and Traverse [2].

EXPERIMENTAL PROCEDURES

In order to demonstrate high temperature transformation toughening in a ZTA-type analogue material, composites containing 20 vol%B-Tb_2O_3 intergranularly dispersed in 80 vol% MgO (20T/80M) were sintered into bend bars and subsequently HIPPed to full density. The microstructures were characterized by x-ray diffraction (XRD), scanning electron microscopy (SEM), transmission electron microspopy (TEM) and energy dispersive X-ray spectroscopy (EDS). Strength and toughness measurements were made from the bend bars up to temperatures of 1400°C. The occurrence of transformation on fracture, polished and ground surfaces was monitored in the context of mechanical properties.

The detailed procedures are described elsewhere [6]. The Tb_4O_7 powder was donated by the Rhone-Poulenc Co., France, while AR grade MgO powder was used. Since Tb_4O_7 consisted of disproportionated mixtures of Tb_2O_3 and TbO_2, the Tb_4O_7 had to be reduced to Tb_2O_3 by annealing in argon (Ar), Ar/H_2 or in vacuum at 1200°C for 15 min.

The processing procedure consisting of dispersing Tb_2O_3 and MgO in isopropyl alcohol with brief ball milling using 12.7 mm diameter ZrO_2 balls, followed by pan drying and sieving through a 200 mesh screen. Pellets and bar shaped (41mm x 4.5mm x 4.5mm) specimens were uniaxially pressed in a die at 34-69 MPa before cold isostatic pressing at 172 MPa. Samples were heated at a rate of 10°C/min to 1750°C for 15 min in a tungsten element furnace in vacuum, or Ar atmosphere. Excess MgO surrounded the specimens during sintering in order to minimize MgO volatilization. After firing, bar-shaped samples were HIPPed at 1750°C under 189 MPa pressure for 15 min. The pressure was also held during cooling to 1000°C which was done by turning off the power to the furnace.

The phase composition was determined by XRD and bulk densities were measured by immersion in hexachloro-1, 3-butadiene. Grain sizes were measured from SEM micrographs of polished specimens thermally etched at 1400°C for 30 to 60 min. Microstructures were examined in a Philips EM400 transmission electron microscope, using selected area diffraction (SAD) and EDS techniques.

High temperature flexure testing was accomplished using a furnace mounted on an Instron testing machine. A large Al_2O_3 tube with stainless steel bellows enabled the system to be operated in an inert argon atmosphere. Flexure tests were conducted at room temperature, 300, 600, 800, 1000, 1075, 1150, 1300 and 1400°C. Flexure specimens were ground to nominal dimensions 3x3x25mm with a 600 grit finish on both tensile and compressive surfaces to assure flatness. Samples were notched using a thin (~150 μm) diamond wafering blade. The notch width varied from ~200 μm at the tip of the notch to 400 μm at the surface of the sample. The notch length was 20% to 25% of the sample thickness.

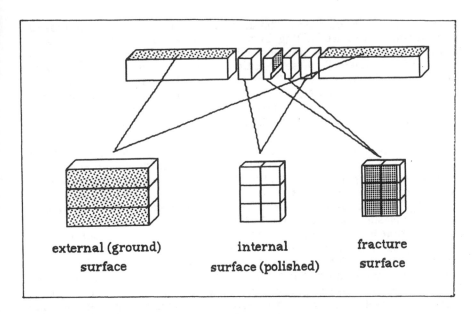

Figure 2. Location of X-ray samples cut from bend bars.

After testing, samples were examined by XRD for transformation. In each case, three samples broken under identical conditions were sectioned, as shown in Fig. 2. External (ground), internal (polished) and fracture surfaces were examined. The room temperature strength of 20T/80M specimens were also measured by breaking un-notched samples in four-point flexure.

RESULTS AND DISCUSSION

Small pellets of pure Tb_2O_3 prepared as described above were sintered at various temperatures. Specimens fired above 1750°C dusted on cooling through the B to C transformation, while those fired below 1700°C remained intact. The former samples were entirely powdered upon removal from the sintering furnace and XRD found entirely C phase. The latter specimens remained in the C phase throughout firing, thereby avoiding the destructive B to C transformation. Even with very rapid furnace cooling (1750°C to 550°C in 7 min) it was not possible to retain pure, bulk Tb_2O_3 in the high temperature B phase. In related work by Sudre, Venkatachari and Kriven [7] however, it was found that the B phase could be retained in pure Tb_2O_3 which was laser melted and roller quenched. The quenching rate in that work was estimated to be 10^7 °C/sec.

In the 20T/80M specimens, the Tb_2O_3 could be retained in the B phase through rapid furnace cooling (as above), or through cooling under pressure in a HIP (also quite rapid). In the former, only partial B phase retention could be obtained. The amount retained was seen to depend critically on the furnace loading, increased thermal mass causing a decrease in the amount of B phase retained. HIPPing (and cooling under pressure) resulted in complete B phase retention. Specimens were commonly 95% dense after sintering and 99% dense after HIPPing. The grains were typically 1 to 10 μm in size, the Tb_2O_3 grains generally being smaller than the MgO (Fig 3.). Attempts to stress-induce the B to C transformation by room temperature grinding at 200 grit, or by liquid nitrogen quenching were not successful.

Fracture toughness was measured from notched-beam, three-point flexure specimens of 20T/80M composites which were broken at ambient and elevated temperatures. The toughness was observed to increase with temperature above 600°C as shown in Fig. 3. The toughness shown for pure MgO was calculated from the fracture surface energy data of Evans et al. [8] and elastic modulus data of Vasilos et al. [9].

Partial B to C transformation of fracture surfaces was observed on samples broken at 1075°C and above. Figure 4 shows XRD patterns taken after fracture at 1150°C, corresponding to the (polished) bulk, fracture surfaces and unconstrained, ground surfaces, respectively. $(222)_c$ and $(400)_c$ peaks of transformed C phase are clearly seen on the fracture surface, indicating a well developed transformation zone on the fracture surface. An equivalent X-ray penetration depth of ~6.4 μm was calculated from 75% attenuation of CuKα radiation, assuming a 20T/80M composition, 2θ=30°, and theoretical density.

Four distinct regions of transformation behavior were observed. Below 1075°C, no B to C transformation was observed. From 1075°C to 1150°C, the B to C transformation occurred only on the fracture surfaces. At 1300°C transformation was also observed on free surfaces, and at 1400°C spontaneous transformation was observed throughout the material.

Increases in toughness prior to the onset of transformation indicated that some other mechanism operated in the intermediate and possibly high temperature range. It is known that slip occurs in B-type sesquioxides above ~800°C [10].

In order to see if surface transformation would lead to strengthening, two sets of specimens (each set containing three samples) were prepared and broken at room temperature in four point flexure. The first set was ground to a 600 grit finish and tested. The second set was ground and then annealed at 1300°C for 20 min in argon before testing. XRD indicated that the diffracting volume on the surface was ~59 % C phase after annealing. The ground and annealed specimens exhibited a 20% increase in strength over the as-ground specimens The values measured were 228 ± 16.0 MPa versus 183 ± 10.7 MPa, respectively.

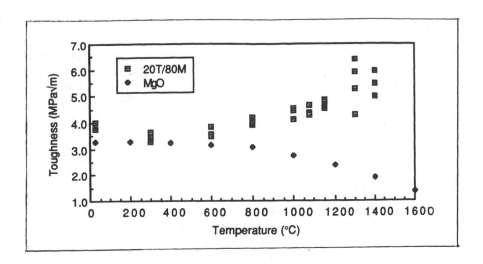

Figure 3. Fracture toughness of 20 vol% Tb$_2$O$_3$-80 vol% MgO composites as a function of temperature.

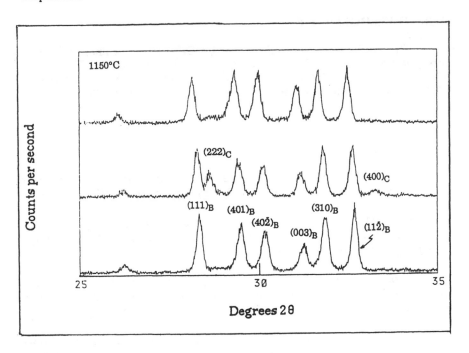

Figure 4. Diffraction patterns from 20T/80M composites broken at 1150°C; (top), polished internal surface, (middle), fracture surface and (bottom), external ground surface, respectively.

195

CONCLUSIONS

In pure Tb_2O_3, the B to C transformation on cooling results in shattering of bulk specimens. Ultra fast quenching and/or a constraining matrix must be used to retain the B phase. Once retained in the B phase at room temperature, transformation cannot be induced by grinding.

B-phase, Tb_2O_3-containing specimens exhibit the B to C transformation at high temperatures, and this is accompanied by an increase in toughness of the composite. Four distinct regions were observed. Firstly, no B to C transformation could be detected below 1000°C. Secondly, localized B to C transformation occurs on the fracture surfaces of specimens broken at temperatures above 1075°C. This could be viewed as the stress-assisted transformation corresponding to traditional transformation toughening as seen in zirconia. Thirdly, transformation of unconstrained ground surfaces (not stress-assisted) is observed above 1300°C. Fourthly, at 1400°C spontaneous bulk transformation occurs.

The formation of a partially transformed surface layer on 20T/80M composites results in improved room temperature strength. This may be due to compressive surface stresses arising from the transformation.

ACKNOWLEDGEMENTS

This work was supported by the U.S. Department of Energy through the Materials Research Laboratory at UIUC, under contract DEA CO2-76ERO-1198. Use of the facilities in the Center for Microanalysis of Materials in the Materials Research Laboratory is gratefully acknowledged.

REFERENCES

1.   Kriven, W.M., 1988. "Possible Alternative Transformation Tougheners to Zirconia: Crystallographic Aspects." J. Am. Ceram. Soc., 71(12):1021-1030.

2.   Foex, M. and J.P. Traverse, 1965. "Étude du Polymorphisme des Sesquioxydes de Terres Rares à Haute Temperatures." Bull. Soc. Franc. Miner. Crist., 89:184-205 (in French).

3.   Foex, M. and J.P. Traverse, 1966. "Remarques sur les Transformations Cristallines Présentees à Haute Température par les Sesquioxydes de Terres Rares." Rev. Int. Hautes Temp. et Refract., 3:429-453 (in French).

4.   Caro, P.E., 1968. "OM_4 Tetrahedral Linkages and the Cationic Group $(MO)_n^{n+}$ in the Rare Earth Oxides and Oxysalts." J. Less Common Metals, 16:367-377.

5.   Lopato, L.M., 1976. "Highly Refractory Oxide Systems Containing Oxides of Rare Earth Elements." <u>Ceram. Int.</u>, 2(1):18-32.

6.   Jero, P.D., 1988. "Investigation of the Lanthanide Sesquioxides as High Temperature Transformation Toughening Agents." Ph.D. Thesis, University of Illinois at Urbana-Champaign, Illinois, USA.

7.   Sudre, O., K.R. Venkatachari and W.M. Kriven, "Kinetics and Crystallography of the Monoclinic (B) to Cubic (C) Transformation in Dysprosia (Dy_2O_3)," this volume.

8.   Evans, A.G., D. Gilling and R.W. Davidge, 1970. "The Temperature Dependence of the Strength of Polycrystalline MgO." <u>J. Mater. Sci.</u>, 5:187-197.

9.   Vasilos, T., J.B. Mitchel and R.M. Spriggs, 1964. "Mechanical Properties of Pure, Magnesium Oxide as a Function of Temperature and Grain Size." <u>J. Am. Ceram. Soc.</u>, 47(12):606-610.

10.   Stobierski, L. and A.M. Lejus, 1982. "The Temperature Dependence of Microhardness of Lanthanide Sesquioxide Single Crystals with A, B and C-Structure." <u>Rev. Int. Hautes Temp. et Refract.</u>, 1:3-8 (in French).

Observation of in-situ Reduction and Oxygen Transport in Zirconia-Tantala Ceramic Alloys

J. R. SELLAR

ABSTRACT

High-resolution electron microscope images are presented of a stabilized anion-excess zirconia alloy (nominally $Ta_2Zr_8O_{21}$) which under electron irradiation undergoes in-situ reduction by the carbon of the electron-microscope support grid. The local oxygen concentration may be related directly to the distance between the modulation fringes which characterize this zirconia-tantala alloy, and the transport of oxygen to the surface of the alloy grains is observed as the fringes move over a period of tens of minutes. Micrographs are presented with examples of the modulation fringes both before and after reduction, and some of the mechanisms suggested whereby the crystals might afterwards relax while maintaining the continuity of the fringes throughout.

INTRODUCTION

The oxygen transport properties of zirconia alloyed with the oxides of lower-valence metals such as CaO and Y_2O_3 make it an excellent candidate for the electrolyte in solid oxide fuel cells. In this paper results are presented of a preliminary electron microscope (EM) study of oxygen transport in an anion-excess zirconia alloy, i.e. when the added metal ion possesses a valency higher than four: in the present case Ta_2O_5. We shall see that oxygen transport is possible under reducing conditions, and that the excess oxygen storage in the alloy is highly ordered. The ordering of the excess oxygen 'pressure' regions of the alloy corresponds to a modulation of the oxygen ion nets which comprise the anion sublattice, in such a way that the modulation is clearly visible in electron micrographs of intermediate magnification. The spatial frequency of the modulation affords a direct measure of the local oxygen ion concentration, which may also be read from the corresponding superlattice diffraction pattern. In all cases to be discussed, the in-situ reduction of the alloy is brought about by a chemical reaction of the ceramic with the amorphous 'holey' carbon electron microscope specimen support grid.

J.R. Sellar, Department of Materials Engineering, Monash University, Clayton, Victoria, Australia

CRYSTALLOGRAPHY AND STOICHIOMETRY IN $Ta_2Zr_{x-2}O_{2x+1}$

The modulated appearance of these alloys in the range $7 \leq x \leq 10.3$ has been linked crystallographically with regions of 'compression' and 'rarefaction' of the oxygen ion network, which is believed to be in very close analogy with zirconia-niobia alloys in the same concentration range [1,2]. X-ray determinations conducted twenty years ago by Galy and Roth of the structures of $Nb_2Zr_6O_{17}$ [3] and $Ta_2Zr_8O_{21}$ [4] were initially interpreted as superstructures of a fluorite - like MO_2 structure. Later, on the basis of the same x-ray data, it was suggested that the extra oxygen atoms were accommodated by transforming ribbons of tetragonal oxygen nets to denser hexagonal nets as in figure 1.

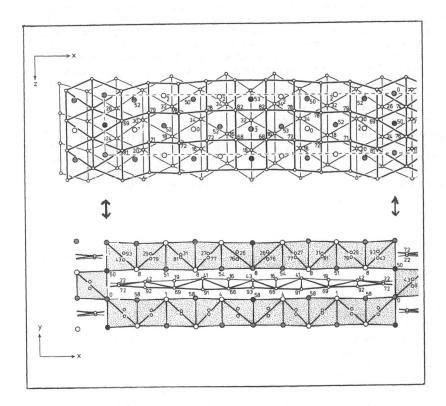

Figure 1   [010] and [001] projections of approximately one-half the unit cell of $Ta_2Zr_8O_{21}[Ta_2O_5.8ZrO_2]$. [010] projection demonstrates the alternating tetragonal and hexagonal oxygen ion arrays and [001] projection displays the antiphase character of the hexagonal nets (here seen side-on) alternating with the oxygen-centered tetrahedra, shown shaded in. The modulation direction is [100]. Region between the overlapping hexagonal nets (arrowed) equals one-half the unit cell. Large circles with vertical hatching are Ta: those without are Zr. Unit cell length in modulation direction is 50.98Å.

Under this model, the structure is considered to be able to accommodate any oxygen stoichiometry in the range by simple topological adjustment of the relative proportions of the locally tetragonal ('rarefied') and hexagonal ('compressed') regions of the oxygen ion network. The more reduction an alloy undergoes, the more closely it approaches MO_2 stoichiometry and the further apart the comparatively oxygen-rich hexagonal regions shift. Accordingly, the spacing between the superlattice diffraction spots grows smaller as reduction proceeds, with a linear relationship between the reciprocal of the modulation wave vector magnitude and composition. More recently, however, this family of structures has been identified in the general case as Type II (or composite) modulated structures, with continuously varying incommensurate super lattice wave-vectors [5].

Throughout the reduction processes observed in these experiments, it is considered that the oxygen ions move much more than the metal ions, so that the metal atoms experience continuously varying co-ordination environments. In common with the zirconia-niobia alloys, however, the orthorhombic metal subcell of the tantala composition series undergoes a small shrinkage in the [100] modulation direction as the relative amount of the more oxygen-rich oxide decreases. Over the corresponding composition range of the niobia alloy for example, the subcell strain amounts to 0.5% [5]. In the present tantala alloy investigation, this often gives rise to microcracking.

In the X-ray investigations already undertaken of alloy specimens prepared in simple molar ratios the unit cell in the [100] direction in $Nb_2Zr_6O_{17}$ is approximately 40Å and in $Ta_2Zr_8O_{21}$ approximately 50 Å. In figure 1, a drawing is shown of $Ta_2Zr_8O_{21}$ in the hexagonal/tetragonal oxygen net model with anion centred tetrahedra [6]. In the [010] projection, we are looking down on the alternating hexagonal and tetragonal layers of the two-dimensional networks. In the [001] projection the antiphase character of the net stacking is evinced, seen here edge-on.

IN-SITU ELECTRON MICROSCOPE OBSERVATIONS

All observations reported in this paper were recorded in a JEOL 200CX electron microscope operating at 200kV with a nominal total pressure near the specimen of approximately 2×10^{-7} torr.

In figure 2 an electron micrograph is presented of a region of tantala-zirconia ceramic alloy with stoichiometry near $Ta_2Zr_8O_{21}$, which has not undergone any reduction (it is not in contact with carbon). Modulations corresponding to ordered variation in oxygen concentration are clearly visible in the thinner parts of the specimen. The modulations are essentially straight and parallel in zirconia-tantala alloys and no disordering was observed where there was no reduction. Similarly, no microcracking was observed in the mechanical preparation of the brittle zirconia-tantala specimens when ground under liquid nitrogen using an agate mortar and pestle: i.e. no microcracks were observed without accompanying reduction.

Figure 2, then, represents the baseline micrograph for all images to follow depicting alloy reduction, and with which they should be compared. For details of specimen preparation, see [5].

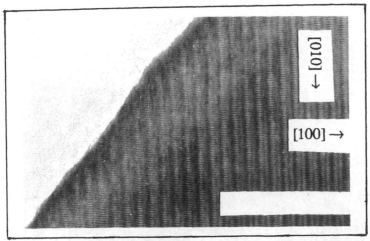

Figure 2   Electron micrograph of $Ta_2Zr_8O_{21}$ in [001] projection, showing characteristics of the modulation due to the excess "oxygen pressure waves", ie. alternating local hexagonal and tetragonal variations of the oxygen ion lattice. Bar is 300Å.

A high resolution electron micrograph of unreduced $Ta_2Zr_8O_{21}$ is presented in figure 3 taken near the thin edge of a grain in [001] projection. Again the antiphase character of the stacking of the alternating hexagonal and tetragonal layer sections of the oxygen ion network in the y-direction is clearly exhibited.

Figure 3   High-resolution image of $Ta_2Zr_8O_{21}$ in [001] projection. The antiphase character of the modulation in this specimen is clear in the thin region near the edge of the specimen. The modulation wavelength may be taken as a measure of the local oxygen concentration. Bar is 30Å.

Figure 4   Thin monocrystalline wedge of $Ta_2Zr_8O_{21}$ after chemical reduction has taken place. Note the heavily pitted and correspondingly depleted holey carbon support grid. The grid was initially touching the specimen everywhere along the alloy crystal's edge. [001] projection. Bar is 300Å.

In figure 4 evidence for chemical reduction as the cause of the bending of the modulation fringes which correspond to the variation of local oxygen composition is adduced by the progressive consumption of the holey carbon support grid bar, causing violent disruption of the formerly straight modulation fringes in those regions of the grain nearest the grid bar. The surrounding unreduced regions of the crystal possess a slightly larger subcell size in the [100] direction which causes tensile stress in the oxygen-depleted region, resulting in the formation of microcracks.

Figure 4 is the result of more than twenty minutes continual motion by the fringes. The reduction process takes place over a period of several minutes before slowing down or ceasing. The fringes in figure 4 were observed to move in a 'jerky' fashion in a direction normal to their length, i.e. in the [100] direction, sometimes swerving suddenly so as to exit the crystal parallel to its surface. Notice the consumption of the carbon grid nearest the thin edge of the crystal. Before the process of reduction took place, the holey carbon continued up to the edge of the alloy specimen, with a regularly-textured surface appearance. After the reduction process, as in figure 4, the consumption of carbon has progressed with a heavy pitting of the 'activated' substrate plainly visible. In some grains the reduction takes place without the formation of cracks, but with widespread disturbance of the oxygen network. Transport of the reactive species over observed distances of up to 1000Å on the support substrate surface is attributable in principle to a Boudouard-type redox reaction (private communication, I.E. Grey, CSIRO). A high-resolution EM image showing the relief of tensile strain in the oxygen-depleted region by a set of parallel, regularly-spaced microcracks is presented in figure 5. A large area of oxygen modulation fringes can be seen traversing the crystal at an angle of approximately 30° to the fringes in the unperturbed regions. The fringes between the cracks are parallel to the cracks.

This result is interpreted to demonstrate two of the mechanisms by which the crystal accommodates reduction. In figure 5, cracks are considered to relieve the tensile strain between the shrunken reduced regions and the unperturbed regions of the crystal. In order to maintain the continuity of the oxygen-related fringes (we consider the tantalum atoms move little) the 30° fringes in the centre of the micrograph appear to move closer together, but there is actually little change in local oxygen concentration. Between the cracks, the fringes are similar in spatial frequency to those of the unreduced regions (eg. in the upper left-hand side of the micrograph) so that the degree of reduction undergone to cause microcracking is not great. If the fringes are tilted, the way in which to 'read' the local excess oxygen concentration is by the distance between fringes in the [100] direction, rather than the direction normal to the fringes.

CONCLUSION

Although preliminary in nature, the foregoing study appears to indicate that under some circumstances zirconia-tantala may be reduced by the carbon of an electron microscope grid while under irradiation in an electron microscope. The clearest indications of this effect are the direct real-time observation of modulation fringes exiting the crystal grain, resulting in a lowering of the oxygen content of the grain, and the corresponding relief by microcracking of the tensile strain arising between the reduced and unreduced regions.

REFERENCES

1.   Hyde, B.G., A.N. Bagshaw, S. Andersson and M. O'Keeffe, 1974. "Some Defect Structures in Crystalline Solids". Ann.Rev.Mat.Sci.,4: 43-92.

2.   Mackovicky, E. and B.G. Hyde, 1981. "Non-commensurate (Misfit) Layer Structures". Struct.Bond.,46: 101-170.

3.   Galy, J. and R.S. Roth, 1973. "The Crystal Structure of $Nb_2Zr_6O_{21}$". J.Solid State Chem., 7: 277-285.

4.   J. Galy (unpublished).

5.   Thompson, J.G., R.L. Withers, J.R. Sellar, P.J. Barlow and B.G. Hyde, 1990. "Incommensurate Composite Modulated $Nb_2Zr_{x-2}O_{2x+1}$ x=7.1 - 10.3". J.Solid State Chem.,88: 465-475.

6.   Hyde, B.G. and S. Andersson, 1989. "Inorganic Crystal Structures", New York, NY, USA: Wiley.

Figure 5   Microcrack relief of strain due to shrinking of metal sublattice due to alloy reduction. Fringes crossing diagonally appear to join together fringes in the reduced and non-reduced areas. Cracks are approximately 150 Angstrom apart, with crack tips approximately equal in size to a fluorite subcell. Bar is 300Å.

PROCESSING AND MICROSTRUCTURAL CONTROL

Hydrothermal Synthesis and Properties of Yttria Fully Stabilized Zirconia Powders

K. HISHINUMA, M. ABE, K. HASEGAWA,
Z. NAKAI, T. AKIBA and S. SŌMIYA

ABSTRACT

Well-crystallized yttria fully stabilized zirconia (Y-FSZ, 8 mol% Y_2O_3 doped ZrO_2) powders were synthesized by homogeneous precipitation under hydrothermal conditions. The mixed aqueous solutions of $ZrOCl_2.8H_2O$, $YCl_3.6H_2O$ and urea were hydrothermally treated at 180°C for 2 hours in a zirconium-lined autoclave. The homogeneous precipitate formed was hydrous ZrO_2 with Y_2O_3, which crystallized into yttria doped zirconia under hydrothermal conditions. The products were cubic fine powders about 10 nm in size. Over 98% theoretically dense 8Y-FSZ ceramics with a homogeneous microstructure were obtained by sintering above 1400°C.

INTRODUCTION

There are several papers related to hydrothermal processing of fine ZrO_2 powders by the authors and others [1-7]. Fine and well crystallized ZrO_2 powders are obtained by hydrothermal treatments. It is not possible to obtain yttria stabilized zirconia under acidic conditions. So an excess amount of ammonia is added to a mixture of yttrium chloride and zirconium oxychloride solutions before hydrothermal treatment [2].

Coprecipitation techniques are used for the preparation of multicomponent powders. In these techniques, it is difficult to obtain a homogeneous distribution of the components in the precipitates because segregation occurs during precipitation. The precipitated powders have to be calcined for crystallization.

Y-FSZ has been widely known to be a good material for the electrolyte of the solid oxide fuel cell (SOFC). Low specific resistivity and good mechanical properties at high temperature are the main requirements for use as an electrolyte. The properties of sintered Y-FSZ ceramics mainly depend on the characteristics of the powders. In this study, we prepared Y-FSZ (8 mol% Y_2O_3 doped ZrO_2) powders by hydrothermal homogeneous precipitation and the properties of these powders will be discussed.

K. Hishinuma, M. Abe, K. Hasegawa, Z. Nakai and T. Akiba
Fine Ceramics Research and Development Division
Chichibu Cement Co. Ltd., 5310 Mikajiri,
Kumagaya, Saitama 360, Japan
S. Sōmiya, The Nishi-Tokyo University,
Uenohara, Yamanashi 409-01, Japan

EXPERIMENTAL PROCEDURE

Figure 1 shows the processing flow chart for the hydrothermal homogeneous precipitation method. The starting materials were $ZrOCl_2.8H_2O$, $YCl_3.6H_2O$ and urea. 120 % of the theoretical urea requirement was used for the homogeneous precipitation. The 600 ml mixed solutions of composition ZrO_2 + 8 mol% Y_2O_3 at 0.5 mol/L total cations concentration, in which the initial pH was <1, were hydrothermally treated at 160°, 180° and 200°C for 2 hours in a zirconium-lined autoclave (1000 mL capacity) and stirred at 500 rpm. The urea decomposed into NH_3 and CO_2 through reaction with H_2O and the solution pH changed to 8. The homogeneous precipitate formed was hydrous ZrO_2 with Y_2O_3 and it crystallized into zirconia under hydrothermal conditions. If urea was not used in this method, yttrium remained in solution. The products were washed with deionized water until the wash water did not become turbid when 1M $AgNO_3$ was added. After washing, these products were dried at 120°C for 12 hours.

The dried powders were calcined at different temperatures from 600° to 1000°C in air for 1 hour, in order to investigate the phase transformation and grain growth phenomena. The products were characterized by X-ray diffraction (XRD), surface area measurement (BET) and thermal analysis (TG/DTA). The crystallite size of zirconia was calculated by using the Sherrer-Warren equation on the half-widths of the cubic (tetragonal) (111) and monoclinic (111) reflections. The product morphology was observed by transmission electron microscopy (TEM). The dried powders were cold isostatically pressed under 196 MPa and sintered at temperatures from 1100° to 1550°C for 2 hours. The density of each sintered body was determined by the Archimedes method. The microstructure of the sintered bodies was observed by scanning electron microscopy (SEM) of their polished and thermally etched surfaces.

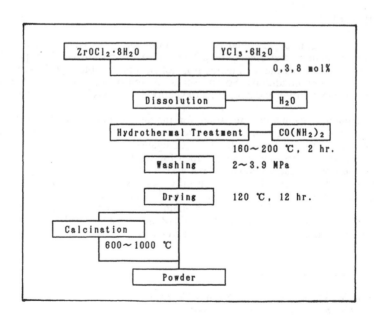

Figure 1. Processing flow chart of the hydrothermal homogeneous precipitation method.

RESULTS AND DISCUSSION

Figure 2 shows the XRD patterns of products prepared by hydrothermal homogeneous precipitation at 180°C for 2 hours. Pure ZrO_2 (0Y) appears to have consisted of tetragonal and monoclinic phases. Adair et al.[6] reported the two mechanisms for crystallization of zirconia particles under hydrothermal conditions. Under acid conditions, crystalline particles formed by precipitation (high temperature hydrolysis). On the other hand, tetragonal zirconia particles crystallize "in-situ" from amorphous, hydrous oxides in more alkaline conditions. Tani et al.[4] reported that tetragonal phase was formed by topotactic crystallization on the nuclei in the amorphous hydrous zirconia. In this method, monoclinic phase was directly prepared from solution by hydrolysis at low pH.

In Y_2O_3 (3 and 8 mol%) doped ZrO_2, only cubic phase was present. ZrO_2 powders were well crystallized and formed solid solutions with Y_2O_3 under hydrothermal conditions. The cubic phase of 3Y was metastable, because it was transformed into the tetragonal phase by calcination [7]. The transformation of metastable cubic to tetragonal was demonstrated by splitting of the (400) line of the XRD patterns. On the other hand, 8Y powder was not transformed by calcination between 600° and 1000°C.

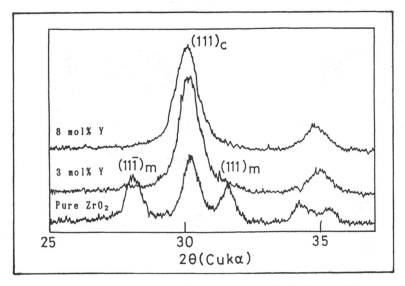

Figure 2. X-ray diffraction patterns of zirconia powders prepared by hydrothermal homogeneous precipitation at 180°C for 2 hours. c:cubic, m:monoclinic.

The results of characterization of products prepared by hydrothermal homogeneous precipitation at 180°C for 2 hours are summarized in Table I. The crystallite size was reduced by the addition of Y_2O_3. This shows that the addition of Y_2O_3 inhibits the crystallite growth of ZrO_2 under the hydrothermal conditions. The crystallite size of 8 mol% Y_2O_3-ZrO_2 (8Y-FSZ) powder was about 10 nm. The specific surface area increased with reduction in the crystallite size. The equivalent particle diameter (D_{BET}) was given by the equation $D_{BET}=6/\rho S$, assuming a spherical shape, where ρ is the powder density measured by helium displacement and S the specific surface area. Since D_{BET} values are close to the crystallite sizes of the products, the products appear to be single-crystal particles .

TABLE I. CHARACTERISTICS OF ZIRCONIA POWDERS PREPARED BY HYDROTHERMAL HOMOGENEOUS PRECIPITATION (180°C, 2 hours).

Y_2O_3 (mol%)	Phase	Crystallite Size (nm)	D_{BET}* (nm)	Surface Area (m^2/g)
0	M+T	M 17, T 13	13	93
3	C′	11	9	118
8	C	10	7	144

M:Monoclinic, T:Tetragonal, C:Cubic, C′:Metastable Cubic
*$D_{BET}=6/\rho S$ (ρ:powder density, S:surface area)

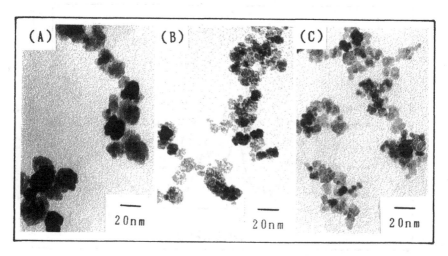

Figure 3. Transmission electron micrographs of zirconia powders prepared by hydrothermal homogeneous precipitation at 180°C for 2 hours.
(A):pure ZrO_2, (B):3 mol% Y_2O_3-ZrO_2, (C):8 mol% Y_2O_3-ZrO_2.

Transmission electron micrographs of the products are shown in figure 3. The pure ZrO_2 particles formed agglomerates 30 nm in size. The Y_2O_3 doped ZrO_2 particles were more dispersed than the pure ZrO_2. The particle size was comparable to the crystallite size determined by XRD.

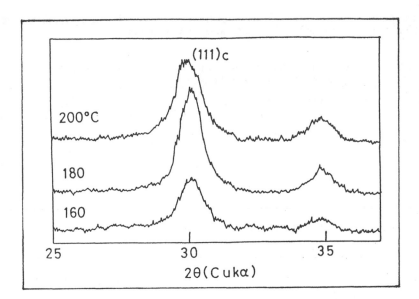

Figure 4. X-ray diffraction patterns of 8Y-FSZ(8 mol% Y_2O_3-ZrO_2) powders prepared by hydrothermal homogeneous precipitation at various temperatures for 2 hours. c:cubic.

Figure 4 shows the XRD patterns of 8 mol% Y_2O_3-ZrO_2 (8Y-FSZ) powders prepared by hydrothermal treatment at various temperatures for 2 hours. The products were cubic ZrO_2 at all temperatures. The crystallite size had a maximum at 180°C. Considering the DTA curve of product prepared at 180°C, the exothermic peak was not observed. The product prepared at 160°C had a large specific surface area and a large weight loss as determined by TG. The DTA curve of product prepared at 160°C shows a small exothermic peak at 255°C. From these results, it seems that complete crystallization of 8Y-FSZ occurs at 180°C under hydrothermal conditions.

Haberko [8] reported that the crystallization temperature is 462°C for the 12 mol% Y_2O_3-ZrO_2 coprecipitate. It is considered that the crystallization temperature of 8Y-FSZ powders under hydrothermal conditions is lower than that of coprecipitated Y-FSZ powders.

Above 180°C, the crystallite size decreased with increase in the hydrothermal temperature. The same phenomenon was observed in 3Y-PSZ powder preparations [7]. It seems that nucleation is dominant rather than crystal growth at higher temperatures in this method. The characteristics of 8Y-FSZ powders prepared by hydrothermal homogeneous precipitation at various temperatures for 2 hours are summarized in Table II.

The variation of crystallite size of zirconia powders with calcination temperature is shown in figure 5. The crystallite size of 0Y (pure ZrO_2) increased significantly above 600°C, whereas those of 3Y and 8Y increased slightly at 800°C. This behavior is considered to have resulted from the rapid "in-situ" sintering of crystallites in the large agglomerates of pure ZrO_2 and confirms that the addition of Y_2O_3 inhibits the crystallite growth of ZrO_2 at low calcination temperatures.

TABLE II. CHARACTERISTICS OF FSZ (8 mol% Y_2O_3) POWDERS PREPARED BY HYDROTHERMAL HOMOGENEOUS PRECIPITATION

Temperature (°C)	Phase	Crystallite Size (nm)	Weight Loss (%)*	Surface Area (m²/g)
160	C	9	17.3	181
180	C	10	9.0	144
200	C	8	7.3	153

C:Cubic, *TG, heating rate is 5°C/h, 20-1100°C, in air

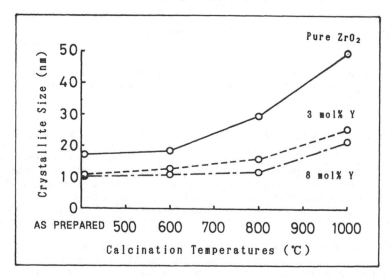

Figure 5. Variation of crystallite size of zirconia powders prepared by hydrothermal homogeneous precipitation at 180°C after 2 hours at various calcination temperatures. (CIP 195 MPa, heating rate:100°C/min, sintered for 2 hours).

Figure 6 shows the variation of the sintered density of 8Y-FSZ prepared under hydrothermal conditions at 180°C for 2 hours. The powder prepared at 180°C was cold isostatically pressed at 196 MPa and sintered at different temperatures for 2 hours. The heating rate was 100°C/hour. We obtained 8Y-FSZ ceramics that were more than 98% of theoretical density (5.971 g/cm³) with a homogeneous microstructure above 1400°C. A scanning electron micrograph of the 8Y-FSZ ceramic sintered at 1500°C for 2 hours is shown in figure 8. The average grain size was 4.2 µm and irregular grain growth and large pores were not observed.

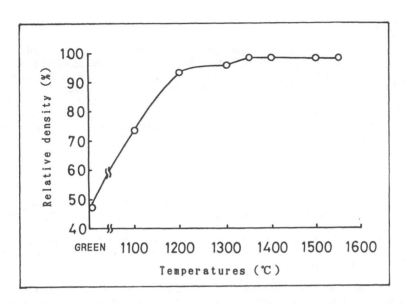

Figure 6. Variation of relative density of 8Y-FSZ (8 mol% Y_2O_3-ZrO_2) with sintering temperature for the powder prepared at 180°C.
(CIP 196 MPa, heating rate:100°C/min, sintered for 2 hours).

Figure 7. Microstructure of 8Y-FSZ (8 mol% Y_2O_3-ZrO_2) sintered at 1500°C for 2 hours.
(Bar = 2.2 μm).

SUMMARY

Well-crystallized 8Y-FSZ powders were obtained by a hydrothermal homogeneous precipitation method. The products were cubic fine powders about 10 nm in size. The crystallite size of ZrO_2 was reduced by the addition of Y_2O_3, which inhibits the crystallite growth of ZrO_2 under hydrothermal conditions.

The 8Y-FSZ powder prepared at 180°C was cold isostatically pressed at 196 MPa and sintered at different temperatures for 2 hours. We obtained over 98% theoretically dense 8Y-FSZ ceramics with a homogeneous microstructure above 1400°C.

REFERENCES

1.   Mitsuhashi, T., M. Ichihara and U. Tasuke, 1979. "Characterization and Stabilization of Metastable ZrO_2 Powders." Am. Ceram. Soc. Bull., 58(8):587-590.

2.   Burukin, A.R., H. Sariçïmen and B.C.H. Steele, 1980. "Preparation of Yttria Stabilized Zirconia (YSZ) Powders by High Temperature Hydrolysis (HTH)." Trans. Br. Ceram. Soc., 79:105-108.

3.   Nishizawa, H., N. Yamasaki, K. Matsuoka and H. Mitsutomo, 1982. "Crystallization and Transformation of Zirconia Under Hydrothermal Conditions." J. Am. Ceram. Soc., 65(7):343-346.

4.   Tani, E., M. Yoshimura and S. Sōmiya, 1984. "Formation of Ultrafine Tetragonal ZrO_2 Powder under Hydrothermal Conditions." J. Am. Ceram. Soc., 66(1):11-14.

5.   Yoshimura, M., S. Kikugawa and S. Sōmiya, 1982. "Preparation of Zirconia Fine Powders by the Reactions between Zirconium Metal and High Temperature-High Pressure Solutions," in High Pressure in Research and Industry, C.M. Backman, T. Johanison and L. Tegner, eds., Uppsala, Sweden: Arkitektkopia ISBN., pp.793-796.

6.   Adair, J.H., R.P. Denkewicz, F.J. Amagada and K. Osseo-Asare, 1988. "Precipitation and In-situ Transformation in the Hydrothermal Synthesis of Crystalline Zirconium Dioxide," in Ceramic Powder Science II, A.G.L. Messing, E.R. Fuller, Jr. and H. Hausner, eds., Columbus, Ohio, USA: The American Ceramic Society, Inc., pp.135-145.

7.   Hishinuma, K., T. Kumaki, Z. Nakai, M. Yoshimura and S. Sōmiya, 1988. "Characterization of Y_2O_3-ZrO_2 Powders Synthesized under Hydrothermal Conditions," in Advances in Ceramics, Vol.24: Science and Technology of Zirconia III, S. Sōmiya, N. Yamamoto and H. Yanagida, eds., Columbus, Ohio, USA: The American Ceramic Society, Inc., pp.201-209.

8.   Haberko, K., 1977. "Some Properties of Zirconia Obtained by Coprecipitation with Different Oxides." Rev. Int. Htes. Temp. et Refract., 14:217-224.

Formation of Ultrafine Multicomponent ZrO$_2$-Based Powders in an RF Plasma

K. A. KHOR and R. MCPHERSON

ABSTRACT

Several ZrO$_2$-based multicomponent ceramic powders have been prepared through the oxidation of the respective metal halide vapour mixture in the tail flame of a radio frequency (rf) thermal plasma. The sub-micron powders were found to have structures that were consistent with co-condensation as the liquid phase to form ultrafine droplet solutions and then subsequent solidification to form a variety of metastable phases.

INTRODUCTION

Ultrafine and ultrapure ceramic oxide powders can be synthesized in a radio-frequency (RF) plasma through the oxidation of the suitable metal halide vapours in the tail flame of the plasma. Pure Al$_2$O$_3$ powders and several binary oxide powders based on Al$_2$O$_3$ have been prepared using this technique [1,2,3,4]. Plasma prepared Al$_2$O$_3$ - TiO$_2$ powders consist of a solid solution of TiO$_2$ in δ-Al$_2$O$_3$ up to ~9 wt.% TiO$_2$ but at higher concentrations each particle consists of a fine dispersion of rutile in δ-Al$_2$O$_3$ [2]. The presence of SiO$_2$ in plasma prepared Al$_2$O$_3$-SiO$_2$ particles tends to stabilize the γ-Al$_2$O$_3$ form up to ~1500°C [3], and the presence of Cr$_2$O$_3$, with a maximum solid solubility of 18wt.%, stabilized the θ-Al$_2$O$_3$ form [4]. Formation of these oxides occurs through co-condensation of two or more oxides to form homogeneous liquid solution droplets. The structure of the powder particles is then determined by the subsequent crystallization details as the droplets cool. Extended solid solutions, metastable phases and extremely fine scale two phase dispersions may be formed in such systems [5].

Zirconia (ZrO$_2$) -based ceramics have attracted considerable attention due to their enhanced mechanical properties, in particular fracture toughness. The toughening and concomitant strengthening are due to microcracking and the volume and shape changes associated with the stress-induced tetragonal to monoclinic (t-->m) transformation, which occurs in the stress field of propagating cracks [6].

K.A.Khor, Nanyang Technological University, School of MPE, Singapore 2263, Singapore.
R.McPherson, CSIRO, Div. Materials Science and Technology, Locked Bag 33, Clayton 3168, Australia.

215

A dispersion of metastable tetragonal ZrO_2 particles, of a critical diameter, thus causes an increase in fracture toughness when they transform to the equilibrium monoclinic structure under stress. Composite powders containing ZrO_2 are therefore of interest for the manufacture of a variety of advanced ceramics.

The control of microstructure in ZrO_2-based ceramics is critical for optimisation of the mechanical properties. In particular, the fracture toughness and fracture behaviour of Al_2O_3 - ZrO_2 are influenced strongly by the size, size distribution, and location of ZrO_2 particles. Small additions (~0.5 mol%) of TiO_2 are found to affect the sintering behaviour of Al_2O_3 - ZrO_2, suggesting the use of TiO_2 as a sintering aid [7].

Chromia (Cr_2O_3) is used as a grain growth inhibitor in Al_2O_3 ceramics, and also forms solid solution with Al_2O_3 which improves the mechanical properties of alumina ceramics [8]. Addition of ZrO_2 to Al_2O_3 - Cr_2O_3 solid solution will increase thermal stress resistance and decrease thermal conductivity since ZrO_2 has a lower thermal conductivity, thus making it a potential material for thermal insulators in which heat loss must be kept to a minimum [9].

Amorphous materials in the Al_2O_3 - ZrO_2 - SiO_2 system have interesting applications for preparation of alkali resistant glasses, chemically stable refractories, glass ceramics and zirconia toughened ceramics [10]. Due to the high melting temperature, it is difficult to prepare zirconia-aluminosilicate glass by conventional melting techniques. Therefore, novel processes such as rapid quenching from the melt and chemical polymerization of metal alkoxides have been employed [11,12,13]. The present paper reports the preparation of ultrafine multicomponent ZrO_2 composite powders in the RF plasma. Characterisation of the powders is carried out using X-ray diffraction and transmission electron microscopy (TEM).

EXPERIMENTAL TECHNIQUES

A radio frequency plasma torch, operated from a 15 kW, 5MHz high frequency generator, was used with the power output controlled via a saturable reactor in the d.c. high tension supply. The plasma gases (argon and oxygen) flowed through a tangential inlet into the confining silica tube (43 mm). The usual operating conditions were : 20 l/min argon, 20 l/min O_2 and 10 l/min cooling O_2. Reaction between the halides and plasma gases took place in a Pyrex tube just below the plasma torch. The powder products were collected by passing the gas stream through an electrostatic precipitator operating at 15 - 20 kV.

The reactants used were AR $ZrCl_4$,$AlCl_3$, $TiCl_4$, CrO_2Cl_2 and $SiCl_4$. Because of difficulties associated with vaporization of the chlorides and maintainence of constant proportion, a system was used in which the required mixture of halide powders and vapours were fed, by argon carrier gas, to a vaporization unit heated to a temperature well above the sublimation temperature of the reactants to achieve flash evaporation. The mixed vapours were then fed into the tail flame of the plasma torch (Figure 1).

The powders produced were examined by X-ray diffraction (XRD) as-prepared and after various heat treatments. Transmission electron microscopy (TEM) was carried out using a Philips EM420 microscope equipped with an EDAX analysis system (model PV 9900) so that individual particles could be analyzed. Quantitative analysis was achieved using 'PV SUPQ' software from the company EDAX, which provides corrections to give high accuracy without the use of standards [14].

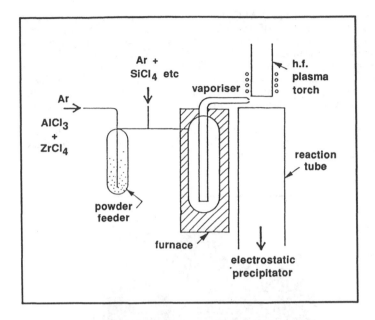

Figure 1 Experimental set up for preparation of powders.

RESULTS & DISCUSSION

ZrO₂-Al₂O₃-TiO₂

The XRD pattern of the as-prepared powders shows the presence of t-ZrO_2 and δ-Al_2O_3. On heat treatment to temperatures above 1100°C, t --> m transformation in ZrO_2 and δ --> α transformation in Al_2O_3 were observed. Aluminium titanate peaks were detected in powders which contained >3.7 wt.% TiO_2 and heat treated above 1300°C.

The particles examined by TEM are found to be spherical with a size range of 0.01 to 0.3 μm. Eutectic-like structures similar to that observed in plasma prepared Al_2O_3 - ZrO_2 powders are found (Figure 2). Electron diffraction suggests that they were two-phase particles. EDAX carried out on individual particles show a concentration of 1 to 8 wt.% TiO_2 believed to be in solid solution. At higher TiO_2 concentrations, the structure changes to reveal TiO_2-rich phases.

ZrO₂-Al₂O₃-Cr₂O₃

The powder collected after plasma reaction is dull green in colour. As the chromium oxide concentration increased, the colour changed to bright green. Powders with less than 15.3 wt.% Cr_2O_3 consist of δ,θ-Al_2O_3 mixture and t-ZrO_2. Above this concentration, α-Cr_2O_3 is formed in the powder. The concentration of ZrO_2 varies from 7.7 to 10.2 wt%. XRD shows that powders heat treated to 1000°C and 1100°C for 1 hour consist of θ-Al_2O_3 and t-ZrO_2. Heat treatment at 1200°C to 1600°C led to the formation of the $(Al,Cr)_2O_3$ solid solution. The t-->m transformation in ZrO_2 occurred after cooling from above 1400°C. The transition temperature to the α phase in Al_2O_3 was found to be ~1250°C. This is slightly lower than the corresponding values in Al_2O_3 - ZrO_2 of the same compositional range (1300° - 1350°C). The lowering of the transition temperature is most likely due to the Cr^{3+} ions occupying the octahedral sites in the Al_2O_3 lattice which will alter the activation energy barrier to the transition and also increase the ordering of the cation vacancies.

TEM observation (Figure 3) shows that the majority of particles contained structures similar to plasma synthesized Al_2O_3 - (10 - 15 wt.%) ZrO_2. At higher Cr_2O_3 concentrations, the particles become more facetted.

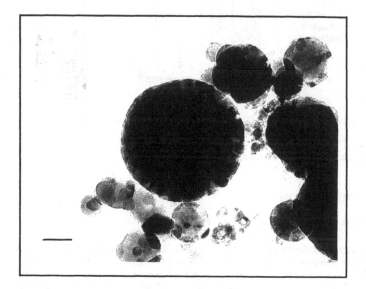

Figure 2   TEM observation of as-prepared Al_2O_3-ZrO_2-TiO_2 particles
(bar = 0.03 μm).

ZrO_2-Al_2O_3-SiO_2

Table I lists the composition and the phases detected by XRD in the as-prepared powders. After heat treatment at 1000°C for 1 hour, all samples consist of tetragonal ZrO_2 and an amorphous phase. The mullite phase ($3Al_2O_3.2SiO_2$) was found in the samples after heat treatment at 1100°C.

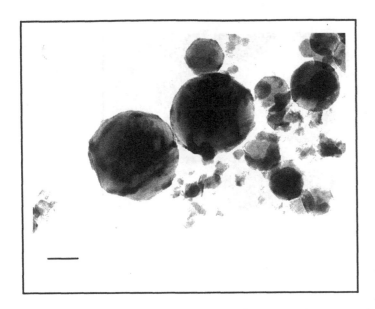

Fig. 3   TEM observation of as-prepared Al_2O_3-ZrO_2-Cr_2O_3 particles
(bar = 0.03 μm).

Table I Phases detected in the as-prepared and heat treated Al_2O_3-ZrO_2-SiO_2 powders

Samples	Wt% Al_2O_3	Wt% ZrO_2	Wt% SiO_2	As-prepared	1100°C
ZAS 1	9.0	7.0	84.0	Amorphous	t-ZrO_2 mullite
ZAS 2	17.3	10.0	73.0	Amorphous t-ZrO_2	t-ZrO_2 mullite
ZAS 3	20.0	15.0	65.0	Amorphous t-ZrO_2	t-ZrO_2 mullite

Electron microscopy and EDAX of individual particles revealed several kinds of structures in the plasma prepared Al_2O_3-ZrO_2-SiO_2 powders. Three groups of amorphous particles have been identified. Type I amorphous particles have no apparent structure. The composition range of these particles is 3 - 10 wt.% Al_2O_3, 82 - 88 wt.% SiO_2 and 0.5 - 5 wt.% ZrO_2. Type II amorphous particles have some apparent sub-structure. Quantitative analysis of these particles showed that they contained 21 - 34 wt.% Al_2O_3, 53 - 68 wt.% SiO_2 and 5 - 12 wt.% ZrO_2. Type III particles contain larger substructures, appearing as dispersed 'spots' or as a network of droplets within the particles (Figure 4). The composition range for Type III particles is 20 - 45 wt.% Al_2O_3, 25 - 45 wt.% SiO_2 and 16 - 45 wt.% ZrO_2.

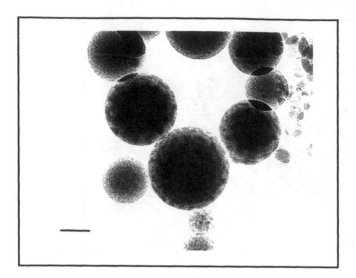

Figure 4   TEM observation of as-prepared Al_2O_3-ZrO_2-SiO_2 particles
(bar = 0.03 μm).

Previous studies on the formation of mixed oxide particles in RF plasma have shown that the oxides are expected to form via condensation in the plasma as fine liquid droplet solutions above the respective homogeneous nucleation temperatures of the crystalline oxides from liquid. Solidification of these liquid solution droplets will then be effectively complete at ~0.8 T_l, where T_l is the equilibrium liquidus temperature. Formation of liquid droplets in the plasma flame will take place when the partial pressure of the reactants in the flame is higher than the partial pressure of the oxide species in equilibrium with the liquid at the particle freezing point.

Thermodynamic data of various Me-Cl-O-Ar systems, where Me=Zr, Al, Ti and Si, have indicated that the oxides will condense fully and form liquid solutions at several hundred degrees above the homogeneous nucleation temperature of crystalline oxide from liquid. Subsequent growth of the droplets will occur through surface reactions and coalescence of the liquid droplets. The microstructure of the plasma sprayed Al_2O_3-ZrO_2-SiO_2 particles showing various degrees of phase separation is consistent with the apparent influence of a metastable miscibility gap which extends across the Al_2O_3-ZrO_2-SiO_2 system. In the case of Type II amorphous particles, the oxides are likely to have condensed fully as a homogeneous liquid oxide solution above the miscibility gap. On the other hand, the microstructure of the particles near the ZrO_2 - SiO_2 side of the ternary system show that the oxides most likely fully condensed within the miscibility gap to form ZrO_2-rich and SiO_2-rich liquids. The ZrO_2-rich liquid subsequently crystallised on further cooling to form a ZrO_2-rich phase within the particle while the SiO_2-rich liquid yields a SiO_2-rich phase with a fine dispersion of ZrO_2 reminiscence of the plasma prepared ZrO_2 - SiO_2 particles in an earlier study [15].

CONCLUSIONS

Ultrafine multicomponent ZrO_2-alloy powders from the systems ZrO_2-Al_2O_3-TiO_2, ZrO_2-Al_2O_3-SiO_2 and ZrO_2-Al_2O_3-Cr_2O_3 have been prepared in a high frequency plasma torch through the oxidation of the respective metal halide vapour mixture in the tail flame of the plasma.

The structures observed are consistent with co-condensation as the liquid phase to form ultrafine droplet solutions and their subsequent solidification to form a variety of metastable phases. TEM of the ZrO_2-Al_2O_3-TiO_2 and ZrO_2-Al_2O_3-Cr_2O_3 powders revealed particles with a two phase microstructure, very similar in appearance to those observed in plasma prepared ZrO_2-Al_2O_3 powders. X-ray diffraction however revealed the presence of θ-Al_2O_3 in the Cr_2O_3 containing powders. Some of the ZrO_2-Al_2O_3-SiO_2 powders were amorphous to X-ray diffraction and, on heat treatment (DTA) crystallised at $1004°$-$1014°C$ to mullite and tetragonal ZrO_2. TEM of the amorphous powders showed that the particles ($\sim 0.1\mu m$) had an extremely fine substructure consistent with spinodal decomposition, which coarsened on heat treatment at temperatures greater than $1200°C$. Several unusual microstructures reminiscent of the plasma prepared ZrO_2-SiO_2 were observed in the particles near the ZrO_2-SiO_2 side of the ternary system. It is likely that, for these particles the oxide solution droplets condensed fully within the miscibility gap to form ZrO_2-rich and SiO_2-rich liquids.

REFERENCES

1. McPherson,R., 1973. "Formation of Metastable Phases in Flame and Plasma Prepared Alumina." J.Mater.Sci., 8: 851-858.

2. Gani,M.S.J. and R.McPherson, 1980. "The Structure of Plasma Prepared Al_2O_3-TiO_2 Powders." J.Mater.Sci., 15: 1915-1925.

3. Gani,M.S.J. and R.McPherson, 1977. "Glass Formation and Phase Transformation in Plasma Prepared Al_2O_3-SiO_2 Powders." J.Mater.Sci., 12: 999-1009.

4. McPherson,R., 1973. "The Structure of Al_2O_3-Cr_2O_3 Powders Condensed from a Plasma." J.Mater.Sci., 8: 859-862.

5. McPherson,R., 1975. "Co-condensation of Refractory Oxides from a RF Plasma." Proc. Int. Round Table on Study and Applications of Transport Phenomena in Thermal Plasmas. Odello, France, IUPAC/CNRS IV-8, edited by Bonet,C.

6. Claussen,N., 1984. "Microstructural Design of Zirconia - Toughened Ceramics (ZTC)." Adv. Ceram., 12 [Sci. Technol. Zirconia 2]: 325 - 351.

7. Osendi,M.I. and J.S.Moya, 1988. "Role of Titania on the Sintering, Microstructure and Fracture Toughness of Al_2O_3/ZrO_2 Composites." J.Mater.Sci. Lett., 7: 15 - 18.

8. Arahori,T. and E.D.Whitney, 1988."Microstructure and Mechanical Properties of Al_2O_3 - ZrO_2 - Cr_2O_3 Composites." J.Mater.Sci., 23: 1605 - 1609.

9. Hasselmann,D.P.H., R.Syed and T.Tien, 1985. "The Thermal Diffusivity and Conductivity of Transformation Toughened Solid Solution of Alumina and Chromia." J.Mater.Sci., 20: 2549 - 2556.

10.   Claussen,N. and J.Jahn, 1978. "Mechanical Properties of Sintered, in Situ-Reacted Mullite - Zirconia Composites." J.Am.Ceram.Soc., 63[3-4]: 228 - 229.

11.   Thorne, D.J., 1969."Glass Refractory Oxide Systems Based on Alumina." J.Brit.Ceram.Soc., 14: 131 - 145.

12.   Yoshimura,M., M.Kaneko and S.Somiya, 1985. "Preparation of Amorphous Materials by Rapid Quenching of Melts in the System Al_2O_3-ZrO_2-SiO_2." J.Mater.Sci.Lett., 4: 1082 - 1084.

13.   Makishima,A., H.Oohashi, M.Wakakuwa, K.Kotani and T.Shoimohira, 1980. "Alkaline Durabilities and Structures of Amorphous Aluminisilicates Containing ZrO_2 Prepared by the Chemical Polymerization of Metal Alkoxides." J.Non-Cryst. Solids 42: 545 - 552.

14.   Sandborg,A.O., 1985."The Edax PV9900: A New Energy Dispersive Microanalysis System." The EDAX Edit 15[3]: 1 - 19.

15.   Khor, K.A., 1988."Plasma Synthesis of Zirconia Alloy Powders." Ph.D. Thesis, Monash University, Australia.

Microstructural Evolution in Spherical Zirconia-Yttria Powders

WALDEMAR PYDA, MARY S. J. GANI and LAP HING CHEN

ABSTRACT

Spray drying of coprecipitated gels was used in the preparation of spherical micropowders of yttria fully stabilized zirconia. Changes in microstructure during spray drying and heat treatment were monitored using scanning electron microscopy and mercury porosimetry. The effect of removing water from the zirconia gel by organic liquid was studied. It has been found that the microstructure and strength of the spherical grains is strongly related to the preparation method of the starting material.

INTRODUCTION

The morphology of ceramic powders strongly influences the microstructure and the properties of the products processed with the powders. Control of the surface area, particle size and particle size distribution, porosity and pore size distribution of the particles and/or particle agglomerates is important in numerous powder applications. They include, powders used for the production of structural ceramics, materials for ultrafiltration, catalysis and chromatography [1-4].

The purpose of this paper is to examine the evolution of the microstructure of porous spherical zirconia powders that are produced by coprecipitation using chlorides as starting materials. The effect of treatment of coprecipitated hydroxides with ethyl alcohol is also investigated.

EXPERIMENTAL

Two zirconia powders containing 8 mol% yttria were prepared as follows:

Salt solutions were made by dissolving zirconium tetrachloride (BDH Chemicals Ltd Poole England - assay 98%) in distilled water and yttrium oxide (99.9% pure, CERAC, Milwaukee, Wisconsin) in HCl. A common solution of the chlorides was added to an intensively stirred ammonia solution. The final pH was kept at 9. The coprecipitated gel was then washed with distilled water until free of Cl⁻ ions. The gel was separated into two samples: W and E. Sample W was dried at 120°C. Sample E was washed additionally with ethyl alcohol and dried at 70°C. The concentration of

Waldemar Pyda, Lap Hing Chen, Centre for Bioprocess Technology, Monash University, Clayton 3168, Vic. Australia

Mary S. J. Gani, Department of Materials Engineering, Monash University, Clayton 3168, Vic. Australia

ethanol in the last filtrate was 86 weight %.

Both W and E were then attrition milled in water using 0.8 mm diameter zirconia balls. Sample W was milled for 3 h while sample E was milled for 1.5 h.

Spray drying was performed using a laboratory scale spray drier (Niro Atomizer - Denmark) with inlet and outlet temperatures of 280±5°C and 100±3°C, respectively. The concentration of the gel particles in the slurries was 17.0±0.2 weight %.

The powders were heat treated at temperatures ranging from 400 to 1400°C in air in a furnace with Super Kanthal heating elements. A heating rate of 50°C/min, a soaking time of 2 h and a cooling rate of 50°C/min was used.

The morphology of the attrition milled gels was observed using transmission electron microscopy (TEM). Scanning electron microscopy (SEM) was used to characterize the morphology of the spray dried and heat treated powders. These methods were also used to determine particle size distributions. X-ray diffractometry was used to identify crystalline phases. The BET technique was used to measure the specific surface area of the powders using a Micromeritics Gemini 2360. The porosity of the powders was determined using mercury porosimetry using a Micromeritic Poresizer 9310. The compressive strength of the particles was found from breakpoints in the density vs logarithm of compaction pressure curves [2, 5]. Uniaxial pressing of the powders was performed in a 10 mm diameter steel mould using the Instron 4505 testing machine. The loading rate was 1mm/min.

RESULTS AND DISCUSSION

CHARACTERISTICS OF THE POWDERS BEFORE HEAT TREATMENT

Transmission electron micrographs of the zirconia gels after attrition milling are shown in figure 1. Electron diffraction patterns (also shown in figure 1) provide evidence that the gels are amorphous. Elementary particles of the gels have size of 5-7nm. These form irregular shaped agglomerates. Additional washing with alcohol results in a more porous microstructure of the agglomerates (Figure 1b). This is in agreement with the studies of Haberko [2, 3], who found that the agglomerates of 6.5 mol% Y_2O_3-ZrO_2 powder derived from the ethanol washed gel have larger porosity (59%) then the agglomerates derived from the water washed one (36%) after calcining at 600°C for 30 min. It was shown [2, 3, 6, 7] that the removal of free water from the accessible surface of the precipitated gels by washing them with suitable organic solvent, prior to drying, reduces the capillary forces which arise in

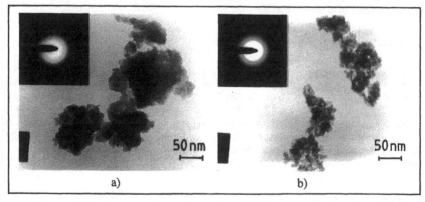

Figure 1. Transmission electron micrographs of 8 mol% Y_2O_3-ZrO_2 gels: a - water washed gel; b - ethanol washed gel.

TABLE I - CHARACTERISTICS OF ATTRITION MILLED GELS AND SPRAY DRIED POWDERS

Property	Sample W	Sample E
Attrition milled gels		
Agglomerate size d_{50}, μm	0.40	0.48
Specific surface area, m^2/g	99.5±3.6	204.6±5.6
d_{BET}, nm	16.3	7.9
Spray dried powders		
Granule size d_{50}, μm	17.8	13.1
Specific surface area, m^2/g	93.5±2.9	207.9±6.9
d_{BET}, nm	17.3	7.8
Compressive strength, MPa	0.153	0.327
Total porosity, vol.%	66.0±1.4	68.2±1.3

± denotes the confidence interval at confidence level of 0.95 in the entire work

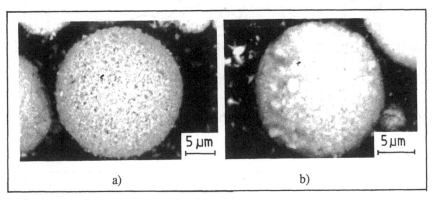

Figure 2. Back scattered SEM micrographs of the powders after spray drying; a - water washed gel; b - ethanol washed gel.

the drying step.

Data from Table I indicate that both gels had a similar degree of fineness before spray drying. The difference in order of magnitude between agglomerate size, d_{50}, and the particle size d_{BET} suggests that the size of the elementary molecules and their area control the high surface area of the gels.

SEM observations (Figure 2) reveal that both powders have a similar morphology and no changes of the surface area and the d_{BET} values of the gels after spray drying are observed (Table I). As prepared spray dried powders show similar porosity and differ only slightly in the granule size.

CHARACTERISTICS OF HEAT TREATED POWDERS

X-ray diffraction analysis of the powders heated to temperatures greater than 500°C revealed only the presence of zirconia solid solution of cubic symmetry. Samples heated at lower temperatures were partially amorphous. Traces of the cubic phase first appeared in the sample heated at 350°C for 2h. A DTA measurement showed the maximum of an exothermic effect at 444°C (a heating rate of 10°C/min).

Heat treatment resulted in a reduction of granule size. The powders prepared from the water washed gel showed a greater shrinkage at temperatures up to 600°C whilst powders prepared by the ethanol route showed the greater rate of shrinkage at temperatures higher than 900°C and reached similar linear shrinkage at 1400°C (36.0

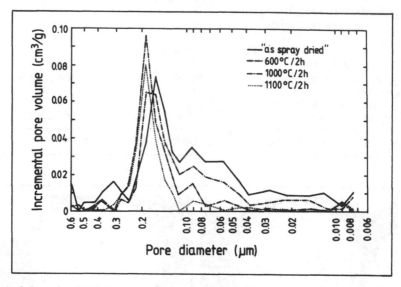

Figure 3. Pore size distribution curves for powders prepared from water washed gel. Conditions of heat treatment indicated.

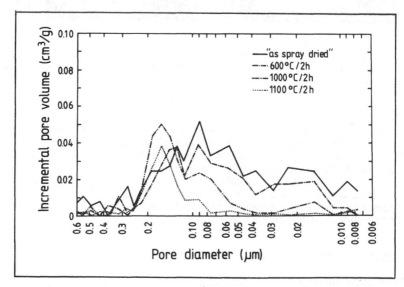

Figure 4. Pore size distribution curves for powders prepared from ethanol washed gel. Conditions of heat treatment indicated.

and 35.7 % for samples W and E, respectively).

A dramatic decrease of the specific surface area of the powders was observed at the temperature range of 375-500°C when the transition of the amorphous gel to the crystalline state occured. It concerns especially the powders prepared from the ethanol washed gel. The value of 207.9±6.9 m^2/g measured for the as spray dried powder fell to 38.5±0.5 m^2/g for the powder heated at 500°C. The values for sample W were 93.5±2.9 and 23.2±0.4.m^2/g, respectively. Nearly constant values (~30 m^2/g - sample E and ~18 m^2/g - sample W) of the specific surface area after heating at temperatures from 600-800°C were observed. A further decrease of the specific surface area occured above 800°C. The powders prepared from the ethanol washed gel had higher surface area then the powders prepared from the water washed gel at temperatures up to 1100°C. This is in agreement with the higher agglomerate porosity of the ethanol washed gel (Figure 1). At higher temperatures they showed a more rapid decrease of their surface area which accompanied the larger shrinkage. The specific surface area of the powders heated at 1200°C was 1.8±0.1 and 2.2±0.1 m^2/g for sample E and W, respectively.

The pore size distributions shown in figures 3 and 4 illustrate the microstructural evolution in the powders during heat treatment. The population of inter-agglomerate pores in the granules of the powders prepared from the water washed gel is shown in figure 3. The "as spray-dried" granules exhibit a maximum population with a pore diameter of about 0.15 μm. (The population of the smaller intra-agglomerate pores is outside the size range shown on the graph. This is indicated by the observation that prior to grinding the maximum in the intra-agglomerate pores was 9 nm.) Heat treatment results in the shrinkage of the smaller inter-agglomerate pores. Increasing the temperature of the heat treatment increases the amount of small pore shrinkage. However the maximum of the pore population is shifted to larger pore size with increasing temperature. The probable reason is an existence of inhomogenities of the pores distribution within the granule. This is a result of inhomogeneous packing of the different in size agglomerates. The driving force of the sintering process is greater within the domains where the pores are smaller in size compared to the domains where the pores are larger. Thus, the former domains shrink with the

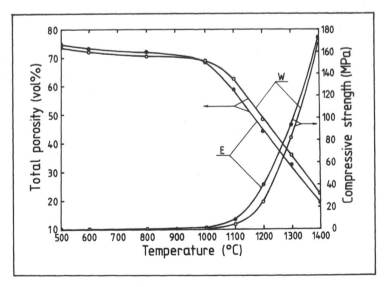

Figure 5. Total porosity of the powders and compressive strength of the granules vs temperature of heat treatment; W - starting gel washed with water; E- starting gel washed with ethanol.

greater rate, and the void space between poorly bonded domains is enlarged. Figures 3 and 4 show that the process of pore enlargement takes place by a redistribution of pore sizes; i.e., pores within tensioning domains redistribute to enlarge inter-domain pores. A similar process was reported by Lange at all [8] in compacts of 6.6 mol% Y_2O_3-ZrO_2 sintered at temperatures from 600-1200°C.

The "as spray-dried" powders prepared from the ethanol washed gel show a different pore size distribution (Figure 4). This distribution is broad and extended towards the smallest pores measured in the system. A maximum corresponds to the pores of about 0.09 μm diameter. The pore size distribution measurements for the ethanol washed gel made before grinding revealed an existence of a broad population of the intra-agglomerate pores which had the maximum at 0.05 μm. This suggests the overlapping of the inter- and intra-agglomerate porosities occurs as a result of the high porosity of the ethanol washed agglomerates. The amount of the smallest pores in the granules prepared from the ethanol washed gel decreases with an increase of the heat treatment temperature similarly to the sample W. Pore enlargement is also observed.

The process of pore enlargement ceases at temperatures between 1000 and 1100°C for both powders, as is shown in figures 3 and 4. The volume of these pores is larger for the powders prepared from the water washed gel.

Figure 5 shows the changes of the total porosity of the studied powders vs temperature. At temperatures below 800°C both powders have similar porosity. Above this temperature, the powders prepared from the ethanol washed gel have smaller porosity. This is a result of their better sinterability which originates from the smaller pore sizes in the starting granules and the overlapping of the inter- and intra-agglomerate porosities.

Figure 5 also illustrates the compressive strength of the powders as a function of temperature. The powders derived from the ethanol washed gel have higher compressive strength than the powders prepared from the water washed gel within the whole range of applied temperatures. This is undoubtedly connected to the lower total porosities presented by these powders and their more uniform microstructure.

CONCLUSIONS

It has been shown that the removal of free water from the coprecipitated zirconia gel by washing with ethanol prior to drying, results in the formation of powders composed of agglomerates with increased porosity. Spherical granules of these powders formed by spray drying have a broad pore size distribution in which the overlapping of the inter- and intra-agglomerate porosities occurs. The granules formed from the agglomerates of the water washed gel show a pore size distribution in which a sharp maximum of the inter-agglomerate porosity can be discerned. The intra-agglomerate pores have sizes out of the range of mercury porosimetry. Enlargement of the inter-agglomerate pores is observed during heat treatment of the powders. The powders prepared from the ethanol washed gel, because of smaller pore sizes and a much smaller difference between inter- and intra-agglomerate pore sizes , show better sinterability. At comparable temperatures they reach smaller total porosities and the higher compressive strength.

ACKNOWLEDGEMENTS

Support was provided under the Generic Technology component of the Industry Research and Development Act 1986. The authors thank A.Hartshorn and W.Braun from ICI Advanced Ceramics for assistance in spray drying and B.C.Muddle from Department of Materials Engineering, Monash University, for assistance in TEM analysis.

REFERENCES

1. Lange, F.F., 1989. "Powder Processing Science and Technology for Increased Reliability." Journal of the American Ceramic Society, 72(1): 3-15.

2. Haberko, K., 1979. "Characteristics and Sintering Behaviour of Zirconia Ultrafine Powders." Ceramics International, 5: 148-54.

3. Haberko, K., 1983. "Scientific Reports of AGH" No. 931, Ceramics z. 47, University of Mining and Metallurgy, Cracow.

4. Ramsay, J.D.F. and R.G.Avery, 1986. "Oxides with Controlled Surface and Porous Properties from Sol-Gel Techniques." Novel Ceramic Fabrication Processes, Brit.Cer.Proc. No.38, Institute of Ceramics, Shelton House.

5. Gasiorek, S., 1980. "Shrinkage as a Result of the Density and Mass Changes During Sintering Process," in Science of Ceramics, Vol.10, H.Hausner, ed., Berlin, Germany: Deutsche Keramische Geselschaft, pp.311-19.

6. Hoch, M. and K.M.Nair, 1976. "Densification Characteristics of Ultrafine Powders", Ceramurgia International, 2(2): 88-97.

7. Dole, S.L., R.W.Scheidecker, L.E.Sheiers, M.F.Berard and O.Hunter, Jr., 1978. "Technique for Preparing Highly-sinterable Powders." Materials Science and Engineering, 32: 277-281.

8. Lange, F.F. and B.I.Davies, 1984. "Sinterability of ZrO_2 and Al_2O_3 Powders; The Role of Pore Coordination Number Distribution", in Advances in Ceramics, Vol.12, Science and Technology of Zirconia II, N.Claussen, M.Rühle, and A.H.Heuer, eds., Columbus, Ohio, USA: the American Ceramic Society, Inc., pp.699-713.

The Structure of Rapidly Quenched Zircon Melts

R. McPHERSON and P. R. MILLER

ABSTRACT

Molten zircon, rapidly quenched by plasma spraying onto a cool substrate, solidifies to a metastable mixture of tetragonal zirconia and silica glass. Transmission electron microscopy reveals a fine dispersion of tetragonal zirconia, particle size <10 nm, in a glass matrix. The microstructure is consistent with spinodal decomposition of the supercooled liquid, within a wide metastable miscibility gap in the ZrO_2-SiO_2 system, and suppression of the equilibrium crystallisation of ZrO_2 from the melt or phase separation by nucleation and growth of droplets at large undercooling. Tetragonal ZrO_2 is retained because of a particle size effect related to a balance between the free energy difference between tetragonal and monoclinic ZrO_2, the ZrO_2-glass interfacial energy and strain energy arising from elastic constraint of thermal expansion mismatch. Heat treatment between 900 and 1100 $^{\circ}C$ results in coarsening of the tetragonal ZrO_2. Above 1100 $^{\circ}C$ the ZrO_2 coarsens to a particle size exceeding the critical dimensions for retention of the tetragonal phase on cooling, ~20 nm, and monoclinic ZrO_2 is observed. Reaction between monoclinic ZrO_2 and SiO_2 to form zircon takes place slowly between 1200 and 1500 $^{\circ}C$ but is catalysed in the presence of a liquid phase.

INTRODUCTION

Zircon decomposes into silica and zirconia on heating to temperatures above approximately 1700°C and complete melting takes place at ~2500°C. Slow cooling of the melt results in a mixture of monoclinic zirconia and silica glass [1] but rapid quenching, by plasma spraying onto a cold substrate, gives a mixture of glass, tetragonal zirconia and monoclinic zirconia [2] or glass and tetragonal zirconia only [3]. A study of plasma dissociated zircon (PDZ), prepared by passing zircon particles through plasma devices in which they melted to form spherical droplets and subsequently solidified, showed that a variety of microstructures were formed which depended on particle size and hence cooling rate [3]. Spherulitic growth of zirconia from the melt dominated at larger particle sizes (~100 μm) but a complex droplet structure of monoclinic and tetragonal zirconia in silica glass was observed at smaller particle sizes (~10μm). A very fine dispersion of tetragonal zirconia in silica glass

R. McPherson and P. R Miller, CSIRO Division of Materials Science and Technology, Locked Bag 33, Clayton, Vic. 3168, Australia.

was observed in some fine particle size spheroidised samples and in thin coatings prepared by spraying onto a cold substrate; it was suggested that this microstructure may have arisen from spinodal decomposition of a metastable ZrO_2-SiO_2 solution liquid. The retention of the tetragonal zirconia form was assumed to be related to a particle size effect, that is, the tetragonal phase is stabilised below a critical diameter because surface and strain energy contributions compensate for the "chemical" free energy difference between the monoclinic and tetragonal forms. The strain energy arises from constraint of thermal expansion mismatch and transformational strain between the zirconia particles and matrix. A critical diameter of ~20 nm has been derived from X-ray diffraction line broadening measurements of plasma-sprayed films [2]. Studies of ZrO_2-SiO_2 heat treated gels have shown a similar effect with a critical size for retention of tetragonal ZrO_2 in a silica glass matrix of ~40 nm [4].

Although PDZ consists of an intimate mixture of silica and zirconia, their reassociation to form zircon occurs only slowly at temperatures around 1500°C, presumably because of nucleation effects and low diffusion rates [5]. There is evidence, however, that the rate of reassociation is increased by oxide additives such as Fe_2O_3 [6] and ZnO [7], an effect probably associated with formation of a liquid phase [8]. The growth of relatively large zircon crystals from a flux containing ZrO_2 and SiO_2 at temperatures of 900-1000°C is well known [9]. The formation of zircon ceramic pigments from ZrO_2-SiO_2 mixtures at temperatures of ~1000°C is also associated with flux additions [10], presumably because material transport is greatly enhanced through a liquid phase compared with solid state diffusion, although it has been suggested that vapour transport is also an important factor [11].

The present paper examines the microstructure of rapidly quenched molten zircon, prepared by plasma-spraying onto a cooled substrate, and structural changes which occur on heat treatment.

EXPERIMENTAL PROCEDURE

A piece of tube, approximately 5 mm thick, formed by spraying molten zircon particles onto a removable mandrel using a PAL 160 kw water stabilised plasma torch*, was used as the source of rapidly quenched molten zircon (PDZ). The microstructure of the deposit was examined by scanning electron microscopy of a transverse fracture surface. Chemical analysis showed that the deposit contained 0.45 wt % Fe, equivalent to ~0.6 wt % Fe_2O_3, which probably arose from contamination from the steel anode of the torch. This material was ground to pass a 53 μm screen and washed with dilute hydrochloric acid to give an Fe content <0.1 wt %.

Samples were heat treated in a platinum crucible at temperatures between 1000 and 1400 °C in air. PDZ samples ground with 20 wt% of a mixture of NaCl-27 wt. % NaF were also heat treated at 1000°C. The structures of the specimens before and after heat treatment were characterised by X-ray diffraction (XRD) and transmission electron microscopy (TEM). The proportion of zircon present after heat treatment was determined from the relative intensities of the (101) zircon and the sum of the intensities of the (111) tetragonal and the (111) and (111), monoclinic reflections using calibration mixtures prepared from zircon and commercial plasma-dissociated zircon. The approximate proportion of zirconia present as the tetragonal form was estimated from the ratio of the intensity of the (111) tetragonal reflection to the sum of the intensities of the {111} tetragonal and monoclinic reflections. TEM samples were prepared by grinding of the appropriate material in ethanol and placing

---

* The sample was supplied by Dr P. Chraska, Institute of Plasma Physics, Czechoslovak Academy of Sciences, Prague, Czechoslovakia.

a drop of the suspension onto a carbon film for examination using a Philips CM-30 scanning transmission electron microscope.

RESULTS

Scanning electron microscopy of a transverse fracture surface of the as sprayed deposit showed that it consisted of the lamellar structure usually observed in plasma-sprayed coatings, with a mean lamellar thickness of ~3µm (figure 1).

XRD showed that the as-prepared sample consisted of tetragonal zirconia with a trace of monoclinic zirconia. Silica in any form was not detected. Heat treatment of the as-prepared PDZ between 1000 and 1200°C resulted in the formation of monoclinic zirconia after cooling. Zircon was first observed after heating at 1200°C and the proportion increased at higher temperatures as shown in Table I. The proportion of zirconia present as the tetragonal form decreased on heat treatment between 1100 and 1300°C, but increased in samples heat treated at 1400°C. The extent of re-association to zircon was greater in the higher Fe_2O_3 material.

The sample mixed with NaCl-27 wt.% NaF and heat treated at 1000 °C for one hour consisted of monoclinic zirconia; after heating for 4 hours a mixture of zircon, monoclinic zirconia and tetragonal zirconia was observed.

TEM of the as-deposited PDZ showed a very fine distribution of tetragonal zirconia crystals and glass. Although the scale of the microstructure was variable, the predominant structure consisted of an interpenetrating distribution of glass and tetragonal zirconia on a scale of 7 - 10 nm as typified in figure 2. Imaging of lattice planes showed that relatively large single crystal regions (~100 nm) extended over the disseminated zirconia phase.

TABLE I - THE EFFECT OF HEAT TREATMENT FOR ONE HOUR ON THE PHASE CONSTITUTION OF PDZ.

Temperature (°C)	% Zircon	% t-ZrO_2	% m-ZrO_2	% SiO_2 glass	Fe content (%)
as-prepared	0	52	15	33	0.45
1200	1	16	50	33	<0.1
1300	5	5	59	31	"
1400	30	16	31	23	"
1300	12	7	52	21	0.45
1400	50	12	22	16	"

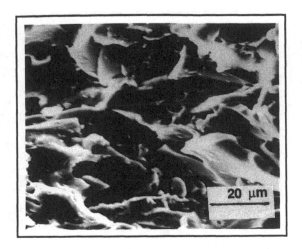

Figure 1. Scanning electron micrograph of transverse fracture surface of plasma-sprayed zircon deposit.

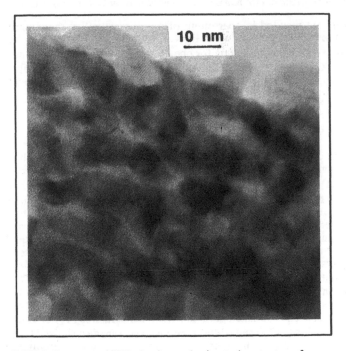

Figure 2. TEM of as-prepared PDZ showing predominant microstructure of tetragonal zirconia and glass.

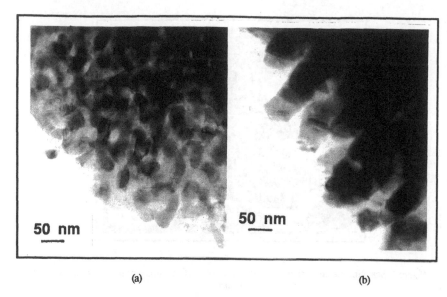

(a)   (b)

Figure 3.TEM of PDZ heat treated for one hour, (a) at 1200°C . (b) at 1300°C.

Heat treatment resulted in coarsening of the microstructure, apparently without change in morphology, as shown in figures 3 (a) and (b). The scale of the structure increased to ~25 nm after heating for one hour at 1200 °C and 50 - 80 nm after one hour at 1300 °C.Twins were observed in some of the zirconia crystals after heat treatment consistent with transformation of the tetragonal phase to monoclinic during cooling.

DISCUSSION

The ZrO_2-SiO_2 equilibrium phase diagram is required as a basis for the interpretation of the solidification of zircon melts. That generally accepted [12] shows a region of stable liquid immiscibility above 2200°C with a consolute point at 2430°C, 70 mole % SiO_2. Experimental data on the miscibility gap is limited to one study [13] and must be regarded as approximate because of uncertainty associated with high temperature measurements and problems with SiO_2 loss by vapourisation at temperatures in the vicinity of the melting point of zirconia. The liquidus of the system calculated from thermodynamic data, which included the reported consolute point, shows flattening over the range 30 - 70 mole % SiO_2 [14]. A wide metastable liquid miscibility gap, estimated to extend to 80 mole % ZrO_2 at 1080°C, would be expected on theoretical grounds [15]. The estimated miscibility gap is shown in figure 4, assuming parabolic shape, together with the spinodal based on the "root three rule" [16]. The estimated glass transition temperatures (T_g) of ZrO_2 and SiO_2 metastable liquids (~1700 and ~1400°C respectively) are also shown [17].

Three types of behaviour may be recognised during cooling of zircon composition melts on the basis of the phase diagram, as discussed previously [3]:

1. Zirconia crystals nucleated from the undercooled melt on relatively slow

cooling grow with spherulitic morphology because unstable interface conditions, arising from a high silica boundary layer, lead to non-crystallographic branching.

2. Direct nucleation of zirconia crystals is suppressed at faster cooling rates, the metastable liquid crosses the miscibility gap boundary, and phase separation into zirconia-rich and silica-rich metastable liquids occurs by nucleation and growth to give a droplet morphology (both the droplet phase and the matrix phase may undergo further phase separation during cooling). Zirconia then crystallises from the high zirconia liquid.

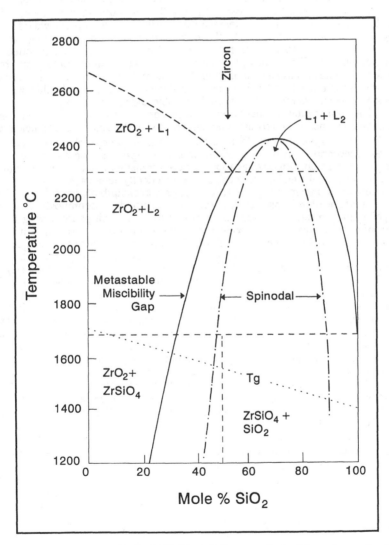

Figure 4. Estimated metastable liquid-liquid miscibility gap and spinodal superimposed on the zirconia - silica equilibrium phase diagram.

3. Droplet nucleation may be suppressed at very high cooling rates so that the the metastable homogeneous liquid is retained to the spinodal. The liquid becomes unstable within the spinodal and spontaneously separates on a very fine scale into zirconia and silica-rich regions by amplification of compositional fluctuations at a characteristic wavelength.

Plasma-spraying onto a substrate results in flattening of the impacting particles, to form discs 2-3 μm thick, and their rapid solidification. The cooling rate, estimated to be $\sim 10^6$ K sec^{-1}, is interface controlled by the heat transfer coefficient between the spreading droplet and underlying material which, in turn, is related to imperfect contact conditions between them [18]. The cooling rate of individual particles within a coating, built up by many successive particle impacts, would not be expected to be strongly influenced by the thermal conductivity of the coating material provided that the deposition surface was cooled, as is the usual practice, by impinging air jets. There would, however, be some variation of cooling rate from particle to particle because not all of them flatten completely on impact. The cooling rate during splat quenching has been shown to depend linearly on splat thickness for a given heat transfer coefficient between the particle and substrate [19].

The suppression of direct crystallisation of zirconia or liquid-liquid phase separation by a nucleation process is not unexpected in the undercooled silicate melt at cooling rates of the order of 10^6 K sec^{-1}. However, the rate of phase separation by spinodal decomposition when a liquid enters the spinodal region of the phase diagram is so high, at temperatures above the glass transition, that it is not possible to suppress it even at the highest quenching rates practically achievable. The maximum rate occurs at an undercooling of $\sim 10\%$ giving a wavelength of, typically, 5-10 nm [20]. A feature of microstructures formed by spinodal decomposition is two interpenetrating phases, with regularity of distribution, each phase having a high degree of connectivity [21]. These features are observed in the present samples of rapidly quenched zircon melts and, although microstructure is not an unequivocal indication of spinodal decomposition, the observations taken together with other evidence from the microstructure of PDZ [3] strongly suggest that the initial stage of solidification has involved spinodal decomposition into two metastable liquids. The much higher diffusion rate to be expected in the low silica regions of the metastable liquid then results in crystallisation of zirconia at large undercooling below the zirconia liquidus to give the observed microstructures of interpenetrating regions of tetragonal zirconia and silica glass. The observation that single crystal zirconia regions are an order of magnitude larger than the scale of the phase separation is consistent with nucleation of tetragonal zirconia throughout the continuous, high zirconia regions of a modulated composition metastable liquid.

Heat treatment of the as-quenched material at temperatures between 1100 and 1300°C results in coarsening of the microstructure without change in morphology, as observed in other spinodally decomposed structures. The tetragonal phase is gradually replaced by the monoclinic structure and the results are consistent with the critical diameter of ~ 20 nm reported by Krauth and Meyer [2]. Heat treatment at 1000 $^\circ$C in the presence of the NaCl-NaF mixture gave a considerable increase in the rate of formation of monoclinic zirconia, that is, an increase in the coarsening rate. The NaCl-NaF composition used corresponds to a eutectic at 680 $^\circ$C [22] suggesting that the increased rate is associated with enhanced diffusion via a liquid phase.

The gradual reassociation to zircon, over the temperature range 1200 to 1400°C, of the low iron PDZ sample, was similar to that observed previously [5]. The increased rate of zircon formation in samples containing ~ 0.6 wt % Fe_2O_3 agrees with the results of Boch and Kapelski [8] and is probably associated with the formation of a liquid phase. The considerable increase in the rate of formation of zircon in the presence of NaCl-NaF liquid is also clearly related to enhanced diffusion via a liquid phase. The increase in proportion of tetragonal phase observed after heat treatment at 1400°C may be explained by the formation of zircon, replacing silica glass as the matrix phase, and the larger critical diameter, estimated to be ~150 nm [23], required for retention of tetragonal zirconia.

CONCLUSIONS

Molten zircon, rapidly quenched by plasma-spraying onto a cooled substrate, has a microstructure consisting of an extremely fine, interpenetrating distribution of tetragonal zirconia and silica glass. The structure is consistent with the spinodal decomposition of an initially homogeneous liquid solution into zirconia-rich and silica-rich phases, within a wide metastable miscibility gap in the zirconia-silica system, followed by crystallisation of tetragonal-ZrO_2 from the zirconia-rich metastable liquid. The tetragonal form of zirconia is retained because of a particle size effect associated with surface energy stabilisation. The critical particle size is ~20nm. Heat treatment at 1000-1200 °C results in coarsening of the structure and transformation of tetragonal-ZrO_2 to monoclinic on cooling. Heat treatment at 1200-1500 °C causes gradual reassociation to zircon. The coarsening and reassociation processes are rapidly accelerated in the presence of a NaCl-NaF eutectic liquid, presumably because of enhanced diffusion via the liquid phase.

REFERENCES

1.   Barlett, H.B., 1931. "X-ray and Microscopic Studies of Silicate Melts Containing Zirconia." J. Amer. Ceram. Soc., 14: 837-843.

2.   Krauth, A. and H.Meyer, 1965. "On Modifications Produced by Chilling and on their Crystal Growth in Systems Containing Zirconia." Ber. Deutsch. Keram. Ges., 42: 61-72.

3.   McPherson, R. and B.V.Shafer, 1984. "Spherulites and Phase Separation in Plasma Dissociated Zircon." J. Mater. Sc., 19: 2696-2704.

4.   Nogami, M. and M.Tomozawa, 1986. "ZrO_2-transformation Toughened Glass-ceramics Prepared by the Sol-gel Process from Metal Alkoxides." J. Amer. Ceram. Soc., 69: 99-102.

5.   McPherson, R., R.Rao and B.V.Shafer, 1985. "The Reassociation of Plasma Dissociated Zircon." J. Mater .Sc., 20: 2597-2602.

6.   Williamson, J.P.H. and D.E.Lloyd, 1981. "The Characterisation of Ceramic Bodies Produced from Plasma Dissociated Zircon." J. Mater. Sc., 16: 1264-1272.

7.   Pepplinkhouse, H.J., 1979. "Synthesis of Zircon by Solid State Reaction." J. Aust. Ceram. Soc., 15: 24-27.

8.   Boch, P. and G.Kapelski, 1986. "Sintering of Zircon Powders." J. Physique, 47: C1-405-409.

9.   Ballman, A.A. and R.A.Laudise, 1965. "Crystallisation and Solubility of Zircon and Phenacite in Certain Molten Salts." J. Amer. Ceram. Soc., 48: 130-133.

10.   Fischer, R. and P.Kleinschmit, 1986. "Modern Pigments for Decoration." Keram. Zeit., 38: 365-367.

11.   Eppler, R.A., 1979. "Mechanism of Formation of Zircon Stains." J. Amer. Ceram. Soc., 53: 57-62.

12.   Butterman, W.C. and W.R.Foster, 1967. "Zircon Stability and the ZrO_2-SiO_2 Phase Diagram." Amer. Mineralogist, 52: 880-885.

13.   Toropov, N.A. and F.Ya.Galakhov, 1956. "Liquation in the System ZrO_2-SiO_2." Bull. Acad. Sci. U.S.S.R., Div. Chem. Sci., 5: 153-156.

14.   Doerner, P., L.J.Gauckler, H.Krieg, H.L.Lukas, G.Petzow and J.Weiss, 1979. "On the Calculation and Representation of Multicomponent Systems." CALPHAD, 3: 241-257.

15.   Galakhov, F.Ya. and G.G.Varshal, 1973. "Causes of Phase Separation in Simple Silicate Systems," in The Structure of Glass, Vol. 8, E.A.Porai-Koshits, ed., New York, NY, USA: Consultants Bureau, pp. 7-11.

16.   Cook, H.E. and J.E.Hilliard, 1965. "A Simple Method of Estimating the Chemical Spinodal." Trans. Met. Soc. AIME, 233: 142-146.

17.   JANAF Thermochemical Tables, 3rd Edition, 1985., J. Phys. Chem. Ref. Data, Supplement No 1, 14: 1693-1677.

18.   McPherson, R., 1981. "The Relationship Between the Mechanism of Formation, Microstructure and Properties of Plasma Sprayed Coatings." Thin Solid Films, 83: 297-310.

19.   Ruhl, R.C., 1967. "Cooling Rates in Splat Quenching." Mater. Sc. Engin., 1: 313-320.

20.   Cahn, J. W. and R.J.Charles, 1965. "The Initial Stages of Phase Separation in Glasses." Physics and Chem. of Glasses, 6: 181-191.

21.   Cahn, J.W., 1968. "Spinodal Decomposition." Trans. Met. Soc. AIME, 242: 166-180.

22.   Sangster, J. and A.D.Pelton, 1987. "Phase Diagrams and Thermodynamic Properties of the 70 Binary Alkali Halide Systems Having Common Ions." J. Phys. Chem. Ref. Data, 16: 509-561.

23.   McPherson, R., B.V.Shafer and Ai Ming Wong, 1982. "Zircon-zirconia Ceramics Prepared from Plasma Dissociated Zircon." J. Amer Ceram. Soc., 65: C57-58.

Preparation of MgO Dispersed with ZrO$_2$ by Hydrolysis of Zirconium Butoxide

TOSHIO SUGIYAMA, AKIKO ICHINOSE, OSAMU YAMAMOTO and YASURO IKUMA

ABSTRACT

A composite powder of ZrO$_2$ dispersed in MgO was formed by a sol-gel method and morphological changes of the ZrO$_2$ particles were studied as the composite powder was calcined. The composite was then sintered at 1450°C and examined by scanning and transmission electron microscopy. The distribution of ZrO$_2$ particles in the composite formed by the sol-gel method was much better than in the case of composites formed by ball milling or coprecipitation. A nanocomposite was formed by the sol-gel method.

INTRODUCTION

Ceramic-based composites have received much attention due to their high strength and their potential as high-temperature structural components [1-7]. Recently, it was found that when the composite contained a second phase of nanometer size, dispersed within the matrix (nanocomposite), it showed high fracture toughness and strength [8,9]. Preparation procedures for nanocomposites have been proposed using various techniques, such as ball milling, solid state reaction, direct synthesis, etc. [9-11]. One of the methods for obtaining nanometer size particles is the sol-gel method in which hydrolysis of metal alkoxides is employed. The technique was used to prepare composites of ZrO$_2$-Al$_2$O$_3$ [12] and mullite-cordierite [13]. The preparation method, however, is restricted to certain experimental conditions because the hydrolysis depends on several factors such as the concentration of alkoxide, kind of solvent, the amount of water, the reaction temperature, etc. In other words, it is essential to control the hydrolysis of alkoxide to get the desired composite.

In the present work, a nanocomposite of ZrO$_2$ dispersed in MgO [4-7] was prepared by controlled hydrolysis of zirconium tetra-butoxide. The distribution of ZrO$_2$ particles with nanometer size and the composite morphology in the powder form and in the sintered bulk are discussed.

Toshio Sugiyama, Akiko Ichinose, Osamu Yamamoto and Yasuro Ikuma
Department of Chemical Technology, Kanagawa Institute of Technology
1030 Shimo-ogino, Atsugi, Kanagawa 243-02, JAPAN

EXPERIMENTAL

The MgO powder (Rare Metallic Co., purity = 99.99%) with an average particle size of 70nm and zirconium tetra-butoxide {Kishida Chemical Co., 85.5% $Zr(OC_4H_9)_4$ and 15% C_4H_9OH} or $ZrOCl_2$ (Kanto Chemical Co., GR) were used as starting materials. Zirconium tetra-butoxide {$Zr(OBu)_4$} was dissolved in n-butanol (C_4H_9OH). The concentration of the $Zr(OBu)_4$ solute was 0.05 mol/l. The $Zr(OBu)_4$ alcoholic solutions were mixed separately with the MgO powder, to give a ZrO_2 content of 10 mol% in the MgO-ZrO_2 composite. Acetone was added at several ratios: acetone/$Zr(OBu)_4$ = 400, 166, and 58 (acetone-1, acetone-2, and acetone-3). In another case, diethylene glycol was added to stabilize $Zr(OBu)_4$ (DEG). The alkoxide solutions were placed in a closed container with a very small opening (about 0.5mm in diameter) to the atmospheric air and were stirred for about 50h. The composite powder was formed by hydrolysis of $Zr(OBu)_4$ with H_2O which was present in the acetone and/or in the air (at 50-80% relative humidity). The hydrolysis reaction was completed by increasing the opening size. After drying, the powders were calcined at 800-900°C for 1h. Figure 1 shows the process in detail. Some composite powders were tested by thermogravimetry-differential thermal analysis (TG-DTA) and examined by transmission electron microscopy (TEM, JEOL 2000EX).

In order to compare the microstructure of specimens prepared by different procedures, the composite powder was prepared by other methods. One was precipitation, in which NaOH solution was added to an aqueous solution containing $MgCl_2$ and $ZrOCl_2$ (coprecipitation). The precipitate was filtered and washed with water three times to remove residual sodium. Another composite powder was prepared by ball-milling the MgO and ZrO_2 (TZ-0, Tosoh, Tokyo) powders.

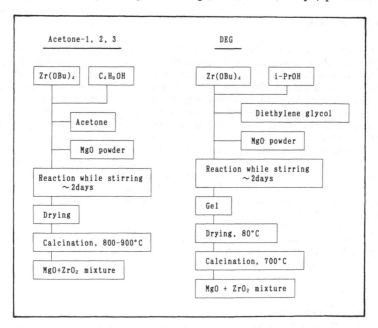

Figure 1. Flow charts of the sol-gel method.

The composite powders were uniaxially pressed into discs at 98 MPa, and then isostatically pressed at 147 MPa. The discs were sintered at 1450°C for 1 h in air. The density of the sintered bodies of MgO containing 10 mol% ZrO$_2$ were determined by an Archimedes method, and the microstructure was examined using scanning (JEOL JXA8600) and transmission electron microscopy.

RESULTS AND DISCUSSION

As a preliminary experiment, ZrO$_2$ powder was made from Zr(OBu)$_4$ without the presence of MgO. A TG-DTA curve of the ZrO$_2$ particles (acetone) is shown in figure 2(a). An endothermic peak was observed below 100°C, which was due to the evaporation of water and solvents adsorbed on the surface of the powder specimen. A strongly exothermic peak caused by the combustion of alkoxy group (C$_4$H$_9$O-) that remained unreacted was observed between 250 and 350°C. The powder was still amorphous at 400°C {Figure 2(b)}. The weak exothermic peak in the temperature range 430-475°C was due to the crystallization of amorphous ZrO$_2$ to the tetragonal phase, which was indicated by the X-ray diffraction pattern of a specimen fired at 600°C {Figure 2(b)}. Increasing temperature up to 1000°C resulted in the transformation of tetragonal phase to monoclinic phase. The former phase could have been either tetragonal or cubic, because (h00) peaks of the cubic phase seen in the range 20°<2theta<80° were not clearly separated into the (h00) and (00h) peaks of the tetragonal phase but were broadened due to the small particle size. Mitsuhashi et al.[14] reported that ZrO$_2$ particles prepared by adding NH$_3$ to a hot 0.2M solution of ZrOCl$_2$ gave cubic-type x-ray profiles at 410°C whereas electron diffraction patterns indicated that the particles were tetragonal with particle size larger than 100nm. In the present study, the ZrO$_2$ particles heated at 600°C were indexed as a tetragonal phase, but they could have been cubic. On the basis of the result shown in figure 2, the mixture of amorphous ZrO$_2$ and crystalline MgO can be fired at 800-900°C to burn off the residual alcohol to produce ZrO$_2$ from Zr(OBu)$_4$.

The TEM micrograph of the MgO powder (as received) is shown in figure 3(a). The surface of the powder is not as clear as for the calcined MgO-ZrO$_2$ composite {Figure 3(c)}. Mg(OH)$_2$ that formed on the surface of the MgO after reaction with H$_2$O in the atmosphere during storage may explain the condition of the surface. However Mg(OH)$_2$, if it existed, must be present in small quantities because X-ray diffraction of the powder indicated that it was MgO. The TEM micrograph of the composite powder (acetone-3) before calcination is shown in figure 3(b). ZrO$_2$ is not clearly seen in the micrograph, although small ZrO$_2$ particles (2-4nm) could be located on the surface of each MgO particle. Figure 3(c) shows the characteristic composite powder after calcination at 900°C for 1h. In this case the ZrO$_2$ particles have grown (dark particles with a diameter of about 10-20nm) and are readily seen on the MgO surface. From the TEM observations, it was found that the fine ZrO$_2$ gel particles with sizes in the range 2 to 4nm were coordinated around the basic MgO powder most probably due to the hydroxyl groups on the surface of the ZrO$_2$ gel particles.

The composite powders prepared by several methods were pressed into discs and sintered at 1450°C. Figure 4 shows the etched surfaces of the sintered composite obtained from the powder prepared by ball milling or coprecipitation from the inorganic Zr and Mg salts. The white grains are ZrO$_2$ while the dark grains are MgO. Some large agglomerates of ZrO$_2$ are seen in the ball milled composite. Coprecipitation from the inorganic Zr and Mg salts improved the microstructure. However, the ZrO$_2$ particles in the sintered body obtained by adding NaOH to ZrOCl$_2$ and MgCl$_2$ aqueous solution are aggregated on grain boundaries of the matrix, as shown in figure 4(b).

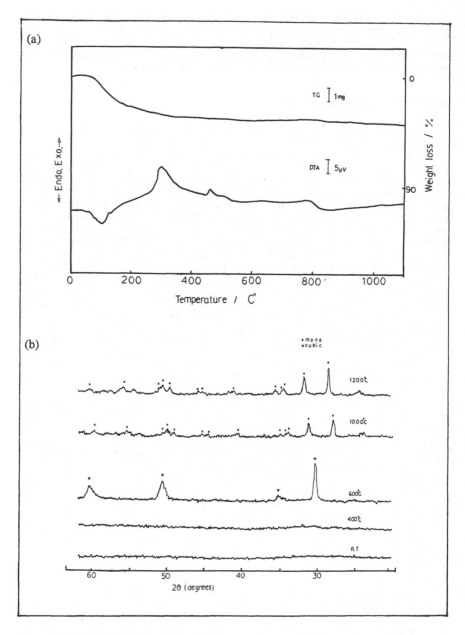

Figure 2. TG-DTA curve (a) and X-ray diffraction patterns (b) of ZrO_2 powder prepared by the sol-gel method.

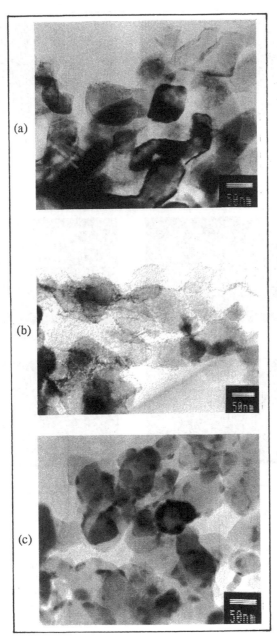

Figure 3. Transmission electron micrographs of powder specimens formed by the acetone-3 method:
(a) MgO as-received, (b) MgO-ZrO$_2$ composite powder before calcination, and
(c) MgO-ZrO$_2$ composite powder after calcination.

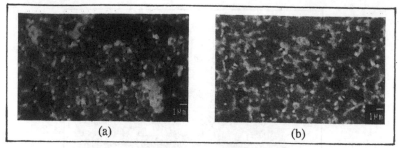

Figure 4. Scanning electron micrographs of sintered composites:
(a) ball milling and (b) coprecipitation.

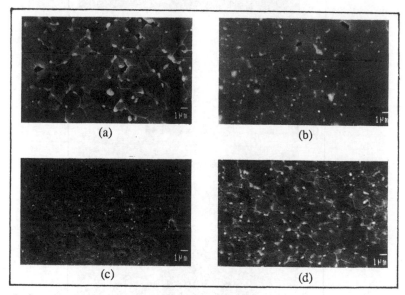

Figure 5. Scanning electron micrographs of sintered composites:
(a) acetone-1, (b) acetone-2, (c) acetone-3, and (d) DEG.

When the NaOH solution was added to the solution of $ZrOCl_2$ and $MgCl_2$ (low pH), zirconium hydroxide precipitated first (pH=2) and then magnesium hydroxide precipitated at high pH(=11). The difference in pH at which these precipitates form might explain the existence of small agglomerates in this composite. However, the coprecipitation technique improved the microstructure significantly compared with the powder prepared by ball milling.

The microstructure of the sintered composites formed by hydrolysis of $Zr(OBu)_4$ (acetone-1, -2, and -3 and DEG) is shown in figure 5. The ZrO_2 particles with diameter of 0.1-1μm were dispersed homogeneously in the matrix and the distribution of ZrO_2 particles is much better than in the case of the composites shown in figure 4. From comparison of the grain size of ZrO_2 after the calcination {10-20nm, figure 3(c)} with that of ZrO_2 after sintering {Figure 5(c), 0.1-0.2μm}, it is obvious that the ZrO_2 grains, formed by the acetone-3 method, grow during sintering. Monodispersed ZrO_2 particles within the MgO grains were found when the composite powder was prepared by hydrolysis of $Zr(OBu)_4$ dissolved in acetone (Figure 5). However, the particles show a heterogeneity due to the difference in the driving force for grain growth. The ZrO_2 particles within the MgO grains are generally small but the ZrO_2 particles on the MgO grain boundaries are large. This is because the grain growth of ZrO_2 takes place by material transport mainly through the MgO grain boundary. From these results, it was found that the ZrO_2 particles were widely spread within the matrix due to the use of the powder composed of nanometer size ZrO_2 coordinated around the MgO particles.

The rate of hydrolysis is proportional to the amount of water in the system during the reaction [15]. Since water is more soluble in acetone than butanol, the hydrolysis rate is faster when there is more acetone: the hydrolysis rate in acetone-1, -2, and -3 decreases in this order. Obviously the reaction of ZrO_2 formation from the inorganic Zr salt is the fastest. Consequently the faster the reaction, the greater the growth of the ZrO_2 particles in the sintered microstructure. In figure 5, we also notice that the slower the hydrolysis reaction, the better the distribution of ZrO_2 in the MgO matrix. This hypothesis is valid also in the microstructure of the DEG sample {Figure 5(d)}. In this case the hydrolysis reaction is very slow because diethylene glycol stabilizes $Zr(OBu)_4$ [16] and it results in uniform distribution of the ZrO_2. When the hydrolysis becomes slower we see a smaller fraction of ZrO_2 particles in figure 5. One might think that the concentration of ZrO_2 is not the same. This is not the case. When we examine the samples under the transmission electron microscope (Figure 6, where dark grains are ZrO_2 and white grains are MgO), very little difference in the concentration of ZrO_2 between these specimens is observed. In figure 5(b), (c) and (d), some of the ZrO_2 particles were too small to be seen.

In the course of the present work, it was found that the hydrolysis of $Zr(OBu)_4$ can be readily carried out with diethylene glycol which stabilizes $Zr(OBu)_4$ or in the presence of acetone which tends to dilute the $Zr(OBu)_4$ solution without enhancing the concentration of hydroxy groups (-OH) in the solution. In other words, the reaction rate becomes lower in acetone and DEG. Obviously, neutralization of $ZrOCl_2$ is extremely fast when compared to the hydrolysis of $Zr(OBu)_4$. Therefore, the MgO nanocomposite dispersed with nanometer size ZrO_2 was considered to be obtained by the hydrolysis reaction of $Zr(OBu)_4$ controlled at the slow rate.

CONCLUSIONS

Composites of ZrO_2 dispersed in MgO were formed by several methods. The distribution of ZrO_2 particles in the composite formed by the sol-gel method was much better than in the case of composites formed by ball milling or coprecipitation. It was found that a nanocomposite was formed if the composite was manufactured using the sol-gel method. In particular, when the hydrolysis reaction of $Zr(OBu)_4$ was very slow ZrO_2, of particle size 2-4nm, was distributed uniformly on the MgO particles.

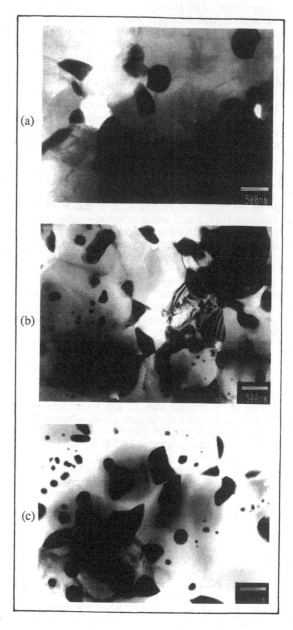

Figure 6. Transmission electron micrographs of sintered composites:
(a) acetone-1, (b) acetone-3, and (c) DEG.

REFERENCES

1.   Claussen, N., 1976. "Fracture Toughness of Al_2O_3 with an Unstabilized ZrO_2 Dispersed Phase." Journal of the American Ceramic Society, 59(1-2):49-51.

2.   Becher, P.F. and G.C. Wei, 1984. "Toughening Behavior in SiC-Whisker-Reinforced Alumina." Journal of the American Ceramic Society, 67(12):C267-269.

3.   Kim, H.E. and A.J. Moorhead, 1991. "Corrosion and Strength of SiC-Whisker-Reinforced Alumina Exposed at High Temperatures to H_2-H_2O Atmospheres." Journal of the American Ceramic Society, 74(6):1354-1359.

4.   Ikuma, Y., A. Yoshimura, K. Ishida and W. Komatsu, 1986. "Phase Transformation and Toughening in MgO Dispersed with ZrO_2." Materials Science Research, 20:295-304.

5.   Mikami, R., Y. Ikuma and W. Komatsu, 1987. "Mechanical Properties of MgO Dispersed with ZrO_2." Gypsum and Lime, No. 209:27-32.

6.   Okamoto, T., Y. Ikuma, M. Shimaoka, T. Shirotori and W. Komatsu, 1989. "Thermal Shock Resistance of ZrO_2-Toughened MgO." Journal of the Ceramic Society of Japan, 97(8):812-817.

7.   Nishida, A., S. Fukuda, Y. Kohtoku and K. Terai, 1992. "Grain Size Effect on Mechanical Strength of MgO-ZrO_2 Composite Ceramics." Journal of the Ceramic Society of Japan, 100(2):191-195.

8.   Nakahira, A., K. Niihara, J. Ohkijima and T. Hirai, 1989. "Al_2O_3/Si_3N_4 Nanocomposites." Funtai oyobi Funmatsu Yakin, 36(2):239-242.

9.   Sasaki, G., H. Nakase, K. Suganuma, T. Fujita and K. Niihara, 1992. "Mechanical Properties and Microstructure of Si_3N_4 Matrix Composite with Nano-Meter Scale SiC Particles." Journal of the Ceramic Society of Japan, 100(4):536-540.

10. Descamps, P., S. Sakaguchi, M. Poorteman and F. Cambier, 1991. "High-Temperature Characterization of Reaction-Sintered Mullite-Zirconia Composites." Journal of the American Ceramic Society, 74(10):2476-2481.

11. Imai, H., J. Sekiguchi and Y. Murakami, 1992. "Characterization of Fine Particles of Ruthenium-Alumina Composites Prepared by Different Methods." Journal of the Ceramic Society of Japan, 100(4):404-407.

12. Hwang, C.S. and W.H. Lin, 1991. "Preparation and Sinterability of Zirconia-Toughened-Alumina Composite Powder." Journal of the Ceramic Society of Japan, 99(4):271-275.

13. Ismail, M.G.M.U., H. Tsunatori and Z. Nakai, 1990. "Preparation of Mullite Cordierite Composite Powders by the Sol-Gel Method: Its Characteristics and Sintering." Journal of the American Ceramic Society, 73(3):537-543.

14. Mitsuhashi, T., M. Ichihara and U. Tatsuke, 1974. "Characterization and Stabilization of Metastable Tetragonal ZrO_2." Journal of the American Ceramic Society, 57(2):97-101.

15. Yoldas, B.E., 1986. "Zirconium Oxides Formed by Hydrolytic Condensation of Alkoxides and Parameters That Affect Their Morphology." <u>Journal of Materials Science</u>, 21:1080-86.

16. Yamamoto, O., T. Sasamoto and M. Inagaki, 1991. "Coating of Carbon Materials with ZrO_2 by Sol-Gel Method." <u>Nippon Kagaku Kaishi</u>, 1991(10):1326-1331.

The Effect of Granule Morphology on the Strength and Weibull Modulus of ZTA Made from Spray Dried Powders

JILL HAYLOCK

ABSTRACT

The strength distribution of ZTA ceramics fabricated from spray dried powders was determined. One of the powders consisted of 'donut' shaped agglomerates, the other of solid spheres. Although the bulk densities of the powders did not differ, the strength of material made from solid spherical granules was considerably higher than for material made from 'donut' shaped granules due to the presence of large crack-like voids resulting from poor interagglomerate sintering. In contrast, the Weibull modulus was very similar for the two materials and it is suggested that the origin of strength limiting flaws may be, in both cases, related to the original spray dried granules.

INTRODUCTION

The influence of granule morphology on the mechanical properties of ceramics made from spray dried powder has been documented on several occasions [1,2]. The main effect of granule morphology is not so much on the green density of the ceramic compacts but rather on the pore size distribution which, in turn, affects the sinter-activity of the compact. Whether a compacted article will achieve full density during sintering depends upon the pore size in the green body rather than the total porosity, since small pores will close up readily whereas larger ones will tend to be retained and may even become enlarged as the matrix shrinks away from the pore [3].

This is particularly important to consider when dealing with high strength ceramics since the strength of ceramics is related to the size of the defects present, large pores being far more detrimental to the strength characteristics than many small pores. In addition, the differential shrinkage within an agglomerate relative to the surrounding powder compact may lead to crack-like voids [4] which may barely affect the overall density of the compact. However, by virtue of their shape, such flaws may severely affect the strength of a ceramic so that measurements of sintered density may not give a reliable indication of its mechanical properties. This is of particular relevance for ceramics made from spray dried powders since the differential drying experienced by a droplet often leads to granules with an inhomogeneous structure.

· Jill Haylock, BHP Research, Melbourne Laboratories, 245-273 Wellington Road, Mulgrave, Victoria AUSTRALIA 3170

The effect of heterogeneities and agglomerates on sintering behaviour of zirconia toughened alumina (ZTA) in particular, has been well documented in the literature. Lange [5] found large crack-like internal surfaces in sintered ZTA produced by differential sintering within dense agglomerates relative to the matrix. Closely packed agglomerates may undergo preferential intra-agglomerate sintering, shrinking away from neighbouring agglomerates creating a lens shaped pore. Ultimately, the pore may shrink somewhat and the agglomerate brought back into contact with the matrix but complete densification is never achieved. According to Lange et. al. [6] since the size of these crack-like voids is proportional to the agglomerate size and strength is inversely proportional to the square root of the crack size, one would expect that the strength would be inversely proportional to the square root of the agglommerate size. The distribution of strength values would also mimic the size distribution of the agglomerates.

The quality of a spray dried powder is most commonly measured by its compressibility. It is assumed that in an easily compressible powder, the agglomerated nature of the powder and hence the heterogeneities will be removed during compaction. Likewise, it has often been assumed that the defects introduced through poor granule morphology would be detectable either through density measurements or observation of the microstructure. Heterogeneities are, however, best observed using an experiment sensitive to them, such as, measuring the strength statistics of the ceramics which gives an indication of the strength limiting flaw population [7].

There have been few studies to date examining the effect of spray dried powder characteristics on the Weibull modulus of the sintered material. In a notable exception, Mosser et. al. [8] determined that alumina pressed from deformable granules generally had a significantly higher strength and Weibull modulus than alumina pressed from less deformable granules. In this case granule deformability was determined by the ductility of the PVA binder.

EXPERIMENTAL

Dispersions of zirconia and alumina were prepared according to a proprietary method developed for this project. The material was spray dried[1] using three different atomizers; a rotary atomizer, a two-fluid nozzle and a pressure nozzle. The spray/air contact was co-current where the rotary atomizer was used and mixed flow for the two-fluid and pressure nozzles.

The median granule size of the spray dried powders was determined using laser particle size analysis[2] of calcined powders. The granule morphology was examined using a scanning electron microscope[3] by mounting calcined powder in epoxy resin and polishing a section to reveal the internal structure. The powders were then fabricated into MOR bars by uniaxial pressing followed by isostatic pressing at 200 MPa. The bars were sintered at 1550°C for 2 hours and diamond ground to 3x4 mm cross-section. The sintered density was measured using a mercury volumeter. The bars were broken in 4 point bending with a 7 mm inner span and a 20 mm outer span and the flexural strength and Weibull modulus determined in the usual way (9).

---

[1] Niro "Production Minor"
[2] Coulter LS
[3] Jeol JSM 6300F

RESULTS

The morphology of the powders produced using the three different atomizing techniques is shown in figure 1. Providing the chemistry of the feed material was optimal, the rotary atomizer consistently produced solid spherical granules. In contrast, powders produced using the two-fluid or pressure nozzles consisted of 'donut' shaped granules, that is, hollow balls with a dimple at one end. Granules produced by the rotary atomizer were also considerably smaller than those produced by the spray nozzles. The granule size distribution of the powders is shown in figure 2.

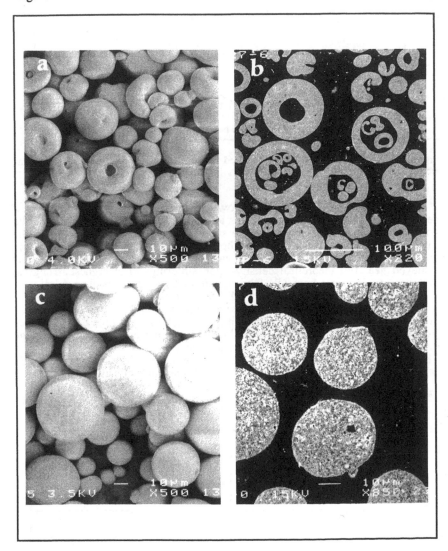

Figure 1. The powder morphology, **a**; *'donut' shaped granules*, external morphology, **b**; internal structure, **c**; *solid spheres*, external morphology, **d**; internal structure.

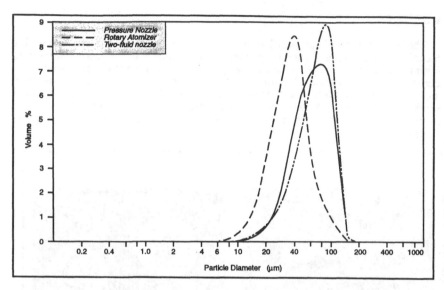

Figure 2. The size distribution of spray dried granules produced using the three atomizing techniques.

The properties of the spray dried powders and the mechanical properties of ceramics fabricated from these powders are shown in Table I. Surprisingly, the powder bulk density was lowest for the solid spherical granules produced by the rotary atomizer, possibly as a result of the smaller granule size. Nevertheless, the green densities of the compacts did not vary considerably and all of the compacts achieved near full theoretical density upon sintering. The highest strengths, however, were obtained from ZTA fabricated from the solid spherical granules.

The Weibull strength distribution for ZTA fabricated from powders made using the rotary atomizer (solid spherical granules) and the two-fluid nozzle ('donut' shaped granules) are shown in figure 3. Although the strength values for the ZTA fabricated from the solid spherical granules are distinctly higher, the strength distributions and hence Weibull modulus, m, are essentially the same.

The microstructure of ZTA fabricated from powder made using the two-fluid nozzle is shown in figures 4a and b. Semi-circular crack-like voids of between 20 and

TABLE I. THE MECHANICAL PROPERTIES OF ZTA CERAMICS FABRICATED FROM THE SPRAY DRIED POWDERS.

Atomizer	Median Granule Diameter (μm)	Powder Bulk Density (g/ml)	Green Density (%)	Sintered Density (%)	Ave. MOR (MPa)
Rotary Atomizer	37	1.3	58	99	850
Pressure Nozzle	63.6	1.4	61	97	600
Two-fluid Nozzle	72.3	1.6	57	98	640

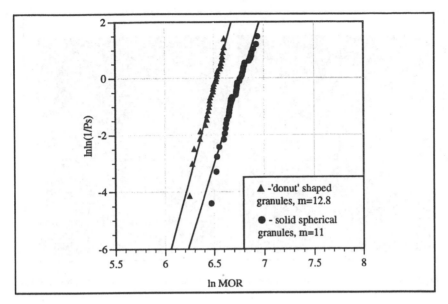

Figure 3. The Weibull strength distributions measured for ZTA fabricated from solid spherical and 'donut' shaped granules.

80µm in diameter are a common feature. Figures 5 a and b show the types of flaws observed in the microstructure of ZTA fabricated from the solid spherical granules The major flaws observed are small (~3 - 10 µm) lenticular crack-like voids but, for the most part, the microstructure is quite homogeneous.

DISCUSSION

The size and shape of the crack-like voids present in the material fabricated from 'donut' shaped granules strongly suggest that they originate from differential sintering between spray dried granules retained in the compacted article. During sintering these agglommerates shrink away from the rest of the matrix resulting in large semi-circular voids. If the agglomerated nature of the powder could be eliminated during pressing and/or sintering, the strength of the material should be greatly improved as these crack-like defects are eliminated. For example, hot pressed and HIP'ed ZTA has often been found to possess flexural strengths up to nearly twice that of cold pressed and sintered ZTA [5]. A powder composed of solid spherical granules would be expected have a more homogeneous structure so that more uniform intra and inter-granular sintering occurs. This should largely eliminate the initial agglomerated structure, producing a ceramic with a smaller defect population. The strength of ZTA produced from such powders should more closely approximate that of hot pressed or HIP'ed ZTA.

The higher strengths measured for ZTA fabricated from solid spherical granules initially appear to confirm the role of granule morphology in determining the mechanical properties of ceramics made from spray dried powders. In addition, SEM observations of the microstructure of ZTA fabricated from solid spherical granules revealed only small lenticular voids compared with those found in ZTA fabricated from 'donut' shaped granules which would seem to indicate better

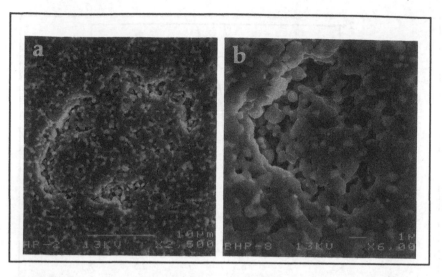

Figures 4a and b. Semi-circular crack-like voids in the microstructure of ZTA fabricated from 'donut' shaped granules.

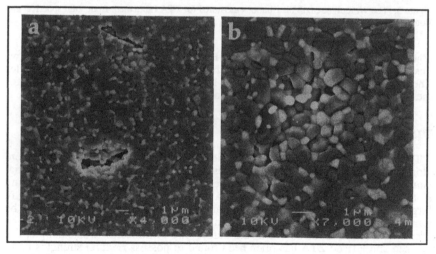

Figures 5a and b. Lenticular flaws in the microstructure of ZTA fabricated from solid spherical granules.

sintering between the spray dried granules. Nevertheless, it could be expected that if solid spherical granules produced more homogeneous sintering behaviour, then the Weibull modulus should also be higher. However the Weibull moduli were very similar indicating no apparent improvement in homogeneity.

Since the strength limiting flaws in the samples made from the 'donut' shaped granules are the result of retention of the granule morphology in the sintered article, the spread of strengths should bear a relationship to the size range of the spray dried granules. As the granule size *range* of the two powders are also very similar, the similar Weibull moduli may indicate that the strength limiting flaws in the material made from solid spherical granules are also related to remnants of the original spray dried granules. The difference in the strength values obtained could then be explained purely in terms of the *size* differences between the 'donut' shaped and the solid spherical granules. If the spray dried granules are smaller, then the defects produced by poor interagglomerate sintering will necessarily also be smaller.

It is possible to test this assertion by determining whether an accurate estimation of the strength distribution can be made from granule size range. The size of fracture initiating defects can be derived from the mechanical properties of the material using the Griffith equation which states that:

$$\sigma = \frac{E\gamma}{\sqrt{\pi c}} \tag{1}$$

where: σ = flexural strength
                  E = Young's Modulus
                  γ = Fracture energy
                  c = Flaw size

The value of E may be reasonably estimated using the rule of mixtures while fracture energy may be estimated from the fracture toughness, K, and Young's modulus using the relationship:

$$\gamma = \frac{K^2}{2E} \tag{2}$$

In the case of semi-circular defects, only part of the flaw would be activated under the stress distribution experienced and result in failure of the specimen. Therefore, the effective flaw size is significantly smaller than the actual flaw size. Nevertheless, it should be possible to derive a relationship between the granule size and strength as theorized by Lange et. al [6]. For instance, for the material fabricated from 'donut' shaped granules, it may be safely assumed that the strength of the weakest ceramic tested results from a defect surrounding one of the largest granules. The strength of this sample then bears a relationship to the size or rather circumference of the largest granules.

Using the Griffith equation, the weakest sample is estimated to have failed from a flaw with an effective length of 34μm. This is approximately 10% of the total circumference of the largest spray dried granules. If this proportion is relatively consistent, then it should be possible to estimate the strength distribution based on the granule size distribution. Such an estimation, however, assumes that the amount of sample tested during the flexural strength test is of a size such that only one of the flaws experiences the maximum stress concentration. A small proportion of samples with very high strength are therefore expected. In reality, a range of flaw sizes are likely to be represented in the volume of material tested, the largest one determining the load at failure. Nevertheless, using the median 90% of the granule distribution range and the estimation that 10% of the circumference contributes to the strength limiting flaw, it is possible to obtain a reasonably accurate estimation of the strength distribution (figure 6).

Similarly, for the ZTA fabricated from solid spherical granules, if a flaw size relationship of 10% of the granule circumference is inferred and a strength distribution is estimated from the granule size distribution, the results show quite a reasonable coincidence with the measured strength distribution (figure 7). In fact, the expected distribution only deviates significantly from the measured distribution with the expectation of a small proportion of samples of a very high strength.

That the strength distribution can be estimated from the size range of the spray dried granules with a reasonable degree of accuracy is compelling evidence that incomplete sintering between granules is the major cause of strength limiting defects in both of the ZTA materials examined. That is, with respect to strength, the solid spherical granules provide no benefit other than that associated with the smaller size of the granules. In this respect, the microstructural appearance is deceptive.

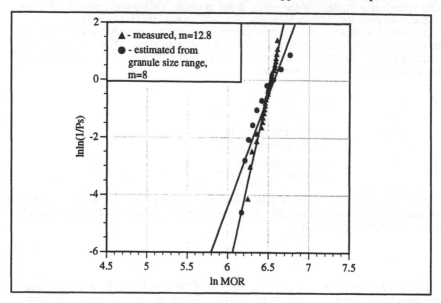

Figure 6. The measured strength distribution (Weibull modulus) and the distribution predicted from the granule size range ('donut' shaped granules).

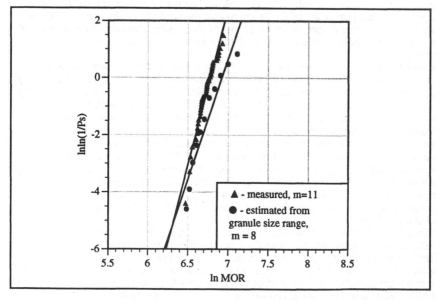

Figure 7. The measured strength distribution versus the distribution predicted from the granule size range (solid spherical granules).

CONCLUSIONS

1. Although the strength of ZTA fabricated from solid spherical granules was considerably higher than that fabricated from 'donut' shaped granules, the Weibull modulus was similar for the two materials.

2. "Donut' shaped granules appear to produce crack-like discontinuities in the microstructure as a result of preferential intra-agglomerate sintering. These defects are likely to be responsible for the relatively low strengths obtained for this material. Using the Griffith equation, the flaw size distribution which gives rise to the strength distribution measured for this material can be estimated with convincing accuracy from the granule size range.

3. Whilst the microstructure of ZTA fabricated from solid spherical granules appears relatively homogeneous, it is apparent that inhomogeneous sintering between granules may give rise to the major flaws determining the strength statistics in these materials. In this case, the high strengths measured for these ceramics results from the small size of the spray dried granules. As with the materials made from 'donut' shaped granules, the strength distribution could be estimated with reasonable accuracy from the granule size distribution.

4. Weibull statistics are a powerful tool, not only for evaluating the strength of a material but also for detecting patterns within the strength data which may be correlated to types of heterogeneities within a material. In the present study, similarities in the Weibull distribution suggest that similar flaw types may be controlling the strengths of the materials, even though the microstructures appear completely different.

ACKNOWLEDGEMENTS

The author would like to thank Mr. Leo Frawly for his assistance with the SEM, Dr. M.V. Swain and Dr. W.J. Sinclair for their helpful comments and BHP Co. Pty. Ltd. for their permission to publish this work.

REFERENCES

1. Lukasiewicz, S.J., 1989. "Spray-drying Ceramic Powders." J. Am. Ceram. Soc., 72(4): 617-624 .

2. Hartshorn, A.J., S.C. Koh and B.E. Reichart, 1988. "Morphology of Spray Dried Particles - The Effect On Ceramics Made By Dry Pressing," in Ceramic Powder Processing Science. Hausner, H., G.L. Messing and S. Hirano. eds., Proceedings of the 2nd international conference held 12-14 October, 1988 in Berchtesgaden, Koln: Deutsche Keramische Gesellschaft .

3. Vasilos, T. and M. Rhodes, 1969. "Solids Processing of Fine Grain Size Ceramics," in Characterization of Ceramic Materials, Saito, S., Y. Kotera and S. Somiya. eds., Proceeding of the First US-Japan Seminar on Basic Science of Ceramics, Tokyo and Kyoto, New York: Elsevier . pp. 93-140.

4. Rhodes, W.S., 1981. "Agglomerate and Particle Size Effects on Sintering Yttria Stabilized Zirconia." J. Am. Ceram. Soc., 64(1): 19-22.

5. Lange F.F., 1983. "Processing Related Fracture Origins: I, Observations in Sintered and Isostatically Hot Pressed Al_2O_3/ZrO_2 Composites." J.Am. Ceram. Soc.,66(6): 396-398.

6.   Lange, F.F. and M. Metcalf , 1983. "Processing Related Fracture Origins II, Agglomerate Motion and Cracklike Internal Surfaces Caused by Differential Sintering". J.Am. Ceram. Soc.,66(6): 398-406.

7.   Lange, F.F., 1989. "Powder Processing Science and Technology for Increased Reliability." J. Am. Ceram. Soc, 72(1): 3-15 .

8.   Mosser, B.D., J.S. Reed and J.R. Varner, 1992. "Strength and Weibull Modulus of Sintered Compacts of Spray-Dried Granules," Am. Ceram. Soc. Bull., 71(1): 105-109.

9.   Richardson, D.W., 1982. Modern Ceramic Engineering. New York: Marcel Dekker Inc.

Preparation and Characterisation of Sol-Gel Derived Zirconia Coatings

MAREE ANAST, BESIM BEN-NISSAN, JOHN R. BARTLETT,
JIM L. WOOLFREY, JOHN M. BELL, TREVOR J. BELL,
DAN R. DE VILLIERS, LEONE SPICCIA, BRUCE O. WEST,
GRAHAM R. JOHNSTON and IAN D. WATKINS

ABSTRACT

The literature reveals significant interest in zirconia coatings prepared using alkoxide-based sol-gel processes. This paper reviews the achievements of our research group in this area. The chemistry of the processes we have used is briefly described. Coatings were prepared on both glass and metallic substrates by dip coating followed by firing under various controlled conditions. Characterisation of all stages of the sol-gel process has been undertaken. Particle size analysis of the species in solution has been performed using static light scattering. The transformation of the gel has been studied by thermal analysis and X-ray diffraction (XRD). The final coatings and coating/substrate interactions have been characterised by

Maree Anast and Besim Ben-Nissan, Department of Materials Science, University of Technology, Sydney, PO Box 123, Broadway, NSW 2007, Australia
John R. Bartlett and Jim L. Woolfrey, Advanced Materials Program, Australian Nuclear Science and Technology Organisation, Private Mail Bag 1, Menai, NSW 2234, Australia
John M. Bell, Department of Applied Physics, University of Technology, Sydney, PO Box 123, Broadway, NSW 2007, Australia
Trevor J. Bell, CSIRO Division of Applied Physics, PO Box 218, Lindfield, NSW 2070, Australia
Dan R. de Villiers, Leone Spiccia and Bruce O. West, Department of Chemistry, Monash University, Wellington Road, Clayton, Victoria 3168, Australia
Graham R. Johnston, Materials Research Laboratory, DSTO Australia, Cordite Avenue, Maribyrnong, Victoria 3032, Australia
Ian D. Watkins, Silicon Technologies Australia Pty Limited, 101 Merton Street, Albert Park, Victoria 3206, Australia

XRD, scanning Auger microscopy (SAM), scanning electron microscopy (SEM), Rutherford backscattering spectrometry (RBS) and precision ultra-microhardness. Information on cracking and adherence of the coating will also be presented. Discussion will be centred on the effect of variation of the solution and processing parameters on the characteristics of the final coatings.

INTRODUCTION

Zirconia coatings have been prepared by most of the available coating techniques, but the recent explosion of research interest in the sol-gel process has led to new opportunities for the preparation of zirconia coatings. This paper reviews the achievements of our research group in this area. The chemistry of the sol-gel process with respect to the preparation of zirconia coatings has been extensively reviewed recently [1], so only the chemistry of the three non-aqueous sol-gel processes investigated by our research group will be summarised here:
1.   Alkoxide/acetic acid solutions,
2.   Alkoxide/acetylacetonate complex solutions,
3.   Alkoxide/chelate complex solutions.
All of these processes involved complexation of the zirconium alkoxides:

$$Zr(OR)_4 + XOH ---> Zr(OR)_3(OX) + ROH \qquad (1)$$

where XOH = complexing agent, followed by partial/complete hydrolysis:

$$Zr(OR)_3(OX) + aH_2O ---> Zr(OR)_{3-a}(OX)(OH)_a + aROH \qquad (2)$$

Tailoring of the properties of the final zirconia coating is primarily achieved by utilising the flexibility of the chemistry of the sol-gel process, including variation of the type of zirconium alkoxide, the type and extent of complexation, and the degree of hydrolysis.

EXPERIMENTAL

SOLUTION FORMULATION

Alkoxide/Acetic Acid Solutions

In this system, a zirconium alkoxide was dissolved in glacial acetic acid and water was added at water:zirconium ratios of up to 32:1. In some solutions nitric acid was also added to stabilise gelation. A polyethylene glycol was also added to several of the solutions to improve the quality of the coating. Various yttrium and cerium compounds were added to some of these solutions to assess the feasibility of producing stabilised zirconia coatings from these solutions. Some of these solutions had only moderate stability, with precipitation beginning one week after preparation.

Alkoxide/Acetylacetonate Complex Solutions

Complexation of a zirconium alkoxide by acetylacetone was explored in these alcoholic solutions in order to ensure hydrolytic stability upon addition of acidified water and a yttrium compound. Addition of a polyethylene glycol to one of the solutions improved the quality of the coating but resulted in poor solution stability, with precipitation beginning within 24 hours of preparation.

Alkoxide/Chelate Complex Solutions

Most of the investigations of our research group have involved alcoholic solutions containing zirconium alkoxide chelates. A variety of zirconium alkoxides were used over a range of concentrations. Water was added to many of these solutions to explore the hydrolytic stability of the solutions and to determine the effect of the extent of hydrolysis on the quality of the coatings. Two polyethylene glycols were also added at various concentrations to some of the solutions to explore the impact of this viscosity modifier on the quality of the coatings. These solutions were usually found to be very stable over long periods of time.

SOLUTION CHARACTERISATION

When the concentration of particles in a solution is low, the monitoring of the static intensity of scattered light can provide information on the size of particles. The scattering intensity (I) is described by the equation:

$$I^{-1} = k[(1/M) + 2Bc + 3Cc + ...] \qquad (3)$$

where k = a constant, M = molecular weight of scattering species, c = concentration of scattering species, and B and C are the associated virial coefficients. If it is assumed that $B = C = 0$ (ideal behaviour), then the scattering intensity is directly proportional to the molecular weight of the scattering species.

In this study, alcoholic solutions of zirconium alkoxide/chelate complexes were combined with water/alcohol solutions in various water:zirconium ratios. All solutions were filtered using Millipore 200nm syringe filter cartridges prior to mixing. The static intensity of light scattered by the zirconium species during hydrolysis was monitored simultaneously at 14 discrete angles (between 26 and 135°) using a Wyatt Technology Dawn-F Static Light Scattering Spectrometer. All data was obtained using HeNe laser irradiation ($\lambda=632.8$nm).

COATING PREPARATION AND CHARACTERISATION

Coating preparation is discussed in more detail elsewhere [2,3,4]. In summary, soda glass microscope slides and polished stainless steel, inconel or titanium alloy pieces were used as substrates after rigorous cleaning. Substrates were withdrawn at controlled speeds of up to 6 cm/sec from the zirconium precursor solutions into an atmosphere of dry nitrogen. The coating was allowed to dry before exposure to atmospheric moisture. The coating was then heated in a controlled manner to various temperatures for specified periods. Further coatings were applied after firing the coating at intermediate temperatures. Experimental details of characterisation utilising Rutherford backscattering spectrometry (RBS) and scanning Auger microscopy (SAM) are available elsewhere [5].

RESULTS AND DISCUSSION

SOLUTION CHARACTERISATION

From a review of the hydrolytic chemistry of zirconium alkoxides and chelates [1], it can be expected that hydrolysis of a zirconium alkoxide/chelate complex will lead to progressive removal of the alkoxide ligands before removal of the chelate, rapidly followed by formation of oxo linkages between zirconium atoms. Turbidity and precipitation can be expected to arise with increasing water:zirconium ratios due to the lower solubility of the condensed species following loss of organic ligands and also due to the growth of more energetically stable zirconium clusters [1]. The growth of such clusters and precipitates can be monitored by light scattering techniques.

The static intensity of light scattered by solutions with water:

zirconium mole ratios (R) of 0.5, 2 and 5 is shown in figure 1. The light was found to be scattered isotropically (*i.e.* the

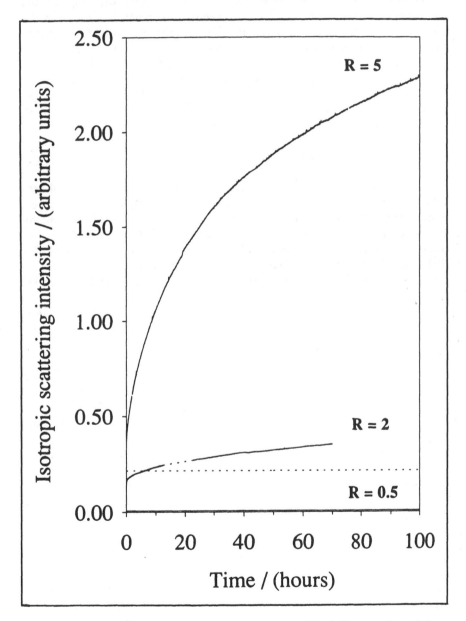

Figure 1. Isotropic scattering intensity from zirconium alkoxide/chelate complex solutions containing water:zirconium mole ratios (R) of 0.5, 2 and 5. The dotted line plotted for R = 0.5 is for a solution left standing for more than 6000 hours prior to analysis.

scattering intensity was independent of scattering angle) for all solutions at all times, indicating that the scattering species had diameters of less than 30nm (λ/20). At no time did any of the solutions display turbidity. The increase in particle size can be seen to have been initially rapid, but decreased with time. Since the rate of hydrolysis is a function of water concentration, the observed increase in particle size with increasing water: zirconium mole ratio was expected. The long term stability of this type of coating solution can be attributed to the very slow growth of condensed zirconium species, even at relatively high water:zirconium ratios.

GEL TRANSFORMATION

A zirconium alkoxide/chelate solution containing no phase-stabilising additives was dried and hydrolysed to a gel. Transformation of this gel by heating in air was monitored using thermal analysis and X-ray diffraction (XRD), with further details reported elsewhere [2]. Thermal effects observed by differential thermal analysis were interpreted as follows:

endotherm 105°C: removal of free water and residual solvent.

exotherm 345°C: condensation and gel rearrangement reactions and loss of residual solvent and additives.

exotherm 550°C: decomposition of unreacted alkoxide and zirconium hydroxide, burn-out of residual organic material and crystallisation of amorphous structure.

Information gained from the thermal analysis was useful for selecting an appropriate firing program for coating preparation. XRD analysis of the gel revealed that it was amorphous up to 400°C, with the first peak appearing at 500°C likely to correspond to a strained tetragonal lattice. Changes observed with further heating were consistent with a relaxation of the strained tetragonal phase to the equilibrium monoclinic phase, although full conversion was not achieved.

COATING CHARACTERISATION

Considerable variation occurred in the quality of the coatings produced from the various sol-gel solutions, but the best quality coatings could usually be prepared by reducing the substrate withdrawal speed to produce thinner coatings. Such coatings appeared smooth and craze-free during SEM analysis. Good quality coatings were obtained on glass from most of the alkoxide/acetic acid solutions. In the absence of polyethylene glycol, coatings from an alkoxide/acetylacetonate complex solution were often crazed and peeling. However, addition of glycol to the solution resulted in thicker coatings of excellent quality. Coatings on glass from alkoxide/chelate complex solutions varied from crazed and peeling to uncrazed for the same withdrawal speed from different solutions. The improvement in the quality of the coatings upon addition of polyethylene glycol to some solutions may be attributed to the improved flexibility of the coating during the shrinkage process which occurs during drying.

Further details regarding zirconia coatings prepared on metal substrates from alkoxide/chelate complex solutions are provided elsewhere [2,3,4]. Variations observed in the coating microstructure and the degree of crazing could be attributed to solution composition and aging. Coating thicknesses, ranging from 50-100nm for single coats, were measured by SEM, RBS and profilometry. The reproducibility of the thickness of each coat for multiply coated samples was confirmed by RBS. With careful control of the hydrolysis and firing conditions, films consisting of up to 15 craze-free coats could be successfully prepared. Zirconia-coated inconel 600 substrates prepared at 800°C were elementally depth profiled using RBS and SAM. Results of the two analyses were consistent, showing the zirconia coatings overlayed a chromia-rich layer which had developed above the bulk inconel. The structure of the zirconia coatings was found to be the same when fired in either air or oxygen atmospheres. XRD analysis of multiple coatings corresponded perfectly with the analysis of the gel described above. A strained tetragonal phase tentatively identified at 500°C was slowly, and only partially, converted to the equilibrium monoclinic phase at higher temperatures. Precision ultra-microhardness measurements were made on multiple coatings which were sufficiently thick to be considered free of influence from the substrate. SEM analysis of high load (200mN) indentations revealed excellent bonding with no de-lamination occuring between coats or from the substrate. Low load (6.5mN) indentation measurements revealed that the coating hardness decreased when the firing temperature was increased above 500°C. This decrease was attributed to the strained tetragonal phase tentatively identified by XRD at 500°C being converted to the equilibrium monoclinic phase at higher temperatures. The hardness value of 10.3GPa for the 500°C coating is very similar to that of bulk zirconia and indicates that such sol-gel derived coatings could find application as wear resistant coatings.

CONCLUSION

Careful control of the type of zirconium alkoxide, the type and extent of complexation, the degree of hydrolysis and the physical processes associated with the non-aqueous sol-gel technique, provide an opportunity to tailor the properties of the final zirconia coating. Through manipulation of the chemistry of the sol-gel solution and control of the coating process, our group has successfully produced both unstabilised and partially stabilised zirconia coatings. Extremely hard, transparent coatings prepared at 500°C are likely to find application as wear resistant coatings for both glass and metal substrates.

ACKNOWLEDGEMENTS

Financial support from the Australian Federal Government Department of Industry Technology and Commerce (GIRD grant) is gratefully acknowledged.

REFERENCES

1.   Watkins, I.D., 1993. "A Review of the Preparation of Zirconia Coatings using Sol-Gel Processes." Journal of the Australian Ceramic Society, to appear.

2.   Ben-Nissan, B.,M.Anast, J.Bell, G.Johnston, B.O.West, L.Spiccia, D.de Villiers and I.Watkins, 1991. "Characterisation of Sol-Gel Derived Zirconia Films," in Proceedings of the Science of Engineered Ceramics 91, S.Kimura and K.Niihara, eds., Tokyo, Japan: The Ceramic Society of Japan, pp. 25-29.

3.   Anast, M., J.M.Bell, T.J.Bell and B.Ben-Nissan, 1992. "Precision Ultra-Microhardness Measurements of Sol Gel Derived Zirconia Thin Films", Journal of Materials Science Letters, to appear.

4.   Bell, J.M., R.W.Cheary, M.Rice, B.Ben-Nissan, J.L.Cocking and G.R.Johnston, 1992. "Crystal Structures of Sol-Gel Deposited Zirconia Thin Films", Austceram 92, Proceedings of the International Ceramic Conference Australia 1992, 2:765-770.

5.   Johnston, G.R., J.L.Cocking, G.Cumming and V.E.Jeleniewski, 1992. Unpublished work.

Slip Casting of Y_2O_3-Stabilized ZrO_2

B. D. BEGG, A. J. RUYS and C. C. SORRELL

ABSTRACT

Since ZrO_2 is a nonplastic material, slip casting of objects using this technique is difficult. The objective of the present work was to develop a slip casting technique for ZrO_2 using water plus a deflocculant/binder that gives a high degree of deflocculation, slow casting rate and adequate green strength.

Y_2O_3-stabilized ZrO_2 (mean particle size d=19.3μm) from ICI Advanced Ceramics was ball milled for 90 hours to produce a powder of average particle size d=1.64μm. Slips containing 65 wt% yttria-stabilized zirconia (Y-TZP) and 35 wt% demineralised water were prepared with 0.05-0.93 wt% sodium carboxymethylcellulose (Na-CMC) added as a deflocculant/binder. Viscosity, thixotropy, casting rate, cast density and diametral shrinkage tests were performed as a function of Na-CMC addition.

Concurrent optimisation of the viscosity, thixotropy, casting rate, cast density and diametral shrinkage was obtained for ~ 0.25 wt% Na-CMC addition. This produced a body with low, yet sufficient green strength for handling as a result of the binding power of the Na-CMC molecules.

INTRODUCTION

The development of zirconia-based ceramics over the past thirty years has transformed zirconia from its refractory beginnings into a ceramic capable of being used in advanced structural engineering applications. The development of stabilized zirconia with superior mechanical properties has led the way for a new field of structural zirconia-based ceramics. The transformation toughening properties of stabilized zirconia give it high strength, excellent fracture toughness and good wear resistance qualities, which make it particularly well suited to high-performance applications [1].

Current forming methods employed in the production of stabilized zirconia are typically dry or isostatic pressing [2,3]. These are well suited to the production of simple and uniformly shaped articles. The formation of complex-shaped ware

B.D. Begg, A.J. Ruys and C.C. Sorrell
Department of Ceramic Engineering, School of Materials Science and Engineering
University of New South Wales, P.O. Box 1, Kensington, NSW 2033

requires costly and labour intensive machining, which has generally made this process uneconomical.

The utilization of wet forming processes, which are more suited to the production of complex-shaped objects than dry forming methods (such as dry or isostatic pressing), may alleviate this problem. Two major wet forming methods that are capable of producing complex-shaped ware are injection molding and slip casting.

Whilst injection molding is suited to the mass production of small complex-shaped articles, unless the production numbers are large, both the initial capital outlay required to purchase the specialized equipment as well as the on-going tooling costs render this forming method nonviable [4,5]. Slip casting, however, represents a simple, effective, reliable and economical means of producing complex-shaped articles.

Fundamental to any slip casting process is the addition of a deflocculant, which ensures the homogeneous suspension of solids in the slip. The slip casting of nonplastic materials, such as Y_2O_3-stabilized zirconia (Y-TZP), further requires the addition of a binder to give sufficient green strength to the cast ware to enable it to retain its shape prior to firing [6,7]. Binders are long macromolecular polymers that provide strength through sufficient wetting and specific adsorption of active groups along the polymer onto the surface of the colloidal particles [6,8].

The long-chain structure and highly adsorbant nature of sodium carboxymethylcellulose (Na-CMC) make it particularly useful as an organic binder [9,10]. The adsorbant anionic nature of Na-CMC serves also to coat the particles with a slight negative charge, which effectively disperses the suspension [6,9,10]. The dual binding/deflocculating properties of Na-CMC, which make it desirable as a sole additive in the slip casting of nonplastic ceramics, have been recognised and reviewed previously by Ruys and Sorrell [11]. Tests carried out to determine the impurities introduced into a body cast with 0.5 wt% Na-CMC reveal a Na concentration of ~ 205 ppm [11].

A BRIEF HISTORY OF THE SLIP CASTING OF ZIRCONIA CERAMICS

The proprietary nature of the slip casting of stabilized zirconia is illustrated not only by the limited number of publications available but also by the fact that many authors have failed to reveal details of their experimental findings [8,12-14]. A summary of the reported data is given in Table I.

The establishment by St. Pierre [21] of the pH-viscosity relationship in the zirconia-water-poly(vinyl alcohol)-hydrochloric acid system provided the groundwork for the slip casting of zirconia. Rempes et al. [16] subsequently undertook a comparative study of the slip casting of twenty nonplastic materials using a variety of casting additives. They reported the destabilising effects of using sodium carboxymethylcellulose (Na-CMC) in acid-deflocculated zirconia casting slips. Low casting times, the unstable nature of the suspension and the presence of cracks in the cast ware led them to recommend other additives for the slip casting of zirconia. Subsequent studies in the use of Na-CMC as a casting additive have shown its instability to pH adjustments [22-25].

Kallinga et al. [15] investigated the viability of slip casting lime-stabilized zirconia under acidic and basic conditions. They concluded that acidic conditions (pH 1.6) were preferable owing to the denser castings and reduced firing shrinkage. This was confirmed by Kwang-Lung and Huey-Chang [26] in their work on the aqueous dispersion of zirconia powders by varying the pH.

TABLE I. SLIP CASTING OF ZIRCONIA AND PSZ

Material	Deflocculant		Binder		Solids	Rel. p_{cast}	Particle Size	Ref.
	(Type)	(pH) [a] (wt%) [b]	(Type)	(wt%)	(wt%)	(%)	(μm)	
6.0 CaO-ZrO_2	Acid	1.60[a]	-	-	-	47	-	[15]
	Base	10.5[a]	-	-	-	42	-	
50 CaF$_2$-ZrO_2	HCl	0.65[a]	Na-CMC+	12	62.6	48.6	-	[16]
50 TiO$_2$-ZrO_2	HCl	0.39[a]	Na-CMC	15	57.2	45.6	-	[16]
5.3 Y$_2$O$_3$-ZrO_2	(NaPO$_3$)$_6$	-	Demol N*	0.25	73	50	0.45	[17]
5.5 Y$_2$O$_3$-ZrO_2	(NaPO$_3$)$_6$	0.1[b]	Demol N	0.05-0.3	65	46-47	< 0.7	[18]
					70	56.4		
4.8 Y$_2$O$_3$-ZrO_2	-	0.4[b]	-	-	75	56.5	0.3	[19]
					80	57.2		
3.7 MgO-ZrO_2	Na-CMC	0.77[b]	Na-CMC	0.77	65	-	-	[20]

+Sodium carboxymethylcellulose
*Demol N; a sodium salt of the polycondensation product of β-napthalenesulfonic acid and formaldehyde

Masson *et al.* [14] later described the slip casting of lime-stabilized zirconia using ethanol-based suspensions as a means of improving the mold life over that obtained from aqueous acidic suspensions. It has also been shown that magnesia-stabilized zirconia may be slip cast from presintered waste material [13].

Taguchi *et al.* [17,18] were the first to carry out an extensive study into the mechanical properties obtained by altering the characteristics of the PSZ casting slip. They reported that the relative density of the cast ware increased linearly with the solids concentration up to a maximum of 73%, where thixotropic behaviour was observed. Firing shrinkage was similarly observed to decrease with increased solids loadings. A solids concentration of 65% was found to give optimum casting control.

The relationship between relative density and firing shrinkage was later confirmed by the same authors in their subsequent work on the effect of milling on the relative density of the cast ware [18]. This investigation showed that the relative density increased with milling time and was independent of the concentration of binder used. The degree of deflocculation was, however, found to play an important role in determining the relative density of the cast specimens.

Moreno *et al.* [19] investigated the rheological and casting parameters of yttria-stabilized zirconia. They illustrated for three suspensions (80%, 75% and 70% solids by weight) that increased solids contents led to greater viscosities and increased casting rates. The relative density of the cast ware was found to be independent of the solids content over this range. It was concluded that, though stable suspensions of up to 80% solids could be obtained, the lower viscosity of the 70% solids suspension was preferred because it gave rise to more controlled casting. The addition of calcium or sodium salts to the PSZ casting slip was found to be detrimental because these are known to have a significant effect on the viscosity of the suspension [27].

The only work known to the authors carried out on the slip casting of zirconia using Na-CMC as the sole casting additive was done by Ruys [20]. He determined that the optimum casting concentration of Na-CMC was 0.77 wt%. A slip containing 65 wt% solids was used and no data were available on the cast density or average particle size.

EXPERIMENTAL PROCEDURE

The objective of the present study was to investigate the feasibility of slip casting of yttria-stabilized PSZ using sodium carboxymethylcellulose as the sole casting additive. The following properties were measured as a function of the level of Na-CMC addition: 1) viscosity, 2) thixotropy, 3) casting rate, 4) casting density and 5) diametral shrinkage.

A 65 wt% Y-TZP[1] and 35 wt% demineralized water slip was prepared to optimise casting control [17]. The slip was milled for 90 hours in a neoprene mill with magnesia-stabilized zirconia balls to reduce the average particle size of the powder. Additions of between 0.05 and 0.93 wt% Na-CMC[2] were then made to the slip, which was milled for 0.5 h in order to achieve adequate mixing.

The viscosity and thixotropy index were measured with a torsional viscometer[3], using a 30 gauge torsion wire (0.32 mm in diameter) and 17.5 mm (11/16 inch) diameter bob. A small-scale laboratory filter press was used in conjunction with a top loading balance to determine the casting rate of the slip.

RESULTS AND DISCUSSION

The average particle size[4] of the milled Y-TZP powder was 1.64 μm, with an average specific surface area[4] of 5.07 m^2/g. The milled powder exhibited a broad range of particle sizes, with 90% of the powder being < 3 μm and 20% < 0.6 μm.

Figure 1 shows the viscosity and three-minute thixotropy index for Y-TZP casting slips as a function of the Na-CMC content. The high viscosities that were observed for low level additions of Na-CMC were the result of the concentration of Na-CMC being insufficient to defllocculate the casting slip fully. Optimal deflocculation occurred between 0.15 wt% and 0.25 wt% Na-CMC, where the resistance to flow was minimized. Higher level additions of Na-CMC caused the viscosity to increase but at a slower rate owing to the formation of a gel by the hydrophyllic Na-CMC molecules around the deflocculated particles [9]. Part of the increase in the viscosity came from the increased concentration of Na-CMC [28].

Thixotropy is a measure of time-dependent viscosity. The broad minimum in the three-minute thixotropy index curve (see figure 1) shows the optimum level of Na-CMC addition to be between 0.20 wt% and 0.50 wt%. The overlapping minima between the viscosity and three-minute thixotropy curves indicate that optimal deflocculation occurs between 0.20 wt% and 0.40 wt% Na-CMC. The resistance to flow of an optimally deflocculated suspension will increase only marginally with time in comparison to that of a flocculated or gelled suspension. Therefore, a corresponding minimum is expected between the thixotropy index and viscosity.

---

1   SY-SUPER 5.2 wt% yttria-stabilised zirconia powder; mean particle size 19.3 μm; 90% < 33.3 μm; specific surface area 0.17m^2/g ; Z-Tech Pty. Ltd., Melbourne, Australia.

2   Grade 1130; >99%; pH 6.5 to 8.0, 1% solution; Daicel Chemical Industries, Ltd., Tokyo, Japan.

3   Model 11940/1 (vs-020) Universal Torsional Viscometer; A. Gallenkamp & Company Ltd., London, England.

4   Horiba Capa-700 Particle Analyser; Horiba Inc., Kyoto, Japan.

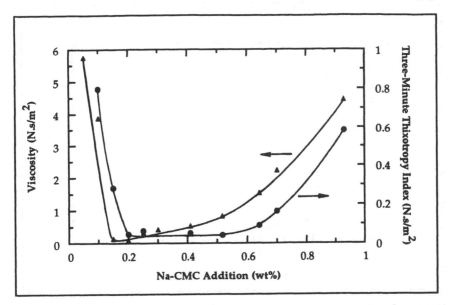

Figure 1.   Viscosity and three-minute thixotropy index versus sodium carboxymethylcellulose
content for Y-TZP slips ball milled for 90 hours.

Casting rates showed an exponential decrease with the concentration of Na-CMC,
as shown in figure 2. Fast casting rates were observed for low level Na-CMC
additions, which were insufficient to deflocculate the slip. The adsorbed molecules
set up a bridging effect that partially opened up the flocculated structure. This is
reflected in the low green densities seen for low level Na-CMC additions (see figure
3) [9]. Higher concentrations reduced the casting rate due to the formation of a gelled
structure. No significant casting rate improvements were obtained for Na-CMC
additions above 0.20 wt%.

The cast density and diametral drying shrinkage results shown in figure 3 illustrate
that, as the level of deflocculation increases with Na-CMC addition to a maximum at
~0.25 wt%, so too does the particle packing efficiency lead to an optimal cast density
at ~0.25 wt% Na-CMC. For higher level additions, void filling by the excess Na-
CMC decreases the packing efficiency and therefore the cast density. The diametral
shrinkage results follow a similar albeit reverse trend to the cast density results, with
the diametral shrinkage decreasing with increasing particle packing efficiency. The
optimally deflocculated castings had sufficient green strength to resist deformation
during handling.

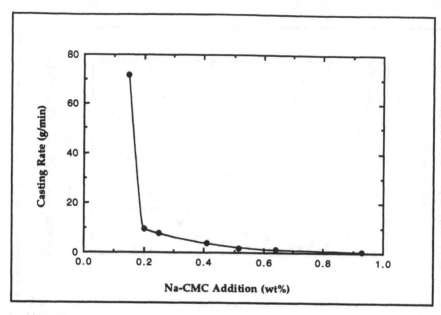

Figure 2. Casting rate versus sodium carboxymethylcellulose content for Y-TZP slips ball milled for 90 hours.

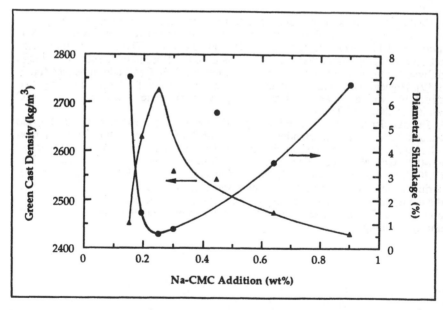

Figure 3. Green apparent solid cast density and diametral shrinkage versus sodium carboxymethylcellulose content for solid cylinders cast from Y-TZP casting slips ball milled for 90 hours .

SUMMARY AND CONCLUSIONS

The present work shows that the slip casting of yttria-stabilized zirconia using Na-CMC as dual deflocculant/binder represents a simple and effective means of fabricating complex articles with a minimum of effort and expense. Slip casting can be achieved using a Y-TZP/water suspension (65/35 wt%) with ~0.25 wt% Na-CMC. In this case, the average Y-TZP particle size was 1.64μm. The concurrent optimization of the viscosity, thixotropy, casting rate, cast density and diametral shrinkage for ~0.25 wt% Na-CMC additions can be achieved while still yielding sufficient green strength for handling. This is possible due to the dual binding and deflocculating qualities of sodium carboxymethylcellulose.

REFERENCES

1.   Butler, E.P., 1985. "Transformation-Toughened Zirconia Ceramics." Mater. Sci. Tech., 1 [6]:417-432.

2.   Garvie, R.C., R.H.J. Hannink and N.A. McKinnon, July 21, 1981. "Partially Stabilized Zirconia Ceramics", U.S. Pat. No. 4,279,655.

3.   Yuan, T.C., G.V. Srinivasan, J.F. Jue and A.V. Virkar, 1989. "Dual-Phase Magnesia-Zirconia Ceramics with Strength Retention at Elevated Temperatures." J. Mater. Sci., 24 [2]:3855-3864.

4.   Creyke, W.E.C., I.E.J. Sainsbury and R. Morrell, 1982. Design with Non-Ductile Materials. London, England: Applied Science.

5.   Richerson, D.W., 1982. Modern Ceramic Engineering: Properties. Processing and Use in Design. New York, NY, USA: Marcel Dekker, Inc.

6.   Onoda, Jr. G.Y., 1978. "The Rheology of Organic Binder Solutions," in Ceramic Processing Before Firing. G.Y. Onoda Jr. and L.L. Hench, eds., New York, NY, USA: Wiley Interscience, pp. 235-251.

7.   Smith, T.A., 1962. "Organic Binders and Other Additives for Glazes and Engobes." Brit. Ceram. Soc. Trans., 61:523-549.

8.   McNamara E.P. and J.E. Comeforo, 1945. "Classification of Natural Organic Binders." J. Am. Ceram. Soc., 28 [1]:25-31.

9.   Höpfner, P., 1981. "Cellulose Ether in Ceramic Slips and Plastic Liquids." Ceram. Forum. Int./Ber. DKG, 58 [4-5]:22-35.

10. Hoechst Kalle AG, Manufacturer's Specifications (Tylose C), Frankfurt, FRG.

11. Ruys A.J. and C.C. Sorrell, 1990. "Slip Casting of High-Purity Al₂O₃ Using Sodium Carboxymethylcellulose as Deflocculant/Binder." Am. Ceram. Soc. Bull., 69 [5]:828-832.

12. Fisher, G., 1986. "Zirconia: Ceramic Engineering's Toughness Challenge." Am. Ceram. Soc. Bull., 65 [10]:1355-1360.

13. Mayer, G.T., 1986. "Mg-PSZ Casting Slip," in Ceramics. The New Era: Proceedings of the Twelfth Australian Ceramic Conference, Melbourne, Victoria, Australia: The Australian Ceramic Society, Melbourne, pp. 69-93.

14. Masson, C.R., S.G. Whiteway and C.A. Collings, 1963. "Slip Casting Calcium Fluoride and Lime Stabilized Zirconia." Am. Ceram. Soc. Bull., 42 [12]:745-747.

15. Kallinga, G.P., V.A. Kolbasova and D.N. Poluboyarinov, 1962. "The Technology of Slip Casting of Zirconium Ware." Ogneupory, 27 [1]:28-34.

16. Rempes, P.E., B.C. Weber and M.A. Schwartz, 1958. "Slip Casting of Metals, Ceramics and Cermets." Am. Ceram. Soc. Bull., 37 [7]:334-339.

17. Taguchi, H., Y. Takahashi and H. Miyamoto, 1985. "Slip Casting of Partially Stabilized Zirconia." Am. Ceram. Soc. Bull., 64 [2]:325.

18. Taguchi, H., Y. Takahashi and H. Miyamoto, 1985. "Effect of Milling on Slip Casting of Partially Stabilized Zirconia." J. Am. Ceram. Soc., 68 [10]:C-264-265.

19. Moreno, R., J. Requena and J.S. Moya, 1988. "Slip Casting of Yttria-Stabilized Tetragonal Polycrystals." J. Am. Ceram. Soc., 71 [12]:1036-1040.

20. Ruys, A.J., School of Materials Science and Engineering, University of New South Wales, personal communication.

21. St. Pierre, P.D.S., 1952. "pH/Viscosity Relationships in the System Zirconia - Water - Poly(Vinyl Alcohol) - Hydrochloric Acid." Brit. Ceram. Soc. Trans., 51 [4]:260-268.

22. Dobrovol'skii A.G., and E.Ya. Popichenko, 1965. "Slip Casting from Titanium Carbide." Sov. Powd. Metall. Met. Ceram., 5 [9]:45-52.

23. Avgustinik A.I. and G.V. Drozdetskaya, 1967. "Conditions for Making Manufactured Objects from Metal-Like Refractories by Dross Casting." Izv. Vyssh. Ucheb. Zaved.. Khim. Khim. Tekhnol., 10 [5]:556-558.

24. Dobrovol'skii A.G., 1969. "Slip Casting of a Zirconium Carbide-Tungsten Composition." Sov. Powd. Metall. Met. Ceram., 9 [2]:94-99.

25. Avgustinik A.I., G.V. Drozdetskaya and G.V. Ordan'yan, 1970. "Structural-Mechanical Properties of Aqueous Suspensions of Titanium Carbide and Zirconium Carbide." Porosh. Metall., 10 [10]:90-94.

26. Kwang-Lung L. and W. Huey-Chang, 1989. "Dispersion of Alkoxide-Hydrolysed Zirconia Powders in Aqueous Suspensions." J. Mater. Sci. Lett., 8 [1]:49-51.

27. Suzuki H. and K. Ozaki, 1987. "Characteristics of Partially Stabilized Zirconia Powder for Slip Casting." Ceram. Powder Sci., 21:627-633.

28. Daicel Chemical Industries, Ltd., Manufacturer's Specifications (CMC Daicel), Tokyo, Japan.

Low Temperature Sinter Forging of Nanostructured Y-TZP

M. M. R. BOUTZ, A. J. A. WINNUBST, A. J. BURGGRAAF,
M. NAUER and C. CARRY

ABSTRACT

Results of sinter forging tests are presented conducted on 2.6Y-TZP powder preforms under constant load at 1100°C. The investigated powder batch has been prepared by a gel precipitation technique, using metal chlorides as precursor chemicals. The dependence of both creep and densification on the applied stress (28 - 84 MPa) has been analyzed by means of interrupted tests. At 93% relative density grain sizes equal to 0.1 μm have been obtained. The value (2.5) of the creep stress exponent found for the sinter forging material is in good agreement with the one determined from creep tests on dense samples of identical composition. The creep behaviour of both porous and dense Y-TZP has been interpreted as interface reaction controlled.

INTRODUCTION

The intrinsic driving force for densification is often represented by the so-called sintering pressure. This sintering pressure is inversely proportional to both particle size and pore radius [1]. A high intrinsic driving force can thus be reached by using nanostructured (crystallite size 8 nm) powders -as done in the present investigation in the case of yttria stabilised tetragonal zirconia polycrystals (Y-TZP)- to produce green compacts, in which the pore radius equals 3 nm after cold isostatic compaction at 400 MPa (see also [2]). During free sintering these compacts reach 95% or more of the theoretical density at a relatively low temperature of 1150°C (10 hrs), the final grain size being equal to 0.2 μm [3]. The application of an external pressure acts as an additional driving force for densification and this can be used to further reduce sintering time and temperature.

Three main types of pressure assisted sintering techniques exist:

M.M.R. Boutz, A.J.A. Winnubst, A.J. Burggraaf, University of Twente, Faculty of Chemical Technology, Laboratory for Inorganic Chemistry, Materials Science and Catalysis, P.O. Box 217, 7500 AE Enschede, The Netherlands.
M. Nauer and C. Carry, Laboratoire de Céramique, EPF Lausanne, MX-D Ecublens, CH 1015 Lausanne, Switzerland.

a.   Hot isostatic pressing (HIPing); here a hydrostatic gas pressure is applied to the specimen and the shrinkage is isotropic, i.e. no shape change (creep) occurs. This technique is mostly applied to remove residual porosity (< 7 vol%) after free sintering in order to improve the mechanical properties.

b.   Hot pressing; a uniaxial pressure is applied to a powder (compact), which is constrained in the lateral direction by the die wall. The radial strain thus equals zero and the unknown parameter is the stress, exerted by the die walls on the specimen, trying to expand. It is commonly used to promote densification of inert, covalent single phase materials, like nitrides and carbides or composite materials, in which large differences in sinterability exist between the different phases.

c.   Sinter forging; here a uniaxial pressure is applied again, but the sample is free to move in the lateral direction. The radial strain can be negative or positive, depending on the relative magnitudes of sintering and creep.

During sinter forging large shear strains can be imposed on the material, which promote the elimination of residual flaws and improve the mechanical properties of the final product if the shear strain surpasses a certain limit as shown by Venkatachari & Raj in the case of alumina [4]. During free sintering and HIP'ing the shear strain is zero, while the shear strain is identical to the axial strain in the case of hot pressing, leading to a fixed ratio between the volumetric strain and the shear strain. Only for sinter forging the processing path is not fixed and the amount of shear strain depends on the magnitude of the applied stress. This makes sinter forging an interesting method for pressure assisted densification.

In this paper the results of sinter forging tests are described, conducted on cylindrical compacts of Y-TZP under constant load at 1100°C. This temperature is 50°C below the sintering temperature (defined as the minimum temperature required to obtain densities $\geq 95\%$) of this particular, non-commercial powder batch.

The dependence of both creep and densification rate on the applied stress has been analyzed using the approach proposed by Raj [5]. For this purpose the density and radial strain have been monitored by means of interrupted tests.

EXPERIMENTAL PROCEDURE

Cylindrical compacts were prepared by cold isostatic compaction at 400 MPa of an ultra-fine (crystallite size 8 nm) yttria-stabilized (2.6 mol% Y_2O_3) zirconia powder, prepared by a gel precipitation technique, using metal chlorides as precursor chemicals (hence 'chloride'-method). Details of the powder synthesis method can be found in [2]. Chemical analysis by AAS of the used powder batch gave the following impurity concentrations (in wt%): .146 Al, .040 Si, .005 K, .004 Na, .001 Ca and < .002 Fe.

Densities were measured by the Archimedes technique (in Hg). Nitrogen adsorption/desorption isotherms were obtained at 77 K using a Micromeritics ASAP 2400 system. Specific surface areas were determined by the BET-method. Pores having widths exceeding 7.5 nm were characterized by Hg penetration assuming a cylindrical pore shape.

The green density after CIP'ing equals 45% of the theoretical one (6.06 g/cm^3). To improve the strength of the green compacts, they were presintered by heating at 120°C/hr to 950°C, immediately followed by cooling down. This resulted in a slight increase in density to 48.5%, but more importantly the accessible surface area decreased dramatically from > 100 m^2/g to 20 m^2/g. Since the porosity is open at such a low density-value, it can be concluded that the neck area increased considerably, thereby improving the mechanical strength and thermoshock resistance. This allowed the use of relatively high heating rates up to 950°C. Dimensions after the presintering-step typically were 15.7 mm (height) x 7.3 mm (diameter).

Sinter forging test were performed at 1100°C in air, using an Instron testing machine. Samples were heated at 600°C/hr to 950°C, followed by 300°C/hr to the end temperature. During heating up a small load (40-50 N≈ 1 MPa) ensured that the piston remained in contact with the shrinking specimen. Once the end temperature was reached, compression was started after a total height reduction of 10.9% for all specimens. This height reduction results from densification during heating from 950 to 1100°C. Three different loads, 1000-2000-3000 N, were used. The time required to increase the load to its final value was kept constant at 200 sec. At the end of the test the furnace was switched off, resulting in a sudden temperature drop.

The axial displacement Δh was taken from the internal LVDT and the radial displacement was calculated from the mass, height and final density of the samples after interrupted tests.

The procedure outlined by Raj [5] has been used to separate creep and densification, which occur simultaneously during sinter forging. The creep or shear strain (ε_e) is defined herein as:

$$\varepsilon_e = 2/3 \mid \varepsilon_z - \varepsilon_r \mid = \mid \varepsilon_z - (\varepsilon_v / 3) \mid \tag{1}$$

Axial (ε_z), radial (ε_r) and volumetric (ε_v) strains have been calculated with the starting dimensions (before testing) as reference values.

The free sintering behaviour of the 2.6Y-TZP compacts was analyzed by means of a Netzsch 402E dilatometer. Density was calculated from the initial one and the observed axial shrinkage. Grain sizes D were determined by the lineal intercept technique from SEM micrographs of polished, thermally etched cuts using D=1.56 L, where L is the average lineal intercept, corrected for the presence of residual porosity using the method of Wurst & Nelson [6].

EXPERIMENTAL RESULTS

SINTER FORGING CHARACTERISTICS

The increase in relative density with time during sinter forging and free sintering at 1100°C is shown in figure 1. The applied loads, 1000-2000-3000 N, resulted in initial stresses equal to 28, 58 and 84 MPa respectively. From figure 1 it is clear that the applied stress has a strong beneficial effect on the densification kinetics. Final densities equal 93-94% for the sinter forged samples, while the free sintering sample reaches only 84% after 10 hours. The time required to reach a certain density value decreases strongly with increasing load. To reach 93-94% 158 min are required under 1000 N, while only 22 min are needed under 3000 N. Prolonged sinter forging (160-185 min) under 2000-3000 N did not increase the density above 94% indicating that pure creep occurs at constant porosity at 1100°C. By raising the temperature to 1150°C densities of 97-99% have recently been obtained by sinter forging for 25 min under an initial stress of 80 MPa.

Upon reaching the final density at 1100°C the grain size equals 0.10 μm and 0.14 μm after respectively 22 min under 3000 N and 158 min under 1000 N. The grain size increases only slightly to 0.15 μm after 185 min under 3000 N. During free sintering the grain size equals 0.15 μm after 200 min. These observations show that grain growth is very sluggish and dynamic grain growth is virtually absent under the investigated experimental conditions.

The observed time dependence of the resulting stress was as follows: under 2000 and 3000 N the effective stress decreases continuously, due to radial expansion of the specimens, while under 1000 N the stress remains essentially constant.

In figure 2 the processing paths ϵ_r vs. ϵ_z of the current experiments are shown. It can be seen that the variation in diameter is only very small under the 1000 N load. Contraction, due to sintering, is balancing with expansion, due to creep. The path for 1000 N lies close to the one, that would result from a hot pressing experiment. The paths for 2000 and 3000 N are typical high shear strain paths [4]. The negative value of ϵ_r in figure 2 is due to the shrinkage during heating up before application of the load.

In figure 3 the creep strain ϵ_e is given as a function of time for the 3 different loads. To reach the final density (93-94%) the specimen has undergone a creep strain equal to 0.26 under 1000 N and this creep strain increases to 0.42 by increasing the load to 3000 N. As mentioned before, from this point on the creep strain increases without any further increase in the volumetric strain. Creep strains equal to 0.63 - 0.97 (not included in figure 3) could be reached under 2000 - 3000 N, while the density remained constant at 93 -94%. To investigate which creep mechanism is operating during sinter forging the stress dependence of the creep rate has been analyzed assuming the relation $\dot{\epsilon}_e \sim \sigma_e^n$ holds, where n is the creep stress exponent. The internal load bearing area changes continuously during sinter forging and with it the stress intensification factor ϕ. This factor is a function of density only; its precise functional dependence however is a matter of debate [7] and depends on the assumed pore and grain geometry. Taking the strain rates at constant density ensures that the stress intensification factor is constant. In figure 4 the stress dependence of the creep strain rate is shown for 85% and 90% relative density. For both density values the stress exponent equals 2.5. Observed creep rates are 10^{-5}-10^{-4} sec^{-1}, typical for superplastic forming, at 1100°C only.

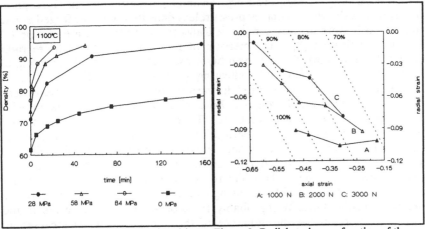

Figure 1 Increase of relative density with time at 1100°C during free sintering (0 MPa) and sinter forging under initial stresses equal to 28, 58 and 84 MPa.

Figure 2 Radial strain as a function of the axial strain of the present sinter forging experiments. Dashed lines represent isodensity-contours.

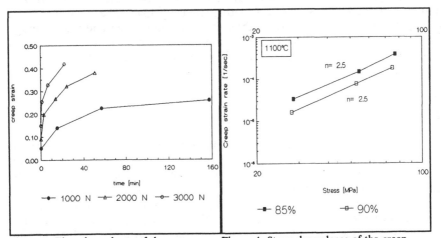

Figure 3 Time-dependence of the creep strain during the present sinter forging experiments.

Figure 4 Stress dependence of the creep strain rate at 85% and 90% relative density.

INFLUENCE OF APPLIED STRESS ON THE POROUS TEXTURE

To investigate if the stress applied during sinter forging affects the evolution of pore size and shape compared to free sintering, samples with a relative density of approximately 80% (2nd stage of sintering with open porosity) were characterized using N_2 physical adsorption and Hg penetration. The N_2 sorption isotherms (not shown) of samples pressureless densified or sinter forged under 1000 - 3000 N were all very similar. All adsorption isotherms are of type IV (BDDT classification [8]), characteristic of mesoporous systems (pore radii between 1 and 25 nm). The observed hystereses are all of type A according to the classification of de Boer [9], indicating a cylindrical pore shape.

Capillary condensation occurred during adsorption at relative pressures near unity in the investigated samples. Pore sizes cannot be calculated with acceptable precision in such a case [10]. Therefore, Hg penetration has been used to measure the pore size distribution (PSD). The pore volumes of the studied samples were too small to accurately determine the pore size with this technique, but some qualitative conclusions can be drawn. The free sintered sample shows a bimodal PSD, while the PSD in the sinter forged materials becomes more unimodal with increasing stress due to the disappearance of the fraction of smaller pores. Most probably, these pores progressively coalesce and form larger pores under the influence of the applied stress. This causes an increase in the mean pore size in the forged samples compared to the free sintered one.

During the 3rd stage of sintering (closed porosity) pores with sizes exceeding the mean grain size are fragmented into smaller ones in sinter forged materials as found by Venkatachari & Raj [4] in the case of alumina. SEM-observations showed that also in our 93 - 94% dense materials the size and number of these 'flaws' decrease with increasing load. The majority of the residual pores have sizes smaller than the mean grain size and their average diameter is observed to decrease with increasing load. Finally, under 2000 N and 3000 N densification takes place homogeneously throughout the sample, while this is not the case under 1000 N.

DISCUSSION

Comparing the density-evolution during free sintering and sinter forging at 1100°C (figure 1) shows that the time needed to reach a certain density value, e.g. 84%, is reduced by a factor 20 - 600 by applying an initial stress equal to 28 - 84 MPa. This drastic reduction in processing time together with the virtual absence of dynamic grain growth at this temperature makes it possible to limit grain growth effectively. In this way the grain size could be kept just within the nanometer-regime ($D \leq 100$ nm), while simultaneously attaining high density (93%) during sinter forging under the initial stress of 84 MPa. To achieve fully dense 0.1 μm Y-TZP from the present powder post-HIPing (1100°C, 200 MPa) 90% dense sinter forged specimens is considered for future experiments.

In reference [3] we have clearly established, that for an identical processing of the green body the sinterability of undoped Y-TZP prepared by the 'chloride'-method is significantly improved compared to commercially available (Tosoh Co., Japan) powders. In recently performed experiments we have noticed that this difference in sinterability decreases if one uses sinter forging as a pressure assisted densification technique. Still, sintering times are reduced by a factor 5 at 1100°C compared to the commercial material and efforts to sinter forge the latter material to densities above 90% failed due to premature fracturing of the specimens.

Shear strains imposed on the material during sinter forging are effective in eliminating residual strength degrading flaws. Consolidation to near-theoretical densities together with a significant reduction in residual flaw concentration leads to improved mechanical properties compared to free sintering as has recently been observed by us in the case of Y-TZP and YCe-TZP.

The creep behaviour of the sinter forging material can be compared with the results from creep tests performed on dense 2.6Y-TZP with a grain size of 0.2 μm and a similar impurity content as reported in [3]. For the dense material a stress exponent equal to 2.2 has been determined at 1100 - 1300 °C in the stress interval 16 - 120 MPa. This value is in fairly good agreement with the creep stress exponent found in this analysis (n= 2.5). The creep behaviour of the dense Y-TZP material has been interpreted as interface reaction controlled grain boundary sliding, in good agreement with the deformation map proposed by Nauer & Carry [11]. The precise character of the interface reaction is still a matter of debate. The value of the creep stress exponent determined for sinter forging suggests that the same mechanism is operating in porous Y-TZP with grain sizes ≤ 0.1 μm (σ= 28-77 MPa, T= 1100°C). It is indeed expected from the deformation map that for 0.1 μm grain size creep is interface reaction controlled under the investigated experimental conditions.

Panda, Wang & Raj [12] have analyzed the sinter forging characteristics of 2.9Y-TZP (Tosoh, Japan) at 1400°C in Ar, using constant piston velocities. They observed a non-linear stress dependence of both densification and creep rate. The creep stress exponent equals 3 at 80 and 95% relative density in their analysis. This value of the creep stress exponent plus the fact, that the volumetric strain seems to be correlated with the effective strain only -irrespective of the applied stress- make them suggest, that a dislocation type of plastic flow is responsible for densification.

Comparing their analysis to ours, we see that the value of their creep stress exponent is somewhat higher: 3 compared to 2.5. More importantly, in our experiments the volumetric strain is not only related to the creep strain, but depends clearly on the applied stress. The sinter forging behaviour of both investigated Y-TZP materials thus differs fundamentally.

CONCLUSIONS

The application of a uniaxial load without any lateral constraints during sintering of Y-TZP at 1100°C leads to a strong reduction of sintering time. Nanostructured (grain size 0.10 μm) Y-TZP could be obtained by sinter forging of the chloride-material under an initial stress of 84 MPa at 1100°C.

The application of a uniaxial load during the 2nd stage of sintering leads to a more uniform pore size distribution, while in the last stage of sintering the average size of flaws and residual pores decreases with increasing load. This offers the possibility to significantly improve final mechanical properties compared to free sintering.

Both densification and creep rate show a non-linear relationship with the applied stress. The creep stress exponent determined in this analysis ($n=2.5$) is in good agreement with the one determined from creep tests on dense samples of identical composition. This seems to indicate, that the same mechanism - interface reaction controlled creep- is rate limiting, irrespective of the presence of porosity (max. 15 vol% here).

ACKNOWLEDGEMENTS

M. Boutz gratefully acknowledges the financial support of Akzo Chemicals BV Amsterdam for this research. R. Olde Scholtenhuis is acknowledged for powder and sample preparation, and M. Smithers for SEM-observations.

REFERENCES

1. Raj, R., 1987. "Analysis of the Sintering Pressure." J. Am. Cer. Soc., 70 (9):C-210-211.

2. Groot Zevert, W.F.M., A.J.A. Winnubst, G.S.A.M. Theunissen and A.J. Burggraaf, 1990. "Powder Preparation and Compaction Behaviour of Fine Grained Y-TZP." J. Mater. Sci., 25: 3449-3455.

3. Boutz, M.M.R., A.J.A. Winnubst, A.J. Burggraaf, M. Nauer and C. Carry, "Low Temperature Superplastic Flow of Yttria Stabilized Tetragonal Zirconia Polycrystals", Proceedings of the 2nd meeting of the European Ceramic Society, Augsburg, Germany, 1991, in print.

4. Venkatachari, K. and R. Raj, 1987. "Enhancement of Strength through Sinter Forging". J. Am. Cer. Soc. 70 (7): 514-520.

5. Raj, R., 1982. "Separation of Cavitation-Strain and Creep-Strain during Deformation". J. Am. Cer. Soc. 65 (3): C46.

6. Wurst, J.C. and J.A. Nelson, 1972. "Lineal Intercept Technique for Measuring Grain Size in Two-phase Polycrystalline Ceramics". J. Am. Cer. Soc. 55 ; 109.

7. Rahaman, M.N., L.C. De Jonghe and R.J. Brook, 1986. "Effect of Shear Stress on Sintering". J. Am. Cer. Soc. 69 (1): 53-58.

8. Gregg, S.J. and K.S.W. Sing, 1982. Adsorption, surface area and porosity, 2nd edn., London, UK: Academic Press, p. 4.

9. de Boer, J.H. and B.C. Lippens, 1964. "Studies on Pore Systems in Catalysts II. The Shapes of Pores in Aluminium Oxide Systems". J. Catal. 3: 38-43.

10. Lecloux, A.J., 1981. "Texture of Catalysts", in Catalysis. Science and Technology, Vol.2, J.R. Anderson and M. Boudart, eds., Berlin, Heidelberg, New York: Springer Verlag, pp. 172-227.

11. Nauer, M. and C. Carry, 1990. "Creep Parameters of Yttria Doped Zirconia Materials and Superplastic Deformation Mechanisms". <u>Scripta Metall. Mater.</u>, 24 (8): 1459-1463.

12. Panda, P.C., J. Wang and R. Raj, 1988. "Sinter-Forging Characteristics of Fine-Grained Zirconia". <u>J. Am. Cer. Soc.</u> 71 (1): C-507-509.

Sinter Forging as a Tool for Improving the Microstructure and Mechanical Properties of Zirconia Toughened Alumina

A. J. A. WINNUBST, Y. J. HE, P. M. V. BAKKER,
R. J. M. OLDE SCHOLTENHUIS and A. J. BURGGRAAF

ABSTRACT

Sinter forging experiments are described of zirconia toughened alumina (15 wt% ZrO_2 / 85 wt% Al_2O_3), starting from composites containing transition (θ-) alumina. Full density (> 99%) was possible within 15 minutes at 1450°C and 40 MPa and resulted in a homogeneous microstructure with an alumina grain size of 0.8 μm and a zirconia grain size of 0.3 μm. Sinter forging gives an increase in strength and fracture toughness by a factor 1.5 - 2 if compared with pressureless sintered powder compacts.

INTRODUCTION

Packing inhomogeneities in powder compacts and differential sintering in a two-phase ceramic system can give rise to the formation of flaws during subsequent (pressureless) sintering. These processing flaws can be removed effectively by applying large shear strains. The strength is optimally improved when shear strains of ± 60% are achieved [1,2]. Pressure-assisted sintering techniques like sinter forging provide the opportunity to reach large shear strains during densification. In sinter forging a powder compact is subjected to a uniaxial pressure without any lateral constraint acting on the compact by die walls. In this way large shear strains are possible. An additional advantage of sinter forging is the opportunity of combining densification and shape forming in a one-step process. Experimental work on shear deformation during sintering of zirconia/alumina composites is limited.

In this paper pressureless and sinter forging experiments are described for powder compacts of alumina doped with 15 wt % zirconia. The starting powders are prepared by a gel-precipitation technique resulting in fine-grained powders containing transition alumina. The effect of sinter-forging on the mechanical properties is also discussed.

A.J.A. Winnubst, Y.J. He, P.M.V. Bakker, R.J.M. Olde Scholtenhuis and A.J. Burggraaf, Laboratories of Inorganic Chemistry, Materials Science and Catalysis, Department of Chemical Technology, University of Twente, P.O. Box 217, 7500 AE Enschede, The Netherlands

EXPERIMENTAL PROCEDURE

The zirconia/85 wt% alumina composite powders were prepared by the hydrolysis of a solution of metal chlorides in NH_4OH (Merck, 25%). A more comprehensive description of this so-called "chloride" method is given in [3,4]. The air-dried gels (120°C) were dry-milled and subsequently calcined in air for 2 hours at 900 or 1100 °C. After isostatically pressing (400 MPa) the pressureless-sintering behaviour was investigated with a Netzsch 402 dilatometer. Nitrogen adsorption/desorption isotherms at -196 °C were measured on green compacts and on compacts after initial sintering using a Micromeritics ASAP 2400 equipment. Bulk densities were determined by the Archimedes technique (in Hg). The ceramic microstructure of polished and thermally etched ceramics was examined using a Jeol JSM-35CF scanning electron microscope.

The sinter forging experiments were performed in static air using an Elatec hot press. The samples were heated to 1450 °C at 2.5 °C/min, while the pressure was linearly increased to 40 MPa in the temperature range of 1150 - 1200 °C. More details are given in [3,4].

Changes in the dimensions of the rectangular samples during sinter forging are a combination of densification and shear deformation. An estimate of the effective shear strain (ϵ_e) was determined on the basis of the final dimensional and density changes of the compacts using the equation according to Raj [5].

The fracture toughness measurements were performed by the 3-point SENB technique with a span of 12 mm and a notch depth of 450 μm. The bending strength was measured by the 4-point bending technique with an inner and outer span of 10 and 20 mm, respectively. More experimental details are given in [4]. Phase composition was analyzed by X-ray diffraction (Phillips PW 1710) using CuKα radiation. The fraction of monoclinic zirconia (V_m) was calculated relative to the total amount of zirconia using the equation of Toraya et al. [6].

RESULTS AND DISCUSSION

POWDER CHARACTERISTICS

The powders calcined at 900 and 1100 °C both contained Θ alumina, while the zirconia phase contains a monoclinic component (75 %). Other powder microstructural parameters are however different. The most important difference between the powders as a function of the calcination temperature is the morphology of the pores within the powder compact. Because of the fact that the main phase is the alumina and zirconia is well dispersed in the final sintered (dense) alumina matrix it is assumed that the stacking of the powder particles during compaction is mainly determined by that of the Θ alumina particles.

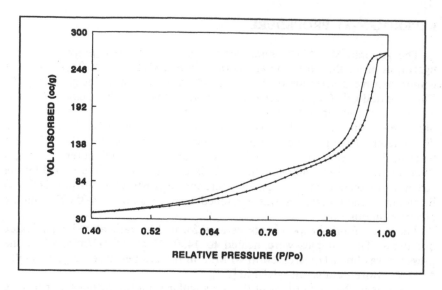

Figure 1:   Nitrogen adsorption/desorption isotherm of a powder calcined at
900 °C and isostatically pressed at 400 MPa

In figure 1 a nitrogen adsorption/desorption isotherm is given of a powder calcined at 900 °C and isostatically pressed at 400 MPa. From this figure it can be seen that two hysteresis loops are present. The hysteresis loop in the relative pressure range of 0.6 to 0.85 is only observed in compacts of powders calcined at 900 °C, while the hysteresis loop in the range 0.9 to ~ 1.0 is found in compacts of both powders. According to De Boer [7] the hysteresis loop observed in the relative pressure range of 0.6 to 0.85 can be built up from a type B hysteresis loop and an isotherm of type II. Such a combination is expected from a capillary system of open slit-shaped pores together with wedge-shaped capillaries with a closed edge at the narrower side. Most transition aluminas have a plate-like structure and therefore this pore structure is ascribed to pores between θ alumina plate-like crystallites, which are partly sintered together forming strong aggregates [3,4]. From the volume of nitrogen adsorbed in the relative pressure range from 0.6 to 0.85 a relative aggregate density of 65 % was calculated, while the individual θ alumina crystallites have a thickness of 6 nm as determined by X-ray line broadening analysis [3,4].

After calcination at 1100 °C these slit shaped pores are not observed in a powder compact. For these powders a θ alumina crystallite size of 35 nm was found [3,4]. It can therefore be concluded that the internal aggregate structure is completely densified and recrystallized to new θ alumina "single" crystals after calcination at 1100 °C.

The hysteresis loop in the relative pressure range of 0.9 to ~ 1, as is observed for both powder compacts, is of a so-called "A"-type [7]. This type of hysteresis loop can correspond to tubular capillaries with slightly widened parts.

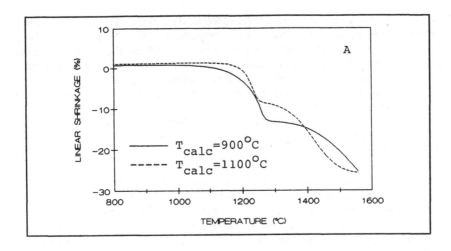

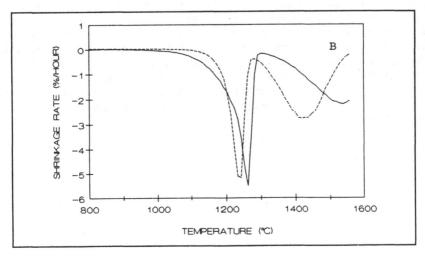

Figure 2: Dilatometer curves (A: densification; B: densification rate)

These pores represent the pores between the spherical-like (dense or porous) aggregates. From these results it can be concluded that the calcination temperature determines the pore morphology of the alumina matrix prior to sintering. This green microstructure strongly influences the pressureless sintering behaviour as will be discussed in the next section.

PRESSURELESS SINTERING

Typical densification (rate) curves of the powder compacts calcined at 900 and 1100 °C are given in figure 2. Macroscopic densification takes place in two distinct temperature regions. The temperature which separates both densification stages resembles the temperature at which the Θ to α alumina phase transformation takes place as is verified by DSC [8]. The drastic decrease in densification rate directly after this phase transformation is probably due to the large increase in grain size during the Θ to α alumina phase transformation at about 1250 °C.

The main difference in densification behaviour between the powder compacts calcined at 900 and 1100 °C occurs in the high temperature densification stage which represents the densification of the α alumina matrix. After the phase transformation the powder compact calcined at 1100 °C already shows macroscopic densification at 1300 °C with a maximum in densification rate at 1420 °C. The powder compact calcined at 900 °C only has the highest densification rate at 1520 °C. This difference in densification behaviour of the α alumina phase as a function of the calcination temperature can be ascribed to a difference in microstructure development during the phase transformation from Θ to α alumina. Just after the phase transformation to α alumina the compact of a powder calcined at 900 °C shows a broad pore size distribution with a large tail in the smaller pore size region. In this case it seems that the irregularly shaped Θ alumina platelets, which are separated by irregularly shaped pores, persist during the transformation to the α alumina phase [3,4]. A compact of a powder calcined at 1100 °C only shows uniform (cylindrical) pores with a narrow distribution after the phase transformation to α alumina. This absence of small pores after the phase transformation is held responsible for the lower temperature necessary for the densification of ZTA in comparison with compacts calcined at lower temperatures.

TABLE I - DENSITY AND MECHANICAL PROPERTIES AFTER
PRESSURELESS SINTERING AND SINTER FORGING AT 1450 °C

T_{calc} (°C)	Sintering conditions		rel. density (%)	σ_f MPa	K_{IC} (MPa√m)
	Pressure (MPa)	time (min)			
900	0	120	91		
900	40	15	100	525±40	9.1±0.8
1100	0	120	98	380±90	5.2±0.1
1100	0	15	87		
1100	40	15	100	800±95	7.4±0.4

The improved sintering behaviour is also observed during isothermal sintering. The compact (T_{calc} = 1100 °C) has a density of 98 % of the theoretical value after 2 hours of sintering at 1450 °C. This ZTA ceramic has an alumina grain size of 1.0 μm while the zirconia grains with a size of 0.4 m are well dispersed in the alumina matrix [3]. For the compact (T_{calc} = 900 °C) the same sintering treatment results in a density of 91 %. This compact densifies to 98 % after two hours of sintering at 1500 °C.

SINTER FORGING

The sinter forging experiments started at a temperature of 1150 °C while the maximum pressure (40 MPa) was present on the sample at 1200 °C (see experimental procedure). During pressureless sintering of these ZTA samples the alumina phase still has the Θ structure at a temperature of 1200 °C.

In Table I results are given of pressureless sintered and sinter-forged samples after a temperature treatment at 1450 °C. From this table it can clearly be seen that sinter-forging gives a strong improvement in densification behaviour. A dense ZTA ceramic is obtained after sinter forging for 15 minutes with a pressure of 40 MPa. Only the ideal powder compact (T_{calc} = 1100 °C) densifies at the same temperature after 2 hours of pressureless sintering.

According to [1] effective flaw elimination requires an effective shear strain of 0.6. After sinter forging for 15 minutes at 1450 °C shear strains of more than 0.75 were measured (according to [5]) while the ceramic was completely dense (99$^+$ %). Both compacts (powders calcined at 900 and 1100 °C) show the same (effective) shear strain and the same density after this treatment. It seems that the powder calcination temperature does not influence the sinter forging behaviour.

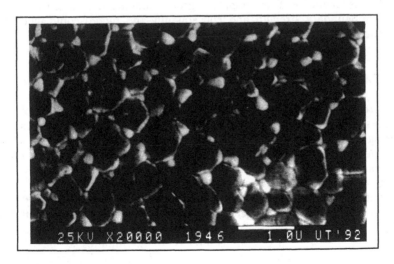

Figure 3: SEM picture of a ZTA compact sinter-forged at 1450 °C

After sinter forging for 15 minutes the microstructures of both compacts show a uniform distribution of the two phases with an alumina grain size of 0.8 μm and a zirconia grain size of 0.3 μm. A scanning electron microscope picture of a sinter-forged sample is given in figure 3. The microstructure has been investigated, both parallel and perpendicular to the direction of the applied load. No morphological texture could be observed in this way.

MECHANICAL PROPERTIES

In Table I also bending strength and fracture toughness data are given for dense, pressureless and sinter-forged compacts (T_{sinter} = 1450 °C). The sinter-forged samples clearly show a higher strength and toughness (a factor 1.5 - 2 higher than pressureless sintered).

X-ray diffraction analysis indicated that about 8 vol. % of the zirconia phase in the bulk of the pressureless and sinter-forged samples has the monoclinic structure. During cooling after sintering the main part of zirconia remains in the tetragonal structure due to the constraint of the alumina matrix acting on the dispersed zirconia particles. The amount of zirconia which transforms to the monoclinic structure during mechanical loading is almost equal for pressureless and sinter-forged samples (± 35 vol. %). The higher strength and toughness indicate that for the sinter-forged samples an extra toughening mechanism must be present beside stress induced phase transformation. One option is the improved perfection of the grain boundary structure, caused by the large effective shear strain during sinter forging. By means of sinter forging micropores will disappear while also amorphous phases at grain boundaries of pressureless sintered samples will move to triple-point junctions of grain boundaries by this technique. These improved grain-boundary structures will improve the grain-boundary strength and consequently the macroscopic mechanical properties [9].

CONCLUSIONS

- ZrO_2/85 wt% Al_2O_3 powders were prepared by the hydrolysis of metal chlorides and subsequent calcination to obtain transition alumina composites for sintering.
- Calcination at low temperature (900 °C) results in aggregates of Θ alumina platelets separated by irregular pores. After calcination at 1100 °C these aggregates are recrystallized to new "single" Θ alumina crystals.
- A compact of a powder calcined at 900 °C densifies at higher temperatures due to the unfavourable aggregate structure and pore morphology (if compared with T_{calc} = 1100 °C).
- Sinter forging of both powder compacts for 15 minutes at 1450 °C at a pressure of 40 MPa results in a complete dense ZTA ceramic. Large (effective) shear strains are observed.

- Strength and fracture toughness increase by a factor of 1.5 - 2 by sinter forging the samples.
- The contribution of stress induced transformation to the strength and toughness is the same in pressureless sintered samples and sinter-forged ones.
- The increase in toughness by sinter forging the samples is due to the improved perfection of the grain boundary structure.

ACKNOWLEDGEMENT

We would like to thank Dr. P. den Exter for helpful discussions. This research was partly supported by the Innovative Research Program on Technical Ceramics (IOP-TK) with financial aid of the Dutch Ministry of Economic Affairs

REFERENCES

1.   Venkatachari, K.R. and R. Raj, 1987. "Enhancement of Strength through Sinter Forging." J. Am. Ceram. Soc., 70: 514 - 520.
2.   Panda, P.C., J. Wang and R. Raj, 1988. "Sinter-Forging Characteristics of Fine-Grained Zirconia." J. Am. Ceram. Soc., 71: C507 - C509.
3.   Exter, P. den, A.J.A. Winnubst and A.J. Burggraaf, "The Pressureless Sintering and Sinter Forging of Zirconia Toughened Alumina" J.Am.Ceram. Soc. submitted.
4.   Exter P. den, 1991. "Synthesis, Microstructure and Mechanical Properties of Zirconia-Alumina Composites". Ph.D. Thesis, University of Twente (Enschede, The Netherlands).
5.   Raj, R., 1970. "Separation of Cavitation-Strain and Creep-Strain during Deformation". J. Am. Ceram. Soc., 65 : C46.
6.   Toraya, H., M. Yoshimura and S. Somiya, 1984. "Calibration Curve for Quantitative Analysis of the Monoclinic-Tetragonal ZrO_2 System by X-ray Diffraction". J.Am.Ceram.Soc., 67:C119-C121.
7.   Boer J.H. de, 1985. "The Shape of Capillaries" pp. 68-94 in The Structure and Properties of Materials, ed. by D.H. Evertett and F.S. Stone, Butterworths Scientific Publications (London).
8.   Exter, P. den, A.J.A. Winnubst and A.J. Burggraaf,1991. "The Sintering of ZTA at Low Temperatures through Transition Alumina Densification", To be published in the proceedings of the second European Ceramic Society Conference (The German Ceramic Society).
9.   Krell A. and P. Blank, 1992. "Inherent Reinforcement of Ceramic Microstructures by Grain Boundary Engineering". J. Eur. Ceram. Soc. 9:309-322.

Sintering of Large TZP Components

H. SCHUBERT, P. W. KOUNTOUROS, J. PROSS, G. PETZOW and A. O. BOSCHI

ABSTRACT

Fast heating rates as normally used in sintering lab samples was found to create difficulties in sintering large ZrO_2-plates. While the outer region of the sample was densified the core showed a pronounced crack pattern. In spite of the low thermal conductivity temperature gradients could be excluded as a reason. Small samples were prepared out of the large green bodies which could be investigated by mercury porosimetry and which could undergo an unconstrained shrinkage in a dilatometer. The core had a 0.7% lower green density but the final density was found to be almost identical. As a consequence, the shrinkage of the core was higher. Being constrained in a large plate this difference in shrinkage causes the appearance of mismatch stresses between core and outside which then cause cracking. The use of a binder containing powder resulted in green bodies with a much better green density distribution, i.e. they could be sintered to almost full density and had a strength difference of only 3 % between outside and core.

INTRODUCTION

It is well established that grain growth and densification are competing processes in sintering [1]. In order to enforce the densification fine grained powders with a very high sinter activity are used which are then rapidly fired to limit the grain growth. For larger components this procedure can be used only to a very limited extent, since high heating rates are expected to cause thermal gradients in materials with a low thermal conductivity. Fine grained powders are also known to cause packing problems which are common in thick cross sections. Hence, high heating rates are expected to cause thermal gradients in materials with a low thermal conductivity.

H. Schubert, P.W. Kountouros, J. Proß, G. Petzow

Max-Planck-Institut für Metallforschung

Heisenbergstr. 5, D 7000 Stuttgart 80

A.O. Boschi, Universidade Federale de Sao Carlos, Brasil

This paper reports the experience in sintering Y–TZP samples of 140 mm x 80 mm x 30 mm (after sintering). The typical powders for TZP processing are co–precipitated materials with 20 – 100 nm crystallite size and particles of approximately 0.3 μm in size. The green density of such fine powders is therefore not very high; typical values are in the order of 50 % t.d.. Furthermore the thermal conductivity of TZP is in the order 2 W/mK at room temperature. Thus, difficulties in sintering larger TZP sample were to be expected.

EXPERIMENTAL PROCEDURE

The starting powders (Tosoh TZ–3Y SB, TZ–2Y B and TZ–2Y, Japan) were tapped in silicone moulds and isostatically pressed at 200 MPa (Weber Pressen, Remshalden Grunbach, Germany). The green density was measured by Archimedes method and the pore radii distribution was investigated by Hg–porosimetry(Carlo Erba, Italy). The samples were sintered at heating rates ranging from 50 to 750 K/h in a box furnace (Nabertherm LHT 17/32, Germany). Shrinkage behavior was tested in a horizontal dilatometer (Baer Gerätebau, Germany). The microstructures and fractured surfaces were investigated by SEM (Cambridge Stereoscan 200, UK).

RESULTS AND DISCUSSION

ESTIMATIONS OF THE TEMPERATURE DISTRIBUTION IN THE SAMPLES

Because of the known low thermal conductivity of zirconia the temperature gradients during fast heat up (15 K/min) were expected to be large. However, a conservative estimation of the temperature gradient can be made using the following equation [2]

$$T = \quad v\,t + (\,v\,(\,\kappa^2 - 1^2\,)\,/\,2\,\lambda\,) + \Sigma\text{–function} \qquad (1)$$

v = heating rate; t = time; κ = distance to the centre,
l = 1/2 sample thickness; λ = thermal diffusivity

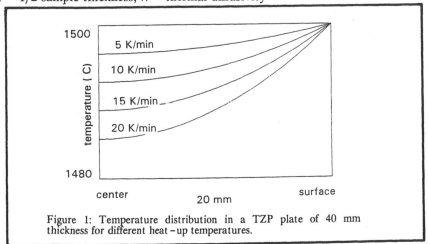

Figure 1: Temperature distribution in a TZP plate of 40 mm thickness for different heat–up temperatures.

This equation describes the temperature in a plate with one dimensional heat transport under heat – up with a constant heating rate. It can be easily shown that for our problem only the periodic part of the equation is important as the contribution of the sum (Σ–term) is negligible small. Estimating the temperature distribution in a 40 mm thick sample shows that the gradients are not very high even for rapid heating rates (Figure 1).

INVESTIGATIONS ON THE POWDERS WITHOUT BINDER

The green bodies pressed from 2Y – TZP powder (without binder) did not show any visual heterogenities at their outside. When these bodies were fractured in the green state a pronounced core structure was then visible (Figure 2). The near surface region showed a comparably flat fracture surface, whereas the core had a pronounced roughness.

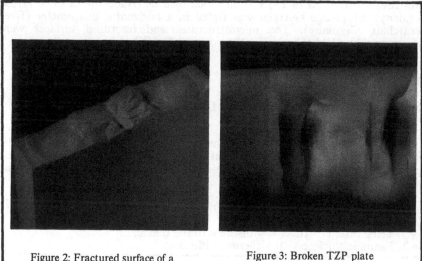

Figure 2: Fractured surface of a green body without binder; core structure.

Figure 3: Broken TZP plate after firing; core cracking.

Samples which contained this type of gradient could not be sintered to full density. Figure 3 shows a plate of approximately 140 mm x 80 mm x 30 mm size showing an extended crack network. The fracture surface of such a part again reveals a relatively dense outside and, in contrast, a crack pattern in the core region. These two positions on the fracture surface have been investigated by SEM. Figure 4a shows the micrograph of a core crack. The grains appear to be rounded, indicating that this part of the fracture surface has been altered by the influence of temperature. This argument is reinforced by the large crack opening. In contrast, the fracture surface of a fresh room temperature crack

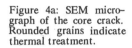

Figure 4a: SEM micro-graph of the core crack. Rounded grains indicate thermal treatment.

Figure 4b: Fresh room temperature crack.

Figure 5: Low magni-fication SEM of the core crack showing relics of granules.

shows sharp edged grains and a typical intergranular crack propagation (Figure 4b). Furthermore, the low magnification micrograph of the crack surface shows a modulation which is typical for relics of granules in green bodies (Figure 5). The presence of hard granules is attributed to the decomposition of binders under the influence of temperature (during spray drying) and usually results in this type of defect. All three observations lead to the conclusion that the cracks have already opened during heat–up.

The crack formation as a consequence of green density gradients was then studied on small samples (5 mm x 10 mm x 30 mm) which were prepared from the large plate by soft grinding. The average pore radius does not differ very much between outside and core, but the final green density is approximately 0.7 wt.% higher in the sample taken from the outside (Figure 6). Another pair of samples was then investigated in the dilatometer (Figure 7). Both samples reached approximately the similar final density (6.1501 g/cm^3 for the core sample and 6.1727 g/cm^3 for the outside sample). Consequently, the shrinkage of the core sample was higher. The largest difference in sample length appears in the temperature range before reaching the soaking temperature as can be seen from the dilatometer curves. It is believed that the cracks in the core are formed in this temperature range. Actually, figure 7 shows that, at any temperature up to about 1420^{0}C, the outside shrinks slightly more than the core. However, from then on there is further shrinkage of the core but none at the outside, so core cracking probably occurs at 1420^{0}C to 1600^{0}C.

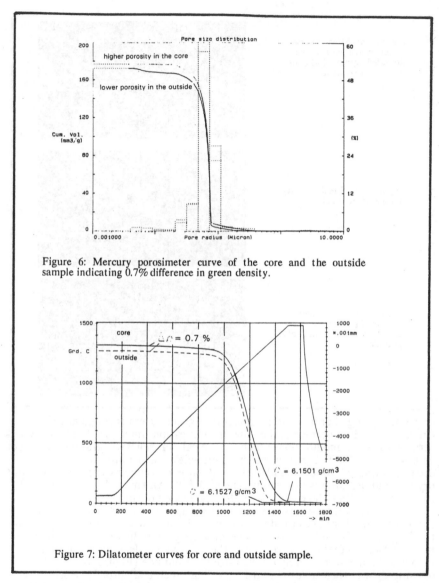

Figure 6: Mercury porosimeter curve of the core and the outside sample indicating 0.7% difference in green density.

Figure 7: Dilatometer curves for core and outside sample.

The porosimeter and dilatometer experiments show quite clearly the detrimental effects of a green density gradient. The main reason for the appearance of a density gradient is friction between the powder particles during pressing. These frictional forces reduce the applied stress from the surface to the inside. Therefore, the effective compaction stress in the inside of the sample is much lower compared to the surface.

BINDER CONTAINING SAMPLES

The use of binders decreases the green density gradient in large samples. There were no measurable differences in green density in the plate made from binder containing powder (TZ–3Y SB). Applying slow heat up (60 K/h) and an approximately 30⁰C higher final sinter temperature (1530⁰C for the large sample compared to 1500⁰C for the small sample) resulted in satisfactory final densities of $\geq$99 % t.d. without fracturing. The microstructure shows only small differences from the outside to the core (Figure 8 a–c). The outside shows a fully densified microstructure without microporosity and the same observation is made for samples taken from a depth of 7 mm. The samples taken from the core region, however, show a microporosity. The grain size of all three samples was very similar, but it was approximately 25% larger compared to a fast fired small lab sample (1 cm x 1 cm; 750 K/h heating rate). Because of the long duration at high temperature the plates have suffered grain growth. In general, the reduction of the frictional forces combined with a sufficiently slow heating rate resulted in dense samples but at the expense of grain growth.

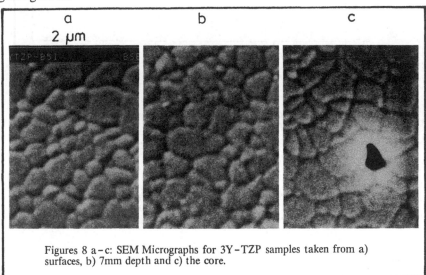

Figures 8 a–c: SEM Micrographs for 3Y–TZP samples taken from a) surfaces, b) 7mm depth and c) the core.

In order to control the quality of the plate, mechanical testing was performed on two sets of samples from the outside and core region prepared by diamond machining. Their strength was investigated by 4 point bend test (40/20 span, 10^4 MPa/s load rate). The strength was slightly higher at the outside (σ_0 = 1042 MPa, $<\sigma>$ = 1006, m = 11) compared to the core (σ_0 = 983 MPa, $<\sigma>$ = 949 MPa, m = 12); see figure 9. The difference in strength was only approximately 3%. Fractography showed the same type of defects in both sets of samples.

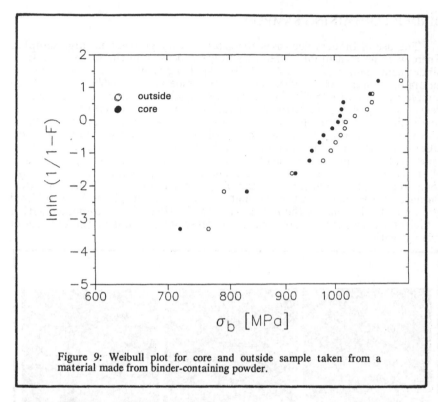

Figure 9: Weibull plot for core and outside sample taken from a material made from binder-containing powder.

ACKNOWLEDGEMENT

This work was supported by the Federal Minister of Science and Technology under contract 03 M 2045 and the state of Baden-Württemberg under KKS-Project.

REFERENCES

1. Kingery, W.D., H.K. Bowen and D.R. Uhlmann, 1976. Introduction to Ceramics, 2nd edition, New York, NY, USA: Wiley.

2. Carslaw, H.S. and J.C. Jaeger, 1959. Conduction of Heat in Solids, 2nd edition, Oxford, UK: Clarendon Press, p.105.

Fabrication and Characterization of Microwave Sintered Zirconia Ceramics

S. A. NIGHTINGALE, R. H. J. HANNINK and S. STREET

ABSTRACT

In the first stage of the project, 3 mol% Y_2O_3-ZrO_2 (Y-TZP) was uniaxially dry pressed at various pressures, and then microwave sintered using a hybrid heating system in a commercial 2.45 GHz microwave chamber. Most samples were successfully sintered to high densities in a relatively short time, although a few were slightly warped and cracked. The results of this work showed that microwave sintering is similar to conventional sintering in that increasing green densities result in increased sintered densities, albeit at shorter hold times.

In the project's second stage, 3 mol% Y_2O_3-ZrO_2 and 9 mol% MgO-ZrO_2 (Mg-PSZ) powders were microwave sintered using a similar technique. For the yttria-zirconia rapid heating and cooling rates resulted in the formation of a dense, basically tetragonal zirconia polycrystal (TZP) material with an average grain size of ~0.2 μm . Fracture toughness of the Y-TZP material was low, and comparison of X-ray diffraction patterns before and after grinding of the surface indicate that the tetragonal to monoclinic transformation did not occur using this stressing method. The Mg-PSZ samples were subjected to a variety of longer heating and cooling cycles. Only one heat treatment cycle resulted in precipitate growth. Ageing of some Mg-PSZ samples in a conventional furnace at 1400°C produced a non-uniform distribution of precipitates whose size was dependent on aging time.

This work has shown that microwave sintering may have some application for production of single phase and TZP zirconia ceramics, but that difficulties may be associated with the development of optimal PSZ material.

S.A. Nightingale and S.Street, University of Wollongong, P.O. Box 8844, South Coast Mail Centre, NSW, AUSTRALIA 2521
R.H.J. Hannink, CSIRO, Division of Materials Science and Technology, Locked Bag 33, Clayton, Victoria, AUSTRALIA 3168

INTRODUCTION

Microwave sintering offers a number of attractive features over conventional heating processes used for ceramic fabrication. The first of these is the volumetric heating which results from the interaction of electromagnetic energy with the material. During exposure to microwaves, dipoles in the material (either permanent or induced) rapidly oscillate in response to the alternating electric field, and friction between dipoles causes rapid, uniform heating. Movement of charge carriers, such as ions, may also contribute thermal energy through ohmic heating. These mechanisms result in a much more uniform distribution of heat in the ceramic than is possible using conventional heating methods which rely on conduction of heat from the surface to the interior. More uniform heating means there is much less danger of damage to the sintering ceramic as a result of thermomechanical stresses, and so very rapid heating is possible if no volatiles are formed. Materials which do contain volatile components must be kept sufficiently small so that gases escape readily, or preheating can be used. It has also been found that for materials which undergo stress induced phase transformations, such as partially stabilized zirconia-alumina composites, there is an increased retention of the tetragonal phase [1].

Another advantage of microwave sintering is that rapid heating to sintering temperatures facilitates very short sintering cycles, making production of dense ceramics with very small grain sizes possible. There is also some evidence to suggest that the activation energy for sintering is lower when using a microwave chamber than in a conventional furnace, making it possible to sinter ceramics to high densities at a lower temperature than would be required to achieve the same density using conventional heating methods [2,3].

Two major problems are associated with microwave sintering of ceramics. The first of these is that most ceramics are not susceptible to microwaves at low temperatures; they must be heated to ~400 to 500°C before they begin to absorb energy from the microwaves. Secondly, once they begin to absorb energy, microwave susceptibility increases exponentially, which can result in an uncontrollable temperature rise, referred to as thermal runaway.

The aims of this project were twofold: firstly, to determine the optimum powder consolidation pressure (maximum green density), and secondly, to determine the effectiveness of microwave sintering on the microstructure and properties of two different zirconia based ceramics. These materials were a 3 mol% Y_2O_3 zirconia (Y-TZP), which would be expected to form single phase tetragonal zirconia polycrystals (TZP), and a 9.3 mol% MgO zirconia, which when conventionally heated to solid solution treat at 1700°C and controlled cooled gives a two phase structure consisting of tetragonal precipitates in a cubic stabilized zirconia matrix and refered to as magnesia-partially stabilised zirconia (Mg-PSZ).

EXPERIMENTAL

EFFECT OF CONSOLIDATION PRESSURE

A spray dried 3 mol% Y_2O_3 zirconia* (Y-TZP), with an average particle agglomerate size of 30μm was used. Samples were prepared by dry uniaxial pressing in a 25.4mm diameter die. Samples were nominally 3.0g in weight and were pressed at approximately 20, 50, 60, 70, 100, 130, 160 and 190MPa, with a dwell time of 20sec. Green densities were determined by direct measurement. Sintering was carried out in a 1.3kW, 2.45 GHz multimode microwave oven containing a mode stirrer and using a technique which had been developed in an earlier project [4]; which uses microwave susceptible hybrid material (or "susceptors") to initiate heating. Samples were subjected to 70% full power for 20 minutes, and then allowed to cool in the oven.# Density and porosity were determined by the Archimedes method using distilled water as the immersion fluid. Scanning electron microscopy (SEM) was used to examine the microstructure.

EFFECT OF HEATING CYCLE

Sintering was performed using the same method as in Part 1. Samples of binder free 3 mol% Y_2O_3 zirconia* were prepared from a spray dried powder by uniaxial pressing into 25.4 mm discs at 188 MPa. The 9.3 mol% MgO zirconia powder** was uniaxially pressed at 40 MPa, and then isopressed at 200 MPa into discs of different thicknesses (nominally 3, 5, and 10 mm) and 24mm in diameter, and bars approximately 50 x 5 x 5 mm.

A variety of sintering cycles were used. These varied from 20 min at 70% power to a sequential series of 10 min at 30%, followed by 20 min at 70%, 10 min at 100% and 10 min at 40%. The aim of the more complex cycle was try to encourage some precipitation during the cooling period.

Temperatures could not actually be measured during sintering, but were assumed to be related to the power cycles when specimen sizes, composition, and arrangement were the same.

Hardness and toughness were measured by Vickers indentation. Bulk density was measured by Archimedes immersion. Microstructures were examined using SEM and transmission electron microscopy (TEM), and grain sizes determined using image analysis techniques.

---

* Z-Tech SY Ultra 5.2 powder supplied by ICI Advanced Ceramics, Monash Business Park, Clayton, Vic. 3168

It must be noted that the microwave oven used is not capable of a true variation in the power. "70% power" means that the magnetron operates cyclically, operating at full power for 70% of the total time, and switching off during intervals comprising 30% of the time.

** 9.3 mol% $MgO-ZrO_2$ powder supplied by CSIRO Division of Materials Science and Technology.

As-sintered, polished and coarsely ground surfaces of the Y-TZP samples were subjected to X-ray diffraction to establish if stress induced transformations would be induced. Three Y-TZP samples were X-rayed before and after aging for 24 hrs. at 240°C in air to give an indication of the stability of the tetragonal phase in the presence of water vapour.

Some of the sintered Mg-PSZ samples were aged in an electric furnace at 1400°C for periods of 0.5, 1, 2, and 4 hours, etched in HF and examined using the SEM. Some Mg-PSZ specimens were also prepared for TEM examination.

RESULTS AND DISCUSSION

EFFECT OF CONSOLIDATION PRESSURE

Sintered densities approaching theoretical density were achieved in 20 minutes using the 2.45GHz microwave hybrid heating system. In total, 72 samples were microwave sintered, all having densities in excess of 98% theoretical. However, some samples were cracked and/or warped after sintering. Cracking was not of the "onion" type observed by Janney et al [3] but was usually a single crack originating from the edge of the sample.

As shown in figure 1, increasing the green density by increasing consolidation pressure resulted in higher sintered densities, as might be expected. Scatter in the sintered densities could be attributed to a number of factors. One of these is the variation in field intensity in the multi-mode heating system that was used. Another factor may have been the degradation of the insulation material. Although it was regularly replaced, reduced efficiency of the insulation may have been enough to lower the temperatures in some trials resulting in lower densities, as shown by Sutton [5] and Holcombe and Dykes [6].

Cracks which formed in some samples may have been the result of excessive heating rates. Although temperature was not measured during sintering, measurements performed at a later stage in this project indicated that the heating rate was in excess of 100°C/min, with a temperature of 1200°C attained after 10 minutes . Janney et al studying similar sized samples of 8 mol% Y_2O_3 zirconia, found that to produce crack free samples slow heating rates were required. Microwave heating rates of 10, 20 and 30°C/min still produced cracked samples, despite the samples having been prefired to remove all volatile components [3].

Localised thermal runaway, and density variations resulting from the uniaxial pressing are also possible causes of cracking.

Microstructural observation showed that microwave sintering was not capable of removing processing defects from the green microstructure. Intact agglomerates and lenticular voids were dispersed throughout sintered samples pressed at 20MPa. These samples had by far the greatest porosity and lowest density of all the sintered samples. At higher consolidation pressures no remnant agglomerate structure was observed. The grain size was found to be similar for all the samples regardless of consolidation pressure, being ~0.2μm, only the density varying.

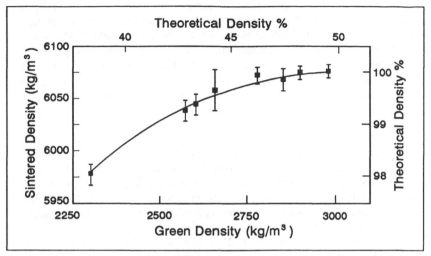

Figure 1: Sintered density as a function of green density.

EFFECT OF HEATING CYCLE

a) Y-TZP

The samples of yttria zirconia were found to be similar to the highest green density samples of Part 1; a dense, single phase tetragonal structure having an average grain size of $0.22\mu m$, and the average bulk density was 6.06 g/cm^3 (99.7% of theoretical). When sintered conventionally, this ceramic would be expected to have a bulk density of 6.07 to 6.08 g/cm^3, and a grain size of approximately 0.5 to $1\mu m$. Accurate measurement of the temperature of the samples during microwave heating has proven to be difficult. However, in the later parts of this work, measurements performed using an infra red pyrometer indicated that a temperature of approximately 1300°C is attained after 15 minutes, and at the end of the 20 minute sintering period the final temperature is approximately 1320°C. It seems that densification has occurred at a temperature lower than would normally be expected for this material, which may account for the smaller grain size in the microwave sintered sample.

Measured hardness results were in the range of 14 to 15 GPa, comparable to conventionally sintered material. The toughness of these ceramics was in the range of 3 to 4 MPa√m, which is low for Y-TZP material. X-ray diffraction of as-sintered, ground and polished surfaces indicate that the zirconia was in the tetragonal form, and that no transformation occurred on grinding. It is well known that the toughness of Y-TZP is related to its grain size; with smaller grain sizes giving lower toughness, as a result of undergoing less transformation to the monoclinic form.

Similarly, aging tests*, which involved exposure to water vapour at 240°C at ambient pressure for 24 hrs, also showed that the material was transformation resistant. No change was observed in the monoclinic phase content, which was less than 4%, as compared to a change from ~5% to 8-10% monoclinic phase for conventionally sintered material [6]. The lack of transformation is attributed to the small grain size. It has also been shown that Y-TZP ceramics with a grain size of less than 0.4μm undergo little degradation in warm, moist environments [7].

Figure 2 compares the microstructures developed as a result of conventional and microwave sintering for Y-TZP material. It is evident from these figures, that while grain size is considerably smaller for the microwave sintered sample, the substructure within grains is very similar. This fact is highlighted by four point DC probe measurements which indicate that the impedance only differs with respect to the conventionally sintered material for the component attributable to porosity (unpublished data Ciacchi and Badwal), the grain boundary and grain matrix components showing only minor variations.

Grain size of microwave sintered zirconia materials has been found to be smaller then the grain size for conventionally heated samples by a number of investigators. Wilson and Kunz [8] sintered bars of a 3 mol% Y_2O_3-ZrO_2 powder at temperatures between 1360 and 1460°C, for a range of times, with total length of the heating cycle varying from 6 to 65 minutes. All samples showed a similar microstructure, with an average grain size of 0.5 μm. This was slightly smaller than the grain size found when the powder was sintered conventionally in the same temperature range. Mechanical properties were slightly inferior for the microwave sintered ceramic .

No work has been reported comparing microwave and conventionally sintered Y-TZP where the thermal cycle has been replicated. Samuels and Brandon [9] compared microwave sintered 12wt% Y_2O_3-ZrO_2 (Y-CSZ) with samples sintered in a SuperKanthal furnace using identical heating profiles and peak temperatures of 1600°C. These workers found that the grains in the microwave sintered samples were larger (~5 μm) than those sintered in the SuperKanthal furnace, with a peak sintering temperature of 1600°C. It appears that to achieve the same density, lower temperatures and shorter times are required for microwave sintering of equivalent material, to that produced by more conventional techniques. However, it seems that if samples which have undergone identical thermal cycles by the two methods are compared, the accelerated diffusion mechanisms which appear to operate in the presence of microwaves also produce greater grain growth. The enhanced grain growth is also consistent with an increase in diffusion parameters. The consolidation and diffusion mechanisms have yet to be identified, but enhanced diffusion during microwave heating has been shown for polycrystalline alumina, zirconia and for ^{18}O in sapphire single crystals [2,3,7,10].

---

*Conducted by ICI Advanced Ceramics.

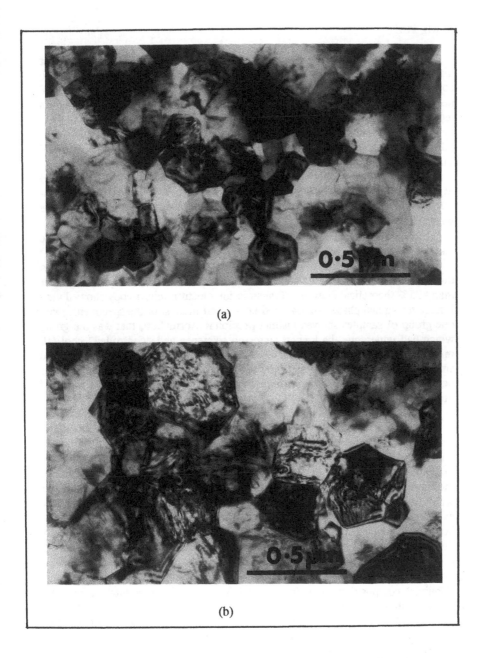

Figure 2: Microstructure of (a) microwave sintered and (b) conventionally sintered Y-TZP (TEM).

b) Mg-PSZ

The temperature required to achieve solid solution for Mg-PSZ is much higher than that required for Y-TZP. A temperature greater than 1700°C is needed to convert the Mg-PSZ material completely to the cubic phase, whereas the Y-TZP need only be heated to just over 1000°C to produce the complete solid solution tetragonal phase. Despite using variations in sample geometry (bar, disc, cylinder) and heating cycle, temperatures attained were not sufficiently high to allow formation of a uniform single phase cubic solid solution in the Mg-PSZ samples.

In general, the following physical properties were obtained for the as-fired Mg-PSZ; the density 5.44-5.60 g/cm^3; grain size 0.8-3.0μm; hardness 8-10GPa and fracture strength 94-240MPa. The hardness and fracture strength were only marginally related to the firing cycle used.

X-ray diffraction scans indicate the samples contained a mixture of monoclinic and cubic phases, no tetragonal phase being detected.

Even though a number of the sintering cycles were considerably longer than those used for the TZP, they were still insufficient to achieve densities greater than ~95% theoretical density. Transmission electron microscopy showed very limited tetragonal phase material, and none at all in most of the specimens. Only one group of samples showed distinct precipitate formation; that was the group which had undergone the longest sintering sequence and "controlled" cool (10 min 30%, 20 min 70%, 10 min 100%, 10 min 40%). These samples contained small precipitates approximately 60 to 100 nm in length. The precipitates occurred in irregularly spread patches throughout some of the cubic stabilized grains.

Some samples which showed no precipitate formation after microwave sintering (30 min, 70% power) were subsequently aged at 1400°C in an electric furnace in order to determine if precipitate growth rate was different to conventionally sintered materials. The material underwent significant precipitation and growth, but the distribution was also non-uniform. The size of these precipitates varied with aging time. After 2 hours of aging, precipitates within the grains were mainly 50 to 100 nm in length, smaller than would be expected when aging conventional Mg-PSZ (~180 nm), but along grain boundaries they were up to 2 μm long. Some regions appeared to be completely devoid of precipitates. It is possible that the cubic grains are solute enriched as a result of the presence of significant amounts of monoclinic zirconia. This would tend to push the composition of the cubic phase towards the fully stabilized composition, lowering the driving force for precipitate formation. Figure 3 shows typical SEM microstructures of the 1400°C aged samples.

Although temperatures have been reached which were sufficient to cause reasonably good densification and formation of some cubic phase, the time at maximum temperature achieved appeared inadequate for formation of a fully dense, complete solid solution single phase microstructure. Recent tests have shown that following a cycle of 20 minutes at 70% power, and 5 minutes at 100% power, the sample temperature has been above 1600°C. This is still not high enough for the complete formation of the cubic solid solution.

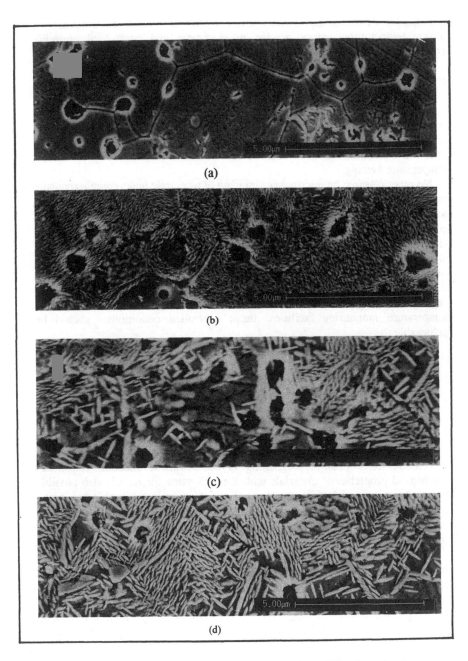

Figure 3: Microwave sintered Mg-PSZ, SEM images (a) as sintered (b) after aging for 1 hr, (c) 2 hrs and (d) 4 hrs at 1400°C.

A new microwave chamber is under construction which will allow attainment of higher temperatures, more precise measurement and control of sintering conditions.

Limitations on the measurement, control and maximum temperatures attainable make production of a uniform microstructure difficult at this stage. Rapid heating does not appear to offer any major advantage, in that the need for a homogeneous distribution of precipitates which are of uniform size requires that complete solid solution is achieved folowed by a controlled precipitation and growth cycle. However, we are confident that further work in the new microwave chamber may show there to be a benefit in the reduced costs for high temperature heating.

No other research on microwave sintering of a ceramic undergoing a requisite precipitation process is known to have been reported. While ceramic composites have been sintered successfully [11,12,13] none of these have required controlled cooling to produce the desired microstructure. By far the majority of microwave sintering experiments have been performed using single phase materials, such as alumina, where small, uniform grains in a dense microstructure are the desired outcome, For such materials the advantages of rapid heating are not counteracted by the disadvantages of a prolonged hold or controlled cooling regime and aging process. Once again, with increased power and more precise temperature monitoring facilities, these controlled conditions should be achievable.

CONCLUSIONS

This work has shown that there is some potential for the development of a process which will enable microwave energy to be used for sintering Y-TZP ceramics to high densities. Materials with a very small, uniform grain size can be produced, which is an advantage if stability in moist environment is required, and where transformation toughening is not of prime concern. Improving mechanical properties of materials with 8 mol% yttria zirconia is also possible by producing materials with smaller grain sizes than is achievable by using conventional sintering methods, which may for example be of value for fuel cell components. The technique could also be of value where thermal conduction, such as for zirconia, is poor.

Dense Mg-PSZ material with the desired microstructure has not yet been achieved with the microwave equipment available. Greater power input, control and temperature measurement has been found to be necessary to produce the required heat treatment cycle. Many of the potential benefits of microwave sintering, such as rapid heating, appear to be countered by the need for a long solid solution hold, controlled cooling cycle or an additional aging process.

ACKNOWLEDGEMENT

The support of CSIRO and the Australian Research Council and the technical assistance of Mr. B. Muldowney for this project are gratefully acknowledged.

REFERENCES

1. Parks, S.S. and T.T. Meek, 1990. "Characterization of ZrO_2-Al_2O_3 Composites Sintered Using 2.45 GHZ Radiation." Ceram. Eng. Sci. Proc., 11(9-10):1395-1404.

2. Swain, M.V. 1988. "Microwave Sintering of Ceramics." Adv. Mat. and Proc., No.9, pp. 76-82.

3. Janney, M.A., C.L. Calhoun and H.D Kimrey, 1992. "Microwave Sintering of Solid Oxide Fuel Cell Materials: 1, Zirconia-8 mol% Yttria". J. Am. Ceram. Soc.,75(2):341-346.

4. Funcik, M., 1990. "Microwave Sintering of Ceramics", B. Eng. Thesis, University of Wollongong.

5. Sutton, W.H., 1988. "Microwave Firing of High Alumina Castables," in Microwave Processing of Materials, W.H Sutton, M.H. Brooks, I.J. Chabinsky, eds., Materials Research Society, Pittsburgh, PA, USA, pp. 287-295.

6. Holcombe, C.E. and N.L. Dykes, 1990. "Importance of "Casketing" for Microwave Sintering of Materials". Journal of Material Science Letters, 9(4):425-428.

7. Unpublished data, ICI Advanced Ceramics.

8. Green, D.J., R.H.J. Hannink and M.V. Swain, 1989. Transformation Toughening of Ceramics, Boca Raton, Florida, USA, CRC Press.

9. Wilson, J. and S.M. Kunz, 1988. "Microwave Sintering of Partially Stabilized Zirconia". J. Am. Ceram. Soc., 71(1):C40-C41.

10. Samuels, J. and J.R. Brandon. "The Effect of Composition on the Enhanced Microwave Sintering of Alumina-Based Composites," Electricity Research and Development Centre, Chester, U.K., paper submitted to the Journal of Materials Science.

11. Janney, M.A. and H.D. Kimrey, 1991. "Diffusion-Controlled Processes in Microwave-Fired Oxide Ceramics". Microwave Processing of Materials II Proceedings, MRS, pp. 215-227.

12. Blake, R.D. and Meek, T.T., 1986. "Microwave Processed Composite Materials". J. Mat. Sci. Let., 5(11):1097-98.

13. Katz, J.D. and R.D. Blake, 1991. "Microwave Sintering of Multiple Alumina and Composite Components". Am. Ceram. Soc. Bull. 70(8):1304-1308.

Densification Behaviour and Microstructural Features of 3Y-TZP with and without Additives

D. D. UPADHYAYA, T. R. G. KUTTY and C. GANGULY

ABSTRACT

High density yttria doped tetragonal zirconia polycrystalline (3Y-TZP) ceramics with fine monomodal grain structure (grain size < 1 micron) have wide applications and immense potential as engineering materials. The components of this material are usually fabricated by cold-compaction of powder, followed by sintering in the temperature range 1200 - 1400°C in air.

In the present investigation, densification studies of green pellets of : (i) 3Y-TZP, (ii) 3Y-TZP with 10 volume percent alumina and (iii) 3Y-TZP with 0.25 mol percent cobalt oxide were carried out up to 1400°C using a high temperature dilatometer. The grain size and distribution of sintered pellets were evaluated by transmission and scanning electron microscopes.

Cobalt oxide was found to increase the shrinkage rate and improve the densification in the temperature range 850-900°C. The maximum shrinkage rate with cobalt oxide was found to be approximately 100 micron/minute at 1060°C (for 10mm long specimens). Alumina was found to retard the shrinkage rate. Two distinct shrinkage peaks were observed in the case of alumina-bearing 3Y-TZP. The maximum shrinkage rate was observed to be 50 micron/minute at 1000°C.

INTRODUCTION

Yttria-containing tetragonal zirconia polycrystalline (Y-TZP) ceramics provide a greater advantage over other types of transformation toughened ceramics in terms of higher strength and toughness. The homogeneous and very fine grain size make Y-TZP amenable to various thermomechanical forming operations common for metals and alloys. In most cases, the presence of a thin grain-boundary amorphous phase, due to impurity pick-up (mainly SiO_2 and Al_2O_3) during processing and subsequent segregation, is observed. It has an important role in tailoring a microstructure which is stable against coarsening during sintering and on deformation at elevated temperatures. Suitable grain boundary chemistry modifications by either incorporating some specific dopants (e.g. MnO or CoO) or composite configuration with other oxides (e.g. with Al_2O_3) are the two attractive approaches usually followed in the Y-TZP system [1-4]. For example, in the MnO-Al_2O_3-SiO_2 system, a liquid phase appears at around 1150°C. A similar phenomenon is expected in the case of CoO-Al_2O_3-SiO_2.

D.D. Upadhyaya, T.R.G. Kutty and C. Ganguly, Materials Group, Bhabha Atomic Research Centre, Bombay 400 085, India.

With the above background information, in the present work the densification kinetics of ZrO_2 - 3 mol% Y_2O_3 (3Y-TZP) and two of its modified versions, viz.: one containing 0.25 mol% cobalt oxide and the other with 10 vol% Al_2O_3, have been studied using a high temperature dilatometer. The microstructures of sintered pellets of these materials are also discussed.

EXPERIMENTAL

The wet-chemical method of co-precipitation was utilised to produce fine TZP powders. The starting materials were high purity salts of $ZrOCl_2.8H_2O$ and $Y(NO_3)_3.5H_2O$. Stock solutions with cation concentration of 0.5M were prepared in distilled water. Mixed solutions were then added dropwise to a vigorously stirred liquor ammonia solution which produced a white, gelatinous precipitate. These gels were filtered, washed, dehydrated with ethanol and dried overnight at 110°C. The dried mass was calcined at 600°C for three hours in air. The resulting product was ball-milled for eight hours using TZP grinding media. For CoO doping, the acetate salt dissolved in acetone (0.25 mol% concentration) was mixed in the ground powder and for 3Y-TZP/Al_2O_3 composites, the co-precipitation route described above was employed.

For dilatometry, the samples (l0mm height x 6mm diameter) were prepared by cold-compaction of powders at 84 MPa without any binder. A high temperature dilatometer (Netzsch Model 402E) was used for evaluation of sintering rate and shrinkage and for studying rate-controlled sintering. For rate-controlled sintering, a heating rate of 10°C/minute was used from room temperature to 1400°C. The densification studies were carried out in both air and argon atmospheres. Coble's exponent model was employed for analysing the mechanism of mass transport and for the estimation of activation enthalpy. The calcined powders and sintered pellets were characterised with a Differential Scanning Calorimeter (DSC) and an X-ray Diffractometer (XRD). Transmission and Scanning Electron Microscopes (TEM & SEM) were employed to examine the microstructural features of samples sintered at 1400°C for two hours.

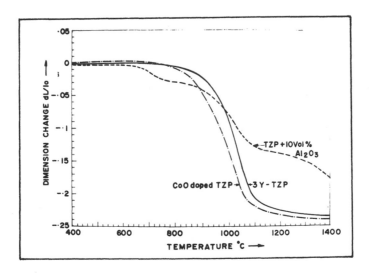

Figure 1. Linear shrinkage curves for the three 3Y-TZP powders (heating rate 10°C/min., air atm.)

RESULTS AND DISCUSSION

DENSIFICATION

Figures 1 and 2 show the linear shrinkage profiles (sintering curves) and shrinkage rate, respectively, for the three 3Y-TZP samples. These two figures reveal that for the pure phase as well as for the CoO doped powders, the onset temperature (800°C) as well as the point of inflection (~1020°C) are unchanged. The total shrinkage was found to be more for CoO doped TZP than the pure phase. The maximum shrinkage rate with CoO was found to be 120 microns/minute at around 1020°C. The difference in behaviour can readily be ascribed to the enhanced grain boundary mobility imparted by the dopant. It has been observed that in Y-TZP subjected to extensive ball-milling, liquid phase sintering is the most probable densification mechanism [2]. In such systems the action of the transition metal oxide dopant is to lower the viscosity of the siliceous glass phase which will facilitate grain boundary transport and thus in turn enhanced grain growth is expected [1]. Recent results [3,5] however, suggest a preferential, space charge segregation of divalent and trivalent solutes. Thus solute drag together with liquid phase sintering promotes the densification of Y-TZP without abnormal grain growth.

A distinct change in densification behaviour of 3Y-TZP/10 vol.% Al_2O_3 composites was observed compared to the other two powders. It essentially showed two shrinkage regions. The lower temperature part starting at around 700°C can mostly be assigned to the amorphous to crystalline transformation of the ZrO_2 powders. This is in agreement with several reports which state that in the presence of Al_2O_3 the crystallisation of ZrO_2 powders is inhibited up to 800°C [6-9]. The exothermic peak at 700°C in the DSC plots in figure 3 further corroborates this interpretation. The change in slope of the linear shrinkage profile at 1100°C is attributed to the γ to α transformation of Al_2O_3.

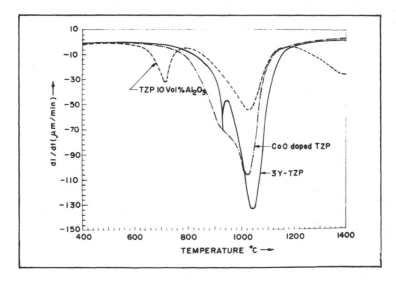

Figure 2. Shrinkage rate curves for the three 3Y-TZP powders.

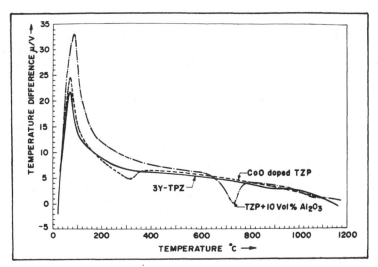

Figure 3. Differential scanning calorimetry of 3Y-TZP compositions (powders were calcined at 600°C).

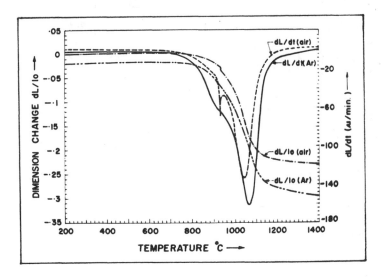

Figure 4. Effect of sintering atmosphere on the densification behaviour of 3Y-TZP.

EFFECT OF ATMOSPHERE

The shrinkage profiles of the samples sintered in air and argon at 1400°C were more or less identical, as shown in figure 4, indicating that the oxygen ion stoichiometry, which could have played a role in controlling mass transport and the densification process, is unaffected up to 1400°C in argon atmosphere.

RATE CONTROLLED SINTERING

Rate controlled sintering is a novel and reliable technique of obtaining the values of kinetic parameters in a densification reaction. This has been made possible by the availability and improved capability of modern instrumental techniques. In this method the shrinkage rate is kept constant between two threshold values. When the dl/dt signal exceeds the first threshold the sample is heated at a constant rate, whereas the temperature is kept constant when dl/dt exceeds the second threshold. This results in isothermal steps in the dl/dt versus T plots.

A typical isothermal step-scan obtained during the rate controlled sintering of 3Y-TZP is shown in figure 5. A plot of dl/l_o versus dl/dt for various steps yields the value of the exponent "n". The value of "n" was found to be ~0.3 in the present material, suggesting that grain boundary diffusion is the rate controlling mechanism [10]. The activation energy obtained from the lnk(T) versus l/T plot has a value of 129 kcal/mol for 3Y-TZP. This value is slightly higher than that reported by other authors (106 kcal/mol) [5]. The difference can be attributed to the segregation energy of associated impurities at the grain boundary [3].

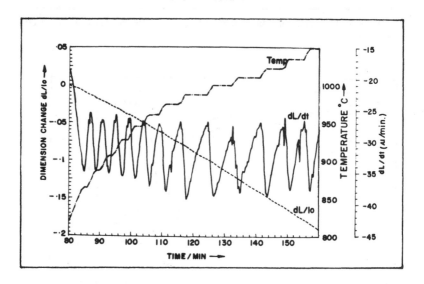

Figure 5. Rate controlled sintering of 3Y-TZP showing isothermal steps.

MICROSTRUCTURAL STUDIES

High density samples of 3Y-TZP without any additive and with either 10 volume percent Al_2O_3 or 0.25 mol percent CoO were utilised for microstructural studies. For TEM the thin sections were made electron transparent by Ar^+ ion milling. Figures 6(a), (b) and (c), respectively, present the typical micrographs for the undoped 3Y-TZP, cobalt and aluminium oxide-containing TZP ceramics. The fine (<0.5 micron) and spherical grains are separated by an apparently continuous thin amorphous phase which is typical for milled materials [1]. The rounded shape of the grains is indicative of the presence of a liquid phase at the sintering temperature. The dense TZP grains possess the fine plate-like, twin related contrast of t-ZrO$_2$. The presence of controlled amounts of glassy phase is beneficial in reducing the residual stresses developed during cooling due to anisotropic thermal contraction of the t-ZrO$_2$ grains [11].

An occasional presence of large size (>1 micron) grains was recorded only for pure TZP. With CoO doping, this tendency of abnormal grain growth was found to be completely eliminated and the average size of the grains is also considerably reduced (~0.3 micron). Similarly, for the alumina dispersed samples, a fine and isometric grain structure was obtained both for the matrix and dispersed phases.

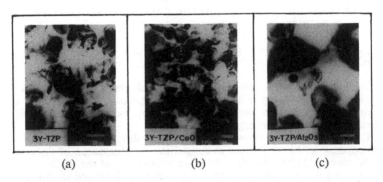

| (a) | (b) | (c) |

Figure 6. Transmission electron micrographs showing the ultrafine grain microstructure of 3Y-TZP ceramics.

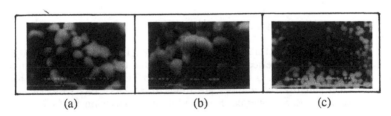

| (a) | (b) | (c) |

Figure 7. Surface grain structure of polished and heat treated 3Y-TZP ceramics : (a) undoped and containing (b) cobalt and (c) aluminium oxide .

Some clustering of discrete Al$_2$O$_3$ particles was obtained. An important feature, however, was the absence of entrapped porosity in the matrix grains and its preferential inclusion (size 10 nm) within the alumina grains. This residual porosity is evidently due to the microporous texture in the low temperature modification of alumina. The other difference observed was with respect to reduction in the amount of intergranular amorphous phase compared to the other two materials. This is due to the gettering effect of alumina in the TZP matrix [12]. Alumina thus plays a dual role during sintering of 3Y-TZP; to control the growth anisotropy due to grain-boundary pinning effects and to refine/clean the boundaries by trapping the impurities.

Microstructural evolution at elevated temperature is of great importance for Y-TZP ceramics for the envisaged superplastic forming applications. For this purpose the polished samples were refired at 1400°C/2 hours and the as-fired surface morphology was studied by scanning electron microscopy. The micrographs of figure 7 show that in the pure as well as cobalt oxide-doped material there is a tendency for abnormal grain growth but in the Y-TZP/Al_2O_3 composite the original homogeneous microstructure remains unaltered. The undesirable grain growth seems to be due to liquid film migration and surface segregation of an extensive glassy phase during high temperature annealing. It leads to the dissolution of the near surface t-ZrO_2 grains and reprecipitation of large Y_2O_3-rich cubic ZrO_2 grains, described as diffusional phase transformation [13]. The alumina dispersion has generated an ultrafine grain microstructure that is stable against coarsening even at high service temperature.

CONCLUSIONS

1.   The present work has demonstrated the usefulness of the dilatometry technique for studying the sintering behaviour of zirconia based ceramics. Rate controlled sintering further enhances the utility in predicting the sintering mechanism and in estimating the activation energy for the process.
2.   The 3Y-TZP/Al_2O_3 showed two distinct shrinkage stages, one at 700°C corresponding to the amorphous to crystalline transformation of the matrix phase and the other at 1100°C for the γ to α transformation of the Al_2O_3 reinforcement. The sintering process was found to be considerably inhibited in this mixed oxide composition.
3.   The use of CoO and Al_2O_3 was beneficial in controlling the grain growth anisotropy and in maintaining a coarsening-resistant microstructure in 3Y-TZP ceramics.

ACKNOWLEDGMENTS

The authors are grateful to the following colleagues of Materials Group, BARC; Messrs T. Jarvis, J. Banerjee and R.K. Bhagat for their assistance in dilatometry, Mr. Sunil Kumar and Dr. G.K. Dey for SEM and TEM support and Mr. K.B. Khan for the XRD pattern. The authors are thankful to Drs. S.K. Roy, Head Ceramic Science Section and S. Banerjee, Head, Metallurgy Division for their keen interest in this work.

REFERENCES

1.   Lange, F.F., H. Schubert, N. Claussen and M. Ruhle, 1986. "Effect of Attrition Milling and Post Sintering Treatment on Fabrication, Microstructure and Properties of Transformation Toughened ZrO_2." Journal of the American Ceramic Society, 69(2):768-774.

2.   Kimura, N., H. Okamura and J. Morishita, 1988. "Preparation of Low-Y_2O_3-TZP by Low-Temperature Sintering," in Advances in Ceramics, Vol.24. Science and Technology of Zirconia III, S. Somiya, N. Yamamota and H. Yanagida, eds., Westerville, Ohio, USA: The American Ceramic Society, Inc., pp.183-191.

3.   Chen, I.W. and L.A. Xue, 1990. "Development of Superplastic Structural Ceramics." Journal of the American Ceramic Society, 73(9):2585-2609.

4.   French, J.D., M.P. Harmer, H.M. Chan and G.A. Miller, 1990. "Coarsening-Resistant Dual-Phase Interpenetrating Microstructure." Journal of the American Ceramic Society, 73(8):2508-2510.

5.   Hwang S.L. and I.W. Chen, 1990. "Grain Size Control of Tetragonal Zirconia Polycrystals Using the Space Charge Concept." Journal of the American Ceramic Society, 73(11):3269-3277.

6.   Pugar, E.A. and P.D.Morgan, 1986. "Coupled Grain Growth Effect in Al_2O_3/10 vol.% ZrO_2." Journal of the American Ceramic Society, 69(6): C120-C123.

7.   Pilate, P. and F. Combier, 1986. "Sintering of Ultra-Rapidly Quenched Al_2O_3-ZrO_2 Mixed Powders." Science of Ceramics, Vol.13, P. Odier, F. Vabannes and B. Cales, eds., pp.C1-255 to C1-259.

8.   Wang, J. and R. Raj, 1991. "Control of the Microstructure of Alumina-Zirconia Alloys Starting from Inorganic Salts." Journal of the American Ceramic Society, 74(7):1707-1709.

9.   Upadhyaya, D.D., P.Y. Dalvi and G.K. Dey, 1992. "Processing and Properties of Y-TZP/Al_2O_3 Nanocomposites." Journal of Materials Science, to be published.

10. O'Hara, M.J. and I.B. Cutler, 1969. "Fabrication Science - 2." Proceedings of the British Ceramic Society, 12:145-154.

11. Lin, Y.J., P. Angelini and M.L. Mecartney, 1990. "Microstructural and Chemical Influences of Silicate Grain-Boundary Phases in Yttria-Stabilized Zirconia." Journal of the American Ceramic Society, 73(9):2728-2735.

12. Rajendran, S., J. Drennan and S.P.S. Badwal, 1987. "Effect of Alumina Addition on the Grain-Boundary and Volume Resistivity of Tetragonal Zirconia Polycrystals." Journal of Materials Science Letters, 6:1431-1434.

13. Chaim, R., D.G. Brandon and A.H. Heuer, 1986. "A Diffusional Phase Transformation in ZrO_2-4 wt.% Y_2O_3 Induced by Surface Segregation." Acta Metallurgica, 34:1933-1939.

MECHANICAL PROPERTIES

A Review of the Mechanics and Mechanisms of Cyclic Fatigue-Crack Propagation in Transformation-Toughened Zirconia Ceramics

M. J. HOFFMAN, R. H. DAUSKARDT, Y.-W. MAI
and R. O. RITCHIE

ABSTRACT

Damage and cyclic fatigue failure under alternating loading in transformation-toughened zirconia ceramics are reviewed and compared to corresponding behavior under quasi-static loading (static fatigue). Current understanding of the role of transformation toughening in influencing cyclic fatigue-crack propagation behavior is examined based on studies which altered the extent of the tetragonal-to-monoclinic phase transformation of Mg-PSZ through sub-eutectoid aging. These studies suggest that near-tip computations of the crack-driving force (in terms of the *local* stress intensity) can be used to predict crack-growth behavior under constant amplitude and variable-amplitude (spectrum) loading, using spatially resolved Raman spectroscopy to measure the extent of the transformation zones. In addition, results are reviewed which rationalize distinctions between the crack-growth behavior of pre-existing "long" (> 2 mm), through-thickness cracks and naturally-occurring, "small" (1 to 100 μm), surface cracks in terms of variations in crack-tip shielding with crack size. In the present study, the effect of grain size variations on crack-growth behavior under both monotonic (R-curve) and cyclic fatigue loading are examined. Such observations are used to speculate on the mechanisms associated with cyclic crack advance, involving such processes as alternating shear via transformation-band formation, cyclic modification of the degree of transformation toughening, and uncracked-ligament (or grain) bridging.

INTRODUCTION

Ceramic materials in their traditional sense have always been thought of as brittle in nature, displaying purely linear elastic deformation behavior; correspondingly, lifetime prediction for ceramic components subjected to cyclic loads is generally based upon the cumulative time under static loading. The development of toughened ceramics designed for use in structural applications, however, relies on energy dissipating processes associated with nonlinear constitutive behavior. These

M. J. Hoffman and Y.-W. Mai, Center for Advanced Materials Technology, Department of Mechanical Engineering, University of Sydney, Sydney, N.S.W. 2006, Australia

R. H. Dauskardt and R. O. Ritchie, Center for Advanced Materials, Lawrence Berkeley Laboratory, and Department of Materials Science and Mineral Engineering, University of California, Berkeley, CA, 94720, USA

321

shielding mechanisms which primarily act behind the crack tip and induce crack-processes result in toughening from fiber and whisker reinforcement, phase-transformation and microcracking, crack deflection and crack bridging, all crack-tip resistance or rising R-curve behavior [1].

In addition to R-curve behavior, toughened ceramics have been shown to suffer mechanical damage under cyclic loading [e.g., 2-18], despite initial claims to the contrary [19]. Consequently, lifetimes are lower than those predicted by simply integrating the effects of environmental damage or static fatigue. While detailed models of the crack-advance mechanisms are currently uncertain, preliminary observations [9,12,16,17] suggest that cyclic fatigue is attributed to a decrease in crack-tip shielding. For example, in monolithic ceramics such as coarse-grained alumina, where toughening is promoted by crack bridging from interlocking grains, the grain bridges can be degraded under cyclic loading due to wear processes from repeated sliding of the interfaces. For many ceramic materials, however, such as transformation-toughened zirconias, the microstructural mechanisms of cyclic fatigue damage remain uncertain.

The objective of this paper is to review what is known about fatigue failure in transformation-toughened zirconia ceramics, with specific reference to partially stabilized zirconia (PSZ). In these materials, toughening is due to the *in situ* stress-induced martensitic transformation of the tetragonal zirconia phase to the monoclinic phase in the vicinity of the crack tip; this phase change involves a ~4% dilation and a small degree of shear strain (~16%) [20-22]. In simple terms, toughening then results as the crack grows into the zone of transformed material, which exerts compressive forces on the crack surfaces, consequently shielding the crack tip from the applied stresses and leading to R-curve behavior (Figure 1).

CYCLIC FATIGUE BEHAVIOR

STRESS/LIFE DATA

Several studies using stress/life (S/N) testing have confirmed that cyclic loads lead to a degradation in strength in PSZ [e.g., 4,8,11,18]. Figure 2, for example, shows failure data in Mg-PSZ from bend samples where cracking was initiated from microhardness indents [18]. Lifetimes for statically loaded samples generally exceed those obtained under cyclic loading; moreover, the strength at failure is generally lower under cyclic loads. As the lifetimes of cyclically loaded samples are significantly less than those predicted from static fatigue models, it is clear that PSZ shows true cyclic fatigue degradation.

Under cyclic loading, there is clear evidence for many ceramics that lifetimes in smooth-bar S/N tests under tension-compression ($R = -1$) loads are distinctly lower than those found in corresponding tests under tension-tension ($R > 0$) loads [4,15]. For Mg-PSZ, this appears to result from the more damaging nature of fully reversed loads in promoting crack initiation as microcrack densities are observed to be far higher after cycling at $R = -1$ compared to $R = 0$ [15].

CRACK-PROPAGATION DATA: LONG CRACKS

The most persuasive evidence for cyclic fatigue degradation, however, has resulted from studies on subcritical crack growth obtained under cyclic loading conditions on "long" (> 2 mm) through-thickness cracks in compact-tension samples of MS-grade Mg-PSZ [1,7]. The rate of crack growth (Figure 3) is several orders of magnitude higher during cyclic loading than under constant loading (static fatigue) conditions at the same (applied) maximum stress intensity (K_{max}); moreover, the threshold stress intensity for crack growth is over 40% lower under cyclic loads.

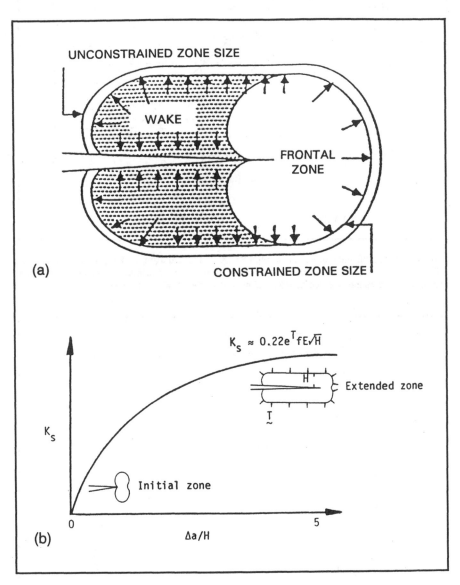

Figure 1. Schematic illustration of a) the transformation zone surrounding a crack in a phase-transforming ceramic, and b) the resulting resistance curve showing the increase in shielding stress intensity K_s with increasing crack extension Δa, normalized with respect to the zone height, H.

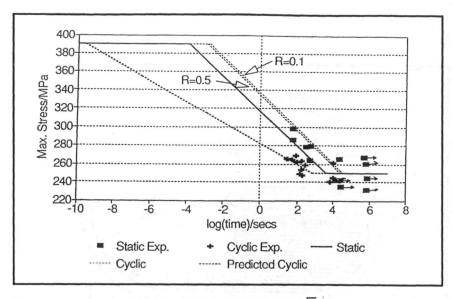

Figure 2. Stress/life data for MS-grade Mg-PSZ ($K_c \approx 11.5$ MPa$\sqrt{m}$) under quasi-static and cyclic ($R = 0.1$ and 0.5) loading, showing a comparison of experimental vs. predicted lifetimes. Horizontal arrows indicate that samples had not failed when tests were stopped [18].

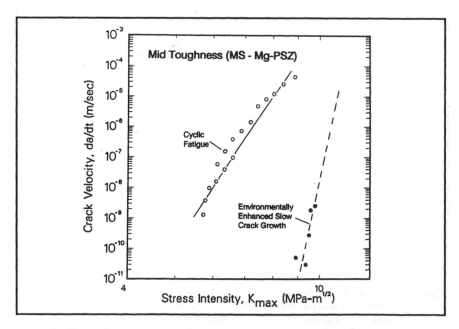

Figure 3. Comparison of the crack velocity vs. maximum stress intensity (K_{max}) curves for static fatigue (slow crack growth) and cyclic fatigue (at $R = 0.1$) in MS-grade Mg-PSZ in a moist air environment, showing cyclic crack velocities up to 7 orders of magnitude faster at equivalent K_{max} levels [7].

324

This is typical of many ceramics at room temperatures; at higher temperatures under creep conditions, conversely, growth rates may be faster under quasi-static conditions [23], although little data exist for zirconia materials.

Effects of Temperature and Environment

Although data are relatively scarce, cyclic fatigue behavior in zirconia ceramics is sensitive to the environment. Cyclic crack-growth rates, shown in figure 4 for as-fired (AF-grade) Mg-PSZ ($K_c \approx 55$ MPa$\sqrt{m}$), display a progressive acceleration in water compared to moist air and dry gaseous nitrogen [7]. Environmental effects have also been observed in rotational fatigue S/N data for PSZ tested in aqueous solutions of HCl and NaOH [4]; these results further indicate that cyclic fatigue lifetimes in PSZ decrease significantly with increasing temperature between 20 and 600°C. However, no data exist to date on the relative susceptibility of the various grades of PSZ (with differing degrees of transformation toughening) to the environment.

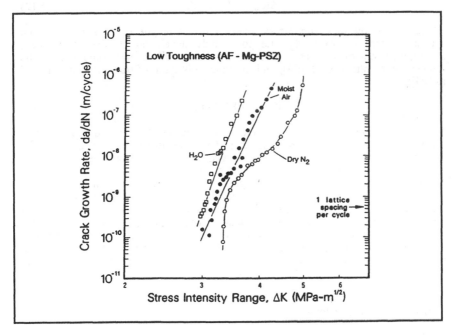

Figure 4. Cyclic crack-growth rates, da/dN, as a function of the stress-intensity range, ΔK, in AF-grade Mg-PSZ ($K_c \approx 5.5$ MPa$\sqrt{m}$) tested at 50 Hz in long-crack C(T) samples, showing accelerated growth-rate behavior in distilled water and moist air compared to dehumidified gaseous nitrogen [7].

Influence of Transformation Toughening

The degree of crack-tip shielding, specifically from transformation toughening, can also have a significant effect on cyclic crack-growth rates [7]. Microstructures which develop larger crack-tip transformation zones and which thus display steeply rising R-curves and high toughness values, are found to show the highest resistance to cyclic crack growth, as shown by the steady-state crack-growth rate vs. applied stress-intensity range (ΔK) data in figure 5a for Mg-PSZ, sub-eutectoid aged to differing (plateau) K_c toughnesses between 2.9 and 15.5 MPa$\sqrt{m}$. In general terms, growth rates (da/dN) appear to follow a simple Paris power-law:

$$da/dN = C \, \Delta K^m \tag{1}$$

where the applied $\Delta K = K_{max} - K_{min}$, and C and m are experimentally determined scaling constants. Note that values of the exponent m in Mg-PSZ range between 21 and 42, values that are far higher than those typically reported for metals. Although the data in figure 5a suggest that the exponent m increases with increasing K_c, in view of the difficulty in defining such large exponents, it is unclear whether this effect is real.

The effect of microstructure on cyclic fatigue can be rationalized in terms of the effect of the crack-tip shielding on the local (near-tip) stress intensity, ΔK_{tip}, which drives the crack [7]:

$$\Delta K_{tip} = K_{max} - K_s \tag{2}$$

where K_{max} is the maximum stress intensity in the fatigue cycle, and where the shielding stress intensity, K_s, due to a fully developed steady-state dilatant transformed zone of width H, is given in terms of the dilatational component of the transformation strain, ϵ^T (~4%), and the volume fraction of transforming phase, f, as [24]:

$$K_s = \alpha \, E' \, \epsilon^T f H^{\frac{1}{2}} \tag{3}$$

E' is the effective Young's modulus and α is a constant that depends upon frontal zone shape (= 0.22 for a hydrostatic stress-field transformation zone and 0.25 for the more semicircular profiles observed in higher toughness Mg-PSZ microstructures [25]). Using Raman spectroscopy to determine the size and shape of the transformation zone surrounding the crack, values of the near-tip stress-intensity range can be computed from equations 2 and 3; by characterizing the crack-growth rate data in terms of ΔK_{tip}, rather than the applied ΔK, crack-growth behavior for the four microstructures can be normalized (Figure 5b) [7]. This implies that the effect of transformation shielding on fatigue-crack growth is essentially identical to its effect on steady-state fracture toughness, i.e., transformation shielding reduces the effective stress intensity actually experienced at the crack tip.

Variable Amplitude Loading

The phase transformation in PSZ also has a significant impact on transient fatigue-crack propagation behavior under variable-amplitude loading [10,11]; examples of the effect of load history on crack-growth rates are shown in figure 6 for MS and TS grades of Mg-PSZ [10]. Similar to behavior in metals, high-low block loading and single tensile overloads result in a transient retardation, whereas low-high block loading results in a transient acceleration. Relatively accurate predictions of the post load-change growth rates have been made by computing the extent of crack-tip shielding in the changing transformation zone following the load changes; these predictions are shown by the dotted lines in figure 6a [10]. The varying size of the transformation zone is measured using spatially-resolved Raman spectroscopy; zone sizes for the loading sequence shown in figure 6b are plotted in figure 7, where it is clear that changes in load significantly affect the degree of phase transformation at the crack tip. Note that significant variations in growth rates are apparent for relatively small changes in ΔK; this follows because of the steep slope of the da/dN-ΔK curves found in ceramics.

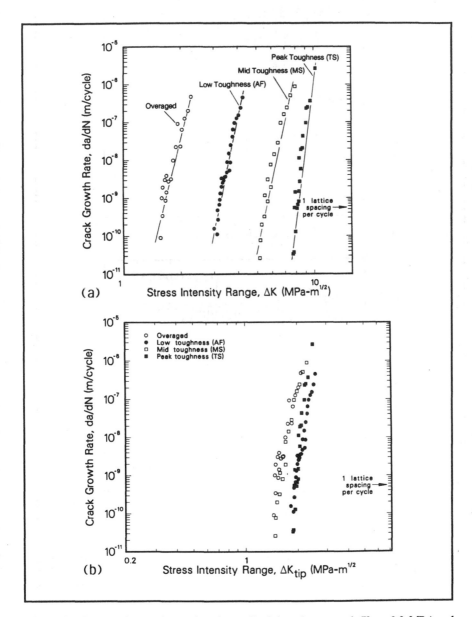

Figure 5. Long-crack growth-rate data for cyclic fatigue in overaged ($K_c \approx 2.9$ MPa) and transformation-toughened Mg-PSZ as a function of a) the applied stress-intensity range, $\Delta K = K_{max} - K_{min}$, and b) the near-tip stress-intensity range, $\Delta K = K_{max} - K_s$ [7].

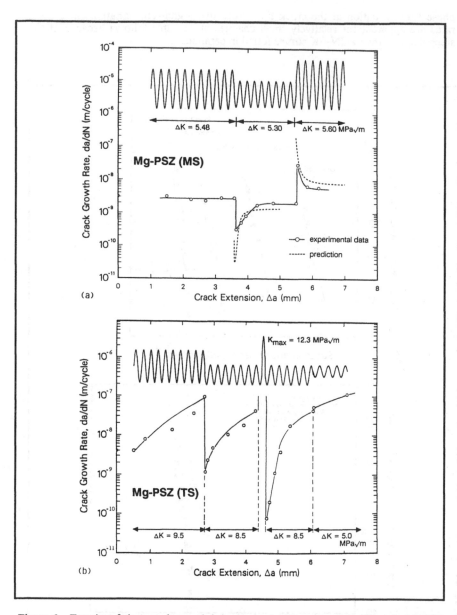

Figure 6. Transient fatigue-crack growth behavior in a) MS-grade and b) TS-grade Mg-PSZ, showing variation in growth rates following various block loading and tensile overload cycling. Predictions in a) rely on steady-state crack-growth rate data (Figure 5b) and transformation-zone size measurements (Figure 7) [10].

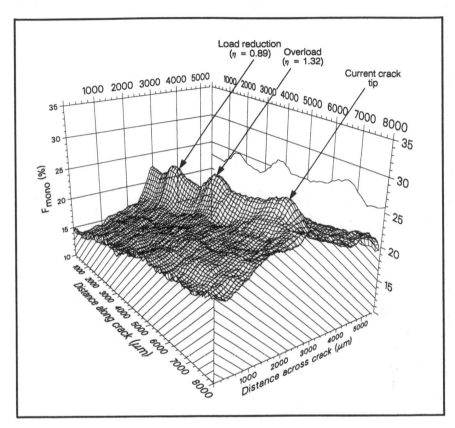

Figure 7. Morphology of the transformation zone, indicated by the volume fraction of the transformed phase, F_{mono}, derived from Raman spectroscopy measurements, surrounding a fatigue crack in TS-grade Mg-PSZ ($K_c \approx 16$ MPa$\sqrt{m}$), following the loading sequence shown in figure 6b [10].

Two models have been developed to predict the effects of variable-amplitude loading in PSZ. One idealizes the transformed zones as a series of steady-state zones with a width depending on the level of applied stress intensity [10]; the other considers a hydrostatic stress profile to cause the transformation to predict the wake effect [11]. Both models provide reasonable predictions of transient growth-rate behavior.

Grain Size Effects

Only limited studies have been performed to date on the role of (cubic) grain size on cyclic fatigue behavior in zirconia ceramics. Shown in figure 8 are R-curve and corresponding fatigue-crack propagation data for MS- and TS-grade Mg-PSZ with grain sizes of ~32, 50 and 100 μm. As noted in previous studies [7], the enhanced degree of transformation toughening in the TS-grade promotes both superior toughness and resistance to fatigue-crack growth. Moreover, in general it is apparent that toughness and fatigue resistance are increased with increase in grain

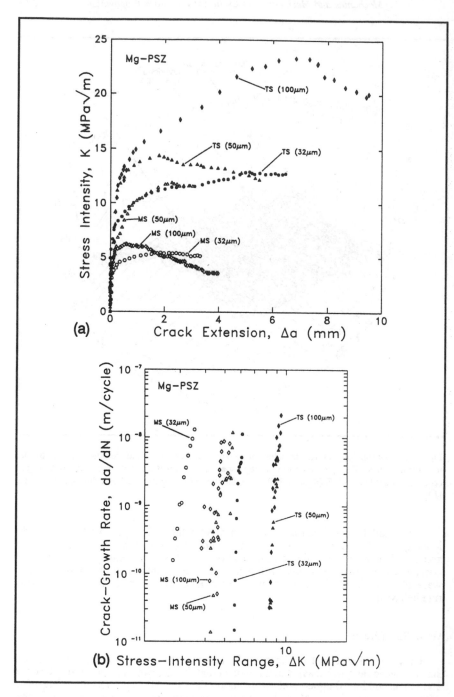

Figure 8. Effect of grain size (32, 50 and 100 μm) on a) R-curve and b) cyclic fatigue-crack growth behavior (long crack) in MS-grade Mg-PSZ.

size (although the 100-µm MS-grade sample shows a surprisingly low toughness); however, despite differences in their R-curve behavior, fatigue-crack growth behavior for 50 and 100 µm microstructures in the MS grade and the TS grade are essentially identical. The reason for such large grain-size effects are as yet uncertain but may be associated with the transformation of large grain-boundary tetragonal particles; preliminary data, however, do not show a consistent scaling of the transformation-zone size with grain size.

CRACK-PROPAGATION DATA: SMALL CRACKS

The question of small cracks (of a size approaching microstructural dimensions) is undoubtedly of special concern in ceramics; because of their low toughness, most ceramic components will not be able to tolerate the presence of physically long cracks in service. Such small cracks may be initiated from inherent processing flaws, interfaces between reinforcement and matrix phases, grain boundaries and other defects in the microstructure.

For Mg-PSZ, experimental data have been generated for naturally-occurring small flaws under both static fatigue [26] (Figure 9a) and cyclic fatigue [15] (Figure 9b) conditions. In both cases, the slopes of the logarithmic crack velocity vs. applied stress intensity curves appear to be negative; this is in direct contrast to long-crack behavior (e.g., Figures 4,5) yet consistent with extensive results on small cracks published for metallic materials [e.g., 27]. Moreover, small-crack growth rates can be seen to be far in excess of those of corresponding long cracks at the same applied stress intensity, and furthermore to occur at applied stress intensities below the long-crack threshold. Although this behavior can be attributed to a number of factors [27], the primary reason is that the extent of crack-tip shielding (in the case of Mg-PSZ from transformation toughening) is diminished with small flaws because of their limited wake. Although the applied stress intensity increases with increase in crack size, the shielding stress intensity is also enhanced as the transformation zone is developed in the crack wake. The near-tip stress intensity which drives the crack (equation 2), and hence the crack-growth behavior is thus the result of a mutual competition between these two factors; initially K_{tip} is diminished with crack extension until a steady-state transformation zone is established, whereupon behavior approaches that of a long crack. Failure ultimately results from the coalescence of such small cracks.

MECHANISMS OF CYCLIC FATIGUE

Detailed models of the crack-advance and microstructural damage mechanisms associated with cyclic fatigue in ceramics are currently not available; such models are integral to material development for optimum fatigue resistance and to the prediction of structural design lifetimes of ceramic components. However, a number of observations have been recently reported for Mg-PSZ which may shed some light on potential mechanisms; these are summarized below.

In situ studies, performed in a scanning electron microscope at temperatures between 20 and 650°C [13], showed only a limited region of stable crack growth in Mg-PSZ above ~450°C; cracking was very unstable (presumably due to the lack of appreciable phase transformation). Extrapolation of these results suggested that no subcritical crack growth by fatigue was possible at all above 750-800°C. At 20°C, however, "slip lines", similar to those seen in metal fatigue, were observed to form at the crack tip; a mechanism of crack advance was proposed involving propagation along these slip bands (which contain transformed precipitates). This mechanism invokes the contentious concept of crack extension in ceramics by alternating crack-tip blunting and resharpening, similar to the well known mechanism for

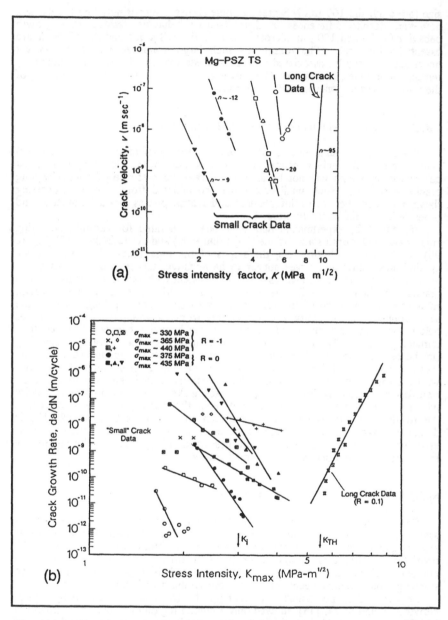

Figure 9. Small-crack data for MS-grade and TS-grade Mg-PSZ under a) static fatigue [26] and b) cyclic fatigue [15] conditions, showing accelerated small-crack growth rates, occurring below the long-crack threshold, with a negative dependency on the applied stress intensity.

fatigue-crack growth in metallic alloys. Whereas this is unlikely in non-transforming ceramics at ambient temperatures, it is conceivable that transformation-band formation at the crack tip in phase transforming ceramics such as Mg-PSZ may provide a feasible mechanism for blunting; in fact, certain authors [13,14] have reported the formation of fatigue striations on fracture surfaces in Mg-PSZ and Y-TZP. In addition, direct observations of blunting in Mg-PSZ have been reported by Davidson *et al.* [13].

The phase transformation itself, however, appears to be relatively unaffected by cyclic loads. *In situ* measurements of the morphology and nature of transformation zones in Mg-PSZ using interference microscopy and spatially-resolved Raman spectroscopy (e.g., figure 10) indicated no change in zone size surrounding monotonically and cyclically loaded cracks at equivalent stress-intensity levels [7,25,28].

In addition to crack-tip shielding due to the transformation, evidence of uncracked-ligament bridging is prevalent during crack growth in Mg-PSZ (Figure 11). Under cyclic loading, progressive wear degradation of the bridges may reduce the bridging capacity of the resulting bridging zone thereby exposing the crack tip to an increased driving force. Similar to behavior in metal-matrix composites [29], such uncracked ligaments most likely result from the formation of microcracks *ahead* of the crack tip.

Evidence of microcracking during fatigue has been reported for Y-TZP [14], where the compliance of samples loaded uniaxially was observed to increase with cyclic loading due to the formation of microcracks in the bulk of the material. These authors proposed that crack initiation occurs from pre-existing surface or bulk flaws, and that the localized stresses surrounding these flaws promote microcrack nucleation and coalescence; this results in significant strength degradation during cyclic loading. Direct observations of microcracks have been made in ZTA, Mg-PSZ and Y-TZP using transmission electron microscopy [30]; the origin of this microcracking was attributed to large residual stresses resulting from the martensitic transformation, or to large localized stresses from the elastic anisotropy [31].

CONCLUSIONS

Experimental evidence clearly shows that phase-transforming zirconia ceramics suffer significant strength degradation and premature failure under cyclic loads. The cyclic fatigue damage appears to be motivated by the inelasticity associated with transformation and enhanced microcracking, and results in the subcritical propagation of cracks by an apparently true cyclic fatigue process; the precise micromechanisms of crack advance, however, are as yet undetermined.

A significant amount of this work has centered on the propagation behavior of "long" (> 2 mm) through-thickness cracks in Mg-PSZ. Based on these studies, cyclic crack-growth rates have been found to show a power-law dependence on the stress intensity with an exponent of above 20, and to be sensitive to such variables as load ratio, crack size, temperature and environment, degree of transformation, grain size and loading spectra. In contrast, only limited studies have addressed the question of fatigue-crack initiation in ceramics; many difficulties, however, exist with the reproducibility and analysis of such data suggesting that statistical approaches will be essential for developing engineering design criteria for these materials.

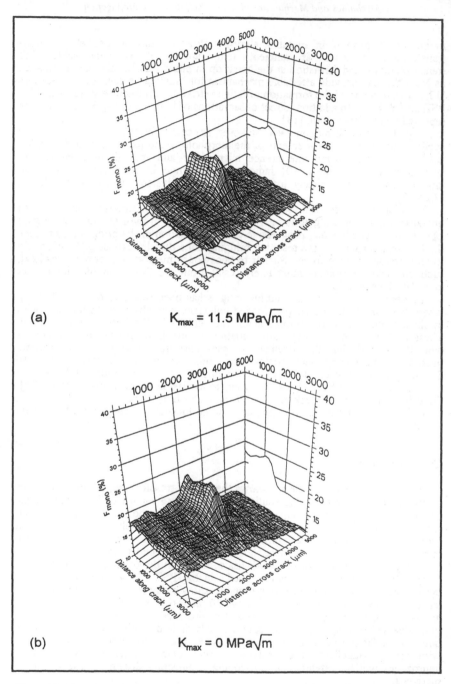

(a) $K_{max} = 11.5\,\mathrm{MPa}\sqrt{m}$

(b) $K_{max} = 0\,\mathrm{MPa}\sqrt{m}$

Figure 10. Transformation zone surrounding a fatigue crack in TS-grade Mg-PSZ *in situ* imaged during a fatigue loading cycle from $K_{min} = 0\,\mathrm{MPa}\sqrt{m}$ to $K_{max} = 11.5\,\mathrm{MPa}\sqrt{m}$ using spatially resolved Raman spectroscopy. F_{mono} is the volume fraction of the transformed phase [28].

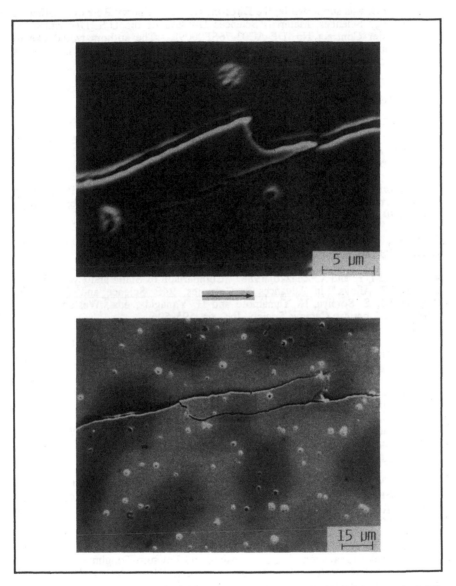

Figure 11. Examples of uncracked ligament bridges just behind the tip of fatigue cracks in TS-grade Mg-PSZ during cyclic loading. Arrow indicates direction of crack growth.

335

ACKNOWLEDGEMENTS

This work was supported by the Director, Office of Energy Research, Office of Basic Energy Sciences, Materials Sciences Division of the U.S. Department of Energy under Contract No. DE-AC03-76SF00098. The authors would like to thank collaborative studies with Drs. W. C. Carter, D. B. Marshall and D. K. Veirs during the course of this work.

REFERENCES

1. Evans, A. G., 1989. "The New High-Toughness Ceramics," in Fracture Mechanics: Perspectives and Directions, ASTM STP 1020, R. P. Wei and R. P. Gangloff, eds., Philadelphia, PA, USA: American Society for Testing and Materials, pp. 267-291.

2. Dauskardt, R. H., W. Yu and R. O. Ritchie, 1987. "Fatigue Crack Propagation in Transformation-Toughened Zirconia Ceramic." Journal of the American Ceramic Society, 70:C248-C252.

3. Ewart, L. and S. Suresh, 1987. "Crack Propagation in Ceramics under Cyclic Loads." Journal of Materials Science, 22:1173-1192.

4. Swain, M. F. and V. Zelizko, 1988. "Comparison of Static and Cyclic Fatigue in Mg-PSZ Alloys," in Advances in Ceramics, 24B Science and Technology Zirconia III, S. Sōmiya, H. Yamamoto and H. Yanagida, eds., Westerville, OH, USA: The American Ceramic Society, Inc., pp. 595-606.

5. Reece, M. J., F. Guiu and M. F. R. Sammur, 1989. "Cyclic Fatigue Crack Propagation in Alumina under Direct Tension-Compression Loading." Journal of the American Ceramic Society, 72(2):348-352.

6. Masuda, M., T. Soma, M. Matsui and I. Oda, 1990. "Cyclic Fatigue of Sintered Silicon Nitride." Journal of the European Ceramic Society, 6:253-258.

7. Dauskardt, R. H., D. B. Marshall and R. O. Ritchie, 1990. "Cyclic Fatigue-Crack Propagation in Magnesia-Partially-Stabilized Zirconia Ceramics." Journal of the American Ceramic Society, 73(4):893-903.

8. Lentz, W., Y.-W. Mai, M. V. Swain and B. Cotterell, 1990. "Time Dependent Strength Behaviour of Mg-PSZ with R-Curve Characteristics." Proceedings of the AUSTCERAM 90 Conference, P. J. Darragh and R. J. Stead, eds., Key Engineering Materials, 18-50:750-753.

9. Ritchie, R. O., R. H. Dauskardt, W. Yu and A. M. Brendzel, 1990. "Cyclic Fatigue-Crack Propagation, Stress-Corrosion, and Fracture-Toughness Behavior in Pyrolytic Carbon-Coated Graphite for Prosthetic Heart Valve Applications." Journal of Biomedical Materials Research, 24:189-206.

10. Dauskardt, R. H., W. C. Carter, D. K. Veirs and R. O. Ritchie, 1990. "Transient Subcritical Crack-Growth Behavior in Transformation-Toughened Ceramics." Acta Metallurgica et Materialia, 38(11):2327-2336.

11. Duan, K., B. Cotterell and Y.-W. Mai, 1991. "Crack Growth of Zirconia Bearing Ceramics under Cyclic Load." Proceedings of Fracture Mechanics of Ceramics Conference, Nagoya, Japan.

12. Lathabai, S., J. Rödel and B. R. Lawn, 1991. "Cyclic Fatigue from Frictional Degradation at Bridging Grains in Alumina." Journal of the American Ceramic Society, 74(6):1340-1348.

13. Davidson, D., J. B. Campbell and J. Lankford, 1991. "Fatigue Crack Growth through Partially Stabilized Zirconia at Ambient and Elevated Temperatures." Acta Metallurgica et Materialia, 39:1319-1330.

14. Liu, S.-Y. and I-Wei Chen, 1991. "Fatigue of Yttria-Stabilized Zirconia: I, Fatigue Damage, Fracture Origins, and Lifetime Prediction;" "II, Crack Propagation, Fatigue Striations, and Short-Crack Behavior." Journal of the American Ceramic Society, 74(6):1197-1205; and 74(6):1206-1216.

15. Steffen, A. A., R. H. Dauskardt and R. O. Ritchie, 1991. "Cyclic Fatigue Life and Crack-Growth Behavior of Microstructurally Small Cracks in Magnesia-Partially-Stabilized Zirconia Ceramics." Journal of the American Ceramic Society, 74(6):1259-1268.

16. Guiu, F., M. J. Reece and D. A. J. Vaughan, 1991. "Cyclic Fatigue of Ceramics." Journal of Materials Science, 26:3275-3286.

17. Hu, X.-Z. and Y.-W. Mai, 1992. "Crack-Bridging Analysis for Alumina Ceramics under Monotonic and Cyclic Loading." Journal of the American Ceramic Society, 75(4):848-853.

18. Hoffman, M. J., W. Lentz, M. V. Swain and Y.-W. Mai, 1992. "Cyclic Fatigue Lifetime Predictions of Partially Stabilized Zirconia with Crack-Resistance Curve Characteristics." Journal of the European Ceramic Society, 8:to be published.

19. Evans, A. G., 1980. "Fatigue in Ceramics." International Journal of Fracture, 16(6):485-498.

20. Hannink, R. H. J. and M. V. Swain, 1982. "Magnesia-Partially Stabilised Zirconia: The Influence of Heat Treatment on Thermomechanical Properties." Journal of the Australian Ceramic Society, 18:53-62.

21. Hannink, R. H. J., 1983. "Microstructural Development of Sub-Eutectoid Aged MgO-ZrO$_2$ Alloys." Journal of Materials Science, 18:457-470.

22. Marshall, D. B., W. L. Morris, B. N. Cox and M. S. Dadkhah, 1990. "Toughening Mechanisms in Cemented Carbides." Journal of the American Ceramic Society, 73:2938-2943.

23. Han, L. X. and S. Suresh, 1989. "High-Temperature Failure of Alumina-Silicon Carbide Composite under Cyclic Loads: Mechanisms of Fatigue Crack-Tip Damage." Journal of the American Ceramic Society, 72(7):1233-1238.

24. McMeeking, R. M. and A. G. Evans, 1982. "Mechanics of Transformation-Toughening in Brittle Materials." Journal of the American Ceramic Society, 65(5):242-246.

25. Marshall, D. B., M. C. Shaw, R. H. Dauskardt, R. O. Ritchie, M. Readey and A. H. Heuer, 1990. "Crack-Tip Transformation Zones in Toughened Zirconia." Journal of the American Ceramic Society, 73(9):2659-2666.

26. Jensen, D., V. Zelizko and M. V. Swain, 1989. "Small Flaw Static Fatigue Crack Growth in Mg-PSZ." Journal of Materials Science Letters, 8:1154-1157.

27. Suresh, S. and R. O. Ritchie, 1984. "The Propagation of Short Fatigue Cracks." International Metals Reviews, 29(6):445-476.

28. Dauskardt, R. H., D. K. Veirs and R. O. Ritchie, 1991. Unpublished work, Lawrence Berkeley Laboratory, University of California, Berkeley.

29. Shang, J.-K. and R. O. Ritchie, 1989. "Crack Bridging by Uncracked Ligaments During Fatigue-Crack Growth in SiC-Reinforced Aluminum-Alloy Composites." Metallurgical Transactions A, 20A:897-908.

30. Heuer, A. H., M. Rühle and D. B. Marshall, 1990. "On the Thermoelastic Martensitic Transformation in Tetragonal Zirconia." Journal of the American Ceramic Society, 73(4):1084-1093.

31. Marshall, D. B. and M. V. Swain, 1988. "Crack Resistance Curves in Magnesia-Partially-Stabilized Zirconia." Journal of the American Ceramic Society, 71(6):399-407.

The Influence of Grinding and Wear on Residual Stress and Strength of Zirconia Ceramics

P. H. J. VAN DEN BERG and G. DE WITH

ABSTRACT

The influence of grinding and wear on strength and residual stress in Mg-PSZ and Y-TZP was studied. The grinding tests were performed with two different diamond wheels on Mg-PSZ and with three different wheels on Y-TZP. The wheels differed mainly in diamond grain size. Strength, residual stress and phase content were measured after the grinding tests. The countermaterial in the wear tests was a hardened steel. Wear tests were performed under ambient conditions and with water as a lubricant under various loads. The strength of worn surfaces was measured. The grinding analysis of Mg-PSZ resulted in a clear and consistent picture of the relation between the surface treatment, and strength, phase content and residual stress. The same analysis of Y-TZP showed that this material is more complex than Mg-PSZ. The results of the wear tests show that the strength of worn Mg-PSZ is influenced by the development of residual stress due to wear testing. The results of the wear tests on Y-TZP show that this material is more complicated, and that more research is required to explain the behaviour of Y-TZP.

INTRODUCTION

The major characteristic of structural zirconia ceramics is the phase transformation [1, 2]. The emphasis of this study was on the relation between mechanical surface interactions and the phase transformation at the surface resulting in residual stresses, see also [3]. Earlier study, [4], showed the correspondence between the amount of monoclinic zirconia and residual stress for Mg-PSZ which is used in this paper. The strength of worn surfaces was measured after wear testing under several conditions. This gives indirect information on the residual stress and the phase transformation in worn samples. These measurements were performed on two types of zirconia ceramics, namely Magnesium Partially Stabilized Zirconia (Mg-PSZ) and Yttria Tetragonal Zirconia Polycrystalline (Y-TZP).

P. H. J. van den Berg, Eindhoven University of Technology, Centre for Technical Ceramics P. O. Box 513, 5600 MB Eindhoven The Netherlands

G. de With, Philips Research Laboratories P. O. Box 80000, 5600 JA Eindhoven The Netherlands, also affiliated to Eindhoven University of Technology, Centre for Technical Ceramics

There are significant differences between Mg-PSZ and Y-TZP. Phenomena like superplasticity, degradation and the re-transformation [5-7], are important to Y-TZP but not to Mg-PSZ. Moreover, the microstructure of both materials is entirely different. The mean maximum grain size of Mg-PSZ is about 60 μm, while the mean maximum grain size of Y-TZP is about 1 μm. Such differences almost certainly result in a different behaviour during grinding or wear testing. Observations on worn surfaces and proposed wear mechanisms are treated in [8].

EXPERIMENTAL

The materials used were the commercially purchased zirconia varieties Mg-PSZ (Nilcra) and Y-TZP (Feldmühle). The hardened steel, denoted as stavax, used in the wear tests was commercially available (Uddeholm).

Often mechanical polishing is the most suitable surface treatment to use as a surface preparation for a ceramic. Mechanical polishing is more reproducible than grinding and the introduction of undesired phenomena during polishing is less than for other surface treatments. A polished surface is, however, not expected to be entirely free of damage caused by the polishing. Polishing was usually done with a diluted soap solution on a tin disk, initially with diamond powder of 4-7 μm, and in the final steps with diamond powder of 2-4 μm. Polishing was also done to remove material from the surface in thin quantities of one and several microns, to obtain depth-information.

Three-point-bend tests to determine strength and fracture toughness were performed on samples of approximately $1 \times 3 \times 15$ mm^3 with a span width of 12 mm in a dry nitrogen gas flow giving a dew point of at least -40 °C. The strength measurements were performed on samples with a polished surface as the surface under tension.

A sample shape of the zirconia materials often used was a shape with ten three-point-bend samples in a row still attached to their common base also denoted as a 'cam'. This particular shape made it possible to perform wear or grinding tests on the surface of the material and to obtain afterwards ten samples of $15 \times 3 \times 1$ mm^3 suitable for a three-point-bend test with the worn or ground surface as the surface under tension. Removal of these ten samples was done by sawing of the sample opposite to the surface tested in tension and has therefore no influence on the measured strength.

The grinding was performed with three diamond wheels that differed mainly in diamond grain size and concentration, whereas the other variables were equal. The grinding was circumference grinding on a Jung machine with a vertical feed of 10 μm, a rotational speed of 2800 rotations/minute with a grinding wheel of 200 mm in diameter. Cooling was done with water and the wheels had a bronze bonding. The maximum diamond grain size of the fine grained wheel, A, was about 64 μm, of the intermediate grained wheel, B, about 107 μm and of the coarse grained wheel, C, about 180 μm. The concentrations for A and B were respectively 1.32 and 1.76 karate/cm^3.

Determination of the various phases in zirconia was performed with X-ray analysis using Cu-Kα. The required peak areas were measured from diffractograms with the help of a digitizer coupled to a personal computer. The quantitative determination of the amount of monoclinic zirconia, V_m, was performed according to [9]. The original data from the quantitative determination are averages over the penetration depth of Cu-Kα. After a discrete integration this results in values averaged over a depth range, determined by the polishing procedure. These values are represented here as step-like profiles.

The residual stress measurements were performed with the conventional bend strip method. The residual stress was calculated from the measured radius of the strip using a slightly adjusted formula correcting for the concentration of stresses at the surface, [4].

Measured for Mg-PSZ were the strength, residual stress and amount of monoclinic zirconia after grinding with A and B. The strength and residual stress were measured as a function of depth. Profiles for the residual stress were derived from bend strip measurements on ground samples, the phase profiles and the results from earlier study which showed the correspondence between residual stress and the amount of monoclinic zirconia [4]. For Y-TZP, strength, residual stress and the amount of monoclinic zirconia were measured after grinding with A, B and C. The depth-analysis for strength and the amount of monoclinic zirconia was performed after all three grinding methods. The residual stress as a function of depth was only determined after grinding with method C.

The wear tests were performed with an instrument especially designed to relate the strength of worn surfaces to the wear test conditions. Ten three-point-bend samples still attached to their common base, are positioned on a rotating disk. The hardness of the disk was 6.5 GPa and the surface of the disk was polished to an R_a of 0.001 µm. The velocity of the disk was constant at about 1.4 m/s. The normal loads used were 14, 19 and 34 N. The tests were performed under ambient conditions and with water as a lubricant. This water was applied to the disk at a rate of 2 ml/min.

RESULTS

The results of the measurements are presented in figures 1-7.

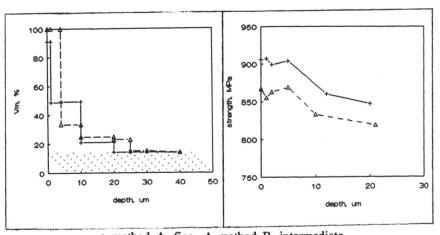

+ method A, fine   Δ method B, intermediate

Figure 1 The calculated amount of monoclinic zirconia, Vm given in %, in Mg-PSZ as a function of depth. The shaded area indicates the possible amount of monoclinic zirconia in the bulk.

Figure 2 The strength-depth curve of Mg-PSZ for methods A and B for the first 20 mm which illustrates the consistent difference in strength between the two grinding methods.

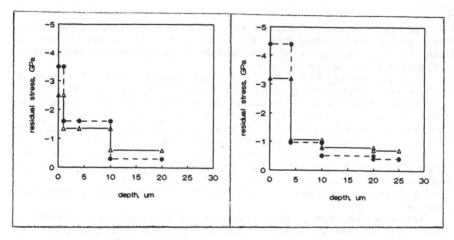

Δ corrected for 0 % • corrected for 15 %

Figure 3: The derived residual stress profile of Mg-PSZ as a function of depth. The amount of monoclinic zirconia present in the bulk of the material is unknown. The profiles for 0 % and 15 % monoclinic zirconia are therefore given.
(left): method A, fine, (right): method B, intermediate.

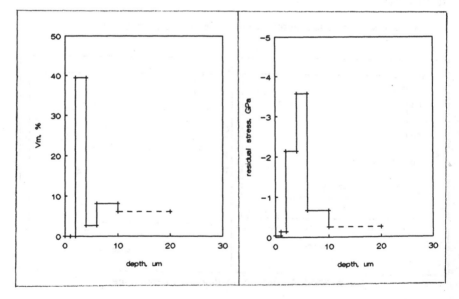

Figure 4a (left): The derived profile for the amount of monoclinic zirconia of Y-TZP after coarse grinding.
Figure 4b (right): The derived profile for the residual stress of Y-TZP after coarse grinding.

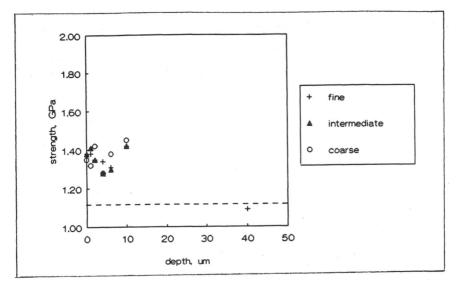

Figure 5: The results of the strength measurements on Y-TZP. Each point is the average of ten measurements. The dotted horizontal line is the strength of Y-TZP of which at least 50 mm has been removed through polishing.

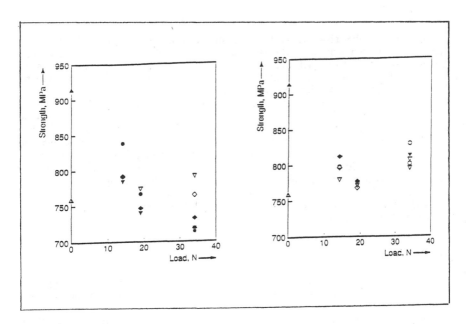

Δ p, 0h, o p, 20h, ∇ p, 70h, □ p, 200h, Δ g, 0h, o g, 20h, ∇ g, 70h, ■ g, 200h

Figure 6: The strength of Mg-PSZ after the wear tests. The p and g stand for the initial surface treatment of respectively polishing and grinding (D46).
(left): with water as a lubricant, (right): ambient conditions

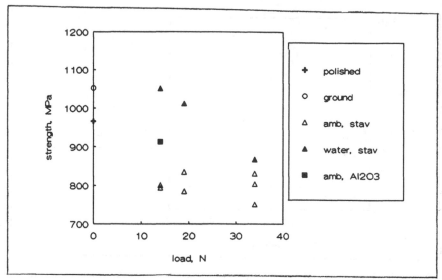

Figure 7: Strength of Y-TZP after wear tests performed under both ambient conditions and with water as a lubricant.

In figure 1 the absolute amount of monoclinic zirconia as a function of depth for the two grinding wheels is given. figure 2 gives the results of the strength measurements, with each point presenting the average of ten three point bend tests. In figure 3 the derived residual stress profiles are shown. These two figures were derived from figure 1 according to [4].

The results from the same type of measurements performed on Y-TZP are given in figures 4 and 5. The strength measurements on Y-TZP in figure 5 show a diffuse scatter of data, showing no correlation with grinding method or depth.

The results of the strength measurements on worn surfaces are shown in figures 6 and 7, with each point indicating the average of ten measurements. The strength of Mg-PSZ and Y-TZP decreases with increasing load after testing with water as a lubricant. The strength of Mg-PSZ after testing under ambient conditions is higher than the strength of polished Mg-PSZ, while the strength of Y-TZP after wear testing under ambient conditions is significantly lower than the strength of polished Y-TZP.

DISCUSSION

The interpretation of the results of the grinding tests performed on Mg-PSZ follows from figures 1-3. There is a clear dependence of the amount of monoclinic zirconia, and thus of residual stress, on diamond grain size. Grinding with a fine diamond grain results in less residual stress then grinding with an intermediate diamond grain, in accordance with the principle that relatively sharp abrasives are accompanied by lower forces operating to lower depth, relative to blunt abrasives [10-13]. The strength after grinding with a fine grain is higher than the strength after grinding with an intermediate grain. This apparent discrepancy, a higher residual stress corresponding to a lower strength, can be explained with the concept of compensating tension beneath the residual stress layer, which is located near inhomogeneities like grain boundaries, and which is larger for larger compressive residual stresses within the residual stress layer [14]. This principle has been illustrated with Finite Element Analysis and it is consistent with fractographic observations [14]. Failure of the material is thus assumed to originate from this subsurface tension. A higher residual stress relates to a lower strength and the strength of polished material is lower than the strength of the material with residual stress. This means that there is a certain value for the residual stress below which the strength increases with increasing residual stress, and above which the strength decreases with increasing residual stress.

The interpretation of the data on Y-TZP is less clear. Grinding with a fine grain does result in the least amount of transformation which is consistent with what is expected from the results on Mg-PSZ. The strength data, however, show a broad band within which most of the data can be plotted, without any significant dependence of strength on diamond grain size, but with values above the strength of polished samples. This implies that the material is strengthened by residual stress, but the precise shape and magnitude of this residual stress profile have no influence on strength. This means that the fracture mechanism of Y-TZP is significantly different from the fracture behaviour of Mg-PSZ. The profiles of the residual stress and the phase content show subsurface maxima and practically no stress or monoclinic zirconia within the first two micrometers from the surface. This can be explained with the re-transformation which has been suggested in [3] for Ce-TZP.

The strength data after wear testing of Mg-PSZ can be reasonably explained with the principles of residual stresses caused by surface interactions. The strength after testing with water as lubricant is less then the strength after testing under ambient conditions for loads of 19 and 34 N. The forces caused by the surface interactions during ambient testing are likely to be higher than during lubricated sliding. A residual stress layer is likely to develop during unlubricated sliding thus strengthening the material. This is consistent with the higher strength after unlubricated sliding at a load of 34 N compared to unlubricated sliding at a load of 19 N.

The data for Y-TZP after lubricated sliding show the same qualitative dependence on load as the results for Mg-PSZ. Lubricated sliding removes material without much mechanical interaction, comparable to polishing, thus resulting in low strength values. The strength values after sliding under ambient conditions are, however, significantly lower. This can be explained with the concept of degradation [5]. The temperatures during unlubricated sliding are high enough to cause degradation of Y-TZP, thus explaining the low strength.

CONCLUSIONS

1. The dependence of the strength of Mg-PSZ on mechanical surface interactions such as wear and grinding with various grain sizes, can be explained by the concepts of the transformation and the associated residual stress.
2. The maximum residual stress and the maximum amount of transformation due to mechanical surface interactions for Y-TZP are located at a few micrometers beneath the surface while for Mg-PSZ both maxima are located at the surface.
3. The strength of Y-TZP ground with various diamond grain sizes is independent of the various amounts of residual stress caused by the grinding, contrary to the behaviour of Mg-PSZ.

4. The concepts of re-transformation and degradation influence the relation between the mechanical surface interaction and the residual stress and strength in Y-TZP. These aspects need further attention.

REFERENCES

1. Hughhan, R. R. and R. H. J. Hannink, 1986. "Precipitation During Controlled Cooling of Magnesia-Partially-Stabilized Zirconia." Journal of the American Ceramic Society, 69(7):556-563.

2. Bansal, G. K. and A. H. Heuer, 1975. "Precipitation in Partially Stabilized Zirconia." Journal of the American Society, 58(5-6):235-238.

3. Swain, M. V. and R. H. J. Hannink, 1989. "Metastability of the Martensitic Transformation in a 12 mol% Ceria-Zirconia Alloy: II, Grinding Studies." Journal of the American Ceramic Society, 72(8):1358-1364.

4. Berg, P. H. J. van den and G. de With, 1992. "Residual Stress and the Stress-Strain Curve for Mg-PSZ." Journal of the European Ceramic Society, 9(4):265-270.

5. Yoshimura, M. 1988. "Phase Stability of Zirconia." Ceramic Bulletin, 67(12):1950-1955.

6. Dominguez-Rodriguez, A., K. P. D. LagerlÖf and A. H. Heuer, 1986. "Plastic Deformation and Solid-Solution Hardening of Y_2O_3-Stabilized ZrO_2." Journal of the American Ceramic Society, 69(3):281-284.

7. Chen, I.-W. and L. A. Xue, 1990. "Development of Superplastic Structural Ceramics." Journal of the American Ceramic Society, 73(9):2585-2609.

8. Berg, P. H. J. van den and G. de With, 1991. "Wear and Strength of Mg-PSZ Worn on Hardened Steel." Journal of the European Ceramic Society, 8:123-133.

9. Evans, P. A., R. Stevens and J. G. P. Binner, 1983. "Quantitative X-ray Diffraction Analysis of Polymorphic Mixes of Pure Zirconia." British Ceramic Transactions, 83: 39.

10. Torrance, A. A. 1987. "An Approximate Model of Abrasive Cutting." Wear, 118:217-232.

11. Abebe, M. and F. C. Appl, 1988. "Theoretical Analysis of the Basic Mechanics of Abrasive Processes Part 1: General Model." Wear, 126:251-266.

12. Abebe, M. and F. C. Appl, 1988. "Theoretical Analysis of the Basic Mechanics of Abrasive Processes Part 2: Model of the Ploughing Process." Wear, 126:267-283.

13. Torrance, A. A. 1990. "The Correlation of Process Parameters in Grinding." Wear, 139:383-401.

14. Berg, P. H. J. van den and G. de With, 1993. "Residual Stress and Strength of Mg-PSZ After Grinding." In print in Wear.

Comparison of Strength and Fracture Toughness of Single and Polycrystalline Zirconia

G. A. GOGOTSI and M. V. SWAIN

ABSTRACT

Sintered polycrystals and single crystals grown by the skull melting technique of ZrO_2 with Y_2O_3 and MgO additions have been studied. The mechanical properties of the materials were investigated over a wide range of temperatures. The stress–strain relationships were measured from -140 to 1400^0C, the hardness from 20 to 900^0C(at 10N) and the critical stress intensity factor, K_{1c}, at -150 and 20^0C. R–curve behaviour, indentation toughness K_c^y, and dynamic elastic moduli E_d were determined at room temperature. It was shown that R–curves for crystals with different additives and ceramics with Y_2O_3 additions were horizontal. At room temperature all these crystals exhibit linear elastic behaviour and those crystals with higher E_d had higher K_{1c}, K_c^y and K_r values. The temperature dependence of the ductile to brittle transition was significantly lower for indentation tests than for flexural tests. Results indicate that the partially stabilised crystals exhibited higher strength and greater creep resistance than polycrystalline materials at high temperatures.

INTRODUCTION

Considerable research over the last decade has shown crystals of partially stabilized zirconium dioxide (PSZC) to be promising materials which have some advantages over ceramics of similar composition (e.g. serviceability at high temperatures [1]). The advances in the application of the direct high–frequency skull melting technique [2] for crystal growth, which allows up to 30 kg of each crystal to be obtained in one heat [3], make it possible to consider their application as structural materials.

As far as one can judge from the known publications, ZrO_2 crystals were studied in parallel: in the former USSR, where the main interest was shown in crystals obtained in Russia from the Ukrainian raw materials, and in the USA and Europe where mainly the crystals of the U.S. company CERES Corp. were investigated. Initially the attention of researchers was directed towards

G.A. Gogotsi, Institute for Problems of Strength, Ukraine Academy of Sciences, Kiev, 252014, Ukraine
M.V. Swain, Department of Mechanical Engineering, University of Sydney, Sydney, NSW 2006, Australia.

fabrication of cubic (fully stabilized) zirconia single crystals (FSZC) for the jewelry industry and laser engineering.

In this paper the mechanical properties of single crystalline and polycrystalline materials of comparable composition are evaluated. The most significant difference between the materials was the high temperature behaviour where it was observed that the brittle to ductile transition temperature was much lower for the polycrystalline materials.

MATERIALS AND EXPERIMENTAL PROCEDURES

To fabricate the crystals [4] for the investigation the typical skull melting equipment designed for commercial production of imitation diamonds (fianite) was used. FSZC crystals fabricated of the same raw material and polycrystalline ceramics of similar composition [5] were also used for the investigation.

The experimental studies were performed using techniques pioneered by Gogotsi [6] for the determination of the mechanical properties for advanced ceramics as well as traditional refractory materials. This approach differs a little from that adopted by other authors [7, 8] who studied crystals using miniature specimens and pursuing physical rather than engineering purposes.

The dimensions of all the specimens were 3.5 × 5 × 50 mm bars which were placed on roller supports, for flexural strength testing on the side width of 5 mm, and for fracture toughness testing on the 3.5 mm edge. In the first case, four–point bending testing was employed, with the distance between the supports being 20/40 mm and in the second case, three–point bending, with the distance between the supports being 20 mm. The fracture toughness (K_{1c}) calculations were made by the formula proposed in reference [9]. All specimens used for the K_{1c} measurements had 2.5 mm deep cuts (blunt cracks). For the R–curve studies a sharp crack was grown to the same depth. It was initiated under conditions of specimen longitudinal compression (fracture surfaces obtained in those tests are shown in Fig. 1). The R–curves were plotted using the compliance data for a bar with a sharp crack subjected to cyclic flexure under strain–controlled loading.

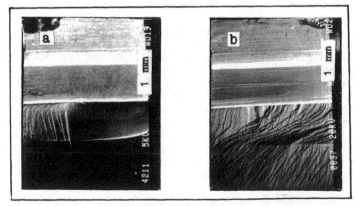

Figure 1. Fracture surfaces of Specimens used for R–curve measurements on a) Y–PSZC–1 and b) Y–PSZC–2. Regions I and II are part of the notch with widths of 1 and 0.15 mm; III is the surface of the initiated sharp crack from the notch and IV is the surface obtained after testing.

The hardness and indentation fracture toughness of materials was measured on the undamaged remnants of polished strength and notched bend fracture bars. The hardness variation with temperature was obtained at an indentation load of 10 N using the test set up designed and manufactured at the Institute for Problems of Strength [6].

Dynamic elastic moduli E_d were determined from the measurements of ultrasound velocity along the specimen axis. Static elastic moduli $, E_{st},$ and the specimen deflections were measured with a deflectometer with the resolution of $\sim 0.1\ \mu m$.

The specimen orientation in relation to the crystal structure was performed using the conventional procedure of the Laue back–reflection X–ray technique.

EXPERIMENTAL RESULTS

At the beginning of the investigation, dynamic elasticity moduli were measured for all specimens (see their average values in Table I). The range (the minimum and maximum values) for 40 specimens of Y–PSZC–1 crystals was 164 to 185 GPa (the specimen axes were oriented approximately in the <111> crystallographic direction), and for 20 specimens of Y–PSZC–2 crystals was 281 to 315 GPa (the specimen axes were oriented in the $\sim$ <001> direction). Almost similar results were observed for the Y–FSZC–10 crystals whose axes were oriented in the $\sim$ <100> direction and the elastic modulus range was 322 to 347 GPa.

TABLE I. CHARACTERISTICS OF THE MATERIALS TESTED

Material	Additive (mol%)	Brittleness	Elastic Modulus (GPa) E_d	E_{st}	Bending Strength (MPa)	Limiting Strain %	Hardness (P=100N) (GPa)	K1c, (MPa.m1/2)
Y-PSZC-1	Y2O3(3)	1	173	149	642	0.43	13.2	9.3
Y-PSZC-2	Y2O3((3)	1	301	297	506	0.17	13.2	12.3
Y-PSZ	Y2O3(3)	1	214	205	830	0.332	-	9.3
Y-TZP-1	Y2O3(3)	1	208	204	760	0.373	13.0	8.9
Y-TZP-2	Y2O3(3)	1	209	205	1021	0.495	13.6	9.5
Y-FSZC-10	Y2O3(10)	1	335	327	221	0.068	-	2.0
Mg-PSZC	MgO(4)	1	-	165	895	0.539	12.7	5.9
Mg-PSZ	MgO(9)	0.44	217	206	521	0.454	13.7	10.4

Room temperature testing of crystals revealed that irrespective of the stabilizing additive all yttria–containing crystals and polycrystalline ceramics exhibited linear elastic behaviour, i.e. they are brittle materials with the brittleness measure* $\chi = 1$ (Table I). But ceramics containing MgO were inelastic, i.e. relatively brittle ($\chi < 1$). Comparative fracture toughness tests were performed, the results of which are listed in Table I. They revealed that all fully stabilized crystals have comparatively low K$_{1c}$. An attempt was also

---

*The brittleness measure $\chi = \sigma_s^2/E\int_0^\epsilon \sigma d\epsilon$. Here σ_s is the ultimate strength, E is the elasticity modulus, ϵ is the ultimate strain, σ are the stresses at the current strain values ϵ (see details in ref. [10]).

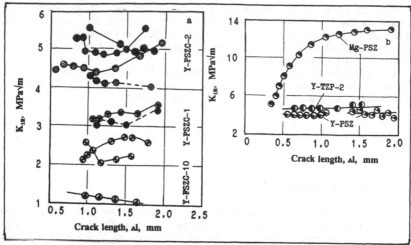

Figure 2.   Crack growth resistance curves (R–curves) of the various materials tested; a) crystalline materials and b) polycrystalline ceramics.

made to evaluate the influence of cooling to −150°C on the specimen fracture toughness. The results for the Y–FSZC–10 and Y–PSZC–1 materials were 2.5 and 12.6 MPa√m respectively.

The crack growth resistance studies revealed that R–curves for Y–PSZC and Y–FSZC crystals, as well as for Y–PSZ and Y–TZP ceramics are actually horizontal (Fig. 2) i.e. no increase in the K_{Ir} with crack propagation is observed, unlike typical Mg–PSZ ceramics (Fig. 2b). The relationship between the ceramics deformation behaviour and R–curves is discussed elsewhere [10]. Horizontal R–curves for similar materials were also observed previously [4, 5, 11].

Vickers indentation caused different modes of fracture of Y_2O_3– and MgO–stabilized ceramics and crystals. Radial cracks initiated from the impression corners in the first ceramics, and a type of buckling of the second material occurred in the area surrounding the impression in the second one (see ref. [11, 13]) and both radial and lateral cracks formed in the crystals (Fig. 3). It was noted that radial cracks did not necessarily nucleate from the corners of the impression in the direction of its diagonals. A more detailed description of the cracking behaviour is presented elsewhere [4, 14].

The characteristic stress–strain diagrams of Y–PSZC crystals tested at various temperatures are shown in Fig. 4 and those for ceramics in Fig. 5. It follows from those figures that partially–stabilized crystals show the least softening at high temperatures. The cubic Y–FSZC–10 crystals (Fig. 6) exhibited a different behaviour: they strengthened at 1400°C. The behaviour of cubic PSZC crystals on heating is considered elsewhere [15].

A decrease in the test temperature down to −140...−150°C induced an increase in the ultimate strengths for all the materials studied (Figs. 4 & 5). A similar increase in strength for the same ceramics was observed in [16].

When studying the hardness variation with temperature (Figs. 7 – 9) the behaviour was found to be alike for all the crystals (Figs. 7 & 9) and for Y_2O_3–based ceramics (Fig. 8); although a decrease in hardness in the temperature range from 20 to 900°C is about the same the temperature zones where the observed curves changed their slopes are different. As is seen from Fig. 7, similar dependences were obtained for crystals with Y_2O_3. At the same

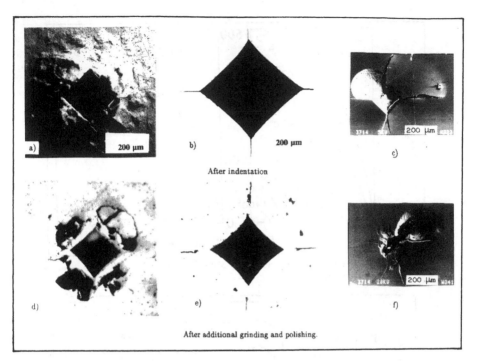

Figure 3.   Vickers indentation impressions; a),b) Y–TZP–2 (load 500N), c),d) Mg–PSZ (load 300N), and e),f) Y–PSZC–1 (load 300N).

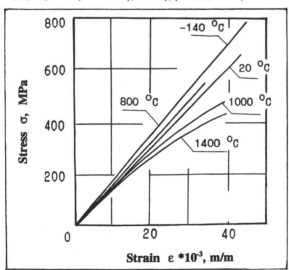

Figure 4.   Stress–strain curves for Y–PSZC–1 as a function of temperature.

time, for the Mg–PSZC ceramics, the slope of the whole curve is somewhat different (Fig. 9). Similar relationships were obtained earlier for Y–PSZC [17, 18] and for Mg–PSZ [13] to those presented here.

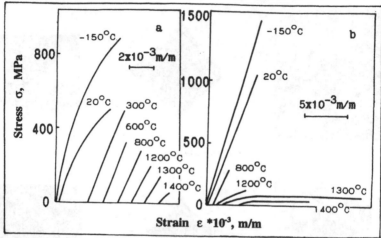

Figure 5.   Stress–strain curves for a) Mg–PSZ and b) Y–TZP–2 at various temperatures.

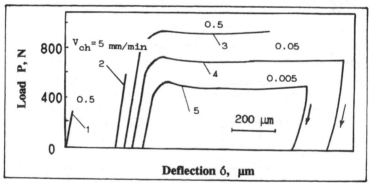

Figure 6.   Load (P) versus deflection (δ) diagrams for Y–FSZC–10 as a function of crosshead speed V_{ch}. 1, T = 20°C; 2 – 5, T = 1400°C.

ANALYSIS OF RESULTS

Before considering the results obtained in detail, let us see how they agree with the published data. If we compare the strength of crystals with the work of Ingel et al [7, 20], who used miniature specimens (for crystals with 3 mol. % of Y_2O_3, the MOR is equal to 1348 MPa), these values are inferior to them and the difference is significant and, probably, cannot be related only to the specimen size effect (only miniature specimens were used in [7]). The cubic crystals (Table I) also had lower strength than in reference [20], where their MOR is 342 MPa. The considered discrepancy in the strength values is probably associated with the details of the testing procedures.

It is interesting to compare the results of the crystals' fracture resistance investigations performed by three independent methods (K_{1c} measurement (SENB), R–curve construction and crack size monitoring after indentation), all of which confirmed a higher fracture resistance of Y–PSZC–2 crystals than that

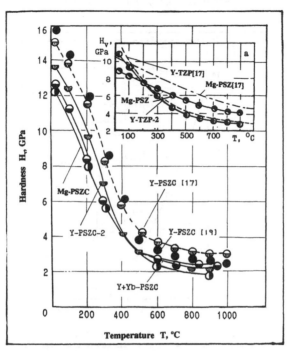

Figure 7.   Hardness versus temperature of crystals and ceramics(insert) with a Vickers
pyramid at 10N load.

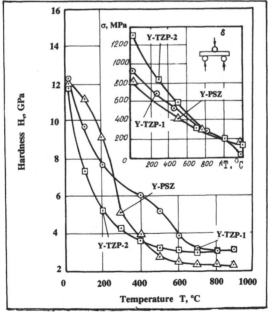

Figure 8.   Vickers hardness versus temperature of ceramics at 10N load, insert
compares the temperature dependence of the MOR.

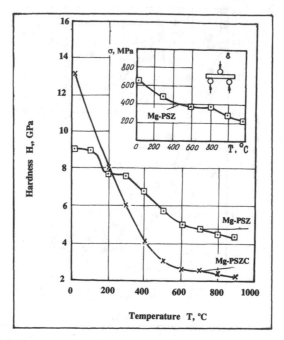

Figure 9.   Hardness versus temperature of Mg–PSZ crystals and ceramics with 10N load, insert shows the temperature dependence of Mg–PSZ strength.

of Y–PSZC–1. However, the scatter of K_{1c} (6.9 to 12.6 MPa√m for Y–PSZC–1 and 8.1 to 13.4 MPa√m for Y–PSZC–2) and crack length values from the impression corners is so large that the data for the two PSZC crystals overlap. At the same time the R–curves are distinctly separated into two groups. The plots in Figure 2 reveal no rising part of such curves and show that the Y–PSZC–2 crystals have an advantage over all the other Y_2O_3–stabilized ceramics. This cannot be said of the crack length about indentations (at a load of 100 N the average crack length for Y–PSZ ceramics is 56 μm, for Y–TZP–2 it is 37 μm). In addition, no lateral cracks appear in the Y_2O_3 ceramics in the indentation zone (the indentation load was 500 N).

Comparison of the K_{1c} and K_{1r} values (Table I and Fig. 3) revealed that the former exceed the latter for crystals (as well as for the Y_2O_3 ceramics), i.e. the results obtained on specimens with a notch (blunt crack) are higher than those for specimens with a sharp crack. This effect (fracture barrier [10]) is characteristic of transformation toughening brittle materials containing ZrO_2 [5, 21] and is of relevance for their sensitivity to macrostress concentration. It was not observed with relatively brittle Mg–PSZ ceramics [5, 10].

It should be noted that if we compare all the data considered above for the Y–PSZC crystals, we can state that there exists direct proportionality between their elasticity moduli and fracture toughness. This relationship was also observed for other crystals stabilized with both Y_2O_3 and $Y_2O_3 + Yb_2O_3$ [11]. The existence of such a relationship was suggested but not verified in [20].

The possibility for the existence of the above effect is also confirmed by the known works (e.g. [22]), which deal with ceramics fracture toughness variation induced by phase transformation during fracture. The relationship we observed follows directly from the expression for the calculation of the transformation–toughening increment of K_{1c} [23]: $K^T = \eta E^x e^t V_f \sqrt{h}$ where η is the coefficient of the phase transformation zone, E^x is the effective elasticity modulus, e^t is the dilational strain, V_f is the volume and h is the size.

However, it is difficult to judge the accuracy of this estimate since we do not know either the relation between the values appearing in it that are characteristic of anisotropic ZrO_2 crystals, or the influence of the crystallographic orientation of the crack, which causes the crystal fracture, on the K_{1c} being determined.

Fractographic investigations revealed that the Y–PSZC–2 crystal fracture surface is rougher than that of Y–PSZC–1 and the higher the specimen toughness, the more pronounced the "steps" on its fracture surface (see Fig. 1).

Having discussed the fracture toughness data, let us again return to the problem of crystal strength based on the relationships presented in reference [12]. If, on the basis of the analysis of numerous experiments, the relationship between strength and fracture toughness obtained there for ZrO_2 ceramics may hold for the case considered here, then it is reasonable that the crystals of our investigation which have higher toughness display lower strength because of transformation plasticity.

In addition to the fracture toughness evaluation from the data on the size of cracks formed on the crystal's surface, the features of the subsurface layer fracture during indentation were investigated. For this purpose, after indentation, the crystal's surface layer was ground off (Fig. 3). Despite the fact that Y–PSZC crystals of different orientation were studied, no "classic" system of semi–circular cracks was observed in either case, though we noticed (Fig. 3f) that sometimes, in crystals, cracks were formed which were similar to Palmqvist's cracks that we observed at all load levels in the ceramics studied. No regularity in the crack propagation pattern was noticed in the PSZC crystals similar to that for Y–FSZC [19, 24].

The above observations reduce the confidence in the possibility of using even the most recognised (e.g. substantiated in [25, 26]) formulas to determine K_c^v for PSZC crystals. For this reason, while there are no standards for the fracture toughness characteristics determination in indentation, it is probably feasible to use the crack lengths and indicate the indentation load when analysing the surface fracture resistance of PSZC crystals (see also [14]).

Considering the results of hardness measurement (e.g. Table I) one can conclude that the dispersion of this data is insignificant and that it is independent of the indentation plane. Having agreed with the analysis of the dependences presented in [17, 18], for the case of high–temperature hardness determination, a reduction in strength and hardness observed in heating (Figs 5 & 8) is almost equal in the ceramics, while in the crystals it differs (see also [5]). This is probably associated with the fact that during indentation high local stresses (which are not observed, say, in bending) appreciably intensify plastic deformation which causes a reduction in the material resistance to penetration of the indenter.

Considering the limited availability of crystal specimens and the nature of their deformation behaviour in the temperature range studied, the authors confined themselves to tests at maximum and minimum temperatures at ambient temperature and 1400°C. At ambient temperature all the crystals (similarly to the ceramics studied) exhibited linear deformation behaviour. At 1400°C the strength of Y–FSZC crystals increased as compared to that at ambient temperature (Fig. 6) at any crosshead speed (see also [7, 15]). The reason for that behaviour may be an effect due to the annealing of the specimens in the course of testing and resulting reduction in the degree of their surface damage caused by machining. The following data obtained when testing annealed specimens at ambient temperature confirmed the above

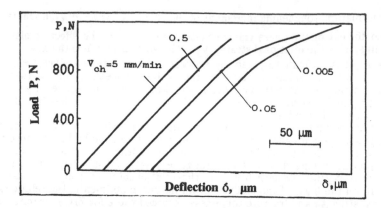

Figure 10.   Load (P) versus deflection (δ) diagrams for Y–PSZC–1 at 1400°C at various crosshead speeds V_{ch}.

assumption; for unannealed specimens the MOR was 134 MPa while for annealed ones it was 172 MPa in Y–FSZC–10 crystals. A similar conclusion was arrived at by Ingel et al [7].

At 1400°C, when the crystals were tested at high V_{ch} values, their load–deflection diagrams, similarly to those at ambient temperature, were linear (2 in Fig. 6). At $V_{ch} = 0.5$ mm/min those crystals exhibited horizontal portions on the diagrams (3 in Fig. 6) while at $V_{ch} = 0.05$ and 0.005 mm/min a pronounced load peak was observed (4 and 5 in Fig. 6), i.e. the upper and lower yield points were monitored. Diagrams illustrating these features in flexure tests were also considered in [7, 15] and in axial compression in [27].

Figure 6 shows that at $V_{ch} = 0.5$ mm/min Y–FSZC–10 crystals sustained higher loads to fracture than at $V_{ch} = 5$ mm/min. At $V_{ch} = 0.5$ mm/min this effect may be anticipated because of sufficient time for the manifestation of microplasticity. It is a well known fact [23] with ceramics that an increase in strength occurs just prior to conditions of significant plasticity or creep beginning to appear.

The behaviour of partially stabilized Y–PSZC–1 crystals at 1400°C and at various V_{ch} values differed from that of ceramics and cubic crystals. Their load deflection diagrams for all speeds V_{ch} (Fig. 10) have linear portions and only at the loads close to fracturing do they exhibit nonlinearity, which with a decrease in speed V_{ch} increases significantly compared to that observed for ceramics and for cubic crystals.

The observations reveal the existence of differences in the mechanical behaviour of the ceramics, cubic and partially–stabilized ZrO_2 crystals studied. But in order to have additional information about the differences in the deformation behaviour · of those materials the authors considered their behaviour under loads close to the fracture limit. For silicon nitride ceramics (at 1200°C) and the cubic crystals (at 1400°C) these are the loads which induce the appearance of a plateau on the diagrams, and for partially stabilized crystals (at 1400°C) fracture occurs prior to the plateau development. Again, for comparison, V_{ch} was chosen equal to 0.5 mm/min.

CONCLUSIONS

The crystals and ceramics are characterised by both similar and different features of mechanical behaviour:

- with a decrease in temperature to -150^0C the strength and fracture toughness of the crystals and ceramics improves by 20 to 25%,
- the crystals and ceramics of higher strength turned out to have lower fracture toughness,
- at ambient temperature the PSZ crystals with an addition of Y_2O_3 and MgO are elastic, ceramics with Y_2O_3 are also elastic while that with MgO is inelastic,
- with an increase in temperature the crystals and ceramics with Y_2O_3 become inelastic while that with MgO first exhibits elasticity and then again inelasticity,
- upon heating to 1200...1300^{0}C the ceramics soften while the cubic and PSZ crystals remain of quite satisfactory serviceability,
- during indentation the PSZ crystals and the ceramics exhibit plasticity at lower temperatures than in bending,
- crystals with high values of elastic moduli caused by their crystallographic orientation have higher fracture toughness,
- Y_2O_3 containing materials exhibited virtually no evidence for rising R–curve behaviour.

ACKNOWLEDGMENTS

The authors would like to thank Dr V.I. Alexandrov and Dr E.E. Lomonova for the crystals and engineer D.Yu. Ostrovoy for help with the testing.

REFERENCES

1. Green, D.J., R.H.J. Hannink and M.V. Swain, 1989. Transformation Toughening of Ceramics, Boca Raton, Fla., USA: Chemical Rubber Company.

2. Alexandrov, V.I., V.V. Osiko, A.M. Prokhorov and V.M. Tatarintsev, 1978. "Fabrication of High-Temperature Materials by the Direct High-Frequency Skull-Melting Technique." Uspekhi Khimii, 13:386-427 (in Russian).

3. Gogotsi, G.A., E.E. Lomonova and V.V. Osiko, 1991. "Studies of Mechanical Characteristics of Zirconia Single Crystals for Structural Application." Ogneupory, 8:14-17 (in Russian).

4. Gogotsi, G.A., E.E. Lomonova and V.G. Pejchev, 1993. "Strength and Fracture Toughness of Zirconia Crystals." Journal of European Ceramic Society, in press.

5. Gogotsi, G.A., A.V. Drozdov, V.P. Zavada and M.V. Swain, 1991. "Comparison of the Mechanical Behaviour of Zirconia Partially-Stabilized with Yttria and Magnesia." Journal of Australian Ceramic Society, 27:37-49.

6. Gogotsi, G.A., 1991. "Test Methods of Advanced Ceramics, Reasonable Approaches for Standardisation of Ceramics". Key Engineering Materials, 56-57:419-434.

7. Ingel, R.P, D. Lewis, B.A. Bender and R.W. Rice, 1982. "Temperature Dependence of Strength and Fracture Toughness of ZrO$_2$ Crystals." Journal of American Ceramic Society, 65:150-152.

8. Lankford, J., R.A. Page, and L. Rabenberg, 1988. "Deformation Mechanism in Yttria-Stabilized Zirconia." Journal of Materials Science, 23:4144-4156.

9. Strawley, J.E., 1979. "Wide Range Stress Intensity Factor Expression for ASTM E-399 Standard Fracture Toughness Specimens." International Journal of Fracture, 12:475-476.

10. Gogotsi, G.A., 1991. "Deformation Behaviour of Ceramics." Journal of European Ceramic Society, 7:87-92.

11. Gogotsi, G.A., A.V. Drozdov, and V.G. Pejchev, 1991. "Mechanical Behaviour of Zirconium Dioxide Crystals Partially Stabilized with Yttrium Oxide." Strength of Materials, 23:86-91.

12. Swain, M.V., 1985. "Inelastic Deformation of Mg-PSZ and its Significance for Strength-Toughness Relationship of Zirconia Toughened Ceramics." Acta. Met., 33:2083-2091.

13. Gogotsi, G.A., M.V. Swain, and J. Davis, 1992. "Partially Stabilized Zirconia and its Mechanical Behaviour under Loading." Ogneupory, 1:2-5 (in Russian).

14. Gogotsi, G.A. and S.N. Dub, "Indentation Fracture Toughness of Zirconia Crystals," to be published in Materials Forum.

15. Gogotsi, G.A., E.E. Lomonova, and D.Yu. Ostrovoy, 1992. "Deformation of ZrO$_2$ Cubic Single Crystals." Ogneupory, 3:15-19 (in Russian).

16. Veitch, S., M. Marmach, and M.V. Swain, 1987. "Strength and Toughness of Mg-PSZ and Y-TZP Materials at Cryogenic Temperature," in Advanced Structural Ceramics, P.F. Becher, M.V. Swain and S. Somiya, eds., MRS Vol. 78, pp. 97-106.

17. Martinez-Fernandes, J., M. Jimenez-Melendo, A. Dominguez-Rodriguez and A.H. Heuer, 1990. "Elevated Temperature Studies of Microindentation of Y-PSZ," in Structural Ceramics-Processing, Microstructure and Properties, Denmark, pp.413-418.

18. Ticare, V. and A.H. Heuer, 1991. "Temperature-Dependent Indentation Behaviour of Transformation-Toughened Zirconia-Based Ceramics." Journal of American Ceramic Society, 74:593-597.

19. Morscher, N.G., P. Pirouz and A.H. Heuer, 1991. "Temperature Dependence of Hardness in Yttria-Stabilized Zirconia Single Crystals." Journal of American Ceramic Society, 74:491-500.

20. Ingel, R., D. Lewis, B.A. Bender and R.W. Rice, 1984. "Physical, Microstructural, and Thermomechanical Properties of ZrO$_2$ Single Crystals." Journal of American Ceramic Society, 67:408-414.

21. Gogotsi, G.A., 1991. "Mechanical Behaviour of ZrO_2-Based Ceramics." Poroshkovaya Metallurgiya, 10:51-56 (in Russian).

22. Evans, A.G. and R.M. Cannon, 1986. "Toughening of Brittle Solids by Martensitic Transformation." Acta. Met., 34:761-800.

23. Swain, M.V., 1989. "Toughening Mechanism for Ceramics." Materials Forum, 13:237-253.

24. Pajares, A., F. Guiberteau, A. Dominguez-Rodriguez, and A.H. Heuer, 1991. "Indentation-Induced Cracks and the Toughness Anisotropy of 9.4 mol% Yttria-Stabilized Cubic Zirconia Single Crystals." Journal of American Ceramic Society, 74:859-891.

25. Niihara, K., R. Morena and D.F.H. Hasselman, 1982. "Evaluation of K_{1c} of Brittle Solids by the Indentation Method With Low Crack-to-Indent Ratios." Journal of Materials Science, 11:13-16.

26. Lawn, B., D.B. Marshall, P. Chautikul and G.R. Antstis, 1990. "Indentation Fracture: Applications in the Assessment of Strength of Ceramics." Journal of Australian Ceramic Society, 16:4-9.

27. Dominguez-Rodrigues, A., K.P.D. Legerlof and A.H. Heuer, 1986. "Plastic Deformation of Solid Solution Hardening of Y_2O_3-Stabilised ZrO_2." Journal of American Ceramic Society, 69:281-284.

Microstructure-Crack Resistance-Fatigue Correlations in Eutectoid-Aged Mg-PSZ

S. LATHABAI and R. H. J. HANNINK

ABSTRACT

Investigations on cyclic fatigue behaviour of Mg-PSZ reported in literature have been exclusively on subeutectoid-aged materials. In this paper we present the results of a study of the crack resistance and fatigue behaviour of Mg-PSZ in the absence of the δ-phase ($Mg_2Zr_5O_{12}$), by using specimens rapidly cooled from the sintering temperature and aged at the eutectoid temperature, 1400°C, for various times. Inert strength testing of specimens with controlled indentation flaws gave a measure of the strength as well as flaw tolerance of the materials. SEM and neutron diffraction were used to characterize the microstructure and phase assemblage. Under-, peak- and over-aged conditions were identified and the latter two conditions selected for R-curve and static and cyclic fatigue testing. The results and their relationship with the microstructure are discussed and compared with those previously obtained for subeutectoid-aged materials.

INTRODUCTION

An attractive feature of magnesia-partially stabilized zirconia (Mg-PSZ) is that by altering thermal treatments, one can tailor the microstructure and thence the mechanical properties of the material for specific applications [1-5]. The primary toughening mechanism in these materials is the stress-induced transformation of metastable tetragonal precipitates, confined within a cubic-stabilized matrix, to the monoclinic phase. The selection of the thermal treatment is dictated by the requirement that tetragonal precipitates of an optimum size, and hence metastability, are formed within the cubic grains.

S. Lathabai and R.H.J. Hannink
CSIRO Division of Materials Science and Technology
Locked Bag 33, Clayton, Vic 3168, Australia

The fabrication process for Mg-PSZ starts with sintering in the cubic phase field of the phase diagram, at temperatures between 1680 to 1720°C depending upon the MgO content [1]. This is followed by a cooling cycle such that a uniform distribution of tetragonal precipitates forms in the grains the majority of which are retained in the tetragonal form to room temperature. The cooling sequence can be a rapid cool with rates of about 500°C/hr, controlled slow cooling, or cooling with isothermal hold steps. The rapid cooling results in fine tetragonal precipitates in the size range 30 to 60 nm which have to be coarsened by aging at or above the eutectoid temperature of 1400°C to bring them to a condition of metastability at room temperature. This is referred to as eutectoid aging. Controlled cooling or isothermal hold cooling result in tetragonal precipitates that are close to the optimum size (~180 nm in diameter). Aging a material cooled by these latter sequences at a subeutectoid temperature of 1100°C produces a ceramic with high fracture toughness and strength primarily due to the precipitation of the δ-phase ($Mg_2Zr_5O_{12}$) at the precipitate-matrix interface, which induces additional strain and increases the metastability of the tetragonal phase [3]. In the commercial fabrication of Mg-PSZ products, this latter approach is generally followed.

It is now accepted that zirconia-based ceramics, particularly those stabilized with magnesia and ceria, constitute some of the toughest monolithic ceramics available, possessing high strength, crack resistance (T-curve, or equivalently, R-curve) behaviour and flaw tolerance. However, it has become apparent over the past 5 years that unlike truly brittle materials, they are susceptible to degradation under cyclic loading [6-11]. Though the mechanisms responsible for the fatigue damage in the zirconia-based materials have not been completely identified, it has been shown that subcritical crack propagation occurs at rates faster than those attained in environmentally-enhanced crack growth under monotonic loading. In the case of Mg-PSZ, cyclic crack velocities have been found to be up to seven orders of magnitude faster and threshold stress intensities almost 40% lower than those measured under sustained loading in identical environments [9]. However, these cyclic fatigue studies have been carried out exclusively on subeutectoid aged materials, heat treated to various toughness levels.

Eutectoid aging at 1400°C presents an alternative approach to promote transformation toughening in Mg-PSZ. However, this approach has not received as much attention as subeutectoid aging because the latter process yields materials with excellent properties and by varying the aging times it is possible to fine tune these properties [1-5]. In this paper, we present the results of a detailed investigation of the effect of eutectoid aging on the microstructure, phase assemblage and mechanical properties of Mg-PSZ. Optimum heat treatments for attaining a combination of high strength and fracture toughness, in the absence of the δ-phase, are identified. A major objective was to study the behaviour of these materials under cyclic loading and compare this with that of subeutectoid-aged Mg-PSZ reported in the literature.

EXPERIMENTAL PROCEDURE

MATERIAL

The material used in this study was 9.9 mole % Mg-PSZ produced by ICI Advanced Ceramics, Australia, supplied in the form of disks (diameter ~ 25 mm and thickness ~ 4 mm) and slabs (~ 50mm x 50mm x 4mm), rapidly cooled from the sintering temperature of about 1700°C (~500°C/hr to 1000°C). Aging at the eutectoid temperature, 1400°C, was carried out for 1, 2, 4, 8, 16 and 32 hours, about 30 disks being aged for each condition. The slabs were aged for two selected durations to be discussed below.

MICROSCOPY AND PHASE ANALYSIS

Specimens representing each aging condition, polished to 1 μm diamond finish and etched with 50% solution of HF were examined using a Leica Stereoscan FE360 scanning electron microscope (SEM).

The phase assemblages resulting from the various aging treatments were analysed by neutron diffraction using the high resolution powder diffractometer at the HIFAR nuclear reactor at the Australian Nuclear Science and Technology Organisation, Lucas Heights, NSW. The diffraction patterns were collected at room temperature over a 2θ range from 0 to 155°, with a step size of 0.05°, and at a wavelength of 0.14925 nm. These patterns were analysed using the Rietveld method. Details of the experimental technique will be published elsewhere [12].

MECHANICAL TESTS

Inert strengths of the materials were assessed by testing the heat treated disks in biaxial flexure using a flat-on-three balls support arrangement. The detrimental effect of moisture in the environment on strength was minimised by using high cross-head speeds on the universal testing machine so that failure times were of the order of ~ 30 ms and by placing a drop of silicone oil on the prospective tensile faces of the specimens prior to testing. The tensile faces of the specimens were polished to 1μm diamond finish and indented with a Vickers pyramid indenter at various loads to introduce controlled flaws of various sizes. A minimum of five specimens was broken at each indentation load and in each case a post-failure examination was carried out to ensure that the failure initiated at the indentation flaw. A few specimens were tested in the unindented state. The tests served to identify optimum aging and over-aging conditions.

The crack resistance behaviour of the materials was evaluated by measuring the toughness as a function of crack extension (T-curves). The procedure used was in general conformance to ASTM E 561-86 [13]. Compact tension specimens with specimen width, w ~35 mm were machined from the slabs alluded to earlier after they had been heat treated to the peak- and over-aged conditions and polished to 1 μm diamond finish. A sharp precrack was introduced at the tip of the machined notch and the specimens were annealed at 1000°C for 30 minutes to retransform any monoclinic phase formed during the notching and precracking. The crack length was monitored using a travelling microscope to a resolution of ±10 μm. A crack opening displacement gauge was attached to the specimen edge by means of knife edges and load-displacement curves were recorded on an x-y plotter.

Static and cyclic fatigue tests were confined to materials heat treated to peak- and over-aged conditions. Compact tension specimens with long cracks (~ 3.5 mm) were used. The tests were carried out under load control using a digitally-controlled servo-hydraulic machine[1]. The cyclic fatigue tests were conducted using sinusoidal loading at a frequency of 10 Hz with a load ratio, $K_{min}/K_{max} = 0.1$ (tension-tension loading). The testing procedure conformed to ASTM E 647-86a for measurement of fatigue crack growth rates in metallic materials [14]. The experimental setups to monitor load, displacement and crack length were identical to those described in the preceeding section.

RESULTS AND DISCUSSION

MICROSTRUCTURE AND PHASE ASSEMBLAGE

Figures 1 a, b and c show the precipitate microstructures of the as-fired material and after aging for 2 hours and 16 hours at 1400°C. It is evident that the precipitates in the as-fired state are small, about 100 nm in size and that with increased aging time, the precipitates coarsen. The grain size of the cubic matrix was about 35 μm and was unaffected by the aging treatment.

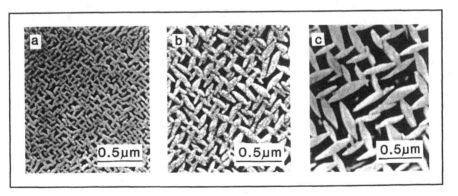

Figure 1. SEM micrographs of (a) as-fired material, (b) after 2 hours aging at 1400°C and (c) after 16 hours at 1400°C (50% HF etch). Note the gradual coarsening of the precipitates with longer aging times.

---

[1] Instron Pty. Ltd, Bayswater, Victoria

The neutron diffraction analysis results are presented in figure 2 in which the proportions of the cubic, tetragonal and monoclinic phases (c-, t- and m-ZrO_2) are plotted against the aging time. The as-fired material consists of cubic and tetragonal phases, indicating that the rapid cooling from the sintering temperature has retained a large proportion of the high temperature c-ZrO_2 down to room temperature and that some t-ZrO_2 has formed. No m-ZrO_2 was detected in this material. On aging at 1400°C for 1 hour, the proportion of t-ZrO_2 has increased at the expense of the c-ZrO_2, and after 2 hours, there is a further increase. However, note that no m-ZrO_2 is as yet detected. The monoclinic phase is first detected in specimens aged for 4 hours; the t-ZrO_2 content has decreased in this material. Aging for 8 hours causes a further decrease in the t-ZrO_2 content and an increase in the m-ZrO_2 content. No further changes in the relative proportions of the tetragonal and monoclinic phases are detected in the materials aged for longer times. It may also be noted that the c-ZrO_2 content remains constant, within experimental error, for aging times longer than 4 hours.

The neutron diffraction analysis together with the microstructural examination provide some insight into the potential for transformation toughening in these materials. It is established that the transformability of t-ZrO_2 is associated with a critical particle size and volume fraction [1,15-18]. If the particle size is smaller than the critical size, toughening is minimal; as

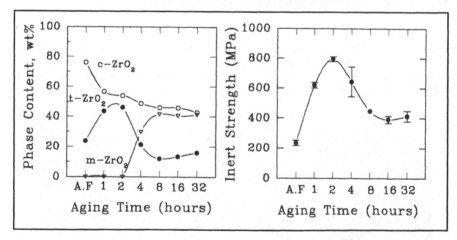

Figure 2. Cubic, tetragonal and monoclinic phase contents as a function of aging time. Aging for 2 hours results in the highest amount of t-ZrO_2.

Figure 3. Inert strength as a function of aging time. Note the similarity to the plot of tetragonal phase content as a function of aging time in Figure 2.

the size increases, the stress-induced transformation becomes easier, and toughness increases. Larger particles transform spontaneously and the amount of t-ZrO_2 available for stress-induced transformation becomes less.

The as-fired material with its fine tetragonal precipitates is stable and no m-ZrO_2 is detected in this material. Apparently, the optimum amount and size of t-ZrO_2 occurs after 2 hours aging. In materials aged for 4 hours and longer, the precipitate sizes exceed the critical size and spontaneous transformation to m-ZrO_2 has occurred on cooling. Further confirmation for this was obtained from the inert strength results and T-curve measurements, as discussed below.

INERT STRENGTH

In figure 3, the inert strength of the unindented specimens is plotted as a function of aging time. The as-fired material has relatively low strength, but even 1 hour at 1400°C has resulted in a three-fold increase in strength. Peak strength is attained on aging for 2 hours. The strength drops on aging for 4 hours; a further drop and levelling off occurs in materials aged for longer duration. Note that this variation in strength with aging time parallels the variation in the t-ZrO_2 content depicted in figure 2. We have observed that the material aged for 2 hours had the optimum size and content of the tetragonal phase; the high strength is a consequence of this.

Figure 4 presents the variation in inert strength in the materials when flaws of different sizes are introduced by indentations at various loads. Although these tests were conducted on all the materials in this study, for the sake of clarity, only results from the as-fired material, and those aged for 2 and 16 hours are shown in figure 4. The results for materials aged 8 and 32

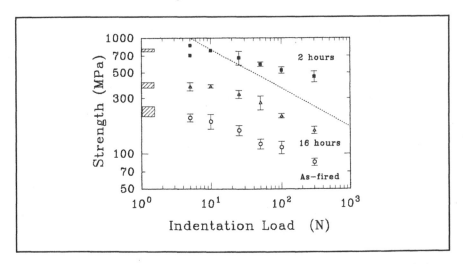

Figure 4. Inert strength-indentation load responses of Mg-PSZ in the as-fired state and after 2 and 16 hours aging at 1400°C. Shaded regions to the left denote failures from natural flaws; these are the strength values plotted in figure 3. Dashed line with -1/3 slope represents prediction from indentation fracture mechanics for a material with single-valued toughness. Note the deviation from the predicted slope in all the three materials, but particularly for the material aged for 2 hours.

hours virtually overlap the data for the material aged for 16 hours; those for the 1 and 4 hours aging follow the trend shown in figure 3, i.e. they fall below the data for the material aged for 2 hours but above the family of data for aging times of 8 hours and more. The dashed line in the plot with a slope of -1/3 represents the variation in strength with indentation load, predicted from indentation fracture mechanics for a material with a single-valued toughness [19]. It can be seen from figure 4 that the strength-indentation load response of the as-fired specimens as well as those aged for 2 hours and 16 hours deviate from this theoretical prediction, the deviation being most pronounced for the material aged for 2 hours. This is an indication that the materials possess crack resistance behaviour, though not to the same extent. It is also evident from figure 4 that even in the presence of an indentation flaw at 500 N, the strength of the peak-aged material (2 hours aging) does not degrade appreciably - i.e. the material is damage tolerant.

Implicit in the theoretical prediction represented by the dashed line is the requirement that well-defined radial cracks form as a result of the indentation. Although cracks formed at the indentation corners even at low indentation loads, the cracks were by no means well-defined and multiple grain boundary cracks were observed around the impression. The analysis also does not consider the stresses associated with the transformation zone around the impression. Despite these observations about the indentation-induced flaws in these materials, the failures always initiated from these sites, indicating that they were the strength-controlling flaws. The assertion of flaw tolerance, especially in the peak-aged material, thus holds.

TOUGHNESS CURVES

The toughness curves for the materials, aged for 2 and 16 hours, representing peak- and over-aged conditions, are shown in figure 5. It can be seen that for both materials the plateau toughness is attained after a crack extension of about 500 μm. However, the plateau toughness of the peak-aged material, $\sim$ 6 MPa√/m, is significantly lower than values reported for the corresponding subeutectoid-aged materials (see, for example ref [9]). Reports of investigations on crack resistance behaviour of eutectoid-aged Mg-PSZ are sparse. However, a recent review on toughening mechanisms for ceramic materials does report a rising R-curve, tending to a plateau value of about 8 MPa√/m for a peak-aged material and about 3 MPa√/m for an over-aged material [20]. The review states that a further annealing of the peak-aged material at 1100°C, however, resulted in significantly tougher material. It would appear that with eutectoid aging, material with high toughness is not as readily achievable as with subeutectoid aging.

STATIC AND CYCLIC FATIGUE

The results of the crack propagation studies under static and cyclic loading on the peak- and over-aged materials are presented in figure 6. The data has been presented in terms of crack extension per second (da/dt) to enable

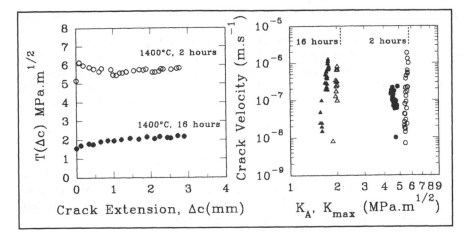

Figure 5. Toughness curves for the materials aged for 2 and 16 hours. Both materials approach the plateau toughness rapidly.

Figure 6. Fatigue crack growth behaviour of peak- and over-aged materials. Open symbols for static loading and closed symbols for cyclic loading. Data plotted as crack velocity da/dt as a function of K_A for static loading and K_{max} for cyclic loading. Dashed lines at the top x-axis are plateau toughness values from figure 5.

direct comparison of crack propagation under cyclic and sustained loading. The dashed lines at the top of the plot denoting the plateau toughness from figure 5 indicate the stress intensity factors where the cracks would be expected to propagate at very high velocity to failure. The solid symbols represent the data for cyclic loading and the open symbols for static loading. In both materials, no crack growth was detected under static loading until the applied stress intensity factor, K_A, was about 90% of the plateau toughness. Under cyclic loading, in both materials, crack propagation occurred only when the peak stress intensity factor, K_{max} was above approximately 75% of the plateau toughness.

Based on these data, it may be concluded that there is a small but significant mechanical fatigue effect in the eutectoid-aged materials, over and above the effect of the environment seen under static loading. However, the degrading effect of cyclic loading in these materials is by no means as dramatic as that reported for subeutectoid-aged materials [9]. The crack resistance exhibited by the peak subeutectoid-aged material is more pronounced than that of the peak-aged material in this study. The nonlinear deformation processes in the crack zone are likely to be more substantial in the former materials and it may be speculated that repeated loading and unloading cause more damage in them. However, Dauskardt and coworkers have observed pronounced cyclic fatigue even in what they refer to as "mid toughness" and "low toughness" subeutectoid-aged materials [9]. These latter materials have plateau toughness values comparable to our peak-aged material.

The mechanisms responsible for damage accumulation have not yet been precisely identified in Mg-PSZ materials, although localized microplasticity at the crack tip, microcracking at grain boundaries and precipitate/matrix interfaces and wedging action at crack surface asperities have all been suggested as possible contributing factors. Using Raman microprobe measurements, it has been shown that cyclic loading does not reduce the degree of transformation toughening [9]. Some recent in situ fatigue experiments in an SEM by Davidson et al indicate that the fatigue crack growth in subeutectoid-aged Mg-PSZ might be associated with "slip line" formation at the crack tip during the repeated loading and unloading and that the fatigue mechanics and mechanisms in Mg-PSZ are very similar to those on some aluminium alloys [21]. Obviously, more such investigations are needed to identify the micromechanisms of fatigue in Mg-PSZ.

SUMMARY

The effect of eutectoid aging on strength, flaw tolerance and fatigue behaviour of Mg-PSZ was investigated. Microscopy and neutron diffraction were used to characterize the microstructure and phase assemblage of the materials. Inert strength and T-curve measurements identified the optimum aging conditions. Crack propagation under sustained and cyclic loading was monitored. The results may be summarized as follows:

(1) The microstructure and phase assemblage varied dramatically with aging time at $1400°C$. The optimum size and proportion of t-ZrO_2 was observed after aging for 2 hours.

(2) The changes in the phase assemblage had direct correlation with the mechanical properties. Under-, peak- and over-aged conditions were identified.

(3) The peak- and over-aged materials both exhibited crack resistance behaviour and flaw tolerance. The plateau toughness of the peak-aged material was considerably lower than reported values for subeutectoid peak-aged material.

(4) The threshold K values in cyclic loading were lower than those in static loading, for both peak- and over-aged materials, but the differences were not as dramatic as for subeutectoid aged materials.

(5) In-situ fatigue experiments on optical and scanning microscopy need to be conducted to identify the fatigue damage mechanisms.

ACKNOWLEDGEMENT

The authors thank A. Hartshorn, Y-W. Mai and B.R. Lawn for invaluable discussions, D. Argyriou and C.J. Howard for providing the data to compile figure 2, I. Fernandez, V. Zelizko and D. Hassard for technical assistance and

M. Boulter for word processing the manuscript. ICI Advanced Ceramics provided the specimens and BHP Materials Research Laboratories allowed access to their Instron 8501. The project was supported by GIRD grant No.15042.

REFERENCES

1. Green, D.J., R.H.J.Hannink, and M.V.Swain, 1989. Transformation Toughening of Ceramics, Boca Raton, Florida, USA:CRC Press, Inc.

2. Hannink, R.H.J. and R.C.Garvie, 1982. "Sub-eutectoid Aged Mg-PSZ with Enhanced Thermal Up-shock Resistance." Journal of Materials Science, 17(9):2637-2643.

3. Hannink, R.H.J., 1983. "Microstructural Development of Sub-eutectoid Aged MgO-ZrO$_2$ Alloys." Journal of Materials Science, 18(2):457-470.

4. Hughnan, R.R and R.H.J.Hannink, 1986. "Precipitation during Controlled Cooling of Magnesia-Partially Stabilized Zirconia." Journal of the American Ceramic Society, 69(7):556-563.

5. Hannink, R.H.J. and M.V.Swain, 1986. "Particle Toughening in Partially Stabilized Zirconia: Influence of Thermal History," in Tailoring Multiphase and Composites Ceramics Conference, G.L.Messing, C.G.Pantano and R.E.Newnham, eds., New York, NY, USA:Plenum Press, pp. 259-279.

6. Dauskardt, R.H., W.Yu and R.O.Ritchie, 1987. "Fatigue Crack Propagation in Transformation-Toughened Ceramics." Journal of the American Ceramic Society, 70(10):C248-C252.

7. Swain, M.V. and V. Zelizko, 1988. " Comparison of Static and Cyclic Fatigue in Mg-PSZ Alloys," in Advances in Ceramics, Vol 24., Science and Technology of Zirconia III, S.Somiya, N.Yamamoto and H.Yanagida, eds., Columbus, Ohio, USA:The American Ceramic Society, Inc., pp. 595-606.

8. Sylva, L.A. and S.Suresh, 1989. "Crack Growth in Transforming Ceramics under Cyclic Tensile Loads." Journal of Material Science, 24(5):1729-1738.

9. Dauskardt, R.H., D.B.Marshall and R.O.Ritchie, 1990. "Cyclic Fatigue Crack Propagation in Magnesia-Partially-Stabilized Zirconia Ceramics." Journal of the American Ceramic Society, 73(4):893-903.

10. Tsai, J.F., C-.S.Yu and D.K.Shetty, 1990. "Fatigue Crack Propagation in Ceria-Partially-Stabilized Zirconia (Ce-TZP)-Alumina Composites." Journal of the American Ceramic Society, 73(6):2993-3001.

11. Grathwohl, G., and T.Liu, 1991. "Crack Resistance and Fatigue of Transforming Ceramics: II, CeO_2-Stabilized Tetragonal ZrO_2." Journal of the American Ceramic Society, 74(12):3028-3034.

12. Lathabai, S., D.Argyriou, C.J.Howard and R.H.J.Hannink, "Microstructure, Phase Assemblage and Mechanical Properties of Eutectoid-Aged Mg-PSZ", under preparation.

13. ASTM Standard E 561-86, 1987. "Standard Practice for R-Curve Determination," in 1987 Annual Book of Standards, Vol. 3.01, Philadelphia, PA, USA:American Society for Testing and Materials, pp. 563-578.

14. ASTM Standard E 647-86a, 1987. "Standard Test Method for Measurement of Fatigue Crack Growth Rates," in 1987 Annual Book of Standards, Vol. 3.01, Philadelphia, PA, USA:American Society for Testing and Materials, pp. 899-926.

15. F.F. Lange, 1983. "Transformation Toughening: Thermodynamic Approach to Phase Retention and Toughening," in Fracture Mechanics of Ceramics, Vol. 5, R.C Bradt, A.G.Evans, F.F.Lange and D.P.H.Hasselman, eds., Plenum Press, New York, pp. 255-274.

16. Lange, F.F. and D.J.Green, 1981. "Effect of Inclusion Size on the Retention of Tetragonal ZrO_2: Theory and Experiments", in Science and Technology of Zirconia, Advances in Ceramics, Vol. 3, A.H. Heuer and L.W. Hobbs, eds., American Ceramic Society, Columbus, Ohio, pp. 217-225.

17. Ruhle, M. and A.H.Heuer, 1984. "Phase Transformations in ZrO_2-containing Ceramics," in Science and Technology of Zirconia II, Advances in Ceramics, Vol. 12, N.Claussen, M.Ruhle and A.H.Heuer, eds., American Ceramic Society, Columbus, Ohio, pp. 14-32.

18. Heuer, A.H., M.Ruhle and D.B.Marshall, 1990. "On the Thermoelastic Martensitic Transformation in Tetragonal Zirconia." Journal of the American Ceramic Society, 73(4):1084-1093.

19. Chantikul, P., G.R.Anstis, B.R.Lawn and D.B.Marshall, 1981. "A Critical Evaluation of Indentation Techniques for Measuring Fracture Toughness: II. Strength Method." Journal of the American Ceramic Society, 64(9): 539-543.

20. Steinbrech, R.W., 1992. "Review: Toughening Mechanisms for Ceramic Materials." Journal of the European Ceramic Society, 10(3):131-142.

21. D.L. Davidson, J.B. Campbell and J. Lankford, 1991. "Fatigue Crack Growth Through Partially Stabilized Zirconia at Ambient and Elevated Temperatures." Acta Metallurgica et Materiala, 39(6):1319-1330.

The Effect of Microstructure on the Plastic Deformation of Mg-PSZ

ANITA GROSS, WARREN BATCHELOR, TREVOR FINLAYSON
and JOHN GRIFFITHS

ABSTRACT

Samples from four batches of Mg-PSZ were plastically deformed under uniaxial tension. The amount of plastic strain that could be induced before fracture was relatively constant for samples from the same batch but varied widely between batches. It is suggested that both grain size and precipitate metastability play an important role in determining the observed creep fracture strains.

INTRODUCTION

Mg-PSZ is a structural ceramic which, at room temperature, incorporates both high toughness and strength [1]. These advantageous mechanical properties are attributable to a microstructure consisting of a cubic (c) matrix containing metastable tetragonal (t) precipitates together with three other phases, monoclinic (m), orthorhombic (o) and the δ-phase ($Mg_2Zr_5O_{12}$). The toughening arises from the 4-5% volume expansion and shape change associated with the stress-induced tetragonal to monoclinic (t$\rightarrow$m) phase transformation. Furthermore, the high strains in and near the transformed precipitates cause microcracking which may further enhance the macroscopic toughness [2].

Under stress zirconia ceramics can plastically deform. The amount of plastic strain that can be induced is large for a ceramic with volume strains of up to 1.8% having been reported in compression for some grades of Mg-PSZ[3]. The mechanisms responsible for the plastic deformation are the volume strains of the tetragonal to monoclinic and tetragonal to orthorhombic transformations,

A. Gross and J. Griffiths, Department of Materials Engineering, Monash University, Clayton, Vic. 3168, Australia

W. Batchelor and T. Finlayson, Department of Physics, Monash University, Clayton, Vic. 3168, Australia

J. Griffiths is now at CSIRO, Division of Manufacturing Technology, P.O. Box 883, Kenmore, QLD 4069, Australia

371

and the strain from microcracking[4, 5]. The study of the plastic deformation of Mg-PSZ remains an area of importance as effective theories of transformation toughening can only be formulated with knowledge of how the strain of transformation is distributed[6].

Previously, plastic deformation measurements on Mg-PSZ have mostly been confined to compression testing[3, 4, 7, 8]. There has been some other work on plastic deformation with continuously increasing tensile stress[9] and with a four-point bending stress[9-13]. The work reported here and elsewhere[13-15] represents the first work on the tensile plastic deformation of Mg-PSZ under constant stress (creep test).

EXPERIMENTAL METHOD

Samples were tested from four separate batches of commercial Mg-PSZ* believed to be fabricated by a process similar to that described by Marshall *et al.* [16]. Both MS and TS grades were supplied. The mechanical and microstructural properties of each batch were determined using XRD, optical microscopy, MOR, tensile and compressive strengths. Each MOR value is from an average of twelve measurements, whereas the tensile and compressive strengths were determined from only two and one tests, respectively. The XRD and MOR measurements were performed by the supplier, ICI Advanced Ceramics.

BATCH CHARACTERIZATION

XRD results of the ground surface monoclinic (GSM) and the polished surface monoclinic (PSM) contents were determined using the Garvie and Nicholson method [17]. The stress transformable tetragonal (STT) content was calculated from the difference between the GSM and the PSM. Optical microscopy of the polished and chemically etched surface was used to determine the grain size. Mechanical strength values were determined from four point bending (MOR), and from uniaxial tension and compression.

SAMPLE PREPARATION AND TESTING

The as-received cylinders were machined by grinding to produce dumb-bell shaped samples with dimensions shown in figure 1. The gauge section was polished with diamond impregnated cloth and then with silicon carbide paper to a final finish of 600 grade SiC paper. The final polish was longitudinal so that no scratches were left transverse to the loading direction. Steel sleeves were glued to the ends of the cylinders as attachment points for Instron *Super–Grips*[†]. The Instron self-aligning *Super Grips* gave a bending stress of less than 5% of the tensile stress provided the sample had been prepared and loaded correctly. Strains were measured by electrical resistance strain gauges glued to the gauge length,

---

*Supplied by ICI Advanced Ceramics, Clayton, Victoria, Australia
†Instron, Bucks, Great Britain

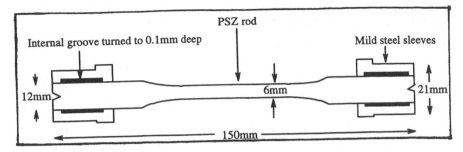

Figure 1. The dimensions of a tensile test sample.

with four gauges used for each test. Three gauges were aligned parallel to the tensile axis and spaced equally around the circumference. These were used to measure the bending moment besides their primary function of measuring the tensile strain. The remaining gauge measured transverse strains.

Throughout each creep test, measurements of Young's modulus (E) and Poisson's ratio (ν), were made on unloading from the creep stress and during loading/unloading cycles in which the maximum stress during a cycle was typically 160-190 MPa, well below the creep stress, so that further creep during cycling was avoided. After each Young's modulus measurement, the sample was reloaded to the creep stress and the experiment continued.

RESULTS AND DISCUSSION

BATCH PROPERTIES

Table I summarizes the properties of each of the four batches of samples used.

Precipitate metastabilities are indicated by the STT values from the XRD data. Batches 1 and 2, having a similar STT value, consist of more stable tetragonal precipitates compared to batches 3 and 4. Batches 1 and 2 have a metastability characteristic of MS grade samples, whereas batches 3 and 4 are TS grade. It should be noted that the PSM value obtained for batch 1 was measured at a later date to the other batches. It is thus possible that the anomalously low value of PSM for this batch was a result of slightly different polishing conditions than those employed in the measurements of the three other batches.

The different measures of strength show that batches 1 and 2 have the highest strengths. This result is consistent with the difference between the MS and TS grades. As a result of the different precipitate metastabilities, the tetragonal precipitates in the TS grade samples transform under lower applied stresses than those in the MS grade samples (see Table II). CREEP DATA

Table II shows typical longitudinal (ε_l) and transverse (ε_t) plastic strains

TABLE I- MICROSTRUCTURAL AND MECHANICAL PROPERTIES OF SAMPLES FROM FOUR DIFFERENT BATCHES OF Mg-PSZ.

Test Procedure		Batch 1	Batch 2	Batch 3	Batch 4
1. XRD	gsm %	23±1	25±1	34 ±1	36±1
	psm %	6±1	10±1	12±1	12±1
	stt %	17±2	15±2	22±2	24±2
2. Grain size (μm)		42±5	50±5	61±6	66±6
3. MOR (MPa)		700±18	692±34	658±17	626±35
4. Tensile strength (MPa)		450±5	434±5	406±5	380±5
5. Compressive strength (MPa)		not determined	1586±5	1513±4	1420±2
Sample type		MS	MS	TS	TS

at fracture of samples from the different batches. It can be seen from Table II that the plastic strains at fracture are similar for samples from the same batch, but vary widely between the batches, with over an order of magnitude difference between batches 1 and 4. For each batch, the range of stresses given in Table II show the stresses at which it is practicable to perform a creep test. Stresses below the indicated range are too small to produce any significant plastic deformation within a time of a few weeks, while stresses above the indicated range tend to produce very rapid fracture. The longitudinal strains in Table II are more than twice as large as the transverse strains. This difference is thought to be partly due to the contributions of both transformation and microcracking in the longitudinal direction, and only transformation strains in the transverse[3, 4, 13, 14, 15, 18, 19].

During each creep test the Young's modulus, E, decreased with increasing strain. The initial measured value of E at 200±5GPa decreased down to a minimum value of 191±5GPa for a batch 4 sample after which it fractured[15]. The progressive decrease in Young's modulus followed by fracture is assumed to be associated with the development of microcracks. The value of ν remained constant at 0.28±0.04 during the test, for strains in the elastic region. However the *effective* Poisson's ratio, ν_{eff}, becomes progressively negative as the material creeps implying an expansion on elongation [15].

When the creep fracture strains from Table II are compared with the microstructural properties of the different batches, it can be seen that the creep fracture strain, ε_l, increases with both the precipitate metastability and the grain size. Since both properties increase with increasing creep fracture strain, it is not possible at the present stage to quantify the contributions of precipitate metastability and grain size to the observed creep fracture strains. While it is clear that precipitate metastability must play a significant role, since reduced precipitate stability will produce a greater volume fraction of transformation

TABLE II- THE VARIATION OF CREEP FRACTURE STRAINS MEASURED IN
A RANGE OF APPLIED STRESSES AMONGST THE FOUR BATCHES

| Batch | σ (MPa) | Creep fracture strains | |
		ε_l	ε_t
1	355-455	.0001-.0005	.0000-.0001
2	291-405	.0010-.0015	.0001-.0002
3	255-340	.0025-.0034	.0006-.0008
4	175-272	.0041-.0053	.0011-.0025

(and hence more plastic deformation), the role of grain size is less obvious.

The tetragonal to monoclinic transformation is thought to proceed by correlated transformation [3, 19]. With this effect, several precipitates within a grain
transform as a result of the applied stress. The transformation of these precipitates then applies an additional stress to neighbouring precipitates, causing
further transformation. This autocatalytic transformation ceases at the grain
boundaries and results in a series of transformation bands within a grain.

Thus, if the volume fraction of transformation within a grain is assumed to
be constant, the plastic strain in the sample from autocatalytic transformation
within that grain will be proportional to the grain size cubed. If the number
of grains within which autocatalytic transformation is nucleated is dependent
only on the test stress and on the metastability of the tetragonal precipitates,
then the creep fracture strain would be expected to scale with the cube of the
grain size provided all other factors were constant. The four different batches
had grain sizes ranging from 42 to 66 μm (see Table I). The ratio of $66^3/42^3$ is
3.88, so such a grain size effect could not explain all of the difference between
the creep fracture strains of the two batches (Table II). However, as has been
previously noted, the precipitates in batch 4 are less stable than those in batch
1 and thus a grain size effect would not be expected to account for all of the
differences between the two batches. Hence while a link between grain size and
creep fracture strain cannot be definitively proven from the available data it can
be seen that such a link is a plausible one.

CONCLUSIONS

Samples from four batches of Mg-PSZ were plastically deformed under uniaxial tension. The amount of plastic strain that could be induced before fracture
was relatively constant for samples from the same batch but varied widely between batches. These creep fracture strains appeared to be correlated with both
grain size and precipitate metastability. While no definitive answer could be
made about the relative importance of these two microstructural properties in
determining the observed creep fracture strains, it is suggested that both play
an important role in determining the observed creep fracture strains.

ACKNOWLEDGEMENTS

Anita Gross was supported by an Australian Postgraduate Research Award (A.P.R.A.), and ICI Advanced Ceramics Pty. Ltd. Warren Batchelor was supported by the Australian Research Council under a postdoctoral research fellowship.

REFERENCES

1.   Green, D.J., R.H.J. Hannink and M.V. Swain, 1989. Transformation Toughening of Ceramics. Boca Raton, Florida, USA: CRC Press.

2.   Evans, A.G., 1984. "Toughening Mechanisms in Zirconia Ceramics," in Advances in Ceramics 12- The Science and Technology of Zirconia II, N. Claussen, M. Rühle, and A.H. Heuer, eds. Westerville, Ohio, USA: The American Ceramic Society, pp.193–212.

3.   Chen, I-W. and P.E. Reyes-Morel, 1986. "Implications of Transformation Plasticity in ZrO_2-Containing Ceramics:I, Shear and Dilatation Effects." Journal of the American Ceramic Society, 69(3):181–189.

4.   Reyes-Morel, P.E. and I-W. Chen, 1990. "Stress-Biased Anisotropic Microcracking in Zirconia Polycrystals." Journal of the American Ceramic Society, 73(4):1026–1033.

5.   Liu, S.-Y. and I-W. Chen, 1990. "Fatigue Deformation Mechanisms of Zirconia Ceramics." Journal of the American Ceramic Society, 75(5):1191–1204.

6.   Chen, I-W., 1991. "Model of Transformation Toughening in Brittle Materials." Journal of the American Ceramic Society, 74(10):2564–72.

7.   Lankford, J., 1983. "Plastic Deformation of Partially Stabilized Zirconia." Journal of the American Ceramic Society, 66(11):c212–c213.

8.   Rogers, W.P. and S. Nemat-Nasser, 1990. "Transformation Plasticity at High Strain Rate in Magnesia-Partially-Stabilized Zirconia." Journal of the American Ceramic Society, 73(1):136–139.

9.   Marshall, D.B., 1986. "Strength Characteristics of Transformation-Toughened Zirconia." Journal of the American Ceramic Society, 69(3):173–180.

10.   Swain, M.V., 1985. "Inelastic Deformation of Mg-PSZ and its Significance for Strength-Toughness Relationship of Zirconia Toughened Ceramics." Acta Metallurgica, 33(11):2083–2091.

11. Swain, M.V., 1988. "Thermoelastic Deformation and Recovery of Mg-PSZ Alloys." in Advances in Ceramics 24, Science and Technology of Zirconia III, S. Sōmiya, N. Yamamoto, and H. Yanagida, eds. Westerville, Ohio, USA: The American Ceramic Society, pp. 439–448.

12. Lim, C.S., T.R. Finlayson, F. Ninio and J.R. Griffiths, 1992. "In-Situ Measurement of the Stress-Induced Phase Transformation in Magnesia Partially Stabilized Zirconia Using Raman Spectroscopy." Journal of the American Ceramic Society, 75(6):1570-1573

13. Finlayson, T.R., J.R. Griffiths, A.K. Gross, and C.S. Lim, 1991. "Creep and Static Fatigue of Magnesia-Partially-Stabilized Zirconia at Room Temperature," in Fracture of Engineering Materials and Structures, S.H. Teoh and K.H. Lee, eds. London: Elsevier Science Publishers, pp 321-326

14. Kisi, E.H., T.R. Finlayson and J.R. Griffiths, 1990. "Phase Determination in Partially Stabilized Zirconia Creep Specimens." Mater. Sci. Forum., 56-58:351–356.

15. Gross, A.K., 1992. "Uniaxial Tensile Creep of Magnesia Partially Stabilized Zirconia." Masters Thesis, Department of Materials Engineering, Monash University, Australia.

16. Marshall, D.B., M.R. James and J.R. Porter, 1989. "Structural and Mechanical Property Changes in Toughened Magnesia-Partially-Stabilized Zirconia at Low Temperatures." Journal of the American Ceramic Society, 72(2):218–227.

17. Garvie, R.C. and P.S. Nicholson, 1972. "Phase Analysis in Zirconia Systems." Journal of the American Ceramic Society, 55(6):303–305.

18. Swain, M.V. and V. Zelizko, 1988. "Comparison of Static and Cyclic Fatigue on Mg-PSZ Alloys," in Advances in Ceramics 24, Science and Technology of Zirconia III, S. Sōmiya, N. Yamamoto, and H. Yanagida, eds. Westerville, Ohio, USA: The American Ceramic Society, pp. 595–606.

19. Heuer, A.H., M. Rühle and D.B. Marshall, 1990. "On the Thermoelastic Martensitic Transformation in Tetragonal Zirconia." Journal of the American Ceramic Society, 73(4):1084–1093.

Preparation and Characterization of Fine-Grained (Mg,Y)-PSZ Ceramics with Spinel Additions

FRANK MESCHKE, GOFFREDO DE PORTU
and NILS CLAUSSEN

ABSTRACT

A fine-grained Mg-PSZ-type material has been produced by adding $MgAl_2O_4$ spinel to a ternary Y_2O_3-MgO-ZrO_2 alloy. A mean cubic grain size below 10 μm was achieved. Aging treatment at 1100°C and 1400°C lead to the formation of transformable tetragonal precipitates and resulted in improved toughness. Transformation toughening is the main toughening mechanism.

INTRODUCTION

Yttria-doped tetragonal zirconia polycrystals (Y-TZP) exhibit very high strength and good toughness [1]. However, owing to an uncontrolled tetragonal (t)-to-monoclinic (m) phase transformation, the low temperature strength decreases dramatically when exposed to humid atmosphere [2]. Mg-partially stabilized zirconia (Mg-PSZ) exhibits much better resistance to low temperature aging, although some strength degradation has also been observed in a few cases [3,4]. The strength of Mg-PSZ is usually lower than that of Y-TZP, but its fracture toughness is much higher [5,6].

The ternary system ZrO_2-MgO-Y_2O_3 has mainly been studied with respect to the effect of Y_2O_3 on the subeutectoid decomposition of Mg-PSZ [7]. In a more recent study on the precipitation behavior in the ZrO_2-MgO-Y_2O_3 system it has been found that the transformable t phase is difficult to precipitate [8]. Consequently, stress-induced transformation does not play an important role in the toughening of these materials.

G. de Portu, CNR-IRTEC, 48018 Faenza, Italy
N. Claussen and F. Meschke, Advanced Ceramic Group, Technische Universität Hamburg-Harburg D-2100 Hamburg 90, FRG

In a previous work it has been demonstrated that fine-grained Mg-PSZ-type materials with a cubic grain size below 10 μm and good low temperature stability can be produced by adding MgO-Al$_2$O$_3$ to ternary Y$_2$O$_3$-MgO-ZrO$_2$ alloys [9]. In this study we examine the precipitation behavior of such a (Mg,Y)-PSZ alloy during aging at 1100°C and 1400°C and its effect on fracture toughness.

EXPERIMENTAL PROCEDURE

A 30:70 ratio mixture of 3 mol% Y$_2$O$_3$- and 9 mol% MgO-containing ZrO$_2$ was prepared using commercial powders (Tosoh TZ-3Y and MEL SCMG3 grade). 3 vol% of Al$_2$O$_3$/MgO at a 1:1 ratio (MgAl$_2$O$_4$) was added to the composition which was attrition milled in ethanol with 3Y-TZP milling-media in a polyethylene liner. After milling for 3 1/2 h the powders were dried. Green bodies were uniaxially pressed into bars of dimensions 60 mm x 6 mm x 4 mm at 47 MPa followed by cold isostatic pressing at 150 MPa. The bars were sintered in a powder bed of the same nominal composition in air at 1700°C for different holding times (5 min, 30 min and 60 min). Samples sintered for 30 min were subsequently aged either at 1100°C (subeutectoid aging) for 2, 5, 8, and 15 h or at 1400°C for 1, 2, and 3 h.

The optimum aging conditions were determined by X-ray analysis comparing the (202) and (220) peak intensities. The phase contents were evaluated using the method proposed by Garvie and Nicholson [10]. Densities of the sintered bodies were measured by Archimedes method and the relative densities were calculated taking the cubic (c), t- and m-ZrO$_2$ as well as the spinel volume content into account.

Microstructures of thermally etched samples were examined by SEM and the grain sizes were estimated by the lineal intercept method. Fracture toughness (K$_{Ic}$) was measured by the indentation-crack-length technique with a load of 588 N [11]. K$_{Ic}$ data were calculated from six indentation tests. Young's modulus was determined by the resonance frequency method [12].

RESULTS AND DISCUSSION

Samples with densities ranging from 5.68 to 5.75 g/cm^3 were obtained after sintering at 1700°C for different holding times (see Table I). As expected, the samples contain MgAl$_2$O$_4$ spinel particles resulting from the Al$_2$O$_3$-MgO additions. Since the m phase content was the lowest for samples sintered at 1700°C for 30 min and the density remained almost constant for longer holding times, only these samples were used for further treatment.

TABLE I. PROPERTIES OF (Mg,Y)-PSZ AFTER SINTERING AT 1700°C

Holding time min	Density g/cm3	m-phase vol%	E-modulus GPa
5	5.68	19.3	n.m.
30	5.75	4.8	203
60	5.74	9.5	n.m.

MICROSTRUCTURE

The typical microstructure of the fine-grained (Mg,Y)-PSZ materials is shown in figure 1. They exhibit a bimodal grain size distribution with small m-ZrO_2 particles and large c-grains. The mean size of m-grains is about 1.5 μm whereas that of c-grains is approximately 5.5 μm. The spinel- and m-ZrO_2 particles are dispersed preferentially at grain boundaries.

The aging procedures lead to the formation of t-precipitates. The content of tetragonal precipitates is represented by the t/c-ratio in Table II and in figure 2. Subeutectoid aging at 1100°C up to 15 h does increase the t/c-ratio while the m-content remain constant, i.e., t precipitates grow, but do not transform into m symmetry. Aging at 1400°C leads to an increase of m phase after aging times > 1 h, while the t/c-ratio exhibits a maximum for a holding time of 2 h. This indicates an overaging as confirmed previously for Mg-PSZ-type materials [5,9,13].

X-ray analysis does not reveal the presence of free MgO, hence a decomposition phenomenon can be excluded. Compositions aged at 1100°C for 8 and 15 h and at 1400°C for 1 and 2 h contain low amounts of m phase and high t/c ratios. The microstructure within the c grains is shown in figures 3 and 4, revealing the presence of two different microstructural morphologies. Besides elongated, lenticular-shaped precipitates as shown in figure 3, there are in the center of some cubic grains one dark spot with a more irregular, brain-like structure, similar to that previously observed [9,14], i.e., possibly domains of tetragonal t' phase. In figure 4 one of these dark spots is presented which is surrounded by large t-precipitates.

TABLE II. PROPERTIES OF (Mg,Y)-PSZ AFTER AGING

Temperature	Holding time min	m-phase vol%	t/c-ratio	Mean grain size μm	K_{Ic} MPa√m
As-sintered	---	4.8	0.70	5.5	5.4±0.3
1100°C	2	5.7	0.87	n.m.	n.m.
	5	6.0	0.92		9.0±0.4
	8	6.5	0.94		11.7±0.3
	15	6.5	0.96		10.9±0.6
1400°C	1	5.7	0.74	n.m.	9.9±0.6
	2	10.7	0.79		10.6±0.7
	3	15.3	0.68		8.1±0.4

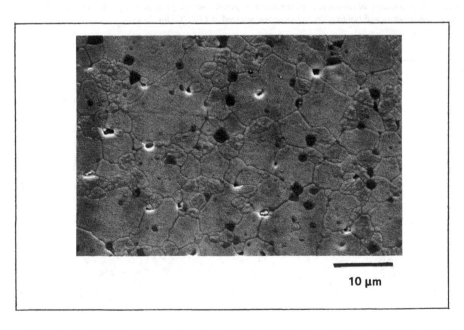

10 µm

Figure 1. Typical microstructure of as-sintered (Mg,Y)-PSZ after sintering at 1700°C for 30 min. Cubic grains have a mean size of 5.5µm, monoclinic ZrO_2- and $MgAl_2O_4$ particles (dark) are preferentially dispersed at grain-boundaries.

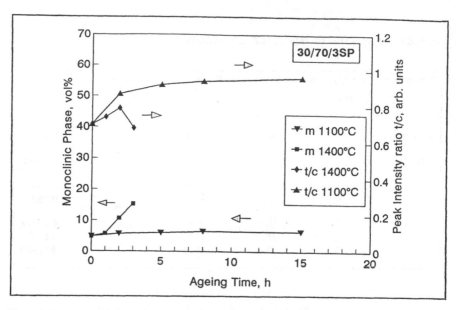

Figure 2. Percentage of monoclinic phase and tetragonal-to-cubic peak intensity ratio as a function of aging temperature and holding time in samples sintered at 1700°C for 30 min.

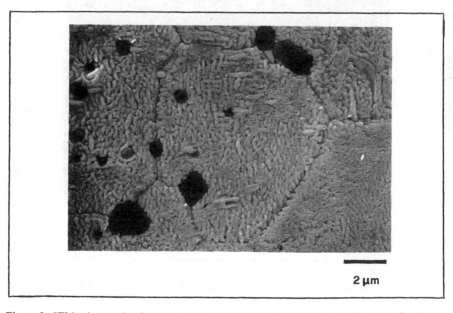

Figure 3. SEM micrograph of a sample aged at 1400°C for 2 h. Spinel particles are distributed preferentially at grain boundaries. Note tetragonal precipitates within cubic grains.

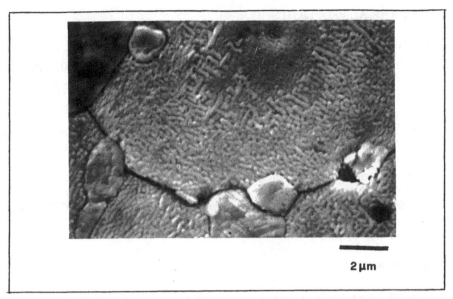

Figure 4. SEM micrograph of a sample aged at 1400°C for 3 h. Dark spot with an irregular morphology in addition to elongated tetragonal precipitates within a cubic grain.

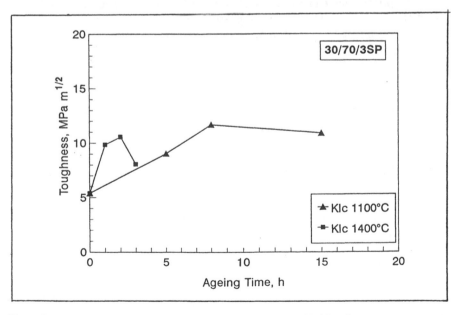

Figure 5. Fracture toughness as a function of aging temperature and holding time.

MECHANICAL PROPERTIES

Fracture toughness results are reported in Table I and figure 5. K_{Ic} values range from 5.4 ± 0.3 MPa$\sqrt{m}$ for the as sintered material to 11.7 ± 0.3 MPa$\sqrt{m}$ and 10.6 ± 0.7 MPa$\sqrt{m}$ for the samples aged at 1100°C and 1400°C, respectively.

The fracture toughness curves coincide very well with those in figure 1 representing the tetragonal volume content. The increase in fracture toughness can be attributed to the presence of a large fraction of transformable t precipitates in the c-grains resulting from the aging treatment.

Considering that the m phase content is low, it is concluded that transformation toughening is the dominant toughening mechanism. However, microcracking could also add to the toughening in samples aged at 1400°C because of increasing m-phase content.

CONCLUSIONS

Mg-PSZ-type ceramics with a cubic grain size < 10 μm can be produced starting from commercial 3 mol% Y_2O_3 and 9 mol% MgO-containing powders if small amounts of $MgAl_2O_4$ spinel particles are added.

Appropriate aging either at 1100°C or 1400°C leads to the formation of a high amount of transformable tetragonal phase. The fracture toughness is increased considerably due to the related transformation toughening mechanism.

ACKNOWLEDGEMENT

F. Meschke is grateful to CEC for providing a grant within the BRITE/EURAM program.

REFERENCES

1. Tsukuma, K., Y. Kubota and T. Tsukidate, 1984. "Thermal and Mechanical Properties of Y_2O_3-Stabilized Tetragonal Zirconia Polycrystals", in Science and Technology of Zirconia II, Advances in Ceramics Vol 12, N. Claussen, M. Rühle and A.H. Heuer, eds, Columbus, Ohio, USA: The American Ceramic Society Inc., pp. 382-390.

2. Masaki, T., 1986. "Mechanical Properties of Y-PSZ After Aging at Low Temperature." Int. J. High Tech. Ceram., 2 : 85-98.

3. Swain, M.V., 1985. "Stability of Mg-PSZ in High Temperature Steam Environment." J. Mat. Sci. Let., 4 :848-850.

4. Sato, T., T. Endo, M. Shimada, T. Mitsudome and N. Otabe, 1991. "Hydrothermal Corrosion of Magnesia-Partially Stabilized Zirconia." J. Mat. Sci., 26 :1346-1350.

5. Hannink, R.H.J., 1983. "Microstructural Development of Subeutoctoidly Aged MgO-ZrO$_2$ Alloys." J. Mat. Sci., 18 :457-470.

6. Claussen, N., 1984. "Microstructural Design of Zirconia-Toughened Ceramics", in Science and Technology of Zirconia II, Advances in Ceramics Vol 12, N. Claussen, M. Rühle and A.H. Heuer, eds, Columbus, Ohio, USA: The American Ceramic Society Inc., pp 325-361.

7. Scott, H.G., 1981. "Phase Relationships in the Magnesia-Yttria-Zirconia System." J. Austr. Ceram. Soc., 17 (1) :16-20.

8. Montross, Ch.S., 1992. "Precipitation and Bulk Property Behaviour in the Yttria-Magnesia-Zirconia Ternary System." Brit. Ceram. Trans. J., 90 :175-178.

9. Meschke, F., G. De Portu and N. Claussen, "Microstructure and Thermal Stability of Fine-Grained Mg-Y-PSZ Ceramics with Alumina Additions." To be published in J. Eur. Ceram. Soc..

10. Garvie, R.C. and P.S. Nicholson, 1972. "Phase Analysis in Zirconia Systems." J. Am. Ceram. Soc., 55 :303-305.

11. Anstis, G.R., P. Chantikul, B.R. Lawn and D.B. Marshall, 1981. "A Critical Evaluation of Indentation Techniques for Measuring Fracture Toghness: I, Direct Crack Measurement." J. Am. Ceram. Soc., 64 :533-538.

12. "Standard Test Method for Young's Modulus, Shear Modulus and Poisson's Ratio for Ceramic Whiteware by Resonance", ASTM, Designation: C 848-78 (Reapproved 1983) pp 268-273.

13. Hannink, R.H.J., R.A. Johnston, R.T. Pascoe and R.C. Garvie, 1981. "Microstructural Changes During Isothermal Aging of a Calcia Partially Stabilized Zirconia Alloy" in Science and Technology of Zirconia , Advances in Ceramics Vol 4, A.H. Heuer and L.W. Hobbs, eds, Columbus, Ohio, USA: The American Ceramic Society Inc., pp 116-136.

14. Yuan, T.C., G.V. Srinivasan, J.F. Jue and A.V. Virkar, 1989. "Dual-Phase Magnesia-Zirconia Ceramics with Strength Retention at Elevated Temperatures." J. Mat. Sci., 24 :3855-3864.

Effect of Stress on Trapped Cracks in Y-TZP

LINDA M. BRAUN and ROBERT F. COOK

ABSTRACT

Two types of radial cracks are generated at indentations in Y-TZP materials. These are "trapped" cracks within the contact-induced transformation zone surrounding the indentation impression and "well-developed" cracks which extend beyond the transformation zone boundary. The probability of generating a well-developed versus a trapped crack is dependent on the transformability of the material (determined by the stabilizer content and grain size) and on the indentation load. The effect of applied stress on well-developed and trapped cracks is investigated by in situ measurements of crack size as a function of a superimposed bending stress. Well-developed cracks grow in stable equilibrium to failure. Weakly trapped cracks escape from the contact-induced transformation zone to well-developed form at high stresses, and exhibit stabilized growth before failure. Strongly trapped cracks do not pop in to well-developed form, but fail spontaneously.

INTRODUCTION

Yttria-stabilized tetragonal zirconia polycrystals (Y-TZP) exhibit high fracture strengths with reported values in excess of 1000 MPa, and fracture toughness values ranging from 5-20 MPa.m$^{1/2}$ [1-6]. These materials use a stress induced tetragonal-to-monoclinic phase transformation to enhance their mechanical behavior. The phase transformation leads to T-curve (R-curve) behavior, i.e. an increase in toughness or crack growth resistance with increasing crack size. There is therefore considerable interest in these materials for structural applications. Retention and transformability of the tetragonal phase are strongly influenced by microstructural parameters such as grain size and stabilizer content. The effect of such parameters on mechanical behavior, specifically strength and toughness, of Y-TZP materials has been extensively investigated [3-7].

A common technique for evaluating ceramic toughness is measurement of the surface crack lengths emanating from indentation contact impressions [8,9].

Linda M. Braun, National Institute of Standards and Technology, Ceramics Division, Gaithersburg, MD 20899 USA

Robert F. Cook, IBM Research Division, T.J. Watson Research Center, Yorktown Heights, NY 10598 USA

In Y-TZP transformation zones have been observed around indentation impressions [4,10,11]. The combined effect of the compressive transformation field surrounding the contact impression and the tensile residual indentation field on indentation fracture mechanics in Y-TZP has recently been considered [12,13]. It was found that a characteristic indentation threshold load exists for crack initiation. Above this (lower) threshold, radial cracks are "trapped" within the contact-induced transformation zone surrounding the indentation impression. Above a second (higher) threshold, the radial cracks are "well-developed", i.e. extend beyond the contact-induced transformation zone boundary. The probability of generating a trapped versus a well-developed crack is found to be strongly dependent on the transformability (microstructure) of the material and on indentation load [12].

The intent of this paper is to consider the effect of a subsequent applied stress on these two regions of radial crack formation in Y-TZP, and to examine the implications of the crack response on strength. We begin with a description of the experimental procedure, including material fabrication and test methodologies. We then present results from observations of indentation measurements of crack lengths and threshold loads. Post-failure fracture surface observations are presented to confirm the geometrical features of the radial cracks. These are followed by results from in situ measurements of crack size as a function of a superimposed bending stress. The discussion incorporates concepts from a previously developed fracture mechanics model [13] of the trapped-crack problem.

EXPERIMENTAL PROCEDURE

MATERIAL FABRICATION

Specimens were fabricated using 2.5 mol% Y_2O_3 (Tosoh)* powder. A slip was prepared containing 20 vol% solids in distilled water, ammonium polyacrylic acid (Darvan)* as a dispersant and NH_4OH to balance the pH. The slip was pressure cast into plates which were subsequently sintered in air at 1300, 1400, 1500, or 1600°C for 1 hr. This yielded specimens 98% dense. Bars were cut from the sintered and surface-ground plates. These were then diamond polished to a 1 μm finish. After polishing, the bars were annealed at 1250°C in air to revert any material that transformed to monoclinic symmetry during machining back to tetragonal, as confirmed by X-ray diffraction [14]. The sintered microstructures were revealed by thermal etching in air to a temperature 50°C lower than the maximum sintering temperature for 10 min. Grain size was measured using the linear intercept technique from scanning electron micrographs using a digital image analyzer.

INDENTATION TESTING

Specimens were indented using a Vickers diamond pyramid with peak loads ranging from 5 to 500 N and a dwell time of 15 s. Contact impression and radial surface crack traces were measured in air by optical microscopy, using bright field, dark field, and Nomarski interference contrast. The indentation hardness, (load/projected contact area), was evaluated directly from measurements of the impression diagonals.

The prospective tensile surface of flexure bars, 3x6x25 mm, were indented with peak loads of either 100 or 300 N with one set of radial cracks aligned perpendicular to the long axis of the bar. In situ observations of indentation-crack growth to failure were made using a 4-pt flexure stressing apparatus attached to an inverted optical microscope. A schematic of the stressing apparatus is shown in figure 1. Load was applied by a stepping micrometer, and was recorded with a standard resistance load cell. Care was taken to ensure that the indentation was within the uniformly stressed region of the 4 pt bend bar. Stable growth (in air) of the surface traces of the radial cracks, aligned perpendicular to the long axis of the bar, was measured directly as a function of load. Load was increased monotonically in increments of $\approx$ 5 MPa in 3 min intervals. Care was taken to ascertain the onset of nonequilibrium moisture-controlled crack growth to rapid failure. No further increase in applied load or crack length measurements were recorded for nonequilibrium growth. All samples failed from the indentation site. Radial crack morphology was determined from optical microscopic examination of fracture surfaces.

RESULTS

MATERIAL CHARACTERIZATION

Table I shows the average grain size for the four different heat treated specimens. Heat treatment at 1300°C yielded specimens with 0.27 μm grain size, which increased to 0.70 μm at 1600°C. All specimens are fine grained (< 1 μm) with equiaxed microstructures. Also given in Table I is the steady-state toughness value, T_∞, as determined from chevron notch-bend tests on specimens cut from the same plates [11,14]. Toughness increased with increasing grain size from 4.2 to 5.9 MPa.m$^{1/2}$.

CONTACT DEFORMATION

Figure 2 is a Nomarski interference micrograph of a 500 N indentation in the material sintered at 1600°C (0.70 μm grain size). Clearly visible is a well-defined contact impression with surface cracks emanating from the corners. Also visible is a contact-induced transformation zone surrounding the contact-impression. This transformation zone is approximately twice the diameter of the contact-impression. Also present are transformation zones surrounding the surface crack traces. The width of these crack-induced transformation zones is approximately 8-10 μm. T-curves measured on this material [11,14] indicate that the toughness, as a result of phase transformation, is saturated at its steady-state value for zones of this magnitude. Similar transformation zones were not as easily detected on those materials with lower steady-state toughness values. Transformation to monoclinic phase was detected, however, using X-ray diffraction [14] because the smaller grain size materials have smaller transformation zone sizes.

A load-invariant hardness was determined for all four materials, shown in Table I, with an average value of 12.9 GPa [12]. This implies that geometrical similarity is maintained with respect to the contact deformation and transformation stress fields.

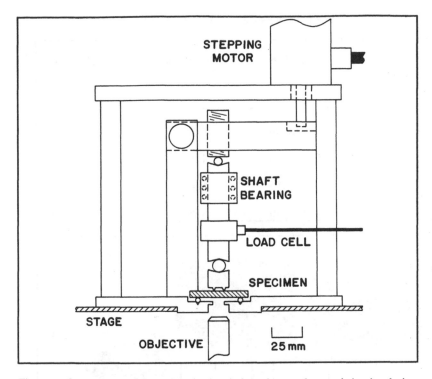

Figure 1. Schematic of fixture for viewing indentation crack growth in situ during specimen stressing to failure. The load is applied by the stepping motor.

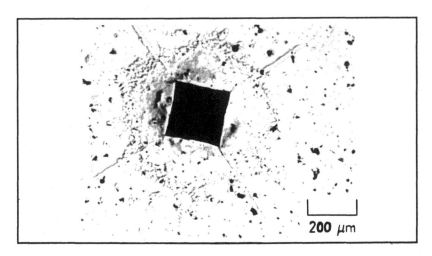

Figure 2. Optical micrograph (Nomarski interference contrast) of a 500 N Vickers indentation with the contact-induced transformation zone surrounding the impression visible (0.70 μm grain size).

TABLE I - PROPERTIES OF Y-TZP MATERIALS

Sintering Temperature (°C)	Grain Size λ (µm)	Steady-State Toughness T_∞ (MPa.m$^{1/2}$)	Hardness H (GPa)
1300	0.27±0.08	4.2±0.2	12.3±0.5
1400	0.36±0.13	4.9±0.1	13.5±0.3
1500	0.42±0.15	5.4±0.1	13.4±0.4
1600	0.70±0.30	5.9±0.3	12.4±0.3

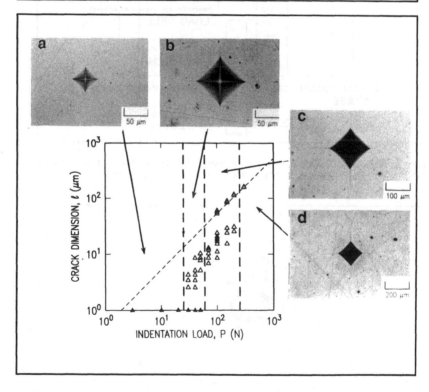

Figure 3. Radial crack length, l, versus indentation load, P, (0.42 µm grain size). Optical micrographs show the four regions of indentation cracking behavior. (a) $P = 20$ N; below the crack initiation threshold. (b) $P = 70$ N; above first threshold for radial cracks trapped within the contact-induced transformation zone. (c) $P = 200$ N; above second threshold for formation of well-developed cracks that have escaped from the transformation zone. (d) $P = 300$ N; above the load required to generate well-developed cracks at all impression corners.

CRACK MORPHOLOGY

The crack morphology of this Y-TZP material was determined as radial from examination of fracture surfaces and serial sectioning of indentations [11-14]. The four radial cracks are generated independently at the corners of the contact impression. Figure 3 is a plot of radial crack-length l, measured from the corner of the contact impression, versus indentation load P for the material sintered at 1500°C (0.42 μm grain size). Four regions of indentation cracking behavior [12] are delineated in figure 3 by the vertical dashed lines.

(a) An indentation threshold load, $P < 30$ N, exists for crack initiation, below which no cracking was observed.

(b) Above the sub-threshold initiation load, short cracks were observed. These cracks are "trapped" within the contact-induced transformation zone surrounding the indentation impression. Note that in this second region of cracking behavior sub-threshold cracking was still observed. As indentation load was increased further, the probability of generating a trapped crack increased until trapped cracks were observed at every impression corner.

(c) A second characteristic threshold load existed, $P \geq 100$ N, above which longer "well-developed" cracks were first observed. Well-developed radial cracks extended beyond the contact-induced transformation zone boundary (south and west corners).

(d) As indentation load was increased, the probability of generating a well-developed crack increased. The fourth region of behavior was characterized by the generation of well-developed cracks at every impression corner.

FRACTURE SURFACE OBSERVATIONS

Post-failure fracture surface observations confirm the radial crack morphology of indentation-generated cracks in Y-TZP. Indentation cracks that are either well-developed or trapped are generated independently of each other during indentation cracking. Figure 4 is an optical micrograph of the fracture surface of a 500 N indentation in the 0.27 μm grain size material. The bright phase-contrast region reveals the radial crack morphology [12,14]. The fracture surface shown in figure 4 is for two well-developed radial crack segments. Previous fracture surface observations on Y-TZP show similar behavior [15-17].

Figure 5 is an optical micrograph of the fracture surface of a 200 N indentation in a 0.42 μm grain size material. The fracture surface is that of an asymmetric indentation with one well-developed crack segment, (W), and one trapped crack segment (T).

Figure 6 is an optical micrograph of the fracture surface from an indentation failure site in the 0.42 μm grain size material. Indentation cracking generated two well-developed cracks that were subsequently stressed to failure. Stable crack growth during applied stressing to failure is visible on the fracture surface indicated by the half-penny trace in figure 6. The outside edge of the half-penny trace delineates the transition to rapid, dynamic crack propagation. Similar traces were observed during fatigue testing of Y-TZP [18]. Figure 7 is an enlargement of the indentation region of figure 6. An escarpment is visible due to the intersection of the two nonplanar radial crack segments.

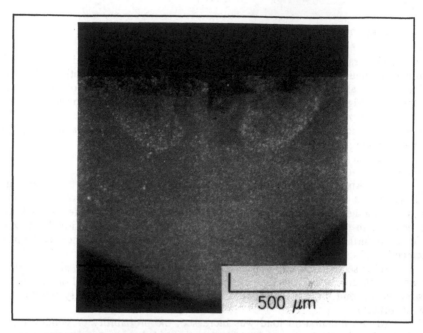

Figure 4. Optical micrograph of the fracture surface trace of two well-developed cracks from a 500 N indentation site in the 0.27 μm grain size material.

Figure 5. Optical micrograph of the fracture surface trace of one well-developed crack, and one trapped crack from a 200 N indentation site in the 0.27 μm grain size material. Arrows denote the independently measured surface trace for the (W) well-developed, and (T) trapped cracks.

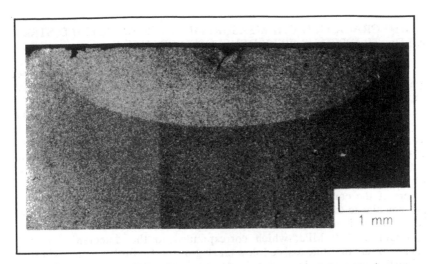

Figure 6. Optical micrograph of the fracture surface trace from an indentation site in the 0.42 μm grain size material. Indentation cracking generated two well-developed cracks that were subsequently stressed to failure. The cracks grew stably in equilibrium as observed by the half-penny trace. The outside edge of the half-penny trace delineates the onset of dynamic crack propagation.

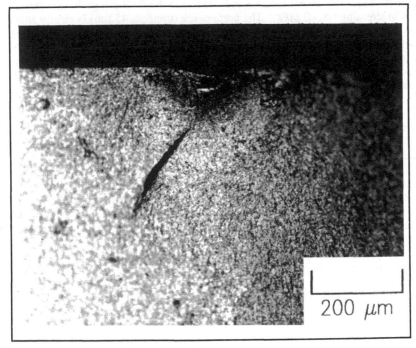

Figure 7. Enlargement of the indentation region of Fig. 6. An escarpment is visible due to the intersection of two different crack planes at failure.

IN SITU CRACK GROWTH MEASUREMENTS IN AN APPLIED STRESS FIELD: TRAPPED VS WELL-DEVELOPED CRACKS

The effect of an applied stress on both trapped and well-developed indentation cracks was determined from in situ crack length measurements as a function of a superimposed bending stress.

Consider first the effect of applied stress on well-developed cracks, (i.e. cracks that extend beyond the contact-induced transformation zone). The cracks have saturated at the steady state plateau toughness value. The effect of increased stress on two such cracks is illustrated in the sequence of optical micrographs for the 0.42 μm grain size material in figure 8. Figure 8a shows a 300 N indentation prior to stressing, $\sigma_a = 0$ MPa. Four well-developed cracks emanate from the corners of the contact impression. With increased stressing, the two cracks perpendicular to the long axis of the bar (north and south cracks) grew in small, stable discrete jumps, figures 8b-c. The applied stress was then increased to a maximum of 100 MPa, which corresponded to the detectable onset of nonequilibrium moisture assisted slow crack growth. No further increase in stress was applied, and the cracks were allowed to propagate unstably to failure at constant stress, figures 8d-e.

Consider now the effect of applied stress on trapped cracks. The effect of stress on a specimen with one well-developed and one weakly trapped crack on opposite indentation corners is shown in the sequence of optical micrographs in figure 9. Figure 9a shows a 300 N indentation in the 0.42 μm grain size material prior to stressing ($\sigma_a = 0$ MPa). The north crack (trapped, 45 μm) is clearly shorter than the three well-developed cracks (180 μm). The length of the well-developed crack normal to the applied stress increased during loading, in small discrete stable jumps, from an initial value of 180 μm to 470 μm, figures 9b-c. During the initial stressing, the weakly trapped crack did not grow (figure 9b). When the applied stress reached a value of 106 MPa, the trapped crack "popped" from its initial length of 45 μm to a value of 480 μm (figure 9c). At this high applied stress level, the trapped crack escaped from the contact-induced transformation zone surrounding the impression, and escaped to a crack length coincident with a well-developed crack. The applied stress was then slightly increased to 108 MPa, at which point both cracks grew in nonequilibrium to failure (figures 9d-e).

Figure 10 shows the effect of stress on a specimen with one well-developed crack and a "well-trapped" crack. Figure 10a shows a 300 N indentation in the 0.42 μm grain size material prior to stressing, with the north crack well-trapped. The length of the well-developed crack prior to stressing was 180 μm and the length of the well-trapped crack 45 μm. Again, the length of the well-developed crack increased during stressing, in small discrete stable jumps, from an initial value of 180 μm to 520 μm (figures 10a-d). The well-trapped crack did not grow at all during stressing (figures 10b-d). This crack was so strongly trapped within the contact-induced transformation zone that it never escaped (even metastably) to well-developed form. The failure condition is shown in figure10e.

Figure 11 is a plot of applied stress, σ_a, versus radial crack length, l, measured from the indentation corner. All data were generated from the crack growth measurements on the 300 N indentations shown in figures 8-10, and are representative of ≈20 crack growth measurements. Filled symbols represent measurements from well-developed cracks and open symbols represent measurements from trapped cracks:

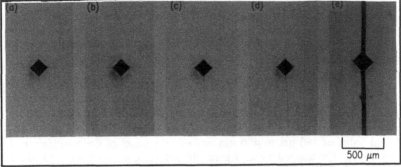

Figure 8. Optical micrographs showing the effect of increased applied stress on two well-developed cracks, P=300 N and 0.42 μm grain size. (a) σ_a=0 MPa, (b-c) stable equilibrium growth, (d) σ_a=100 MPa nonequilibrium growth, (e) failure.

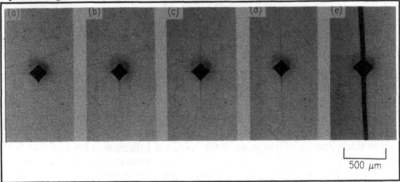

Figure 9. Optical micrographs showing the effect of increased applied stress on one well-developed crack and one weakly trapped crack, P=300 N and 0.42 μm grain size. (a) σ_a=0 MPa, (b) stable equilibrium growth of well-developed crack, trapped crack did not grow, (c) stable equilibrium growth of well-developed crack, trapped crack escapes into well-developed form, σ_a=106 MPa, (d) nonequilibrium growth of both cracks, σ_a=108 MPa, (e) failure.

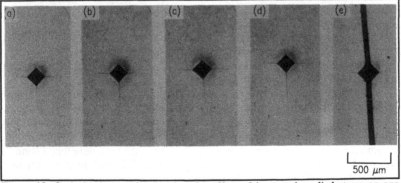

Figure 10. Optical micrographs showing the effect of increased applied stress on one well-developed crack and one well-trapped crack, P=300 N and 0.42 μm grain size. (a) σ_a=0 MPa, (b-c) stable equilibrium growth of well-developed crack, trapped crack did not grow, (d) nonequilibrium growth of well-developed crack, σ_a=125 MPa, (e) failure.

(a) The triangles represent data from the two well-developed, symmetric cracks on opposite indentation corners in figure 8. The applied stress was increased from 0 MPa to a maximum of 100 MPa. Prior to the onset of nonequilibrium slow crack growth, the crack length increased monotonically and stably during stressing, from 180 µm to 440 µm.

(b) The data represented by circles correspond to the well-developed crack (filled circle) and the trapped crack (open circle) on opposite indentation corners in figure 9. The applied stress was increased from 0 MPa to a maximum of 108 MPa. The length of the well-developed crack increased during this stressing from an initial value of 180 µm to 470 µm prior to the onset of nonequilibrium slow crack growth. The trapped crack (figure 9b) did not grow until the applied stress was increased to 106 MPa, at which point it "popped" from its initial length of 45 µm to 480 µm.

(c) The data represented by squares show the effect of stress on the well-developed (filled square) and well-trapped (open square) cracks in figure 10. The applied stress was increased from 0 MPa to a maximum of 125 MPa. The length of the well-developed crack increased from 180 µm to 520 µm. The well-trapped crack did not grow at all during this stressing.

IN SITU CRACK GROWTH MEASUREMENTS IN AN APPLIED STRESS FIELD: EFFECT OF GRAIN SIZE

Figure 12 is a similar plot of applied stress, σ_a, versus radial crack length, l, for 100 N indentations in different grain size materials (0.27 µm grain size), (0.42 µm grain size), and (0.70 µm grain size). A fixed indentation load of 100 N was chosen because it generated three types of crack patterns in the different grain size materials [12]; i.e. two symmetric well-developed cracks in the 0.27 µm grain size material, a well-developed plus a well-trapped crack in the 0.42 µm grain size material, and two well-trapped cracks in the 0.70 µm grain size material.

The effect of stress on the two well-developed cracks in the 0.27 µm grain size material is represented by the squares in figure 12. Crack length increased monotonically during stressing from 65 µm to 150 µm in stable equilibrium with a maximum applied stress of 120 MPa. These well-developed cracks are shorter, and the stress at failure higher, than for the corresponding data set in figure 11 because the indentation load is smaller, (100 N versus 300 N).

The data represented by triangles in figure 12 corresponds to the 0.42 µm grain size material with one well-developed and one well-trapped crack. The well-developed crack grew stably with increased stress from 65 µm to 140 µm, and the well-trapped crack did not grow at all.

Consider now the case of two symmetric cracks trapped within the contact-induced transformation zone surrounding the indentation impression, represented by diamonds in figure 12. The trapped cracks did not grow during stressing (< 5 µm). These cracks were strongly trapped within the contact-induced transformation zone. Therefore, they never escaped into well-developed form but rather failed spontaneously at a critical applied load. The failure stress for these trapped cracks was relatively high, 265 MPa.

DISCUSSION

A fracture mechanics model has been developed to explain the observed

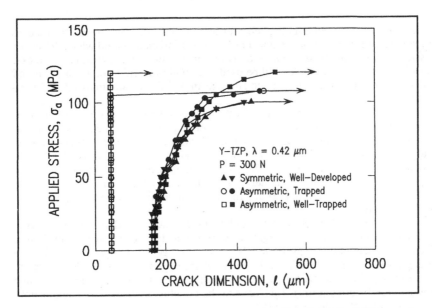

Figure 11. Plot of applied stress, σ_a, versus crack dimension, l, for the 0.42 μm grain size material, $P=300$ N, showing the effect of applied stress on well-developed, trapped and well- trapped cracks.

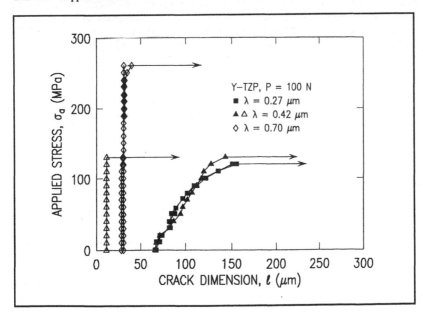

Figure 12. Plot of applied stress, σ_a, versus crack dimension, l, for the 0.27 μm, 0.42 μm, and 0.70 μm grain size material, $P=100$ N, showing the effect of applied stress on two well-developed cracks, one well-developed and one trapped crack and two trapped cracks.

transformation-induced trapping behavior in Y-TZP [13]. That model is based on the superposition of the indentation tensile residual mismatch field which drives the cracks, and the compressive phase transformation field, which retards the cracks. Conclusions from that model will be used to explain the effect of a superimposed bending stress on crack growth and strength in Y-TZP, as summarized in figures 11 and 12.

The first conclusion is that all of the well-developed cracks, regardless of the type of crack on the opposite indentation corner or grain size of the material, start at approximately the same initial crack length, 180 μm for the 300 N indentation load, and 65 μm for the 100 N load. Note that as grain size increases, the steady-state toughness increases in Table I. There does not appear to be a direct correlation between the toughness and the initial length of the well-developed cracks, and this is considered in some detail in [13] in terms of a competition between toughness and crack geometry.

The trapped crack systems fail at higher stresses than the symmetric, well-developed systems. The degree of trapping influences the maximum failure stress and extent of crack growth to failure; the stronger the trapping, the higher the strength and amount of stable crack growth of the opposing well-developed cracks. Interestingly, both the weakly-trapped and the well-trapped crack have the same initial crack length (45 μm). This leads to the conclusion that crack length is not a good indicator of degree of trapping.

The well-developed cracks propagate at the steady-state toughness because there is a fully saturated transformation zone surrounding these indentation cracks, figure 2. Growth of the trapped cracks, however, is influenced by the strong contact-induced compressive transformation field. The trapped crack systems fail at higher stresses and at longer escaped crack lengths than the symmetric (initially well-developed) systems, because they remain pinned within the transformation zone at high stress values. The initially well-developed crack in these asymmetric systems has its driving force diminished by a geometrical interaction with the opposing trapped crack.

These results have implications for impact damage in Y-TZP. Impact damage is similar in many respects to indentation produced cracking. These results are relevant to impact-induced transformation zones and the subsequent effect of an applied stress on impact damage.

CONCLUSIONS

1. Well-developed cracks were observed to grow in stable equilibrium to failure.

2. Trapped and well-developed cracks can grow independently of one another during applied stressing to failure, and can influence each other's growth.

3. Weakly trapped cracks pop into well-developed form at high stresses, and can grow in a stabilized manner prior to failure.

4. Strongly trapped cracks do not pop into well-developed form, but fail spontaneously.

5. Growth of trapped and well-developed cracks is strongly influenced by the

contact-induced transformation zone surrounding the indentation impression.

ACKNOWLEDGEMENTS

The authors wish to thank Eric Liniger for experimental assistance, and Brian Lawn for a critical review of the manuscript. Funding, for LMB, was provided by the U.S. Air Force Office of Scientific Research.

*Certain trade names and products of companies are identified in this paper to adequately specify the materials and equipment used in this research. In no case does such identification imply that the products are necessarily the best available for the purpose or that they are recommended by NIST.

REFERENCES

1. Tsukuma , K., Y. Kubota and T. Tsukidate, 1984. "Thermal and Mechanical Properties of Y_2O_3-Stabilized Tetragonal Zirconia Polycrystals," in Science and Technology of Zirconia II, Vol.12, Columbus, OH, USA: The American Ceramic Society, Inc., pp.382-390 .

2. Matsui, M., T. Soma and I. Oda, 1984. "Effect of Microstructure on the Strength of Y-TZP Components," in Science and Technology of Zirconia II, Vol.12, Columbus, OH, USA: The American Ceramic Society, Inc., pp.371-381.

3. Masaki, T.,1986. "Mechanical Properties of Toughened ZrO_2-Y_2O_3 Ceramics." J. Am. Ceram. Soc., **69** [8]: 638-640 .

4. Swain, M.V., 1986."Grain-Size Dependence of Toughness and Transformability of 2 mol% Y-TZP Ceramics." J. Mater. Sci., **5** [11]: 1159-1162 .

5. Masaki, T., and K. Sinjo, 1987. "Mechanical Properties of Highly Toughened ZrO_2-Y_2O_3." Ceramic International, **13**: 109-112 .

6. Green, D.J., R.H.J. Hannink and M.V. Swain,1989. Transformation Toughening of Ceramics, Boca Raton, FL, USA: CRC Press, Inc.

7. Wang, J., M. Rainforth and R. Stevens, 1989. "The Grain Size Dependence of the Mechanical Properties of TZP Ceramics." Br. Ceram. Trans. J., **88**: 1-6 .

8. Evans, A.G., and E.A.Charles, 1976. "Fracture Toughness Determinations by Indentation." J. Am. Ceram. Soc., **59** [7-8]: 371-372 .

9. Anstis, G.R., P. Chantikul, B.R. Lawn and D.B. Marshall, 1981. "A Critical Evaluation of Indentation Techniques for Measuring Fracture Toughness: I, Direct Crack Measurements." J. Am. Ceram. Soc., **64** [9]: 533-538.

10. Lai, T.R., C.L. Hogg and M.V. Swain, 1989. "Evaluation of Fracture Toughness and R-curve Behaviour of Y-TZP Ceramics." Trans. ISIJ, **29** [3]: 240-245.

11. Anderson, R.M., and L.M. Braun, 1990. "Technique for the R-Curve Determination of Y-TZP Using Indentation-Produced Flaws." J. Am. Ceram. Soc., **73** [10]: 3059-3062.

12. Cook, R.F., L.M. Braun and W.R. Cannon, submitted. "Trapped Cracks at Indentations: I, Experiments on Y-TZP." J. Mater. Sci.

13. Cook, R.F., and L.M. Braun, submitted. "Trapped Cracks at Indentations: II, Fracture Mechanics Model." J. Mater. Sci.

14. Braun, L.M., 1990. Evaluation of Microstructural Effects on the Mechanical Behavior of Yttria Stabilized Tetragonal Zirconia Polycrystals. Ph.D., Rutgers University, New Jersey.

15. Orange, G., L.M. Liang and G. Fantozzi, 1988. in Science of Ceramics, Vol.14, D. Taylor, ed., Stoke-on-Trent,UK: Institute of Ceramics, pp.709.

16. Jones, S.L., C.J. Norman and R. Shahani, 1987. "Crack-Profile Shapes Formed Under a Vickers Indent Pyramid." J. Mater. Sci. Lett., **6**: 721.

17. Akiyama, S., Y. Kimura and Sekiya, 1989. "Evaluation of Thermal Shock Resistance in Ceramic Materials." J. Soc. Mater. Sci. Jpn, **38**: 1415.

18. Liu, S.-Y., and I.-W. Chen, 1991. "Fatigue of Yttria-Stabilized Zirconia: II, Crack Propagation, Fatigue Striations, and Short-Crack Behavior." J. Am. Ceram. Soc., **74** [6]: 1206-1216.

Relation between Flexural Strength and Phase Transformation in Ground Tetragonal Zirconia

T. MIYAKE, F. ASAO and C. O-OKA

ABSTRACT

The transformation on the ground surface of Y_2O_3-ZrO_2 was investigated by Raman microprobe analysis. The amounts of monoclinic phase observed on two sections intersecting perpendicularly differ and it was proved that transformation induced by grinding shows anisotropy according to grinding condition or grinding method. The relation between the two components of the anisotropic transformation, one is parallel and the other is perpendicular to the applied tensile stress, and strengthening was examined. The linear relation predicted by theoretical analysis, between strengthening and transformed volume fraction was valid when the monoclinic fraction along the direction which shields the crack opening against the applied stress was taken as the transformed volume fraction.

INTRODUCTION

Evaluation of the strength of ceramics after machining is inevitable because during machining flaws are induced at the surface and the strength is reduced because of these flaws. On the other hand strengthening of surface-ground ceramics has been reported [1]. Particularly tetragonal zirconia and its composites show remarkable strengthening [2-4], because the transformation from the tetragonal to monoclinic form (t→m), which involves volumetric dilatation, is induced by stresses during grinding and biaxial compressive residual stresses are generated at the surface. This surface toughening by machining is one of the main toughening mechanisms in ZrO_2 ceramics. Also from an industrial viewpoint it is important because machining is usually required for engineering ceramics used as structural components. For these reasons theoretical and experimental investigations have been made to clarify the relation between transformation and strengthening in tetragonal zirconia and its composites [2-4].

Takushi Miyake, Electoronics Department,Nagoya Municipal Industrial Research Institute, 3-4-41 Rokuban, Atsuta-ku, Nagoya 456, Japan

Fumihiro Asao, Mechanical Engineering Department, Nagoya Municipal Industrial Research Institute

Chihiro O-oka, Inorganic Materials Department, Nagoya Municipal Industrial Research Institute

TABLE I - PROPERTIES OF Y_2O_3-ZrO_2 SPECIMENS

Composition Y_2O_3 (mol%)	Density (g/cm^3)	Young's Modulus (GPa)	Flexural Strength (MPa)	Fracture Toughness (MPa•m$^{1/2}$)	Grain Size (µm)
4.5	5.76	203	630	5.2	0.9

TABLE II - GRINDING CONDITIONS OF SPECIMENS

Type of grinding	Wheel	Wheel speed (m/min.)	Work-feed speed (m/min.)	Depth of cut (µm)
Peripheral	140 grit straight type	1,600	15 (constant)	5 - 20
			5 - 15	10 (constant)
Face	140 grit cup type	600 or 1,800	0.125 or 0.5	50 or 200

The residual stress at the ground surface of a sintered ceramic exhibited planar anisotropy [5]. It is expected that transformation and transformation-strengthening by grinding will exhibit similar anisotropy.

X-ray diffraction is commonly used to measure monoclinic concentration, but it is inadequate to determine the extent of the transformation at the ground surface because of limited spacial resolution [6,7]. The Raman microprobe has high ability to distinguish between monoclinic and tetragonal polymorphs in micro spatial resolution and it is useful for investigating the distribution of transformed particles.

In the present work, the transformation behavior along two directions perpendicular to each other on ground Y_2O_3-ZrO_2 were measured by Raman microprobe spectroscopy and the transformation anisotropy induced by grinding was investigated. The relation between the directional component of transformation which induces a crack shielding stress and a strengthening effect, was examined.

EXPERIMENTAL PROCEDURE

ZrO_2^*, containing 4.5 mol% Y_2O_3 sintered at atmospheric pressure, was used for testing. The mechanical properties, density and grain size of this material are summarized in Table I. The material was cut and ground to bend specimens, 4mm by 3mm by 40mm in size. The specimens were ground on one face with two different types of 140-grit diamond wheel, one a cup type and the other a straight type. The grinding conditions are listed in Table II. To change the amount and depth of transformation, grinding conditions are varied over a wide range.

To investigate the relation between transformation and flexural strength, the

---

*NGK Spark Plug Co., Ltd., Nagoya, Japan

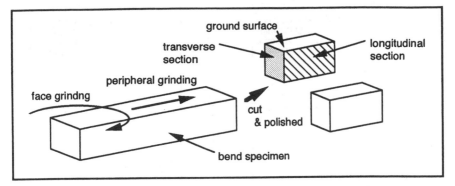

Figure 1.   Schematic illustration of specimen preparation for Raman measurement.

distribution of monoclinic phase in the ground surface layer was measured and the strength of the same sample was determined by a four-point-bend test with the ground surface being in tension. Bending was performed over the outer span 30mm with a crosshead speed 0.5mm/min.

Raman microprobe[**] with a 488 nm line from an argon ion laser was used. The optical beam diameter on the sample surface and the output power were 1 μm and 500mW, respectively. Monoclinic fraction is calculated by the Raman intensities ratio using the equation proposed by Clark and Adar [7].

RESULTS AND DISCUSSION

ANISOTROPY OF TRANSFORMATION ON GROUND SURFACE

To examine the anisotropy of transformation induced by grinding, the monoclinic fractions on two sections perpendicular to each other, one parallel to and the other perpendicular to the tensile stress, i.e. the longitudinal direction of the bend specimen, were measured. These sections and their locations within a bend specimen are illustrated in figure 1. The specimen for Raman measurement was cut from the bend specimens and over 35μm was removed by polishing to be free from the damage by cutting. Polishing was done initially with a 15μm diamond disk, sequentially with a 6μm diamond disk and finally with 0-1μm diamond powder .

Monoclinic fractions in the surface layer of these two cross sections were measured by Raman microprobe spectroscopy from the ground surface to the inner region. These data are plotted in figures 2 and 3, from which it can be seen that grinding influences the monoclinic fraction to a depth greater than 30μm.

The variation of monoclinic fraction with distance from the ground surface on two sections intersecting perpendicularly, are shown in figure 2 for the case of peripheral grinding (a) and face grinding (b). Monoclinic fraction on both cross

---

[**]NR-1100, Japan Spectroscopic Co., Ltd., Tokyo, Japan

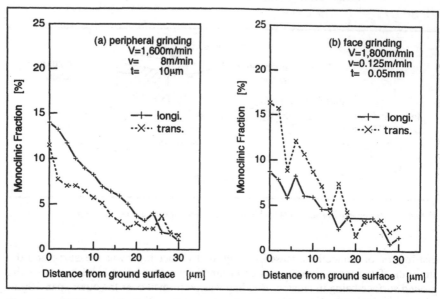

Figure 2. Difference of anisotropy in ground surface layer due to the difference of grinding method. (a) is monoclinic fraction vs. depth from the ground surface for peripheral grinding and (b) is that for face grinding. For (a) a 140 grit straight wheel was used, while (b) a cup-type wheel of the same grit size was employed.

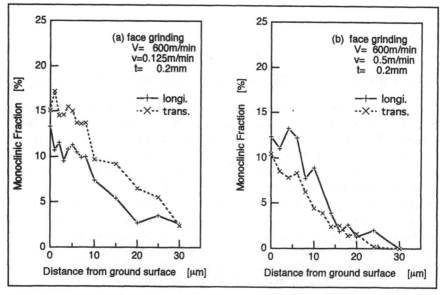

Figure 3. Difference of anisotropy in ground surface layer due to the difference of grinding condition in case of face grinding. Between (a) and (b), the ratio of work-feed speed to wheel speed differs.

404

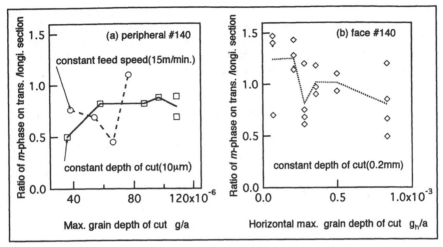

Figure 4. Dependence of anisotropy on grinding conditions. (a) is the ratio of monoclinic fraction on the transverse section to that on the longitudinal section vs. maximum grain depth of cut in the case of peripheral grinding. It includes data for the condition of constant depth of cut and for the condition of constant work-feed speed. (b) is for the case of face grinding with constant depth of cut. Horizontal maximum grain depth of cut is used.

sections decreased similarly with depth, but the fractions on the two cross sections do not coincide with each other in the region near the ground surface. Monoclinic fraction resulting from grinding shows anisotropy. The degree of anisotropy is largest at the ground surface and became smaller with depth. For peripheral grinding, there was a greater monoclinic fraction on the longitudinal section than at the corresponding distance from the ground surface on the transverse section. But the relation of monoclinic fraction in the two sections was reversed for face grinding. The anisotropy varies with grinding method.

The anisotropy also varies with grinding conditions. For the data in figure 3 the same grinding method, face grinding, was used but the work-feed speed was increased by a factor of three for (b) by comparison with (a). The amount of monoclinic fraction varied with the change of work-feed speed but, at the same time, the relation of monoclinic fraction on the two sections was reversed throughout most of the surface region measured.

Further dependence of anisotropy on grinding condition is shown in figure 4, for the case of peripheral grinding (a) and face grinding (b), respectively. Grinding conditions are represented by maximum grain depth of cut, $g = 2a(v/V)\sqrt{t/D}$, in the case of peripheral grinding and horizontal maximum grain depth of cut, $g_h = a(v/V)$, in the case of face grinding, where a is the mean distance between grains, v is work-feed speed, V is wheel speed, t is the depth of cut and D is the wheel diameter. In the case of peripheral grinding with a 140-grit wheel, changing the depth of cut with constant work-feed speed resulted in a wide variation in the ratio of the monoclinic fraction on the transverse section to that on the longitudinal section. On the other hand, change in the work-feed speed at constant depth of cut gave a small change in this ratio. In the case of face grinding the variation with change in speed ratio at constant depth of cut showed a wide variation. Unlike the data of peripheral grinding at constant depth of cut, the ratio of the monoclinic fraction on the transverse section to that on the longitudinal section exceeds 1.0 for some grinding conditions (as observed in figure 2).

RELATION BETWEEN TRANSFORMATION AND FLEXURAL STRENGTH

From the above experimental results the transformation induced by grinding shows anisotropy, it is necessary to examine the relation between strength and each component of the anisotropic transformation and make clear which component contributes to strengthening. Figure 5 shows the relation between the flexural strength

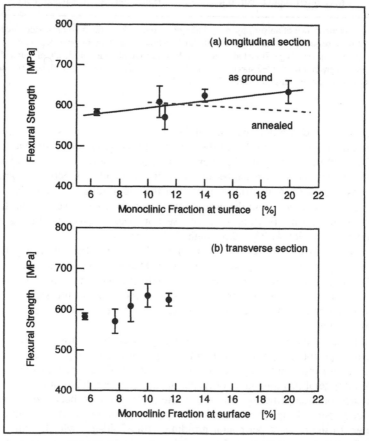

Figure 5.   Relation between flexural strength and monoclinic fraction at the ground surface in the case of peripheral grinding with a 140 grit straight wheel.   (a) is the relation on the longitudinal section and (b) is that on the transverse section.

and the monoclinic fraction in the ground surface layer of the two perpendicular sections for the case of peripheral grinding. Figure 6 shows the same relation for the case of face grinding.

From figure 5 it is seen that the monoclinic fraction observed at the surface on the longitudinal section increased, as did the flexural strength. This is because the compressive stress along the longitudinal direction, induced by the volume increase in transformed particles on the longitudinal section, shielded any crack opening against the applied bending stress. On the transverse section flexural strength increased with the monoclinic fraction but not in the same systematic manner as that observed on the longitudinal section.

For face grinding, shown in figure 6, flexural strength also increased linearly with the monoclinic fraction on the longitudinal section, although figure 6 contains results ground with different depths of cut. However, there is a less systematic

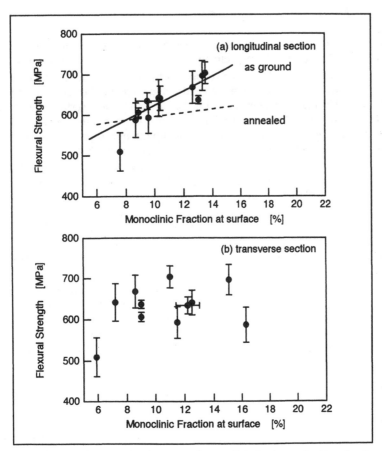

Figure 6.   Relation between flexural strength and monoclinic fraction at the ground surface in the case of face grinding with a 140 grit cup wheel. (a) is the relation on the longitudinal section and (b) is that on the transverse section.

dependence of flexural strength on monoclinic fraction measured on the transverse section. It is supposed that relation between flexural strength and the monoclinic fraction on the transverse section is not essential in face grinding and also in peripheral grinding, taking the fact that the monoclinic fraction on the transverse section by peripheral grinding had an almost constant relation with that on the longitudinal section as observed in figure 4, into account.

According to the theoretical analysis for a surface-toughened ceramic with the assumption of linear variation of monoclinic fraction with depth, the fracture strength is given by [3,8,9]

$$\sigma_f = \frac{K_{IC}}{y\sqrt{\pi a}} + \frac{MvE}{1-v}\frac{\Delta V}{3V} \tag{1}$$

and

$$M = 1 - (2a/\pi s) \qquad 0 \leq a \leq s \tag{2}$$

where K_{IC} is the fracture toughness, y is a geometrical factor, E and v are Young's modulus and Poisson's ratio respectively, v is the volume fraction of transformed material at the surface, $\Delta V/V$ is the relative volume expansion accompanying the tetragonal to monoclinic transformation and a and s are the flaw size and the depth transformed. The first term of equation 1 is the strength without surface strengthening and the second term is the strengthening effect due to transformation. According to equation 1, strengthening shows a linear relation with transformed volume fraction, assuming the constant ratio of flaw size a to transformed depth s.

To know the net strengthening effect by transformation, expressed by the second term in equation 1, retransformation by annealing at 1350°C for 30min. was done. The flexural strength after annealing was plotted vs. the pre-anneal surface monoclinic fraction in figures 5(a) and 6(a). From figures 5(a) and 6(a), the drop of strength after annealing showed a nearly linear increase with the pre-anneal monoclinic fraction on the longitudinal section. Equation 1 is valid for the strengthening by grinding when the monoclinic fraction on the longitudinal section is used for the transformed volume fraction v.

Assuming $s \approx a$ in equation 2 resulting in a minimum M, gives a minimum increase in fracture stress in equation 1. From the experimental results, the increase of fracture stress is far less than the minimum value predicted by equation 1. The gradient of linear relation between the strength and the monoclinic fraction on the longitudinal section differs between   peripheral and face grinding. Quantitative estimation by equation 1 was not successful for these systems.

CONCLUSION

The Raman microprobe has been a useful technique for local characterization of the transformation behavior in ZrO_2 ceramics. The Raman microprobe is used to show the existence of anisotropy of transformation in the ground surface layer and the contribution of the anisotropic transformation to strengthening by grinding was investigated. Although the transformation behavior shows plane anisotropy, a linear relationship was observed between the strengthening and the monoclinic fraction along the direction in which the volumetric dilatation shielded the crack against applied stress .

REFERENCES

1.   Suzuki, K., K.Tanaka, Y.Yamamoto and H. Nakagawa, 1989. "Effect of Grinding Residual Stress on Bending Strength of Ceramics." Journal of the Society Materials Science, Japan., 38 (429): 582-588.

2.   Swain, M.V., 1980. "Grinding-induced tempering of ceramics containing metastable tetragonal zirconia." Journal of Materials Science Letters, 15:1577-1579.

3.   Claussen, N. and M.Rühle, 1981. "Design of transformation-toughened ceramics," in Advances in Ceramics Vol.3, A.H.Heuer and L.W.Hobbs, eds., Columbus, OH, USA:The American Ceramic Society, pp.137-163.

4.   Claussen, N., 1985. "Transformation toughening of ceramics," in Fracture of Non-Metallic Materials, K.P.Herrmann and L.H.Larsson. eds., Brussel and Luxembourg:ECSC, EEC, EAEC, pp.137-156.

5.   Kisimoto, H.,A.Ueno, H.Kawamoto and S.Kondou, 1987. "X-Ray Residual Stress Measurement of Sintered Si_3N_4 ." Journal of the Society Materials Science, Japan., 36 (407): 810-816.

6.   Garvie, R.C. and P.S. Nicholson,1972. "Phase Analysis in Zirconia Systems." Journal of the American Ceramic Society, 55(6):303-305.

7.   Clarke, D.R. and F.Adar, 1982. "Measurement of the Crystallographically Transformed Zone Produced by Fracture in Ceramics Containing Tetragonal Zirconia." Journal of the American Ceramic Society, 65(6):284-288.

8.   Richmond, O., W.C.Leslie and H.A.Wriedt, 1964. "Theory of Residual Stresses Due to Chemical Concentration Gradients." Transactions of ASM Q, 57(1): 294-300.

9.   Lawn, B.R. and D.B.Marshall, 1977. "Contact Fracture Resistance of Physically and Chemically Tempered Glass Plates." Physics and Chemistry of Glasses, 18 (1):7-18.

Subcritical Crack Growth and R-Curve Behavior in Al$_2$O$_3$-Toughened Y-TZP

DANIEL E. GARCÍA, JÜRGEN RÖDEL and NILS CLAUSSEN

ABSTRACT

Effects of Al$_2$O$_3$ addition made as either fine Al$_2$O$_3$ particles or Al$_2$O$_3$ platelets on the room temperature mechanical properties of Y-TZP were studied in composites of Al$_2$O$_3$/3Y-TZP, which were hot-isostatically pressed. Alumina additions to the 3Y-TZP matrix, irrespective of particulate morphology, result in an increase in fracture toughness. In situ crack propagation studies have shown that Y-TZP and Al$_2$O$_3$/Y-TZP composites are strongly susceptible to subcritical crack growth. An R-curve in the Al$_2$O$_3$/Y-TZP composite was obtained with the crack resistance rising by 2 MPa.m$^{1/2}$ during 80-120 μm crack extension.

INTRODUCTION

The properties of transformation toughened materials [1-4] are now well described and the basic principles of the transformation toughening mechanism [5-8] are laid out. There are still numerous pathways open, however, to elaborate on newly found material improvements or possible synergistic mechanisms between various toughening approaches. Specifically, this is evident in the case of alumina toughened zirconia, which will be the focus of this report. Irrespective of precise content of (yttria) stabilizer or zirconia grain size, we initially attempt to establish some salient points in this composite material.

The original result by Tsukuma et al.[9,10], that the fracture strength in 3-pt. bending of Y-TZP can be raised with additions of Al$_2$O$_3$ to more than 2GPa, has been confirmed by Rajendran et al.[11]. There is also agreement that additions of alumina improve the stability of Y-TZP under hydrothermal conditions [11-13]. Some inconsistencies in the results of the fracture toughness measurements in these composites still need to be addressed. Tsukuma et al.[9] described an increase or decrease with alumina content in the fracture toughness values depending on testing technique. Tu et al.[14] have recently reported an increase of K$_{Ic}$ with an addition of 5 wt% Al$_2$O$_3$ to Y-TZP.

It was thus deemed pertinent to initiate further investigations with controlled fracture toughness determinations, including measurements of subcritical crack

Daniel E. García, Jürgen Rödel and Nils Claussen
Advanced Ceramics Group
Technische Universität Hamburg-Harburg
D-2100 Hamburg 90
Federal Republic of Germany

growth of Y-TZP and Al$_2$O$_3$/Y-TZP as well as measurements of the R-curve behavior of Al$_2$O$_3$/Y-TZP. In situ crack propagation studies were also undertaken in order to find clues for the mechanistic effect of Al$_2$O$_3$ on strength and toughness of Y-TZP.

A final issue concerns the influence of particulate shape of the Al$_2$O$_3$ additions on mechanical properties. Recent work of Heussner and Claussen [15] established that additions of 5 vol% platelets of Al$_2$O$_3$ into 3Y-TZP resulted in a strong drop in strength, but an increase in fracture toughness. The intent is thus to give some guidelines for the toughening effect of Al$_2$O$_3$ platelets as compared to Al$_2$O$_3$ particulates in Y-TZP.

Our approach will be twofold. First, we will establish the general trends in fracture toughness and four-point (4-pt.) bending strength of Y-TZP containing either platelets, broken platelets or particulates and platelets as a function of initial platelet size. Second, we will report the results of detailed fracture mechanics studies on compact tension specimens, designed to contrast the mechanical properties of Y-TZP with the most promising Al$_2$O$_3$/Y-TZP composite.

EXPERIMENT

SAMPLE PREPARATION AND CHARACTERIZATION

Zirconia powder containing 3 mol% Y$_2$O$_3$ (3Y-TZP) (Tosoh Tokyo, Japan) was mixed with 15 vol% Al$_2$O$_3$ platelets (Atochem, Paris, France) and attrition milled in ethanol for either 20 min or 4 h. The first treatment lead to a dispersion of unbroken platelets in the ZrO$_2$ matrix and was designated 20min-TZP, while the samples resulting from the second treatment contained broken platelets and were termed 4h-TZP. Some compositions containing TZP, 10 vol% Al$_2$O$_3$ powder (Ceralox 0.8 µm) and 15 vol% Al$_2$O$_3$ platelets were attrition milled for 20 min (20min-ATZP). The property data given by the supplier of the Al$_2$O$_3$ platelets used in the present work are as follows (in wt%): Al$_2$O$_3$: >97, F: 1 - 2, Li: 0.6, Na: 0.15, Si: 0.05 and S: 0.025. Three different platelet sizes with medium diameter 4.7, 5.3 and 11 µm were used. After milling, the platelet-powder mixtures were dried in a rotational vacuum drier and then screened through a 0.2 mm sieve, uniaxially pressed in a steel die at 15 MPa into rectangular bars of 6 mm x 6 mm x 60 mm and cold isostatically pressed at 800 MPa. Comparative studies with Y-TZP, prepared as the Al$_2$O$_3$/ZrO$_2$ composites, were also performed. All compositions were presintered at 1400°C for 2 h and then hot-isostatically pressed at 1500°C and 200 MPa for 10 min in Ar, using a graphite heating element. Sintered densities were measured gravimetrically using Archimedes' principle and the fracture toughness was determined using the chevron-notch technique (CNB) [16] in 4-point bending (span 30/10 mm). The specimen halves were then tested in 4-point bending (span 12/6 mm) to determine the flexural strength. The phase composition of the bulk and fracture surfaces was determined by x-ray diffraction (XRD). Fracture surfaces and microstructure were analysed by scanning electron microscopy (SEM). The average grain size was determined using the linear intercept method.

CRACK PROPAGATION STUDIES

Samples from two powder compositions, as received Y-TZP (designated rTZP) and an alumina platelet/TZP composition attrition milled for 4 h with the initial platelet size being 4.7 µm (4h-TZP4.7) were chosen for detailed crack propagation studies. These were performed on compact tension specimens (CT) [17] with a distance of 20 mm between load point and back edge of the sample and a thickness of 2 mm. Crack propagation was studied both with optical microscopy as well as scanning electron microscopy with an improved version of a device as introduced by Rödel et al.[18,19].

CT specimens were precracked by first cutting a half chevron notch into the sample and then placing a Vickers indent of 49 N at the thin end close to the notch tip. The specimen was then placed into the testing device and the radial crack emanating away from the notch was extended under load application on the stage of the optical microscope. After about 300 μm crack extension, the specimen was removed and the area of the elastic plastic zone cut out. The sample was then loaded again and the crack allowed to grow through the half chevron region into the area of constant sample thickness.

Crack propagation was qualitatively studied both in the regime of the half-chevron notch as well as in the constant thickness region. Subcritical crack growth data were taken at the plateau of the R-curve (more than 500 μm after initial crack extension) with the K-value determined from the ASTM calibration [17] and the measured load and crack length at crack propagation. The crack velocity was calculated by taking the time for a certain degree of crack extension with a stop watch and measuring crack growth optically. Subcritical crack growth was also followed on a monitor and partly recorded for further control measurements. R curves were recorded after annealing samples containing cracks of several millimeter length for 0.5 h at 1300°C in air.

RESULTS

GENERAL CHARACTERISATION OF COMPOSITES

Attrition milling for 20 min left platelets essentially unbroken and allowed comparison to broken platelet or particulate reinforced TZP of identical chemistry, a composition which was available with the powders being attrition milled for 4 h.

The densities obtained for the sintered composites were between 98% and 99.7% for all 10 powder compositions with no clear functional dependences on platelet size or milling treatment. Grain size, G, was only determined for the two compositions which were studied in greater detail. An average grain size of 0.3 μm was found for rTZP while G for the composite 4-hTZP4.7 was 0.37 μm.

Typical microstructures of a TZP sample as contrasted to a composite containing broken platelets (4h-TZP4.7) are given in figure 1 with the photographed areas both containing the tip of a radial indentation crack.

The content of monoclinic zirconia at the fracture surface as determined by XRD [20] was found to lie between 15 and 25% for all compositions studied. No clear correlations between these results and the fracture toughness values measured could be made. The same method showed that the bulk material consisted of tetragonal zirconia and α-alumina only.

Fracture strengths are given in figure 2. TZP without additions of Al_2O_3 (here plotted as the composite with initial platelet size 0 μm) had a strength of about 1600 MPa. Additions of alumina platelets led to a decrease in strength with increase in initial platelet size. This effect was least noticeable when the powder composition was milled for 4 h, since most platelets were broken during this process. In particular, compositions containing alumina with the smallest initial platelet size, experienced no strength reduction compared to the TZP material.

In contrast, as shown in figure 3, additions of Al_2O_3, irrespective of particle morphology, led to an increase in CNB fracture toughness in almost all cases. Fracture toughness of TZP without alumina was found to lie between 7 and 8 $MPa.m^{1/2}$. Additions of 10 vol% alumina powder gave values of 10 $MPa.m^{1/2}$. The highest fracture toughness was obtained with the composite containing broken platelets of initial size of 4.7 μm. Initial platelet size, whether broken or not, had

a)

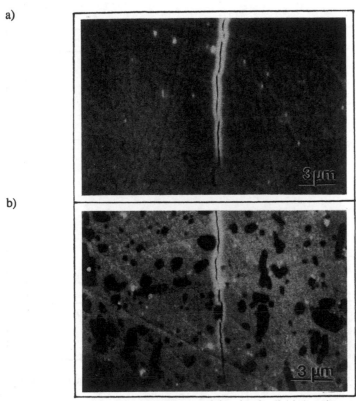

b)

Figure 1. Representative microstructures containing tip of radial indentation crack in (a) 3Y-TZP (TZP), and (b) 3Y-TZP/15 vol% Al_2O_3 (4h-TZP4.7) composite.

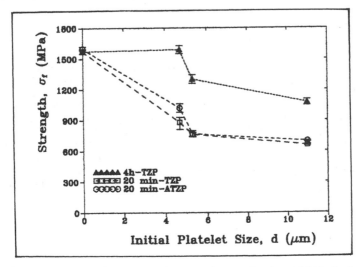

Figure 2. Fracture strength of 3Y-TZP/Al_2O_3 composites as a function of initial platelet size.

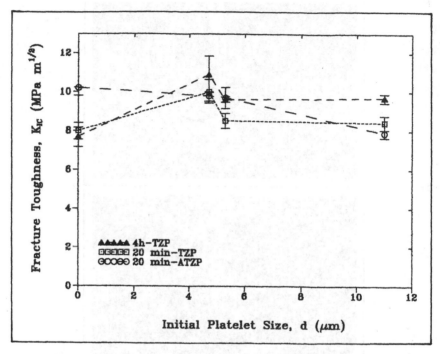

Figure 3. CNB fracture toughness results for 3Y-TZP/Al_2O_3 composites of varying initial platelet size.

only small effects on K_{Ic}, with a trend to smaller values with increasing initial platelet size.

CRACK PROPAGATION STUDIES

Crack propagation in both composites investigated occurred in discrete jumps of 10-50 µm, when studied in the electron microscope under a vacuum of 10^{-5} Torr. Crack advancement in the laboratory atmosphere, as studied under the optical microscope appeared continuous as observed on the video screen with a magnification of 1100X. Rather, crack velocity could be controlled very well by changing the applied load and crack velocities over several orders of magnitude could be obtained in a stable manner. Microcracking, except for microcrack formation in front of the crack tip with subsequent linkup to the main crack could not be observed for either material. Crack paths in rTZP were very straight, whereas crack deflection and formation of small elastic bridges were apparent in 4h-TZP4.7. The latter is depicted and marked by arrows in figure 4.

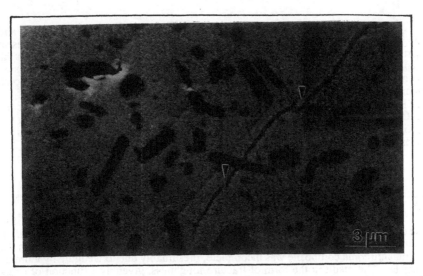

Figure 4. Crack tip region in 3Y-TZP/Al₂O₃ (4h-TZP4.7) composite showing elastic bridges marked by arrows.

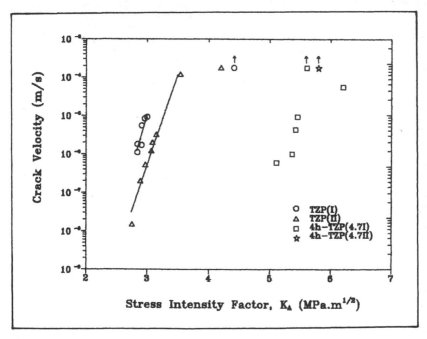

Figure 5. Crack velocity data in 3Y-TZP (rTZP) and 3Y-TZP/15 vol% Al₂O₃ (4h-TZP4.7) platelet composite in air with ~70% relative humidity at room temperature.

Some data for subcritical crack growth for two samples of rTZP (I and II) and one sample of 4h-TZP4.7 are given in figure 5, together with instability values in inert atmosphere, which are marked with arrows to designate high velocities in the graph. Most strikingly, the stress intensity factors obtained for the compact tension geometry are considerably lower than those obtained by using the chevron notch technique. K_{Ic} computed with the values supplied by CNB was 6.7 MPa.m$^{1/2}$ for rTZP. K-values for comparable velocities are higher by about 2 MPa.m$^{1/2}$ for the Al_2O_3/ZrO_2 composite when compared to rTZP. Though the data taken are really not sufficient to calculate crack growth rates with confidence, approximate numbers are given for rTZP by casting the region I into the general crack growth equation [21]:

$$v = A \, K_I^n \tag{1}$$

The values for A and n are (for two samples I and II): YTZPI: A=2.6x10^{-23}, n=37; for YTZPII: A=2.2x10^{-23}, n=34.

Attempts to measure an R-curve for the rTZP material were not successful, since the crack healed perfectly close to the crack tip during the described annealing treatment and it proved difficult to locate the original crack tip. Crack healing in the alumina/zirconia composite was incomplete, presumably due to wedging by alumina particles. The result in the form of fracture toughness, given as a function of crack extension, is provided in figure 6. Subcritical crack growth again influenced the measurements and attaches a certain degree of uncertainty to the results. The obtained fracture toughness values are best described by determining a crack growth rate of about 1 x 10^{-6} m/s. Since this velocity has to be regarded as approximate, the K-values carry an uncertainty of about 0.3 MPa.m$^{1/2}$ when each data point is normalized to the same velocity. Nevertheless, the result appears reproducible. An R curve in this composite can be measured, with the fracture toughness rising from about 3.5 MPa.m$^{1/2}$ to about 5.5 MPa.m$^{1/2}$ after 80 - 120 mm crack extension. The latter value still lies far below all the values as measured by the CNB method.

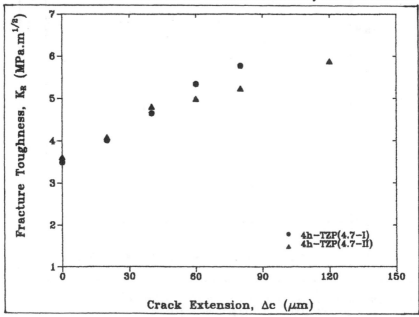

Figure 6. R-curve behavior of 3Y-TZP/15 vol% Al_2O_3 (4h-TZP4.7) platelet composites.

DISCUSSION

GENERAL CHARACTERIZATION

Before we discuss the effect of alumina particulate additions on microstructure development of TZP, we need to separate the effect of alumina doping (in the range of about 1 mol%) on microstructural parameters. Small amounts of alumina additions to TZP are known to increase the densification rate [13,22], the grain growth rate [22], fracture toughness [13] and fracture strength [13]. The fact that the grain size in our rTZP material (G = 0.3 μm) is nearly unaffected by the presence of broken platelets in 4h-TZP4.7 (G = 0.37 μm) is felt to be a combined effect of increased grain growth due to the alumina dissolved in the lattice [13,22] and reduced grain growth due to particle drag [23] of alumina particulates.

The amount of monoclinic zirconia found at the fracture surface correlates well with results obtained by Rajendran et al.[11], where their results lay between 15 and 25% for the sintered bodies, but 25 to 35% for the hipped specimens. As in their work, no clear correlation between fracture toughness and content of monoclinic zirconia at the fracture surface could be found.

Extended attrition milling times led to platelet fracture resulting in a reduced size of possible fracture origins and therefore enhanced fracture strength compared to materials prepared after brief milling times. Incorporation of platelets into the matrix has now generally been recognized to cause low fracture strengths, notably so in platelet reinforced TZP [15]. Fracture strengths measured were not as high as obtained by other workers before [9-11]. This is in part due to the higher fracture strengths found if using 3-pt. bending as done by Tsukuma et al. [9,10] and Rajendran et al.[11]. Fracture toughness results appear to be indifferent to particulate shape, but depend slightly on particulate size. Optimum materials in terms of high strength and toughness can be obtained with broken alumina platelets with a small starting size.

CRACK PROPAGATION STUDIES

Fracture toughness values as determined on CT specimens are about 40% lower than the equivalent data taken with the chevron notch technique. It is not yet clear why this discrepancy arises, but both the creation of compressive surface stresses as well as overloading the sample to initiate the crack at the chevron notch tip could explain this tendency. Clearly, the toughness values determined by crack propagation of well formed cracks in compact tension specimens is felt to be the more reliable technique. In this context, we refer again to the work by Tsukuma et al.[9], who showed that differences in K_{Ic} up to 10 MPa.m$^{1/2}$ can be found when comparing the microindentation technique with the chevron notch technique, with the latter again giving the higher values. Nevertheless, the tendency, that additions of alumina increase the fracture toughness by about 40%, is verified with both techniques.

All TZP materials investigated proved to be very susceptible to subcritical crack growth with stress exponents between 34 and 37 in region I. A plateau region (region II) is at least indicated. Again, we reiterate that the number of data points is insufficient to give reliable estimates for these n values, but the fact that subcritical crack growth can be stably followed over 4 orders of magnitude in velocity and a range of 1.5 MPa.m$^{1/2}$ in the stress intensity factor remains as factual evidence. This is in fair agreement with previous measurements by Ashizuka et al.[24] who found a stress exponent n of 41 using dynamic fatigue data. More important, however, we find indications of an extended region II, which leads to the observation that subcritical crack growth in Y-TZP occurs at about 2 MPa.m$^{1/2}$ below the instability value in inert atmosphere.

The crack resistance in the alumina/zirconia composite studied was found to increase from 3.5 MPa.m$^{1/2}$ to about 5.5 MPa.m$^{1/2}$ for crack extensions from 80 to 120 μm. The base value of 3.5 MPa.m$^{1/2}$ represents the crack tip toughness of the material (estimated [25] to lie at 2.0 MPa.m$^{1/2}$) plus contributions from crack bridging in a zone behind the crack tip amounting to several millimeters crack length and contributions from microcracks, which did not heal during the high temperature treatment. Since it is difficult to determine exactly the point of first crack propagation, it is likely that the base value of the R-curve lies somewhat lower than 3.5 MPa.m$^{1/2}$. Irrespective of the uncertainties of possible contributions due to microcrack toughening, we surmise that no more than 1 MPa.m$^{1/2}$ (the difference between the value of the measured R-curve and the crack tip toughness of Y-TZP) toughening is due to crack bridging. Therefore, most of the toughening effect by the alumina inclusions is felt to be due to process zone shielding, probably transformation toughening, but possibly also microcrack toughening.

Finally, the limited amount of data available, particularly the absence of Raman spectroscopy, allows us only to speculate on the exact toughening mechanism of the alumina additions. Three factors are deemed relevant:
1. Thermal expansion mismatch of zirconia and alumina creates residual stresses which superimpose on the crack tip stresses and lead to a broadening of the transformation zone [26].
2. Crack deflection leads to deviation from a straight crack path and therefore leads effectively to a wider transformation zone.
3. Formation of elastic crack bridges can be seen as an overlapping of multiple crack tips, which again causes a broadening of the transformation zone.

These factors are in contrast to a reduction in fracture toughness owing to incorporation of alumina due to the fact that the fraction of transformable phase (tetragonal zirconia) is reduced.

While we are not yet in the position to balance all these factors against each other, we still propose that a widening of the transformation zone is the likely candidate for the increased toughness due to incorporation of alumina into a Y-TZP matrix.

CONCLUSIONS

1. Additions of alumina to Y-TZP, irrespective of particulate morphology, result in an increase in fracture toughness.
2. Detrimental effects on fracture strength are negated if the alumina is present in the form of particulates or small platelets.
3. Y-TZP and an Al$_2$O$_3$/Y-TZP composite were found to be strongly susceptible to subcritical crack growth.
4. An R-curve in the Al$_2$O$_3$/Y-TZP could be obtained with the crack resistance rising by more than 2 MPa.m$^{1/2}$ during 80-120 μm crack extension.
5. It is speculated that additions of Al$_2$O$_3$ to Y-TZP lead to increased process zone shielding and a small contribution by crack bridging.

ACKNOWLEDGEMENTS

Daniel E. Garcia thanks RHAE-CNPq (Brazil) for a two-year scholarship.

REFERENCES

1. Garvie, R. G., R. H. J. Hannink and R. T. Pascoe, 1975. "Ceramic Steel?" Nature, 258 (5537):703-704.

2. Claussen, N., 1978. "Stress-Induced Transformation of Tetragonal ZrO_2 Particles in Ceramic Matrices." Journal of the American Ceramic Society, 61(1-2):85-86.

3. Gupta, T. K., F. F. Lange and J. H. Bechtold, 1978. "Effect of Stress-Induced Phase Transformation on the Properties of Polycrystalline Zirconia Containing Metastable Tetragonal Phase." Journal of Materials Science, 13(7):1464-1470.

4. Porter, D. L., A. G. Evans and A. H. Heuer, 1979. "Transformation-Toughening in Partially Stabilized Zirconia (PSZ)." Acta metall, 27:1649-1654.

5. McMeeking, R. M. and A. G. Evans, 1982. "Mechanics of Transformation-Toughening in Brittle Materials." Journal of the American Ceramic Society, 65(5):242-246.

6. Lange, F. F., 1982. "Transformation Toughening, Part 1-5." Journal of Materials Science, 17(1):225-263.

7. Evans, A. G. and R. M. Cannon, 1986. "Toughening of Brittle Solids by Martensitic Transformations." Acta metall, 34(5):761-800.

8. Green, D. J., R. H. J. Hannink, and M. V. Swain, 1989. Transformation Toughening of Ceramics, Boca Raton, FL., USA,: CRC Press, Inc.

9. Tsukuma, K. and K. Ueda, 1985. "Strength and Fracture Toughness of Isostatically Hot-Pressed Composites of Al_2O_3 and Y_2O_3-Partially-Stabilized ZrO_2." Journal of the American Ceramic Society, 68(1):C4-C5.

10. Tsukuma, K. and T. Takahata, 1987. "Mechanical Property and Microstructure of TZP and TZP/Al_2O_3 Composites." Materials Research Society Symposium Proceedings, 78:123-135.

11. Rajendran, S., M. V. Swain and H. J. Rossell, 1988. "Mechanical Properties and Microstructures of Co-Precipitation Derived Tetragonal Y_2O_3-ZrO_2-Al_2O_3 Composites." Journal of Materials Science, 23:1805-1812.

12. Tsukuma, K., K. Ueda, K. Matsushita and M. Shimada, 1985. "High-Temperature Strength and Fracture Toughness of Y_2O_3-Partially-Stabilized ZrO_2/Al_2O_3 Composites." Journal of the American Ceramic Society, 68(2):C56-C58.

13. Kihara, M., T. Ogata, K. Nakamura and K. Kobayashi, 1988. "Effects of Al_2O_3 Addition on Mechanical Properties and Microstructure of Y-TZP." Journal of the Ceramic Society of Japan, International Edition, 96:635-642.

14. Tu, G. F., Z. T. Sui, Q. Huang and Q. Z. Wang, 1992. "Sol-Gel Processed Y-PSZ Ceramics with 5 wt.% Al_2O_3." Journal of the American Ceramic Society, 75(4):1032-1034.

15. Heussner, K-H. and N. Claussen, 1989. "Yttria- and Ceria-Stabilized Tetragonal Zirconia Polycrystals (Y-TZP, Ce-TZP) Reinforced with Al_2O_3 Platelets." Journal of the European Ceramic Society, 5:193-200.

16. Munz, D. G., J. L. Shannon, Jr. and R. T. Bubsey, 1980. "Fracture Toughness from Maximum Load in Four Point Bend Tests with Chevron Notch Specimens." International Journal of Fracture, 16:137-141.

17. Annual Book of ASTM Standards, Vol. 3.01, E-399-83, 480-504, American Society for Testing and Materials, Philadelphia, USA, 1979.

18. Rödel, J., J. F. Kelly and B. R. Lawn, 1990. "In Situ Measurements of Bridged Crack Interfaces in the Scanning Electron Microscope." Journal of the American Ceramic Society, 73(11):3313-3318.

19. Rödel, J., J. F. Kelly, M. R. Stoudt and S. J. Bennison, 1991. "A Loading Device for Fracture Testing of Compact Tension Specimen in the SEM." Scanning Microscopy, 5(1):29-35.

20. Garvie, R. C. and P. S. Nicholson, 1972. "Phase Analysis in Zirconia Systems." Journal of the American Ceramic Society, 55(6):303-305.

21. Wiederhorn, S. M., 1973. "Subcritical Crack Growth in Ceramics," in Fracture Mechanics of Ceramics, Vol.2, R.C. Bradt, D.P.H. Hasselman and F.F. Lange, eds., New York, NY, USA: Plenum Press, pp. 613-646.

22. Buchanan, R. C. and D. M. Wilson, 1984. "Role of Al_2O_3 in Sintering of Submicrometer Yttria-Stabilized ZrO_2 Powders," in Advances in Ceramics, Vol.10, W.D. Kingery, ed., Columbus, OH, USA: The American Ceramic Society, Inc., pp. 526-538.

23. Brook, R. J., 1976. "Controlled Grain Growth," in Treatise on Materials Science and Technology, Vol.9, F.F.Y.Wang, ed., New York, NY, USA: Academic Press.

24. Ashizuka, M., H. Kiyohara, E. Ishida, M. Kuwabara, Y. Kubota and T. Tsukidate, 1986. "Fatigue Behavior of Y_2O_3-Partially Stabilized Zirconia." Yogyo-Kyokai-Shi, 94(4):432-439.

25. Anderson, R. M. and L. M. Braun, 1990. "Technique for the R-Curve Determination of Y-TZP Using Indentation-Produced Flaws." Journal of the American Ceramic Society, 73(10):3059-3062.

26. Heussner, K-H., 1991. "Mechanische Eigenschaften von ZrO_2 (TZP)/Al_2O_3-Platelet-Verbundwerkstoffen (Ph.D. Thesis at TUHH, Hamburg), VDI Verlag, 5 (225), Düsseldorf.

Cr$_2$O$_3$ Particulate Reinforced Y-TZP Ceramics with High Fracture Toughness and Strength

ZHENXIANG DING, RAINER OBERACKER
and FRITZ THÜMMLER

ABSTRACT

The improvement of fracture toughness of TZP materials is accompanied by a loss of strength in many cases. The intention of the present work was the simultaneous improvement of fracture toughness and strength of Y$_2$O$_3$ (2 and 3 mol% Y$_2$O$_3$) stabilized TZP. The mechanical properties are characterized by bending strength, fracture toughness, hardness and elastic modulus. X-ray, SEM, TEM and HRTEM were used to characterize the phase composition and microstructure. The maximum 4-point bending strength and fracture toughness were about 1470 MPa and 7 MPa$\sqrt{\text{m}}$, respectively, compared with 1200 MPa and 4.8 MPa$\sqrt{\text{m}}$ of pure 2Y-TZP. The main reinforcement mechanisms are stress induced martensitic transformation and microcracking.

INTRODUCTION

It is well known that Y$_2$O$_3$ stabilized tetragonal zirconia polycrystals (Y-TZP) are suitable engineering ceramics due to their good mechanical properties. Lange [1], Evans [2] and McMeeking [3] studied the toughening mechanisms systematically and found that the stress-induced martensitic phase transformation of ZrO$_2$ from tetragonal to monoclinic symmetry is the dominating mechanism. In spite of the high fracture toughness due to transformation toughening, brittle fracture is still the character of Y-TZP ceramics and limits their practical applications. Further toughening of these ceramics is therefore necessary. The improvement of fracture toughness of TZP materials, however, is accompanied by a loss of strength in many cases. On the other hand, the development of the nanometer powder technique, in which ultrafine (nanometer size) second phase particles are dispersed mainly in the matrix particles, had remarkably enhanced strength and fracture toughness of many ceramics simultaneously [4, 5]. The dispersion of suitable

Zh. Ding, R. Oberacker, F. Thümmler: Institut für Keramik im Maschinenbau, Universität Karlsruhe, Haid-und-Neu-Str. 7, W-7500 Karlsruhe 1, Germany

ultrafine second phase particles in Y-TZP matrix could probably have a similar effect. Cr_2O_3 is a potential candidate. The grain size of commercially available Cr_2O_3 powders is about the same as that of Y-TZP powders. Its partial solubility in ZrO_2 (about 0.7 mol%) [6] and high elastic modulus could improve the strength of Y-TZP due to a solid solution effect and the favourable distribution of mechanical load on the two phase components of the composite [7]. The mismatch between thermal expansion coefficients of the matrix and the second phase benefits the t-m transformation [8] and therefore microcrack formation, which could contribute to the enhancement of fracture toughness.

Only little attention was paid to the ZrO_2-Cr_2O_3 system to date. This is perhaps mainly due to the poor densification ability in air of ZrO_2-Cr_2O_3 mixtures [9]. Cr_2O_3 is very easily oxidized to higher oxides such as gaseous CrO_3 [10, 11] in air at the sintering temperature needed (about 1500°C), which generate pores in the sintering bodies. Thus, the sintering of ZrO_2-Cr_2O_3 in air results in little or no densification [9]. In order to avoid the volatilization of higher oxides of chromium many investigators have sintered zirconia-chromia mixtures in reducing atmospheres and obtained >98% of the theoretical density [9]. Leistner and Elstner have received the highest density of a 60%ZrO_2-40%Cr_2O_3 mixture after sintering under an O_2-partial pressure close to that of the Cr-Cr_2O_3 equilibrium. The enhanced density of ZrO_2-Cr_2O_3 composites in reducing atmospheres was attributed to the avoidance of the oxidation of Cr_2O_3 and the formation of metallic chromium, which contributes to a liquid phase sintering mechanism [12]. The aim of the present work is to develop a new processing technique for the densification of zirconia-chromia composites and to improve the mechanical properties of Y-TZP by Cr_2O_3 particulate additions.

EXPERIMENTAL

The starting materials used in this work were commercial ZrO_2 powders stabilized by 2 and 3 mol% Y_2O_3 (TZ2Y, TZ3Y)[*] and pigment Cr_2O_3 powder[+] with an average grain size of 0.26, 0.27 and 0.27 μm, respectively. The required quantities of the powders were mixed in an attritor using ZrO_2 mill balls in a medium of isopropanol for 1.5 hours. The slip was then wet filtered with a polymer gauze with a mesh size of 6 μm, in order to eliminate hard, large agglomerates which were not comminuted during milling. After drying these mixtures in a rotary evaporator, they were die pressed with a pressure of 10 MPa and subsequently cold isostatically repressed at 400 MPa. The densification behavior in reducing and Ar atmosphere was studied in a temperature range of 1250°C - 1550°C. The pure TZP powders were also treated with the same procedure. Specimens for the study of mechanical properties were presintered in Ar up to pore closure at a temperature range of 1450°C - 1550°C, 2h. They were then densified by HIPing at 1450°C, 150 MPa for 0.5h. The density of sintered and HIPed specimens was determined by the Archimedes' principle. X-ray, SEM, TEM and HRTEM were used to

---

*Toyo Soda Manufacturing Co., Ltd., Kanayawa, Japan
+Bayer AG, Leverkusen, Germany

characterize phase composition and microstructure. Phase analysis of ZrO_2 was performed by the method developed by Garvie et al [13]. The hardness was measured according to Vickers (HV10) with a load of 98 N. Specimens with a geometry of 45x4.5x3.5 mm³ were used to determine Young's modulus by an ultrasonic method, 4-point bending strength σ_c (span 40/20 mm, loading speed 100 N/s), and fracture toughness K_{IC} (3-point bending, cross head speed 10 μm/s). The incipient crack was generated by the bridge method.

RESULTS

DENSIFICATION BEHAVIOR

ZrO_2 materials containing Cr_2O_3 can be easily densified in reducing atmospheres. However, the following disadvantages are possible. First, the metallic chromium, which is formed during sintering, changes the phase composition and may degrade the mechanical properties, especially hardness and strength, of the composite. Second, tetragonal ZrO_2 can be destabilized. For instance, on the as-sintered surface of a TZ3Y-30vol%Cr_2O_3 composite, which was sintered in reducing atmosphere, 60% monoclinic ZrO_2 was detected.

The poor sinterability of ZrO_2-Cr_2O_3 composites is suggested mainly to be due to the volatilization of higher oxides of chromium [9, 10]. In a neutral, but not reducing atmosphere, such as Ar, with very low O_2 content this problem must be avoided. This hypothesis was confirmed by sintering of Y-TZP/Cr_2O_3 composites under flowing Ar in a furnace with Mo-heating elements. No m-ZrO_2 or metallic Cr were detected on the as-sintered surface by x-ray phase analysis. Figure 1 shows the influence of temperature and Cr_2O_3 content on the sintered density of 3Y-TZP in Ar atmosphere. Cr_2O_3 strongly impeded the densification of 3Y-TZP, and the density of the composites with more than 10 vol% Cr_2O_3 was not satisfactory. Higher sintering temperature

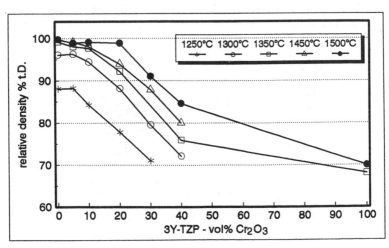

Figure 1   Influence of temperature and Cr_2O_3 content on the sintered density of 3Y-TZP

and longer sintering time can improve the sintered density, but the problem of grain growth can then not be avoided. In order to raise the density and to realize an ultrafine microstructure, which is very important for the simultaneous improvement of strength and fracture toughness, a two-stage densification route was utilized: if the specimens are pressureless sintered at a temperature range of 1450°C - 1550°C up to pore closure, that is about 90% of the theoretical density, they can be fully densified by a containerless HIPing process. This route was successful for Y-TZP with up to 30 vol% Cr_2O_3. It was very difficult to reach pore closure without raising the sintering temperature, if the Cr_2O_3 content exceeded 30 vol%. The composites of $2Y\text{-}TZP/Cr_2O_3$ were more difficult to sinter than $3Y\text{-}TZP/Cr_2O_3$, so that the necessary sintering temperature had to be 50°C higher than that of the latter.

MICROSTRUCTURE

Figure 2 illustrates the microstructure of a $TZ3Y/30Cr_2O_3$ composite. A homogeneous distribution of Cr_2O_3 was reached. Cr_2O_3 particles were dispersed mainly at the grain boundaries of ZrO_2; only a few were observed inside ZrO_2 matrix particles. This is probably attributed to the ultrafine grain size of the ZrO_2 phase.

The grain growth of the ZrO_2 matrix was not influenced by Cr_2O_3 additions. The mean grain size of ZrO_2 in the 3Y- and 2Y-TZP was about 0.35 and 0.42 μm, respectively, independent of the chromia content, while the average grain size of Cr_2O_3 was about 0.6μm. This is because of the relatively wide grain size distribution of the Cr_2O_3 starting powder and stronger grain growth of the Cr_2O_3 phase. Oversize Cr_2O_3 particles up to 20 μm were observed. Such over size second phase particles are detrimental for the mechanical properties because they are potential fracture origins. No grain boundary phases were detected in $Y\text{-}TZP/Cr_2O_3$ composites by the microstructural studies using HRTEM. Figure 3 shows the $ZrO_2\text{-}Cr_2O_3$ phase boundary. The atomic planes of the two phases contact directly with each other.

In as-HIPed specimens microcracks were detectable only in a very limited number, which probably are formed during the preparation of TEM foils.

Figure 2 Microstructure of a $3YTZP\text{-}30Cr_2O_3$ composite after HIPing (TEM, bright: Cr_2O_3)

After bend testing the microstructure of the composites was studied again. In this case TEM foils were taken from fractured specimens at a location between the inner span, where the bending moment is constant. Much more intensive microcracking was observed in these loaded specimens. The microcrack formation took place mainly along the phase boundaries of ZrO_2-Cr_2O_3, at which the ZrO_2 grains are transformed into the monoclinic modification. An example is demonstrated in figure 4.

TRANSFORMATION AND MECHANICAL PROPERTIES

On the as-HIPed surface of 3Y-TZP containing Cr_2O_3 no m-ZrO_2 was detected by x-ray phase analysis, while an increasing tendency towards m-ZrO_2 was observed on the as-HIPed surface of 2Y-TZP/Cr_2O_3 composites. The transformability of the 2Y-TZP matrix (about 60 - 65%) was twice as high as that of the 3Y-TZP material. The t-m transformation of ZrO_2 was surprisingly not reduced by Cr_2O_3, despite the higher elastic modulus of Cr_2O_3.

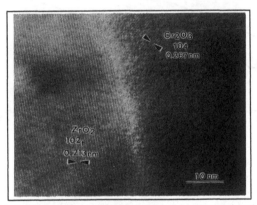

Figure 3 HRTEM image showing grain boundary of ZrO_2-Cr_2O_3 without boundary phases

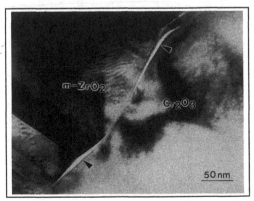

Figure 4   Microcracks along the grain boundary of ZrO_2-Cr_2O_3 (TEM)

The measured Young's modulus and hardness of Cr_2O_3 containing Y-TZP are shown in figure 5. Both were enhanced by Cr_2O_3 additions. The change of the Young's modulus of Y-TZP with Cr_2O_3 content follows the rule of mixture accurately, using a Young's modulus of 311 GPa for pure Cr_2O_3. The hardness of 2Y-TZP composites is lower than that of 3Y-TZP. This could be the result of the slightly larger grain size and the higher tendency of the 2Y-TZP matrix to transform from tetragonal to the monoclinic modification during indentation, as suggested by Sakuma et al [14] and Tsukuma et al [15].

The dependence of bending strength (σ_c) and fracture toughness (K_{IC}) on Cr_2O_3 and Y_2O_3 content are shown in figure 6. In spite of the reducing volume fraction of tetragonal ZrO_2 with increasing Cr_2O_3 content, the

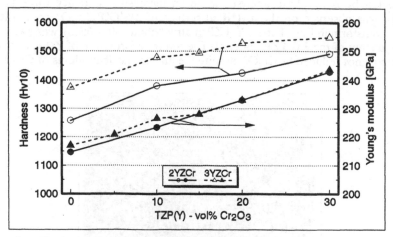

Figure 5   Hardness and Young's modulus of Y-TZP/Cr_2O_3 composites

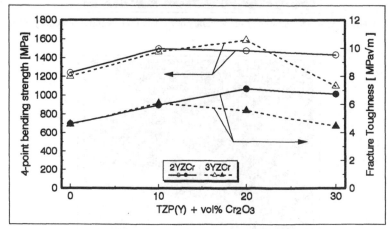

Figure 6   Simultaneous enhancement of bending strength and fracture toughness of Y-TZP by Cr_2O_3 particulate additions

bending strength and fracture toughness of Y-TZP were simultaneously enhanced. Maximum mean strength of 1600 MPa (3Y-TZP) and 1470 MPa (2Y-TZP) was measured at 20 vol% Cr_2O_3. There was no evident difference of the dependence of bending strength on Y_2O_3 content. This is consistent with the results in the system ZrO_2-Al_2O_3 [16]. However, the fracture toughness of Y-TZP/Cr_2O_3 composites with more than 10 vol% Cr_2O_3 was quite different between the two stabilizer contents of 2 and 3 mol% Y_2O_3.

FRACTOGRAPHY

The fracture behavior of Y-TZP/Cr_2O_3 ceramics was examined using scanning electron microscopy. Figure 7 shows the predominant intergranular fracture mode in these ceramics. The oversize Cr_2O_3 particles, however, appeared to fracture transgranularly (arrowed in figure 7). The fracture origins of Cr_2O_3 containing Y-TZP were mostly agglomerates, Cr_2O_3 clusters and oversize Cr_2O_3 particles. No pores were detected as fracture origins due to the full densification. In figure 8 typical fracture origins are shown. Fracture origins in more than 70% of the Cr_2O_3 containing specimens were volume flaws. In comparision to that, mainly surface flaws were detected in pure Y-TZP in spite of the same fabrication procedure. This indicates a reduced flaw concentration in Cr_2O_3 containing Y-TZP. The relationship between the measured bending strength and flaw size (a_c) of pure Y-TZP and Y-TZP/Cr_2O_3 composites is given in figure 9. It is suggested that the improvement of the flaw tolerance by Cr_2O_3 additions is attributed to the enhanced fracture toughness.

DISCUSSION

In comparision to traditional densification procedures of Cr_2O_3 and Cr_2O_3 containing materials in reducing atmospheres, the phase change during sintering was avoided by using Ar gas as the sintering atmosphere. The two-stage consolidation route allowed full densification of Y-TZP with up to 30 vol% Cr_2O_3. This procedure is also suitable for the fabrication of other engi-

Figure 7   Fracture surface of a 2Y-TZP-20 vol% Cr_2O_3 composite (SEM) showing transgranular fracture of large Cr_2O_3 grains (arrowed)

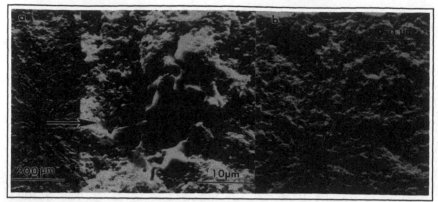

Figure 8   Typical fracture origins of Y-TZP/Cr$_2$O$_3$ composites (SEM). a) Cr$_2$O$_3$ cluster;
            b) oversize Cr$_2$O$_3$ particle

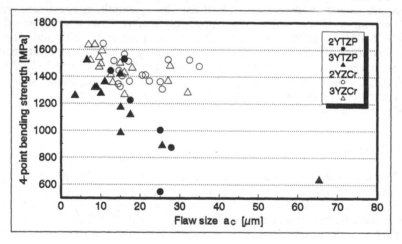

Figure 9   Relationship of measured bending strength and flaw size a$_c$ of Y-TZP/Cr$_2$O$_3$
            composites

neering ceramics, if a fine and homogeneous microstructure is needed.

Y-TZP has been found to always contain some residual glassy phase [17].
This glassy phase is SiO$_2$ rich and distributed in grain junctions and grain
boundaries continuously with a thickness of about 1.5 - 2 nm. The main
sources of this glassy phase are the impurities (SiO$_2$) of the powder. The Y-
TZP powders used in this work contain <0.2 wt% SiO$_2$ (data provided by the
manufacturer). In the grain boundaries of Y-TZP/Cr$_2$O$_3$ composites,
however, no traces of any amorphous phases were detected. This means, that
Cr$_2$O$_3$ has a grain boundary cleaning function. SiO$_2$ has perhaps a higher
chemical affinity for Cr$_2$O$_3$ than for ZrO$_2$ and forms a chemical compound
with the former one, as in the case of Y-TZP/Al$_2$O$_3$ composites [18, 19]. Or

SiO_2 could be soluble in Cr_2O_3 and form a SiO_2/Cr_2O_3 solid solution. The details of this phenomenon are not yet clear and should be studied further.

The increasing tendency of m-ZrO_2 on the as-HIPed surface of 2Y-TZP and the almost constant transformation on fracture surfaces could be attributed to the reaction of Cr_2O_3 with Y_2O_3. There is only one compound in the system Cr_2O_3-Y_2O_3, $YCrO_3$, which was found to occur at about 1250°C in a ZrO_2-33 mol% Y_2O_3 composite containing Cr_2O_3 [20]. This reaction results in a reduced Y_2O_3 concentration in the ZrO_2/Cr_2O_3 phase boundaries and increases the transformation tendency of tetragonal ZrO_2 particles. The fact that microcracks occured predominantly along ZrO_2-Cr_2O_3 phase boundaries, supports this argument. Another possible reason for the enhanced transformation could be the residual stresses resulting from the mismatch of the thermal expansion coefficients of ZrO_2 ($\alpha \approx 11 \cdot 10^{-6}$ /K) and Cr_2O_3 ($\alpha \approx 7.5 \cdot 10^{-6}$/K). Due to its higher thermal expansion, tensile residual stresses should result in the ZrO_2 matrix phase during cooling from the sintering temperature. It is believed, that tensile residual stresses enhance the phase transformation of ZrO_2 [8]. Evidently, these destabilizing effects have compensated for the physical stabilization due to the higher Young's modulus of Cr_2O_3, so that the stress-induced phase transformation is not reduced by Cr_2O_3 additions.

The high bending strength of Y-TZP and Y-TZP/Cr_2O_3 composites can be attributed to the minimization and elimination of processing defects and the realization of a fine microstructure. The careful powder processing, especially the wet filtration, separated the hard, large agglomerates. The use of the two-stage fabrication route allowed the elimination of residual porosity, which is decisive for the strength improvement of ceramic materials. The relatively small flaw size, shown in figure 9, indicates that the upper limit of mechanical mixing was probably already reached. In order to minimize processing defects and to further improve the strength, new fabrication processes, such as sol-gel techniques, must be applied. The grain size distribution of Cr_2O_3 should be also improved.

The strengthening of Y-TZP by Cr_2O_3 additions is mainly due to the increased transformation reinforcement and the higher Young's modulus of Cr_2O_3. In spite of the reduction of the absolute volume fraction of tetragonal ZrO_2 with increasing Cr_2O_3 content the transformability of t-ZrO_2 is enhanced by the formation of $YCrO_3$. So the effectiveness of the transformation reinforcement is not reduced. Furthermore, the clean grain boundaries of Y-TZP/Cr_2O_3 composites must also contribute to strengthening.

The condition of toughening of brittle materials by residual stresses is the creation of compressive stress zones in the matrix [21, 22], which requires a lower thermal expansion of the matrix phase. This is in contrast to the case of Y-TZP/Cr_2O_3 composites. The fact that the fracture toughness of Y-TZP was enhanced by Cr_2O_3 additions in spite of the "wrong" thermal mismatch is because of the transformable matrix. The enhanced transformation led additionally to microcrack toughening. The higher fracture toughness of 2Y-TZP materials is consistent with the theory of transformation toughening.

CONCLUSIONS

Y-TZP/Cr_2O_3 ceramic composites with up to 30 vol% Cr_2O_3 were fully densified by pressureless sintering and subsequent post HIPing in Ar atmosphere. Hardness, bending strength and fracture toughness of Y-TZP were simultaneously enhanced by Cr_2O_3 additions. The main reinforcement mechanisms in Y-TZP/Cr_2O_3 ceramics are stress induced martensitic phase transformation of ZrO_2 and microcracking. Flaw tolerance of the Y-TZP was improved by Cr_2O_3 additions due to enhanced fracture toughness. The relatively small flaw size shows that the upper limit of mechanical mixing was already reached. In order to reduce flaw size and enhance the strength of this material further, new fabrication processes must be applied.

REFERENCES

1. Lange, F. F, 1982. "Transformation Toughening Part 3: Experimental Observations in the ZrO_2-Y_2O_3 System." J. Mater. Sci., 17: 240-248.

2. Evans, A. G., 1984. "Toughening Mechanisms in Zirconia Alloys," in Adv. in Ceramics, vol. 12: Science and Technology of Zirconia II, N. Claussen, M. Rühle and A. Heuer, eds., Columbus, OH, USA: the American Ceramic Society, pp193 - 212.

3. McMeeking, R. M. and A. G. Evans, 1982." Mechanics of Transformation Toughening in Brittle Materials." J. Am. Ceram. Soc., 65[5]:242-246.

4. Niihara, K. and A. Nakahira, 1988. "Strengthening of Oxide Ceramics by SiC and Si_3N_4 Dispersions." in Proceedings of the Third International Symposium on Ceramic Materials & Components for Engines, V. J. Tennery, ed., Las Vegas , NV, USA: the American Ceramic Society, pp919 - 926.

5. Niihara, K. and A. Nakahira, 1991. "New Design of Structural Ceramics: Nanocomposites". Lecture at the "Second Conference of the European Ceramic Society", 11-14 Sept. 1991, Augsburg, Germany.

6. Yoshimura, M., M. Jayaratna, and S. Somiya, 1982. "Phase Equilibria in the Pseudoternary System ZrO_2-Y_2O_3-Cr_2O_3 at 1600°C." J. Am. Ceram. Soc., 65[10]:C-166-C-168.

7. Eigenmann, B., 1992. "Röntgenographische Analyse inhomogener Spannungszustände in Keramiken, Keramik-Metall-Fügeverbindungen und dünnen Schichten." Ph.D. Thesis, University Karlsruhe, Germany.

8. Marshall, D. B. and M. V. Swain, 1988. "Crack Resistance Curves in Magnesia-Partially-Stabilized Zirconia." J. Am. Ceram. Soc., 71[6]: 399-407.

9. Yamaguchi, A., 1981. "Densification of Cr_2O_3-ZrO_2 Ceramics by Sintering." J. Am. Ceram. Soc., 64[4]: C-67.

10. Grimley, R. T., R. P. Burns and M. G. Ingraham, 1960. "Thermodynamics of the Vaporization of Cr_2O_3: Dissociation Energies of CrO, CrO_2, and CrO_3." Chem. Phys., 34: 664-7.

11. Caplan, D. and M. Cohen, 1961. "The Volatilization of Chromium Oxide." J. Electrochem. Soc., 108[5]: 438-442.

12. Leisner, H. and I. Elstner, 1983. "Sintering Behaviour of Refractories in the Ternary System ZrO_2-Al_2O_3-Cr_2O_3 in Dependence of the Firing Atmosphere." Report BMFT-FB-T-83-105, Bundesministerium für Forschung und Technologie, Germany.

13. Garvie, R. C., R. H. Hannik and M. V. Swain, 1982. "X-Ray Analysis of the Transformed Zone in Partially Stabilized Zirconia (PSZ)." J. Mater. Sci. Letters, 1: 437-440.

14. Sakuma, T., Y. Yoshizawa and H. Suto, 1985. "The Microstructure and Mechanical Properties of Yttria-Stabilized Zirconia Prepared by Arc-Melting." J. Mater. Sci., 20: 2399-2407.

15. Tsukuma, K. and M. Shimada, 1985. "Strength, Fracture Toughness and Vickers Hardness of CeO_2-Stabilized ZrO_2 Polycrystals (Ce-TZP)." J. Mater. Sci., 20[4]: 1178-1184.

16. Tsukuma, K., K. Ueda and M. Shimada, 1985. "Strength and Fracture Toughness of Isostatically Hot-Pressed Composites of Al_2O_3 and Y_2O_3-Partially-Stabilized ZrO_2." J. Am. Ceram. Soc., 68 [1]: C4-C5.

17. Rühle, M., N. Claussen and A. H. Heuer, 1984. "Microstructural Studies of Y_2O_3-Containing Tetragonal ZrO_2 Polycrystals (Y-TZP)." in Adv. in Ceramics, vol. 12: Science and Technology of Zirconia II, N. Claussen, M. Rühle and A. Heuer eds., Columbus, OH, USA: the American Ceramic Society, pp352 - 370.

18. Drennan, J., and S. P. S. Badwal, 1988. "The Influence of Al_2O_3 Additions to Yttria-Containing Tetragonal Zirconia Polycrystals (Y-TZP): A Microstructural and Electrical Conductivity Study," in Adv. in Ceramics, vol. 24: Science and Technology of Zirconia III. S. Somiya, N. Yamamoto and H. Yanagida, eds., Westerville, Ohio, USA: the American Ceramic Society, pp807 - 817.

19. Mecartney, M. L., 1987. "Influence of an Amorphous Second Phase on the Properties of Yttria-Stabilized Tetragonal Zirconia Polycrystals (Y-TZP)." J. Am. Ceram. Soc., 70[1]: 54-58.

20. Strakhow,V. I. and V. K. Novikov, 1976. "Reaction of Cr_2O_3 with ZrO_2 Stabilized by Rare Earth Oxides." Inorg. Mater. (Engl. Transl.), 12[2]: 247-250.

21. Wei, G. C. and P. F. Becher, 1984. "Improvements in Mechanical Properties in SiC by the Addition of TiC Particles." J. Am. Ceram. Soc., 67[8]:571-74.

22. Taya, M., S. Hayashi, Albert S. Kabayashi, and H. S. Yoon, 1990. "Toughening of a Particulate-Reinforced Ceramic-Matrix Composite by Thermal Residual Stress." J. Am. Ceram. Soc., 73[5]: 1382-1391.

Observations of Transitions in Creep Behaviour of a Superplastic Yttria-Stabilized Zirconia

DAVID M. OWEN and ATUL H. CHOKSHI

ABSTRACT

The phenomenon of superplasticity has been observed in several single and dual phase ceramics. Typically, the dependence of creep rate ($\dot{\epsilon}$) on stress (σ) is characterized by $\dot{\epsilon} \propto \sigma^n$, where n is the stress exponent. Previous studies on superplastic ceramics have reported values of n ranging from 1 to 3 for nominally identical materials, and the discrepancy has been attributed partly to differences in the content of minor impurities. This paper describes the results of a detailed creep study on fine-grained superplastic zirconia. The deformation behavior was characterized under constant stress in uniaxial compression over the temperature range of 1600 to 1750 K. Two distinct creep behaviors were observed: at higher stresses n~2, whereas at lower stresses n~3. These results are compared to the available theoretical creep models.

INTRODUCTION

Superplasticity refers to the ability of many fine grained ceramics and metallic alloys to achieve tensile elongations on the order of several hundreds of percent. This phenomenon is well established in many metals and is utilized commercially to form complex shapes. Current interest in superplasticity of ceramics follows the study by Wakai et al. [1] that reported an elongation to failure of ~170% in a yttria-stabilized tetragonal zirconia polycrystal (YTZP); the maximum elongation recorded to date in a ceramic is 800% in YTZP [2]. More recently, Wittenauer et al. [3] have demonstrated the ability to form superplastic YTZP-20%Al$_2$O$_3$ into a spherical cap from a flat sheet using gas pressure and their results open up the exciting possibility of forming complex shapes with superplastic ceramics. Although large strains have been reported for other single phase ceramics such as alumina and hydroxyapatite, and ceramic composites such as YTZP-20%Al$_2$O$_3$ and Si$_3$N$_4$/SiC, monolithic YTZP has been the model

D.M. Owen and A.H. Chokshi, AMES Department, 0411, University of California, San Diego, La Jolla, Ca, 92093-0411, USA.

material for numerous investigations of superplasticity in ceramics [1-7].

Typically, the high temperature deformation behavior of ceramics is characterized using an equation of the form

$$\dot{\varepsilon} = \frac{ADGb}{kT} \left(\frac{\sigma}{G}\right)^n \left(\frac{b}{d}\right)^p \tag{1}$$

where $\dot{\varepsilon}$ is the steady state strain rate, σ is the true stress, d is the grain size, T is the absolute temperature, n is the stress exponent, p is the inverse grain size exponent, G is the shear modulus, b is the magnitude of the Burgers vector, k is Bolztmann's constant, and A is a dimensionless constant. The diffusion coefficient, D, of the species controlling the rate of deformation may be expressed as $D = D_0 \exp(-Q/RT)$, where Q is the activation energy for diffusion, D_0 is the frequency factor and R is the gas constant.

Previous studies [1-7] on the deformation behavior of superplastic zirconia have been limited typically to strain rates greater than 10^{-5} s^{-1}, and they have yielded disparate values for the rate controlling parameters: the reported stress exponents and inverse grain size exponents have ranged from 1 to 3 [7]. The observed differences have been attributed in part to variations in the concentrations of glass forming impurities [4] and the failure to account for the occurrence of concurrent grain growth [2].

The present study was undertaken to characterize the constant stress compressive creep behavior of superplastic YTZP for $\dot{\varepsilon}$ ranging from 10^{-7} to 10^{-2} s^{-1} under conditions minimizing microstructural instability.

EXPERIMENTAL

A tetragonal zirconia stabilized with 3 mol% yttria (3YTZP) was the material chosen for the present creep study. Cylindrical specimens 5 mm in diameter and 10 mm in length were obtained from Nippon Kagaku Togyo Co. Ltd., Japan. The mean linear intercept grain size, L, of the as-received specimens was measured as 0.41 μm.

Experiments were conducted in uniaxial compression using a constant stress creep apparatus [8]. A closed loop feedback system, consisting of a linear variable capacitance (LVC) transducer, a load cell and a stepper motor, was controlled by a personal computer to maintain a constant stress. The applied stresses ranged from 3 to 200 MPa and the experiments were conducted in air at temperatures ranging from 1623 to 1723 K. Prior to testing, some specimens were annealed at 1823 K for 5 and 40 hours to obtain grain sizes of 0.66 and 1.3 μm, respectively.

The deformed specimens were sectioned and polished metallographically to a 1 μm finish. The polished sections were thermally etched at 1673 K for 1 hour and examined using scanning electron microscopy. The average linear intercept grain size was measured in directions parallel, L_1, and perpendicular, L_2, to the compression axis. The average intercept grain size, $\bar{L}$, is defined as $(\bar{L}_1\bar{L}_2{}^2)^{1/3}$ and the grain aspect ratio, GAR, as $\bar{L}_1/\bar{L}_2$.

RESULTS

All tests were conducted at single stresses and temperatures and terminated prior to failure at true strains of ≤ 50%; no significant barrelling was noted in the deformed specimens. The feedback loop described above maintained true stresses constant to within 3%.

The results from individual creep tests are presented in the form of $\dot{\varepsilon}$ versus ε, Figure 1, for specimens with an initial grain size of 0.41 μm, deformed at a temperature of 1723 K and two stresses. Two distinct types of creep behavior are noted in Figure 1: deformation at lower stresses is characterized by a brief primary creep region of decreasing strain rate, followed by a region of well defined steady state. At higher stresses, following a short primary and brief steady state region, the strain rate increases slightly with increasing strain.

Figure 2 illustrates the variation in true strain rate, at a true strain of 15%, with stress on a logarithmic scale for tests conducted at T=1723 K with an initial grain size of $\bar{L}_O$=0.41 μm. The stress exponent, n, defined as

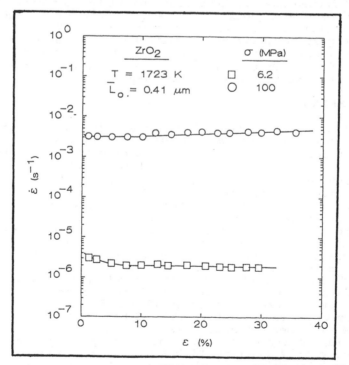

Figure 1 The variation of $\dot{\varepsilon}$ with ε for specimens with $\bar{L}_O$=0.41 μm deformed at T=1723 K and stresses of 6.2 and 100 MPa.

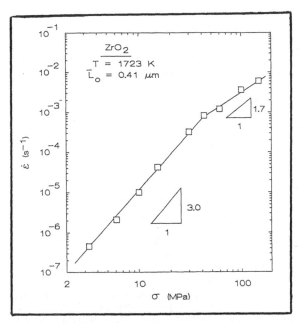

Figure 2 The variation of $\dot{\varepsilon}$ with σ for specimens with $\overline{L}_0 = 0.41$ μm deformed at T=1723 K. The stress exponent, n, changes from ~3 at low stresses to ~2 at high stresses.

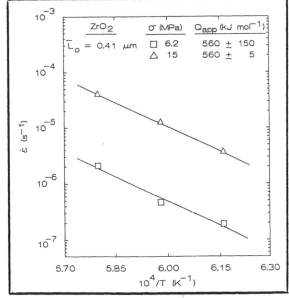

Figure 3 The dependence of $\dot{\varepsilon}$ on 1/T for specimens with $\overline{L}_0 = 0.41$ μm deformed at stresses of 6.2 and 15 MPa where n~3. The activation energy is 560 kJ mol^{-1} for both stresses.

435

TABLE I – MICROSTRUCTURAL EVOLUTION DURING CREEP DEFORMATION OF
3YTZP

σ (MPa)	ε (%)	$\overline{L}_1$ (μm)	$\overline{L}_2$ (μm)	$\overline{L}_1/\overline{L}_2$	$\overline{L}_f/\overline{L}_o$
6.1	33	0.58	0.59	0.98	1.43
15	32	0.39	0.43	0.91	1.02
100	37	0.40	0.44	0.93	1.04
150	43	0.44	0.48	0.90	1.14

$\partial \log \dot{\varepsilon}/\partial \log \sigma$ is seen to decrease from 3.0 ± 0.2 at lower stresses to 1.7 ± 0.2 at higher stresses: the transition appears to occur reasonably sharply at $\sigma \approx 40$ MPa.

The effect of temperature on strain rate is illustrated in Figure 3 by an Arrhenius plot of $\dot{\varepsilon}$ versus $1/T$ for specimens with $L_o=0.41$ μm tested at two different stresses and temperatures of 1623, 1673 and 1723 K. The activation energy for deformation was calculated to be 560 ± 150 and 560 ± 5 kJ mol^{-1} at stresses of 6.2 and 15 MPa, respectively. The data at the two stresses and all temperatures reveal stress exponents of ~ 3.

Limited creep experiments at 1723 K and specimens with grain sizes of 0.66 and 1.3 μm suggest that the creep rate is significantly dependent on the grain size in both regions with n~3 at low stress and n~2 at high stress.

Linear intercept grain sizes of the creep deformed specimens were determined in the directions parallel and perpendicular to the loading axis. Sufficient measurements were made to limit the error for 95% confidence to below 5%. Some of the data obtained are listed in Table I: stresses of 6.1 and 15 MPa correspond to the n~3 region whereas stresses of 100 and 150 MPa correspond to the n~2 region. Inspection of the data reveals clearly that there is no significant change in the grain aspect ratio, thereby indicating that the grains remain essentially equiaxed after deformation in both the high stress and low stress regions. The experimental results in Table I also demonstrate that there was very little concurrent grain growth during creep, except for the test at 6.2 MPa which lasted ~2 days.

DISCUSSION

The stress-strain rate data on superplastic metallic alloys are usually divisible into three regions, termed regions I, II and III in order of increasing stress. Maximum tensile ductility is observed in the superplastic region II where n ≤ 2, and the ductility is reduced in regions I and II where n ≥ 3. Although there is no theory that can completely rationalize the mechanical and microstructural characteristics of superplasticity in region II [9], the significant contribution of grain boundary sliding to deformation is well recognized and accepted [10]. Deformation at high stress in region III is often attributed to intragranular dislocation creep with some contribution from grain boundary sliding.

The origin of region I, with $n \geq 3$, is not as well understood. The observation of region I has been attributed variously to excessive concurrent grain growth [11], the operation of a threshold stress [12] and the occurrence of an as-yet unidentified mechanism [13].

The present experimental results on the superplastic zirconia demonstrate clearly an apparent transition from a region at high stresses with $n \approx 2$ to a region at low stresses with $n \approx 3$, in a manner consistent with data on superplastic metallic alloys.

Rai and Grant [11] have shown that excessive concurrent grain growth during long term experiments at low strain rates in an Al-Cu eutectic alloy may lead to an apparent increase in the value of n and a spurious region I. Nieh and Wadsworth [2] reported n=3 for a superplastic 3YTZ tested under conditions where the microstructure was unstable: however, by accounting for the effect of concurrent grain growth, a corrected value of n=1.5 was obtained. In the present study, there was very limited grain growth; the stress exponents reported in Figure 2 were not significantly modified when the data were corrected for the observed grain growth.

Mohamed [12] re-evaluated data from studies on the superplastic Zn-22%Al and Pb-62%Sn metallic alloys exhibiting region I behavior by considering the possibility of a threshold stress. The applied stress, σ in equation (1), was replaced by an effective stress σ_e $(=\sigma-\sigma_o)$, where σ_o is the threshold stress. The data from both regions fell on a single straight line when plotted as $\dot{\varepsilon}^{(1/n)}$ versus σ on a linear scale, thereby indicating the operation of a single deformation mechanism; σ_o is the stress obtained when the data are extrapolated to $\dot{\varepsilon}=0$ and n is the stress exponent in region II. A similar analysis suggested that the present experimental results cannot be interpreted in terms of a simple threshold stress.

The above discussion suggests that the transition in stress dependence of creep rate in 3YTZP may be a result of two sequential deformation mechanisms in which the slower process will determine the total creep rate.

Studies of grain boundaries in YTZP have revealed that yttria segregates to the boundaries and may inhibit grain growth [14-17]. Creep processes may be similarly affected by the segregation of yttria to grain boundaries in YTZP: at higher stresses associated with deformation where $n \approx 2$, the impurities may not significantly hinder grain boundary sliding, while at lower stresses where $n \approx 3$ the limited mobility of dislocations over obstacles in grain boundaries may become rate controlling. It is important to note that the grains remain equiaxed following creep deformation in both regions: this observation suggests that there are similar microstructural processes involved in creep deformation in both the high and low stress regions. In this context, it is to be noted that the microstructure retains its equiaxed nature even after tensile elongations of up to 700% [17].

It is interesting to note that Nauer and Carry [5] have reported limited creep data on YTZP with varying yttria contents: the data were interpreted in terms of two regions with n~2 and p~1 at lower stresses, and n~1 and p~3 at higher stresses.

CONCLUSIONS

A superplastic 3 mol% yttria-stabilized zirconia exhibits a transition in creep behavior from a region with n~3 at low stresses to a region with n~2 at high stresses. These results are consistent with the transition between superplastic regions I in II observed in many superplastic metals. Concurrent grain growth or the operation of a threshold stress could not account for the observed differences in rate controlling parameters in regions I and II, and the regions may represent the operation of two distinct sequential creep processes.

ACKNOWLEGMENTS

We are grateful to Dr. T.G. Nieh for providing the material used in this study. The research was supported by the National Science Foundation under Grant no. DMR-9023699. One of us (DMO) acknowledges a fellowship from the Army Research Office under contract no. DAAL86-K-0169.

REFERENCES

1. Wakai, F., S. Sakaguchi and Y. Matsuno, 1986. "Superplasticity of Yttria-Stabilized Tetragonal ZrO_2 Polycrystals." Adv. Ceram. Mater., 1: 259-263.

2. Nieh, T.G. and J. Wadsworth, 1990. "Superplastic Behavior of a Fine-Grained,Yttria-Stabilized, Tetragonal Zirconia Polycrystal (Y-TZP)." Acta Met., 38: 1121-1133.

3. Wittenauer, J., T.G. Nieh, J. Wadsworth, 1992. "A First Report on Superplastic Gas-Pressure Forming of Ceramic Sheet." Scripta Met. Mater., 26: 551-556.

4. Carry, C. 1989. "High Ductilities, Superplastic Behaviors and Associated Mechanisms in Fine Grained Ceramics," in Superplasticity, M. Koybayashi and F. Wakai, eds., Proceedings of the MRS International Meeting on Advanced Materials, vol. 7, Pittsburgh, PA, USA: Materials Research Society, pp. 199-215.

5. Nauer, M. and C. Carry, 1990. "Creep Parameters of Yttria Doped Zirconia Materials and Superplastic Deformation Mechanisms." Scripta Met. Mater., 24: 1459-1463.

6. Chen, I.-W. and L.A. Xue, 1990. "Development of Superplastic Ceramics." J. Am. Ceram. Soc., 73: 2585-2609.

7. Chokshi, A.H., 1992. "Superplasticity in Fine Grained Ceramics and Ceramic Composites: Current Understanding and Future Prospects." <u>Mater. Sci. Engg.</u> (in press).

8. Owen, D.M. and A.H. Chokshi, 1991. "A Comparison of the Tensile and Compressive Creep Behavior of a Superplastic Yttria-Stabilized Tetragonal Zirconia-20 wt% Alumina Composite," in <u>Superplasticity of Advanced Materials</u>, S. Hori, M. Tokizane, and N. Furushiro, eds., Osaka, Japan: JSRS, pp.215-220.

9. Arieli, A. and A.K. Mukherjee, 1982. "The Rate Controlling Deformation Mechanisms in Superplasticity - A Critical Assesment." <u>Met. Trans. A</u>,13A: 717-732.

10. Lin, Z.-R., A.H. Chokshi and T.G. Langdon, 1988. "An Investigation of Grain Boundary Sliding in Superplasticity at High Elongations." <u>J. Mater. Sci.</u>, 23: 2712-2722.

11. Rai, G. and N.J. Grant, 1975. "On the Measurements of Superplasticity in an Al - Cu Alloy." <u>Met. Trans.</u>, 6A: 385-390.

12. Mohamed, F.A., 1983. "Interpretation of Superplastic Flow in Terms of a Threshold Stress." <u>J. Mater. Sci.</u>, 18: 582-592.

13. Mohamed, F.A. and T.G. Langdon, 1975. "Creep Behavior in the Superplastic Pb-62%Sn Eutectic." <u>Phil. Mag.</u>, 32: 697-709.

14. Nieh, T.G., D.L. Yaney and J. Wadsworth, 1989. "Analysis of Grain Boundaries in a Fine-Grained, Superplastic, Yttria-Containing, Tetragonal Zironia." <u>Scripta Met</u>. 23: 2007-2011.

15. Hwang, S.L. and I.W. Chen, 1990. "Grain Size Control of Tetragonal Zirconia Polycrystals Using the Space Charge Concept." <u>J. Am. Ceram. Soc.</u>, 73: 3269-3277.

16. Stoto, T., M. Nauer and C. Carry, 1991. "Influence of Residual Impurities on Phase Partitioning and Grain Growth Processes of Y-TZP Materials." <u>J. Am. Ceram. Soc.</u>, 74: 2615-2621.

17. Schissler, D.J., A.H. Chokshi, T.G. Nieh and J. Wadsworth, 1991. "Microstructural Aspects of Superplastic Tensile Deformation and Cavitation Failure in a Fine-Grained Yttria Stabilized Tetragonal Zirconia." <u>Acta Met. Mater.</u>, 39: 3227-3236.

Cyclic Fatigue Crack Propagation Behaviour of 9 Ce-TZP Ceramics with Different Grain Size

TIANSHUN LIU, YIU-WING MAI, MICHAEL V. SWAIN
and GEORGE GRATHWOHL

ABSTRACT

Cyclic fatigue crack growth behaviour has been investigated in 9 mole % Ce-TZP ceramics with grain sizes varing from 1.1 to 3.0 μm. To ascertain the interaction between crack resistance behaviour and cyclic fatigue crack growth, cyclic fatigue tests were conducted with short double-cantilever-beam (s-DCB) specimens in two conditions: (a) with a sharp precrack without pre-existing t-m transformation and (b) with a sharp crack after R-curve measurements, i.e. with preformed t-m transformation in the crack region. Fatigue crack propagation occurs at applied stress intensity factors as low as about 40% of the plateau K_I values measured in the R-curves. The size and shape of the t-m transformation zones are different for specimens in R-curve measurements and in cyclic fatigue tests. For the specimens without pre-existing t-m transformation the overall crack growth behaviour can be described by the Paris power law relation: $da/dN = A\Delta K_I^m$ with m values of 15 for the 1.1 μm grain size and between 8 and 9 for the other materials with larger grain sizes. For the specimens with the preformed transformation zone a "V" shape da/dN versus $\Delta K_{I,app}$ relation is obtained. Explanations for these different results in the two conditions are discussed in terms of crack tip shielding effects.

INTRODUCTION

For nearly two decades there has been a growing interest in studying static and especially cyclic fatigue in ceramic materials [1,2]. True cyclic fatigue effects have been observed in Mg-PSZ [3,4], alumina [5,6], Y-TZP(A) [7-9] and Ce-

---

T Liu, Y-W Mai and M V Swain, Department of Mechanical Engineering, University of Sydney, New South Wales 2006, Australia
G Grathwohl, Institut für Keramik im Maschinenbau, University Karlsruhe, Haid-und-Neu-Str 7, 7500 Karlsruhe 1, Germany

TZP(A) [10,11] ceramics with processing flaws, indentation cracks and long sharp cracks. Efforts have been made to understand the micromechanisms for this cyclic fatigue effect and the interactions between mechanical fatigue and the crack shielding effects caused by stress induced t-m transformation, grain and fibre bridging. It is realised that the same toughening mechanisms that impart a crack-resistance curve in a ceramic are also responsible for its weakness to resist cumulative damage due to cyclic fatigue. For coarse-grained aluminas the failure mechanisms have now been indentified as mainly due to the frictional degradation of the grain-bridges at the crack wake due to repeated abrasions [12]. In transformation toughened zirconia based ceramics attempts have been made to establish the connections between the R-curve and cyclic fatigue [13], though the failure mechanisms are not yet fully worked out.

In the present paper both the stress induced t-m transformation behaviour under cyclic loading and the cyclic fatigue crack propagation behaviour in 9 mole % Ce-TZP ceramics with different grain sizes are reported. An attempt is made to understand the interrelations between crack shielding and cyclic fatigue crack growth.

EXPERIMENTAL PROCEDURE

Cyclic fatigue crack propagation behaviour was studied in four 9 mole % Ce-TZP ceramics with average grain sizes of 1.1, 1.6, 2.2 and 3.0 μm. The transformation behaviour and the crack resistance curves of these same ceramics have been investigated previously [14]. Short double-cantilever-beams (s-DCB) were used to measure the transformation and the crack growth under cyclic loading. The dimensions of the s-DCB were $6.5 \times 32 \times 40$ mm and a sharp precrack was initiated from a Chevron-like prenotch. The specimens were then annealed in air at 800°C to annihilate the t-m transformation which had taken place during precracking. Initial crack length to sample width ratio, a/w, was chosen at about 0.3. Crack growth was measured on the polished surface with an optical travelling microscope with an accuracy of 0.01 mm.

Cyclic fatigue tests were conducted on an Instron machine model 8501 under load control with a frequency of 30 Hz and a stress ratio, K_{min}/K_{max}, of 0.2. To assess the interaction between crack-resistance curve behaviour and cyclic fatigue crack growth, cyclic fatigue tests were conducted in two conditions: (a) with a sharp precrack without pre-existing transformation, and (b) with a sharp crack after R-curve measurements, i.e. with a preformed t-m transformation zone. The applied stress intensity factor $K_{I,app}$ was calculated from:

$$K_{I,app} = \frac{P}{tw^{1/2}} \cdot \frac{2+a/w}{1-(a/w)^{3/2}} \cdot f(a/w) \tag{1}$$

where P is applied load, t specimen thickness, w specimen width, a crack length and f(a/w) is the geometry function taken from Srawley and Gross [15].

EXPERIMENTAL RESULTS

Figure 1 shows the crack growth behaviour of four 9 mole % Ce-TZP ceramics with sharp precracks without pre-existing transformation zones under both monotonic (R-curve measurement) and cyclic loading. For the Ce-TCP-I ceramic with an average grain size of 1.1 μm crack growth took place in a steady

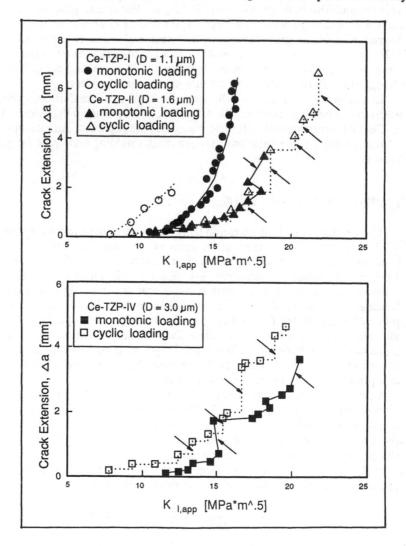

Figure 1. Crack propagation behaviour of Ce-TZP-I, II, and IV ceramics under monotonic and cyclic loading. Arrows indicate pronounced autocatalytic transformation.

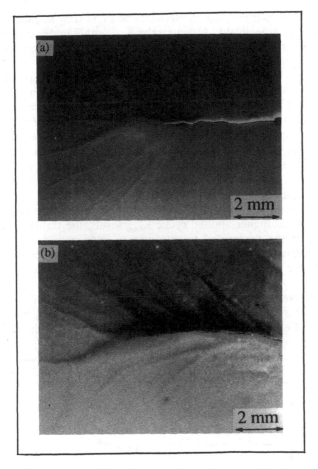

Figure 2. Stress induced t-m transformation zone of the Ce-
TZP-II (grain size = 1.6 μm) after (a) quasi-stable crack
extension and (b) cyclic fatigue test.

and continuous fashion under both monotonic and cyclic loading. In contrast the
Ce-TZP's with larger grain sizes showed a stepwise crack growth behaviour
caused by "autocatalytic" transformation [14]. At higher applied $K_{I,app}$ levels the
increased "autocatalytic" transformation led to unstable crack extension, the
amount of which increased with increasing grain size. Crack growth under cyclic
loading occurred at lower $K_{I,app}$ levels than under monotonic loading. Figure 2
shows the stress induced t-m transformation zones of the Ce-TZP-II ceramic with
a grain size of 1.6 μm after both quasi-stable crack extension
(figure 2-a) in the R-curve measurement [14] and cyclic fatigue
tests (figure 2-b). The maximum applied stress intensity factors

$K_{I,max}$ corresponding to these pictures were 20 and 21.8 MPa$\sqrt{m}$ under monotonic and cyclic loading respectively. Although the $K_{I,max}$ value under cyclic loading is slightly higher than under monotonic loading, the transformation zone (width) is smaller under cyclic loading than under monotonic loading.

Figure 3 shows the crack growth behaviour as a function of applied $K_{I,app}$ for Ce-TZP-II specimen under both monotonic and cyclic loading. Here the specimen was first loaded monotonically in the R-curve measurements to a crack extension of about 4 mm. The same specimen with the already preformed t-m transformation zone was then subjected to cyclic loading. Although the applied cyclic $K_{I,app}$ values were lower than the maximum $K_{I,app}$ value in the monotonic loading test, cyclic fatigue crack growth can still be observed. At low applied cyclic $K_{I,app}$ values continuous cyclic crack propagation within the preformed transformation zone was observed, and there were no marked changes of the transformation zone both in size and geometry. However, with further fatigue crack growth and at higher applied cyclic $K_{I,app}$ levels, the transformation zone size grew out of the preformed transformation zone with a stepwise (auto-catalytic) transformation behaviour and partially unstable crack extension. Similar behaviour was also observed in the Ce-TZP-III ceramic with the grain size of 2.2 μm.

Figure 4 shows the fatigue crack propagation da/dN versus $\Delta K_{I,app}$ curves of the four Ce-TZP's measured in specimens without pre-existing transformation zones. For the Ce-TZP-I ceramic with a grain size of 1.1 μm the fatigue crack

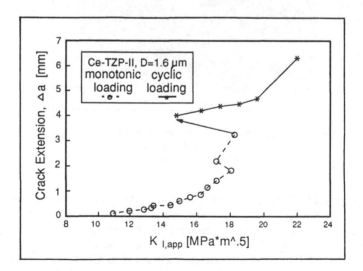

Figure 3. Crack propagation behaviour under monotonic and cyclic loading. The specimen was first loaded monotonically in the R-curve measurement with a crack extension of about 4 mm, after which the same specimen with the already pre-formed transform-ation zone was subjected to cyclic loading.

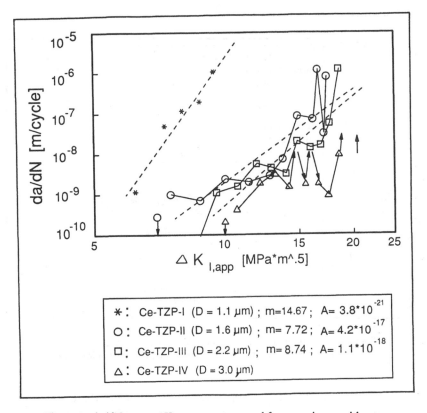

Figure 4. da/dN versus $\Delta K_{I,app}$ curves measured from specimens without a pre-existing transformation zone.

growth rate (FCGR), da/dN, increased with increasing applied stress intensity factor range $\Delta K_{I,app}$. Cyclic fatigue crack propagation occurred at a maximum applied $K_{I,max}$ value as low as 40% of the maximum $K_{I,app}$ value in the R-curve [14]. The fatigue crack growth can be described by the Paris power law equation: da/dN = $A \times \Delta K_{I,app}^{m}$ with m = 15 and A = 3.8×10^{-21}. For the Ce-TZP's with larger grain sizes and hence larger transformation zones "zigzag" type da/dN versus $\Delta K_{I,app}$ curves were observed, where unstable crack extension associated with pronounced "burst-like" autocatalytic transformation occurred at higher applied $\Delta K_{I,app}$ values. If the power law equation is used to fit these experimental data m is about 8 to 9 and A is given in the inset of the figure. The extent of the unstable crack extension and the autocatalytic transformation were both dramatically increased for the Ce-TZP-IV with the grain size of 3 μm leading to highly unstable fatigue crack growth behaviour.

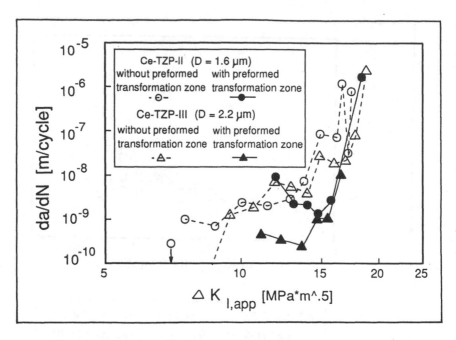

Figure 5. Comparison of da/dN versus $\Delta K_{I,app}$ curves measured
from specimens with and without preformed transformation zone.

Figure 5 shows the fatigue crack growth behaviour of the Ce-TZP-II and III ceramics with grain sizes of 1.6 and 2.2 μm measured in specimens with a pre-existing transformation zone due to early R-curve measurements (solid lines). A V-shape da/dN versus $\Delta K_{I,app}$ curve is observed here. At the lower applied $\Delta K_{I,app}$ levels (left side of the minimum FCGR) da/dN decreased with increasing applied $\Delta K_{I,app}$. However, da/dN increased with further increase in the applied $\Delta K_{I,app}$ beyond the minimum FCGR da/dN. Figure 5 also compares the FCGR of the Ce-TZP-II and III ceramics measured with (solid lines) and without (dashed lines) the pre-existing transformation zone.   Cyclic fatigue crack propagation occurs at higher applied $\Delta K_{I,app}$ levels for specimens with the pre-existing transformation zone.   At the same applied $\Delta K_{I,app}$ levels da/dN in specimens without the pre-existing transformation zone is much higher than in the specimens with the transformation zone. However, it is noted that at much higher $\Delta K_{I,app}$ values the da/dN data for both materials with and without the pre-existing transformation zone seem to merge together. This indicates that at large crack growth the crack tip shielding effect of specimens without the preformed t-m transformation zone becomes similar to that of specimens with the preformed t-m transformation zone.

DISCUSSION

For specimens without a preformed transformation zone the crack tip stress intensity factor ΔK_{tip} can be described by the following equation:

$$\Delta K_{tip} = \Delta K_{app} - \Delta K_{s} + \Delta K_{d} \qquad (2)$$

where ΔK_{s} is the shielding effect due to the stress induced t-m transformation developed under cyclic loading and ΔK_{d} is the cyclic fatigue induced damage.

Figure 6-a shows schematically the dependence of the shielding ΔK_{s} and the cyclic fatigue damage ΔK_{d} on the applied ΔK_{app} under cyclic loading. Since the transformation zone under cyclic loading is smaller than under monotonic loading as obtained in R-curve measurement, the shielding ΔK_{s} under cyclic loading is smaller than under monotonic loading. For the Ce-TZP-II, III and IV ceramics the autocatalytic transformation behaviour results in a stepwise increasing R-curve behaviour under both cyclic and monotonic loading.

Although the micromechanisms for the cyclic fatigue damage have not been fully established and understood yet, it is expected that ΔK_{d} will increase with applied ΔK_{app}. From figure 6-a and equation 2 we can obtain the ΔK_{tip} versus ΔK_{app} and hence da/dN versus ΔK_{app} relations as shown schematically in figure 6-b.

For specimens with preformed t-m transformation zone after R-curve measurement, the crack tip stress intensity ΔK_{tip} is also described by equation 2. However, the shielding ΔK_{s} and its dependence on crack extension Δa and applied ΔK_{app} are quite different from specimens without a preformed transformation zone. As shown schematically in figure 7-a the shielding ΔK_{s} is initially very high because of the large preformed transformation zone which

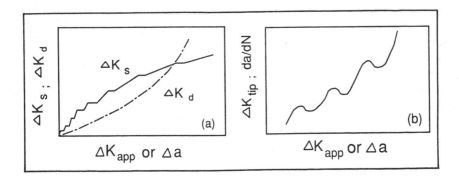

Figure 6. (a) Dependence of shielding ΔK_{s} and cyclic fatigue damage ΔK_{d} on applied ΔK_{app} and crack propagation (Δa) under cyclic loading for specimens without a pre-existing transformation zone. (b) Schematics of the ΔK_{tip} versus ΔK_{app} or Δa and the da/dN versus ΔK_{app} relations for Ce-TZP-II, III and IV ceramics without preformed transformation zone.

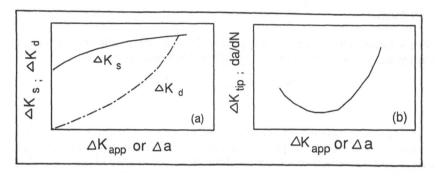

Figure 7. (a) Dependence of shielding ΔK_s and cyclic fatigue damage ΔK_d on applied ΔK_{app} and crack propagation (Δa) under cyclic loading for specimens with a pre-existing transformation zone. (b) Schematics of the ΔK_{tip} versus ΔK_{app} or Δa and the da/dN versus ΔK_{app} relations for Ce-TZP-II, III and IV ceramics with the preformed transformation zone.

increases also with cyclic fatigue crack growth due to the shape of the transformation zone of these ceramics [16]. From figure 7-a and equation 2 we can obtain the ΔK_{tip} versus ΔK_{app} and hence da/dN versus ΔK_{app} relations as shown schematically in figure 7-b. Quantitative prediction of these experimental fatigue data is difficult as it is mathematically involved to calculate ΔK_s, and ΔK_d is virtually unknown for these Ce-TZP materials.

For the Ce-TZP-I ceramic with a grain size of 1.1 μm and a limited transformation zone size as described earlier [19], the crack propagation relation can be described very well by the Paris power law equation da/dN = $A\Delta K_{I,app}^{m}$ with m = 15. For the Ce-TZP-II, III and IV ceramics with larger grain size, however, a "zigzag" type of da/dN versus $\Delta K_{I,app}$ curve has been observed, which is caused by the interaction between the "autocatalytic" shielding effect and the cyclic fatigue damage. Here the specimen size is the crucial factor in deciding the equivalent m value to describe the crack propagation relation. For small specimens the individual autocatalytic transformation associated with the unstable crack extension may result in complete failure of the specimen and m is difficult to obtain. For very large specimens, however, m values obtained from the regression of all the fatigue data can be safely used for approximate life-time prediction even though the crack growth behaviour is really discontinuous or periodic.

CONCLUSION

Cyclic fatigue crack propagation has been investigated in this work showing a strong interaction between the cyclic fatigue damage and the crack shielding effect. The size (width) of the stress induced t-m transformation zone under cyclic loading was smaller than that measured under monotonic loading in R-curve tests, although the applied stress intensity factor $K_{I,app}$ was nearly the same. For the Ce-TZP-I ceramic with a grain size of 1.1 μm the crack growth rate can be described by a power law function: $da/dN = A\Delta K_{I,app}^{m}$ with $m = 15$. For the Ce-TZP-II, III and IV ceramics with larger grain size the stress induced t-m transformation under cyclic loading was characterized by "autocatalytic" behaviour. This "autocatalytic" transformation behaviour leads to a stepwise and/or partially unstable crack propagation under cyclic loading. Cyclic fatigue crack propagation relations for these materials were then characterized by a "zigzag" da/dN versus $\Delta K_{I,app}$ curve. For the Ce-TZP-II and III specimens with t-m transformation zones already preformed before the cyclic fatigue tests a lower fatigue crack growth rate compared to specimens without the transformation zone was measured because of the larger crack shielding effect. A "V"-type da/dN versus $\Delta K_{I,app}$ curve was observed in this situation.

REFERENCES

1.   Evans, A.G. and E.F.Fuller, 1974. "Crack Propagation in Ceramic Materials Under Cyclic Loading Conditions." Metal. Trans. A, 5:27-33.

2.   Evans, A.G., 1980. "Fatigue in Ceramics." Int. Journ. of Fracture, 16:485-498.

3.   Dauskardt, R.H., D.B. Marshall, and R.O. Ritchie, 1990. "Cyclic Fatigue-Crack Propagation in Magnesia-Partially-Stabilized Zirconia Ceramics." J.Am.Ceram.Soc., 73(4):893-903.

4.   Hoffman, M.J., W. Lentz, M. Swain and Y-W. Mai, 1992. "Cyclic Fatigue Lifetime Predictions of Partially Stabilized Zirconia With Crack-Resistance Curve Characteristics." J.European Ceram.Soc., in press.

5.   Lathabai, S., Y-W. Mai and B.R. Lawn, 1989. "Cyclic Fatigue Behaviour of an Alumina Ceramic with Crack-Resistance Characteristics." J.Am.Ceram.Soc., 72(9):1760-63.

6.   Fett, T., G. Martin, D. Munz and G. Thun, 1991. "Determination of da/dN-ΔK_I Curves for Small Cracks in Alumina in Alternating Bending Tests." J.Mater.Sci., 26:253-257.

7.   Grathwohl, G. and T. Liu, 1991. "Crack Resistance and Fatigue of Transforming Ceramics: I, Materials in the ZrO_2-Y_2O_3-Al_2O_3 System." J.Am.Ceram.Soc., 74(2):318-325.

8.   Liu, S-Y and I-W. Chen, 1991. "Fatigue of Yttria-Stablized Zirconia: I, Fatigue Damage, Fracture Origins and Lifetime Prediction." J.Am.Ceram.Soc., 74(6):1197-1205.

9.   Liu, S-Y and I-W. Chen, 1991. "Fatigue of Yttria-Stablized Zirconia: II, Crack Propagation, Fatigue Striations and Short-Crack Behaviour." J.Am.Ceram.Soc., 74(6):1206-1216.

10.   Tsai, J-F, C-S. Yu and D.K. Shetty, 1990. "Fatigue Crack Propagation in Ceria-Partially-Stabilized Zirconia (Ce-TZP)-Alumina Composites." J.Am.Ceram.Soc., 73(10):2992-3001.

11.   Grathwohl, G. and T. Liu, 1991. "Crack Resistance and Fatigue of Transforming Ceramics: II, CeO_2-stabilized Tetragonal ZrO_2." J.Am.Ceram.Soc., 74(12):3028-3034.

12.   Lathabai, S., J. Rodel and B.R. Lawn, 1991. "Cyclic Fatigue from Frictional Degradation at Bridging Grains in Alumina." J.Am.Ceram.Soc., 74(6):1340-1348.

13.   Mai, Y-W, X.Z. Hu, K. Duan and B. Cotterell, 1992. "Crack-Resistance Curve and Fatigue in Ceramic Materials," in Fracture Mechanics of Ceramics, R.C.Bradt, D.P.H.Hasselmann and F.F.Lange, eds., NY, NY, USA:Plenum Press, in press.

14.   Liu, T., Y-W. Mai, M.V. Swain and G. Grathwohl, 1993. "Transformation and R- Curve Behaviours of 9Ce-TZP Ceramics: Grain Size and Specimen Geometry Effect," to be published.

15.   Srawley, J.E. and G. Gross, 1974. "Compendium," J.Engng.Fracture Mech., 4:587.

16.   Liu, T., G.Y. Lu and Y-W Mai, 1992. "Calculation of the Crack Shielding Effect in 9Ce-TZP Ceramics," unpublished work.

J_R-Δa Curves of 9 mol % Ce-TZP Ceramics with Different Grain Size and Specimen Geometry

TIANSHUN LIU, YIU-WING MAI, MICHAEL V. SWAIN
and GEORGE GRATHWOHL

ABSTRACT

Transformation and R-curve behaviours have been investigated in 9 mol % Ce-TZP ceramics with different grain sizes. Both single edge notched beam (SENB) and short double-cantilever-beam (s-DCB) specimens were tested to measure the crack-resistance curves. The size and shape of the transformation zone do not only depend on grain size, but are also strongly influenced by the specimen geometry. This different transformation behaviour has led to different crack-resistance curves. These experimental results are discussed in terms of the thermodynamics of transformation, the autocatalytic effect and fracture mechanics.

INTRODUCTION

One of the most important features of transformation toughened zirconia-based ceramics is the increasing crack-resistance (R-curve) behaviour caused by the crack shielding effect due to the stress induced t-m transformation around the crack tip. This toughening behaviour has been successfully investigated both theoretically and experimentally for the Mg-PSZ, Y-TZP and ZTA ceramics. Some recent studies [1-3] have demonstrated that Ce-TZP ceramics exhibit pronounced transformation plasticity leading to high fracture toughness in the range 12 to 18 MPa$\sqrt{\text{m}}$. However, the shapes of the transformation zones surrounding cracks in the Ce-TZPs are very different from those in other zirconia-based ceramics. In Mg-PSZ ceramics the zone extends approximately equal distance ahead and to the side of the crack, whereas in

Tianshun Liu, Yiu-Wing Mai and Michael V Swain, Centre for Advanced Materials Technology, Department of Mechanical Engineering, The University of Sydney, NSW 2006, Australia
George Grathwohl, Institut fuer Keramik im Maschinenbau, University Karlsruhe, Haid-und-Neu Str. 7, W-7500 Karlsruhe 1, Germany

Ce-TZPs the zone is very elongated. The grain size dependence of the transformation behaviour in Ce-TZPs has also been investigated showing a marked increase of the t-m transformation zone size with increasing grain size [3].

In the present paper the transformation and the crack-resistance (R-curve) behaviours of a 9 mol % Ce-TZP ceramic and the influence of grain size and specimen geometry are investigated. Very different transformation zone size, zone shape and, consequently, different R-curve behaviour have been found for different grain size and specimen geometries.

EXPERIMENTAL WORK

The materials for testing were prepared from a single batch of CeO_2-ZrO_2 powder (Ce-14, Unitec Ceramics Ltd., Stafford, England) containing 9 mol % CeO_2. The powder was first uniaxially pressed at 6 MPa in a hard metal die to produce rectangular plates which were isostatically pressed at 200 MPa. The plates were then sintered at temperatures between 1450° and 1550°C with heating and cooling rates of 3°C/min.

Single edge notched bend (SENB) and short-double-cantilever-beam (s-DCB) specimens were cut and ground from the sintered plates. The sizes of the SENB and the short-DCB specimens were $5 \times 10 \times 45$ ($t \times w \times L$) and $6.5 \times 32 \times 40$ ($t \times h \times w$) mm^3, respectively, which are shown schematically in figure 1(a). One side of both the SENB and short-DCB specimens was polished. For the three-point bend test on the SENB specimens a natural sharp crack was generated by the so-called "bridge method". In the short-DCB test the samples were precracked by initiating a sharp crack from a Chevron-like prenotch. These specimens were then annealed in air at 800°-1200°C to annihilate the t-m transformation which had taken place during precracking. In both specimen types the initial crack length to sample width ratio, a_o/w, was chosen at about 0.3 and the crosshead rate of the Instron testing machine was kept at about 1 μm/s. The load-point displacement was measured by a clip gauge and a LVDT in the s-DCB and SENB geometries respectively. Crack growth was measured from the polished surface optically with a travelling microscope with an accuracy of ± 0.01 mm. The crack-resistance curve was calculated using both stress intensity factor (K) and fracture energy (J-integral) approaches.

For the SENB geometry K_I is given by:

$$K_I = \frac{3PL}{2tw^2} \, a^{1/2} f(a/w) \tag{1}$$

where P is applied load, L (40mm) span width, t thickness, w specimen width, a crack length, and f(a/w) is a function of a/w and can be obtained from stress

Materials	Sintering conditions	Average grain size [µm]	Rel. density [% TD]
Ce-TZP-I	$1450\,^{\circ}$C, 1h	1.62	97
Ce-TZP-II	$1500\,^{\circ}$C, 1h	2.22	98
Ce-TZP-III	$1550\,^{\circ}$C, 1h	3.00	97

TABLE II - SPECIMEN THICKNESS t, CRACK LENGTH a AND
LIGAMENT LENGTH (w-a) OF THE SENB AND s-DCB SPECIMENS.

specimen	t [mm]	a [mm]	w-a [mm]
SENB	5	3-7	7-3
s-DCB	6.5	15-20	25-20

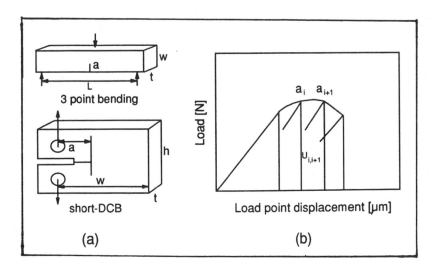

Figure 1. (a) Schematics of the SENB and s-DCB specimens; and (b) typical
load-load point displacement curves.

453

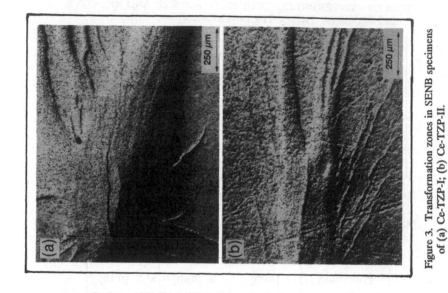

Figure 3. Transformation zones in SENB specimens of (a) Ce-TZP-I; (b) Ce-TZP-II.

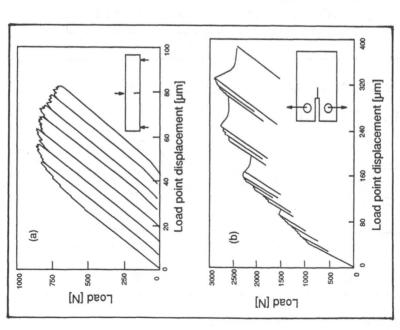

Figure 2. Load versus load-point displacement curves of Ce-TZP-I. (a) in three-point bending; (b) in short-DCB test.

intensity factor handbooks.

The J-integral is calculated using the following equation [5]:

$$J_{i+1} = J_i \frac{w - a_{i+1}}{w - a_i} + \frac{2U_{i,i+1}}{t(w - a_i)} \tag{2}$$

where $U_{i,i+1}$ for the crack growth interval $(a_{i+1} - a_i)$ is shown in figure 1(b).

For short-DCB geometry K_I is given by

$$K_I = \frac{P}{tw^{1/2}} \cdot \frac{(2 + a/w)}{1 - (a/w)^{3/2}} \cdot f(a/w) \tag{3}$$

The J-integral is obtained from [6]:

$$J_{i+1} = \left[J_i + \frac{\eta_i}{w - a_i} \frac{U_{i,i+1}}{t} \right] \left[1 - \frac{\gamma_i}{w - a_i} (a_{i+1} - a_i) \right] \tag{4}$$

where $U_{i,i+1}$ is already shown in figure 1(b), η_i and γ_i are given in [7].

Using the K_I values calculated from equations 1 and 3 for the SENB and s-DCB geometries, the elastic part of the J-integral, J_{el}, was also calculated by:

$$J_{el} = \frac{K_I^2(1 - v^2)}{E} \tag{5}$$

where v is Poisson's ratio and E is Young's modulus. At equilibrium crack growth $K_I = K_R$ and $J = J_R$ [8].

EXPERIMENTAL RESULTS

TRANSFORMATION BEHAVIOUR

Table I shows the sintering conditions, the grain sizes obtained and the relative density of the Ce-TZP materials. With increasing sintering temperature the grain sizes were increased. Figure 2 shows the typical load versus load-point displacement curves for the 9 Ce-TZP ceramics in the SENB and short-DCB geometries. It can be seen clearly, that the 9 Ce-TZP ceramics investigated here undergo stepwise plastic deformation and localised intensity. Similar behaviour was also observed by Grathwohl and Liu [3] which was believed to be caused by the autocatalytic effect of the stress induced t-m transformation. Figures 3 and 4 show the t-m transformation zones of both the SENB and short-DCB geometries, respectively, with different grain sizes. (Note that these transformation zones correspond to the plateau values of the J_R - Δa curves in

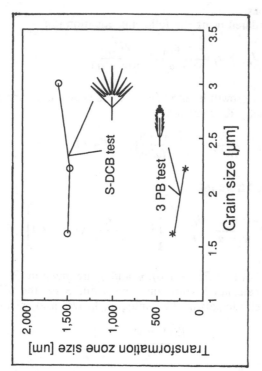

Figure 5. Grain size and specimen geometry dependence of the size and shape of the transformation zone.

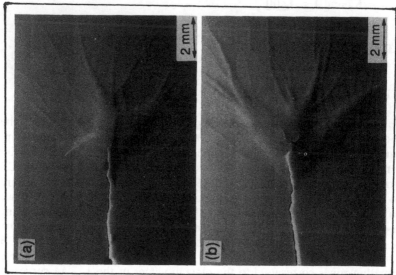

Figure 4. Transformation zones in s-DCB specimens of (a) Ce-TZP-I; (b) Ce-TZP-III.

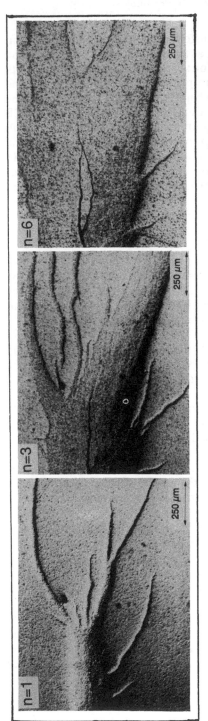

Figure 6. Development of the transformation zone during repeated loading/unloading of a Ce-TZP-I SENB specimen in three-point bending. n indicates the number of cycles.

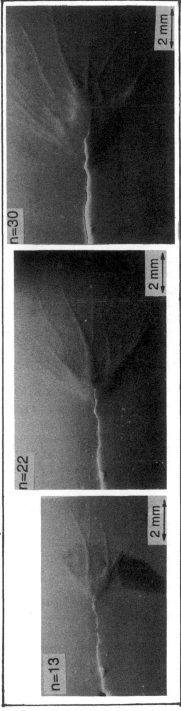

Figure 7. Development of the transformation zone during repeated loading/unloading of a Ce-TZP-II s-DCB specimen. n indicates the number of cycles.

457

figure 9.) These results show that the transformation behaviour is dependent both on grain size and specimen geometry as shown in figure 5. Figures 6 and 7 show the development of the transformation zones with crack extension in both the SENB and short-DCB geometries, respectively, for the Ce-TZP-I and Ce-TZP-II materials. The elongated zone shape in the SENB geometry and the large zone volume with long and slender transformation "tails" in the s-DCB geometry are clearly visible. Crack deflection and crack branching as a result of microcracking in the t-m transformation zone were also observed with optical microscopy in the Ce-TZP-III material with a grain size of 3.0 μm (figure 8). However, with the other Ce-TZP materials with smaller grain size such crack branching and crack deflection due to microcracking in the transformation zone cannot be observed optically or by SEM.

CRACK-RESISTANCE CURVES

Figure 9 shows the crack-resistance (J_R) curves for the 9 Ce-TZPs. Since the transformation zone size is large relative to the specimen dimensions the J-integral approach has been used to characterize the crack-resistance behaviour. Three important characteristics can be summarized in these J_R - Δa curves:

(1)   In the SENB specimens the J-integral values are overall much higher than the J_{el} values derived from the K_I values. This indicates that the specimen size is too small for the K concept to apply.

(2)   In the s-DCB specimens the J-integral values agree well with the J_{el} values for the crack extension of less than 1.5 mm, where the transformation zone size is still small compared to the specimen size. With increasing crack extension and larger transformation zones, however, the J-integral values become much higher than the J_{el} values. Thus the K concept is only valid for the s-DCB specimens, if the crack extension as well as the transformation zone size are small.

(3)   The J_R-Δa curves obtained from the SENB geometries show very different trends. The J_R-Δa curves for the SENB geometry show relatively higher starting values than the s-DCB geometries and tend to a plateau after a crack growth of 1-2 mm. However, the J_R-curves for the s-DCB specimen rise rapidly with crack growth with no apparent plateau value.

Figure 10 shows the increase of the transformation zone size in the lengthwise direction for both the SENB and s-DCB specimens as a function of crack extension. Quite clearly, for the same amount of crack extension, the transformation zone size is much larger and the zone size increases much faster for the s-DCB than the SENB specimens.

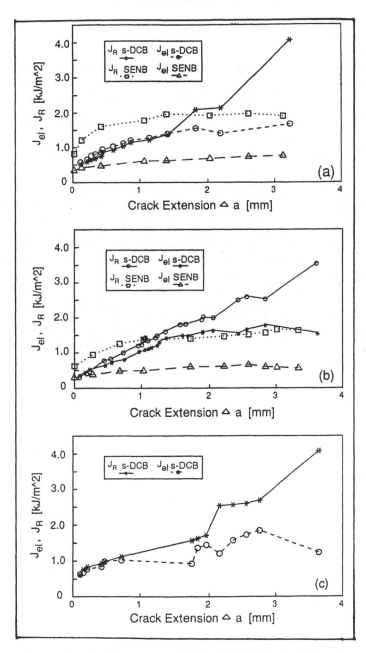

Figure 9. Crack-resistance curves (J_R) for (a) Ce-TZP-II, (b) Ce-TZP-II and (c) Ce-TZP-III measured in SENB and s-DCB specimens. Note that "autocatalytic transformation" and stepwise crack growth are demonstrated by the discontinuous segments of the J_R-Δa curves for the s-DCB specimens.

DISCUSSION

Both the transformation zone size (see figure 5) and crack-resistance curves (see figure 9) for the 9 mol % Ce-TZPs investigated here show a strong dependence on specimen geometry. Two factors have to be considered to explain the difference in the transformation zone size with different specimen geometry and size. Firstly, as shown in Table II, the state of stress in these two specimen types is quite different. While the SENB specimens do not satisfy the full plane strain conditions, they are not far from it. The s-DCB specimens, on the other hand, are hardly near plane strain and are more close to plane stress. These differences lead to different thickness constraint effects [9]. By analogy with crack tip plastic deformation fields the relatively lower thickness constraint in the s-DCB geometry produces a larger transformation zone size. Secondly, these two different specimen geometries and sizes and hence the different boundary conditions will result in different "in-plane constraint" effects [10]. The higher "in-plane constraint" effect in the s-DCB specimen allows the development of a larger transformation zone size before any constraint relaxation due to the free-surface boundary effects occurs. However, further theoretical analysis is needed to understand the correlations between the transformation zone size, the stress state, the constraint effect and the elastic-plastic stress field. In particular the formation and size of the "tails" due to the "autocatalytic" effect in the SENB and s-DCB geometries have to be fully explained. The through-thickness transformation zones also have to be investigated further.

For the evaluation of valid plane strain K_I values, the crack length a, thickness t and the ligament length w-a must satisfy the following requirements:

$$t, a, w - a > 50h \qquad (6)$$

where h is the lengthwise transformation zone size. By comparing Table II with figures 5 and 10 it can be seen that for the 9 mol % Ce-TZP ceramics, except for the first couple of mm of crack extension, the specimen size and thickness are too small for valid K_I measurements in the s-DCB geometry. The SENB geometry is not suitable to measure the K_R-Δa curve in these materials. It is more appropriate to determine the J_R-Δa curves for the 9 mol % Ce-TZP ceramics using equations 2 and 4. These J_R-curves have been shown in figure 9 and only the valid data points are shown here.

The J_R-Δa curves measured from the SENB and s-DCB specimens show that for identical materials the s-DCB geometry has a greater slope dJ_R/da than that for the SENB geometry, caused by a much larger transformation zone size at a given crack growth (see also figure 10). Although the crack initiation J-values for the SENB geometry are larger than for the s-DCB geometry we believe this may be associated with the accuracy in identifying the precise crack initiation points on the load versus load-point displacement curves in the SENB test. Nevertheless, the different specimen geometry and size can indeed lead to different transformation and crack resistance behaviours.

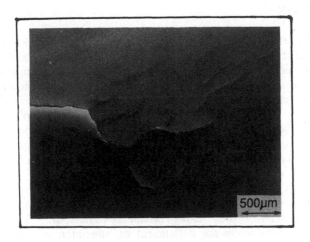

Figure 8. Microcracking in the t-m transformation zone of the Ce-TZP-III material with a grain size of 3 μm.

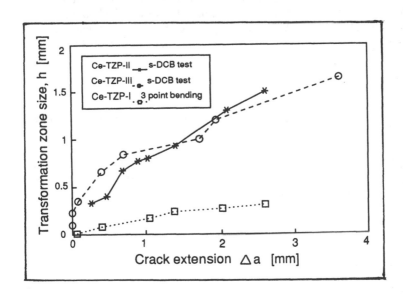

Figure 10. Lengthwise transformation zone size in SENB and s-DCB geometries as a function of crack extension.

CONCLUSION

A very large transformation zone associated with a "burst-like" (autocatalytic) transformation behaviour was observed in the 9 Ce-TZP ceramics investigated here which leads to a discontinuous stepwise "transformation plasticity" and load versus load-point displacement curve. There is also a pronounced specimen geometry effect on the t-m transformation zone in these materials both in shape and in size. For example the transformation zone size of the short-DCB geometry is about 3-5 times larger than that of the SENB geometry at identical crack extension. This is thought to be caused by the differences in the stress state, thickness constraint and the plastic stress field between the two specimen types.

For the 9 Ce-TZP ceramics with the very large transformation zones and hence with pronounced transformation plasticity the J_R-curve approach has to be used to characterize their crack-resistance behaviour. The J_R-Δa curves measured experimentally are dependent on specimen geometry because the transformation zones in the SENB and s-DCB specimens are dissimilar.

ACKNOWLEDGEMENTS

The authors wish to thank the Australian DITAC Industry Research Development Board, Manufacturing and Materials Technology Grant #15042 and the MWK Baden-Wuerttemberg under the KKS-funding, Keramikverbund Karlsruhe-Stuttgart, Germany for the financial support of this work.

REFERENCES

1. Rose, L. R. F. and M. V. Swain, 1988. "Transformation Zone Shape in Ceria-Partially-Stabilized Zirconia." Acta Metall., 36:955-962.

2. Yu, C.-S. and D.-K. Shetty, 1989. "Transformation Zone Shape, Size, and Crack-Growth-Resistance (R-Curve) Behaviour of Ceria-Partially-Stabilized Zirconia Polycrystals." J. Am. Ceram. Soc., 72: 921-928.

3. Warren, R. and B. Johannesson, 1984. "Creation of Stable Cracks in Hardmetals Using 'Bridge' Indentation." Powder Metallurgy, 27:25-29.

4. Garwood, S. J., J. N. Robinson and C. E. Turner, 1975. "The Measurement of Crack Resistance Curves (R-Curve) Using the J-Integral." Int. J. Fract., 11:528-530.

5. "Standard Test Method for J_{IC}, a Measure of Fracture Toughness," Designate E813 817, Assoc. Bd. of ASTM Standards, pp.686-700, 1988.

6. Inghels, E., A. H. Heuer and R. Steinbrech, 1990. "Fracture Mechanics of High-Toughness Magnesia-Partially-Stabilized Zirconia." <u>J. Am. Ceram. Soc.</u>, 73:2023-2031.

7. Mai, Y-W. and B. Lawn, 1986. "Crack Stability and Toughness Characteristics in Brittle Materials." <u>Ann. Rev. Mater. Sci.</u>, 16:415-439.

8. Atkins, A. G. and Y-W. Mai, 1985. <u>Elastic and Plastic Fracture: Metals, Polymers, Ceramics, Composites</u>, Chichester, UK:Ellis Horwood and New York, NY, USA:Halstend Press.

9. Wu, S-X., Y-W. Mai and B. Cotterell, 1992. "Ductile-Brittle Fracture Transition Due to Increasing In-Plane Constraint in a Medium Carbon Steel" presented at the 24th National Symposium on Fracture Mechanics, June 30-July 2, 1992, Gatlinburg, Tennessee, USA.

Scratch Experiments on Ceria Stabilised Tetragonal Zirconia

ANOOP K. MUKHOPADHYAY, TIANSHUN LIU,
MICHAEL V. SWAIN and YIU-WING MAI

ABSTRACT

Single pass scratch experiments were conducted with a Vickers indenter on ceria stabilised tetragonal zirconia polycrystals having grain sizes from 0.5 to 2.7 μm to understand the friction and wear characteristics. For a given grain size, the frictional force and volume of the wear groove and transformation zone size (h) increased with applied load ($\approx$ 8 N - 20 N). At a given load h increased with grain size. Plastic ploughing, and to a small extent localised microcracking, grain boundary microfracture and chipping were involved in the wear damage process.

INTRODUCTION

Because of their exceptionally high strength, hardness and fracture toughness zirconia (ZrO_2) based ceramics offer a potential possibility for wear resistant applications [1-5]. However, despite the exceptional toughness properties of Ce-TZP (ceria stabilised tetragonal zirconia), previous work is mainly focussed on the friction and wear behaviours of PSZ (partially stabilised zirconia) and Y-TZP (yttria stabilised tetragonal zirconia) ceramics [2,4,5]. Friction and wear of zirconia based ceramics are sensitive to microstructural parameters, e.g. porosity and grain size, contact stress, stress induced tetragonal (t) to monoclinic (m) phase transformation, atmosphere and temperature [2,4,5]. Depending on the extent of transformation, plastic deformation, delamination and grain fracture have been found on the wear surface [2,4-6]. These studies, however, fail to identify clearly the wear mechanisms because there is too much damage on the wear surface due to repeated abrasion. Thus, single

A K Mukhopadhyay, T Liu, M V Swain and Y-W Mai, Centre for Advanced Materials Technology, Department of Mechanical Engineering, University of Sydney, NSW 2006, Australia

pass scratch experiments are preferred to elucidate the mechanisms of wear initiation in abrasive wear.

The purpose of this paper was to identify the influence of grain size vis a vis phase transformation on the friction and wear characteristics in single pass scratch tests at a constant speed of 6 mm/s and under moderate normal loads of about 8 N to 20 N on 9 mol % Ce-TZP ceramics the grain size of which varied between 0.5 μm and 2.7 μm.

EXPERIMENTAL WORK

Details of the material synthesis and microstructural characterisation have been published elsewhere [7]. Briefly, the CeO_2-ZrO_2 powder containing 9 mol % CeO_2 and trace impurities (HfO_2, SiO_2) was obtained from commercial supplier (Ce-14, Unitec Ceramics Limited, Stafford, UK) and conventionally sintered. Density measurements by the water immersion technique gave values between 97.57%-99.83% of the theoretical value. Grain size measurements by the linear intercept technique gave average values of 0.5 μm, 1.0μm, 1.4 μm, 1.5 μm, and 2.7 μm for the Ce-TZP 1, Ce-TZP 2, Ce-TZP 3, Ce-TZP 4, and Ce-TZP 5 materials, respectively. The monoclinic phase contents in the different Ce-TZP materials were measured on ground as well as fractured surfaces using X-ray diffractometry.

Detailed descriptions of the experimental methods used for wear tests have been reported elsewhere [8]. In short, the scratch experiments were conducted in a reciprocating sliding machine. A sharp Vickers indenter of apex angle 136° and tip radius less than 0.5 μm was used on polished and annealed (500°C, 0.5 h in air) rectangular test specimens measuring $10 \times 6 \times 4$ mm^3. The friction force was measured using a LVDT and the scratching speed was 6 mm/s. The dead weight of the loading assembly produced an effective force of 8 N at the indenter and in the scratch experiments the normal load was varied between 8 N and 20 N using an adjustable lever arm. In the single pass tests the loaded indenter scratched the sample surface once with a leading plane. Care was taken to assure the vertical alignment of the indenter on the sample surface. After each single test the indenter was cleaned with ethanol to remove any adhered wear debris. The force ratio (f) was calculated as the ratio of the tangential (F) to normal (P) forces; and the wear volume determined from the measured values of groove width and depth (using both scanning electron microscopy and a surface profilometer assuming a triangular cross-section and the length of the groove (6 mm)). The wear damage process was studied with the aid of optical as well as scanning electron microscopy (SEM).

EXPERIMENTAL RESULTS

SCRATCH CHARACTERISTICS

Figures 1a to 1c show the effects of increasing the normal load on the scratch characteristics in the single pass tests. For a given grain size of Ce-TZP, the tangential force (F) increased with normal load (P), but the force ratio (f) seemed to be insensitive to variation in applied normal load as well as grain sizes, figure 1b. The effect of increasing the normal load on wear volume is shown in figure 1c. Generally, wear volume exhibits a power law dependence on normal load, an observation which has been previously reported for Y-TZP, PSZ, alumina and other ceramics [2,3,5,6,8,9].

Figure 2 shows the effect of increasing the normal load on the transformation zone size (h) as a function of grain size. For a given grain size, the transformation zone size surrounding the scratch increased with increasing applied normal load. For a given load, the transformation zone size showed an increasing trend with respect to grain size. For the purpose of comparison the monoclinic phase content [{(m)/(m + t + c)%}] on the ground surface as well as on the fracture surface of the five Ce-TZP materials are plotted in the same figure as a function of grain size. On the ground surface the monoclinic phase content increased slowly with grain size. The trend was similar for the fracture surface except that the relative magnitude of monoclinic phase was higher. As a typical example, figures 3a and 3b show the Nomarski interference optical micrographs of scratches produced under loads of 8 N and 18 N respectively, in the Ce-TZP 1 material of grain size 0.5 μm. Notice that the width of transformation zone beside the scratch is indeed very small. It appears that there is a small surface uplift in these regions. In contrast, the Ce-TZP 5 material with the coarsest grain size of 2.7 μm shows an appreciable transformation zone at both 8 N and 18 N normal loads, figures 3c and 3d, during the single pass scratch experiments. Here the surface uplift is more obvious. Such surface uplift has been observed also during static indentation studies using a Vickers indenter and has been related to the transformation plasticity.

DAMAGE PROCESSES

The microstructure of brittle materials, particularly those which exhibit a rising crack growth resistance characteristic with respect to crack extension can exert a significant influence on their wear behaviour [10]. The microstructures of the five Ce-TZP materials had been studied previously [9]. It was observed that the extent of intergranular fracture was more in the coarse grained materials than in the fine grained material, i.e. Ce-TZP 1. Typical examples of scratches made at relatively low loads of 8 N or 10 N are shown in figures 4a and 4b for the Ce-TZP 1 and 3

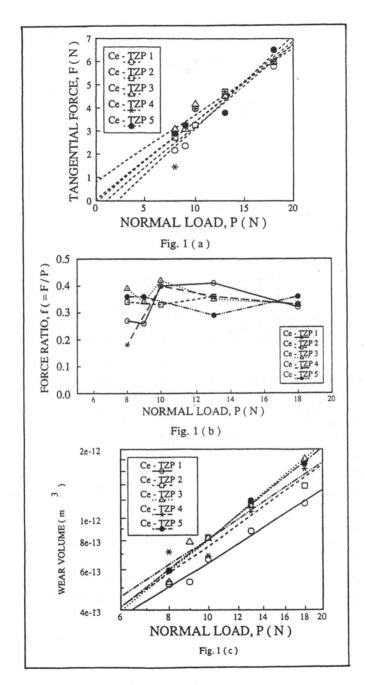

Figure 1. Variation of (a) tangential force (F), (b) force ratio (f), and (c) wear volume with normal load (P) for Ce-TZP.

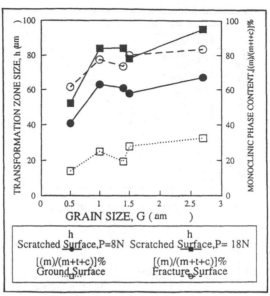

Figure 2. Relationship between the transformation zone size (h), grain size (G) and monoclinic phase content [(m)/(m + t + c)] % on the scratched, ground and fractured surfaces of Ce-TZP.

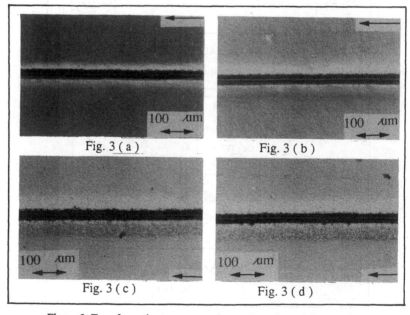

Figure 3. Transformation zone around scratch made on 0.5 μm grain size Ce-TZP for (a) P = 8 N and (b) P = 18 N, and on 2.7 μm grain size Ce-TZP for (c) P = 8 N and (d) P = 18 N.

ceramics, respectively. In these and the subsequent micrographs the arrows indicate the sliding direction of the indenter. Ce-TZP 1 and 3 had grain sizes of 0.5 μm and 1.4 μm, respectively. The main mode of deformation for these materials appears to be plastic ploughing. Also, there is no significant difference between the deformation modes of very fine grain sized material and the relatively coarse grain sized material. Increasing the load usually resulted in small amount of additional grain boundary microfracture near the edge of the plastically deformed groove, figure 5a. At the highest applied load of 18 N, typically the groove was characterised by a heavily deformed region near the middle of the groove and a slightly damaged region near the edge of the groove, figure 5b. In the case of 1.0 μm grain size Ce-TZP material, there appears to be some small amount of very localised chipping which has caused removal of thin layers of material in isolated areas. The presence of microcracks in the groove is also very noticeable. At high magnification, figure 5c, these microcracks seem to be interconnected in a few regions and the entire region where these cracks form appears heavily deformed.

DISCUSSION

In the single pass scratch experiments the tangential force of friction varies linearly with the normal load on the pyramidal indenter. Similar behaviour has been reported from low load, low speed scratch experiments on hot-pressed alumina [3]. The ratio of tangential to normal forces, f, however does not vary appreciably with load, figure 1b. The present force ratio values span a range from about 0.2 to about 0.4. These values are comparable to the friction coefficient data for PSZ [5] and Y-TZP [6], although the experimental conditions are very much different. According to the ploughing theory [11] of friction, the force ratio, f, is given by:

$$f = [\{Cot\,\Theta + \mu(1 + 2\,Cosec\,\Theta)\}/\{3 - \mu\,Cot\,\Theta\}]\tag{1}$$

where Θ is the semi apex angle of the indenter and μ is the Coulomb friction coefficient. As a typical example, if we consider the f values of Ce-TZP 2 from figure 2, with $2\Theta = 136°$, the μ values are estimated using equation 1 as 0.19, 0.17, 0.21 and 0.17 corresponding to normal loads of 8 N, 10 N, 13 N and 18 N, respectively. This implies a load independent average Coulomb friction coefficient of 0.19 $\pm$ 0.02. Thus, it appears that ploughing was the major component of deformation in the present materials.

The increase in wear volume, figure 1c, with increasing normal load is possibly a consequence of increased plastic deformation. Notice that there is no appreciable variation in wear volume with respect to changes in the grain size.

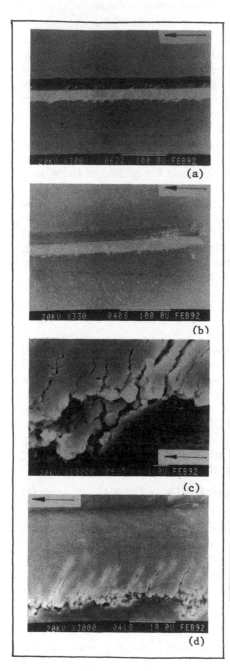

Figure 4. Scratch made at (a) P = 10 N
on 0.5 μm Ce-TZP; and at
(b) P = 8 N on 1.4 μm Ce-TZP.

Figure 5. (a) Microcracking and grain
boundary microfracture in the scratch
made at 13 N load on 0.5 μm Ce-TZP.
(b) Microcracking and chipping in
the scratch made at 18 N load on
1.0 μm Ce-TZP. (c) Details of
microcrack interaction shown in (b).

470

The increase in transformation zone widths with grain size and applied normal load, figures 2 and 3, should be viewed in terms of autocatalysis of the stress induced t → m transformation [7,12]. The current understanding of the t → m transformation supports that transformation is nucleation controlled. It has been appreciated that the autocatalysis may be initiated by a multiple nucleation event [12]. The probability of such nucleation would depend on availability of a nucleating defect. This probability, in turn, is a sensitive function of the thermodynamic potential that acts as a driving force in terms of either temperature or stress. Now, it may be assumed that under similar conditions of applied normal plus tangential loading, the number of such thermodynamically potent defects which could nucleate t → m transformation would be much smaller in a fine grained matrix, than in a relatively coarse grained one. In other words, the autocatalytic effect may be expected to be activated much more easily in a coarse grained matrix than in a fine grained one [12]. This means that the number of transformable grains may be relatively smaller in a fine grained matrix than in a coarse grained matrix of Ce-TZP. Consequently, the transformation zone would be rather narrow in the fine grained matrix as observed in the figures 3a and 3b for 0.5 μm grain sized Ce-TZP. Using similar logic the relatively larger transformation zone sizes in the Ce-TZP material of the largest grain size, e.g. 2.7 μm may be anticipated, figures 3c and 3d. Also, it is interesting to note that the width of the transformation zone changes much more rapidly in going from 0.5 μm to 1.0 μm grain size than in going from 1.0 μm to 2.7 μm grain size, at any given applied load (e.g. 8 N or 18 N) during the single pass scratch experiments, figure 2. Unfortunately, it is not possible to draw a definitive conclusion on the relationship between wear and transformation zone size from these experiments.

At a given grain size, the load dependence of the measured transformation zone size (figure 2) may be a consequence of the stress induced t → m transformation process itself. It has been noted [12] that once the autocatalysis process is initiated, multiple nucleation events may follow in a stimulated fashion to enable further transformation to propagate quickly over an extended region. At higher applied loads, better assistance being externally provided, the nucleation probability is effectively enhanced also. If this were the scenario, one would expect higher amount of t → m transformation to be involved in the relatively coarse grained materials compared to the fine grained material given the similar amount of enhancement in externally applied stress system. A more detailed examination of figure 2 indicates that increasing the applied load in the single pass scratch experiments from 8 N to 18 N enhances the transformation zone size by about 10 μm only in Ce-TZP 1 material of grain size 0.5 μm whereas in the relatively coarse grained materials the zone size enhancement is by about 20-30 μm.

Finally, the main deformation mechanism was ploughing by the pyramidal indenter through the material. The micrographs presented in figures 4 and 5 seem to support this contention. However, ploughing was not the sole component of deformation. Some very localised microcracking particularly near the edges of the groove always occurred. Large scale brittle fracture induced material removal was essentially absent in these materials. These observations are in sharp contrast to our observations in the case of sintered alumina of grain size larger than 1 μm, where brittle fracture controlled wear mechanism dominated in similar scratch tests as reported here [9].

CONCLUSIONS

The single pass scratch experiments conducted with a Vickers pyramid indenter moving at a constant speed of 6 mm/s under applied normal loads of 8 to 18 N on 9 mol % Ce-TZP materials with grain sizes of 0.5 μm, 1.0 μm, 1.4 μm, 1.5 μm and 2.7 μm enabled the following two major conclusions to be drawn:

1. For a given grain size the wear volume of the scratch groove and the width of transformation zone (h) surrounding the scratch increased with normal load. There was no significant influence of the grain size on volume of the wear groove. However, at a given load, the width of transformation zone increased with an increase in grain size. Particularly in the coarse grained materials the transformation zone was characterised by considerable surface uplift indicating significant influence of transformation plasticity in the process. These observations could be rationalised in terms of stress induced t → m transformation and autocatalytic effect of the phase transformation.

2. Predominantly plastic ploughing and very small amount of localised microcracking, grain boundary microfracture and chipping characterised the wear mechanism of the present Ce-TZP ceramics. These observations are suggested to be a consequence of stress induced phase transformations in the contact zone.

ACKNOWLEDGEMENTS

Y-W Mai wishes to thank the Australian Research Council (ARC) for the continuing support of this project and A K Mukhopadhyay acknowledges the award of an ARC Postdoctoral Fellowship tenable at the University of Sydney.

REFERENCES

1. Green, D.J., R.H.J. Hannink and M.V. Swain, 1989. Transformation Toughening of Ceramics, Florida, USA:CRC Press Inc.

2. Bundschuh and K-H Zum Gahr, 1991. "Influence of Porosity on Friction and Sliding Wear of Tetragonal Zirconia Polycrystal." Wear, 151(1):175-191.

3. van Groneou, A.B., N. Maan and J.D.B. Veldkamp, 1975. "Scratching Experiments in Various Ceramic Materials," Philips Research Report, 30:320-359.

4. Birkby, I., P. Harrison and R. Stevens, 1989. "The Effect of Surface Transformation on the Wear Behaviour of Zirconia TZP Ceramics." J. Eur. Ceram. Soc., 5:37-45.

5. Hannink, R.H.J., M.J. Murray and H.G. Scott, 1984. "Friction and Wear of Partially Stabilised Zirconia: Basic Science and Practical Applications." Wear, 100:355-366.

6. Stachowiak, G.W. and G.B. Stachowiak, 1989. "Unlubricated Friction and Wear Behaviour of Toughened Zirconia Ceramics." Wear, 132(1):151-171.

7. Grathwohl, G. and T. Liu, 1991. "Crack Resistance and Fatigue of Transforming Ceramics: II, CeO_2-Stabilised Tetragonal ZrO_2." J. Am. Ceram. Soc., 74(12):3028-3034.

8. Mukhopadhyay, A.K. and Y-W Mai, 1992. "Deformation Characteristics in Scratching of Fine Grained Alumina Ceramic." Submitted to J. Mater. Sci.

9. Mukhopadhyay, A.K. and Y-W Mai, 1993. "Grain Size Effect on Abrasive Wear Mechanisms in Alumina Ceramics," to be published in the Proceedings of the 9th International Conference on the Wear of Materials, April 13-17, San Francisco, CA, USA.

10. Cho, S.J., B.J. Hockey, B.R. Lawn and S.J. Bennison, 1989. "Grain Size and R-Curve Effects in Abrasive Wear of Alumina." J. Am. Ceram. Soc., 72(7):1249-1252.

11. Goddard, J. and H. Wilman, 1962. "A Theory of Friction and Wear During the Abrasion of Metals." Wear, 5:114-135.

12. Reyes-Morel, P.E. and I-W Chen, 1988. "Transformation Plasticity of CeO_2-Stabilised Tetragonal Zirconia Polycrystals: I, Stress Assistance and Autocatalysis." J. Am. Ceram. Soc., 71(5):343-353.

Ceria-TZP with Intergranular Alumina

B. MUKHERJEE and G. BANERJEE

ABSTRACT

Ce-TZP alumina composites with alumina as the minor phase have excellent mechanical properties. Alumina has been dispersed as an intergranular phase up to 50 volume percent. Their sintering, mechanical properties and microstructure have been studied. The formation of β-alumina from impurities present in the powder was noticed in the microstructure. This did not substantially affect the properties of the materials.

INTRODUCTION

Since the seventies, it has been known that partially stabilized zirconia (PSZ) shows exceptional mechanical properties namely fracture toughness and flexural strength. This is due to the presence of three allotropic modifications of zirconia - cubic, tetragonal and monoclinic - and to the martensitic transformation of tetragonal to monoclinic [1]. Initially it raised very high hopes that in the near future PSZ might replace a large number of metallic components. This expectation was tempered subsequently by the knowledge that the mechanical properties of partially stabilized zirconia deteriorate substantially at high temperature. Yttria-TZP which exhibits high strength and toughness degrades rapidly at around 200°C in the presence of water vapour [2,3]. Ceria-TZP shows certain advantages as far as degradation and fracture toughness are concerned. Studies on Ce-TZP [4] show that it can exhibit very high fracture toughness but relatively lower strength. PSZ has a coarse grain structure in which coherent precipitates of a second phase make up the transformable material, whereas TZP has a fine grain uniform structure made up of tetragonal phase only [5]. It has been reported that Y-TZP with around 20 wt% Al_2O_3 formed by HIP [6] shows very high strength. In this paper the microstructure and mechanical properties of Ce-TZP with intergranular alumina have been studied.

B. Mukherjee and G. Banerjee, Central Glass & Ceramic Research Institute, Calcutta, India

EXPERIMENTAL DETAILS

The raw materials, zirconia, alumina and ceria were attritor milled. The ratio of the charge to the grinding media was maintained at 1:7. Isopropyl alcohol was used as the medium. The surface area of the attritor milled powder was around $20m^2/g$. Powder was initially pressed into bars at a pressure of 90 MPa and subsequently isopressed at 270 MPa. The bars were sintered in air in a super kanthal furnace using the same heating and cooling rate for all samples. The bars were sintered at different temperatures from 1400°C to 1700°C and soaking was given for 1,2 and 4 hours. The bars were tested at room temperature for flexural strength. Bulk densities were measured by Archimedes Principle. Mechanical properties were tested at three different temperatures. The first was at 800°C, below the M_S temperature. The second was at 1000°C, between M_S and M_f. The third was at 1250°C, above M_f. M_s is the temperature at which the monoclinic to tetragonal transformation takes place under equilibrium conditions. M_f is the temperature at which the transformation from monoclinic to tetragonal is complete. Phase analysis was done by X-ray diffraction using the Garvie and Nicholson [7] technique. Room temperature flexural strength was measured in a three point bending machine. High temperature modulus of rupture was measured in a hot MOR apparatus. The loading rate for high temperature modulus of rupture was one Newton per second, the span being 43 mm. Fracture toughness was measured in 3 point bending after notching the specimen by a diamond saw. All mechanical property data reported here represent the average of measurements on 5 samples. Microstructures of the samples were examined with the help of a scanning electron microscope.

RESULTS AND DISCUSSION

The batch compositions of the materials are shown in Table I. Initially only Ce-TZP samples with 8.6, 12 and 14 mole% CeO_2 were prepared. Out of these 12 mol% CeO_2 was selected because of its flexibility in terms of processing

TABLE I - BATCH COMPOSITIONS

No.	ZrO_2 (mol%)	CeO_2 (mol%)	Al_2O_3 (mol%)	Sample No.
1	91.4	8.6	0	-
2	88	12	0	-
3	86	14	0	-
4	77.15	10.52	12.33	A1
5	66.83	9.12	24.05	A2
6	57.04	7.78	35.18	A3

TABLE II - DENSITY OF A1, A2 AND A3 SAMPLES SINTERED AT DIFFERENT TEMPERATURES (g/cc)

Sample	1400°C	1450°C	1500°C	1600°C
A1	5.756	5.802	5.830	5.850
A2	5.350	5.476	5.517	5.555
A3	4.508	4.987	5.138	5.308

parameters and ease of retaining the tetragonal phase. Subsequently batches were prepared with 10, 20 & 30 wt% alumina. Table II indicates that much higher densities were achieved for 10 (A1) and 20 wt% (A2) alumina samples at a much lower temperature than for the 30 wt% alumina (A3). There was little effect on the bulk density by increase in the soaking time at 1500°C.

Flexural strength of the samples showed higher values with increasing sintering temperature (Figure 1). This may be due to the different grain sizes of ZrO_2 and Al_2O_3 or better densification of the samples. The high temperature flexural strengths of A1 samples sintered at different temperatures are plotted against the soaking time in figure 2. Their strength increased from one to two hours soaking and subsequently decreased at 4 hours. The high temperature strength tended to fall for higher sintering temperatures, which seems to be related to changes in grain size. In general for A1, A2 and A3 samples the high temperature flexural strength increased from 850 to 1000°C and then fell at 1250°C. We are carrying out further experiments, which will be reported in the future, to help explain these results.

The monoclinic content of the samples plotted against sintering temperature for four hours soaking time (Figure 3) shows that the higher the temperature of sintering the greater is the monoclinic content. However, after a soaking period of one hour, the monoclinic content was considerably less (Figure 4).

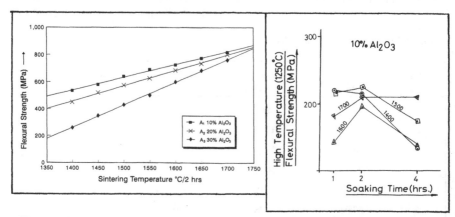

Figure 1. Flexural strength versus sintering temperature

Figure 2. Flexural strength of 10% Al_2O_3/Ce-TZP sintered at different temperatures

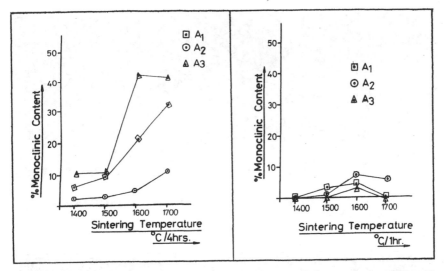

Figure 3. Monoclinic content after soaking 4 hrs   Figure 4. Monoclinic content after soaking 1 hr

This difference can be attributed to the effect of grain coarsening at higher soaking times, which increased their probability of conversion from tetragonal to monoclinic.

The fracture toughness K_{1c} (Figure 5) of A1, A2 and A3 samples was measured for the different sintering temperatures - 1400, 1450, 1550, 1600 and 1650°C. All samples, but particularly A2 and A3 i.e. with 20% and 30% Al_2O_3 respectively, showed increasing K_{1c} values up to 1600°C and then a sharp fall at higher temperature. The fall is more pronounced for higher alumina content samples, whose maximum K_{1c} values were much higher.

The SEM micrographs of A1, A2 and A3 sintered at 1500 and 1600°C are shown in figures 6(a,b,c) and 7(a,b,c) respectively. It can be seen that the zirconia grain size increased with increase in sintering temperature for all the samples, from about 0.5-1.0μm at 1500°C to 2-3μm at 1600°C. The average zirconia grain size decreased with increasing alumina content, e.g. at 30%

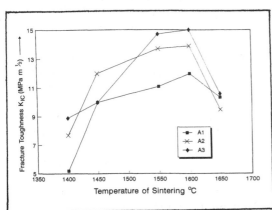

Figure 5. Fracture toughness for different sintering temperatures

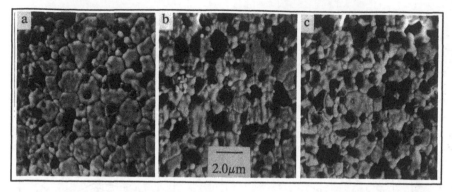

Figure 6. Scanning electron micrographs of Ce-TZP/Al$_2$O$_3$ sintered at 1500°C for 2 hrs.
a) 10% Al$_2$O$_3$ b) 20% Al$_2$O$_3$ c) 30% Al$_2$O$_3$

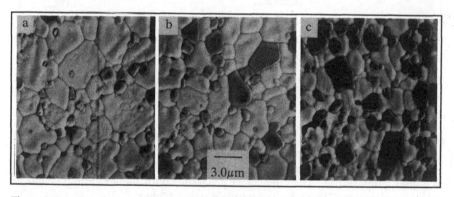

Figure 7. Scanning electron micrographs of Ce-TZP/Al$_2$O$_3$ sintered at 1600°C for 2 hrs.
a) 10% Al$_2$O$_3$ b) 20% Al$_2$O$_3$ c) 30% Al$_2$O$_3$

alumina the average zirconia grain size was almost half that in 10% alumina samples. At lower sintering temperatures (e.g. 1450°C) the zirconia grain size was even smaller and porosity became evident, especially at 30% Al$_2$O$_3$.

SEM observation revealed both transgranular and intergranular fracture at fracture surfaces (Figure 8). XRD of the fracture surface showed that a considerable amount of monoclinic had been formed (Figure 9). XRD on ground and polished surfaces showed t → m transformation in samples sintered at 1600°C (Figure 10) but not in samples sintered at 1500°C (Figure 11) or at 1450°C. This may be due to the smaller zirconia grain size in the latter materials, as shown in figures 6 and 7. The grain size needs to attain a critical value for stress-induced conversion from the tetragonal to the monoclinic phase.

The microstructures of samples sintered at 1500°C and 1600°C showed occasional long, large grains of alumina (Figure 12a and b). These were identified as ß-alumina [Na$_2$O.MgO.15Al$_2$O$_3$] which was confirmed by energy dispersive X-ray spectra and by high angle x-ray diffraction. The occurrence of ß-alumina in similar materials has been reported earlier [8-11]. The mechanical properties did not seem to have been adversely affected by its presence.

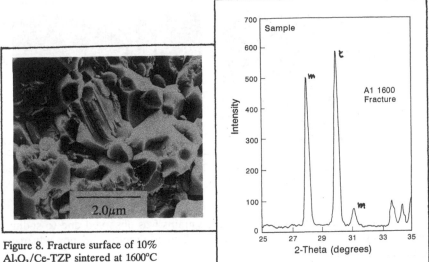

Figure 8. Fracture surface of 10% Al₂O₃/Ce-TZP sintered at 1600°C

Figure 9. X-ray diffractometer trace of fracture surface of 10% Al₂O₃/Ce-TZP sintered at 1600°C

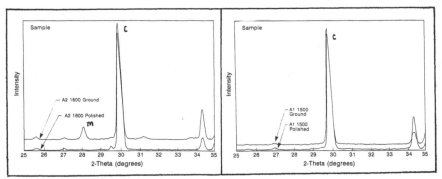

Figure 10. X-ray diffractometer traces of ground and polished surfaces for 20% Al₂O₃/Ce-TZP sintered at 1600°C

Figure 11. X-ray diffractometer traces of ground and polished surfaces for 10% Al₂O₃/Ce-TZP sintered at 1500°C

SUMMARY

(1) Ceria-TZP samples with 10-30% alumina reach similar high levels of flexural strength at high sintering temperatures.

(2) Fracture toughness values were also high, and peaked at a sintering temperature of 1600°C.

(3) The microstructure of the samples showed that alumina inhibited the grain growth of zirconia. Both the alumina and zirconia grains increased in size with increasing sintering temperature.

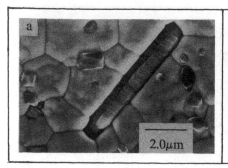

Figure 12. Scanning electron micrographs of β-Al₂O₃ grains in Ce-TZP/Al₂O₃ specimens sintered at 1600°C for 2 hrs. a) 10% Al₂O₃ b) 20% Al₂O₃

(4) The occasional presence of ß-alumina grains was observed due to sodium and magnesium impurities in the powder. The β-alumina did not adversely affect the mechanical properties.

(5) The m → t transformation occurred for samples sintered at 1600°C but was not observed in samples sintered at 1500°C and 1450°C.

ACKNOWLEDGEMENT

The authors would like to express sincere thanks to the Director of Central Glass & Ceramic Research Institute for permitting them to present this paper. The authors would also like to thank Dr. R.H.J. Hannink and Dr. S. Lathabai of CSIRO, Division of Materials Science and Technology, Clayton, Victoria, Australia for rendering valuable guidance and assistance during work done at their Laboratory.

REFERENCES

1. Garvie, R.C., R.H.J. Hannink and R.T. Pascoe, 1975. "Ceramic Steel." Nature (London), 258(5537):703-704.

2. Matsui, M., T. Soma and I. Oda, 1986. "Stress-Induced Transformation and Plastic Deformation for Y₂O₃-Containing Tetragonal Zirconia Polycrystals." Journal of the American Ceramic Society, 69(3):198-202.

3. Sato, T. and M. Shimada, 1985. "Transformation of Yttria-Doped Tetragonal ZrO₂ Polycrystals by Annealing in Water." Journal of the American Ceramic Society, 68(6):356-359.

4. Tsukuma, K. and M. Shimada, 1985. "Strength, Fracture Toughness and Vickers Hardness of CeO₂-Stabilized Tetragonal ZrO₂ Polycrystals (Ce-TZP)." Journal of Materials Science, 20(4):1178-1184.

5.   Chen, I-W. and P.E. Reyes-Morel, 1987. "Transformation Plasticity and Transformation Toughening in Mg-PSZ and Ce-TZP," in <u>Advanced Structural Ceramics</u>, P.F. Becher, M.V. Swain and S. Somiya, eds., Pittsburgh, Penn., USA: Materials Research Society Symposium Series Vol 78, pp. 75-88.

6.   Tsukuma, K., K. Ueda and M. Shimada, 1985. "Strength and Fracture Toughness of Isostatically Hot-Pressed Composites of Al_2O_3 and Y_2O_3-Partially-Stabilized ZrO_2." <u>Journal of the American Ceramic Society</u>, 68(1):C4-C5.

7.   Garvie, R.C. and P.S. Nicholson, 1972. "Phase Analysis in the Zirconia System." <u>Journal of the American Ceramic Society</u>, 55(6): 303-305.

8.   Green, D.J., 1985. "Transformation Toughening and Grain Size Control in $ß''$-Al_2O_3/ZrO_2 Composites." <u>Journal of Materials Science</u>, 20(7):2639-2646.

9.   Drennan, J., 1985. "The Observation of ß-Alumina Type Phases in Zirconia-Toughened Alumina (ZTA)." <u>Journal of Materials Science Letters</u>, 4(6):725-727.

10.   Green, D.J., R.H.J. Hannink and M.V. Swain, 1989. <u>Transformation Toughening of Ceramics</u>, Boca Raton, Fla., USA:CRC Press, Inc., pp. 144-152, 196.

11.   Bevan, D.J.M., B. Hudson and P.T. Moseley, 1974. "Intergrowth of Crystal Structures in β-Alumina." <u>Materials Research Bulletin</u>, 9(8):1073-1084.

Resistance to Crack Propagation in an Al$_2$O$_3$/ZrO$_2$ Ceramic

KAI DUAN, YIU-WING MAI and BRIAN COTTERELL

ABSTRACT

The equilibrium crack growth behaviour of an Al$_2$O$_3$/ZrO$_2$ ceramic composite AS8 (alumina with 8 wt % monoclinic in the form of 15-20 μm polycrystalline clusters) developed by Garvie et al. at the CSIRO Division of Materials Science and Technology, Australia and intended for refractory applications was studied. The crack-resistance (R) curves for three specimen geometries including double-cantilever-beam (DCB), compact tension (CT) and single edge notched beam (SENB) were obtained and their crack toughening mechanisms discussed. In-situ crack growth observations and microscopic studies identified the co-existence of a fracture process zone around the crack tip and crack bridges at the crack wake. Bridging was shown to be a significant toughening mechanism responsible for the pronounced R-curves. Because of the large scale bridging, the R-curves depend on specimen geometry and size. The equivalent crack wake bridging stresses were also determined based on a new crack compliance theory developed in this laboratory.

INTRODUCTION

For many Al$_2$O$_3$/ZrO$_2$ (ZTA) composites, the improvement of mechanical properties is strongly dependent on the microstructures of both matrix and additives, Al$_2$O$_3$/ZrO$_2$ ratio and the stability of zirconia [1]. A well known success is alumina matrix with partially stabilised zirconia

K Duan and Y-W Mai, Centre for Advanced Materials Technology, Department of Mechanical Engineering, The University of Sydney, NSW 2006, Australia
B Cotterell, Department of Mechanical and Production Engineering, National University of Singapore, 10 Kent Ridge Crescent, Singapore 0511

(PSZ) produced by adding stabilisers (such as Y_2O_3 and CeO_2, etc.) where both improved toughness and strength can be achieved because of the transformation toughening mechanism [2,3]. However, in some alumina-unstable zirconia composites the improved crack tolerance which originates from a microcracking mechanism is accompanied by a reduction in the strength of the material due to the increased crack density and size [4]. Additionally, both transformation and microcracking toughening mechanisms have also been found to operate simultaneously in some ZTA's [5,6].

We have recently measured and analysed the mechanical properties of a sintered Al_2O_3/ZrO_2 composite, designated AS8, developed by Garvie et al. at the CSIRO Division of Materials Science and Technology, Australia. This material is not the same as the "conventional" Al_2O_3/ZrO_2 composites (e.g. ZTA's studied by Hori et al. [7]) because its strength and modulus are considerably lower. It is originally designed as an advanced refractory with a good thermal shock resistance due to a microcracking toughening mechanism [8]. As a part of a larger study on its mechanical properties, we present in this paper a detailed analysis of its crack-resistance (R)-curve behaviour.

MATERIAL AND EXPERIMENTS

The material used in the experimental work was made in the CSIRO Division of Materials Science and Technology. The powder of alumina (Al_2O_3) (RC 172DB, Reynolds Metals Co., Bauxite AR 72011, U.S.A.) with 8 wt % ZrO_2 (S-grade, Magnesium Elektron Ltd., Manchester, U.K.) and 0.3 wt % PVA was formed with ~ 10 MPa and isopressed with 210 MPa afterwards. The green bodies were then sintered at 1600°C for 1 hour in air. These specimens were ground with a 150-220 mesh diamond wheel and polished with 1 μm diamond paste.

The microstructure of this ZTA is shown in figure 1 for a thermally etched and carbon coated specimen taken with a 35C scanning electron microscope (SEM). As described by Garvie and Goss [8], the material consists of an Al_2O_3 matrix with aggregates of ~ 13 μm containing individual monoclinic ZrO_2 grains with size ~ 1-2 μm.

Crack-resistance (R) curves for the ZTA have been measured in three specimen geometries, i.e. grooved double-cantilever-beam (DCB), compact tension (CT) and single edge notched beam (SENB) specimens (see figure 2). All experiments were carried out in an Instron 4302 testing machine. Crack lengths were measured directly with an optical travelling microscope after the specimen was unloaded. Dye penetrants were used to make the observation easier. A linear displacement transducer was placed between two knife edges mounted on the front faces of the specimen to measure the Crack-Mouth-Opening-Displacement (CMOD). During loading, the CMODs were plotted against the load P. A typical P-CMOD curve for a DCB specimen is shown in figure 3 where the slopes of the unloading lines

Figure 1. The microstructure of AS8

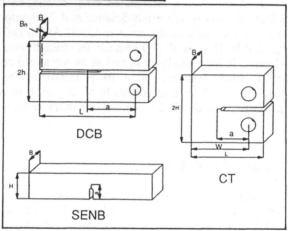

Figure 2. Geometry and dimensions of the DCB (L=52 mm, h= 9 mm, B=3.5 mm, B$_n$=2.5 mm and L-a>2h), CT (W=23.9mm, B=5.9 mm, 2H=28.6 mm and a/W=0.39~0.88) and SENB (B=4.0 mm, H=7.4 mm and a/H=0.33~0.51) specimens.

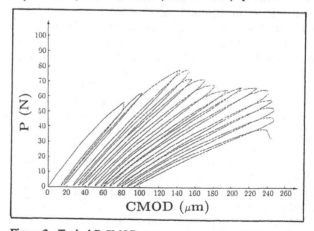

Figure 3. Typical P-CMOD curves measured in a DCB specimen.

give the compliances of the propagated crack. The fracture resistance K_R can be obtained by substituting the crack length a and corresponding fracture load P into the appropriate stress intensity factor expressions. For the DCB geometry the K-expression derived by Freiman et al. [9] was used; the K-expression for the CT geometry has been given in ASTM-E399 and for the SENB specimens under 4-point bending this can be obtained from the stress intensity factor handbook [10]. From these K-expressions the corresponding CMOD-P equations are also obtained.

To examine the toughening mechanisms associated with the rising R-curve, polished surfaces of some CT specimens were observed with both optical and scanning electron microscopy. The specimen surfaces for observations under SEM have been coated with a thin carbon layer. The specimens were also wedged open and dye penetrant used to reveal the damage around the propagating crack with a stereo optical microscope. In addition a SENB specimen was loaded in a four-point fixture to allow in-situ crack growth to be monitored while being observed under a Zeiss microscope. Crack-bridging stresses were evaluated using a CT specimen based on the successive-cutting technique developed by Hu and Wittmann [11] and modified by Hu and Mai [12].

EXPERIMENTAL RESULTS AND ANALYSIS

K_R-CURVES

The K_R-curves of AS8 measured with the DCB, CT and SENB geometries are compared in figure 4 where the fracture resistance K_R has been plotted against crack extension Δa. It is obvious that the material exhibits a pronounced R-curve behaviour.

Whilst there is a fair bit of experimental scatter in the $(K_R, \Delta a)$ data in the DCB and CT geometries, their K_R-curves are similar and commence at about 2.5 to 3 MPa√m rising to a plateau value approximately $6 \sim 7$ MPa√m over a crack extension of about 5 mm. The DCB crack-resistance curve obtained here is similar to that reported by Garvie and Goss [8]. The SENB geometry, however, gives a crack-resistance curve which is quite different from those of the other two geometries. Here, the fracture resistance increases rapidly over a short distance of crack extension to values above the plateau of the DCB and CT geometries. This geometric effect on the R-curve has been observed in other ceramics and explained in terms of the large scale bridging at the crack wake [13,14] relative to the ligament size of the SENB specimen as the crack approaches the back face. Therefore, to measure the full K_R-curve, the CT and DCB geometries are preferred to the SENB geometry.

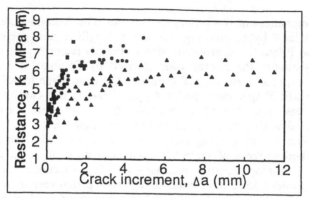

Figure 4. K_R-curves of AS8 measured from DCB ($\triangle$), CT ($\bullet$) and SENB ($\square$) specimens.

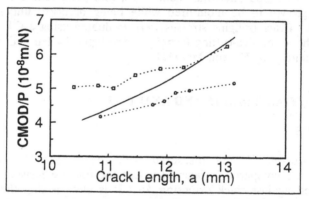

Figure 5. Compliance measurements for a CT specimen. Theoretical curve (—); during crack growth with developing bridges (O); and crack-bridges removed in crack wake by sawcutting ($\square$).

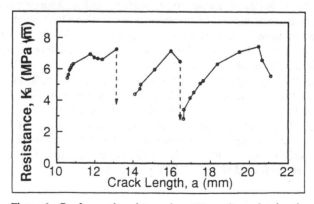

Figure 6. Crack growth resistance in a CT specimen showing the effect of removing the crack-wake bridges by sawcutting.

COMPLIANCE VARIATION AND CRACK-BRIDGING

The theoretical compliance (CMOD/P) of a CT specimen as a function of crack length can be calculated easily. These values are shown in figure 5. When a crack is loaded the compliance initially increases before any crack extension can be detected because of the formation of a damage zone ahead of the crack tip. This effect causes the compliance to be larger than the theoretical value. However, as the crack propagates further, the effect from crack wake bridging gradually overtakes that of the crack tip damage zone so that the measured compliances are smaller compared to the theoretical values. Figure 5 demonstrates this effect of crack-bridges quite clearly as the crack extends for about 4 mm. At this point, when the load is removed and the crack bridges are successively removed in the wake region, the compliances slowly increase until when the whole bridging zone is removed the experimental compliance falls on the theoretical compliance curve. Figure 6 shows the K_R-curve behaviour as the crack extends. Renotching removes all the crack bridges developed along the crack length and this causes the crack-resistance to drop back to its initial value before rising again with further crack growth. This test convincingly proves that crack bridging is a predominant mechanism of toughening in this AS8 material.

IN-SITU OBSERVATION OF R-CURVE MECHANISMS

To confirm the existence of bridges at the crack wake SEM photo-micrographs were taken on the polished and carbon coated surfaces of an AS8 CT specimen which was cracked and then wedged open upon unloading, figure 7. It can be seen that the crack growth is always along an irregular path due to the irregular shapes and sizes of the bridging particles distributed along its length. Crack branches are also seen resulting in cracks overlapping at their ends and some of these are connected underneath the overlaps. Crack bridges are also shown in this photograph. Further evidence of overlapping cracks and ligament crack-bridging along the fracture path is also demonstrated in a SENB specimen loaded in a specially designed jig for observation under a Zeiss optical microscope. These results along with the compliance measurements after renotching all point to crack bridging as a significant toughening mechanism.

ESTIMATION OF CRACK BRIDGING STRESS

Toughening due to bridging can be modelled by smearing closure stresses along the crack face so that the bridging stress $\alpha_b(x)$ at a distance x from the crack tip is a function of the crack face open displacement w(x). For crack-bridging in ceramic materials the σ_b-w relation obeys a strain softening law approximated by [15]

$$\sigma_b = \sigma_m \left[1 - \frac{w}{w_c} \right]^n \tag{1}$$

where w_c is a critical crack open displacement at which point the bridging stress is zero. Once this relationship is determined the theoretical crack-resistance (K_R) curve can be predicted based on procedures described earlier [15]. However, an iterative technique is needed to determine the maximum bridging stress σ_m and softening exponent n to give the best fit to the R-curve data. For ceramic materials, it may be assumed that the crack surfaces of common fracture mechanics specimen geometries (SENB, CT and DCB) remain straight during crack propagation. Therefore, equation 1 can be written as [11,12]

$$\sigma_b = \sigma_m \left[1 - \frac{x}{X} \right]^n \tag{2}$$

A simplified method to evaluate crack bridging stresses has been proposed recently [11,12] based on the compliance CMOD/P approach. Let C(a) be the compliance of a specimen with a crack length a with no bridging and $C_u(x)$ the compliance measured in an experiment for the same a but with a bridging zone x. It is expected that

$$C(a) \geq C_u(x) \tag{3}$$

The function $C_u(x)$ represents the compliance with an unsaturated bridging zone developed during crack extension or that with a bridging zone which has been fully developed and then partially removed by sawcutting [11,12]. The changes in compliance with crack extension can be related to the bridging stress $\sigma_b(x)$ at the distance x from crack tip by [11]

$$\frac{\sigma_b(x)}{\sigma_m} = - \frac{C^2(a)}{C'(a)} \frac{C_u'}{C_u^2} \tag{4}$$

Substituting equation 2 into the above equation and integrating both sides leads to a ϕ-function [11,12,16]:

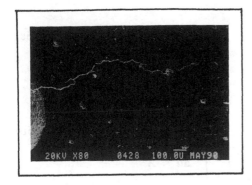

Figure 7. Crack growth pattern of AS8 obtained using a 35C SEM.

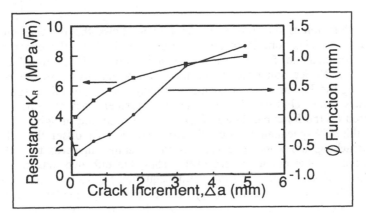

Figure 8. The Φ-function curve and K_R-curve of a CT specimen.

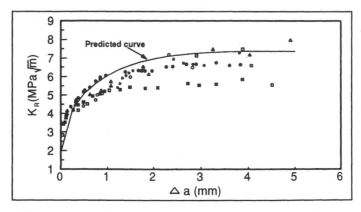

Figure 9. Comparison of predicted and experimental K_R-curves in the CT specimen. (Different symbols represent K_R-curves for different specimens).

$$\phi = \frac{C(a)}{C'(a)} \left\{ \frac{C(a)}{C_u} - 1 \right\} \tag{5a}$$

$$= \begin{cases} \dfrac{X}{n+1} & \Delta a \geq X \\[2ex] \dfrac{X}{n+1} \left\{ 1 - \left[1 - \dfrac{\Delta a}{X} \right]^{n+1} \right\} & \Delta a < X \end{cases} \tag{5b}$$

This ϕ function is dependent on Δa and it allows the n value and the saturated bridging zone X to be determined easily from the experimental data. For example, figure 8 shows the ϕ function and K_R-curve for a CT specimen. Notice that both curves have similar features with the ϕ-curve mirroring the K_R-curve. ϕ is negative initially because of microcracking at the crack front. However, it rises again with further crack-bridging.

Based on the K_R-curve data in figure 8, the saturated bridging zone size X of this CT specimen is approximately 5 mm. To offset the effect of microcracking, the whole ϕ curve has to be shifted upwards to make the minimum ϕ value equal to zero [12]. Thus, the effective maximum $\phi \approx$ 1.80 mm and substituting this value into equation 5a, we have n = 2.33. A theoretical crack resistance (K_R) curve can be calculated with an approximate method and to give the best fit to the experimental data of the CT geometry shown in figure 9 we have used σ_m = 96 MPa.

Since equations 1 and 2 are material properties σ_m and n are independent of specimen geometries. Therefore, provided a full K_R-curve can be developed, such as in the DCB and CT geometries, X and n must be the same. In this respect the SENB geometry will not be suitable to determine n using the above analysis, equation 5a, because there is no saturated K_R-curve.

CONCLUSIONS

It has been unequivocally demonstrated that AS8, a sintered Al_2O_3/ZrO_2 refractory consisting of an agglomerate of individual monoclinic zirconia particles dispersed in an alumina matrix, exhibits a pronounced R-curve behaviour. K_R curves can be measured with both CT and DCB geometries. It is shown that crack-wake bridging is a significant toughening mechanism using the technique of crack compliance measurements and in-situ fracture experiments in a testing rig observed with an optical microscope. The crack bridging stresses are estimated using a new compliance analysis developed by Hu et al. [11,12] and they give a theoretical K_R-curve in good agreement with experimental data.

ACKNOWLEDGEMENTS

We wish to thank the Australian Research Council (ARC) for the continuing support of this work. K Duan was supported by an ARC Junior Research Fellowship.

REFERENCES

1.   Becher, P. F. and V. T. Tennery, 1983. "Fracture Toughness of Al₂O₃-ZrO₂ Composites," in Fracture Mechanics of Ceramics, R. C. Bradt, A. G. Evans, D. P. H. Hasselman and F. F. Lange, eds., New York, NY, USA: Plenum Press, pp. 383-389.

2.   Lange, F. F., 1982. "Transformation Toughening, Part 4 Fabrication, Fracture Toughness and Strength of Al₂O₃-ZrO₂ Composites." J. Mater. Sci., 17:247-254.

3.   Maschio, S., O. Sbaizero and S. Meriani, 1989. "Structure and Fracture Behavior of Al₂O₃/(Ce)-Stabilised ZrO₂ Composites," in Euro-Ceramics '89, G. de With, R. A. Terpstra and R. Metselaar, eds., London, UK: Elsevier Applied Science Ltd., pp. 3.377-3.383.

4.   Claussen, N., J. Steeb and R. F. Pabst, 1977. "Effect of Induced Microcracking on the Fracture Toughness of Ceramics." Am. Ceram. Soc. Bull. 56:559-562.

5.   Rühle, M., 1988. "Microcrack and Transformation Toughening of Zirconia-Containing Alumina." Mater. Sci. Eng., A105/106:77-82.

6.   Duan, K., Y-W. Mai, M. V. Swain and B. Cotterell, 1990. "Mechanical Properties of a Sintered Al₂O₃/ZrO₂ Composite," in Proceedings Austceram '90, P. J. Darragh and R. J. Stead, eds., Perth, Australia, pp. 130-136.

7.   Hori, S., M. Yoshimura and S. Sōmiya, 1986. "Strength-Toughness Relations in Sintered and Isostatically Hot-Pressed ZrO₂-Toughened Al₂O₃." J. Am. Ceram. Soc., 69:169-172.

8.   Garvie, R. C. and M. F. Goss, 1988. "Thermal Shock Resistant Alumina/Zirconia Alloys," in Advanced Ceramics II, S. Sōmiya, ed., London: Elsevier Applied Science Ltd., pp. 69-87.

9. Freiman, S. W., D. R. Mulville and P. W. Mast, 1973. "Crack Propagation Studies in Brittle Materials." J. Mater. Sci., 8:1527-1533.

10. Tada, H., P. C. Paris and G. R. Irwin, 1985. Stress Analysis of Crack Handbook, 2nd edition, St. Louis, Missouri: Paris Production, Inc.

11. Hu, X. Z. and F. H. Wittmann, 1991. "An Analytical Method to Determine the Bridging Stress Transferred within the Fracture Process Zone: I, General Theory." Cement and Concrete Research, 21:1118-1128.

12. Hu, X. Z. and Y-W. Mai, 1992. "A General Method for Determination of Crack-Interface Bridging Stresses." J. Mater. Sci., 27:3502-3510.

13. Cotterell, B. and Y-W. Mai, 1987. "Crack Growth Resistance Curve and Size Effect in the Fracture of Cement Paste." J. Mater. Sci., 22:2734-2738.

14. Cotterell, B. and Y-W. Mai, 1988. "Modelling Crack Growth in Fibre-Reinforced Cementitious Materials." Mater. Forum, 11:341-351.

15. Foote, R. M. L., Y-W. Mai and B. Cotterell, 1986. "Crack Growth Resistance Curves in Strain-Softening Materials." J. Mech. Phys. Solids, 34:593-607.

16. Hu, X. Z., Y-W. Mai and S. Lathabai, 1992. "Compliance Analysis of a Bridged Crack Under Monotonic and Cyclic Loading." J. European Ceram. Soc., 9:213-217.

Flank Wear Mechanisms of Zirconia Toughened Alumina Cutting Tools during Machining of Steel

XING SHENG LI, IT-MENG LOW and DAN PERERA

ABSTRACT

Two zirconia toughened alumina (ZTA) cutting tool grades SN60 (<10% ZrO_2) and AZ5000 (between 10-20% ZrO_2, fine grained) were used for the evaluation of wear mechanisms when machining a high tensile steel (AISI 4340). Experimental studies were carried out at various cutting speeds (300-600 m/min), feed rates (0.1-0.4 mm/rev) and depths of cut (0.5-2.0 mm), in the dry condition. The worn cutting edge surfaces were examined by optical and scanning electron microscopies (SEM). The flank wear was determined by a toolmaker microscope. The AZ5000 grade was found to provide better flank wear resistance under the conditions investigated. The wear behaviour and wear mechanisms of the tools are discussed in relation to the cutting parameters (i.e. cutting speed, feed rate and depth of cut) and microstructures.

INTRODUCTION

Alumina based cutting tools attract keen interest in production engineering due to their ability to display high material removal rates with the concomitant reduction in production costs [1]. However, a major obstacle for a more general acceptance of these ceramic tools is their relatively low strength, poor toughness and high susceptibility to thermal cracking [2]. In order to increase the fracture toughness of these alumina ceramic tools, two innovations have been made; one is SiC whisker-reinforced alumina, the other is zirconia toughened alumina (ZTA). The latter is chemically compatible with steel while the former reacts readily with Fe during machining [3].

It has been well established that flank wear on the tool is the limiting factor in material removal operations during machining of metals. This problem constitutes a

Xing Sheng Li and It-Meng Low, Department of Applied Physics, Curtin University of Technology, GPO Box U 1987, Perth, W.A. 6001, Australia
Dan Perera, Australian Nuclear Science and Technology Organisation, Private Mail Bag 1, Menai, N.S.W. 2234, Australia

493

major concern for the engineering application of ceramic tools [4]. The wear on the tools may be attributed to either mechanical (athermal) or chemical (thermal) mechanisms, or some combination of the two [5]. Therefore, the appropriate material design and economic application of ceramic cutting tools would certainly require a good understanding of the tool wear mechanisms as well as the identification of the dominant wear mechanisms on the flank land.

This paper describes the evaluation of wear rates and flank wear mechanisms of ZTA tools when machining high tensile steel (AISI 4340). The wear behaviour and wear mechanisms of these tools are discussed in relation to the cutting parameters (i.e. cutting speed, feed rate and depth of cut) and the microstructures.

EXPERIMENTAL TECHNIQUES

CUTTING TOOL MATERIALS

Two grades of commercial ZTA cutting tools namely SN60 and AZ5000 were used in this investigation. They were supplied by the Kyocera Corporation (Japan). Less than 10 wt% ZrO_2 was present in SN60 while AZ5000 contained more than 10 but less than 20 wt% ZrO_2. The inserts were 12.7 mm square and 4.76 mm thick and conformed to the ISO specification SNGN 120408. The mechanical properties and microstructure for these inserts are shown in TableI and figure 1 respectively.

TABLE I. PROPERTIES OF CERAMIC CUTTING TOOL MATERIALS USED IN STUDY

Manufacture Code	Hardness (HV_{10})	Fracture Toughness (K_{IC}, MPa.m$^{1/2}$)	Average Grain Size (µm)
SN60	1650	4.0	5.0
AZ5000	1800	6.0	1.0

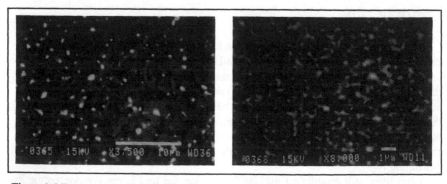

Figure 1. Microstructure of (a) SN60 (left) and (b) AZ5000 (right) (backscattered scanning electron micrographs)

WORKPIECE MATERIAL

Machining tests were conducted on an AISI 4340 steel bar (Rockwell hardness

HRB=95±2). The initial diameter was 150 mm and the length 660 mm. The alloying composition of the steel is shown in TableII.

TABLE II. ALLOYING COMPOSITION (WT%) OF WORKPIECE MATERIAL (AISI 4340)

C	Mn	P	S	Si	Ni	Cr	Mo
0.40	0.70	0.035	0.04	0.25	1.85	0.80	0.25

CUTTING TEST

Machining was performed on a MACSON lathe. This machine was powered by an 11 kW motor which provided stepwise speed control throughout the range of 47-1600 r.p.m. The cutting conditions are shown in Table III. The tool holder used was CSBNR2525N43 (NTK) and the tool angles were (-6, -6, 6, 6, 15, 15, 0.8).

TABLE III. CUTTING CONDITIONS DURING MACHINING OPERATION

Cutting speed (m/min)	300, 400, 500, 600
Feed rate (mm/rev)	0.1, 0.2, 0.3, 0.4
Depth of cut (mm)	0.5, 1.0, 1.5, 2.0
Environment	dry

The orthogonal cutting force components were measured during turning using a 3-axis piezo-electric dynamometer. The signals from the dynamometer were fed through a charge amplifier and recorded on a personal computer. The three force components, viz feed force (Fx), thrust force (Fy) and principal force (Fz) were measured.

The tool life of an insert was determined by either the total tool failure or the critical flank wear (VB = 0.3 mm) criterion. The width of the flank wear land, VB, was measured perpendicular to the major cutting edge, and was measured from the position of the original major cutting edge. The typical wear pattern of cutting tools is shown in figure 2. The flank wear was measured with a toolmaker's microscope (magnification: 40). The worn cutting edge surfaces were examined by both optical and scanning electron microscopy.

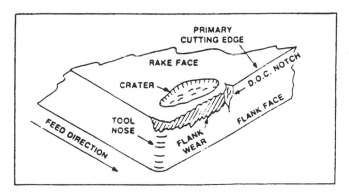

Figure 2. Schematic representation of cutting tool wear

RESULTS AND DISCUSSION

WEAR RATES AND TOOL LIFE

Figures 3 and 4 show the development of flank wear of tested tools at the cutting speeds of 300 and 600 m/min respectively. Figure 5 shows the influence of cutting speed on the flank wear of inserts after 1 minute cutting. The flank wear at the lower cutting speed (300m/min) was approximately the same for both ZTA inserts, but the flank wear at the higher cutting speed (600 m/min) for SN60 was more than twice that of AZ5000. From this result it can be deduced that the influence of composition and microstructure of cutting tools on the flank wear rate is rather important, especially under high cutting speeds.

Figures 6 and 7 show the flank wear as a function of cutting time at the feed rates of 0.1 and 0.4 mm/rev respectively. Figure 8 shows the influence of feed rate on the flank wear of inserts after 1 minute cutting. The tool wear rates were approximately the same for both grades at the lower feed rates, but AZ5000 showed better flank wear resistance at the higher feed rates.

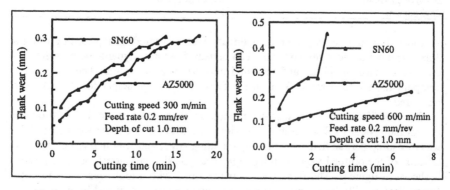

Figure 3. Flank wear versus cutting time   Figure 4. Flank wear versus cutting time

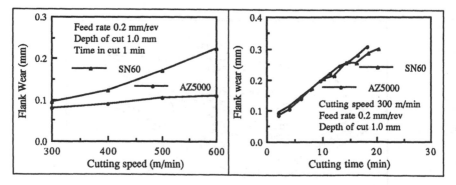

Figure 5. Flank wear versus cutting speed   Figure 6. Flank wear versus cutting time

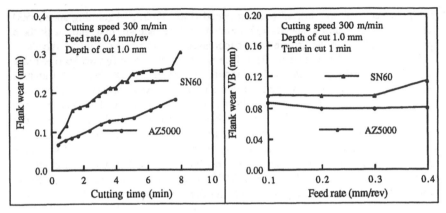

Figure 7. Flank wear versus cutting time   Figure 8. Flank wear versus feed rate

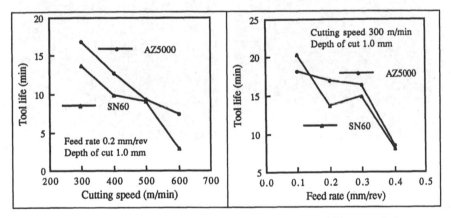

Figure 9. Tool life versus cutting speed   Figure 10. Tool life versus feed rate

Figure 9 shows the tool life of tested cutting tools at different cutting speeds at a feed rate of 0.2 mm/rev. The tool life of AZ5000 was clearly better than that of SN60. Figure 10 shows the tool life of tested cutting tools at different feed rates at a cutting speed of 300 m/min. It can be seen that the tool life decreases with increasing feed rates.

FLANK FACE WEAR

The appearance of the worn flank face is shown in figure 11. The surface was characterised by very large ridges parallel to each other separated by grooves. The ridges normally had a smooth appearance whereas the grooves showed evidence of wear tracks. The extent of ridge formation was accentuated some way back from the cutting edge. When comparing flank faces at different cutting speeds in figures 12 (cutting speed=300 m/min), 13 (cutting speed=400 m/min), 14 (cutting speed=500 m/min) and 15 (cutting speed=600 m/min), it was shown that the flank face appeared rougher and the groove widths increased with increase of cutting speeds.

The influences of feed rate on flank face wear are shown in figures 16 (feed=0.1 mm/rev), 12 (feed=0.2 mm/rev), 17 (feed=0.3 mm/rev) and 18 (feed=0.4 mm/rev) respectively. With increase of the feed rate (from 0.1 to 0.3 mm/rev) the flank face became rougher and the groove width increased. However the flank face became smoother and groove width decreased with further increase in the feed rate (from 0.3 to 0.4 mm/rev). Comparing figures 16 and 19, as well as figures 17 and 11, it was shown that under the light cutting condition (feed=0.1 mm/rev) the flank faces of AZ5000 and SN60 looked nearly the same, but under the heavy cutting condition (feed=0.3 mm/rev) the flank face of AZ5000 looked smoother than that of SN60 which indicates that AZ5000 possessed better flank wear resistance.

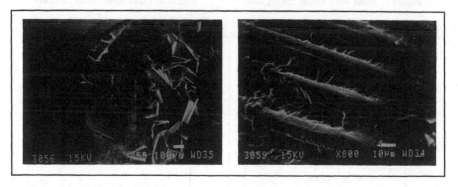

(a)   low magnification   (b)   high magnification

Figure 11. Scanning electron micrographs showing flank wear of AZ5000 after cutting at 300 m/min, 0.3 mm/rev and DOC 1.0 mm until VB=0.3 mm

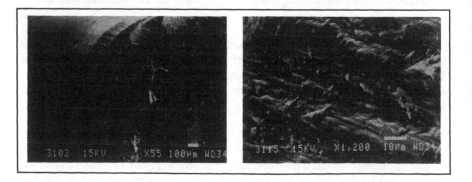

(a)   low magnification   (b)   high magnification

Figure 12. Scanning electron micrographs showing flank wear of SN60 after cutting at 300 m/min, 0.2 mm/rev and DOC 1.0 mm until VB=0.3 mm

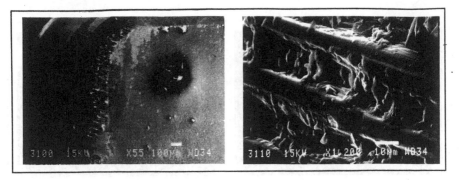

(a)   low magnification   (b)   high magnification

Figure 13. Scanning electron micrographs showing flank wear of SN60 after cutting at 400 m/min, 0.2 mm/rev and DOC 1.0 mm until VB=0.3 mm

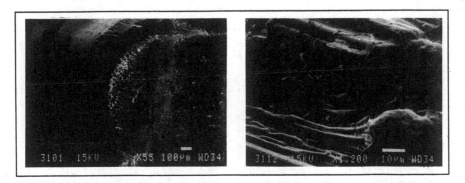

(a)   low magnification   (b)   high magnification

Figure 14. Scanning electron micrographs showing flank wear of SN60 after cutting at 500 m/min, 0.2 mm/rev and DOC 1.0 mm until VB=0.3 mm

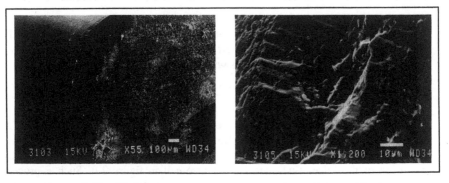

(a)   low magnification   (b)   high magnification

Figure 15. Scanning electron micrographs showing flank wear of SN60 after cutting at 600 m/min, 0.2 mm/rev and DOC 1.0 mm until VB=0.3 mm

 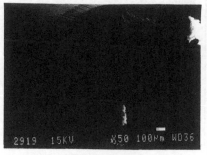

Figure 16. Scanning electron micrograph shows flank wear of SN60 after cutting at 300 m/min, 0.1 mm/rev and DOC 1.0 mm until VB=0.3 mm

Figure 17. Scanning electron micrograph shows flank wear of SN60 after cutting at 300 m/min, 0.3 mm/rev and DOC 1.0 mm until VB=0.3 mm

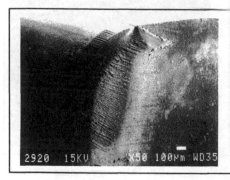 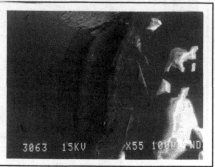

Figure 18. Scanning electron micrograph shows flank wear of SN60 after cutting at 300 m/min, 0.4 mm/rev and DOC 1.0 mm until VB=0.3 mm

Figure 19. Scanning electron micrograph shows flank wear of AZ5000 after cutting at 300 m/min, 0.1 mm/rev and DOC 1.0 mm until VB=0.3 mm

When comparing the flank faces of both ZTA cutting inserts tested (Figures 11-19), no distinct difference between their wear topography can be discerned thus indicating that similar wear mechanisms were involved.

FLANK WEAR MECHANISMS

Intergranular fracture appears to be the dominant flank wear mechanism for ZTA inserts . Examination of the flank face showed visible cracks on the ridges (indicated by the arrow a in figure 20). These micro-cracks led to partial intergranular fracturing of the ridges (indicated by the arrow b in figure 20). Such a process would be temperature- and stress-dependent. In a machining process, cutting tools are used under severe high temperature and high stress environments. Under these circumstances, micro-cracks may be initiated at grain boundaries due to dislocation pile-ups. With progress of the cutting process, the length of the micro-crack will extend. When the micro-crack length reaches a certain extent, intergranular fracturing will occur. The dependence of cutting temperature on cutting speed is rather large, since an average temperature increase of two hundred degrees (from 1100 to 1300 °C) has been recorded when changing the cutting speed from 400 to 600 m/min during machining steel with an alumina ceramic tool [3]. The cutting forces were approximately the same when changing the cutting speed from 300 to 600 m/min (figure 21), so the stress may be the same. The flank wear increased notably with increasing cutting speed from 300 to 600 m/min (figure 5). This indicates that flank wear is more temperature-dependent than stress-dependent. When changing the feed rate from 0.1 to 0.4 mm/rev, the cutting force increased remarkably (figure 22), so the stress would increase. But the flank wear did not display much difference with the increase of feed rate from 0.1 to 0.4 mm/rev (figure 8). This also indicates that stress is not the main controlling factor for flank wear.

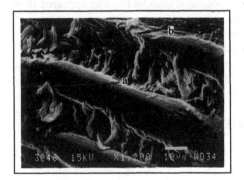

Figure 20. Scanning electron micrograph showing flank wear of AZ5000 after cutting at 300 m/min, 0.2 mm/rev and DOC 1.0 mm until VB=0.3 mm

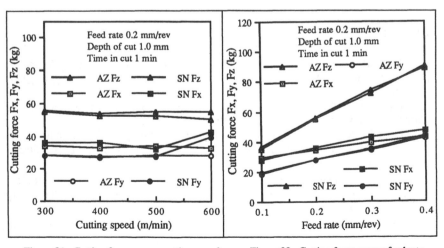

Figure 21. Cutting force versus cutting speed   Figure 22. Cutting force versus feed rate

CONCLUSIONS

The following conclusions are based on turning tests carried out with commercial ZTA inserts on high tensile steel (AISI 4340) at cutting speeds ranging from 300 to 600 m/min, feed rate from 0.1 to 0.4 mm/rev and depth of cut from 0.5 to 2.0 mm.

(1)   With increase in cutting speed, the flank wear of the tools increased significantly. The influence of feed rate on the flank wear was smaller than that of cutting speed.

(2)   For the two zirconia-toughened alumina cutting inserts investigated, under light cutting conditions (i.e smaller cutting speed, feed rate and depth of cut), flank wear resistance was similar, but under heavy cutting conditions (i.e greater cutting speed, feed rate and depth of cut) AZ5000 demonstrated better flank wear resistance than SN60.

(3)   Micro-cracks induced intergranular fracture appeared to be the dominant flank wear mechanism.

(4)   The two zirconia-toughened alumina cutting inserts showed similar flank wear mechanisms under the cutting conditions investigated.

(5)   The higher hardness, better fracture toughness and finer grain structure of AZ5000 resulted in this material being more resistant to flank wear, especially under heavy cutting conditions.

ACKNOWLEDGMENTS

X S Li wishes to thank the Commonwealth Government of Australia and AINSE for financial support in the form of scholarship and research supplement respectively. I M Low is grateful to the Australian Research Council for funding under the Mechanism B scheme. The authors would also like to thank the Kyocera Corporation for providing the ceramic cutting inserts.

REFERENCES

1.   Tonshoff, H.K., E.Brinksmeier and S.Bartsch, 1987. "Notch Wear and Chemically Induced Wear in Cutting with Al_2O_3-Tools." Annals of the CIRP, 36(2):537-541.

2.   Brandt, G., 1986. "Flank and Crater Wear Mechanisms of Alumina-Based Cutting Tools when Machining Steel." Wear, 112:39-56.

3.   Narutaki, N., Y.Yamane and K.Hayashi, 1991. "Cutting Performance and Wear Characteristics of an Alumina-Zirconia Ceramic Tool in High-Speed Face Milling." Annals of the CIRP, 40(1):49-52.

4.   Kim, S., D.R.Durham and V.F.Turkovich, 1991. "Microscopic Studies of Flank Wear on Alumina Tools." Transactions of the ASME Journal of Tribology, 113(1):204-209.

5.   Tonshoff, H.K. and S.Bartsch, 1988. "Wear Mechanisms of Ceramic Cutting Tools." American Ceramic Society Bulletin, 67(6):1020-1025.

Potentially Strong and Tough ZrO_2-Based Ceramic Composites $\geq 1300°C$ by Electrophoretic Deposition

PATRICK S. NICHOLSON, PARTHO SARKAR
and XUE N. HUANG

ABSTRACT

Electrophoretic deposition is used to synthesize laminates of YPSZ/Al_2O_3. The technique is shown to result in dense composites with layers as thin as 3 µm. A protocol for tough laminates $\geq 1300°C$ strong enough to withstand quenching to room temperature is suggested, based on the morphology of natural composites. A thin YPSZ surface layer is strengthened by a subsurface layer of MgO/$MgAl_2O_4$ followed by a functionally-gradiented 100% Al_2O_3 to 100% PSZ layer. The interior consists of alternate layers of Al_2O_3/YPSZ with La-aluminate at their interfaces. Al_2O_3 fibres with 10 nm of gold on their surface serve as the deposition electrode, resulting in central curved layers of YPSZ and Al_2O_3. Microstructures of the resultant composites are presented and Vickers indentation used to demonstrate crack paths therein.

INTRODUCTION

Y-PSZ ceramics have excellent room-temperature strength (>1000 MPa) and high toughness (>5 MPa $\sqrt{m}$). The strength and toughness decrease markedly with increasing temperature $>500°C$ because of the decreasing contribution of the ZrO_2 transformation. The addition of 10 to 40 wt% Al_2O_3 enhanced the bend strength of PSZ (2 to the 3 mol% Y_2O_3) composites at T < 1000°C [1-4], but at temperatures $>1200°C$, YPSZ deforms superplastically [5] and/or viscously via liquid phase because of SiO_2 and other impurities. This paper describes a protocol that turns this "deformability" of YPSZ at high temperatures to good advantage.

There are two schools of thought as to the source of the "ductility/viscosity" of YPSZ at 1300°C. The material is ultrafine grained

Patrick S. Nicholson, McMaster University, Hamilton, Ontario L8S 4L7
Partho Sarkar, McMaster University, Hamilton, Ontario L8S 4L7
Xue N. Huang, McMaster University, Hamilton, Ontario L8S 4L7

and reported to be superplastic [5]. It also contains residual silica and in fact will not sinter without it [7]. Electrical conductivity evidence [4-6] supports the latter source of deformation and a "web" of silicate can be detected on the platelets of Al_2O_3-platelet/ZrO_2 composites (Figure 1).

The "strength" of YPSZ drops to 63 ± 10 MPa at 1300°C (down from 1170 ± 106 MPa at 25°C). In fact, flexure samples do not fracture but strain until the beam centre touches the lower fixture (Figure 2).

The load-deflection curves for α-Al_2O_3 and YPSZ at 1300°C are shown in Figure 3. Clearly the α-Al_2O_3 is "stiff" and the YPSZ "compliant". This combination of mechanical response is the basis of the toughness of natural ceramic composites such as mollusk shells. These have an aragonite ($CaCO_3$-stiff)/conchiolin (protein-compliant) laminar structure [8]. The softer protein induces crack deflection and interfacial propagation (Figure 4) substantially increasing the toughness of the composite. This rationale suggests that combining the stiffness of Al_2O_3 and the compliance of ZrO_2 should toughen laminar ZrO_2/Al_2O_3 composites ≥ 1300°C providing both phases are theoretically dense.

LAMINAR YPSZ/Al_2O_3 COMPOSITES BY ELECTROPHORETIC DEPOSITION

This section describes the synthesis of YPSZ/Al_2O_3 laminates by electrophoretic deposition (EPD)[9].

Electrophoretic deposition (EPD) is the process whereby powder particles are deposited from a suspension onto a shaped electrode of opposite charge, on application of a dc electrical field. The rate of deposition is controllable via applied potential and can be very fast.

EPD is actually two processes: electrophoresis – the motion of particles in a stable suspension under the influence of an electrical potential gradient; and deposition – the coagulation of a dense layer of particles on the electrode. Stability is introduced electrostatically for when the particle moves in a liquid, some of the ions in an associated diffuse region (double-layer), "shear-off" leaving an imbalance of charge and a shear – or zeta – potential (ζ) on the particle. This induces interparticle repulsion and the higher the ζ the more stable is the suspension against coagulation (coagulation is due to the London-van der Waals (LVDW) attractive forces which operate on very close particle approach). The sign of the charge on the particle/double layer system is the same as the particle as this is now incompletely balanced by the double layer.

A dense deposit is formed during EPD so the charges that served to maintain the particles apart must be reduced to a level conducive to coagulation. The particles move in a stationary liquid so their diffuse double-layer will thin ahead and to the equator of the electrophoresing particles, reducing the ζ threat. The distortion is further promoted by the ions in the double layer being attracted to the other electrode in the system.

A second ζ reduction mechanism is due to free ions of the same sign as the particles moving to the same electrode. Some discharge or are involved in electrode reactions to give the current but a number could

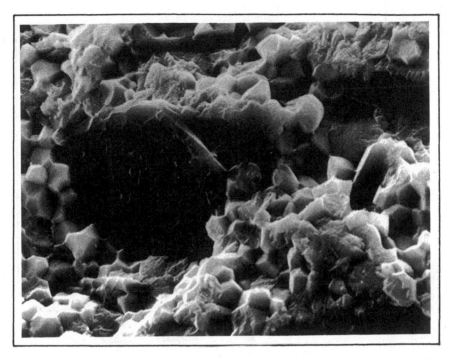

Figure 1   A "web" of silicate grain-boundary phase on the surface of an Al_2O_3 platelet on the fracture surface of an Al_2O_3/YPSZ composite.

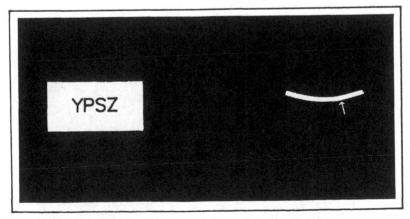

Figure 2   A 1300°C bend specimen of YPSZ. The sample bent until its centre touched the lower fixture.

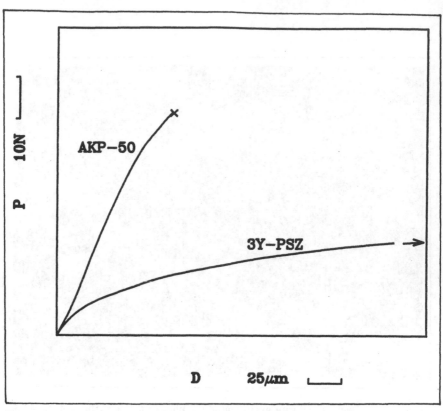

Figure 3   Load (units of 10N) vs. deflection (µm) for AKP5O Al_2O_3 and 3Y–PSZ at 1300°C.

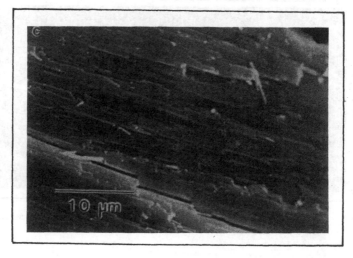

Figure 4   A stepped crack path between the layers of aragonite and along the layers of conchiolin of a strombus-gigas mollusk shell (after Reference [8]).

react with the ions of opposite sign in the particle-associated double-layers to form neutral molecules. This process will also thin the double-layers and allow sufficiently close particle approach for coagulation. This mechanism is also enhanced by the double-layer distortion as the ions in the double-layer "tail" are weakly held so will react with the counter-ions more readily.

The role of charges in the EPD process requires the dielectric constant (e) of the suspension be maximum. Ideal would be water (e = 80) but electrolysis occurs so organic liquids (methanol (33), ethanol (24), propanol (18) and acetone (21) are used. The present investigations utilized ethanol and the natural ions in the Al_2O_3 and ZrO_2 powders.

The mode of transformation from deflocculated to coagulated particles is an advantage of the EPD process. The solids concentration of the suspension is low (<3 v/o) so there is minimal particle interaction. The deposition rate is, however, comparable with slip or pressure casting because it is controlled by the applied voltage. Adjustment of the latter can control the deposition rate to be constant, a degree of control not available in slip-casting; green densities of 60% theoretical can be achieved. At such low concentrations of particles there are no thixotropic tendencies and, as the deflocculated-coagulated transformation is induced electrically, the process is controllable. The elimination of soft agglomerates by mixing and hard agglomerates and inclusions by settling is also much more effective in the low volume-percent-solid suspensions of EPD and, as the greenware contains no organics, no burn-out procedures are required.

Two suspensions were formulated (10 w/o of Al_2O_3 (Sumitomo, Tokyo, AKP-50) and 10 w/o of 3% yttria stabilised zirconia (YPSZ) (Tosoh, Tokyo, TZ-3Y)) in ethanol, sequentially electrophoretically deposited and the composites sintered at 1550°C for 6 hrs. with heating and cooling rates of 300°C/hr.

Figure 5 (a) is a low magnification micrograph of an 80 layer microlaminate of total thickness ~1.5 mm. Figure 5(b) is a higher magnification of 5(a) showing the thickness of the YPSZ layer is ~2 μm i.e. 3 grain thicknesses. The interface uniformity is in the submicron range and it is well bonded. The thickness of the Al_2O_3 layer in these micrographs is ~12μm. The overall volume percentage YPSZ is ~15% in keeping with the low level of conchiolin between the aragonite layers of mollusk shell. The composite has a porosity <0.5 v/o. The Al_2O_3 grain size is large but can be reduced by inclusion of ZrO_2 (Figure 6). Figure 7 displays the controllability of the EPD process; the YPSZ layers have uniform thickness ~7 to ~18 μm. Composites of total thickness >3 mm have been synthesised.

The perfection of the Al_2O_3/ZrO_2 interface in these laminates means orthogonal cracks traverse unimpeded. Inclusion of lanthanium aluminate [10] layers has thus been explored to control the propagation of such cracks. La_2O_3/Al_2O_3 powder mixtures in suspension were deposited between the Al_2O_3 and ZrO_2 layers. The La_2O_3 and Al_2O_3 react at sintering temperatures to produce platey lanthanium aluminate, porous intervening layers (Figure 8). The porosity can be eliminated by

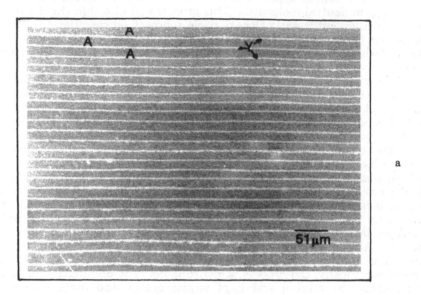

Figure 5   (a)   an 80-layer Al$_2$O$_3$ (grey)-YPSZ (white) electrophoretically deposited,
                   1550°C-sintered laminar composite.
            (b)   higher magnification micrograph of 5(a).

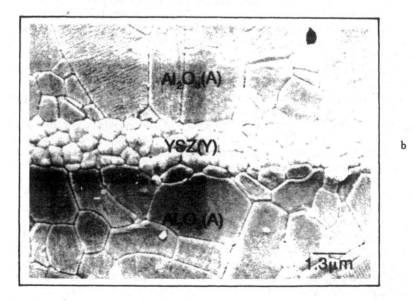

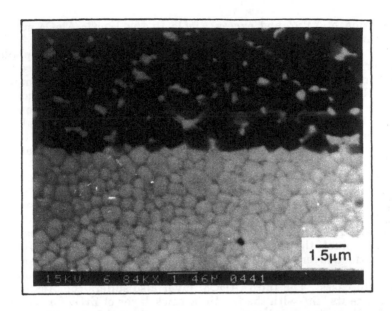

Figure 6   A ZrO_2/Al_2O_3 layer and a YPSZ layer of an electrophoretically deposited laminate composite.

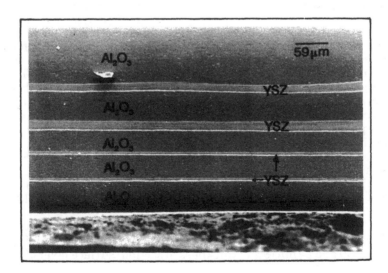

Figure 7   Layers of YPSZ (5, 8, 20 and 15 μm thickness) in an Al_2O_3/YPSZ laminar composite.

inclusion of YPSZ and excess Al_2O_3 in the La_2O_3-Al_2O_3 suspension (Figure 9).

The effort to reproduce the stiff/compliant layer morphology of natural ceramic composites overlooks one important difference between ceramics and mollusks; ceramics operate at elevated temperature ($\geq 1300°C$) and must be cooled during normal use. Such a quench induces large tensile stresses in the surface and laminate design must make allowance therefor. This is accomplished by inclusion of a high thermal expansivity layer beneath the surface layer, eg. MgO ($\alpha = 13.7 \times 10^{-6}/°C$) under a thin ZrO_2 ($10.0 \times 10^{-6}/°C$) surface layer. The latter offers the added advantage of a "plastic" surface layer at operating temperatures. A layer of Al (8% the composite thickness) was found to markedly increase the toughness of a brittle metal-matrix composite (Al/SiC) [11]. Simple stress calculations for a 1550°C quench suggest the MgO would suffer 1200 MPa tension. To moderate this value, spinel ($MgAl_2O_4$: $\alpha = 7.7 \times 10^{-6}/°C$) could be added to the MgO suspension. At 30 v/o spinel, the tension is reduced to ~600 MPa. The MgO of the subsurface layer will locally dissolve in the outer YPSZ layer and also form spinel with the Al_2O_3 of the third layer. Good bonding would result in both cases. The third (Al_2O_3) layer also suffers compression and to enhance its "fit" with the fourth (a thick layer of ZrO_2 for toughness at 1300°C), an intervening functionally-gradiented layer (100% Al_2O_3 to 100% ZrO_2) will be deposited (Figure 10).

THE TOTAL CERAMIC-CERAMIC COMPOSITE FOR $\geq 1300°C$ APPLICATION

The ideal ceramic/ceramic composite has a strong surface and a tough interior. Most components have two surfaces thus most economic is to deposit from the centre in both directions simultaneously. In this case the deposition electrode is located at the centre of the section. Two such electrodes have been explored; ie.; graphite cloth and Al_2O_3 fibres with 10 nm of gold evaporated onto their surfaces. A graphite cloth electrode covered in ZrO_2 and sintered at 1550°C is shown in Figure 11. Subsequent layers can be laid onto this substrate (prior to sintering) as required. The sintered result of the alternate deposition of Al_2O_3 and ZrO_2 on 5 µm diameter, Au-coated, Al_2O_3 fibres* is shown in Figure 12 (fibres shown in section).

Lanthanium aluminate (LA), LA-Al_2O_3-ZrO_2, Al_2O_3 and ZrO_2 layers were subsequently deposited and crack paths in the laminate layers explored by Vickers indentation cracks.

The component deposition is completed by a functionally-gradiented layer, an Al_2O_3 layer, a 70 : 30, MgO : $MgAl_2O_4$ layer (Figure 13) and a thin (8% of thickness of section) surface layer of ZrO_2.

---

*   Almax-C

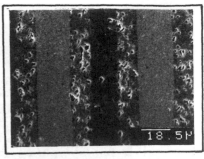

Figure 8 EPD Al_2O_3/YPSZ laminates with porous layers of La-aluminate at each interface.

Figure 9 A "dense" La-aluminate (platey-mineral) layer containing YPSZ and Al_2O_3.

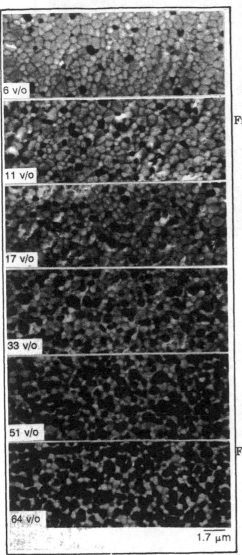

6 v/o

11 v/o

17 v/o

33 v/o

51 v/o

64 v/o

1.7 μm

Figure 10 A functionally-gradiented layer of YPSZ (white) and Al_2O_3 (black) synthesized by electrophoretic deposition and sintered at 1550°C (v/o indicates Al_2O_3 level).

511

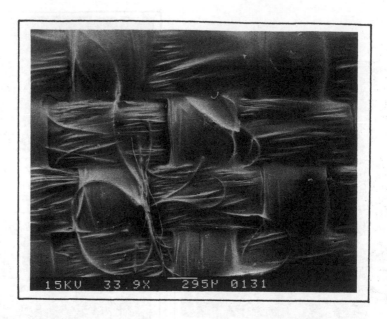

Figure 11   YPSZ electrophoretically deposited on graphite cloth and sintered at 1550°C.

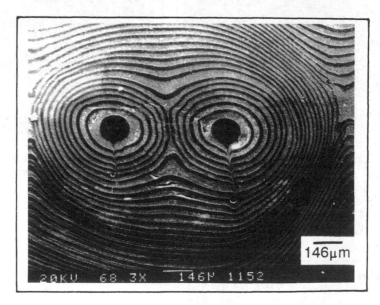

Figure 12   Alternate layers of Al_2O_3 (dark) and YPSZ (white) electrophoretically deposited on Al_2O_3 fibres coated with 10 nm of evaporated gold.

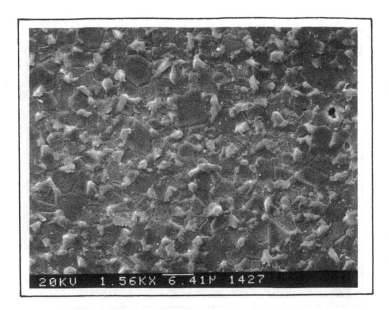

Figure 13　A dense 70 : 30 MgO (dark grey) : MgAl$_2$O$_4$ (light grey) electrophoretically deposited layer fired at 1550°C.

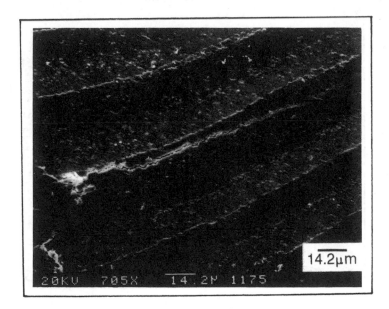

Figure 14　The curved interface between Al$_2$O$_3$ and YPSZ layers deposited on Al$_2$O$_3$-fibres guides a corner crack from a Vickers indent.

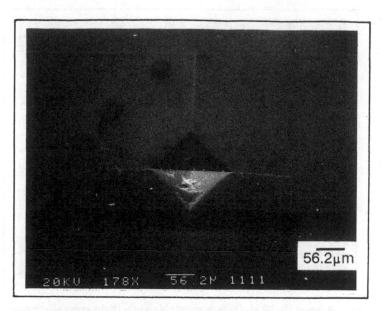

Figure 15   A Vickers indent in a layer of YPSZ. (Note the termination of the bottom corner crack by the La-aluminate layer.)

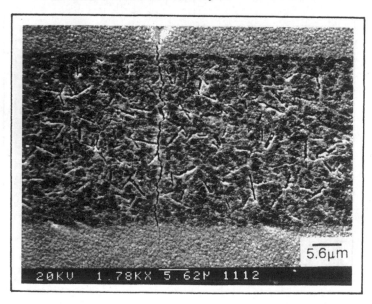

Figure 16   A high magnification micrograph of Figure 15 showing the corner crack terminates in the next YPSZ layer beyond the La-aluminate.

CRACK PATHS IN THE INTERIOR OF THE ELECTRO-PHORETICALLY DEPOSITED ZrO₂/Al₂O₃/La-ALUMINATE, LAMINAR COMPOSITES

The curved ZrO_2/Al_2O_3 interfaces associated with the interior Al_2O_3-fibre electrode tend to confine the indent cracks to the layers (Figure 14). The dense lanthanium aluminate layer is a very effective crack stopper as shown in Figure 15   The indent is located in a layer of YPSZ and three of the corner cracks therein can clearly be seen. The fourth crack is considerably slowed by the La-aluminate-Al_2O_3-ZrO_2 layer and stops just within the next ZrO_2 layer (Figure 16). The porous (100%) La-aluminate layer also considerably disturbs the crack path (Figure 17) and offsets the crack before re-entry into a ZrO_2 layer.

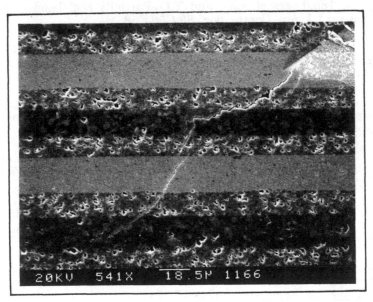

Figure 17   Crack-path perturbation by porous La-aluminate layers along the interfaces of Al_2O_3/YPSZ layers.

CONCLUSIONS

The stiffness of Al_2O_3 and the compliance of ZrO_2 render dense laminates of the two oxides potentially tough ≥1300°C. Al_2O_3/ZrO_2 laminates have been synthesised by electrophoretic deposition followed by sintering at 1550°C. To successfully quench these composites to room temperature it is proposed to prestress their surface layer of ZrO_2 by a subsurface one of $MgO/MgAl_2O_4$. Interspersed layers of porous and dense La-aluminate have also been explored to toughen the deposits. Crack paths from Vickers indents suggest these layers increase the composite toughness.

REFERENCES

1.   Esper, E.J., K.H. Friese and H. Geier, 1984. "Mechanical, Thermal and Electrical Properties in the System of Stabilised $ZrO_2(Y_2O_3)/Al_2O_3$," in <u>Science and Technology of Zirconia II</u>, N. Claussen, M. Rhüle and A.H. Heuer, eds., Columbus, Ohio, USA: The American Ceramic Society, Inc., pp. 528-536.

2.   Tsukama, K. and K. Ueda, 1985. "Strength and Fracture Toughness of Isostatically Hot Pressed Composites of Al_2O_3 and Y_2O_3-Partially-Stabilised ZrO_2." <u>J. Am. Ceram. Soc.</u>, 68:C4-C5.

3.   Tsukama, K. and K. Ueda, 1985. "High Temperature Strength and Fracture Toughness of Al_2O_3 and Y_2O_3-Partially Stabilised ZrO_2/Al_2O_3 Composites." <u>J. Am. Ceram. Soc.</u>, 68:C56-C58.

4.   Kulczycki, A. and M. Wasiucionek, 1986. "Mechanical and Electrical Properties of $6.5Y_2O_3ZrO_2$ Electrolytes Doped with Alumina." <u>Ceram. Int.</u>, 12:181-187.

5.   Chen, I-W., 1990. "Superplastic Ceramics," in <u>Ceramic Powder Science III</u>, G.L. Messing, S. Hirano and H. Hausner, eds., Columbus, Ohio, USA: The American Ceramic Society, Inc., pp.607-617.

6.   Miyayama, M., H. Yanagida and A. Asada, 1986. "Effects of Al_2O_3 Additions on Resistivity and Microstructure of Yttria-Stabilised Zirconia." <u>Bull. Am. Ceram. Soc.</u>, 65:660-664.

7.   Shackleford, J., P.S. Nicholson and W.W. Smeltzer, 1974. "Influence of SiO_2 on Sintering of Partially Stabilised Zirconias." <u>Bull. Am. Ceram. Soc.</u>, 53:865-867.

8.   Laraia, V.J., M. Aindow and A.H. Heuer, 1990. "An Electron Microscopy Study of the Microstructure of the Strombus Gigas Shell." <u>Matls. Res. Sci. Symp. Proc.</u>, 174:117-124.

9.   Sarkar, P., X. Huang and P.S. Nicholson, 1992. "Structural Ceramic Microlaminates by Electrophoretic Deposition." <u>J. Am. Ceram. Soc.</u>, 75:2907-2909.

10.   Tsukuma, K. and T. Takahata, 1987. "Mechanical Properties and Microstructure of TZP and TZP/Al_2O_3 Composites." <u>Mat. Res. Soc. Symp. Proc.</u>, 78:123.

11.   Yost Ellis, L. and J.J. Lewandowski, 1991. "Laminated Composites with Improved Bend Ductility and Toughness." <u>J. Matls. Sci. Lett.</u>, 10:461-463.

Crack Resistance Curves in Layered Ce-ZrO$_2$/Al$_2$O$_3$ Ceramics

D. B. MARSHALL and J. J. RATTO

ABSTRACT

Crack resistance curves have been measured in layered Ce-ZrO$_2$/Al$_2$O$_3$ composites with various layer thicknesses and with Ce-ZrO$_2$ materials of two different initial fracture toughnesses. Increasing the transformability of the Ce-ZrO$_2$ phase shifts the R-curves to higher values of stress intensity factor. However, the slopes of the R-curves were relatively insensitive to layer thickness over the range examined.

INTRODUCTION

A recent study has shown a strong enhancement of transformation toughening in Ce-ZrO$_2$ materials containing layers of either Al$_2$O$_3$ or a mixture of Al$_2$O$_3$ and ZrO$_2$ [1]. The presence of these layers modified the shape and size of the crack tip transformation zone in two ways, both of which increased the degree of crack shielding: the transformation zone spread along the region adjacent to the layers, thus increasing the zone width of a normally incident crack by up to a factor of 10, and the elongated portion of the transformation zone that forms ahead of a crack in uniform Ce-ZrO$_2$ was truncated by the layers. The toughening was evaluated using controlled crack growth experiments in composites containing multilayered regions embedded within uniform Ce-ZrO$_2$. The critical stress intensity factor needed to grow a crack was found to increase from ~5 MPa.m$^{1/2}$ within the Ce-ZrO$_2$, to more than 18 MPa.m$^{1/2}$ after growing through ~18 layers of 30 μm thickness.

In this paper, the effects of varying the dimensions of the layered regions, increasing the number of layers, and increasing the starting fracture toughness of the Ce-ZrO$_2$ material are investigated.

EXPERIMENTS

Composites of Ce-ZrO$_2$ with layers containing a mixture of 50% by volume of Al$_2$O$_3$ and Ce-ZrO$_2$, were fabricated using a colloidal technique described previously [2]. The technique involved sequential centrifuging of solutions containing suspended particles to form the layered green body, followed by drying and sintering at 1600°C for 3h. Use was made of a technique described recently by Velamakanni et al.[3], and Chang et al.[4], in which an aqueous electrolyte (NH$_4$NO$_3$) was used to produce short range repulsive hydration forces and to reduce the magnitudes of the longer

D.B. Marshall and J.J. Ratto, Rockwell International Science Center, 1049 Camino Dos Rios, Thousand Oaks, CA 91360 USA

range electrostatic forces between the suspended particles. Such conditions produce a weakly attractive network of particles which prevents mass segregation during centrifugation, but, because of the lubricating action of the short range repulsive forces, allows the particles to pack to high green density. Two ZrO_2 powders containing 12 mole% CeO_2 were used, one from Tosoh (TZ12Ce) and the other an experimental powder from Ceramatec. The Ceramatec powder was chosen because other studies have shown that high toughness (~14 $MPa.^{1/2}$) can be achieved. In our previous work with the Tosoh powder, the fracture toughness of the Ce-ZrO_2 was ~5 $MPa.m^{1/2}$. The alumina powder was from Sumitomo (AKP30). The composites contained multilayered regions sandwiched between regions of uniform Ce-ZrO_2 as illustrated in figure 1, in order to allow continuous measurement of crack growth, first through the Ce-ZrO_2 then through the layered region.

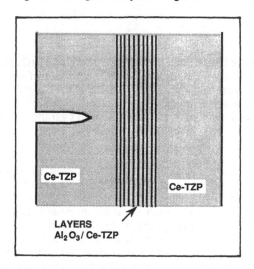

Figure 1. Layered composite test specimen.

Crack growth experiments with notched beams were done in two steps, using two different loading fixtures, which operated on the stage of an optical microscope and allowed high magnification observation of the side of the beam during loading [1]. All experiments were done in a dry nitrogen atmosphere. The dimensions of the beams were approximately 28 x 6 x 1 mm, with the initial notch of 170 µm width and approximately 2 mm depth. First, a stable crack was initiated from the root of the notch under monotonic loading, using a very stiff fixture that made use of WC/Co flexure beams in series with the test specimen [1]. The beams were equivalent to very stiff springs in <u>parallel</u> with the specimen and thus acted as a crack arrester. The initial crack growth was induced without use of a load cell, in order to stiffen the loading system further. After thus growing the crack for ~500 µm, the loading system was changed to include a load cell with conventional four-point loading through rollers in order to allow measurement of the fracture toughness (or crack growth resistance). As the cracks grew stably through the layered composites, the stress intensity factors were evaluated from the measured loads and crack lengths (obtained from optical micrographs) using the expression from [5].

RESULTS

Composites were fabricated using the Tosoh powder with the layer arrangement of figure 1 with four different layer thicknesses, as shown in figure 2. The critical stress intensity factor, K_R, measured as a function of crack growth up to and through the layered regions, is shown in figure 3, with the crack position being measured relative to the beginning of the layers to allow direct comparison of the responses of the different layer thicknesses (crack growth in the Ce-ZrO₂ up to the layers thus corresponds to negative position in this plot). Three of the specimens contained 20 layers and one contained 40 layers. In all cases, K_R increased almost linearly as the cracks grew through the layers. The peak value of K_R coincided with the onset of unstable growth as the crack approached the end of the layered region and the transformation zone penetrated the Ce-ZrO₂ beyond. Therefore, since there is no indication in these data of K_R saturating to a constant value, the maximum toughnesses are limited by the finite extent of the layered regions. The slopes of the R-curves are all similar, although there is an indication that the slope is maximum at a layer thickness of ~35 μm.

Figure 2. Optical micrographs of multilayered composites fabricated with Tosoh powder: darker phase is Al₂O₃, lighter phase is Ce-ZrO₂.

The cracks used for the quasi-static resistance curve measurements of figure 3 were all surrounded by transformation zones that increased in width from ~15 μm

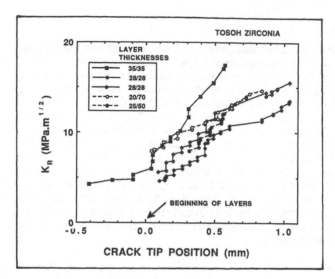

Figure 3. Crack resistance curves from the composites shown in figure 2. Data labelled by layer thicknesses of Al_2O_3/ZrO_2 in microns.

in the $Ce-ZrO_2$ to ~200 µm at the end of the layered region. These transformation zones were observed by optical interference microscopy, which detects out-of-plane surface distortions due to the transformation strain. Increasing zone widths have also been observed for cracks that were growing unstably through the layers. An example is shown in figure 4 for a crack that grew unstably during initiation from the initial notch and arrested after penetrating the first three layers of Al_2O_3/ZrO_2 (the fringes in this micrograph correspond to contours of constant surface uplift).

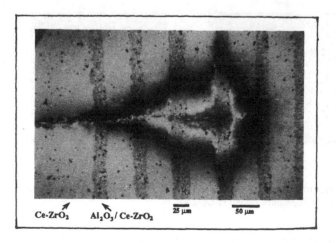

Figure 4. Optical interference micrograph showing transformation zone around crack that initiated unstably in the $Ce-ZrO_2$ and arrested in the layered region.

Resistance curves from two composites fabricated with Ceramatec powder are shown in figure 5. One of these composites was sintered almost to full density and contained 21 layers, whereas the other contained 10% porosity and 41 layers. In both composites, K_R increased almost linearly, with similar slope, as the cracks grew through the layers. However, in the fully dense composite, the R-curve was displaced to larger values of K_R by an amount (~4 MPa.m$^{1/2}$) equal to the difference in fracture toughnesses of the Ce-ZrO₂ materials (≈5-6 MPa.m$^{1/2}$ in the low density material, c.f., 9-10 MPa.m$^{1/2}$ in the higher density material). The peak values of K_R were limited by unstable crack growth as the cracks approached the ends of the layered regions and transformation zones penetrated the uniform Ce-ZrO₂.

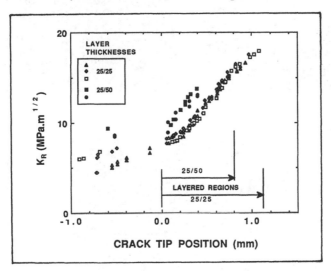

Figure 5. Crack resistance curves for multilayered composites fabricated from Ceramatec Ce-ZrO₂ powder. Data labelled by Al₂O₃/ZrO₂ layer thickness in microns.

The widths of the transformation zones in the composite containing the higher toughness Ceramatec Ce-ZrO₂ were much larger than in the other materials. The initial zone width within the Ce-ZrO₂ was ~200 μm (Figure 6a, c.f.,~15 μm in the Tosoh material). After the crack had penetrated ~10 of the layers, the transformation zone had spread to a width of ~3 mm (Figure 6b). At this stage, the front of the zone penetrated the uniform Ce-ZrO₂ beyond the layered region. The crack then grew unstably and arrested after extending ~1 mm into the Ce-ZrO₂. The zone width within the Ce-ZrO₂ beyond the layered region (Figure 6c) was much larger (~1.5 mm) than the steady-state zone width for a crack grown in this Ce-ZrO₂ material (such as in front of the layers, where the zone width is ~200 μm), but smaller than the zone width within the layers. This result suggests that the zone width of the unstable crack as it exits the layered region is determined by the applied stress intensity factor, as it would be for a quasi-statically growing crack (zone width approximately proportional to K^2). The measured increase in zone width from 200 μm to 1.5 mm would correspond to an increase in K by a factor of approximately 2.7, roughly consistent with the observed increase in K at the point of instability.

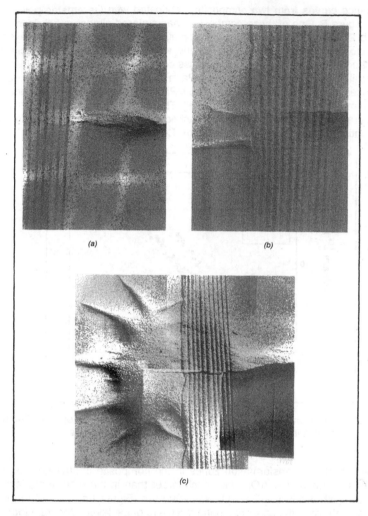

Figure 6. In situ optical micrographs (Nomarski interference) from specimen of figure 5 (25 μm Al_2O_3/ZrO_2 layer thickness, 50 μm $Ce-ZrO_2$) at several stages of crack growth:
    (a) crack within uniform $Ce-ZrO_2$,
    (b) crack extended partly through multilayered region, and
    (c) after crack grew unstably out of multilayered region.

DISCUSSION

All of the composites tested exhibited crack resistance curves that increased almost linearly as the cracks grew through the layered regions. The slopes of the R-curves were similar in composites with various layer thicknesses, although the slope of the R-curve from the composite with the intermediate thickness layer (35 μm) was significantly higher than the others. Although the range of layer thicknesses tested was not large (~25 μm to 70 μm), these results do show that the previously observed toughening effect is not limited to a narrow range of layer thicknesses. The peak toughnesses measured were in the range 14-18 $MPa.m^{1/2}$, limited in all cases by the finite extent of the layered regions (there was no evidence of an approach to steady state in the R-curves).

The large transformation zones observed in the composite with the higher toughness Ceramatec ZrO_2 suggest that very high fracture toughnesses could be achieved in a composite containing more layers. With the observed zone width (which had not saturated) of 3 mm and the relation $K_s \propto \sqrt{w}$ [6,7], where K_s is the steady-state shielding stress intensity factor and w is the zone width, a fracture toughness of more than 30 $MPa.m^{1/2}$ would be predicted.

ACKNOWLEDGEMENT

Funding for this work was provided by the U.S. Air Force Office of Scientific Research under Contract No. F49620-89-C-0031. The authors are grateful to Prof. F.F. Lange of the University of California, Santa Barbara, for discussions that led to development of the colloidal technique used to consolidate the multilayered composites, and R. Cutler of Ceramatec for supplying the Ce-ZrO_2 powder used to fabricate the highly transformable material.

REFERENCES

1.   Marshall,D.B., J.J. Ratto and F.F. Lange, 1991. "Enhanced Fracture Toughness in Layered Microcomposites of Ce-ZrO_2 and Al_2O_3." <u>J. Am. Ceram. Soc.</u>, 74(12): 2979-2987.

2.   Marshall,D.B., 1992. "The Design of High Toughness Laminar Zirconia Composites." <u>Bull. Amer. Ceramic Soc.</u>, 71(6):969-973.

3.   Velamakanni,B.V., J.C. Chang, F.F. Lange and D.S. Pearson, 1990. "New Method for Efficient Colloidal Particle Packing Via Modulation of Repulsive Lubricating Hydration Forces." <u>Langmuir</u>, 6:1323-1325.

4.   Chang,J.C., F.F. Lange and D.S. Pearson, "Centrifugal Consolidation of Al_2O_3 and Al_2O_3/ZrO_2 Composite Slurries vs Interparticle Potentials: Particle Packing and Mass Segregation." <u>J. Am. Ceram. Soc.</u>, in press.

5.   Tada,H., 1985. <u>The Stress Analysis of Cracks Handbook,</u> 2nd Edition, St. Louis, USA:Paris Prod.Inc.

6.   McMeeking,R.M. and A.G. Evans, 1982. "Mechanics of Transformation Toughening in Brittle Materials." <u>J. Am. Ceram. Soc.</u>, 65(5):242-246.

7.   Marshall,D.B., A.G. Evans and M. Drory, 1983. "Transformation Toughening in Ceramics," in <u>Fracture Mechanics of Ceramics,</u> Vol. 6, R.C. Bradt, A.G. Evans, D.P.H. Hasselman and F.F. Lange, eds., New York, NY, USA:Plenum Press, p289.

Bonding of Zirconia Ceramics and Strengthening of Adhesion of Zirconia Coatings by Electric Field Assisted Treatment

AKIRA OHMORI, KATSUYUKI AOKI and SABURO SANO

ABSTRACT

The phenomena occurring during the electric-field-assisted bonding of zirconia ceramics to copper and nickel plate and during strengthening of adhesion of plasma sprayed zirconia coatings on copper are clarified. It was found that a Zr-Cu-O layer formed at the cathodic interface between the zirconia and copper, causing adhesion. Similar results were obtained in the bonding of zirconia ceramic to nickel plate.

In the case of a zirconia coating sprayed onto a copper substrate, a similar reaction layer also formed at the cathodic interface, significantly increasing the adhesive strength.

INTRODUCTION

Zirconia ceramics are very attractive engineering materials because of their high toughness [1]. One of the problems encountered in the successful use of ceramics is how to satisfactorily bond them together or to a metal. Generally, this is achieved by brazing, solid-state diffusion bonding or fusion-welding by laser beam [2]. A strong bonded joint is usually achieved through the formation of a reaction layer at the bonded interface.

In solid-state diffusion bonding of solid electrolyte ceramics, an applied electric field can force the mobile ion in the ceramic to transport to one electrode. In the case of bonding of a glass to a metal Na^+ ion in the glass is forced to move from the anode side to the cathode side and a diffusion layer is formed at the anode interface to bond the glass and metal [3]. Zirconia ceramics are solid electrolytes in which oxygen-ion transportation occurs under an applied electric field [4]. The transport of oxygen-ions in zirconia leads to deficiency at the cathode

Akira Ohmori, Research Center for High Energy Surface Processing (Thermal Spraying Center), Welding Research Institute, Osaka University, 11-1 Mihogaoka, Ibaraki, Osaka 567, Japan
Katsuyuki Aoki, Graduate Student (Present address: Ebara Corporation, 4-2-1 Fijisawa, Fujisawa, Kanagawa 251, Japan)
Saburo Sano, Graduate Student (Present address: Government Industrial Research Institute Nagoya, 1-1 Hirate, Kita-ku, Nagoya, Aichi 462, Japan)

producing zirconium metal which may assist the formation of a bond.

Plasma-sprayed zirconia coatings are well known as thermal barrier coatings but the adhesion of the coating to the metal substrate is very limited because it depends mainly on mechanical interlocking at the interface. The formation of a chemical diffusion layer at the interface usually leads to a remarkable improvement of adhesion [5] and may be expected through heat-treatment under the assistance of an electric field.

In the present report, the phenomena occurring at the interface during electric-field-assisted bonding were investigated in order to clarify the effectiveness of applying an electric field in bonding of zirconia ceramics to metal and improving the adhesive strength of plasma-sprayed zirconia coatings.

MATERIALS AND EXPERIMENTAL PROCEDURE

MATERIALS

Commercially available 15mol%MgO-PSZ, Y_2O_3-PSZ and industrial grade pure copper and nickel were used. Cylindrical ceramic specimens 3mm long and 10mm diameter were used.

Zirconia powders stabilized with 8wt%Y_2O_3, 24wt%MgO and 25wt%CeO_2-3wt%Y_2O_3, respectively were plasma-sprayed onto industrial grade pure copper plate of 10mm diameter. The surface of the copper was polished with grade 600 emery paper and grit-blasted before spraying. Plasma spraying was performed with a Plasmadyne SG-100 spray gun. Zirconia coatings had a thickness of 0.15-0.20mm.

EXPERIMENTAL PROCEDURE

The specimen arrangement for bonding is shown in figure 1(a). To prevent metal oxidation, the experiment was carried out in a vacuum of 1.33×10^{-3}Pa. The chamber was evacuated and the sample heated to the desired temperature. An external pressure of 0.5MPa for copper and 20MPa for nickel was applied to the electrode in order to achieve better electrical contact. After the desired temperature was reached, a DC voltage was applied and the current kept constant for the required

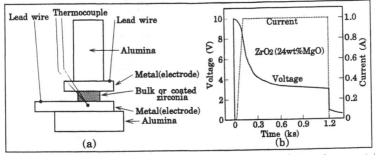

Figure 1 a) Arrangement of sample for field assisted bonding of zirconia to metal and b) typical voltage and current curves during field assisted bonding.

time despite the decrease of electric resistance. Figure 1(b) shows a typical record of change in current and voltage during electric-field-assisted treatment. During the treatment of zirconia coatings, the sample was set so that the coating was on the cathode side of the sample. Table I shows typical conditions used in electric-field-assisted bonding and treatment.

Cross-sections of bonded joints and treated coating-substrate interfaces were examined with optical microscopy, scanning electron microscopy (SEM) and analyzed by energy-dispersive X-ray (EDX) analysis.

RESULTS

ELECTRIC-FIELD-ASSISTED BONDING OF ZIRCONIA CERAMICS TO COPPER AND NICKEL

Bonding of Y_2O_3-PSZ Plate

Figure 2 shows the microstructure of a cross section of a $Cu/PSZ(Y_2O_3)/Cu$ specimen bonded at 973K for 3.6ks under an electric field of 100V at a current of 1A (current density : 2550mA/cm^2). Reaction layers formed at both interfaces. At the cathodic interface the reaction layer consisted of a homogeneous Cu/Zr alloy as shown by

TABLE I ELECTRIC FIELD ASSISTED TREATMENT CONDITIONS

Parameters	Electric field assisted treatment of coating	Electric field assisted bonding of ceramics
Temperature	~1073K	~1073K
Pressure	~0.5MPa	~20MPa
Voltage	10V	~100V
Current	~2A	~2A
Atmosphere	1.33×10^{-3}Pa	1.33×10^{-3}Pa

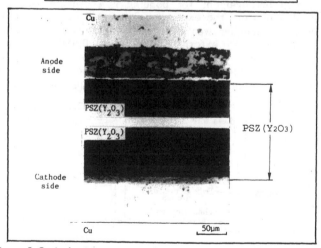

Figure 2 Optical microstructure of anode and cathode interfaces of a $Cu/PSZ(Y_2O_3)/Cu$ joint.

the EDX traces in figure 3. At the anodic interface, a thick layer of Cu_2O was formed, with a thin metallic copper layer between it and the zirconia.

A $Ni/PSZ(Y_2O_3)/Ni$ joint was bonded at 1073K for 3.6ks under an electric field of 100V at a current of 1A. Figure 4 shows the EDX line analysis for Ni and Zr across the cathodic interface. A Ni-Zr reaction layer is formed which is Ni-rich near the nickel side and Zr-rich at $PSZ(Y_2O_3)$ side. However, no clear formation of nickel oxide was found at the anodic interface.

Bonding of MgO-PSZ Plate

A $Cu/PSZ(MgO)/Cu$ joint was bonded at 1073K for 1.2ks under an electric field of 50V at a current of 1A. Reaction layers again formed at both interfaces, although that at the cathodic interface was discontinuous, as shown in figure 5. A $Ni/PSZ(MgO)/Ni$ composite formed at 1073K for 3.6ks under an electric field of 100V at a current of 1A also had a discontinuous alloy layer at the cathodic interface but no nickel oxide layer was observed at the anodic interface.

The experiments in which zirconia was bonded to metals without an applied electric field resulted in no reaction layer at the interface between the ceramic and metal. On the other hand under the influence of an electric field, the bonding of a zirconia ceramic to a metal can be realized by the formation of a metallurgical reaction layer at the cathodic interface. In the case of copper, the bonding can also be achieved by the formation of an oxide layer at the anodic interface. Tests showed that the strength of a bonded $Ni/PSZ(Y_2O_3)/Ni$ joint reached over 15MPa and increased with bonding temperature.

ELECTRIC-FIELD-ASSISTED TREATMENT OF SPRAYED ZIRCONIA COATINGS

Reaction at The Interface

Figure 6 shows typical microstructures from cross sections of zirconia coatings sprayed on copper. These have been heat-treated at 1073K for 1.2ks with an applied 1A electric current and are compared with those

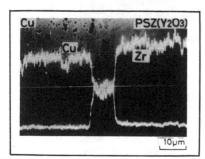

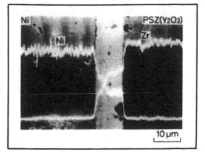

Figure 3 SEM microstructure and EDX element analysis results for the cathode interface of a $Cu/PSZ(Y_2O_3)/Cu$ joint.

Figure 4 SEM microstructure and EDX element analysis results for the cathode side interface of a $Ni/PSZ(Y_2O_3)/Ni$ joint.

of an as-sprayed coating. It can be clearly seen that the reaction layers at both anodic and cathodic interfaces formed in the same way as those described above.

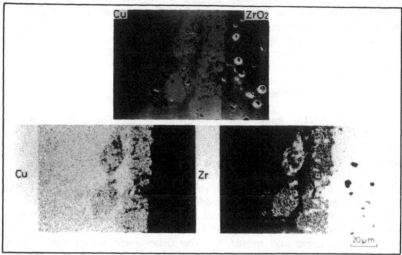

Figure 5 SEM microstructure and EDX element analysis results for the cathode interface of a Cu/PSZ(MgO)/Cu joint.

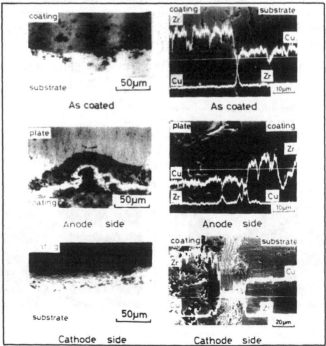

Figure 6 Microstructure and EDX element analysis results of an electric-field-assisted treated $ZrO_2(8wt\%Y_2O_3)$ coating applied on a copper substrate.

Figure 7 shows the dependency of the thickness of reaction layers formed at both anode and cathode side interfaces on time. Growth of both reaction layers is parabolic, suggesting that diffusion is the rate controlling process. Figure 8 shows that an Arrhenius relationship exists between the thickness of the reaction layer and the temperature of treatment.

Effect of Coating Materials on The Formation of Reaction Layers

$ZrO_2(25wt\%CeO_2-3wt\%Y_2O_3)$ coating sprayed on copper, heat-treated at a temperature of 1073K for 1.2ks with a 10V and 1A electric current, forms a Cu_2O layer at the anode interface with a clear thin metallic layer between the Cu_2O and the zirconia coating. However, only a thin reaction layer was formed at the cathode interface. A $ZrO_2(24wt\%MgO)$ coating was sprayed on copper and heat-treated at 1093K for 1.2ks with a 10V and 1A electric current. Similar reaction layers were formed, although they showed some discontinuity.

Under the same treating conditions, in particular, a similar quantity of the electricity, the thickness of the Cu_2O layer formed at the anode interface is about the same. However, the reaction layer at the cathode interface between the coating and substrate was greatly affected by the

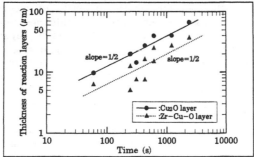

Figure 7 Dependency of the growth of reaction layers at the anode and cathode interfaces on time during electric field assisted treatment of $ZrO_2(8wt\%Y_2O_3)$ coatings.

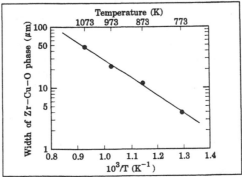

Figure 8 Dependency of the growth of Zr-Cu-O phase at the cathode interface on temperature during electric field assisted treatment of $ZrO_2(8wt\%Y_2O_3)$ coatings.

coating materials used. In the case of a $ZrO_2(8wt\%Y_2O_3)$ coating, a homogeneous reaction layer is more easily formed.

Adhesive Strength of Sprayed Zirconia Coatings

Table II shows the adhesive strength of zirconia coatings sprayed on copper after electric-field-assisted treatment compared with those of as-sprayed coatings and heat-treated coatings. The adhesive strength of the coatings after electric-field-assisted treatment were over 2 to 3 times that of the as-sprayed ones except for $ZrO_2(25wt\%CeO-3wt\%Y_2O_3)$ coatings. Almost all samples fractured in the zirconia coating after electric-field-assisted treatment, suggesting that the adhesive strength of the coatings was larger than the observed values.

DISCUSSION

Present results clearly show that the interfacial reaction products formed during electric-field-assisted bonding of zirconia ceramics to metals are similar regardless of whether the ceramic is present as a coating or as bulk material. When copper was used as anode, Cu_2O was formed and since the current is mainly carried by oxygen ions, the amount of oxide formed depends on the quantity of electricity. Therefore when copper was used as substrate, the thickness of Cu_2O formed was almost constant. With nickel, no clear oxide layer was formed, suggesting that oxygen ions were evolved as oxygen gas.

At the cathodic interface, the transport of oxygen ions will cause a deficiency of oxygen ion, leading to the production of zirconium metal. Formation of the reaction layer is controlled by diffusion. The layered structure formed at the interface between copper and $PSZ(Y_2O_3)$ suggest the occurrence of mutual diffusion. Figure 9 shows the microstructure of a cross section of a $PSZ(Y_2O_3)/Cu$ joint bonded by field assisted bonding, in which a tungsten mark wire was originally pressed into the copper. The surface of the tungsten wire appears in the middle of the joint after bonding confirming the mutual diffusion. The apparent activation energy for diffusion in the zirconia-copper system was estimated from figure 8 to be about 112kJ/mol.

The formation of a reaction layer can be affected by the type, amount and distribution of stabilizer in zirconia, because it may inhibit diffusion. With increase in the amount of stabilizer, formation of a homogeneous reaction layer along the interface would become difficult. Therefore, the reaction layer at the interface between the $ZrO_2(8wt\%Y_2O_3)$ coating and copper substrate was more homogeneous

TABLE II ADHESIVE STRENGTH OF ELECTRIC FIELD ASSISTED TREATED ZIRCONIA COATINGS

Coatings	As-sprayed	Heat-treated	Electric-field assisted treated	(Quantity of electricity)
$ZrO_2(25wt\%CeO_2-3wt\%Y_2O_3)$	4.3MPa	4.4MPa	5.5MPa	(900C)
$ZrO_2(24wt\%MgO)$	3.0MPa	1.7MPa	6.7MPa	(116C)
$ZrO_2(8wt\%Y_2O_3)$	2.9MPa	3.0MPa	9.0MPa	(360C)

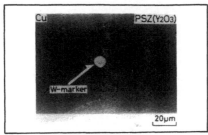

Figure 9 SEM microstructure of the cathode interface of a Cu/PSZ(Y$_2$O$_3$)/Cu joint with a tungsten marker.

than with other coatings.

CONCLUSIONS

The phenomena occurring at the interface between zirconia ceramics and metals under an applied electric field were investigated using sandwiched bulk zirconia and coatings sprayed on copper substrates. It was found that the transportation of oxygen-ions from the cathode towards the anode in the zirconia ceramics takes place under the electric field. The oxygen ions react with the metal contacting the ceramic at the anode side to form an oxide. The transportation of oxygen ions causes a deficiency of oxygen in the zirconia and leads to the formation of a reaction layer at the cathode interface. The formation of such reaction layers can contribute to strong bonding between zirconia ceramics and metals such as copper and nickel and an effective improvement of the adhesive strength of zirconia coatings sprayed on copper substrates.

REFERENCES

1. Evance,A.G., 1991."Perspective on the development of High-Toughness Ceramics." Journal of American Ceramic Society, 73(2):187-206.

2. Japanese Welding Society, 1990. Welding-Bonding Handbook, Tokyo, Japan: Maruzen Publishing Co. Ltd., pp.1088-1103.

3. Arata,Y., A.Ohmori, S.Sano and I.Okamoto, 1984. "Pressure and Field-assisted Bonding of Glass to Aluminum." Transactions of JWRI, 13(1):35-40.

4. Arata,Y., A.Ohmori and S.Sano, 1987."Interfacial Phenomena During Field Assisted Bonding of Zirconia to Metal." Transactions of JWRI, 15(2):215-218.

5. Tucker,R.C., 1974."Structure Properties Relationships in Deposits Produced by Plasma-Spray and Detonation Gun Technique." Journal of Vacuum Science and Technology, 11(4):725-734.

CONCLUSION

REFERENCES

ELECTROLYTE BEHAVIOUR AND DEFECTS

Atomic Defects in Yttria Stabilized and Calcia Stabilized Zirconia Determined by Mechanical Loss Measurements

M. WELLER

ABSTRACT

Mechanical loss (internal friction) experiments were applied to study the local crystallographic structure of atomic defects in cubic ZrO_2. Specimens with different longitudinal axis ([100], [110], [111]) were prepared from crystals of ZrO_2-10 mol% Y_2O_3 and ZrO_2-16 mol % CaO. The orientation dependence of the loss maxima indicates that defects of trigonal ([111]) symmetry are present consisting of oxygen vacancies on nearest neighbour sites to the dopant cations. The shape factor of the dipoles was determined as $\delta\lambda \approx 0.12 - 0.14$ for Y_2O_3 and ≈ 0.05 for CaO doped ZrO_2.

INTRODUCTION

Atomic defects in cubic zirconia are a direct consequence of doping with lower valent cations which are required for stabilization of the cubic phase (about 8 mol % Y_2O_3 or about 16 mol % CaO). For charge compensation oxygen vacancies are created, i.e. one vacancy for every two Y^{3+} ions or for every Ca^{2+} ion. Stabilized zirconia exhibits excellent ionic conduction at high temperatures because of migration of oxygen ions via these vacancies. Oxygen ion conduction is also observed in other oxides with the fluorite structure as, e.g., ThO_2 and CeO_2. Both exhibit the cubic fluorite phase as pure oxides, thus allowing study of the influence of doping on electrical conductivity in a rather dilute system. Several reviews exist on this subject [see e.g. 1-3].

Ionic conduction occurs by migration of freely mobile oxygen vacancies thus leading to superior ionic conduction. At lower temperatures the oxygen vacancies (2+ charged) are associated with the lower valent (negatively charged) dopant cations due to Coulomb attraction. Ionic conduction, which requires free oxygen vacancies, is possible only by dissociation of these complexes. As a consequence ionic conductivity at lower temperatures is determined by a higher activation enthalpy than that at higher temperatures

M. Weller, Max-Planck-Institut für Metallforschung, Institut für Werkstoff- wissenschaft
Seestraße 92, D-W-7000 Stuttgart, Germany

since an additional binding energy is involved.

Ionic conductivity is connected with migration of isolated (freely mobile) oxygen vacancies representing *isotropic* atomic defects. Mechanical loss (internal friction), though, gives complementary information, especially at lower temperatures, where association effects prevail. Such mechanical experiments indicate atomic defects which are *anisotropic*, i.e. elastic dipoles (which may constitute electric dipoles as well). Thermally activated reorientation of (anisotropic) dipoles in alternating stress fields (mechanical oscillations) gives rise to loss maxima. Such maxima were observed in several oxides. In early experiments Wachtmann [4] reported loss maxima for CaO doped ThO_2 and Lay and Whitmore [5] for CaO doped CeO_2 which they assigned to defect complexes of oxygen vacancies with dopant atoms. Additional support for this interpretation came from the adjoint existence of dielectric loss maxima. Nowick and coworkers [6-8] later combined measurements of internal friction (anelastic relaxation), dielectric relaxation and electrical conductivity on doped CeO_2. The atomistic model proposed by these authors (e.g. [4,5,8,9]) assumes that oxygen vacancies are located at sites which are nearest neighbours to the dopant atoms and may jump around them. For higher dopant levels additional loss maxima were observed, e.g., in $CeO_2-Y_2O_3$ [6] or tetragonal zirconia (Y-TZP) [9] which were assigned to larger agglomerates of dopant cations with vacancies.

Additional information on defect properties may be obtained from internal friction measurements on *monocrystalline* specimens [10]. Such experiments require single crystals with dimensions of a few cm, from which specimens with different crystallographic orientations may be prepared. Such crystals are now available for yttria and calcia stabilized zirconia. The orientation dependence of the height of each loss maximum gives information on the atomic symmetry of the defect undergoing relaxation. Furthermore, the components of the dipole tensor describing the anisotropy of the local elastic distortion may be estimated. This paper reports on first measurements on yttria and calcia stabilized monocrystals.

EXPERIMENTAL

SPECIMENS

The crystals of cubic zirconia were supplied by the Ceres Co (North Billerica, Ma, USA). They were prepared by the skull melting technique. The ZrO_2-10 mol% Y_2O_3 crystal was about 60mm long and had a 30x30 mm^2 basal plane. This allowed preparation of rectangularly shaped samples with dimensions of about 40x5x1 mm with three different orientations of the longitudinal axis [100], [110] and [111]. The specimens of calcia stabilized zirconia had to be prepared from two crystals with smaller size (about 40 mm long and a 15x15 mm^2 basal plane). This only permitted preparation of two shorter specimens each about 25 mm long. The longitudinal faces of all specimens were ground and polished after sawing. The crystal orientations were determined using Laue back reflexion.

INTERNAL FRICTION MEASUREMENTS

The internal friction measurements were performed with two types of apparatus operating in different frequency ranges.
(i) Low frequency measurements were carried out with an automatic, computer controlled torsion pendulum. The samples were clamped at their ends in special

jaws in which wedges were pressed against two faces of the rectangular bars. Torsional (shear) oscillations may be excited around their longitudinal (vertically oriented) axis in the range of 2 to 15 Hz by suitable adjustment of the inertia of the pendulum. For thermal exchange He-gas with a pressure of $\simeq$ 10 mbar was used.

(ii) High frequency measurements in the kHz range were performed by exciting the horizontally held specimens in flexure eigenvibrations. One (longitudinal) side of the crystals was coated with conductive silver paint required for electrostatic excitation of the oscillations. The samples were supported by two pairs of thin wires (acting simultaneously as thermocouples) which were placed at nodes of oscillation. Sample oscillations were detected by a 60 MHz FM system (for details see [11]). The kHz-apparatus requires a minimum specimen length of 40 mm. As a consequence only the longer specimens from yttria stabilized zirconia could be measured with this apparatus.

The internal friction, Q^{-1}, (representing the mechanical loss angle) was determined with both types of apparatus from the logarithmic decrement, δ, of the freely decaying amplitudes as $Q^{-1} = (\delta/\pi)(1-\delta/2\pi+...)$. Internal friction and oscillation frequency, f, were measured with increasing temperature with constant mean heating rates between 1 and 2 K/min. Additionally, the kHz apparatus allows determination of Youngs modulus, E, from the frequency of the eigenvibrations according to

$$E = 0.9464 \, \rho \, l^4 \, f^2/d^2 \tag{1}$$

(ρ = density, l = length, d = thickness of the specimen) [11]. From the torsion pendulum, absolute values of the shear modulus, which (as E in eq.(1)) is proportional to f^2, cannot be obtained easily.

EXPERIMENTAL RESULTS

Results for *yttria stabilized zirconia* are shown in figures 1 and 2 for three specimens with [100], [110] and [111] orientation of the longitudinal axis. Figure 1 shows the internal friction, Q^{-1}, versus temperature T obtained with the kHz

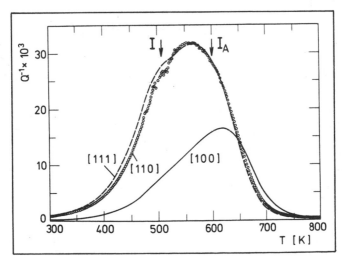

Figure 1: Q^{-1} vs. T of ZrO_2-10 mol % Y_2O_3. Flexure oscillations (f $\approx$ 3 kHz).

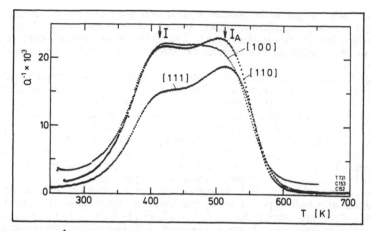

Figure 2: Q^{-1} vs. T of ZrO_2-10 mol % Y_2O_3. Torsional oscillations ($f \approx 3$ Hz).

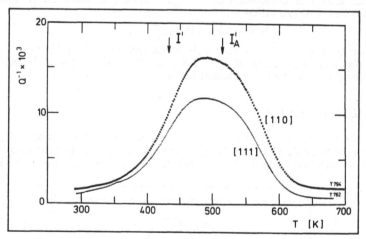

Figure 3: Q^{-1} vs. T of ZrO_2-16 mol % CaO. Torsional oscillations ($f \approx 3$ Hz).

apparatus ($f \simeq 2$ Hz). Figure 2 represents the corresponding measurements with the low frequency torsion pendulum. With the kHz apparatus the following values of Young's modulus were obtained for the three orientations at T = 300 K, according to equation 1

$$E_{100} = 343 \text{ GPa}; \quad E_{110} = 179 \text{ GPa}; \quad E_{111} = 156 \text{ GPa}. \tag{2a}$$

The shear modulus may then be calculated by assuming a mean value of C_{44} ($\cong G$) = 58 GPa from literature [12] as

$$G_{100} = 58 \text{ GPa}; \quad G_{110} = 84 \text{ GPa}; \quad G_{111} = 98 \text{ GPa}. \tag{2b}$$

Measurements on *calcia stabilized zirconia* obtained with low frequency torsion oscillations are shown in figure 3 for two crystals with [110] and [111] orientation of the longitudinal axis.

DATA ANALYSIS

THEORETICAL BACKGROUND

The internal friction measurements on yttria and calcia stabilized zirconia in figures 1 to 3 show a prominent loss maximum between 400 and 600 K (depending on frequency). Apparently, the maximum is composed of two overlapping maxima indicating that more than one species of anisotropic atomic defects are present. The resulting mechanical loss maximum is usually described by a superposition of several Debye maxima

$$Q^{-1}(\omega\tau) = \Delta \frac{\omega\tau}{1+(\omega\tau)^2} \qquad (3)$$

Equation 3 describes a loss maximum centered at $\omega\tau = 1$ with height $Q_m^{-1} = \Delta/2$. Δ is the relaxation strength. The relaxation time τ generally obeys an Arrhenius equation

$$\tau = \tau_\infty \exp(H/kT) \qquad (4)$$

with the activation enthalpy, H, of thermally activated reorientation of the elastic dipoles.

In our experiments, $Q^{-1}(\omega\tau)$ is measured, as usual, at $\omega \approx$ const. as a function of temperature by which τ is varied with temperature according to equation 4. The temperature variation, $Q^{-1}(T)$, may be calculated by combining equations 4 and 5. This gives, if the maximum is positioned at $T = T_p$ (for $\omega\tau = 1$), and the weak temperature dependencies $\Delta(T)$ and $\omega(T)$ are neglected (i.e. $\omega \simeq$ const., $\Delta \simeq$ const.)

$$Q^{-1}(T) \simeq \frac{\Delta}{2} \operatorname{sech} \left[\frac{H}{k} \left(\frac{1}{T} - \frac{1}{T_p} \right) \right] \qquad (5)$$

A more refined treatment including the temperature dependence of $\Delta(T)$ and $\omega(T)$ is given, e.g., in [13].

An analysis of the loss maxima in zirconia shows that the submaxima are substantially broader than simple Debye maxima. This indicates that relaxation occurs by atomic jumps in a multitude of somewhat different environments (due to interaction of defects). This may be taken into account by a continuous distribution of relaxation times. More specifically we consider a Gaussian distribution of relaxation times according to

$$\psi(z) = \frac{1}{\beta\sqrt{\pi}} \exp\left[- \left(\frac{z}{\beta}\right)^2\right] \qquad (6)$$

where $z \equiv \ln(\tau/\bar{\tau})$. We may also assume that the most probable relaxation time, $\bar{\tau}$, obeys the Arrhenius equation given by equation 4.

The magnitude of the relaxation, which determines the peak height, is usually expressed as the relaxation amplitude. By changing from moduli to compliances E^{-1} and G^{-1} the corresponding relaxation amplitudes δE^{-1} and δG^{-1} may then be expressed by the following equations [10]:

$$\delta E^{-1} = \Delta \cdot E^{-1} \qquad (7a)$$

$$\delta G^{-1} = \Delta \cdot G^{-1} \qquad (7b)$$

(E = Young's modulus, G = shear modulus). The relaxation amplitudes are proportional to the concentration of defects (c_0) and their dipole strength (anelastic shape factor) $\delta\lambda$. For simple defects in cubic crystals the following equations hold

$$\delta E^{-1} = A \cdot \frac{c_o v_o}{kT} (\delta\lambda)^2 \cdot F_E (\Gamma) \qquad (8a)$$

$$\delta G^{-1} = B \cdot \frac{c_o v_o}{kT} (\delta\lambda)^2 \cdot F_G (\Gamma). \qquad (8b)$$

Here v_o is the molecular volume and A and B are numerical parameters near unity (see e.g. [10]). The dependence on crystal orientation is expressed in the functions $F_E(\Gamma)$ or $F_G(\Gamma)$. Γ is the orientation parameter for the longitudinal axis

$$\Gamma = \cos^2\alpha\cos^2\beta + \cos^2\beta\cos^2\gamma + \cos^2\gamma\cos^2\alpha \qquad (9)$$

(α, β, γ = angles between longitudinal axis and cubic directions), which is, for example, $\Gamma = 0$ for a [100] orientation, and $\Gamma = 1/3$ for a [111] orientation. In case of polycrystalline samples Γ has to be estimated by averaging over all grains. For a random polycrystalline grain aggregate $\Gamma = 0.2$ is a good average (assuming uniform stress distribution). In other cases, the texture is the primary structural factor determining Γ.

For simple defect symmetries both $F_E(\Gamma)$ and $F_G(\Gamma)$ are linear functions of Γ [10]. For trigonal ([111]-oriented) and tetragonal ([100]-oriented) defects we have

$$F_G = 1-2\,\Gamma; \; F_E = \Gamma, \; \text{trigonal} \qquad (10a)$$

$$F_G = \Gamma; \; F_E = 3\,\Gamma-1, \; \text{tetragonal} \qquad (10b)$$

ANALYSIS OF LOSS SPECTRA

The experimental loss spectra as represented in figures 1 to 3 were decomposed into two submaxima by applying a nonlinear regression procedure. This method is described in [13] for a superposition of Debye maxima and was extended by including lognormal distributions of relaxation times according to equation 6 for every maximum [14] similarly as in [15].

The low frequency measurements on yttria stabilized zirconia of figure 2 were decomposed into two maxima. For the three crystal orientations calculations gave the following results:

Maximum I: T_p = 407 to 420 K; $\beta \approx 4$ (w = 2.8)

Maximum I_A: T_p = 505 to 518 K; $\beta \approx 3$ (w = 2.2)

(w = relative peak width at half maximum). The relaxation strengths allow calculation of the relaxation amplitudes, δG^{-1}, for different orientations using equations 7b and 2b. The results for maximum I are shown in figure 4 as a function of the orientation factor Γ. The loss spectra of kHz flexural vibrations were decomposed in a similar way with the following results:

Maximum I: T_p = 504 to 510 K; $\beta \approx 3$ (w = 2.2)

Maximum I_A: T_p = 590 to 613 K; $\beta \approx 2$ (w = 1.7).

The relaxation amplitudes, δE^{-1}, calculated from the Δ values using equations 2a and 7a are included in figure 4. The quality of the fits was noticeably better for the kHz-flexure oscillations than for the low frequency torsion oscillations, especially in the range of maximum I. In all calculations the mean activation enthalpy was assumed as $H = 1.0$ eV for maximum I and $H = 1.3$ eV for

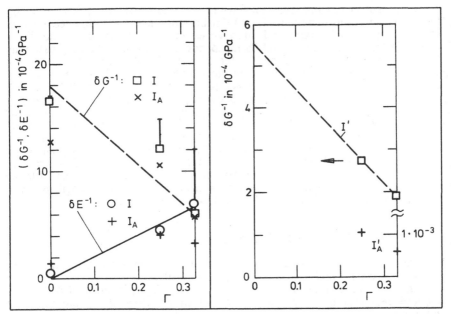

Figure 4: Variation of relaxation amplitudes δE^{-1} and δG^{-1} with orientation parameter Γ for ZrO_2–10 mol % Y_2O_3.

Figure 5: Variation of relaxation amplitude δG^{-1} with Γ for ZrO_2–16 mol % CaO.

maximum I_A (corresponding to $\tau_\infty \simeq 10^{-14}$s) and fixed in the fitting procedures.

The measurements on calcia stabilized zirconia (figure 3) were decomposed into two maxima (I' and I_A') which are positioned closer together than those in yttria stabilized zirconia. The fitting procedure gave the following results:

Maximum I': T_p = 432 and 434 K, β = 2.6 and 2.8 (w ≈ 2)

Maximum I'_A: T_p = 511, and 519 K; β = 5.3 and 4.3 (w ≈ 3)

The fitting quality is much better than for ZrO_2-Y_2O_3. The relaxation amplitude, δG^{-1}, for both maxima is plotted in figure 5, versus Γ. Maxima I' and I_A' correlate well with two peaks reported by Pelaud et al [20] for ZrO_2 – 17.5 % CaO with low frequency (isothermal) internal friction measurements.

ORIENTATION DEPENDENCE OF RELAXATION AMPLITUDES

The orientation dependence of the relaxation amplitudes as represented in figures 4 and 5 leads to the following conclusions:
(a) $\underline{ZrO_2\text{-}Y_2O_3}$. – (i) The relaxation amplitudes $\delta G^{-1}(\Gamma)$ and $\delta E^{-1}(\Gamma)$ of *maximum I* (see figure 4) may well be approximated by linear relations as expected for anelastic relaxation of atomic defects. (ii) The orientation dependence of both relaxation amplitudes agrees well with that for *trigonal* ([$\bar{1}$11] – oriented) defects given by equation 10a). For a tetragonal ([100]-defect)

we would expect just the reverse behaviour. Consequently this defect symmetry may be ruled out for maximum I. (iii) Furthermore, the orientation dependences of δG^{-1} and δE^{-1} of maximum I allow an estimate of the values of the dipole tensor $\delta\lambda$ from equations 8a and 8b independent from each other. By assuming that all oxygen vacancies (5 mol % for 10 mol % Y_2O_3) are associated with yttria ions (representing an upper limit for c_o) we arrive (with $A = B = 4/9$ for trigonal defects) at $\delta\lambda = 0.12$ from $\delta G^{-1}(\Gamma)$ and $\delta\lambda = 0.14$ from $\delta E^{-1}(\Gamma)$. This excellent agreement from two independent types of measurement strongly indicates trigonal defect symmetry. *Maximum I_A* does not exhibit a linear relation between either δG^{-1} or δE^{-1} and Γ. We suppose, then, that defects of lower symmetry and interaction of defects may contribute to this relaxation.

(b) ZrO₂ - CaO. - The result for maximum I' for two crystals is shown in figure 5. The dashed line in figure 4 corresponds to $\delta G^{-1}_{111} / \delta G^{-1}_{100} = 1/3$, which is expected for trigonal symmetry according to equation 10a, and fits the measured data well. This gives with $c_0 = 8$ mol % a dipole strength of $\delta\lambda = 0.05$, i.e. about one half of that for yttria doped zirconia.

DISCUSSION AND CONCLUSIONS

The mechanical loss spectra of yttria and calcia stabilized zirconia exhibit two loss maxima, I and I_A, or I' and I_A', respectively.

The low temperature loss maximum I in $ZrO_2-Y_2O_3$ is interpreted as stress induced reorientation of elastic dipoles which are created by the following reactions

$$Y_2O_3 \xrightarrow{ZrO_2} 2Y'_{Zr} + V_O^{\cdot\cdot} + 3\,O_O^x \tag{11a}$$

$$Y'_{Zr} + V_O^{\cdot\cdot} \longrightarrow (Y_{Zr}'V_O^{\cdot\cdot})^{\cdot} \tag{11b}$$

The assignment of maximum I to complexes of oxygen vacancies and yttrium ions $[Y'_{Zr}V_O^{\cdot\cdot}]^{\cdot}$ is derived from the following sequence of arguments obtained from experiments on tetragonal zirconia [16, 17]. (i) In tetragonal zirconia with 2-3 mol % yttria a mechanical loss maximum is observed at about 380 K for 1 Hz ($H \approx 0.95$ eV). (ii) The height of the maximum increases with yttria contents indicating participation of yttrium ions. (iii) The mechanical loss maximum correlates with a dielectric loss maximum with about equal activation enthalpy. The relaxation times of both maxima differ by a constant factor of about 2. Maximum I in Y-TZP was therefore assigned [16, 17] to thermally activated

reorientation of elastic and electric dipoles consisting of $(Y_{Zr}'V_O^{\cdot\cdot})^{\cdot}$ pairs. The tetragonality of TZP (c/a ratio) is only about 1.02, thus allowing to compare directly results from cubic and tetragonal zirconia. In tetragonal zirconia the peak temperature (and thus the activation enthalpy for reorientation of dipoles) increases with yttria content [17]. This may explain that loss maximum I in cubic zirconia is positioned at a slightly higher temperature than in TZP.

In calcia stabilized zirconia the equivalent defect reaction by doping is

$$CaO \xrightarrow{ZrO_2} Ca''_{Zr} + V_O^{\cdot\cdot} + O_O^x \tag{12a}$$

$$Ca'' + V_O^{\cdot\cdot} \longrightarrow (Ca''_{Zr} V_O^{\cdot\cdot})^x \tag{12b}$$

Quite naturally maximum I' in Ca-doped zirconia is assigned to reorientation of $(Ca''_{Zr} V_O^{\cdot\cdot})^x$ dipoles.

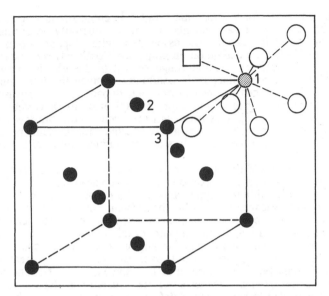

Figure 6: Atomic model for defect in zirconia doped with a lower valent cation
(position 1); □ = oxygen vacancy

The orientation dependence of the relaxation amplitude of both maxima as represented in figures 4 and 5 points to a trigonal symmetry of the dipoles. This leads to the atomic model presented in figure 6. It shows the cubic fluorite cell with position of the Zr ions in a fcc lattice. The sites of oxygen ions corresponding to 8-fold coordination are depicted around one cation (site 1) which may be occupied by a dopant ion (Y^{3+} or Ca^{2+}). The oxygen vacancy (□) may be positioned at eight nearest neighbour sites around the dopant ion with <111> orientation of the dipole axis in each case. This corresponds to a trigonal defect symmetry. Thermally activated jumps of the vacancy (induced by mechanical stress) corresponding to reorientation of the [111]-dipole axis leads to loss maxima I or I', respectively. This eight position model was already elaborated earlier (e.g. [4-6]). One specific prediction that the relaxation time for mechanical and dielectric relaxation of trigonal defects should differ by a factor of two was verified for several (polycrystalline) oxide ceramics with fluorite structure [4-6, 16, 17].

In the present experiments the trigonal atomic site symmetry comes directly from experiments with monocrystals. This strongly indicates that the oxygen vacancies are predominantly associated with dopant ions, even in the highly defective cubic phase of yttria and calcia stabilized zirconia. The shape factor $\delta\lambda$ describing the anisotropy of the distortion pattern created by the dipoles comes out twice as large for yttria stabilized ZrO_2 ($\delta\lambda \simeq 0.12 - 0.14$) as for calcia stabilized ZrO_2 ($\delta\lambda \simeq 0.05$). From the ionic radii of the dopant atoms one might expect the reverse behaviour since Ca^{2+} has a larger ionic radius (0.112 nm) than Y^{3+} (0.102 nm) [19]. From EXAFS measurements on yttria stabilized zirconia [21] it was concluded that oxygen vacancies should be preferentially sited adjacent to Zr ions. This is hardly understandable within the concept of defect associates, i.e. dipoles, which are formed by Coulomb attraction of oppositely charged defects (oxygen vacancy plus dopant cation).

The atomic model for loss maxima I and I' in cubic zirconia agrees well with concepts deduced from measurements of ionic conductivity according to which the oxygen vacancies should be associated with dopant atoms at lower temperatures [1-3]. The activation enthalpy for (localized) reorientation jumps of the dipoles (vacancies), in the range of 1.2 to 1.3 eV, is expected to be different from that controlling ionic conductivity since the latter one requires dissociation of the oxygen vacancies from their dopant atoms. Surprisingly, both activation enthalpies agree within experimental error for a variety of ZrO_2 - Y_2O_3 compounds [16, 17]. This leads us to the conclusion that low temperature ionic conductivity and dielectric loss are connected with the same process, which is supposed to be a displacement or loss current induced by local reorientation of the dipoles [9, 17].

The existence of additional maxima (I_A, I_A') at higher temperatures may be quite naturally assigned to relaxation of oxygen vacancies within Y- or Ca-clusters from which different types and configurations may exist. Interactions with other oxygen vacancies as well as the existence of ordered or partly ordered domains [22] may contribute to the broadening. Atomistic calculations of Baumard and Abelard [18] predict clusters of different size and configuration for heavily doped zirconia. Indications for agglomeration of Y ions in tetragonal zirconia become also apparent from past sinter annealing treatments with tetragonal zirconia [9]. We assume that maxima I_A, I_A' correspond to different configurations of oxygen vacancies within clusters of dopant atoms. Such clusters are expected to have a lower defect symmetry (e.g. orthorhombic). This is also indicated in the orientation dependence of the relaxation amplitudes (Figures 4 and 5).

SUMMARY

Yttria and calcia stabilized zirconia exhibit a highly defective structure due to the high dopant concentrations required for stabilization. Agglomeration and defect interaction lead to two overlapping loss maxima. By separation of the maxima we arrive at specific informations on the atomic configuration of defects (dipoles), i.e. their atomic symmetry and local distortion. Apparently, internal friction measurements are a valuable tool for studying atomic defects in ceramic materials thus enabling a sort of mechanical loss spectroscopy.

Acknowledgements - The author is indepted to the Deutsche Forschungsgemeinschaft, Bonn-Bad Godesberg, for financial support. Thanks are due to Dr. J. Diehl for his support and continuous interest and to Prof. C.A. Wert for helpful discussions and valuable comments.

REFERENCES

1. Kilner, J.A. and B.C.H. Steele, 1981. "Mass-Transport in Anion-Deficient Fluorite Oxides," in Nonstoichiometric Oxides, O.T. Sørensen, ed., New York, NY, USA: Academic Press, pp 233-269.

2. Nowick, A.S., 1984. "Atom Transport in Oxides of the Fluorite Structure," in Diffusion in Crystalline Solids, G.E. Murch and A.S. Nowick eds., New York, NY, USA: Academic Press, pp 143-188.

3. Weller, M., 8-14 September 1991. "Defects in Oxide Ceramics", International Summer School on Mechanical Spectroscopy, Cracow, Poland, to be published.

4. Wachtman, J.B., 1963. "Mechanical and Electrical Relaxation in ThO_2 containing CaO." Phys.Rev., 131 (2): 517-527.

5. Lay, K.W. and D.H. Whitmore, 1971. "Dielectric and Anelastic Relaxation in Ca-Doped Cerium Dioxide." phys.stat.sol.(b), 43: 175-190.

6. Andersen, M.P. and A.S. Nowick, 1981. "Relaxation Peaks Produced by Defect Complexes in Cerium Dioxide Doped with Trivalent Cations." J. de Physique, 42, C5: 823-828.

7. Wang, Da Yu, D.S. Park, J. Griffith and A.S. Nowick, 1981. "Oxygen Ion Conductivity and Defect Interactions in Yttria-Doped Ceria." Solid State Ionics, 2: 95-105.

8. Wang, Da Yu, and A.S. Nowick, 1983. "Dielectric Relaxation in Yttria-Doped Ceria Solid Solutions." J.Phys.Chem.Solids, 44: 639-646.

9. Weller, M., H. Schubert and P. Kounturos, this conference.

10. Nowick, A.S. and B.S. Berry, 1972. Anelastic Relaxation in Crystalline Solids, New York, USA, and London, UK: Academic Press.

11. Weller, M. and E. Török, 1987. "Automatic System for Measurements of Internal Friction and Elastic Moduli." J. de Physique, 48, C8: 371-376.

12. Ingel, R.P. and D.L. Lewis III, 1988. "Elastic Anisotropy in Zirconia Single Crystals." J.Am.Ceram. Soc., 71: 265-271.

13. Haneczok, G. and M. Weller, 1990. "Analysis of Internal Friction Spectra Caused by Snoek-Type Relaxations." J. Less-Comm. Metals, 159: 269-276.

14. Haneczok, G. and M. Weller, to be published.

15. Berry, B.S. and W.C. Pritchet, 1987. "Computer Modelling of Relaxation Peaks Involving a Gaussian Distribution of Activation Energies." phys.stat.sol. (b), 144: 375-384.

16. Weller, M. and H. Schubert, 1986. "Internal Friction, Dielectric Loss, and Ionic Conductivity of Tetragonal ZrO_2 - 3% Y_2O_3 (Y-TZP)." J. Am. Ceram. Soc., 69: 573 - 577.

17. Weller, M. and H. Schubert, 1992. "Defects in ZrO_2-Y_2O_3 studied by Mechanical and Dielectric Loss Measurements", in Solid State Ionics, M. Balkanski, T. Takahashi and H.L. Tuller, eds., North-Holland, Elsevier Science Publ., Amsterdam, NL, pp 569 - 574.

18. Baumard, J.F. and P. Abelard, 1984. "Defect Structure and Transport Properties of ZrO_2-Based Solid Electrolytes." Advances in Ceramics, 12: 555-571.

19. Shannon, R.D., 1976. "Revised Effective Ionic Radii and Systematic Studies of Interatomic Distances in Halides and Chalcogenides." Acta Cryst.A, 32: 751-767.

20. Pelaud, S., P. Mazot and J. Woirgard, 1990. "Etude par frottement intérieur et impédance complexe d'un zircone calciée cubique ZrO_2 - CaO." J. de Phys., 51: 1979 - 1985.

21. Catlow, C.R.A., A.V. Chadwick, G.N. Greaves and L.M. Moroney, 1986. "EXAFS Study of Yttria Stabilized Zirconia." J.Am.Ceram.Soc., 69: 272-277.

22. Pascual, C. and P. Duran, 1983. "Subsolidus Phase Equilibria and Ordering in the System ZrO_2-Y_2O_3." J. Am. Ceram. Soc., 66: 23-27.

Mechanical and Dielectric Loss Measurements in Y_2O_3-ZrO_2 and TiO_2-Y_2O_3-ZrO_2 Ceramics

M. WELLER, H. SCHUBERT and P. KOUNTOUROS

ABSTRACT

Tetragonal zirconia with varying contents of Y_2O_3 (2-4 mol %) and (additional) TiO_2 (0 - 15 mol %) were prepared by coprecipitation. Mechanical loss (internal friction) measurements show a distinct loss maximum at about 380 K (1 Hz); it correlates with a dielectric loss maximum at about the same temperature. The loss maximum is attributed to thermally activated (localized) reorientation of pairs of oxygen vacancies and yttrium ions constituting elastic and electric dipoles. For higher Y_2O_3 contents ($\gtrsim$ 4 mol %) a satellite maximum appears at $\simeq$ 500K (1 Hz); its height increases by post sinter annealing (8 to 36h at $\approx$ 1700K). The 500 K maximum is assigned to larger agglomerates of yttrium ions with oxygen vacancies. Comparative mechanical and electrical impedance measurements on 3Y-TZP (by variation of frequency) indicate that the localized reorientation of the dipoles contributes to the "bulk semicircle."

INTRODUCTION

Defects in metal oxides possessing the fluorite type crystal structure are introduced by aliovalent substitution in the cation sublattice and the corresponding creation of charge compensating oxygen ion vacancies. Mechanical and dielectric loss measurements are well suited for studying of defects in doped oxides as was, e.g., demonstrated for ceria and zirconia [1-4].

Detection of a particular defect implies for the defect of being *anisotropic* thus constituting an elastic (or electric) dipole. Thermally activated reorientation of the dipoles gives rise to a loss maximum at a distinct temperature (or frequency) its height being proportional to the concentration and to the dipole strength of the defects. In tetragonal zirconia (3Y-TZP) a pronounced loss maximum was observed at about 380 K (1 Hz) [3] and assigned to pairs of yttrium ions and oxygen vacancies constituting elastic and electric dipoles. In stabilized cubic zirconia requiring higher (10 mol %) yttria contents

M. Weller, H. Schubert and P. Kountouros, Max-Planck-Institut für Metallforschung, Institut für Werkstoff-wissenschaft, Seestraße 92, D-W-7000 Stuttgart 1, Germany

the loss spectrum is more complicated [5] with several (overlapping) loss maxima indicating agglomeration and interaction of defects.

In this paper the influence of doping on defect configurations in tetragonal zirconia is demonstrated for a larger range of yttria contents (2-4 mol %). The configuration of defect aggregates does not only depend on the dopant level but also on the distribution of the substitutional cations. This may be modified by additional post-sinter annealing treatments. Substitution of Zr^{4+} by Ti^{4+} increases the stability range of the tetragonal phase [6]. The influence of Ti^{4+} on the defect pattern is studied in 3Y-TZP with 5 and 15 mol % TiO$_2$.

Isolated (freely mobile) oxygen vacancies are *isotropic* and thus not directly detectable by mechanical or dielectric loss measurements. However, just the free migration of oxygen vacancies represents the basic mechanism of ionic conduction at elevated temperatures and is usually measured with electrical impedance measurements. For comparing mechanical loss and electrical impedance measurements (both should give complementary information) a new type of apparatus is applied allowing to measure mechanical loss at constant temperature as a function of frequency.

EXPERIMENTAL DETAILS

The investigated tetragonal ZrO$_2$ specimens with various yttria content (2-4 mol %) were prepared by coprecipitation. The coprecipitation technique used for Ti-3Y-TZP powders (5 and 15 mol % TiO$_2$) is described in detail elsewhere [7]. The powders (Tosoh Co., Shimanyo, Yamaguchi 746, Japan) were isostatically pressed at 200 and 630 MPa and subsequently sintered in air between 1600 and 1750 K up to 4h. X-ray diffraction patterns show only the tetragonal phase except in the 4 mol % Y$_2$O$_3$-doped ZrO$_2$ specimen which contained about 15% of the cubic phase.

Thin bar shaped samples with dimensions of 40x5x1 mm^3 were prepared for the mechanical loss (internal friction) experiments. The greater part of measurements was carried out with an inverted torsion pendulum in the Hz-range. The samples were mounted at their ends (in special gripping jaws) and excited to torsional oscillations around the longitudinal axis (see also [5]). Complementary experiments in the kHz-range were carried out with an apparatus in which the same specimen are excited to eigenvibrations in bending [8]. The mechanical loss angle, Q^{-1}, was determined from the logarithmic decrement, ϑ, of the freely decaying amplitudes according to Q^{-1}=(ϑ/π)(1-$\vartheta/2\pi$). In addition, a new type of torsion apparatus was used for measurements of the loss angle at constant temperature in which the frequency can be varied between 10^{-4} Hz and 10 Hz [9]. The specimen is excited to forced vibrations and the loss angle is determined from the phase shift between applied stress (τ) and strain (ϵ) by means of an impedance analyser (Schlumberger 1260).

Disc shaped samples with 10-16 mm ϕ and 1 mm thickness were produced for dielectric loss measurements. The two measuring electrodes were painted as colloidal suspension of silver onto the two opposed faces. The dielectric loss angle, ϕ, was determined from electrical impedance measurements as tan ϕ = 1/(ωRC) = 1/(2πfRC). (R = resistance, C = capacitance of an equivalent circuit).

Mechanical or dielectric loss resulting from thermally activated reorientation of anisotropic atomic defects (dipoles) is usually determined by (one or superposition of several) Debye maxima

$$\tan \delta = \Delta \ \frac{\omega\tau}{1+(\omega\tau)^2}.\tag{1}$$

Equation (1) describes a peak centred at $\omega\tau = 1$ with height $\Delta/2$ ($\Delta =$ relaxation strength). In case of dielectric relaxation δ represents the loss angle ($\delta \equiv \phi$), while in the anelastic case δ is known as the internal friction (mechanical loss) ($\delta \equiv Q^{-1}$). The relaxation time, τ, describing the kinetics of defect reorientation usually obeys an Arrhenius equation

$$\tau^{-1} = \tau_\infty^{-1} \exp(-H/kT)\tag{2}$$

(H = activation enthalpy; τ_∞^{-1} = attempt frequency). According to equations 1 and 2 the Debye maximum may be measured in two ways: (i) as a function of temperature at constant frequency by variation of τ with T; (ii) at constant temperature in forced vibrations by variation of the applied frequency (impedance measurements).

RESULTS

LOSS SPECTRA

(a) *Variation of Loss Spectra with Y_2O_3 Content.* - Figure 1 shows the temperature dependance of the mechanical loss angle as obtained with the torsional pendulum (f = 3 Hz) for polycrystalline specimens with different yttria contents. The loss spectrum for cubic stabilized zirconia (CSZ) with $\simeq$ 10 mol % Y_2O_3 ([110] oriented monocrystal from ref. [5]) is included for comparison. For lower Y_2O_3 contents (2-3 mol %) a single loss maximum appears around 370 K designated as maximum I. With increasing yttria contents maximum I is shifted to slightly higher-temperatures (corresponding to an increase of the activation enthalpy). The 4 mol % sample exhibits an additional satellite peak around 500 K. For the 10 mol % sample this maximum has grown to larger height than maximum I. Figure 2 shows the corresponding measurements for 3 kHz bending oscilations. The loss spectrum exhibits similar variations with yttria contents as observed for low frequencies (temperatures) in figure 1.

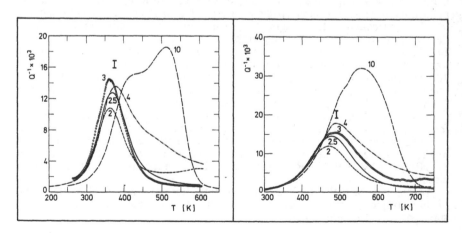

Figure 1: Q^{-1} vs. T (f $\approx$ 3 Hz) for ZrO_2 with different Y_2O_3 contents (2−10 mol%).

Figure 2: Q^{-1} vs. T (f $\approx$ 3 kHz) for $ZrO_2-Y_2O_3$.

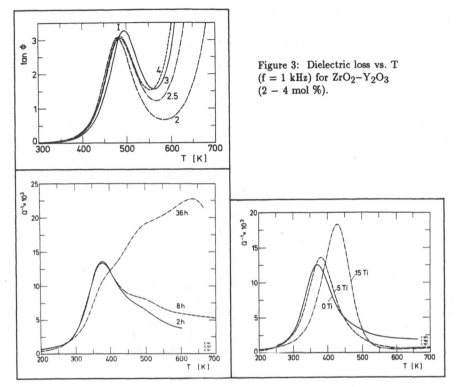

Figure 3: Dielectric loss vs. T (f = 1 kHz) for ZrO_2–Y_2O_3 (2 – 4 mol %).

Figure 4: Influence of post–sinter–annealing treatments on mechanical loss of ZrO_2–4 mol % Y_2O_3 (f ≈ 3 Hz).

Figure 5: Q^{-1} vs. T (f ≈ 3 Hz) of 3Y–TZP with various TiO_2 contents (0, 5 and 15 mol%).

The results of dielectric loss measurements obtained with a measuring frequency of f = 1 kHz are shown in figure 3 for various Y_2O_3 contents. In contrast to the mechanical measurements, the Y_2O_3 content has only a weak influence on the dielectric loss maximum I. Only the peak temperature increases slightly with yttria contents. The steep increase at higher temperatures belongs to a second dielectric loss maximum at about 820 K (1 kHz) which has no mechanical counterpart. The increase in peak temperature of maximum I with Y_2O_3 content corresponds to a slight increase of the activation enthalpies [10].

(b) *Influence of Post–Sinter Annealing Treatments.* - Some of the specimens were subjected to additional annealing treatments after sintering. Figure 4 shows mechanical loss measurements (3 Hz) for specimens with 4 mol % Y_2O_3 after additional annealing for 8 h and 36 h at 1400 °C. The satellite maximum which is only weakly developed after 2 h sintering at 1400 °C is increased by 8 h post–sinter annealing and has grown to larger height than maximum I by 36 h annealing.

(c) *Influence of Doping with TiO_2.* - Figure 5 shows mechanical loss measurements for TZP–3mol % Y_2O_3 containing 5 and 15 mol % TiO_2. The influence of doping with TiO_2 is reflected by an increase of maximum I and a shift to higher temperatures. This corresponds to an increase of the activation enthalpy from 0.9 to 1.1 eV.

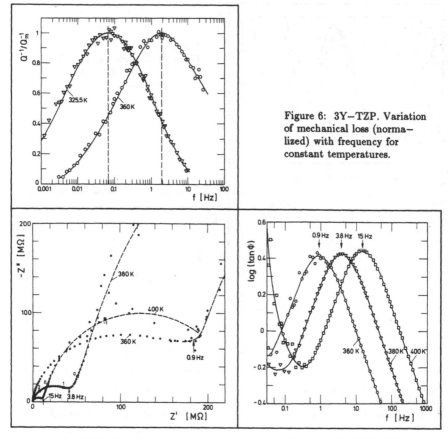

Figure 6: 3Y−TZP. Variation of mechanical loss (normalized) with frequency for constant temperatures.

Figure 7: 3Y−TZP. Complex impedance plots for different temperatures.

Figure 8: 3Y−TZP. Dielectric loss vs. frequency calculated from data of Fig. 7.

CORRELATION OF MECHANICAL AND DIELECTRIC LOSS

Measurements of loss maximum I at constant temperature by variation of frequency were carried out with 3Y-TZP specimens. Figure 6 shows measurements of the mechanical loss angle for T = 325.5 K and 360 K as a function of frequency. Electrical impedance measurements as a function of frequency at different temperatures were performed using the same type of impedance analyser as for the mechanical impedance measurements. Figure 7 depicts impedance plots (−Z''vs Z') (Z' = real part, Z'' = imaginary part) for three temperatures. The loss angle was calculated from these data as $\tan \phi = |Z'/Z''|$. The result is ploted in figure 7 as log (tan ϕ) vs f. Obviously the peak frequencies of the loss maxima in figure 8 correspond to the minima in the impedance plots characterizing the transition from the first semicircle ("bulk") to the second one ("grain boundary").

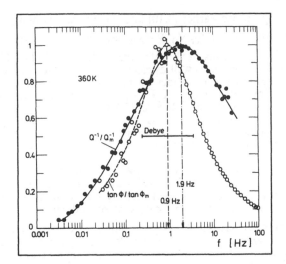

Figure 9: Comparison of mechanical and dielectric loss at 360K of TZP−3Y.

In figure 9 both mechanical and dielectric loss measurements (normalized maxima) are compared directly for one temperature (360 K). As can be seen clearly the dielectric loss maximum is positioned at lower frequency (f = 0.9 Hz) than the mechanical loss maxima (f = 1.9 Hz) corresponding to a ratio of peak frequencies of 2.1. The dielectric loss maximum is 1.4 times broader than a simple Debye maximum (extending over 1.14 decades in f). The mechanical loss maximum is considerably broader (2.4 times a Debye maximum). The evaluation of the temperature and/or frequency position of the mechanical (ml) and dielectric (dl) loss maxima for which ($\omega\tau=1$) gives the following relaxation parameters: H_{ml} = 0.93 eV, $\tau_{\infty(ml)}$ = 8.4·10^{-15} s; H_{dl} = 0.83 eV, $\tau_{\infty(dl)}$ = 4.4·10^{-13} s.

DISCUSSION AND CONCLUSIONS

Tetragonal zirconia, one of the high temperature phases of zirconia, can be stabilized by doping with various amounts of yttria. By addition of yttria, Zr^{4+} cations are replaced by lower valent Y^{3+} cations, and, as a consequence, oxygen vacancies are formed as charge compensating defects

$$Y_2O_3 \xrightarrow{ZrO_2} 2\, Y'_{Zr} + V_O^{..} + 3\, O_O^x \tag{3a}$$

For every two dopant cations one oxygen vacancy is generated. On cooling from the sintering temperature the mobility of the cations is rapidly frozen in and only oxygen vacancies ($V_O^{..}$) are mobile. Coulomb interaction leads then to the formation of associates according to the reaction

$$2Y'_{Zr} + V_O^{..} \longrightarrow (Y'_{Zr}\, V_O^{..})^{.} + Y'_{Zr}. \tag{3b}$$

The net charge of the complexes is compensated by unassociated Y'_{Zr} located elsewhere in the lattice. Further agglomeration may be described by

$$(Y'_{Zr}\, V_O^{..})^{.} + Y'_{Zr} \longrightarrow (Y'_{Zr}\, V_O^{..} Y_{Zr})^x. \tag{3c}$$

As was already proposed earlier [3,10] loss *maximum I* is assigned to thermally activated reorientation of $(Y_{Zr}'V_O^{..})'$-associates by application of an electric or elastic field. This is consistent with the following facts. The $(Y'_{Zr}V_O^{..})'$ complexes represent elastic as well as electric dipoles. Their reorientation occurs by localized jumps of oxygen vacancies around yttrium ions with about the same activation enthalpy ($H_{ml} \approx H_{dl} = 0.9$ eV). The attempt frequencies (τ_∞^{-1}) for reorientation are in the range of 10^{13} to 10^{14} s^{-1} as expected for atomic defects. The height of loss maximum I is supposed to be proportional to the concentration of dipoles, that is (according to equations 3a and 3b) the yttria contents. This indeed observed by the mechanical loss experiments (see figures 1 and 2).

An atomistic model for defects in oxides with the fluorite structure was proposed earlier on the basis of experiments on doped ThO_2 [11] and CeO_2 [12,13]. This model could be verified directly by mechanical loss experiments with single crystals of yttria and calcia stabilized cubic zirconia [5]. These experiments indicate that the $(Y_{Zr}'V_O^{..})'$ complexes causing maximum I exhibit trigonal <111> symmetry. This corresponds to oxygen vacancies positioned on one of (eight) nearest neighbour sites around the dopant atoms. According to Nowick [14] the relaxation times for mechanical and dielectric loss for this defect configuration should differ by a factor of 2 ($\tau_{dl}/\tau_{ml} = 2$).

Tetragonal zirconia exhibits a slightly distorted fluorite structure. The $(Y_{Zr}'V_O^{..})'$ dipoles causing loss maximum I are oriented parallel to the space diagonal of the tetragonal cell corresponding to monoclinic symmetry. However, since the tetragonal distortion in TZP is only small (c/a $\simeq$ 1.02) one may expect that the ratio τ_{dl}/τ_{ml} is close to that in cubic crystals. As shown, e.g., in figure 8 this is indeed observed ($\tau_{dl}/\tau_{ml} = 2.1$). In principle, the activation enthalpy for reorientation of the (same) dipoles by application of mechanical stress or an electrical field should be the same. This is in fact observed in other cubic oxides with the fluorite structure, as e.g. CeO_2 or ThO_2 [11–13] with low doping content. In TZP which requires higher dopant levels for stabilization of the tetragonal phase the slight difference between H_{ml} and H_{dl} (increasing with Y_2O_3 content [10] may be due to different interactions of the dipoles with elastic and electric fields. Elastic (mechanical) fields couple with all defects which are elastically anisotropic, whereas electrostatic fields apply only to defects carrying dipole moment. Interaction of the defects (dipoles) with each other at higher dopant levels may also be different in elastic and electric fields and thus lead to (slightly) different reorientation enthalpies.

For higher yttria contents more complicated (larger) agglomerates may be formed (see, e.g., equation 3c). Consequently the satellite maximum occurring in ZrO_2 with 4 mol % Y_2O_3 at about 500 K for 3 Hz (see figure 1) is assigned to some configuration of $(Y_{Zr}'V_O^{..}Y_{Zr}')^x$ agglomerates. (The 500K-maximum is probably identical with maximum I_A in cubic zirconia [5]). For higher dopant levels we may expect that a certain fraction of two Y^{3+} ions are close together [15] with vacancies somewhere in between. Such complexes may exhibit considerable elastic anisotropy, but a negligable small electrical dipole moment (and thus be electrically inactive, see figure 3).

This interpretation of the high temperature maxima with defect agglomerates is supported by the observation in figure 4 that post–sinter annealing at 1700 K increases this maximum. Agglomeration of Y^{3+} ions, being mobile at that temperature (diffusion enthalpy $\simeq$ 4.5 eV [16]) may occur. The 500 K maximum (I_A) is also observed in non transformable tetragonal (t') phase [17]. Thus we may exclude the possibility that this maximum is connected with the small fraction (15%) of cubic phase present in ZrO_2–4 mol% Y_2O_3.

The assignment of loss maximum I to dipoles leads to conclusions regarding the interpretation of electrical (ionic) conductivity measurements. Reorientation of the dipoles is apparently controlled by jumps of *localized* (bound) oxygen vacancies around yttrium ions. Ionic conductivity, on the contrary, is understood as migration of isolated (dissociated) oxygen vacancies over macroscopic distances. At lower temperatures the oxygen vacancies are bound in dipoles which must be dissociated for long range migration. At high temperatures the dipoles are dissociated. As a consequence ionic conductivity at high temperatures is expected to occur with a lower activation enthalpy H_σ^{HT} than at lower temperatures (H_σ^{LT}). This general behaviour of two regimes in the temperature dependence of ionic conductivity is reported for zirconia (e.g. [18, 19]) and several other doped oxides [20, 21]. At first sight, it might be expected that the activation enthalpy for dipole reorientation (H_{ml}, H_{dl}) corresponding to localized jumps is smaller than that for low temperature conductivity (requiring dissociation of the complexes). There is now evidence from several electrical impedance measurements on 3Y-TZP [3, 22] that at low temperatures H_{dl} and H_σ^{LT} are equal within experimental error ($\simeq 0.8 \pm 0.05$ eV). This was earlier considered by us as accidental [3].

This observation and the correlation between dielectric loss and electrical impedance measurements as presented in figures 7 and 8 leads us to another consideration. Obviously the dielectric loss maxima in figure 8 are connected with the first semicircle of the impedance diagram in figure 7. Originating back to Bauerle [22] the first (high frequency) semicircle is usually interpreted as "bulk" conductivity that means free migration of oxygen vacancies. From the direct correlation of this ("bulk") semicircle with the corresponding dielectric loss maximum and from the equality of the activation enthalpies ($H_\sigma^{LT} \approx H_{dl} \approx H_{ml}$) we conclude that the bulk semicircle is connected with the reorientation of dipoles. Their reorientation in an alternating electrical field causes a displacement (or polarization) current which gives rise to dielectric loss. Further progress in understanding of the underlying atomistic processes is expected from four point DC measurements which are in progress.

Acknowledgements - Thanks are due to Deutsche Forschungsgemeinschaft, Bonn-Bad Godesberg, for financial support.

REFERENCES

1.   Andersen, M.P. and A.S. Nowick, 1981. "Relaxation Peaks Produced by Defect Complexes in Cerium Dioxide Doped with Trivalent Cations." J. de Physique (Paris), 42, C5: 823–828.

2.   Wang, D.Y. and A.S. Nowick, 1983. "Dielectric Relaxation in Yttria-Doped Ceria Solid Solutions." J.Phys.Chem.Solids, 44: 639–646.

3.   Weller, M. and H. Schubert, 1986. "Internal Friction, Dielectric Loss, and Ionic Conductivity of Tetragonal ZrO_2 - 3%Y_2O_3 (Y-TZP)." J.Am.Ceram.Soc., 69: 573–577.

4.   Weller, M., "Defects in Oxides." International Summer School on Mechanical Spectroscopy, Sept. 9–14, 1991, Cracow, Poland, to be published.

5.   Weller, M., "Atomic Defects in Yttria Stabilized and Calcia Stabilized Zirconia Determined by Mechanical Loss Measurements," this conference.

6.   Kountouros P. and G. Petzow, "Defect Chemistry, Phase Stability and Properties of Zirconia Polycrystals," this conference.

7.   Kountouros, P. and H. Schubert, 1991. "Bloating Effect During Sintering of TZP," to be published in the Proceedings of the Second Conference of the European Ceramic Society, Augsburg, Germany.

8.   Weller, M. and E. Török, 1987. "Automatic System for Measurements of Internal Friction and Elastic Moduli." J. de Physique, 48: C8, 371–376.

9.   Weller, M., "Torsion Apparatus for Mechanical Loss Measurements in the Frequency Range of 10^{-4} Hz up to 10 Hz," to be published.

10. Weller, M. and H. Schubert, 1992. "Defects in ZrO_2-Y_2O_3 Studied by Mechanical and Dielectric Loss Measurements." Solid State Ionics, M. Balanski, T. Takahashi and H.L. Tuller eds., Elsevier Science Publ.: pp.569–574.

11. Wachtmann, J.B., 1963. "Mechanical and Electrical Relaxation in ThO_2 containing CaO." Phys.Rev., 131 (2): 517–527.

12. Lay, K.W. and D.H. Whitmore, 1971. "Dielectric and Anelastic Relaxation in Ca-Doped Cerium Dioxide." phys.stat.sol. (b), 43: 175–190.

13. Andersen, M.P. and A.S. Nowick, 1981. "Relaxation Peaks Produced by Defect Complexes in Cerium Dioxide Doped with Trivalent Cations." J. de Physique, 42, C5: 823–828.

14. Nowick A.S., 1985. "The Combining of Dielectric and Anelastic Relaxation Measurements in the Study of Point Defects in Insulating Crystals" J. de Physique, 46, C10: 507–511.

15. Baumard, J.F. and P. Abelard, 1984. "Defect Structures and Transport Properties of ZrO_2 - Based Solid Electrolytes." Adv. Ceramics, 12: 555–571.

16. Oishi, Y., Ken Ando and Y. Sakka, 1983. "Lattice and grain boundary diffusion coefficients of cations in stabilized zirconias." Advances in Ceramics, 7, pp 208–219.

17. Weller, M., A. Foitzik, M. Stadtwald-Klenke and M. Rühle, "Mechanical Loss Measurements on Ferroelastic t'-ZrO_2," to be published.

18. Badwal, S.P.S. and M.V. Swain, 1985. "ZrO_2-Y_2O_3: electrical conductivity of some fully and partically stabilized single grains." J.Mat.Sci.Lett., 4: 487–489.

19. Badwal, S.P.S., 1983. "Electrical Conductivity of Sc_2O_3-ZrO_2 compositions by 4-probe d.c. and 2-probe complex impedance techniques." J.Mat.Sci, 18: 3117–3127.

20. Kilner, J.A. and B.C.H. Steele, 1981. "Mass Transport in Anion-Deficient Fluorite Oxides," in Nonstoichiometric Oxides, O.T. Sorensen, ed., New York, NY, USA: Academic Press, pp. 233–269.

21. Nowick, A.S., 1984 "Atom Transport in Oxides of the Fluorite Structure," in Diffusion in Crystalline Solids, G.E. Murch and A.S. Nowick, eds., New York, NY, USA: Academic Press, pp. 143–188.

22. Leibold B. and N. Nicoloso, 1989. "Impedance and Voltage Relaxation Studies of the Oxygen Sensor Systems $Pt/O_2/YSZ$, $Pt/O_2/TiO_2$ and $Pt/O_2/\delta$-Bi_2O_3." Non Stoichiometric Compounds, J. Nowotny and W. Weppner eds., Kluwer Acad. Publ., pp. 557–579.

23. Bauerle, J.E., 1969. "Study of Solid Electrolyte Polarization by a Complex Admittance Method." J. Phys. Chem. Solids, 30: 2657–2670.

Ionic Conductivity of Tetragonal- and Cubic-ZrO$_2$ Doped with In$_2$O$_3$

L. J. GAUCKLER, K. SASAKI, H. HEINRICH,
P. BOHAC and A. ORLIUKAS

ABSTRACT

The ionic conductivity of tetragonal (t' and t) and cubic ZrO$_2$ doped with 15-45 mol% InO$_{1.5}$ has been investigated by impedance spectroscopy and DC 4-probe measurements up to 1000°C. Both t'- and t-phase polycrystals with the same InO$_{1.5}$ concentration were prepared via cooling from the cubic phase field and low-temperature sintering, respectively. The t-phase reveals higher intragrain ionic conductivity and a lower activation energy, compared to the t'-phase of the same solute concentration. On the other hand, the total conductivity of the fine polycrystalline t-phase is lower than that of the coarser-grained t'-phase because of its larger grain boundary resistivity. The total ionic conductivity of the polycystalline cubic phase near the eutectoid composition is comparable to the total conductivity of 8 mol% Y$_2$O$_3$-doped ZrO$_2$.
Tetragonality and domain size of the t'-phase are strongly dependent on cooling conditions from high temperature.

INTRODUCTION

The dependence of ionic conductivity on solute concentration in zirconia-based systems has been intensively studied [1,2]. Fully-stabilized 8 mol% Y$_2$O$_3$-ZrO$_2$ (FSZ) has higher ionic conductivity in the temperature range of 800-1000°C than tetragonal (TZP) and partially stabilized zirconias (PSZ). It has been shown previously that, in several zirconia-based binary systems, the cubic phase near the low solubility limit transforms to the t'-phase on cooling [3-11]. Good mechanical properties have been observed for t'-phase polycrystals due to its ferroelastic behavior [8,9]. However, only little is known about the electrical properties of the t'-phase [10,11].

The phase diagram of the system ZrO$_2$-InO$_{1.5}$ has been recently revised

L.J.Gauckler, K.Sasaki, P.Bohac, A.Orliukas
Swiss Federal Institute of Technology (ETH-Zürich),
Nichtmetallishe Werkstoffe, CH-8092 Zürich, Switzerland.
H.Heinrich, Institut für Angewandte Physik, CH-8093 Zürich, Switzerland.

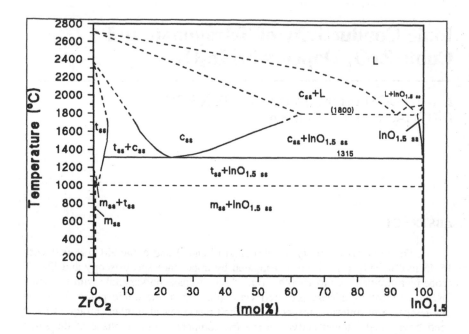

Figure 1: Phase diagram of the system ZrO_2-$InO_{1.5}$ [7]: m_{ss}: monoclinic ZrO_2 solid solution, t_{ss}: tetragonal ZrO_2 solid solution, c_{ss}: cubic ZrO_2 solid solution, $InO_{1.5\ ss}$: bcc $InO_{1.5}$ solid solution, and L: liquid.

[7]. The system possesses an extended cubic phase field, ranging between 12 and 48 mol% $InO_{1.5}$. The cubic phase near the lower solubility limit transforms to the t'-phase. On the other hand, after coprecipitation and calcination, the starting powder in the system consists only of the tetragonal (t) phase up to 50 mol% $InO_{1.5}$. The supersaturation of the solute in this phase remains up to 1000°C because of the slow phase partitioning of tetragonal grains into t+c. Therefore, in the system, it is possible to prepare both t- and t'-phases with the same solute concentration in order to study the effect of crystal structure on ionic conductivity.

In the present study, ionic conductivity is studied in tetragonal (t and t') and cubic ZrO_2 single phase polycrystals with the $InO_{1.5}$ solute concentrations between 15 and 45 mol%. An emphasis is put on the ionic conductivity in the t'-phase and the relations between ionic conductivity and crystalline phases.

EXPERIMENTAL

The starting zirconia powders with $InO_{1.5}$ concentrations of 15, 25, 35 and 45 mol% were prepared via the coprecipitation of hydroxides with ammonia solution. The precipitates were then washed twice with distilled water and once with ethanol, vacuum-dried and subsequently calcined at 800°C for 1 h. For further experimental details see [7]. The powders were uni-axially pressed at 50 MPa, either into pellets of 16 mm diameter and 2 mm thickness for impedance spectroscopy or into bars of dimensions $5\times1.5\times55$ mm^3 for 4-probe DC conductivity measurements. The powder compacts were sintered at 1600°C for 6 h and then furnace-cooled at the rate of 10°C/min. Other powder pellets (15 mol% $InO_{1.5}$) were air-quenched ($\approx$800°C/min) after sintering. In order to prepare t-phase polycrystals, green specimens of 15 mol% $InO_{1.5}$ were isostatically pressed at 300 MPa, sintered at 1000°C for 3 h and furnace-cooled.

Phase compositions of the specimens were analyzed by X-ray diffraction in the 2Θ range 20 to 80° by Cu Kα radiation, after polishing with SiC paper up to grit #4000. The reflections of $(400)_t$ and $(004)_t$, $(400)_{t'}$ and $(004)_{t'}$, and $(400)_c$ between 72 and 76° were used to distinguish between the tetragonal (t and t') and cubic ZrO_2 phases, and to determine lattice parameters calibrated against the Si (331) reflection.

Microstructures of 15 mol% $InO_{1.5}$-ZrO_2, air-quenched and furnace-cooled were characterized by transmission electron microscopy (Philips, CM-30, 300 kV)

DC conductivity was measured by the 4-probe method in the temperature range from 500 to 1000°C using a micro-ohm meter (Keithley, No. 580). AC conductivity measurements were carried out in the frequency range 40 Hz to 1 MHz using an LCR meter (Hewlett Packard, HP 4284A). From the impedance spectroscopy, relaxation frequencies were derived.

RESULTS AND DISCUSSION

PHASE RELATIONS

The cubic phase region in the system ZrO_2-$InO_{1.5}$ extends from 14 to 49 mol% $InO_{1.5}$ at 1600°C, but is thermodynamically stable only above 1315°C (figure 1). However, after cooling from 1600°C at a rate of 10°C/min, the specimens with 25, 35 and 45 mol% $InO_{1.5}$ consist of the cubic phase only, due to the slow kinetics of the eutectoid decomposition.

Lattice parameters and preparation conditions of 15 and 25 mol% $InO_{1.5}$ specimens are summarized in Table I. XRD peaks in the 2Θ range between 72 and 76° are shown in figure 2. The air-quenched 15 mol% $InO_{1.5}$ specimen consisted of the t'-phase only. No peak of the monoclinic phase was detected. Identification as t' rather than c was based on (a) the presence of the $(004)_{t'}$ peak, and (b) the smaller cell dimensions of t' compared with c (figure 2, and table I), which has a constant lattice parameter up to 45 mol %$InO_{1.5}$ [5,7]. Figure 3a shows a TEM dark-field micrograph, in [110] orientation imaged with a (001) reflection, of the air-quenched 15 mol% $InO_{1.5}$ specimen . The presence of domains can be verified. In the specimen, twins associated with the c→t' phase transformation have also been observed. Selected-area diffraction has revealed the {112} tetragonal reflection in the <111> direction.

The t'-phase could be obtained not only after air-quenching but also after furnace-cooling from 1600°C. Figure 2 shows the XRD peaks. In the furnace cooled sample, both the $(400)_{t'}$ and $(004)_{t'}$ peaks have been observed like in the air-quenched sample. The $(400)_{t'}$ peak of the furnace-cooled specimen is lower and broader compared to that of the quenched specimen due to the small domain size of 3 nm.

TABLE I. LATTICE PARAMETERS AND PREPARATION CONDITIONS OF THE
SAMPLES WITH 15 AND 25 MOL% $InO_{1.5}$

Samples	Phase	Sintering condition	Cooling rate -(K/min)	Grain size (μm)	Domain size (nm)	Lattice const. a (nm)	Lattice const. c (nm)	(c/a)
15mol% air-quenched	100% t'	1600°C 6h	800	10-15	20	0.51044	0.51528	1.0095
15mol% furnace cooled	100% t'	1600°C 6h	10	10-15	3	0.51040	0.51636	1.0117
15mol% low T sintering	100% t	1000°C 3h	3	0.1-0.2	---	0.50940	0.51840	1.0177
25mol% furnace cooled	100% c	1600°C 6h	10	10-15	---	0.51201	---	1

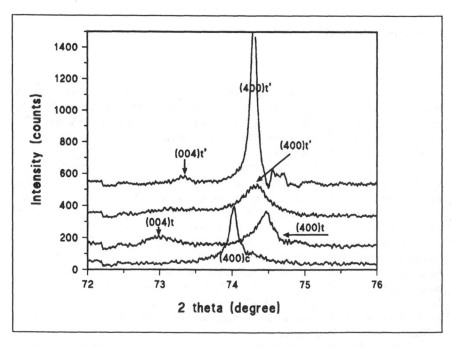

Figure 2: XRD reflections of 15 mol% $InO_{1.5}$-ZrO_2 air-quenched (t'), furnace-cooled at 10°C/min (t'), low-temperature-sintered (t), and 25 mol% $InO_{1.5}$-ZrO_2 (cubic), furnace-cooled.

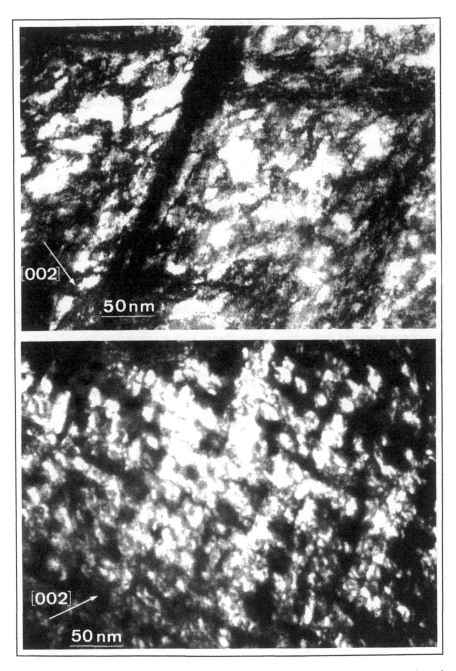

Figure 3: TEM dark-field micrograph in the [110] orientation imaged with a (001) reflection, of 15 mol% $InO_{1.5}$-ZrO_2; (a) air-quenched and (b) furnace-cooled.

A TEM dark-field micrograph of the furnace-cooled specimen is shown in figure 3b. The domain size in the furnace-cooled specimen is 3 nm with high tetragonality (c/a-ratio). The same composition quenched rapidly from the cubic phase field shows a much larger domain size of 20 nm average diameter and a much lower tetragonality than the slow cooled material. These results show that domain size and tetragonality of t' material in this system is depending on thermo-mechanical stresses during cooling.

After the coprecipitation and calcination process, ZrO_2-$InO_{1.5}$ powders from 15 mol% up to 50 mol% $InO_{1.5}$ consist of the t-phase only [7]. This supersaturation of $InO_{1.5}$ in the t-phase remains after sintering at 1000°C, because the segregation kinetics of the solute are very sluggish. To produce a specimen with the t-phase, a powder compact of 15 mol% $InO_{1.5}$-ZrO_2 was sintered at 1000°C for 3h. The XRD peaks of the t-phase in the 2Θ range between 72 and 76° are shown in figure 2 and the lattice parameters are shown in Table I.

IONIC CONDUCTIVITY

Figure 4 shows the intragrain conductivity in the t-, t'- and c-phases, measured by impedance spectroscopy. as a function of reciprocal temperature.Up to 500°C, the t-phase has a higher intragrain conductivity than t', even though the solute concentration is the same in both phases. Both tetragonal phases

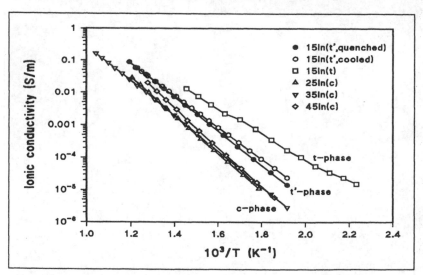

Figure 4: Intragrain conductivity in 15, 25, 35 and 45 mol% $InO_{1.5}$-ZrO_2 with tetragonal (t, t') and cubic phases.

TABLE II: ACTIVATION ENERGY OF INTRAGRAIN CONDUCTIVITY IN $InO_{1.5}$-ZrO_2
(T>800 °C)

Composition mol% $InO_{1.5}$	phase	Conditions	Activation energy (eV)
15	t	furnace cooled	0.8
15	t'	furnace cooled	1
15	t'	quenched	1.1
25	c	furnace cooled	1.2
35	c	furnace cooled	1.15
45	c	furnace cooled	1.25

have higher conductivities compared to the cubic phase, in the low temperature range. The activation energies of the ionic conductivity are shown in Table II.

The t-phase has an activation energy of 0.80 eV, which is consistent with the values of 0.84 eV reported by Sellars et al. [13] and 0.77-0.82 eV in 5.5-9.0 mol% $InO_{1.5}$-ZrO_2 (TZP) by Turrillas et al. [15].

The results of the total DC ionic conductivity measurements up to 1000°C are given in figure 5. The total ionic conductivity consists of contributions from the grains (intragrain conductivity) and grain boundaries. The latter suppress the conductivity in the 15 mol % $InO_{1.5}$ material at temperatures below 700 -800 °C. Above this temperature the resistivities of the grain boundaries become negligible.

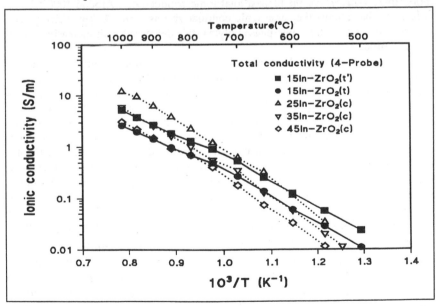

Figure 5: Total conductivity of 15, 25, 35 and 45 mol% $InO_{1.5}$-ZrO_2 measured by the 4-probe method between 500 and 1000°C.

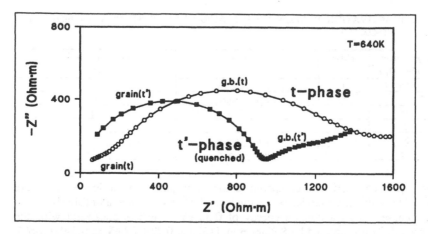

Figure 6:   The Cole-Cole plot at 640°C of 15 mol% InO$_{1.5}$-ZrO$_2$ as the t'-phase (grain size 10-15 μm) and the t-phase (grain size 0.1-0.2 μm).

This causes the deflections of the Arrhenius curves corresponding to the intersections of the intragrain and total conductivity lines in figures 8 and 9. Above 650°C, the cubic phase polycrystal with the lowest solute concentration (25 mol% InO$_{1.5}$-ZrO$_2$) has the highest total ionic conductivity. Around 1000°C, this polycrystal has almost the same ionic conductivity as 8 mol% Y$_2$O$_3$-ZrO$_2$. The ionic conductivity in the cubic phase polycrystals decreased with increasing solute concentration.

The fine grained t-phase polycrystal has a lower total conductivity than the t' or the cubic ZrO$_2$ polycrystals at temperatures higher than 700 °C. Figure 6 shows the Cole-Cole plot of 15 mol% InO$_{1.5}$-ZrO$_2$ prepared by quenching from 1600°C, and by sintering at 1000°C. The first specimen consists of the t'-phase and the latter of the t-phase. The t-phase polycrystal has a small semicircle (on the lower Z' side) with low intragrain resistivity, which is followed by a large semicircle (on the higher Z' side) with high grain boundary resistivity. On the other hand, the t'-phase grain has a higher resistivity in grains but lower resistivity in grain boundaries than that of the t-phase polycrystal. The grain boundary resistivity in the t'-phase polycrystal after air-quenching was one third of that in the t-phase polycrystal.

In the Cole-Cole plot of the t'-phase polycrystal in figure 6, only three semicircles were observed. The relaxation frequency, $\nu_R = 1/\tau = 2\pi f$, can be determined from the frequency, f, of an applied electrical field at the top of a semicircle [16]. The first semicircle with the highest relaxation frequency is due to the ionic conduction in the crystals [1,17-19]. The third semicircle corresponds to the interfacial polarization at the Pt electrodes. The relaxation frequency versus T of the second semicircle is shown in figure 7. This relaxation frequency is the same as in the c-phase both having the same grain size but the latter no domains. Therefore this second semicircle in the t'-phase polycrystals corresponds to the grain boundary relaxation being the same in the t' and c. The absence of a fourth semicircle suggests that the antiphase boundaries in the t' have little effect on the ionic conductivity.

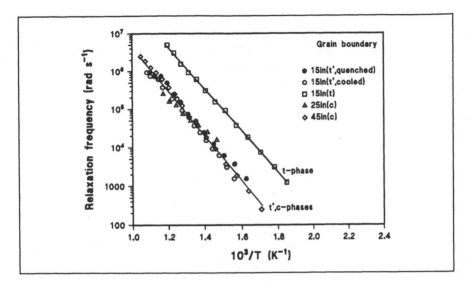

Figure 7: Relaxation frequency at grain boundaries in 15, 25 and 45 mol% $InO_{1.5}$-ZrO_2 t-, t'- and c-phase polycrystals.

In order to compare the intragrain and intergrain contributions to the total conductivity in the t- and t'-phases, the ionic conductivities measured by 4-probe method and impedance spectroscopy are plotted in figures 8 and 9. Total conductivities measured by both methods were consistent, which also supports the argument that the second semicircle is due to the grain boundary relaxation. In the t'-phase polycrystal, the grain boundary contribution to the total resistivity is rather

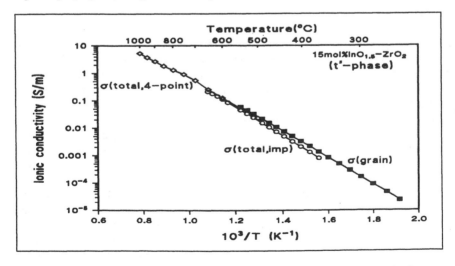

Figure 8: Total and intragrain conductivities in 15 mol% $InO_{1.5}$-ZrO_2 with t'-phase after furnace-cooling from 1600°C.

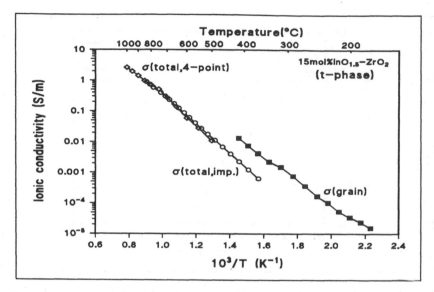

Figure 9:   Total and intragrain conductivities in 15 mol% InO$_{1.5}$-ZrO$_2$ of the t-phase after sintering at 1000°C for 3 h.

small so that the total conductivity of the polycrystal is dominated by the intragrain conductivity. On the other hand, the high grain boundary resistivity of the fine grained t-phase causes its total conductivity to be below the intragrain conductivity.

SUMMARY

1) In the system ZrO$_2$-InO$_{1.5}$, both the 100% t- and t'-phase polycrystals with the same solute concentration were prepared. The t'-phase was obtained via the diffusionless transformation of the cubic phase (c→t'), and the t-phase via the coprecipitation process followed by calcination and low temperature sintering at 1000°C.

2) The tetragonality of the t'-phase in the ZrO$_2$-InO$_{1.5}$ system depends on cooling conditions. The t'-phase with a low tetragonality was obtained after rapid quenching, leading to a large domain size in the t'-grains. The t'-phase with a higher tetragonality is obtained after slow cooling, resulting in t' microstructure with a smaller domain size.

3) The ionic conductivities of the t- and t'-phases differ. Below 500°C, the intragrain conductivity of the t-phase is higher than that of the t'-phase. On the other hand, due to the smaller grain size of the t-phase polycrystal compared to the t'-phase and the c-phase, the total conductivity of the t-phase polycrystal is depressed by grain boundary resistivity and is lower than that of the t'-phase polycrystal between 500 and 1000°C.

4) No extra relaxation process due to the domain structure of the t'-phase was observed.

5) The cubic phase polycrystal near the eutectoid concentration has the highest total ionic conductivity at 700°C and higher. The total conductivity is almost the same as that in ZrO_2 with 8 mol% Y_2O_3 near 1000°C. Between 500 and 650°C, the t'-phase polycrystal has the highest total ionic conductivity in zirconias doped with In_2O_3.

ACKNOWLEDGEMENTS

This study was supported by the Swiss Federal Office of Energy. We thank Dr. Dubal from this office for his support, continuous interest and encouragement.

REFERENCES

1. Orliukas, A., K. Sasaki, P. Bohac and L.J. Gauckler, 1991. "Ionic Conductivity of ZrO_2-Y_2O_3 Prepared from Ultrafine Coprecipitated Powders," in Proc. 2nd. Intl. Symp. Solid Oxide Fuel Cells,   F. Grosz, P. Zegers, S.C. Singhal and O. Yamamoto, eds, Athens, Greece: Commission of European Communities, pp. 377-385.

2. Nowick, A.S., 1984. "Atom Transport in Oxides of the Fluorite Structure," in Diffusion in Crystalline Solids, G.E. Murch and A.S. Nowick eds., N.Y. USA: Academic Press, pp. 143-188.

3. Scott, H.G., 1975. "Phase Relationships in the Zirconia-Yttria System."   J. Mater. Sci., 10(9):1527-1535.

4. Heuer, A., R. Chaim and V. Lanteri, 1988. "Review: Phase Transformations and Microstructural Characterization of Alloys in the System Y_2O_3-ZrO_2", in Advances in Ceramics, Vol. 24, Science and Technology of Zirconia III, S. Somiya, N. Yamamoto and H. Yanagida eds., Columbus, Ohio, USA: The Am. Ceram. Soc. Inc., pp. 3-20.

5. Sheu, T.S., T.Y. Tien and I.W. Chen, 1992. "Cubic-to-Tetragonal (t') Transformation in Zirconia-Containing Systems." J. Am. Ceram. Soc., 75(5):1108-1116.

6. Yoshimura, M., 1988. "Phase Stability of Zirconia." Ceram. Bull., 67(12):1950-1955.

7. Sasaki, K., P. Bohac and L.J. Gauckler, 1993. "Phase Equilibria in the System ZrO_2-$InO_{1.5}$." J. Am. Ceram. Soc., in print.

8. Jue, J.F., J. Chen and A.V. Virkar, 1991. "Low-Temperature Aging of t'-Zirconia: The Role of Microstructure on Phase Stability." J. Am. Ceram. Soc., 74(8):1811-1820.

9. Jue, J.F., and A.V. Virkar, 1990. "Fabrication, Microstructural Characterization, and Mechanical Properties of Polycrystalline t'-Zirconia." J. Am. Ceram. Soc., 73(12):3650-3657.

10. Ciacchi, F.T. and S.P.S. Badwal, 1991. "The System Y_2O_3-Sc_2O_3-ZrO_2: Phase Stability and Ionic Conductivity Studies." J. Europ. Ceram. Soc., 7:197-206.

11. Badwal, S.P.S. and J. Drennan, 1992. "Microstructure/Conductivity Relationship in the Scandia-Zirconia System." Solid State Ionics, 53-56:769-776.

12. Sasaki, K., A. Orliukas, P. Bohac and L.J. Gauckler, 1992. "Electrical Conductivity of the ZrO_2-In_2O_3 System," in Proc. Europ. Ceram. Soc. 2nd. Conf., in print.

13. Sellars, A.P. and B.C.H. Steele, 1988. "Factors Affecting Ionic Conductivity in Doped Polycrystalline Tetragonal Zirconia (TZP)." Mater. Sci. Forum, 34-38:255-260.

14. Hohnke, D.K., 1980. "Ionic Conductivity of $Zr_{1-x}In_{2x}O_{2-x}$." J. Phys. Chem. Solids, 41(7):777-784.

15. Turrillas, X., A.P. Sellars and B.C.H. Steele, 1988. "Oxygen Ion Conductivity in Selected Ceramic Oxide Materials." Solid State Ionics, 28-30:465-469.

16. Hench, L.L. and J.K. West, 1990. Principles of Electronic Ceramics, New York, NY, USA: John Wiley & Sons, pp. 185-236.

17. Abelard, P. and J.F. Baumard, 1982. "Study of the dc and ac Electrical Properties of an Yttria-Stabilized Zirconia Single Crystal $(ZrO_2)_{0.88}$-$(Y_2O_3)_{0.12}$." Phys. Rev., B26(2):1005-1016.

18. Orliukas, A., P. Bohac, K. Sasaki and L.J. Gauckler, "Relaxation Dispersion of Ionic Conductivity in a $Zr_{0.85}Ca_{0.15}O_{1.85}$ Single Crystal." Submitted to J. Europ. Ceram. Soc.

19. Kleitz, M., H. Bernard, E. Fernandez and E. Schouler, 1981. "Impedance Spectroscopy and Electrical Resistance Measurements on Stabilized Zirconia", in Advances in Ceramics, Vol. 3, Science and Technology of Zirconia, A.H. Heuer and L.W. Hobbs, eds., Columbus, Ohio, USA: The Am. Ceram. Soc. Inc., pp. 310-336.

Ionic and Electronic Conductivity of TiO$_2$-Y$_2$O$_3$-Stabilized Tetragonal Zirconia Polycrystals

A. KOPP, H. NÄFE, W. WEPPNER,
P. KOUNTOUROS and H. SCHUBERT

ABSTRACT

The system TiO$_2$-Y$_2$O$_3$-ZrO$_2$ was studied to establish the stability range of the tetragonal phase. The materials were prepared by the coprecipitation method in order to obtain fine, homogeneous and sinterative powders. 1 - 30 mol%TiO$_2$ was added systematically to 1, 2, 3 and 4 mol%Y$_2$O$_3$-doped ZrO$_2$. The ionic conductivity of the ternary tetragonal ZrO$_2$ was investigated using impedance spectroscopy at temperatures between 250 and 700 °C. The ionic conductivity decreases with increasing TiO$_2$ content. Hole and electron transport properties as a function of temperature between 500 and 700 °C and oxygen partial pressure between 10^2 and 10^5 Pa using the Hebb-Wagner polarization technique show an increase of the electron conductivity with increasing TiO$_2$ content whereas the hole conductivity remains unchanged. As in the case of undoped samples, the hole conductivity follows a 1/4 power law dependence on the oxygen partial pressure which indicates a large concentration of ionic defects.

INTRODUCTION

Solid solutions of the Y$_2$O$_3$-ZrO$_2$ system are well known for their good oxygen ion conductivity at high temperatures. Especially the tetragonal phase has recently attracted much interest because of its combination of high ionic conductivity and outstanding mechanical stability. The addition of titania is known to have a beneficial effect on the sintering behaviour of the zirconia polycrystals. The influence of titania on the mechanical stabi-

A. Kopp, H. Näfe and W.Weppner, Max-Planck-Institut für Festkörperforschung, Heisenbergstr. 1, 7000 Stuttgart 80, Germany

P. Kountouros and H. Schubert, Max-Planck-Institut für Metallforschung, Institut für Werkstoffwissenschaft, PML, Heisenbergstr. 5, 7000 Stuttgart 80, Germany

lity was studied thoroughly by Haberko and coworkers [1] and they reported a high fracture toughness of the samples doped with more than 10 mol% TiO_2.

The addition of dopants, such as titania, is also of practical interest because of a possible impact on the electronic conductivity. Mixed conduction of electronic and ionic species may be favourable for the electrode reaction of oxygen gas sensors to allow a lowering of the working temperature and for kinetically less impeded electrode reactions in fuel and water electrolysis cells [2].

The present experiments were conducted in order to establish the stability range of the tetragonal phase and to examine the ionic and electronic conductivities in titania doped yttria stabilized zirconia. The ionic conductivity was measured by impedance spectroscopy. For the determination of the electronic conduction the Hebb-Wagner polarization method [3,4] was used.

EXPERIMENTAL ASPECTS

SAMPLE PREPARATION

The coprecipitation method was used to prepare the powders with compositions of 0 - 30 mol% titania and 0 - 4 mol% yttria, respectively. The procedure is described in detail in [5]. The powders were isostatically pressed (200 MPa) into cylindrical bars and sintered at 1450 °C for 2 h in air. Pellets were cut from these bars, ground and polished to a surface roughness of 1 μm. The phases were examined by X-ray diffraction. The notation of the samples is the following: ZnYmT, where n and m indicate the yttria and titania content in mol%, respectively.

For the impedance and polarization measurements samples with a thickness of 2 - 3 mm and 50 - 100 μm, respectively, were used. The electrodes were prepared by painting platinum paste (No. 6204 0210 Demetron, Hanau, Germany) onto both sides of the samples and firing at 800 °C.

ELECTRICAL MEASUREMENTS

The ionic conductivity was measured by an impedance bridge (HP 4192a, Hewlett Packard) over a frequency range from 5 Hz to 13 MHz at temperatures between 250 and 700 °C.

The electron and hole conductivities were determined by the Hebb-Wagner technique [3,4]. Applying a dc voltage below the decomposition voltage, the ionic flux decreases and in steady state the current is only carried by electrons and holes. For the current voltage dependence the following relation [4,6] holds:

$$i = \frac{ART}{LF} \left\{ \sigma_e \left(\exp\left[\frac{EF}{RT} \right] - 1 \right) + \sigma_h \left(1 - \exp\left[-\frac{EF}{RT} \right] \right) \right\} \quad (1)$$

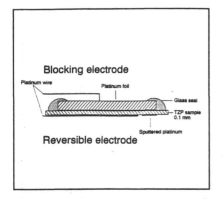

Figure 1: Experimental arrangementof the Hebb-Wagner polarization cell using a thin sample sealed by glass at one side.

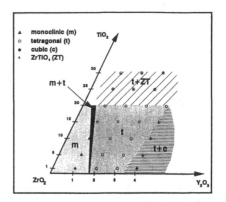

Figure 2: Phase composition of the TiO₂-Y₂O₃-ZrO₂ samples sintered between 1300 and 1450 °C for 4 h.

where i, E, σ_e, σ_h, A, L, R, T and F are the current, applied voltage, electron and hole conductivity at the oxygen partial pressure of the reference electrode, geometrical area, length of the sample, gas constant, temperature and Faraday´s constant, respectively. For low applied voltages the first term in equation (1) may be neglected. The current reaches a plateau and from its height the hole conductivity may be determined. At higher voltages the first term dominates and the current increases exponentially. From this exponential increase the electron conductivity may be calculated.

The polarization measurements were performed between 500 - 700 °C using Ar/O₂ mixtures to adjust the reference oxygen partial pressure from 10^2 to 10^5 Pa. The blocking electrode was made by sealing a platinum foil onto one of the electrodes with a specially synthesized glass [7] as shown in figure 1. The samples were placed in a spring-loaded sample holder within a heated quartz tube. The gas composition was measured by a potentiometric zirconia oxygen gauge. The details of the polarization measurements are described elsewhere [8].

RESULTS

The phase regions of the ZrO₂-rich part of the TiO₂-Y₂O₃-ZrO₂ compositions are shown in figure 2 . The density observed for the sintered samples was > 98.5 % of the theoretical density.

The impedance spectra of Z2Y10T are given in figure 3 for different temperatures. Three semicircles may be distinguished. In accordance with literature [9] these are related to the bulk, grain boundary and electrode resistance. The grain boundary resistance is seen to change more strongly with increasing temperature than the bulk resistance. Hence, the activation energy for the transport across the grain boundary is higher than that for the grains (cf. figure 6).

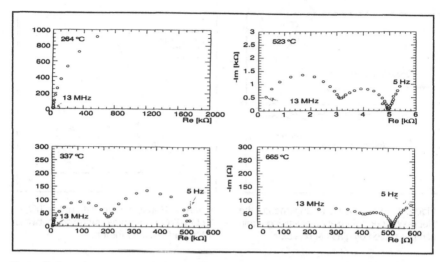

Figure 3: Impedance spectra at various temperatures of 10 mol% TiO$_2$-doped ZrO$_2$ (+ 2 mol% Y$_2$O$_3$).

Figure 4 shows the bulk and total conductivity of the samples with 2 mol% Y$_2$O$_3$ as a function of the titania content. At small TiO$_2$ doping concentrations the bulk conductivity increases whereas at larger dopant contents the conductivity decreases for all temperatures. The same behaviour is observed with the samples stabilized with 3 mol% yttria.

Figure 5 presents the dependence of the activation energy on the titania content for 2 and 3 mol% yttria stabilized zirconia. The activation energies of the bulk and the total conductivity both increase with increa-

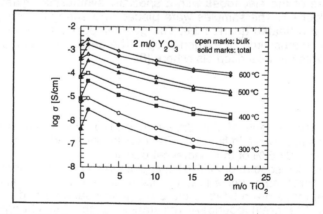

Figure 4: Bulk and total resistances of titania doped ZrO$_2$ (+ 2 mol% Y$_2$O$_3$) as a function of the titania content at 300, 400, 500 and 600 °C, shown as effective conductivities in terms of the macroscopic dimensions of the samples.

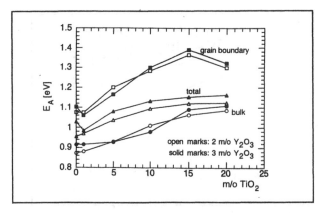

Figure 5: Activation energies of ZrO_2 (+ 2 and 3 mol% Y_2O_3) as a function of the titania content.

sing titania content. The activation energy of the grain boundaries shows a maximum at about 15 mol% titania.

Figure 6 shows typical current-voltage polarization curves for titania free TZP (tetragonal zirconia polycrystals), Z2Y1T, Z2Y5T, and Z2Y10T. In these measurements the voltages had to be corrected for the instantaneous voltage drop due to ohmic polarization which was determined by voltage interruption as described in the literature [10,11].

Figure 6 shows the current-voltage curves with accordingly corrected values. Doping with 1 mol%TiO_2 keeps the plateau current unchanged while the exponential increase starts at a lower voltage compared to undoped material. In the case of the sample containing 5 mol% TiO_2 the exponential increase appears at such low voltages that no current plateau is seen anymore. The current-voltage curve for Z2Y10T has a normal cur-

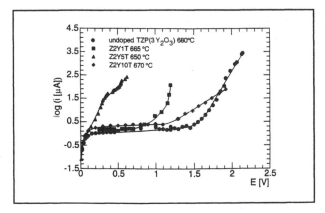

Figure 6: Current-voltage curves of titania free TZP, Z2Y1T, Z2Y5T and Z2Y10T (the numbers indicate the amount of yttria and titania, respectively, in mol%).

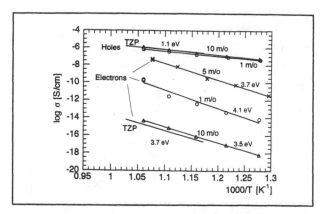

Figure 7: Arrhenius diagram of the electron and hole conductivities of ZrO_2 (+ 2 mol% Y_2O_3) doped with 1, 5 and 10 mol% TiO_2. For comparison the curves for titania-free TZP are shown.

rent plateau but the slope of the exponential increase at higher voltages is much smaller than the theoretically expected one even after the correction for the instantaneous voltage drop.

The temperature dependence of the electron and hole conductivity is given in figure 7. The magnitude and the activation energy (1.1 eV) of the hole conductivity are independent of the titania content and comparable to the values obtained for the sample without titania. The activation energy of the electron conductivity is found to be independent of the titania content. It is about 3.7 eV which agrees with the value for titania-free TZP [7]. The magnitude of the electron conductivity, however, depends strongly on the titania content. With the exception of the sample with 10 mol% TiO_2 the electron conductivity increases with increasing titania concentration by up to seven orders of magnitude.

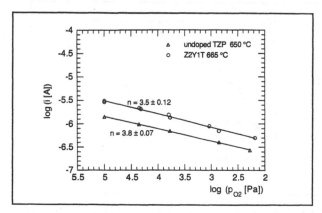

Figure 8: Polarization current at 500 mV (against air) as a function of the oxygen partial pressure.

The polarization current was measured as a function of oxygen partial pressure at 500 mV polarization voltage (against air) at which predominantly hole conduction occurs. The results in figure 8 show a 1/4 power law dependence.

DISCUSSION

For *cubic* zirconia with a higher yttria content of 10 mol% Liou and Worrell [12] reported an increase of the total conductivity between 400 and 800 °C for increasing titania content (0, 1 and 5 mol% TiO_2). These data were confirmed by Matsui [13] using samples with compositions of 0, 2, 4 and 6 mol% TiO_2 and 9 mol% Y_2O_3. Naito and Arashi [14] reported, however, that the conductivity decreases with increasing titania content (5, 7.5 and 10 mol% TiO_2, 10 mol% Y_2O_3). The present results on *tetragonal* zirconia show the same dependence.

Electron and hole conductivity investigations on the *cubic* material were performed by Liou and Worrell [15] and Arashi and Naito [16]. They used the blocking electrode (800 - 1000 °C, 5, 7.5 and 10 mol% TiO_2, 10 and 12 mol% Y_2O_3) and the oxygen permeation technique (1300 - 1500 °C, 7.5 and 10 mol% TiO_2, 10 mol% Y_2O_3), respectively. Both reported, at much higher operating temperatures, an increasing electronic conductivity with increasing titania content, which agrees with the results of the present work on *tetragonal* zirconia.

In the case of the highest titania content (10 mol% TiO_2) an unexpected decrease of the electronic conductivity was found compared to moderate titania contents. This result might be interpreted by a decrease of the effective number of electrons available for the transport at doping levels ≥ 5 mol% TiO_2 but is more likely due to interfacial processes such as minute amounts of $ZrTiO_4$ at the grain boundary, since the polarization technique only provides an integral information on the overall electronic conductivity. Further work is required to interpret the observed results.

As in cubic zirconia the $p_{O_2}^{1/4}$ power law dependence of the hole and electron conductivity may be explained by a large concentration of oxy-

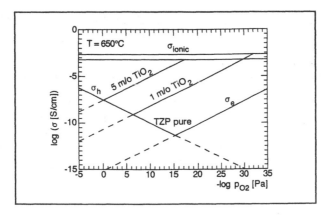

Figure 9: Brouwer diagram of pure TZP and titania doped TZP with 1 and 5 mol% TiO_2 at 650 °C.

gen vacancies which may be considered to be independent of the oxygen activity. This obviously remains valid in the case of lower yttria dopant concentration and in the presence of small amounts of titania.

From the present results a Brouwer diagram may be constructed which is shown in figure 9. For that purpose a $p_{O_2}^{-1/4}$ power law dependence of the electron conductivity is assumed to be fulfilled. It is seen that with increasing titania content the intrinsic point, i.e. the oxygen partial pressure of equal hole and electron concentration, shifts to higher oxygen partial pressures.

REFERENCES

1. Haberko, K., W. Pyda, M. M. Bucko, M. Faryna, 1991. "Study on Preparation of Tetragonal Polycrystals in the TiO_2-Y_2O_3-ZrO_2 System." Proc. 2nd. Europ. Ceram. Soc. Conf., Augsburg, in press.

2. Badwal, S. P. S., F. T. Ciacchi, 1986. "Performance of Zirconia Membrane Oxygen Sensors at low Temperatures with Nonstoichiometric Oxide Electrodes." J. Appl. Electrochem., 16: 28-40.

3. Hebb, M., 1952. "Electrical Conductivity of Silver Sulfide." J. Chem. Phys., 20: 185-190.

4. Wagner, C., 1957. "Galvanic Cells with Solid Electrolytes Involving Ionic and Electronic Conduction." Proc. 7th Intern. Com. Electrochem. Thermodyn. Kin., Lindau, Butterworth, London, pp. 361-377.

5. Kountouros,P., H. Schubert, 1991. "Bloating Effect During Sintering of TZP." Proc. 2nd. Europ. Ceram. Soc. Conf., Augsburg, in press.

6. Patterson, J. W., E. C. Bogren, R. A. Rapp,1967. "Mixed Conduction in $Zr_{0.85}Ca_{0.15}O_{1.85}$ and $Th_{0.85}Y_{0.15}O_{1.925}$ Solid Electrolytes."J. Electrochem. Soc.,114: 752-758.

7. Näfe, H., 1990. "High-Vacuum Tight, Liquid Sodium Resistant Joint between ThO_2 Ceramic and Metal."J. Nucl. Mater.,175: 67-77.

8. Kopp, A., H. Näfe, W. Weppner, 1992. "Characterization of the Electronic Charge Carriers in TZP." Solid State Ionics,53-56: 853-858.

9. Bauerle, J. E., 1969, "Study of Solid Electrolyte Polarization by a Complex Admittance Method." J. Phys. Chem. Solids,30: 2657-2670.

10. Burke, L. D., H. Rickert, R. Steiner, 1971. "Elektrochemische Untersuchungen zur Teilleitfähigkeit, Beweglichkeit und Konzentration der Elektronen und Defektelektronen in dotiertem Zirkondioxid und Thoriumdioxid." Z. Phys. Chem., NF 74: 146-167.

11. Swinkels, D. A. J., 1970. "Rapid Determination of Electronic Conductivity - Limits of Solid Electrolytes." J. Electrochem. Soc., 117: 1267-1269.

12. Liou, S. S., W. L. Worrell, 1989. "Electrical Properties of Novel Mixed-Conducting Oxides." Appl. Phys., A49: 25-31.

13. Matsui, N., 1990. "Effects of TiO_2 Addition on Electrical Properties of Yttria-Stabilized Zirconia." Denki Kagaku, 58: 716-722.

14. Naito, H., H. Arashi, 1992. "Electrical Properties of ZrO_2-TiO_2-Y_2O_3 System." Solid State Ionics, 53 - 56: 436-441.

15. Liou, S. S., W. L. Worrell, 1989. "Mixed-Conducting Oxide Electrodes for Solid Oxide Fuel Cells." Proc. of First Int. Symp. on Solid Oxide Fuel Cells, Pennington N. J., Ed. S. C. Singhal.

16. Arashi, H., H. Naito, 1992. "Oxygen Permeability in ZrO_2-TiO_2-Y_2O_3 System." Solid State Ionics, 53 - 56: 431-435.

Mixed Conduction and Oxygen Permeation of ZrO_2-$Tb_2O_{3.5}$-Y_2O_3 Solid Solutions

G. Z. CAO, X. Q. LIU, H. W. BRINKMAN,
K. J. DE VRIES and A. J. BURGGRAAF

ABSTRACT

$(ZrO_2)_{1-x-y}(Tb_2O_{3.5})_x(Y_2O_3)_y$ solid solutions with a fluorite-type structure, where $x = 0.3$-0.5 and $y = 0$-0.1, were prepared using citrate synthesis. The solid solutions are mixed oxygen ion and electronic conducting materials. The total electrical conductivity reaches 3.8 S/m and the oxygen permeation is about 2.6 x 10^{-7} mol/m^2.s, at 1173 K. The influence on the electrical conductivity and oxygen permeation of the valence change of $Tb^{3+/4+}$ ions and of the addition of yttria to the solid solutions is discussed.

INTRODUCTION

Mixed conducting materials, in which both oxygen vacancies and electrons are mobile at elevated temperatures, have potential applications, such as gas separation membranes, dense cathodes for solid oxide fuel cells and in electro-catalytic reactors. Zirconia with a fluorite type structure stabilized by doping with rare-earth oxides, such as Y_2O_3, has been widely investigated for its high oxygen ionic conductivity. However, these materials show, in general, only very limited electronic conductivity, which limits their oxygen permeability. Recently Iwahara et al. [1] reported that zirconia stabilized by terbium oxide demonstrates high mixed conductivity. As a consequence improved oxygen permeability may be expected. In this system, the presence of three-valent terbium ions occupying the zirconium ion sites and introducing an effective negative charge on each site, results in the formation of oxygen vacancies in the lattice. These oxygen vacancies are mobile, resulting in ionic conductivity. The coexistence of both four and three valent terbium ions introduces the possibility for electron

G.Z. Cao, X.Q. Liu, H.W. Brinkman, K.J. de Vries and A.J. Burggraaf, Laboratory of Inorganic Chemistry, Materials Science and Catalysis, Department of Chemical Engineering, University of Twente, POB 217, 7500 AE Enschede, Netherlands

hole conductivity due to the hopping of electron holes between three and four valent terbium ions.

Due to the influence of temperature and/or oxygen partial pressure the defect structure of ZrO$_2$-Tb$_2$O$_{3.5}$ solid solutions, i.e. the concentration of the mobile oxygen vacancies and electron holes, may change. For example in the electrochemical permeation process, driven by an oxygen partial pressure gradient across the material, at the gas solid interface three valent terbium ions can be oxidized to the four valent state causing the related oxygen ion vacancy concentration to be reduced drastically to such a level that the oxygen permeation process is limited. To limit such an oxygen partial pressure influence the defect structure of ZrO$_2$-Tb$_2$O$_{3.5}$ solid solutions has to be stabilized.

In the present study, we introduced a small amount of yttrium oxide into the ZrO$_2$-Tb$_2$O$_{3.5}$ solid solutions in order to have a minimum oxygen vacancy concentration. Thus the oxygen permeation will not be limited by an oxygen vacancy concentration correlated only with the valence change of terbium ions.

EXPERIMENTAL

PREPARATION OF THE SPECIMENS

The citrate synthesis method was applied for the preparation of the mixtures [2]. The raw materials used are ZrOCl$_2$.H$_2$O (> 99 wt%, Merck), Y$_2$O$_3$ (99.99 %, Aldrich Chemical Company Inc.) and Tb$_2$O$_{3.5}$ (3N, Highways Int.). The compositions of the starting mixtures are presented in Table I. The synthesized powders were calcined at 1273 K in air for 10 hours, and then pressed isostatically (at 400 MPa) into pellets of 25 mm diameter and 5 mm thickness. The pellets were sintered at 1773 K for 3 hours with a heating and cooling rate of 1.0 K/min.

MEASUREMENT OF CONDUCTIVITY AND PERMEATION

To ensure compatibility with the data presented by Iwahara et al. [1] the total electrical conductivity of the bulk of the material was measured using a frequency response analyzer (SOLARTRON 1255) at a fixed frequency of 10 kHz. The specimens used were 12 mm in diameter and between 1.9 mm and 2.9 mm thick. Platinum electrodes of 300 nm thickness and 10 mm diameter were deposited on both sides of the specimen by DC-sputtering for 30 min followed by a heat treatment at 1223 K in air for 60 min.

Because this type of single frequency measurement only produces reliable data for materials with high electronic transference numbers, also impedance measurements were performed (frequency from 1 MHz to 10 mHz) to check the results from the single frequency measurements. For these measurements 300 nm thick gold electrodes with a diameter of 10 mm were DC-sputtered on both sides of the specimen. The electrodes were given the same heat treatment as the platinum electrodes.

TABLE I - COMPOSITION AND DENSITY OF THE SPECIMENS

Specimens No.	Composition (mol%)			Density (g/cm^3)	Density %
	ZrO_2	$Tb_2O_{3.5}$	Y_2O_3		
TYZ-1	70.0	22.8	7.2	5.18	76
TYZ-2	70.0	25.0	5.0	6.33	92
TYZ-3	70.0	27.5	2.5	6.54	93
TYZ-4	70.0	30.0	0	6.47	92
TYZ-5	50.0	40.0	10.0	6.39	89
TYZ-6	50.0	44.9	5.1	6.51	88
TYZ-7	50.0	47.6	2.4	7.00	94

The ionic transference numbers were determined using an oxygen concentration cell [3]. The specimens used were 12 mm in diameter and 2 mm thick. Platinum electrodes were sputtered on both sides and heat-treated in the same way as for the conductivity measurement.

The oxygen permeation was determined in an oxygen permeation reactor as described by Bouwmeester et al. [4]. The specimens were 12 mm in diameter and 1-2 mm thick and were sealed with glass to a quartz permeation cell, which was then heated to 1173 K. Air (P_{O2} = 0.21 atm) was flushed along one side while helium gas (P_{O2} = 10^{-4} atm) was flushed along the other side. The total pressure was maintained at 1 atm on both sides of the specimens and the flow rate for both gases was 20 ml(STP)/min. The amount of oxygen gas which permeated through the specimen was determined by a gas chromatograph. The oxygen permeability was calculated from the latter data.

RESULTS AND DISCUSSION

PREPARATION OF THE SPECIMENS

The sintered specimen had a relative density of about 90 % (except TYZ-1, also see Table I) and were gas tight. Due to small cracks, specially in the samples with higher terbium concentrations, higher densities could not be

TABLE II - CONDUCTIVITY AND OXYGEN PERMEATION AT 1173 K.

Specimens No.	σ_t (S/m)	t_i	J_{O2} $(mol/m^2.s)$
TYZ-2	1.2	0.72	2.3×10^{-7}
TYZ-4	1.2	0.37	2.6×10^{-7}
TYZ-6	1.8	0.046	
TYZ-7	3.8	0.038	

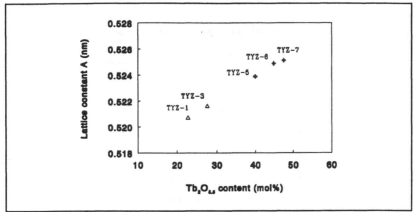

Figure 1.   The lattice parameter a of (Tb,Y)-ZrO$_2$ solid solutions
as a function of Tb$_2$O$_{3.5}$ content.

achieved. This crack formation is caused by the valence change of terbium ions, which is accompanied by a change in the size of ions and of the unit cell (Tb^{3+}, r = 1.04 Å and Tb^{4+}, r = 0.88 Å), as the temperature changes [5-6].

XRD analysis of (ZrO$_2$)$_{1-x-y}$(Tb$_2$O$_{3.5}$)$_{1-x}$(Y$_2$O$_3$)$_y$ indicates that all compositions with x = 0.3-0.5 and y = 0.0-0.1, after sintering at 1773 K for 3 hours formed zirconia solid solutions with a fluorite-type structure. The lattice constant of stabilized zirconia increased with the amount of terbium oxide doped as shown in figure 1. The influence of the yttrium oxide on the lattice constant is seen to be rather small.

CONDUCTIVITY AND OXYGEN PERMEATION

The total electrical conductivity of the bulk σ_t and the ionic transference number t_i at 1173 K of the materials studied are summarized in Table II. The total electrical conductivity and ionic transference numbers as functions of temperature are plotted in figures 2 and 3, respectively. Comparison of data obtained by the single frequency measurements and impedance spectroscopy showed that for the composition TYZ-2 the single frequency data were not reliable (see also figure 2). Therefore the conductivity data for this composition should be taken from the impedance spectroscopy measurements.
It is seen that the zirconia solid solutions doped with a combination of in total 30 mol% of yttria and terbia have a relatively low total electrical conductivity in comparison with that of those doped with a combination of 50 mol% of yttria and terbia. Although the electrical conductivity increases with a total amount of dopant, the ionic transference number decreases strongly, indicating that the ionic conductivity decreases with the increasing amount of dopant.

Some preliminary oxygen permeation measurements were carried out. The oxygen permeability at 1173 K for specimens TYZ-2 and TYZ-4 were 2.3 x 10^{-7}-

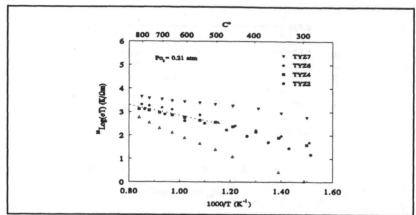

Figure 2.   An Arrhenius plot of the total conductivity of (Tb,Y)-ZrO$_2$ solid solutions (closed symbols). Data obtained for TYZ-2 by single frequency measurements (open symbols) and literature data [1] (dash line) for TYZ-4 are added.

mol/m^2.s and 2.6 x 10^{-7} mol/m^2.s, respectively, i.e. not much influenced by the addition of yttria into the system.

DISCUSSION

Under the condition of thermodynamic equilibrium, the ZrO$_2$-Tb$_2$O$_{3.5}$ solid solutions can be formulated as:

$$Zr_{1-x}(Tb^{4+}_{1-z} Tb^{3+}_z)_x O_{2-\delta} V_\delta \qquad (1)$$

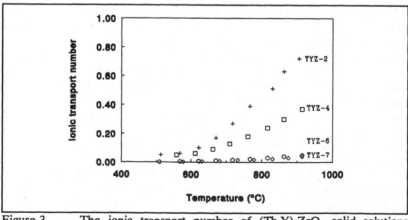

Figure 3.   The ionic transport number of (Tb,Y)-ZrO$_2$ solid solutions determined by using an oxygen concentration cell.

with oxygen deficiency $\delta = zx/2$. V is the oxygen vacancy. Using Kröger-Vink notation, the system undergoes the following reactions

$$O_O^x + 2h° \rightleftharpoons V_O^{°°} + \frac{1}{2}O_2 \tag{2}$$

The electron holes are provided by the reaction

$$Tb_{Zr}^x \rightleftharpoons Tb_{Zr}' + h° \tag{3}$$

The electroneutrality condition becomes:

$$[Tb_{Zr}'] = [h°] \tag{4}$$

Applying the law of mass action to reaction 3 and combining with equation 4, we have:

$$[h°] = (K_{(3)} * [Tb_{Zr}^x])^{\frac{1}{2}} \tag{5}$$

where $K_{(3)}$ is the equilibrium constant of reaction 3. Combining reactions 2 and 3 we have the overall reaction:

$$2Tb_{Zr}^x + O_O^x \rightleftharpoons 2Tb_{Zr}' + V_O^{°°} + \frac{1}{2}O_{2(g)} \tag{6}$$

The oxygen vacancies are responsible for the ionic conduction, while the hopping of the electron holes between Tb_{Zr}' and Tb_{Zr}^x ions contributes to the electronic conduction. The total electrical conductivity is then given by:

$$\sigma_t = \sigma_e + \sigma_i = [h°] * \mu_h * q + 2[V_O^{°°}] * \mu_O * q$$

$$= (K_{(3)} * [Tb_{Zr}^x])^{\frac{1}{2}} * \mu_h * q + 2[V_O^{°°}] * \mu_O * q \tag{7}$$

where μ_h and μ_O are the mobilities of electron holes and oxygen ion vacancies, respectively; q is the unit charge; $[Tb_{Zr}^x]$ and $[V^{°°}]$ are the concentrations of the Tb^{4+} ions and oxygen ion vacancies, respectively.

The ionic transference number is given by:

$$t_i = \frac{\sigma_i}{\sigma_e + \sigma_i} = \frac{1 - (K_{(3)} * [Tb_{Zr}^x])^{\frac{1}{2}} * \mu_h}{(K_{(3)} * [Tb_{Zr}^x])^{\frac{1}{2}} * \mu_h + 2[V_O^{°°}] * \mu_O} \tag{8}$$

One can see from equations 7 and 8 that the total electrical conductivity and ionic transference number are dependent on the concentration and mobilities of both electron holes and oxygen vacancies.

The dopant of yttria into the system is expected to result in an increasing amount of oxygen vacancies. They will cause a decrease in the concentration of the three valent terbium ions according to equation 6. The increasing amount of oxygen vacancies is expected to increase both the ionic conduction and the ionic transference number according to equations 7 and 8, assuming that the mobilities of the electron holes and oxygen vacancies remain unchanged.

The ionic conductivities at 1173 K of samples TYZ-2 and TYZ-4, calculated from the total electrical conductivity and the ionic transference number (Table II), are 0.86 and 0.44 S/m, respectively. The ionic conductivity of sample TYZ-2 is twice as large as that of sample TYZ-4. This is ascribed to an increasing concentration of oxygen vacancies caused by the addition of three valent yttrium ions to the $Tb_2O_{3.5}$-ZrO_2 system. However, the oxygen permeation at 1173 K of samples TYZ-2 and TYZ-4 are almost the same, which implies that the oxygen permeation here is not rate-limited by the ionic conductivity.

When the zirconia solid solutions contain a relatively large total amount of terbia and yttria (50 mol%), the electronic conductivity is predominant. The ionic transference number measured by using an oxygen concentration cell shows no significant difference. Thus a sharp increase of the total conductivity, from sample TYZ-6 to TYZ-7, implies a sharp increase of the electronic conductivity, which corresponds with an increased amount of terbia in the system. The electronic conductivity at 1173 K of samples TYZ-2 (0.34 S/m) and TYZ-4 (0.84 S/m) is also seen to increase with the amount of terbia in the specimen. An interpretation for this observation may be that when the total amount of terbia in the system increases, according to literature [5] the fraction of the four valent terbium ions will increase (at the oxygen partial pressure 0.21 atm). So the sharp increase in the electronic conductivity may be ascribed to the increase in the concentration of the Tb^{4+} (electron holes), which according to literature [5], determines the electronic conductivity.

From figure 2 it is observed that the total electrical conductivity values obtained in the present study for our sample TYZ-4 are not very different from those of the same composition reported by Iwahara et al.[1] (who determined the electrical conductivity using a digital impedance meter with a 10 kHz signal).

Further work on the influence of the addition of yttria on the stabilization of the concentration of oxygen vacancies will be published elsewhere.

CONCLUSIONS

In the ZrO_2-$Tb_2O_{3.5}$-Y_2O_3 system, zirconia solid solutions with a fluorite type structure can be formed in the range of total dopant between 30 to 50 mol% of which 0 to 10 mol% is yttria. The solid solutions are mixed conducting materials. Their total electrical conductivity is in the same order as that of YSZ (8 mol% yttria stabilized zirconia), but the oxygen permeation is about 2 to 3 orders higher.

The electronic conductivity increases with the amount of terbia present in the system. The addition of yttria into the $Tb_2O_{3.5}$-ZrO_2 material leads to an increasing concentration of oxygen vacancies and thus results in a sharp increase of the ionic conductivity, but has no detectable influence on the oxygen permeability. This probably indicates that the ionic conductivity is not the rate-limiting factor for oxygen permeation.

Single frequency impedance measurements can only be used for samples with ionic transference numbers considerably smaller than unity, otherwise impedance spectroscopy has to be applied.

ACKNOWLEDGEMENTS

The authors would like to acknowledge Dr. I.C. Vinke, Ir. C.S. Chen and Mr. H. Kruidhof for their experimental assistance and fruitful discussion.

REFERENCES

1. Iwahara, H., T. Esaka and K. Takeda, 1988. "Mixed Conduction and Oxygen Permeation in Sintered Oxides of a System ZrO_2-Tb_4O_7", in Science and Technology of Zirconia III, S. Somiya et al. ed., The Amer. Ceram. Soc., Inc., Columbus, OH, pp. 907-916.

2. Van de Graaf, M.A.C.G. and A.J. Burggraaf, 1984. "Wet-Chemical Preparation of Zirconia Powders: Their Microstructure and Behaviour", in Science and Technology of Zirconia II, N. Claussen et al. ed., The Amer. Ceram. Soc., Inc., Columbus, OH, pp. 744-765.

3. De Vries, K.J., T., van Dijk and A.J. Burggraaf., 1979. "Electrical Conductivity in Ceramic Solid Solutions of ZrO_2-Ln_2O_3", in Fast Ion Transport in Solids, Electrodes and Electrolytes, P. Vashista, J.N. Mundy and G.K. Shenoy ed., Elsevier North Holland, Amsterdam, pp. 679-682.

4. Bouwmeester, H.J.M., H. Kruidhof, A.J. Burggraaf and P.J. Gellings, 1992. "Oxygen Semipermeability of Erbia-Stabilized Bismuth Oxide". Solid State Ionics, 53-56: 460-468.

5. Arashi, H., S. Shin, H. Miura, A. Nakashima and M. Ishigame, 1989. "Investigation of Valence Change of Tb Ions in ZrO_2-Tb_4O_7 Mixed Conductor Using XANES Measurements". Solid State Ionics, 35: 323-327.

6. Van Dijk, M.P., K.J. de Vries and A.J. Burggraaf, 1985. "Electrical Conductivity and Defect Chemistry of the System $(Tb_xGd_{1-x})_2Zr_2O_{7+y}$ ($0 \leq x \leq 1$; $0 \leq y \leq 0.25$)". Solid State Ionics, 16: 211-224.

Luminescence of Anion Vacancies and Dopant-Vacancy Associates in Stabilized Zirconia

ERIC D. WACHSMAN, FRANCOIS E. G. HENN,
NAIXIONG JIANG, PIETER B. LEEZENBERG,
RICHARD M. BUCHANAN, CURTIS W. FRANK,
DAVID A. STEVENSON and JOSEPH F. WENCKUS

ABSTRACT

The desired properties of solid-oxide ionic conductors depend upon the concentration and local environment of mobile ionic species. These species are lattice-oxygen vacancies and vacancy associates. The identity and concentration of these species are implied from the influence of oxygen stoichiometry, dopant concentration, and structure on electrolytic properties. We have recently reported a dependence of the optical properties (absorption and luminescence) on the oxygen stoichiometry and dopant in yttria-stabilized zirconia (YSZ) and erbia-stabilized bismuth oxide (ESB) [1-3]. Absorbance, excitation and emission spectra were ascribed to electronic transitions of F-center type defects (electron-occupied oxygen-vacancies). We proposed a tentative band diagram and defect energy levels for YSZ based upon these results. The spectroscopic characteristics of these materials can provide a better understanding of the behavior of dopants and anion vacancies on the material, and the influence of these species on conductivity and electrocatalysis. In the present paper we present additional luminescence spectra of stabilized zirconia as a function of dopant and dopant concentration.

INTRODUCTION

Zirconia is stabilized in the fluorite phase by the addition of calcia (CaO) or rare earth oxides such as yttria (Y_2O_3) or scandia (Sc_2O_3). These dopants not only stabilize the fluorite phase, they also create anion vacancies ($V_O^{\bullet\bullet}$) in order to preserve charge neutrality. As the dopant concentration increases, dopant-vacancy associates form (e.g., $Ca_{Zr}''-V_O^{\bullet\bullet}$) both due to coulombic attraction and lattice strain energy. Defect association reduces the ionic conductivity, becoming more prevalent at large dopant concentrations (≥ 10 mol%). This has been the subject of several studies in order to explain the observed maximum in conductivity with dopant concentration [4-7]. However, these studies have overlooked the interaction of $V_O^{\bullet\bullet}$ with electronic species present in these oxides. Characterization of the nature of these point defects is needed in order to understand the conductivity mechanism in these materials and to provide information for the development of more highly conductive materials.

E.D. Wachsman, Materials Research Center, SRI International, Menlo Park, California, USA 94025

F.E.G. Henn, N. Jiang, P.B. Leezenberg, R.M. Buchanan, C.W. Frank and D.A. Stevenson, Departments of Materials Science and Chemical Engineering, Stanford University, Stanford, California, USA 94305

J.F. Wenckus, Ceres Corporation, North Billerica, Massachusetts, USA 01862

Oxygen-stoichiometric yttria-stabilized zirconia (YSZ), similar to most ionic crystals and ceramics, is transparent in the single crystalline form and is white in the polycrystalline form. This lack of absorption in the visible range is due to its large band gap and an insignificant population of defects that absorb in the visible. However, upon partial reduction, either by annealing in a reducing atmosphere or vacuum, or by application of a d.c. potential, polycrystalline zirconia blackens and single crystal zirconia becomes highly colored (yellow-orange) indicating the presence of color-centers. Color-centers are defects with energy levels lying within the band gap of the host material and having characteristic visible absorption, hence color, as well as the potential for luminescence.

A particular type of color-center, the F-center, is important for understanding ionic conduction in solid-oxide electrolytes. The F-center is an anion vacancy with trapped electrons in order to maintain local charge neutrality. There are numerous perturbations of the F-center based on their local environment and electron occupation. In YSZ, for example, a single oxygen vacancy with only Zr^{4+} nearest neighbors and two trapped electrons (V_O^x) is an F-center. This same vacancy with either zero or one trapped electron would be an F^{2+}-center ($V_O^{\bullet\bullet}$) or an F^+-center ($V_O^\bullet$), respectively. If the single electron occupied vacancy had as a nearest neighbor Y^{3+} the associated defect, $(Y_{Zr}'V_O^\bullet)^x$, is charge neutral (as signified by the x superscript) and the complex is an F_A-center (where the subscript, A, indicates that the anion vacancy has an extrinsic cation in a nearest neighbor position). Numerous other combinations are possible upon association with either additional extrinsic cations or neighboring anion vacancies, as well as with either more or fewer electrons.

Due to the large number of potential defect structures in stabilized zirconia, the exact nature of these defects is still a subject of controversy. EPR has been used to identify single-electron trap defects in YSZ (paired electrons are not paramagnetic). The observed EPR signals have been attributed to F_A centers [8,9], impurities [10], and variations of the F^+-center where the electron is in a cation $4d^1$ level (rather than the 1s level of the anion vacancy) localized at a Zr^{3+} associated with a single anion vacancy $((Zr_{Zr}'V_O^{\bullet\bullet})^\bullet)$ [11-13], and two anion vacancies $((V_O^{\bullet\bullet}Zr_{Zr}'V_O^{\bullet\bullet})^{\bullet\bullet\bullet})$ [13].

In the later study the two EPR signals were intensity correlated with specific optical spectra (absorption at 480 nm and 375 nm, respectively) indicating a common electronic defect [13]. The absorption spectra in this study are similar to results we have previously presented for YSZ [1,2] and similarly attributed to intrinsic Zr_{Zr}-V_O defects. However, our interpretation differs in the electronic occupancy of these defects. We have used uv-visible absorbance and fluorescence spectra in 10 mol% YSZ and 20 mol% erbia-stabilized bismuth oxide (ESB) to identify F-center type defects in these materials, and proposed a tentative band diagram for YSZ based upon these studies [2,3]. We expand this study here to include the effects of dopant type and concentration on excitation and emission spectra of zirconia based electrolytes.

EXPERIMENTAL

Single crystal samples of 15.2 mol% calcia- (CSZ) and several nominal compositions of yttria-stabilized zirconia (YSZ) were prepared at Ceres Corporation by skull-melting (an RF-heated crystal growth process). These samples and a polycrystalline sample of 7.5 mol% scandia-stabilized zirconia (SSZ) were examined using fluorescence spectroscopy. An electron microprobe (EM) equipped with energy dispersive x-ray analysis (EDAX) was used to quantify the dopant concentration and determine the presence of any impurities. Compositional analysis of the YSZ samples is shown in Table I. The oxygen non-stoichiometry (δ) is calculated assuming full compensation of Y' with $V_O^{\bullet\bullet}$. No impurities were identified in any of the samples with the EDAX analysis (sensitivity of <0.5 wt%).

Table I - COMPOSITION OF $Zr_{1-c}Y_cO_{2-\delta}$

EM Analysis, c (atom%)*	Y_2O_3 Concentration (mol%)	$\delta (= c/2)$
17.5±0.3	9.6	0.088
22.0±0.4	12.4	0.110
27.7±0.3	16.1	0.139
32.0±0.3	19.0	0.160
35.4±0.5	21.5	0.177
40.7±0.9	25.5	0.204

* ±1σ

Fluorescence spectra were taken with a Spex Fluorolog 212 spectrophotometer; all experiments were at room temperature in air. Details of the experimental apparatus are presented elsewhere [1,2]. Excitation and emission spectra were taken as a function of wavelength, and peaks are identified by the energy (eV) corresponding to the maximum emission intensity. For broad emission bands a filter was used to limit secondary harmonics of the excitation ($2\lambda_{ex}$). This filter was not used in our previous investigations [1-3] and results in a ~0.1 eV shift for one of the YSZ peaks. Therefore, accuracy of the broad band peak positions is given to 0.1 eV. For sharp peaks, accuracy is given to 0.01 eV.

The effect of crystal lattice orientation on luminescence was evaluated using <100>, <110>, and <111> oriented single crystals of YSZ. No measurable anisotropic effect was observed.

EFFECT OF DOPANT TYPE

In our previous investigations, we identified two different luminescent chromophores with characteristic excitation and emission band shapes and relative energies [1,2]. The first chromophore exhibits broad featureless Gaussian excitation and emission bands (typical of F-centers). The excitation band ($h\nu_1 = 4.1$ eV) was attributed to excitation of an electron in a single electron occupied intrinsic zirconia oxygen-vacancy ($V_O^\bullet$) to the conduction band.

$$V_O^\bullet + h\nu_1 \rightarrow V_O^{\bullet\bullet} + e' \tag{1}$$

The resulting emission band ($h\nu_2 = 2.1$ eV) was attributed to electronic recombination with a $V_O^{\bullet\bullet}$:

$$V_O^{\bullet\bullet} + V_O^x \rightarrow 2V_O^\bullet + h\nu_2 \tag{2}$$

These excitation and emission bands were observed in both YSZ and monoclinic ZrO_2.

In Figures 1 and 2 can be seen similar excitation ($h\nu_1 = 4.3$ eV) and emission ($h\nu_2 = 2.2$ eV) bands, respectively, for CSZ. The 0.1-0.2 eV spectral difference between YSZ and CSZ, for a transition to or from the conduction band, can be explained by the difference in band gap energy (E_G) between these two materials. The E_G of YSZ, measured with electron energy loss spectroscopy [14], is 5.2 ± 0.2 eV while that for CSZ, calculated from the temperature dependence of the minority carriers [15], is 5.6 eV. The E_G of these materials and the effect of doping, however, has not yet been fully determined. It has not been established whether an increase in Y_2O_3 ($E_G = 5.6$ eV [16]) or CaO ($E_G = 7.7$ eV [16]) concentration increases the E_G of ZrO_2 ($E_G = 4.99$ eV [16]) in some proportional rule of mixing or whether it reduces E_G due to

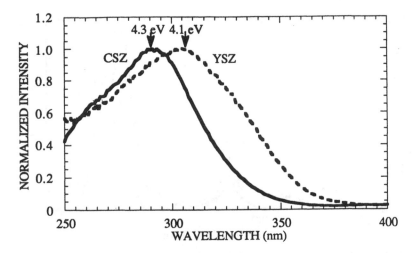

Figure 1. Broad band excitation spectra of 15.2 mol% calcia- (CSZ) and 9.6 mol% yttria- (YSZ) stabilized zirconia. Emission intensity at 2.1 eV and 2.2 eV for YSZ and CSZ, respectively, as a function of excitation wavelength.

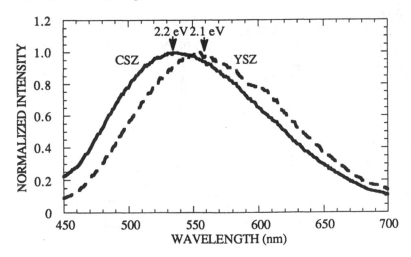

Figure 2. Broad band emission spectra of 15.2 mol% calcia- (CSZ) and 9.6 mol% yttria- (YSZ) stabilized zirconia. Intensity as a function of emission wavelength for 4.1 eV (YSZ) and 4.3 eV (CSZ) excitation.

the defect energy states introduced by these dopants into the band gap.

The second chromophore exhibits multiple narrow excitation and emission peaks. A 2.54 eV excitation peak was attributed to local (intra-defect) excitation of a lanthanide-associated vacancy $((Ln_{Zr}'V_O^\bullet)^x)$. The resulting emission doublet at 2.22 and 2.27 eV (as well as neighboring peaks) was attributed to internal relaxation of the local excited state in $(Ln_{Zr}'V_O^\bullet)^x$. These excitation and emission peaks were observed in both YSZ and ESB. Figures 3 and 4 show identical excitation and

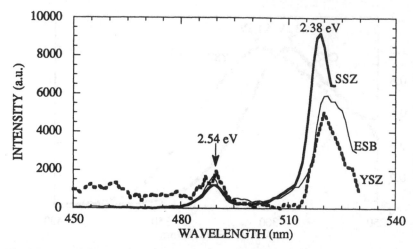

Figure 3. Intra-defect excitation spectra of 7.5 mol% scandia- (SSZ) and 9.6 mol% yttria- (YSZ) stabilized zirconia, and 20 mol% erbia-stabilized bismuth oxide (ESB). 2.27 eV emission intensity as a function of excitation wavelength.

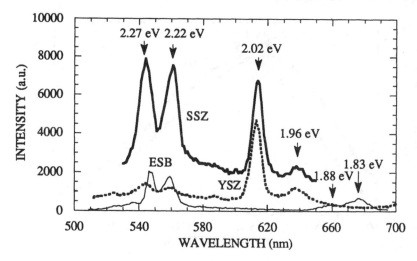

Figure 4. Intra-defect emission spectra of of 7.5 mol% scandia- (SSZ) and 9.6 mol% yttria- (YSZ) stabilized zirconia, and 20 mol% erbia-stabilized bismuth oxide (ESB). Intensity as a function of emission wavelength for 2.54 eV excitation.

emission peaks, respectively, for SSZ. The neighboring emission doublet at 1.96 and 2.02 eV was observed in YSZ at approximately the same energy and relative intensities. In ESB a similar neighboring doublet was observed at lower energy (1.83 and 1.88 eV) with reversed relative intensities. None of these peaks were observed either previously in the monoclinic ZrO_2 or in the present study with CSZ. These new findings are still consistent with our aforementioned chromophore assignments.

EFFECT OF DOPANT CONCENTRATION ON EMISSION SPECTRA

The effect of yttrium concentration on emission spectra from direct intra-defect excitation (2.54 eV) was observed as a change in relative intensity of the intra-defect emission doublet (2.22 and 2.27 eV), with a negligible energy shift. The ratio of intensities of the 2.02 eV to the 2.27 eV emission peaks ($I_{2.0}/I_{2.3}$) is plotted in Figure 5 as a function of yttrium concentration. The relative emission intensity decreases from 1.3 to 0.2 over the yttrium concentration investigated. This decrease in $I_{2.0}/I_{2.3}$ with yttrium concentration is further evidence for the assignment of the 2.22 and 2.27 eV emission doublet to a Y'-associated chromophore.

Broad band emission spectra were obtained at the excitation wavelength (near 4.1 eV) corresponding to maximum intensity for each of the YSZ samples and are shown in Figure 6. With the exception of the 12.4 mol% sample, there is an approximately linear shift in emission band maximum toward higher energy with increasing dopant concentration. Spectra from the 12.4 mol% sample generally deviated from the trends observed with the other samples (e.g., Figures 5 and 6). However, as no impurities were found in this or any of the samples we cannot discount the spectra of this sample at this time.

The blue shift of the emission band maximum can be described in terms of a change in relative intensity of two overlapping emission bands. As the concentration of yttria increases, the relative population of yttrium associated oxygen-vacancies increases. Therefore, the intensity of emission due to yttrium associated oxygen-vacancies (2.27 eV) increases relative to that of the un-associated oxygen-vacancies (2.1 eV).

This model is consistent with the expected effect of increasing dopant concentration on population of associated dopant-vacancy defect pairs. Moreover, the two observed changes in relative emission intensity are self consistent by this model if the 2.02 eV emission peak from direct intra-defect excitation and the 2.1 eV emission band from conduction band excitation arise from optical relaxation in the same un-associated defect, Eq. 2.

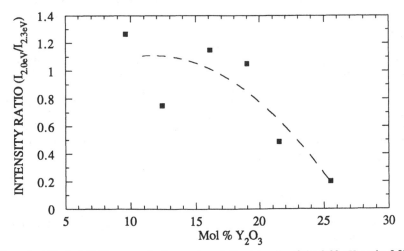

Figure 5. Effect of Y_2O_3 concentration on the relative intensity of the 2.02 eV to the 2.27 eV emission peaks ($I_{2.0}/I_{2.3}$).

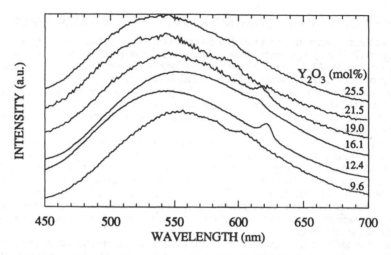

Figure 6. Effect of Y_2O_3 concentration on broad band emission spectra. Spectra are offset vertically to show peak location. Apparent peak at ~620 nm is due to secondary reflection of excitation ($2\lambda_{ex}$).

EFFECT OF DOPANT CONCENTRATION ON EXCITATION SPECTRA

A more complicated effect of yttrium concentration is observed with the broad band excitation spectra. In Figure 7 is plotted the excitation energy, corresponding to maximum emission intensity (near 2.1 eV), as a function of Y_2O_3 concentration and the resulting fraction of vacant-anion sites ($\delta/2$). A minimum is observed at ~ 13 mol% Y_2O_3 ($\delta/2 = 0.06$). At higher dopant concentrations a nearly linear increase in excitation energy with increasing Y_2O_3 concentration and δ is observed.

A similar trend was obtained by Hohnke [4] when comparing the conductivity activation energy as a function of dopant concentration in a number of doped fluorite oxides. This effect of dopant concentration on activation energy is due to the increase in Y_{Zr}'-$V_O^{\bullet\bullet}$ association with increasing Y_2O_3 concentration.

No direct correlation between conductivity activation energy and fluorescence excitation energy can be made at this time. However, if dissociation of a dopant-vacancy pair requires electron-hole recombination to overcome coulombic attraction, then two mechanistic steps are possible depending on whether a dopant-vacancy pair has a trapped electron (e.g., $(Y_{Zr}'V_O^{\bullet})^x$) or not (e.g., $(Y_{Zr}'V_O^{\bullet\bullet})^{\bullet}$):

$$(Y_{Zr}'V_O^{\bullet})^x + h^{\bullet} \rightarrow Y_{Zr}' + V_O^{\bullet\bullet} + h\nu_3 \tag{3}$$

$$(Y_{Zr}'V_O^{\bullet\bullet})^{\bullet} + h^{\bullet} \rightarrow (Y_{Zr}'h^{\bullet})^x + V_O^{\bullet\bullet} + h\nu_4 \tag{4}$$

These steps as shown would result in luminescence and may explain the similar dependence on doping concentration.

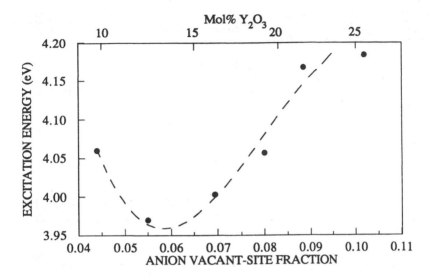

Figure 7. Broad band excitation energy as a function of Y_2O_3 concentration and fraction of vacant anion sites ($\delta/2$).

CONCLUSION

Fluorescence spectra have been used to elucidate electronic transitions in stabilized zirconia. These transitions involve excitation and relaxation of electronic species in oxygen vacancy defects. Observed changes in relative intensity and excitation energy with Y_2O_3 concentration may provide a method to determine relative populations of associated and un-associated oxygen vacancies. This information provides a basis for the understanding of the conductivity mechanism in these materials and the development of more highly conductive materials.

ACKNOWLEDGMENT

E. D. Wachsman, the Stanford authors, and F. E. G. Henn respectively acknowledge the Gas Research Institute, the Electric Power Research Institute, and the French Foreign Ministry (Lavoisier Award) for their support.

REFERENCES

1.   Wachsman, E.D., N. Jiang, C.W. Frank, D.M. Mason and D.A. Stevenson, 1990. "Spectroscopic Investigation of Oxygen Vacancies in Solid Oxide Electrolytes." <u>Applied Physics A,</u> 50:545-549.

2.   Wachsman, E.D., 1990. "Photophysics of Solid State Structures as Related to: A. Conformation and Morphology of Polyimide; B. Conductivity and Reactivity of Solid Oxide Electrolytes." Ph.D. Dissertation, Stanford University.

3.   Wachsman, E.D., G.R. Ball, N. Jiang and D.A. Stevenson, 1992. "Structural and Defect Studies in Solid Oxide Electrolytes." <u>Solid State Ionics,</u> 52:213-218.

4.   Hohnke, D.K., 1979. "Ionic Conduction in Doped Zirconia," in Fast Ion Transport in Solids, P. Vashishta, J.N. Mundy and G.K. Shenoy, eds., New York, NY, USA:Elsevier North Holland, Inc., pp.669-672.

5.   Subbarao, E.C., and T.V. Ramakrishnan, 1979. "Ionic Conductivity of Highly Defective Oxides," in Fast Ion Transport in Solids, P. Vashishta, J.N. Mundy and G.K. Shenoy, eds., New York, NY, USA:Elsevier North Holland, Inc., pp.653-656.

6.   Kilner, J.A. and R.J. Brook, 1982. "A Study of Oxygen Ion Conductivity in Doped Non-Stoichiometric Oxides." Solid State Ionics, 6:237-252.

7.   Catlow, C.R.A., 1984. "Transport in Doped Fluorite Oxides." Solid State Ionics, 12:67-73.

8.   Thorp, A.S., A. Aypar and J.S. Ross, 1972. "Electron Spin Resonance in Single Crystal Yttria Stabilized Zirconia." Journal of Materials Science, 7:729-734.

9.   Shinar, J., D.S. Tannhauser and B.L. Silver, 1986. "ESR Study of Color Centers in Yttria Stabilized Zirconia." Solid State Ionics, 18&19:912-915.

10.   Genossar, J., and D.S. Tannhauser, 1988. "The Nature ESR Centers in Reduced Stabilized Zirconia." Solid State Ionics, 28-30:503-507.

11.   Azzoni, C.B. and A. Paleari, 1989. "EPR Study of Electron Traps in X-Ray-Irradiated Yttria-Stabilized Zirconia." Physical Review B, 40(10):6518-6522.

12.   Azzoni, C.B. and A. Paleari, 1989. "Effects of Yttria Concentration on the EPR Signal in X-Ray-Irradiated Yttria-Stabilized Zirconia." Physical Review B, 40(13):9333-9335.

13.   Orera, V.M., R.I. Merino, Y. Chen, R. Cases and P.J. Alonso, 1990. "Intrinsic Electron and Hole Defects in Stabilized Zirconia Single Crystals." Physical Review B, 42(16):9782-9789.

14.   Weimhöfer, H.D., S. Harke and U. Vhorer, 1989. "Electronic Properties and Gas Interaction of LaF_3 and ZrO_2," in Proceedings of the 7th International Conference on Solid State Ionics, manuscript No. 6pb-15.

15.   Heyne, L. and N.M. Beekmans, 1971. "Electronic Transport in Calcia-Stabilized Zirconia." Proceedings British Ceramic Society, 19:229-263.

16.   Strehlow, W.H. and E.L. Cook, 1973. "Compilation of Energy Band Gaps in Elemental and Binary Compound Semiconductors and Insulators." Journal Physical Chemical Reference Data, 2(1):163-199.

Impedance Spectroscopy of Microstructure Defects and Crack Characterisation

M. KLEITZ, C. PESCHER and L. DESSEMOND

ABSTRACT

Recent results have shown that impedance spectroscopy can be used for crack characterisation in electric ceramics. The capacitive component of the blocking effect and its relaxation frequency appear to be as significant as the resistive term. Appropriate combinations of these parameters allow us to evaluate the average width of the cracks. Measurements performed on welds between single crystals and porous materials have confirmed this statement. They have also revealed a serious risk of error in the determination of the specific conductances in highly porous and composite materials.

INTRODUCTION

Since the pioneering work by Bauerle, it has been recognized that grain boundaries in sintered stabilized zirconia result in an additional resistance, at least at relatively low temperature [1]. Various interpretations and models have been put forward for this detrimental effect which can more than double the ohmic drop in an electrochemical cell. Most of them refer to two types of microscopic observations :
- traces of a second phase frequently precipitated at the grain boundaries
- dopants and dissolved impurities which appear to segregate into the grain sub-surfaces [2-9].
Both phenomena are viewed as resulting in local resistance increases. The precipitated phase, most frequently containing silica, is less conducting than stabilized zirconia. The solute enrichment shifts the zirconia doping away from its maximum conductivity concentration.
In agreement with these observations, the electric behavior of the material has frequently been modeled by a series circuit [1]. In terms of experimental results, the data are presented on impedance diagrams.
With Schouler and Bernard we have examined an alternative interpretation sketched in figure 1 [10,11]. The mobile ions (oxide vacancies) are divided into two

M. Kleitz, C. Pescher and L. Dessemond
Laboratoire d'Ionique et d'Electrochimie des Solides de Grenoble (associé au C.N.R.S.)
I.N.P.G., Domaine Universitaire, B. P. 75 38402 Saint Martin d'Hères Cédex - France.

593

groups : those which permeate through the grain boundaries without suffering any trapping or slowing down and those which, for any reason, are blocked at certain grain boundary locations forming sorts of space-charges resulting from the current line distribution around the blocked contact. The "blocker", which can be the precipitated phase, for example, is modeled by a capacitance C_g in series with the conductance Σ_g representing the blocked ions (Figure 1a). Because the natures of the blocked and permeating ions are the same, their mobilities are identical and the populations of both categories are proportional to the corresponding conductances (in figure 1a, respectively Σ_p and Σ_g for the permeating and blocked ions). Because this description is more independent with respect to the chemical nature of the blocking agent, we will call it a "geometrical model" as opposed to the other descriptions which will be called "chemical models". The geometrical model is also called a "parallel model" and the others "series models" in reference to the equivalent circuits.

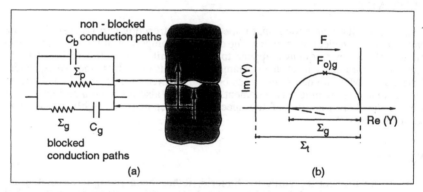

Figure 1. Grain boundary effect. Basic diagram of the parallel model.

At sufficiently high frequency, the blocking capacitance C_g is "short-circuited" and the measured conductance Σ_t is equal to the sum of both conductances :

$$\Sigma_t = \Sigma_p + \Sigma_g \qquad (1)$$

That conductance characterizes the total conductance of the material in the absence of a blocking effect, as expected.

For the parallel model, the experimental results are better represented on admittance diagrams with the parameters determined as shown in figure 1b.

To characterize the blocking effect, we have defined the blocking coefficient α_R [12]:

$$\alpha_R = \frac{\Sigma_g}{\Sigma_t} \qquad (2)$$

It simply gives the fraction of ions being blocked under the measuring conditions over the total number of mobile ions in the sample. This blocking coefficient can in fact be calculated both in terms of admittance or impedance diagram parameters. In terms of impedance parameters it obeys the formula :

$$\alpha_R = \frac{R_g}{R_b + R_g} \tag{3}$$

where R_b and R_g are the resistance parameters describing the intragrain behavior and the grain boundary effect. A parameter defined by a similar equation was also used by Inozemtsev et al. [13].

This blocking coefficient is of interest for another reason : it is dimensionless. It is independent with respect to the geometrical factor of the measuring cell. This is important when comparing results from different sources and when using cells with rather inaccurate geometrical factors such as the point electrode cells used in the investigation reported here.

To take full advantage of this situation we have also introduced a dimensionless capacitive coefficient similarly defined by the equation :

$$\alpha_C = \frac{C_g}{C_b} \tag{4}$$

where C_b is the capacitance associated with the true dielectric properties of the material as determined by the high frequency semicircle in impedance or the high frequency straight line in admittance.

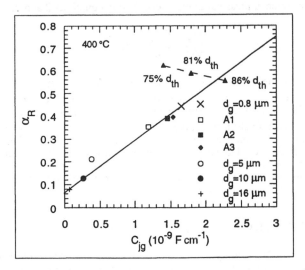

Figure 2. Original Bernard's diagram [11]. Blocking factor versus grain boundary capacitive effect in YSZ (9 mole%) for : d_g nominally pure samples with variable grain sizes as indicated ; A1, A2, A3 small grain samples, with variable amounts of Al_2O_3 partly substituted for Y_2O_3, respectively 0.44, 0.93 and 1.73 mole% Al_2O_3 ; d_{th} porous samples of various relative densities as indicated.

Two other features will be made use of in presenting our results :

a) Schouler pointed out that the relaxation frequencies can be regarded as identification parameters, i.e. as "signatures" of the observed phenomena [10]. On an Arrhenius diagram, not only the dielectric specific relaxation frequencies but also the grain boundary relaxation frequencies (and the electrode relaxation frequencies) are found on well defined lines or rather narrow bands [10].

The usage of such a type of diagram (Schouler's diagram) provides an easy way to identify a semicircle on an isolated diagram recorded at a single temperature.

b) Bernard working on nearly pure YSZ with grain sizes varying from approximately 0.8 to 16μm observed that α_R and C_g decrease proportionally as the grain size increases (Figure 2) [11].

These two results are described by the basic relaxation equation for the parallel circuit of figure 1 which is :

$$\frac{C_g}{\Sigma_g} 2\Pi F_{o)g} = 1 \tag{5}$$

If $F_{o)g}$ is a characteristic parameter for the grain boundaries, constant at a given temperature, C_g and Σ_g are proportional at this temperature :

$$\frac{\Sigma_g}{C_g} = 2\Pi F_{o)g} \tag{6}$$

In terms of dimensionless parameters, combining equations 2 and 4, gives :

$$\frac{\alpha_R}{\alpha_C} = \frac{F_{o)g}}{F_{o)b}} = \alpha_F \tag{7}$$

where $F_{o)b}$ is the dielectric relaxation frequency of the material (measured on the intragrain response). We will call α_F the relaxation frequency ratio $F_{o)g} / F_{o)b}$.

Accordingly, we will frequently present our results on a relaxation frequency Arrhenius diagram (Schouler's diagram) and on an α_R versus α_C diagram (Bernard's diagram) to compare them to the conventional grain boundary responses.

Let us finally mention that according to our selection of the "parallel model", most of our data are deduced from admittance diagrams. With respect to the conventional approach from the impedance diagrams, this can introduce slight differences, particularly for the capacitance parameters.

CRACK DETECTION AND CHARACTERISATION

The finding summarized by the d_g points on figure 2 provided evidence that the blocking in nearly pure zirconia was a specific property of the grain contact and not predominantly influenced by the impurities (which were maintained at a constant and low level of concentration in these investigated samples) [11]. At this stage, we suggested that the grain boundary blocking could be due to simple mismatches between the grains with nanovoids prohibiting the jumps of the oxide ions. From this conclusion it was natural to deduce that cracks which certainly generate voids along them, should be observable by impedance spectroscopy.

The cell used for the measurements is sketched in figure 3 (for more details see [14]). The cracks were generated by Vickers indentation at room temperature. The measurements were done locally with point electrodes of approximately $0.1mm^2$ area, located 0.25 mm away from an indentation corner. The relative reproducibility of the results was found to be better than 15%.

A first series of impedance and admittance diagrams were plotted with sintered YSZ (9mole% Y_2O_3, average grain size 13μm, relative density 95%, main impurity level : silica 220ppm).

As the Vickers indentation load increased we did observe an alteration of the grain boundary semicircle both in size and in depression.

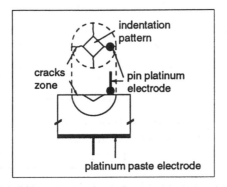

Figure 3. Diagram of the electrochemical cell used for local impedance spectroscopy characterisation of cracks.

Typical results are shown in figure 4 in terms of the blocking factor α_R. At first, the blocking factor slightly decreases. Then the expected monotonic increase in blocking is observed with a large jump around an indentation load of 200N.

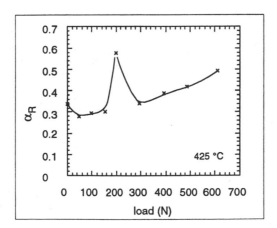

Figure 4. Blocking factor versus indentation load for cracks in sintered YSZ, at 425°C.

The depression angle β_g of the blocking effect semicircle that we interpreted in terms of heterogeneity of the property, shows a marked increase (Figure 5) at least up to a load of 400N. This can be interpreted in terms of a crack response sufficiently close to that of the grain boundaries to overlap with it, but significantly different to introduce a heterogeneity in the overall response.

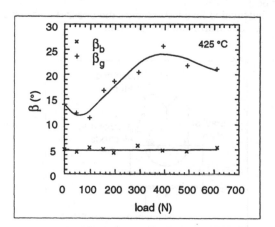

Figure 5. Semicircle depression angles in sintered YSZ : β_b bulk response and β_g (grain boundary + crack) response.

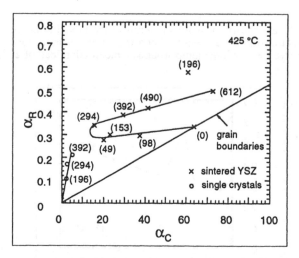

Figure 6. Bernard's diagram of the grain boundary and crack characteristic parameters. Indentation loads are given in brackets. The straight lines represent the different load domains. For regular grain boundaries, the proportionality between α_R et α_C is supposed to be the response of a grain boundary density increase.

On Bernard's diagram (Figure 6), the response did show a change as a function of the load. Under low load, the experimental point evolution is characterized mostly by a decrease of the capacitive coefficient with a relative slow variation of the blocking factor. Around a load of 290N, a change occurs towards a simultaneous increase of α_R and α_C which is supposed to be the regular response of a crack density increase (in fact, the capacitive coefficient increases more rapidly than expected from the strict proportionality between the two coefficients). On this diagram, the peculiar point at 196N appears comparable to that obtained under an indentation load of about 600N.

Similar measurements were performed on single crystals of YSZ doped with 10mole% Y_2O_3 (kindly provided by Crismatec*). Typical impedance diagrams are shown in figure 7 (for more details see [15]). In the absence of any interference with a grain boundary response, the cracks are here directly observed as a specific additional semicircle. On this material it was impossible to go higher in load than 390N because of chipping effects.

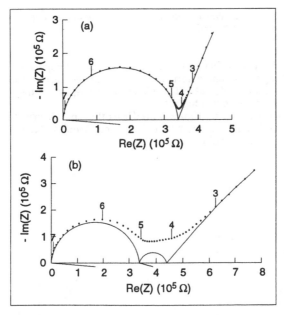

Figure 7. Typical impedance diagrams obtained with single crystal reference sample (a) and with a single crystal indented under a 392 N load (b), at 400°C. Measurements were performed with point electrodes in air. The numbers indicate the a.c. frequency logarithm.

Figure 8 shows the variation of α_R which regularly increases with the indentation load. The depression angle β_g (Figure 8) shows an increasing heterogeneity which however remains relatively small and close to the usual heterogeneity of the grain

---

* Crismatec (single crystal manufacturer), subsidiary of St. Gobain, 2 rue des Essarts, 38610 Gières (F). Fax : (33) 76.44.17.53.

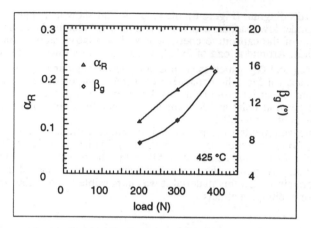

Figure 8. Variation with indentation load of the blocking factor and depression angle. YSZ single crystals (10 mole% Y_2O_3).

boundaries. On Bernard's diagram (Figure 6) the results appear to follow a rather well defined straight line characteristic of a regular increase in the density of the observed microstructure defects. The slope, with respect to the reference grain boundary line is however much higher.

The Arrhenius diagram on figure 9 confirms that the observed cracks relax at frequencies higher than regular grain boundaries and therefore that their characteristic parameters are significantly different.

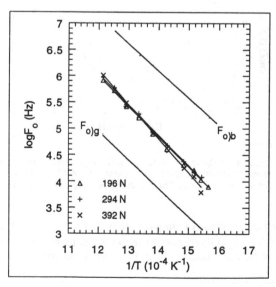

Figure 9. Arrhenius diagram of the crack relaxation frequencies (admittance) for single crystals.

At this stage, the essential conclusion is that cracks can actually be observed using impedance spectroscopy and that their response is different than that of the grain boundaries in terms of relaxation frequency.

The interpretation we propose for the differences in behavior between cracks and grain boundaries is the following :

- The blocked conductance Σ_g and therefore the blocking factor α_R are proportional to the area density S_g of the blocker (the voids along the grain boundaries and cracks in our present interpretation).

- The blocking capacitance C_g and therefore the capacitive coefficient α_C obey an equation of the type :

$$\alpha_C = A \times \frac{S_g}{E_g} \qquad (8)$$

where A is a constant and E_g is average width of the blocker.

This gives a α_R / α_C ratio proportional to the average microstructure width :

$$\frac{\alpha_R}{\alpha_C} = B \times E_g \qquad (9)$$

where B is a constant.

- The results obtained by Bernard which can be summarized in the statement "α_R remains proportional to α_C when the grain boundary area varies" means that under the experimental conditions the average width E_g of the voids along the grain boundaries remained constant.

- The fact that the slope of the α_R versus α_C line for the single crystal (Figure 6), in other words, for a pure crack distribution, is higher than that of the regular grain boundaries indicates that the crack width is bigger than that of the grain boundaries (by a factor 10).

- The heterogeneity observed with cracks in sintered materials appears be due to a variation of the average width E_g of the observed microstructure defects. The cracks being wider than the grain boundaries, the average width increases with the crack density. Under low load, the behavior observed on Bernard's diagram corresponds to an opening of the voids along the grain boundaries, probably due to a crack propagation along them. Then, at higher load, the marked increase in α_R and therefore in the density of blocking internal surfaces indicates a generation of new cracks probably across the grains. The more rapid increase in α_C than in α_R means that under high load the relative density of narrow cracks is higher.

If the interpretation is correct one can conclude that the α_R / α_C ratio is really proportional to the width of the observed microstructure defect and therefore that the relaxation frequency ratio α_F which is proportional to the previous ratio (Equation 7), can be used as a measurement of the blocker width.

CONTACT AND WELDING BETWEEN SINGLE CRYSTALS

To check this conclusion, we performed measurements on contacts between single crystals. Pieces were cut from the same bar (kindly provided by Crismatec*) of YSZ doped with 10mole% Y_2O_3. Their bases were oriented perpendicular to the [111] direction and polished with a fine diamond buff wheel. Two of these pieces were further polished with a 1μm diamond paste and simply pressed in contact. Two others were welded by curing at 1700°C for 20 hours and two others at 2000°C for 7 hours.

In each case a reference piece was submitted to the same temperature cycle. It was used to measure the resistivity of the material and to simulate the semicircle of the bulk parts of the assemblies. After the measurements, small pieces were also cut from the welded crystals. Their responses were also measured and compared to those of the reference samples. The agreement was always within 10%.

Impedance measurements were carried out on the assemblies with symmetrical electrochemical cells. For greater accuracy, the semicircle responses of the bulk parts were simulated from the data on the reference samples and used to deconvolute the high frequency responses which are in fact sums of two semicircles with close relaxation frequencies. By using a least square fitting the accuracy of the experimental deconvolutions was found better than 10%.

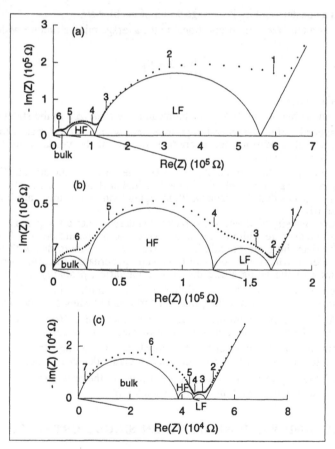

Figure 10. Impedance diagrams of various contacts and "welding" between single crystals : (a) Bi1 : polished surfaces, k = 2.731 cm⁻¹, 400 °C ; (b) B17 : welding at 1700°C (20h), k = 2.285 cm⁻¹, 401 °C ; (c) B20 : welding at 2000°C (7h), k = 2.731 cm⁻¹, 400 °C. The numbers indicate the a.c. frequency logarithm.

In the case of the welding at 2000°C (Figure 10c), a resolution of the high frequency response into a single semicircle would result in a relaxation frequency and a conductivity half the size of the associated parameters of the corresponding reference sample. Furthermore the accuracy of such a deconvolution is only 15%.

Typical deconvoluted impedance diagrams are shown in figure 10. They are composed of three semicircles :

- The dielectric response of the material (which relaxes at a frequency slightly higher than 10^6 Hz under the experimental conditions of figure 10)

- An intermediate response, referenced HF, which relaxes between 10^4 and 10^5 Hz under the same experimental conditions.

- A low frequency one, referenced LF, with relaxation frequencies close to 10^3 Hz.

The LF semicircle is enormous with the simple contact between the single crystals.

The HF semicircle becomes predominant after the welding at 1700°C and markedly decreases after the welding at 2000°C.

On Schouler's diagram (Arrhenius diagram) the relaxation frequencies are located as shown in figure 11. These signatures of the observed microstructure defects are in fact significantly different from those of conventional grain boundaries. Microscopic observations of the welding at 2000°C indicate a predominant concentration of pores of about 10µm width which correspond to the HF semicircle.

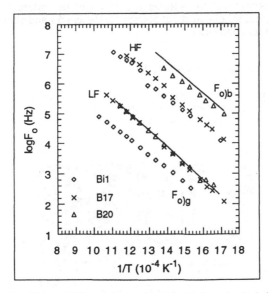

Figure 11. Arrhenius diagram of the various relaxation frequencies (admittance) observed on contacts and welding between single crystals (symbols defined in Figure 10 caption).

Based on these data and assuming that the α_F relaxation frequency ratio is proportional to the width of the blocker, we calculated the widths shown in Table I.

For the simple contact between the single crystals, it is reasonable to find a width of about 0.6µm associated with the HF relaxation, because the average roughness of the surfaces in contact is of this order of magnitude.

TABLE I - CALCULATED WIDTHS (E_g) OF THE INVESTIGATED MICROSTRUCTURE DEFECTS.
Assumption : the relaxation frequency is proportional to the width. Calibration of the
scale with the pores observed on welded single crystals.
Their average width was evaluated to be about 10 μm.

RESPONSE	$F_o / F_o)_b$	E_g
HF circle (welding at 2000 °C)	0.932	10 μm
HF circle (polished contact)	$6.22 \ 10^{-3}$	0.6 μm
LF circle (welding at 1700 °C)	$1.21 \ 10^{-3}$	13 nm
LF circle (polished contact)	$2.89 \ 10^{-4}$	3 nm
cracks in single crystals	$4.33 \ 10^{-2}$	500 nm
cracks in sintered samples : - 49 N	$1.38 \ 10^{-2}$	150 nm
- 612 N	$6.72 \ 10^{-3}$	70 nm
grain boundaries in regular sintered YSZ	$9.86 \ 10^{-4}$	10 nm

Using this "calibration" to evaluate the width of the microstructure defects previously examined, we obtain about 500nm for the cracks and about 10nm for the voids along the grain boundaries.

PORES IN SINTERED CERAMICS

As further confirmation of our conclusion we can quote recent results by Guindet [16]. The measurements were carried out on ZrO_2 - Y_2O_3 - CeO_2 fully stabilized cubic solutions which exhibit a narrow spectrum of pore sizes (Figure 12).

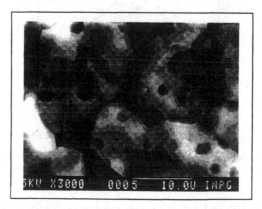

Figure 12. Micrograph of $(ZrO_2)_{0.883}$ - $(Y_2O_3)_{0.098}$ - $(CeO_2)_{0.019}$ samples [16].

Typical diagrams are shown in figure 13. Here an intermediate semicircle is clearly observable. The derivation developed for the welded crystals gives here an

α_F frequency ratio of about 0.2 and therefore a pore diameter in the range 1.4 - 2.6µm. The microscopic observations (Figure 12) indicate a diameter of about 2µm.

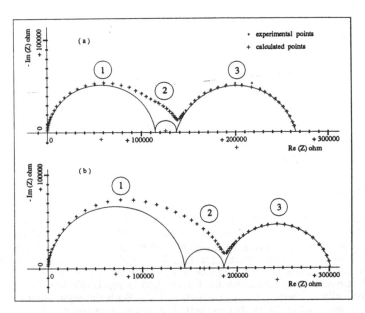

Figure 13. Impedance diagrams of porous ZrO_2 - Y_2O_3 - CeO_2 samples [15] : (a) $(ZrO_2)_{0.883}$ - $(Y_2O_3)_{0.098}$ - $(CeO_2)_{0.019}$, (b) $(ZrO_2)_{0.830}$ - $(Y_2O_3)_{0.092}$ - $(CeO_2)_{0.078}$; 340 °C, air, k = 4.074 cm^{-1}.

BULK CONDUCTIVITY MEASUREMENTS

Extrapolating the previous derivation leads to an important conclusion : the relaxation frequency of pores slightly larger than 10µm is likely to be very close to that of the dielectric specific relaxation of the material. Then, the corresponding semicircles will overlap to such an extent that the presence of a pore response will not be visible. This is a serious risk of error in the measurements of the bulk characteristics of ceramic materials.

This was checked with samples prepared as follows : a sample of sintered YSZ of about 95% theoretical density was ground and sieved to select appropriate aggregates of about 40 to 250µm average diameter. New samples were prepared by pressing these aggregates together and sintering at 1750°C for 2 hours. The micrographs of the samples show irregular pores of about 15µm. A typical impedance diagram is shown in figure 14. The intermediate HF semicircle is no longer visually separable from the bulk response. The evaluation we made of the true bulk response is significantly smaller than the observed semicircle. Under such a situation, if the pore interference is not taken into account, significant error in the determination of the bulk properties will result.

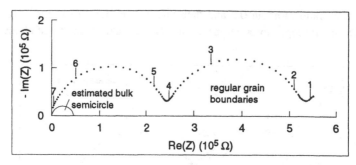

Figure 14. Impedance diagram of a highly porous YSZ (9 mole%) sample. $d = 0.54\ d_{th}$, average pore diameter : 15 μm, $k = 1$ cm^{-1}, 380 °C (the numbers indicate the measuring a.c. frequency logarithm).

CONCLUSIONS

Cracks are observable in impedance spectroscopy. Measurements can be made on bulk parts by using large area electrodes on both sides of the zone to be examined (as was done with the welded crystals) or locally with a point electrode (0.1mm^2 area electrodes were used in the reported results). The required technology is not overly complex and it should be even possible to carry out experiments "in vivo" under variable stress. We have checked that the method is applicable to other materials either ionically or electronically conducting [17]. With the equipment presently available our conclusion is that measurements are possible provided that the conductivity of the material satisfies the following conditions :

$$10^3 < \sigma \text{ (in Scm}^{-1}) < 10^7 \qquad (10)$$

With thin layers the upper limit can be pushed forward up to 10^{10} Scm^{-1}.

The technique allows the average width of the examined microstructure defects to be determined. Measurements were performed on widths varying from a few nm up to 10μm. In fact the technique has been found more accurate or at least more convenient in terms of semicircle overlapping for relatively small widths.

Measurements carried out on materials with widely open defects revealed a serious risk of error in the determination of the specific properties because of a possible interference of the defect response with the specific response. This is especially important in highly porous materials and in composites.

ACKNOWLEDGEMENTS

The authors sincerely thank M. Hénault for the preparation of the samples.

REFERENCES

1. Bauerle, J. E., 1969. "Study of Solid Electrolyte Polarization by a Complex Admittance Method." Journal of Physics and Chemistry of Solids, 30 : 2657-2670.

2. Beekmans, N. M. and L. Heyne, 1976. "Correlation between Impedance Microstructure and Composition of Calcia-Stabilized Zirconia." Electrochimica Acta, 21 : 303-310.

3. Badwal, S. P. S., 1984. "Electrical Conductivity of Single Crystal and Polycrystalline Yttria-Stabilized Zirconia." Journal of Materials Science, 19 : 1767-1776.

4. Hughes, A. E., and B. A. Sexton, 1989. "XPS Study of an Intergranular Phase in Yttria-Zirconia." Journal of Materials Science, 24 : 1057-1061.

5. Badwal, S. P. S. and A. E. Hughes. 1991. "Modification of Cell Characteristics by Segregated Impurities," in Proceedings of the Second International Symposium on Solid Oxide Fuel Cells, F. Grosz, P. Zegers, S. C. Singhal and O. Yamamoto, eds., Brussels, Belgium : Office for Official Publications of the European Communities, pp. 445-454.

6. Winnubst, A. J. A., P. J. M. Kroot and A. J. Burggraaf, 1983. "AES / STEM Grain Boundary Analysis of Stabilized Zirconia Ceramics." Journal of Physics and Chemistry of Solids, 44 (10) : 955-960.

7. Whalen, P. J., F. Reidinger, S. T. Correale and J. Marti, 1987. "Yttria Migration in Y-TZP during High-Temperature Annealing." Journal of Materials Science, 22 : 4465-4469.

8. Hughes, A. E. and S. P. S. Badwal, 1990. "Impurity Segregation Study at the Surface of Yttria-Zirconia Electrolytes by XPS." Solid State Ionics, 40 / 41 : 312-315.

9. Hughes, A. E. and S. P. S. Badwal, 1991. "Impurity and Yttrium Segregation in Yttria-Tetragonal Zirconia." Solid State Ionics, 46 : 265-274.

10. Schouler, E., April 1979. "Study of Solid Electrolyte Cells by Complex Impedance Method. Applications to the Measure of Conductivity and to the Study of Oxygen Electrode Reaction (in Fr.), " Ph. D. Thesis, Grenoble National Polytechnic Institute, Grenoble, France.

11. Bernard, H., July 1980. "Microstructure and Conductivity of Sintered Stabilized Zirconia (in Fr.), " Ph. D. Thesis, Grenoble National Polytechnic Institute, Grenoble, France.

12. Kleitz, M., H. Bernard, E. Fernandez and E. Schouler, 1981. "Impedance Spectroscopy and Electrical Resistance Measurements on Stabilized Zirconia," in Advances in Ceramics, vol. 3, Science and Technology of Zirconia, A. H. Heuer and L. W. Hobbs, eds. Columbus, OH, USA : The American Ceramic Society, pp. 310-336.

13. Inozemtsev, M. V., M. V. Perfil'ev and A. S. Lipilin, 1974. "Effect of Temperature and Sintering Time of Solid Oxide Electrolytes on their Electrical Properties." Elektrokhimiya, 10 (10) : 1471-1476.

14. Dessemond, L. and M. Kleitz, 1992. "Effects of Mechanical Damage on the Electrical Properties of Zirconia Ceramics." Journal of the European Ceramic Society, 9 : 35-39.

15. Dessemond, L., J. Guindet, A. Hammou and M. Kleitz, 1991. "Impedancemetric Characterization of Cracks and Pores in Zirconia Based Materials," in Proceedings of the Second International Symposium on Solid Oxide Fuel Cells, F. Grosz, P. Zegers, S. C. Singhal and O. Yamamoto, eds., Brussels, Belgium : Office for Official Publications of the European Communities, pp. 409-416.

16. Guindet, J., September 1991. "Contribution to the Study of Anode Materials for Solid Oxide Fuel Cell (in Fr.), " Ph. D. Thesis, Grenoble National Polytechnic Institute, Grenoble, France.

17. Pescher, C., July 1980. "Characterization of a New Technique of Crack Detection : Impedance Spectroscopy (in Fr.), " D. E. A. Diploma, Grenoble National Polytechnic Institute, Grenoble, France.

SECTION V

OXYGEN SENSORS

Amperometric Tetragonal Zirconia Sensors

B. Y. LIAW, J. LIU and W. WEPPNER

ABSTRACT

This paper discusses recent progress made on amperometric oxygen sensors based on tetragonal zirconia polycrystals (TZP) operated at intermediate and near-ambient temperatures. Sensor characteristics such as ac and dc conductivities, complex impedance responses and dc polarization behaviors are described. TZP has favorable properties for lower-temperature sensor application in spite of the lower defect concentration compared to conventional cubic stabilized zirconia (CSZ). The sensors measured oxygen partial pressure from almost 1 atm down to ppm range. Cross-sensitivity can be eliminated by considering the shape of the i-V curves. At high oxygen partial pressure ($\geq$ 0.1%), the low-biased polarization indicates that the sensor was operating under the limitation of the bulk and grain boundary transport of the TZP electrolyte. At low oxygen partial pressure, the catalytic behavior at the electrode/electrolyte interface dominates in such a polarization.

INTRODUCTION

Zirconia-based electrochemical oxygen sensors are often used to transduce oxygen partial pressures or concentrations into electrical signals [1]. This type of sensor has found useful applications for monitoring gas quality, controlling the air/fuel ratio in internal combustion engines, controlling oxidation and reduction processes in steel making, or detecting impurities and hazardous materials in special environments. The sensing function is generally categorized into either potentiometric or amperometric modes of operation.

The potentiometric mode of operation employs the Nernst equation, in which the measured cell emf represents a logarithm of the oxygen partial pressure in the environment versus a known reference such as air. A well-known example is the "λ-probe" used in automobiles to monitor exhaust gas composition to control the air/fuel

B. Y. Liaw, Hawaii Natural Energy Institute, School of Ocean and Earth Science and Technology, University of Hawaii at Manoa, 2540 Dole Street, Holmes Hall 246, Honolulu, HI 96822, USA
J. Liu and W. Weppner, Max-Plank-Institut für Festkörperforschung, W-7000 Stuttgart 80, Germany

ratio for fuel economy. This type of operation is insensitive to small oxygen partial pressure change and has to rely on a reliable reference for correct measurements.

Similar to polarographic analysis, several amperometric sensors were later introduced to overcome these drawbacks. These amperometric sensors became interesting alternatives in providing better sensitivity with a simpler device configuration that eliminates the use of a reference electrode. The sensors measure limiting currents that are related linearly to oxygen concentration. To date sensors of this type have been reported to detect H_2O [2, 3], CO [4], CO_2 [5], NO/NO_2 [4] and alcohols [6].

In the early report by Dietz [7], a porous diffusion barrier was used for the gas-diffusion-controlled type sensing application. The diffusion mechanism can be either ordinary gas diffusion or Knudsen diffusion control, depending on the type and geometry of the diffusion barrier, the desired operating range of oxygen partial pressure and temperature. The sensor requires an in-situ planar heater to control the operating temperature above 600° C.

A limiting-current polarographic-type oxygen sensor (Figure 1) was later reported by Usui, Asada, Takeuchi and their co-workers [2, 3, 6, 8-11]. Their sensor uses CSZ as the solid electrolyte. An encapsulated cavity with a long, narrow gas diffusion barrier was provided to ensure the ordinary gas diffusion mechanism to be the rate limiting step in the sensing device. This configuration offers many advantages in sensing characteristics, including: an operating temperature below 500° C; elimination of the reference electrode; good detectability over a wide range of oxygen concentration, typically from 0.1 to over 95%; small size and weight; and a long lifetime without caring much about the degradation problem at the electrolyte/electrode interface.

Weppner and his co-workers [12-14] have reported the use of tetragonal zirconia polycrystal (TZP) electrolyte for amperometric limiting-current sensor applications, aiming to improve the operation characteristics. The use of TZP has additional advantages over CSZ; namely, an even lower operating temperature due to improved ionic conductivity, and higher resistance to thermal shock. Liaw and Weppner [12]

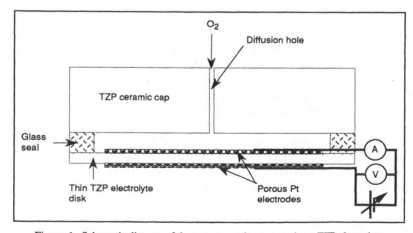

Figure 1. Schematic diagram of the amperometric sensor using a TZP electrolyte

have reported sensors working above 200° C in high oxygen concentrations (≥ 0.1%). Liu and Weppner [14] later developed similar sensors for detecting trace oxygen concentrations in argon. This paper summarizes the sensor operation characteristics.

PRINCIPLES

The amperometric sensor measures oxygen partial pressures as limiting currents by imposing a constant bias over the Pt electrodes and the zirconia electrolyte, as shown in Figure 1. A typical amperometric sensor consists of a thin disk of zirconia electrolyte and two porous Pt electrodes housed in a diffusion barrier structure typically made of a ceramic disk with a thin, long diffusion hole and a seal material such as glass that has a thermal expansion coefficient compatible with the ceramic materials. The anode is exposed to the gas environment. The cathode resides in the cavity of the diffusion barrier structure, which has to be gas-tight so the gas diffusion occurs solely through the long diffusion hole of controlled geometry.

As expected from Fick's (first) law, a diffusion flux should be linearly proportional to the concentration (chemical potential) gradient; thus, a limiting current can be measured according to [7]

$$i_L = 4 \, q \, D_{O_2\text{-}X} \, A \, P_{O_2} \, / \, k \, T.l \qquad (1)$$

where q is the charge transferred in the dissociation of oxygen, $D_{O_2\text{-}X}$ is the ordinary oxygen diffusion coefficient in the gas mixture O_2-X, A is the cross section of the diffusion hole, P_{O_2} is the oxygen partial pressure in atm, k is the Boltzmann constant, T is the absolute temperature in Kelvin, and l is the length of the diffusion hole.

When the oxygen concentration is high, the limiting current does not follow a linear relationship with the concentration. Under the polarographic process with a sufficiently high bias, a limiting current i_L can be measured. Such a limiting current is a function of total pressure p according to [11]

$$i_L = - \, 4 \, q \, D_{O_2\text{-}X} \, A \, p \, \ln(1 - P_{O_2}/p) \, / \, k \, T \, l \qquad (2)$$

The difference between (1) and (2) is due to the consideration of both chemical-potential and pressure gradients in the case of high oxygen content. Under limiting-current operation, the oxygen diffused into the cavity is constantly evacuated by the sensor, leaving almost zero oxygen concentration at the cathode. This process creates a pressure difference across the diffusion hole, which leads to the term $p \, \ln(1 - P_{O_2}/p)$ in equation 2.

The proper function of the amperometric sensor solely relies on the gas diffusion through the diffusion hole. The diffusion-hole geometry decides the diffusion mechanism. The sensor geometry, including the diffusion-hole dimensions, cavity size and the electrochemical cell dimensions, determines sensor output characteristics. Demands for lower operating temperatures, higher sensitivity, wider detection range, better reliability, and faster response time require additional engineering considerations.

The diffusion mechanism depends on the diameter of the diffusion barrier, ϕ, and the mean free path of the oxygen diffusion, λ. When $\phi \gg \lambda$, ordinary gas diffusion dominates. If $\lambda \gg \phi$, Knudsen diffusion becomes predominant. The dependence on pressure and temperature are different in the two mechanisms, which leads to different sensor configurations for sensing requirements.

Dietz [7] has discussed the dependence on pressure and temperature in both diffusion mechanisms. The limiting current is proportional to $p(T)^{0.5}$ in Knudsen diffusion but only to $(T)^{0.7}$ in ordinary diffusion. The Knudsen diffusion also depends on the nature of the barrier surface, which is difficult to control. Sensors based on ordinary diffusion are thus often desired because of their simple configurations. This paper will therefore concentrate on the ordinary-diffusion type sensors.

We consider the ordinary diffusion applies to our sensors. The diameter of the diffusion hole is of the order of 0.1 mm, which is much longer than the oxygen mean free path in the operating temperatures. Taking 673 K as an example, we can calculate the oxygen diffusion coefficient in argon using the Fuller-Schettler-Giddings model:

$$D_{O_2\text{-}Ar} = 1.00 \times 10^{-3} \, T^{1.75} \, (1/M_{O_2} + 1/M_{Ar})^{1/2} / p[V_{O_2}^{1/3} + V_{Ar}^{1/3}]^2 \qquad (3)$$

where M_{O_2} (32.0 g/mol) and M_{Ar} (39.95 g/mol) are the molecular and atomic weights of oxygen and argon, respectively, and V_{O_2} (16.3 cm^3) and V_{Ar} (16.2 cm^3) are "diffusion volumes" of the oxygen molecule and argon atom, respectively. The diffusion coefficient at 673 K is thus calculated to be 0.822 cm^2/s. The mean free path of oxygen diffusion in argon-mixed stream is related to the diffusion coefficient by

$$D_{O_2\text{-}Ar} = (3\sqrt{2} \, \pi / 64) \, c \, \lambda \qquad (4)$$

where c is the velocity of oxygen in the mixed stream as depicted by the Maxwell-Boltzmann distribution, namely

$$c = (8 \, k \, T / \pi \, M_{O_2})^{1/2} \qquad (5)$$

The value of λ at 673 K is therefore about 1.86×10^{-3} cm, compared to $\phi \approx 0.01$ cm. The condition of $\phi \gg \lambda$ is thus valid and will guarantee that ordinary diffusion predominates.

RESULTS

OPERATION IN THE HIGH-OXYGEN-CONCENTRATION REGIME

Limiting-current sensors using the TZP electrolyte have been used [12] to detect oxygen partial pressures at high concentrations over 0.1%. The gas stream used was oxygen-argon mixtures. A temperature range from 600 to 200° C (873–473 K) was investigated. The diffusion hole dimensions were about 0.1 mm in diameter and about 0.5–1.0 cm long. The electrolyte disk had a thickness of about 0.1 mm.

Both ac impedance and dc polarization measurements were conducted on open

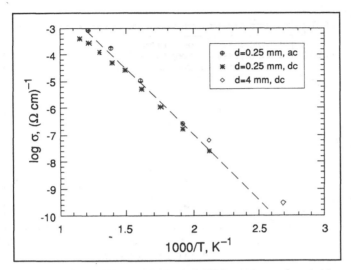

Figure 2. Arrhenius behavior of the TZP electrolytes of different thickness. An activation enthalpy of 0.94 eV was determined from the slope of the dashed line.

electrolytes and amperometric sensors. The ionic conductivity as a function of reciprocal temperature was obtained with an activation enthalpy of 0.94 eV. Figure 2 displays such an Arrhenius behavior. This enthalpy value represents both inter- (bulk) and intragranular (grain boundary) contributions and agreed with values reported by others [15-17].

Figure 3 shows a series of polarization curves from sensors operating at different temperatures but similar oxygen concentrations. An interesting feature in Figure 3 is the decline in slope of the initial polarization before the limiting-current plateaus. This feature is expected from the decrease of ionic conductivity in the electrolyte as a function of temperature, reducing the sensitivity. A second feature reflects the relationship between temperature and limiting current. A temperature dependence can be derived from equations 2 and 3, and it comes close to $T^{0.75}$. A third feature indicates that the limiting-current plateau ends at a lower voltage as temperature increases. The end of the plateau and the increasing current represent the decomposition of the TZP electrolyte. The higher the temperature the easier the TZP decomposes. Therefore, the sensor should operate at a lower temperature to minimize the degradation of the electrolyte.

OPERATION IN THE LOW-OXYGEN-CONCENTRATION REGIME

More recently, Liu and Weppner [14] have investigated sensor behavior in the low-oxygen-concentration regime, below 0.1% (1000 ppm), to detect trace oxygen in inert gases. The results show that the polarization behavior is different from that found in the high-concentration studies.

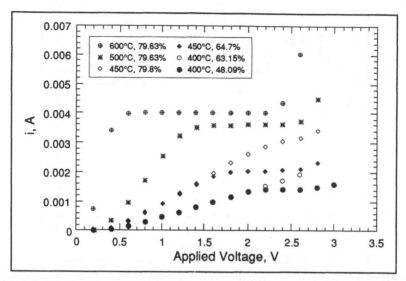

Figure 3. Polarization i-V curves of sensors operating at several temperatures and oxygen concentrations

When the oxygen concentration is low, equation 2 will converge to the form described in equation 1. The limiting current is then directly proportional to the oxygen partial pressure. This provides a unique property for oxygen sensing under amperometric operations. However, since the oxygen concentration is low, the magnitude of limiting current will be low, sometimes, of the same order as the leakage current of the instruments or of the minority-carriers. Therefore, additional electronic circuitry design and engineering of the sensor geometry are required for adequate sensor operation in low concentrations.

Figure 4 shows the variations in the complex impedance signals under different oxygen concentration and bias. Explanations of these dependencies are not trivial but deserve great attention. A typical curve in the Nyquist plot consists of three semicircular components. The first two semicircles represent the inter- (bulk) and intra-granular (grain boundary) conductions, respectively. The corresponding conductivity values agree with data previously reported by others. The third semicircle-like component is attributed to the interfacial behavior at the electrode/electrolyte boundaries.

Further demonstration of the polarization effect upon the influence of oxygen concentration in the low-concentration regime ($\leq 0.1\%$ or 1000 ppm) is shown in Figure 5 for results in the 400–500° C range. At 9.7 ppm, limiting-current plateaus at about 1 µA level can be detected. A strong temperature dependence of part of the polarization curve was observed. When the oxygen concentration increased to 86 ppm, such a dependence became even more profound. This effect almost impaired the measurement of the limiting-current plateaus as expected in the 10 µA range. These behaviors indicate a strong effect on the overall transport property of the sensor at the electrode/electrolyte interface.

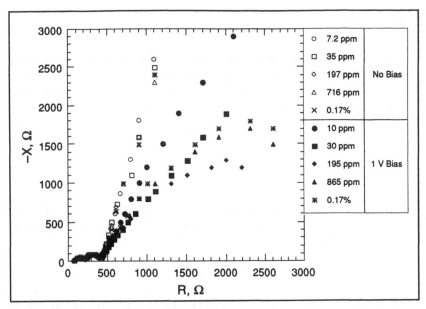

Figure 4. A Nyquist plot showing the oxygen concentration and bias dependence of the complex impedance of the sensor operating at 395° C

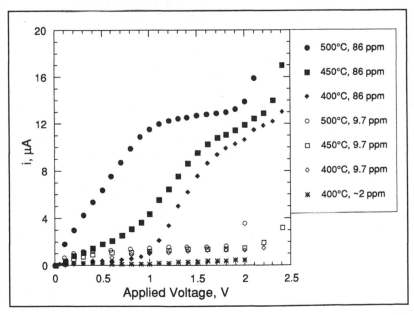

Figure 5. Polarization i-V curves of a sensor working with low oxygen concentrations between 400 and 500° C

DISCUSSION

The overall transport mechanism can be considered through more subtle distinction of microscopic steps. Considering the amperometric sensor configuration, we may divide the transport phenomena into: (1) gas diffusion through the diffusion hole, in which the ordinary chemical diffusion coefficient D_{O_2-x} and the diffusion-hole geometry (as defined by the constant, A/l) are important; (2) molecular oxygen adsorption by the Pt electrode; (3) chemical dissociation of molecular oxygen into atomic species;

$$O_{2(ad)} = 2\, O_{ad} \qquad (6)$$

(4) surface diffusion of oxygen adatoms to the Pt, $O_{2(g)}$|TZP three-phase-boundary line (TPBL), where a surface diffusion coefficient D_{surf} is specified; (5) charge transfer reaction as depicted by

$$O_{ad} + V_{\ddot{O}} + 2\, e^{\cdot} = O_O^x, \qquad (7)$$

in which the Kröger-Vink notation is used, and a reaction constant k_{ct} and an exchange current density I_{ex} are specified; (6) ionic and electronic defect diffusion through the electrolyte; and, subsequently, (7) the reversed sequence of reactions and transport processes from (5) to (2) that follows.

Each step described above can be the rate-limiting factor, although step (1) has to be the only rate-limiting factor to enable the proper function of the sensor. It is therefore important to understand the i-V characteristics of the dc polarization curves and the results from the ac impedance measurements. Through the variation of oxygen concentration, bias, temperature, and device geometry, we attempt to understand the rate-limiting factors in the sensor operation for future optimization.

Figure 6 shows a typical i-V polarization curve of the sensor. The whole curve can be divided into four regions. Region (I) describes the initial polarization caused by a small bias, which results from the presence of an activation overpotential. The work by Liaw and Weppner [12] indicated that the activation seems to relate with the charge transfer reaction (5) (equation 7) with a Tafel slope corresponding to $\alpha = 0.591$. Region (II) represents a different rate-limiting process that becomes dominating ahead of the one in (I) as the bias increases. Liaw and Weppner reported the case limited by the ionic transport through the TZP electrolyte when the oxygen concentration was high. They derived the dc conductivity from the slope of the i-V curves in this region and showed a consistency with those measured by ac impedance techniques (Figure 2). However, Liu and Weppner [14] recently reported different behaviors in this region if the oxygen concentration became low. If the rate-limiting step in this region is strictly due to ionic transport, it seems unlikely that the oxygen concentration would cause such a profound effect. It is known that defect concentration in the doped zirconia electrolyte remains essentially constant and invariant over a wide oxygen partial pressure range [18]. The different results observed between high- and low-concentration regimes might reflect the different states of gas|Pt interface, Pt catalytic reactivity and surface diffusion

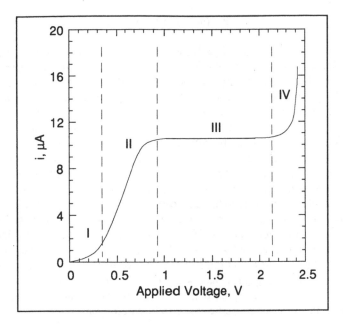

Figure 6. Schematic illustration of a typical polarization i-V curve in which four discrete regions are labelled for their specific rate-limiting mechanism

among the sensors. Region (III) exhibits the limiting-current plateau under the polarographic process of the sensor, and the ordinary diffusion through the barrier has become the predominant rate-limiting step. Region (IV) is where the electrolyte begins to decompose through reduction reactions, and the resulting electronic leakage becomes the primary source of current.

We have shown that the polarization behavior in Region (II) is critical in sensor operation. A discussion of possible processes involved in this region warrants an opportunity of future improvements. We will focus on steps (2) to (4) and correlate them to the results reported in the previous section. Steps (1) and (5) have been identified in other regions and should not be responsible for the behavior in this region.

The Langmuir isotherm was commonly used to describe oxygen adsorption on Pt through steps (2) and (3), which allows the establishment of the equilibrium relationship between the oxygen partial pressure at the gas|Pt interface, $P_{O_2,i}$, and the oxygen activity (at the active sites) on the Pt surface, a_O. The isotherm can be expressed as [19]

$$\theta = K_A P_{O_2,i} / (1 + K_A P_{O_2,i}) \tag{8}$$

in correspondence to equation 6, where θ is the surface coverage fraction of the active sites on Pt and K_A is the adsorption equilibrium constant of oxygen on the Pt surface:

$$K_A = (a_O)^2 / P_{O_2,i} \tag{9}$$

Thus, $$\theta = (a_O)^2 / [1+(a_O)^2]$$ (10)

Upon application of the bias, K_A will change accordingly; subsequently, it changes a_O and θ. When the oxygen partial pressure is high, the bias will not affect the value of θ much. As the oxygen activity approaches unity, θ will be close to 0.5. However, if the oxygen partial pressure is low, θ will be a strong function of a_O, as equation 10 will become $\theta \approx (a_O)^2$. This would explain the results reported in Figure 4, where the bias strongly affects the shape of the third impedance semicircle, inasmuch as the bias changes the θ value that will significantly vary the interface resistance and capacitance. On the other hand, the Langmuir adsorption isotherm is a thermodynamic property of the gas⏐Pt interface, often depending on temperature and oxygen partial pressure. The significant temperature dependence of the slopes in the polarization curves, as reported in Figure 5, could be attributed to this adsorption process; although, surface diffusion is also a possible explanation. More careful studies and modelling are needed.

It is known from several studies [20-24] that Pt morphologies on the electrolyte surface exhibit different natures, which could affect the Langmuir isotherm, the surface diffusion and the charge-transfer rates. There are no adequate studies in our work or by others in characterizing this behavior on TZP. However, this behavior seems to affect the sensor operation to a great degree. To improve the reliability and sensitivity of the sensor, the interface properties have to be carefully specified.

CONCLUSION

The amperometric limiting-current sensor uses the polarographic principles and offers a simple oxygen sensing device potentially applicable to many domestic and industrial processes and environments that require the control or monitoring of the oxygen concentration. This type of sensor can operate over a wide range of oxygen concentration (from ppm to essentially 1 atm) with better sensitivity than the conventional potentiometric types. A lower-temperature operation is feasible in sensors using the TZP materials than those using cubic stabilized electrolytes. The operating characteristics of amperometric sensors using TZP electrolytes in high- and low-oxygen-concentration regimes are presented and discussed. The sensor performance seems to be strongly influenced by the electrode/electrolyte interface properties particularly at low oxygen partial pressures. Further optimization of the sensor will depend on the improvement of the interface properties and the ionic conductivity of the TZP electrolytes.

REFERENCES

1. Dietz, H., W. Haecker and H. Jahnke, 1977. "Electrochemical Sensors for the Analysis of Gases," in <u>Advances in Electrochemistry and Electrochemical Engineering, Vol. 10</u>, H. Gerischer and C.W. Tobias, eds., New York, NY, USA: Wiley Interscience, John Wiley and Sons, pp.1-90.

2. Usui, T., Y. Kurumiya, K. Nuri and M. Nakazawa, 1989. "Gas-Polarographic Multifunctional Sensor: Oxygen-Humidity Sensor." <u>Sensors and Actuators</u>, 16:345-358.

3. Usui, T., A. Asada, K. Ishibashi and M. Nakazawa, 1991. "Humidity-Sensing Characteristics in Wet Air of a Gas Polarographic Oxygen Sensor Using a Zirconia Electrolyte." <u>J. Electrochem. Soc.</u>, 138(2):585-588.

4. Jahnke, H., B. Moro, H. Dietz and B. Beyer, 1988. "Electrochemische Grenzstromsensoren für Partialdruckmessungen." <u>Ber. Bunsenges. Phys. Chem.</u>, 92:1250-1257.

5. Mari, C.M., D. Narducci and L. Facheris, 1992. "Solid State Limiting Current Sensor for CO_2 Determination." <u>ISSI Lett.</u>, 3(1):1-2.

6. Usui, T., A. Asada, K. Ishibashi and M. Nakazawa, 1991. "Output Characteristics in an C_2H_5OH-N_2 Gas Mixture of a Gas Polarographic Oxygen Sensor Using a Zirconia Electrolyte." <u>Jap. J. Appl. Phys.</u>, 30(7):1496-1497.

7. Dietz, H., 1982. "Gas-Diffusion-Controlled Solid-Electrolyte Oxygen Sensors." <u>Solid State Ionics</u>, 6:175-183.

8. Takeuchi, T. and I. Igarashi, 1988. "Limiting Current Type Oxygen Sensor," in <u>Chemical Sensor Technology, Vol. 1</u>, T. Seiyama, ed., Amsterdam:Elsevier, pp.79-95.

9. Usui, T., K. Nuri, M. Nakazawa and H. Osanai, 1987. "Output Characteristics of a Gas-Polarographic Oxygen Sensor Using a Zirconia Electrolyte in the Knudsen Diffusion Region." <u>Jap. J. Appl. Phys.</u>, 26(12):L2061-L2064.

10. Usui, T. and A. Asada, 1988. "Operating Temperature of a Limiting-Current Oxygen Sensor Using a Zirconia Electrolyte," in <u>Advances in Ceramics, Vol. 24, Science and Technology of Zirconia III</u>, S. Somiya, N. Yamamoto and H. Yanagida, eds., Columbus, OH, USA:American Ceramic Society, pp.845-853.

11. Usui, T., A. Asada, M. Nakazawa and H. Osanai, 1989. "Gas Polarographic Oxygen Sensor Using an Oxygen/Zirconia Electrolyte." <u>J. Electrochem. Soc.</u>, 136:534-542.

12. Liaw, B.Y. and W. Weppner, 1991. "Low Temperature Limiting-Current Oxygen Sensors Based on Tetragonal Zirconia Polycrystals." J. Electrochem. Soc., 138:2478-2483.

13. Liaw, B.Y., J. Liu, A. Menne and W. Weppner, 1992. "Kinetic Principles for New Types of Solid State Ionic Gas Sensors." Solid State Ionics, 53/56:18-23.

14. Liu, J. and W. Weppner, "TZP-Based Sensors for Measuring Trace Oxygen in Inert Gases." to be published.

15. Badwal, S.P.S. and M.V. Swain, 1985. "ZrO_2-Y_2O_3: Electrical Conductivity of Some Fully and Partially Stabilized Single Grains." J. Mat. Sci. Lett., 4:487-489.

16. Bonanos, N., R.K. Slotwinski, B.C.H. Steele and E.P. Butler, 1984. "High Ionic Conductivity in Polycrystalline Tetragonal Y_2O_3-ZrO_2." J. Mat. Sci. Lett., 3:245-248.

17. Bonanos, N. and E.P. Butler, 1985. "Ionic Conductivity of Monoclinic and Tetragonal Yttria-Zirconia Single Crystals." J. Mat. Sci. Lett., 4:561-564.

18. Casselton, R.E.W., 1970. "Low Field DC Conduction in Yttria-Stabilized Zirconia." Phys. Stat. Sol. (a), 2:571-585.

19. Smith, J.M., 1956. Chemical Engineering Kinetics, New York, NY, USA:McGraw-Hill Book Co., pp.206-214.

20. Badwal, S.P.S. and F.T. Ciacchi, 1986. "Microstructure of Pt Electrodes and Its Influence on the Oxygen Transfer Kinetics." Solid State Ionics, 18/19:1054-1059.

21. Badwal, S.P.S. and F.T. Ciacchi, 1988. "Differential Response Rates of Electrode/Y-FSZ/Y-TZP/Electrode Cells," in Materials Science Forum, Vol. 34-36, Ceramic Developments, C.C. Sorrell and B. Ben-Nissan, eds., Switzerland:Trans Tech Publications Ltd., pp.231-235.

22. Wang, D.Y. and A.S. Nowick, 1979. "Cathodic and Anodic Polarization Phenomena at Platinum Electrodes with Doped CeO_2 as Electrolyte, I. Steady-State Overpotential." J. Electrochem. Soc., 126(7):1155-1165.

23. Wang, D.Y. and A.S. Nowick, 1981. "Diffusion-Controlled Polarization of Pt, Ag, and Au Electrodes with Doped Ceria Electrolyte." J. Electrochem. Soc., 128(1):55-63.

24. Verkerk, M.J., M.W.J. Hammink and A.J. Burggraaf, 1983. "Oxygen Transfer on Substituted ZrO_2, Bi_2O_3, and CeO_2 Electrolytes with Platinum Electrodes, I. Electrode Resistance by D-C Polarization." J. Electrochem. Soc., 130(1):70-78.

A Probe for Measuring Oxygen Activity in Molten Glasses

J. J. JI and M. P. BRUNGS

ABSTRACT

A novel probe has been developed for measuring oxygen activity in molten glasses. The probe consists of a water-cooled metal support, an yttria-stabilised zirconia reference electrode and an alumina indicator electrode. Both the indicator and reference electrode can be quickly and easily attached to the support, providing a simple solution to the problem of replacing electrodes which have been corroded by the molten glass. The experimental results from both gas and glass trials show that the accuracy and signal stability of the water-cooled probe are the equal of a conventional probe.

INTRODUCTION

Oxygen activity (or "oxidation state") in glass influences both glass colour and rate of refining. Traditionally, the activity has been determined spectrophotometrically on solid samples taken at the end of the manufacturing process. With growing market demand to improve glass quality, it has become important to develop a technique capable of measuring the oxygen activity at an earlier state, that is, while the glass is still molten. The most promising technique to date is using an electrochemical concentration cell commonly known as an oxygen probe.

Currently most research effort is being directed towards the development of oxygen probes which use zirconia electrolytes as part of the reference electrode. A major problem in the development of commercial oxygen probes based on zirconia is the relatively rapid corrosion of the zirconia in the molten glasses. One solution to the problem has been to use a larger piece of zirconia, in the form of a long, solid, yttria stabilised zirconia cylinder attached to an alumina tube [1,2]. In operation, the end of the zirconia cylinder just touches the glass

J. J. Ji & M. P. Brungs   School of Chemical Engineering & Industrial Chemistry
University of New South Wales   P.O.Box 1, Kensington, N.S.W. 2033, Australia.

surface. As the zirconia is corroded, the cylinder is lowered to maintain contact. There is a problem with the reference electrode being above the molten glass at a lower temperature, which leads to measurement errors. Baucke proposed [3] to use a heater to control the temperature of the reference chamber so that the temperature of the reference electrode is equal to that of molten glass. This probe is complicated by the need for a mechanism to continually lower the zirconia cylinder and keep it in contact with the glass surface. The probe signal may be affected by furnace atmosphere.

On the laboratory scale, an original oxygen probe consisting of a zirconia tube reference electrode and an alumina encased platinum indicator electrode has been shown to operate reliably, respond quickly and be unaffected by furnace atmosphere [4,5,6]. It has also proved to be a valuable reference electrode in subsequent electrochemical studies of glass [7,8,9]. However, this probe is not practical in commercial use because of corrosion problems and the difficulty of joining the zirconia electrode to an alumina support tube. This project was aimed at overcoming these limitations by developing a support in which small expendable zirconia (reference) and alumina (indicator) elements might be firmly held and easily replaced.

DESIGN OF THE SUPPORT

We believed that the best electrical and mechanical connection of the replaceable probe elements could be obtained by using a water cooled support. This raised the problem of thermal shock, since the "hot end" of the ceramic elements would be at temperature well above 1000^0C. To overcome the thermal shock problem, a shield called a "thermal gradient crown" was developed (Figure 1). The progressive nature of the shielding reduces the thermal gradient in the ceramic to reduce the resultant stress to avoid failure. Mathematical analysis of the temperature and thermal stress distribution using a simplified model of the shield indicates that the optimum length of crown is about 70 mm. This result has been verified experimentally [10].

The reference electrode consisted of a standard 4.5 mol% yttria (partially) stabilised zirconia tube (diameter 10 mm, length 150 mm). The inside bottom of the tube was coated with platinum paste to provide a contact for the lead wire. The indicator electrode was fabricated in the laboratory by calcining alumina powder which had been isostatically pressed around the 0.5 mm diameter platinum indicator wire. To increase the strength of the indicator electrode, a thicker platinum (2 mm diameter) wire was used on the exposed zone. Both the support and thermal gradient crown were made of copper (because of its high thermal conductivity) plated with nickel to decrease absorption of thermal radiation.

Mechanical attachment of the ceramic elements to the support presented another problem. Being brittle, the ceramics could be damaged by any mechanical shock during replacement. Moreover, a ceramic confined in a water cooled metal support would expand more than the metal under operating conditions, causing failure of the ceramic. We therefore decided to put an elastic material between the ceramic and the metal. A commercial silicone sealant,

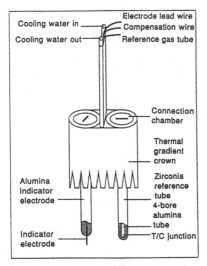

Figure 1. Schematic of the oxygen probe.

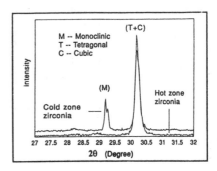

Figure 2. Water induced phase change of zirconia. Results have been displaced on vertical axis for the sake of clarity.

which adhered to both metal and ceramic was found suitable for this application. To ensure ease of replacement, a hollow bolt configuration was designed such that the ceramic elements were first fixed inside the bolt with silicone sealant and then the bolt was screwed into the holder.

In the reference electrode, a 4-bore alumina tube was used for introducing and insulating the thermocouple (Pt/Pt+13%Rh) and electrode (Platinum) wires and also for introducing reference gas to the end of the zirconia tube. Consequently three electrical contacts and one hole had to be placed in a small area. This was accomplished by increasing the connection area with a teflon "collar" which fitted over the top of the alumina tube. Electrical and gas connections were made via this collar when it was slipped into a mating connector in the top of the probe body. Once in place, a spring loaded cap was screwed down on top of the collar. This arrangement ensured good contact between the reference electrode and the zirconia electrolyte while allowing for some movement due to the different thermal expansion of zirconia and alumina. Base metal compensating wires were used from the connector in the probe body to the outside. It was assumed that little error would be introduced by this arrangement since all the connections were at the same low temperature [11].

RESULTS

After 100 hours in a gas fired muffle furnace at a ceramic tip temperature of $1300^{0}C$, the probe was disassembled and examined. The silicone sealant holding the ceramic in the hollow bolt connecter and the teflon collar and mating connector were all in good order, as were the electrical connections. It was found that the ceramic elements in the probe could be replaced in less than two minutes.

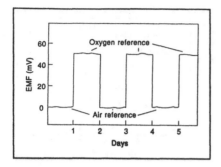

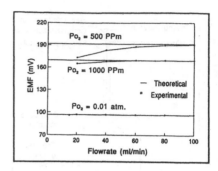

Figure 3. Probe signal as a function of time and reference gas at 1200°C.

Figure 4. Effect of flowrate on EMF at 1200°C.

Probes were operated with ceramic tip temperature in the range 1100-1300°C for a period totalling 600 hours with no sign of thermal shock in the alumina indicator electrode. However after approximately 50 hours, microcracks were found in a cool zone of the zirconia electrolyte. XRD analysis (Figure 2) indicates that this microcracking is the result of a water induced phase change from tetragonal to monoclinic zirconia. This problem can be overcome by the use of 6-8 mole% yttria partially stabilised zirconia tubes which have been shown [12] not to undergo the phase change.

In the initial trials, the indicator and reference electrodes were shorted together so that the probe could operate as a gas probe. Trials were conducted using air as the measured gas and a variety of carbon monoxide/carbon dioxide and oxygen/nitrogen mixtures as reference gases. As indicated by figure 3, the probe signal was found to be very stable (± 0.5mV) throughout all experiments. When the reference gas was changed from air to oxygen a new stable signal was obtained within one minute. Signal stability was maintained even after the probe had been used to stir the molten glass in a crucible, indicating that there should be no problem when the probe is used to measure the oxygen potential of flowing glass.

To determine the accuracy of the probe, flow meters were used to produce a number of air/nitrogen mixtures with a broad range of oxygen activities. It was found that at higher oxygen activities, the probe signal was independent of gas flow rate, but when the oxygen activity dropped below 10^{-3} atm, permeation of oxygen into the reference caused an error (Figure 4). This phenomenon is well documented [13].

The results obtained using a CO/CO_2 mixture ($P_{O_2}=1.26*10^{-11}$ at 1200°C) as the reference gas were in close agreement with those obtained with the original probe [14] and were independent of gas flow rate above 1000°C (Figure 5). Below 1000°C the measured oxygen activity was more than predicted (due to the presence of some contaminating oxygen in the initial gas mixture) and increased further still when the gas flow rate was increased. These results indicate that at lower temperatures, because of the relatively short length of the zirconia tube, the CO/CO_2 gas mixture does not achieve equilibrium. Thus when gas flow rate is increased, there is even less time for equilibrium to be achieved and the signal

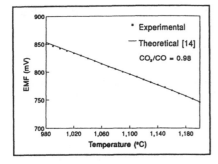

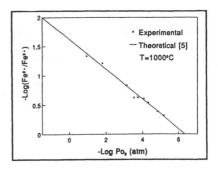

Figure 5. Accuracy of probe measurements in gases.

Figure 6. Accuracy of probe measurements in molten glass.

moves farther from the predicted (equilibrium) value.

Over the temperature range 1000-1200°C, the oxygen activities of high-iron content (2.4 wt%) sodium disilicate glasses were measured with the probe, using air as the reference gas. Actual oxygen activities for the glasses were obtained by determining the Fe^{2+}/Fe^{3+} ratio of quenched glass samples [15]. Again the performance of the water cooled probe was very similar to that of the original probe (Figure 6).

Because the reaction tube (ID=70 mm) in the furnace used had a relatively low volume and thermal capacity, insertion of the probe caused a 30°C drop in furnace temperature. Glass temperature decreased approximately 50°C because of the additional conductive heat loss via the probe ceramics. However, it is believed that in a glass furnace or feeder, the thermal effect of the water-cooled probe on the flowing mass of glass would be negligible.

CONCLUSION

A novel probe has been developed for measuring oxygen activity in molten glasses. The probe consists of a water-cooled support, a zirconia reference electrode, an alumina indicator electrode and a thermal gradient crown around the ceramics. Both the indicator and reference electrodes can be quickly and easily attached to the support, offering a viable solution to the corrosion problem encountered when measuring oxygen activity in molten glasses. The experiment results indicate that there is no significant difference between the performance of the water-cooled probe and a conventional oxygen probe.

REFERENCES

1. Baucke, F.G.K., 1988. "Electrochemical Cells for On-line Measurement of Oxygen Fugacity in Glass-forming Melts." Glastech. Ber., 61(4):87-90.

2. Muller-Simon, H., K.W.Mergler and H.A.Schaeffer, 1989. "Oxygen Activity

Measurement of Melts in Glass Tanks Using Electrochemical Sensors." Glass 89, XV International congress: Nauka, 150-155.

3. Baucke, F.G.K., 1989. German Pat. No.3811864.

4. Tran, T. and M.P.Brungs, 1980. "Applications of Oxygen Electrodes in Glass Melts. Part 1. Oxygen Reference Electrode." Phys. Chem. Glasses, 21(4):133-140.

5. Tran, T. and M.P.Brungs, 1980. "Applications of Oxygen Electrodes in Glass Melts. Part 2. Oxygen Probes for the Measurement of Oxygen Potential in Sodium Disilicate Glass." Phys. Chem. Glasses, 21(5):178-183.

6. Tran, T. and M.P.Brungs, 1983. UK Pat. No.2057695.

7. Maric, M., M.P.Brungs and M.Skyllas-Kazacos, 1988. "Anodic Voltametric Behaviour of a Platinum Electrode in Molten Sodium Disilicate Glass Containing Fe_2O_3." J. Non-Cryst. Solids, 105:7-16.

8. Maric, M., M.P.Brungs and M.Skyllas-Kazacos, 1989. "Voltametric Studies of the Fe^{2+}/Fe^{3+} Species in Molten Sodium Disilicate Glass." Phys. Chem. Glasses, 30(1):5-11.

9. Maric, M., M.P.Brungs and M.Skyllas-Kazacos, 1989. "Qualitative and Quantitative Chronopotentiometric Determinations of Fe^{2+} Species in Molten Silicates." J. Non-Cryst. Solids, 108:80-86.

10. Ji, J.J., November 1992. "Development of Oxygen Probe for Molten Glasses." Second Progress Report, Department of Industrial Chemistry, University of New South Wales, Sydney, Australia.

11. ASTM, 1981. "Manual on the Use of Thermocouples in Temperature Measurement," 3rd ed. Philadelphia, PA, USA :p73.

12. Sato, T. and M.Shimada, 1985. "Transformation of Yttria-doped Tetragonal ZrO_2 Polycrystals by Annealing in Water." J. Am. Ceram. Soc., 68(6):356-359.

13. Etsell, T.H. and S.N.Flengas, 1972. "The Determination of Oxygen in Gas Mixtures by Electromotive Force Measurements Using Solid Oxide Electrolytes." Met. Trans. 3:27-36.

14. Tran, T. and M.P.Brungs, 1980. "Application of Oxygen Electrodes in Glass Melts. Part 3: An Oxygen Concentration Cell for Thermodynamic Studies of CO/CO_2 and Ni/NiO Systems." Phys. Chem. Glass, 21(5):184-188.

15. Maxwell, C.J. and M.P.Brungs, 1984. "Simple and Rapid Spectrophotometric Methods for the Determination of the Ferrous-ferric Ratio in Glass and Geological Materials." Glass Technology, 25(5):244-246.

SOLID OXIDE FUEL CELLS

Recent Progress in Zirconia-Based Fuel Cells for Power Generation

S. C. SINGHAL

ABSTRACT

High temperature solid oxide fuel cells based upon yttria-stabilized zirconia electrolyte offer a clean, pollution-free technology to electrochemically generate electricity at high efficiencies. This paper reviews the designs, materials and fabrication processes used for such fuel cells. Most progress to date has been achieved with tubular geometry cells. A large number of tubular cells have been electrically tested, some to times up to 30,000 hours; these cells have shown excellent performance and performance stability. In addition, successively larger size electric generators utilizing these cells have been designed, built and operated since 1984. Two 25 kW power generation field test units have recently been fabricated; these units represent a major milestone in the commercialization of zirconia-based fuel cells for power generation.

INTRODUCTION

The high oxygen ion conductivity over wide ranges of temperature and oxygen pressures in stabilized zirconia has led to its use as a solid oxide electrolyte in a variety of electrochemical applications. Zirconia-based oxygen sensors are widely used in combustion control, especially in automobiles, atmosphere control in furnaces, and as monitors of oxygen concentration in molten metals. Other applications include electrochemical pumps for control of oxygen potential, steam electrolyzers and high temperature solid oxide fuel cells (SOFC's).

High temperature fuel cells utilizing yttria-stabilized zirconia electrolyte offer a clean, pollution-free technology to electrochemically generate electricity at high efficiencies. These fuel cells provide many advantages over traditional energy conversion systems; these include high efficiency, reliability, modularity, fuel adaptability, and very low levels of NO_x and SO_x emissions. Furthermore, because

S. C. Singhal, Science and Technology Center, Westinghouse Electric Corporation, 1310 Beulah Road, Pittsburgh, PA 15235, U.S.A.

of their high temperature of operation ($\sim 1000^{\circ}$C), these cells can be operated directly on natural gas eliminating the need for an expensive, external reformer system. These fuel cells also produce high quality exhaust heat which can be used for process heat or a bottoming electric power cycle to further increase the overall efficiency. This paper reviews the current status of the solid oxide fuel cell technology for power generation.

OPERATING PRINCIPLE OF A SOLID OXIDE FUEL CELL

A solid oxide fuel cell essentially consists of two porous electrodes separated by a dense, oxygen ion conducting electrolyte. The operating principle of such a cell is illustrated in figure 1. Oxygen supplied at the cathode (air electrode) reacts with incoming electrons from the external circuit to form oxygen ions, which migrate to the anode (fuel electrode) through the oxygen ion conducting electrolyte. At the anode, oxygen ions combine with H_2 (and/or CO) in the fuel to form H_2O (and/or CO_2), liberating electrons. Electrons flow from the anode through the external circuit to the cathode. The reactions at the two electrodes can be written as follows:

$$\text{Cathode:} \quad 1/2 \; O_2 + 2 \; e^- = O^{2-} \qquad (1)$$

$$\text{Anode:} \quad H_2 + O^{2-} = H_2O + 2 \; e^- \qquad (2)$$

$$CO + O^{2-} = CO_2 + 2 \; e^- \qquad (3)$$

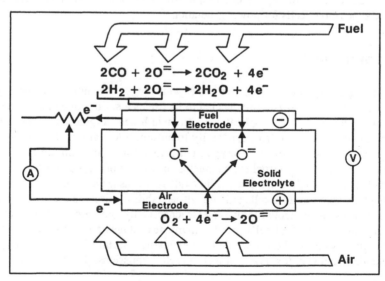

Figure 1. Operating principle of a solid oxide fuel cell.

The overall cell reaction is simply the oxidation of fuel (H_2 and/or CO) and the open circuit voltage, E, of the fuel cell is given by the Nernst equation:

$$E = \frac{RT}{4F} \ln \left\{ \frac{P_{O_2} \ (\text{oxidant})}{P_{O_2} \ (\text{fuel})} \right\} \tag{4}$$

where R is the gas constant, T is the cell temperature, F is the Faraday constant, and P_{O_2}'s are the oxygen partial pressures. When a current passes through the cell, the cell voltage (V) is given by:

$$V = E - IR - \eta_A - \eta_F \tag{5}$$

where I is the current passing through the cell, R is the electrical resistance of the cell, and η_A and η_F are the polarization voltage losses associated with irreversibilities in electrode processes on the air side and the fuel side, respectively. To keep the IR loss low, the electrolyte in various solid oxide fuel cell designs is fabricated in the form of a thin film.

DESIGNS OF SOLID OXIDE FUEL CELLS

Solid oxide fuel cells of several different designs are presently under development; these include planar, monolithic and tubular geometries [1-3]. The materials being considered for cell components in these different designs are either the same or very similar in nature. In the planar design, illustrated in figure 2, the

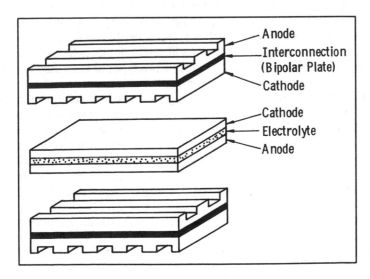

Figure 2. Cross-flow planar solid oxide fuel cell design.

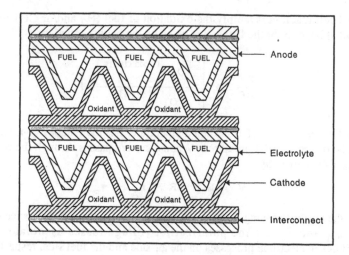

Figure 3. Co-flow monolithic solid oxide fuel cell design.

cell components are configured as thin, flat plates. The interconnection having ribs on both sides forms gas flow channels and serves as bipolar gas separator contacting the anode and the cathode of adjoining cells. The dense electrolyte and interconnection are fabricated by tape casting, powder sintering or CVD [4,5], whereas the porous electrodes are applied by slurry methods, screen printing, or by plasma spraying [6]. The planar cell design offers improved power density but requires high temperature gas seals at the edges of the plates to isolate oxidant from fuel. Difficulties in successfully developing such high temperature seals have limited the development and use of planar design cells for SOFC generators.

In the monolithic design also, the different cell components are fabricated as thin layers [7]. In the co-flow version of the monolithic solid oxide fuel cell, illustrated in figure 3, the cell consists of a honeycomb-like array of adjacent fuel and oxidant channels. Such a cell is made of two types of laminated structures, anode/electrolyte/cathode and anode/interconnect/cathode. The anode/electrolyte/cathode composite is corrugated and stacked alternately between flat anode/interconnect/cathode composites. A cross-flow version of the monolithic solid oxide fuel cell, in which the fuel and oxidant flows are at right angle to each other, has also been investigated. Even though the monolithic solid oxide fuel cells offer potentially the highest power density of all SOFC designs, their fabrication involving co-sintering of the cell components at elevated temperatures has proven to be a formidable task. As a result, the development of large scale stacks utilizing monolithic solid oxide fuel cells has not progressed very far.

The most progress to date has been achieved with the tubular geometry fuel cell. Figure 4 illustrates the design of the Westinghouse tubular geometry cell. In this design, the active cell components are deposited in the form of thin layers on a ceramic support tube. The remaining sections in this paper relate to this tubular design solid oxide fuel cell.

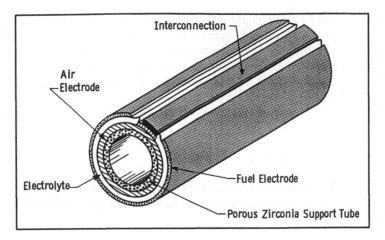

Figure 4. Tubular solid oxide fuel cell design.

MATERIALS AND FABRICATION PROCESSES

The materials for different cell components have been selected based on the following criteria:

(a) Suitable electrical properties required of different cell components to perform their intended cell functions.

(b) Chemical stability at high temperatures during cell operation as well as during cell fabrication.

(c) Minimum reactivity and interdiffusion among different cell components.

(d) Matching thermal expansion among different cell components.

In addition to above materials selection criteria, the fabrication processes have been chosen so that every sequential component fabrication process does not affect those components already fabricated onto the support tube. The materials and fabrication processes used for different cell components are discussed below.

SUPPORT TUBE

The support tube acts both as a structural member onto which the active cell components are fabricated and as a functional member to allow the passage of air (or oxygen) to the air electrode (cathode) during cell operation. To perform both these functions, the support tube must have adequate mechanical integrity with sufficient porosity for gas permeation, must have matching thermal expansion with the air electrode, and should not react to any significant extent with the air electrode. To satisfy these requirements, a porous support tube (PST) made of zirconia stabilized with 15 mole percent calcia is used. The tube is made by extrusion of calcia-stabilized zirconia powder followed by sintering in air. The mechanical properties, e.g., rupture strength and elastic modulus, of these support tubes vary as a function of porosity [8]. To obtain sufficient strength such that the tubes can undergo all processing and operational conditions without structural failure and to provide adequate gas permeability, the tubes have about 35 percent porosity.

AIR ELECTRODE (CATHODE)

With operating temperature of around $1000^{\circ}C$ in air or oxygen atmosphere, the material for the air electrode in solid oxide fuel cells has to meet the following requirements:

(a) High electronic conductivity.

(b) Chemical and dimensional stability in environments encountered during cell operation and during fabrication of interconnection, electrolyte and fuel electrode layers.

(c) Thermal expansion match with other cell components.

(d) Sufficient porosity and good adherence at the surface of the electrolyte.

(e) Compatibility and minimum reactivity with the support tube, the electrolyte and the interconnection with which air electrode comes into contact.

To satisfy these requirements, strontium-doped lanthanum manganite $(La_{0.9}Sr_{0.1}MnO_3)$ is used as the air electrode material. Lanthanum manganite is a p-type perovskite oxide and shows reversible oxidation-reduction behavior. The material can have oxygen excess or deficiency depending upon the ambient oxygen partial pressure and temperature. Although, it is stable in air and oxidizing atmospheres, it dissociates at $1000^{\circ}C$ at oxygen pressures $\leq 10^{-14}$ atm. The electronic conductivity of lanthanum manganite is due to small polaron hopping which is enhanced by doping with a divalent ion such as strontium. The defect chemistry, electrical conduction, and cathodic polarization behavior of doped lanthanum manganites have been studied in detail [9,10], and strontium-doped lanthanum manganite has been found to satisfy all the requirements to be an effective air electrode. Furthermore, the reactivity and interdiffusion studies [11,12], between strontium-doped lanthanum manganite and yttria-stabilized zirconia electrolyte have shown any interactions between these two materials at $1000^{\circ}C$ to be minimal.

The strontium-doped lanthanum manganite air electrode is deposited onto the calcia-stabilized zirconia support tube in the form of approximately 1.4 mm thick layer by a slurry filtration technique followed by sintering. To provide for adequate gas permeability, sintering conditions are controlled to achieve about 35 percent porosity in the air electrode.

ELECTROLYTE

Solid oxide fuel cells are based on the concept of an oxygen ion conducting electrolyte through which the oxygen ions migrate from the air electrode (cathode) side to the fuel electrode (anode) side where they oxidize the fuel (H_2, CO, etc.) to generate an electrical voltage. Zirconia doped with about 10 mole percent yttria is used as the electrolyte. Electrical properties and defect structure of yttria-stabilized zirconia have been thoroughly reviewed [13,14]. This material exhibits cubic fluorite structure in which yttrium (Y^{3+}) substitutes for the zirconium (Zr^{4+}) cations generating oxygen vacancies. The high ionic conductivity of yttria-stabilized zirconia is attributed to these oxygen ion vacancies along with low activation energy for oxygen ion migration. The conductivity of yttria-stabilized zirconia at 1000°C is maximum (about 0.1 $S \cdot cm^{-1}$) at about 10 mole percent yttria as shown by Strickler and Carlson [15]; the activation energy is also least near this composition. The thermal expansion of 10 mole percent yttria-stabilized zirconia is about $10 \times 10^{-6}/\,^\circ$C; materials for all other cell components are chosen to have thermal expansion near this value.

For optimum cell performance, the yttria-stabilized zirconia electrolyte must be free of porosity so as not to allow gases to permeate from one side of the electrolyte to the other, it should be uniformly thin to minimize ohmic loss, and it should have high oxygen ion conductivity with transport number for oxygen ions close to unity and a transport number for electrons as close to zero as possible. Electrolyte with these desired properties is deposited in the form of about 40 μm thick layer by an electrochemical vapor deposition process. In this process, discussed in detail by Isenberg [16] and Pal and Singhal [17], chlorides of zirconium and yttrium are volatilized in a predetermined ratio and passed along with hydrogen and argon over the outer surface of the porous air electrode. Oxygen mixed with steam at a predetermined ratio is passed inside the porous calcia-stabilized zirconia tube over which the porous air electrode is deposited. In the first stage of the reaction, designated as the chemical vapor deposition (CVD) stage, molecular diffusion of oxygen, steam, metal chlorides, and hydrogen occurs through the porous air electrode and these react to fill the pores in the air electrode with the yttria-stabilized zirconia electrolyte according to the following reactions:

$$2\ MeCl_y + y\ H_2O = 2\ MeO_{y/2} + 2y\ HCl \qquad (6)$$

$$4\ MeCl_y + y\ O_2 + 2y\ H_2 = 4\ MeO_{y/2} + 4y\ HCl \qquad (7)$$

where Me is the cation species (zirconium and yttrium); and y is the valency associated with the cation. The temperature, the pressure and the different gas flow

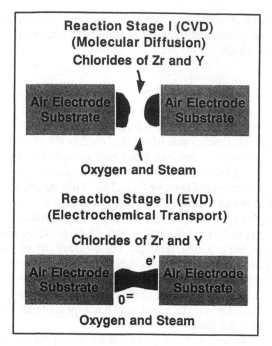

Figure 5. Two stages of reaction occurring during the deposition
of the yttria-stabilized zirconia electrolyte.

rates are so chosen that the above reactions are thermodynamically and kinetically favored.

During the second stage of the reaction after the pores in the air electrode are closed, electrochemical transport of oxygen ions maintaining electroneutrality occurs through the already deposited electrolyte in the pores from the high oxygen partial pressure side (oxygen/steam) to the low oxygen partial pressure side (chlorides). The oxygen ions, upon reaching the low oxygen partial pressure side, react with the metal chlorides and the electrolyte film grows in thickness. This second stage of the reaction is termed the electrochemical vapor deposition (EVD) stage.

The two stages of reaction, viz., the CVD stage and the EVD stage, responsible for the growth of the yttria-stabilized zirconia electrolyte film, are schematically shown in figure 5. In this deposition process, the flows of the metal chloride vapors are maintained above a critical level to eliminate any gas-phase control of the EVD reaction. Furthermore, the ratio of yttrium chloride to zirconium chloride is so chosen that the electrolyte deposited contains about 10 mole percent yttria.

The growth of the electrolyte film is parabolic with time, as shown in figure 6, and occurs by the oxygen ions diffusing through yttria-stabilized zirconia from the oxygen/steam side to the chlorides side. The rate controlling step in this process is the electronic transport (diffusion of electrons) through the electrolyte film; the average electronic transport number of the electrolyte during EVD at 1473 K is

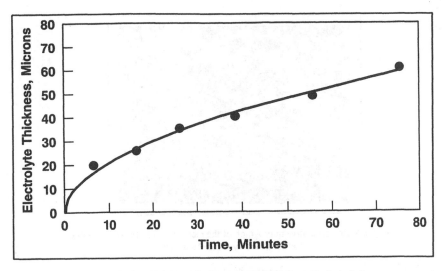

Figure 6. Growth kinetics of the yttria-stabilized zirconia electrolyte
by the electrochemical vapor deposition process at 1473 K.

about 0.6×10^{-4} [17]. The electrochemical vapor deposition process ensures the
formation of a pore-free, gas-tight, uniformly thick layer of the electrolyte over
porous air electrode. A representative micrograph of the electrolyte layer over
porous air electrode is shown in figure 7.

Figure 7. Representative micrograph of the electrochemically
vapor deposited yttria-stabilized zirconia electrolyte
over porous air electrode.

FUEL ELECTRODE (ANODE)

The fuel electrode of a solid oxide fuel cell must be electronically conducting,
stable in the reducing environment of the fuel, and must have interconnected

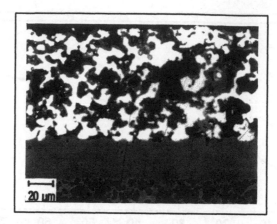

Figure 8. Representative micrograph of the nickel/yttria-stabilized zirconia
fuel electrode over the electrolyte.

porosity to allow the passage of the fuel to the electrolyte surface. Due to the reducing atmosphere of the fuel gas, a metal such as nickel (or cobalt) can be used for SOFC anodes. However, the thermal expansion coefficient of nickel is considerably larger than that of the yttria-stabilized zirconia electrolyte; this large thermal expansion mismatch can cause delamination of the anode from the electrolyte surface. Also, nickel can sinter at the cell operating temperature resulting in a decrease in anode porosity. To circumvent these problems, a skeleton of yttria-stabilized zirconia is formed around the nickel particles. This skeleton of yttria-stabilized zirconia supports the nickel particles, inhibits sintering of the nickel particles during cell operation, provides adherence to the electrolyte, and also provides an anode thermal expansion coefficient closer to that of the electrolyte. Such nickel/yttria-stabilized zirconia anodes show very low diffusion and activation polarization losses during cell operation. Such anodes have also been shown to possess sufficient catalytic activity at 1000°C to reform natural gas and other hydrocarbons in situ.

A 100-150 μm thick layer of nickel/yttria-stabilized zirconia is applied over the electrolyte by a two-step process. In the first step, nickel powder is applied over the electrolyte by dipping in a nickel slurry. In the second step, yttria-stabilized zirconia is grown around the nickel particles by the same electrochemical vapor deposition process as used for depositing the electrolyte. A representative micrograph of the fuel electrode is shown in figure 8. The fuel electrode contains 40 to 45 percent porosity.

INTERCONNECTION

Interconnection serves as the electric contact to the air electrode and also protects the air electrode material from the reducing environment of the fuel on the fuel electrode side. The requirements of the interconnection are most severe of all cell components and include:

(a) It should have nearly 100 percent electronic conductivity.

(b) Since it is exposed to air (or oxygen) on one side and fuel on the other, it should be stable in both oxidizing and reducing atmospheres at the cell operating temperature.

(c) It should have low permeability for oxygen and hydrogen to minimize direct combination of oxidant and fuel during cell operation.

(d) It should have a thermal expansion close to that of the air electrode and the electrolyte.

(e) It should be non-reactive with the air electrode, electrolyte and the electric contact material (e.g., nickel).

To satisfy these requirements, doped lanthanum chromite is used as the interconnection material. Lanthanum chromite is a p-type conductor; its conductivity is due to small polaron hopping from room temperature to $1400^{0}C$ at oxygen pressures as low as 10^{-18} atm. The conductivity is enhanced as lower valence ions (e.g., Ca, Mg, Sr, etc.) are substituted on either the La^{3+} or Cr^{3+} sites. The defect chemistry, oxidation-reduction behavior, and the thermal expansion behavior of these chromites have been extensively studied [9,18].

In the tubular solid oxide fuel cell, Mg-doped lanthanum chromite interconnection is deposited in the form of about 40 μm thick, 9 mm wide strip on the porous air electrode along the cell length by an electrochemical vapor deposition process [16,19], similar to that used for depositing the electrolyte. This process ensures the deposition of a gas-tight, dense interconnection; a representative micrograph of the electrochemically vapor deposited magnesium-doped lanthanum chromite interconnection over porous air electrode is shown in figure 9. This

Figure 9. Representative micrograph of the electrochemically vapor deposited Mg-doped lanthanum chromite interconnection over porous air electrode.

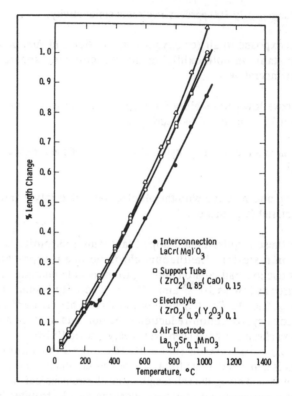

Figure 10. Thermal expansion of materials in the tubular solid oxide fuel cell.

material has good electronic conductivity, low permeability for oxygen and hydrogen, and is stable in both oxidizing and reducing environments [20].

Figure 10 shows the thermal expansion characteristics of the different cell components [8]. The thermal expansion of $La_{0.9}Sr_{0.1}MnO_3$ air electrode is closely matched to that of the calcia-stabilized zirconia support tube and the yttria-stabilized zirconia electrolyte. However, the thermal expansion of the Mg-doped lanthanum chromite interconnection is about 10-12 percent lower than that of other cell components. Better thermal expansion matching can be obtained if strontium-doped lanthanum chromite is used as the interconnection material.

Development efforts to reduce the cost of the cells utilizing alternate fabrication techniques and to increase the power output per cell are currently in progress [21]. Longer cells and increased power output per unit cell length are desirable to improve SOFC power plant economics. Over the past few years, the active length of the tubular SOFC has been increased from 30 cm (in pre-1986 cells) to 100 cm (currently in production) with the result being a corresponding increase in power output per cell. In parallel with increasing the length, the thickness of the calcia-stabilized zirconia porous support tube (PST) has been reduced. Early technology tubular solid oxide fuel cells had a 2 mm wall thickness PST (the thick-wall PST).

Although sufficiently porous to allow air flow to the air electrode, there was an inherent impedance to air flow toward air electrode. Efforts were successful in first reducing the PST wall thickness to 1.2 mm (the thin-wall PST) and ultimately in eliminating the PST altogether. As the PST wall thickness was reduced and the PST eventually eliminated from the cell, the air electrode thickness was increased. The present technology tubular solid oxide fuel cells have an air electrode tube onto which the other cell components are deposited. The combined effect of increasing the cell length from 30 cm to 100 cm and eliminating the PST has resulted in a six-fold (from 20 to 120 W) increase in power output per cell [21].

OPERATION AND PERFORMANCE OF A SOLID OXIDE FUEL CELL

The ceramic porous support tube is closed at one end. For cell operation, oxidant (air or oxygen) is introduced through a ceramic injector tube positioned inside the cell. The oxidant is discharged near the closed end of the cell and flows through the annular space formed by the cell and the coaxial injector tube. Fuel flows on the outside of the cell from the closed end and is electrochemically oxidized while flowing to the open end of the cell generating electricity.

At the open end of the cell, the oxygen-depleted air exits the cell and is combusted with the partially depleted fuel. Typically, 50 to 90 percent of the fuel is utilized in the electrochemical cell reaction. Part of the depleted fuel is recirculated in the fuel stream and the rest combusted to preheat incoming air and/or fuel to the fuel cell. The exhaust gas from the fuel cell is at 600 to 900°C depending on the operating conditions and can be used in a cogeneration system for producing process steam or in a steam turbine bottoming unit for an all-electric system.

A large number of single solid oxide fuel cells of the tubular design have been electrochemically tested [22] at 875 to 1200°C, some to times up to 30,000 hours, either on humidified hydrogen, a mixture of hydrogen and carbon monoxide, or natural gas. These cells have shown excellent performance and performance stability during long-term operation. A voltage-current plot at different temperatures is shown in figure 11, and the power output as a function of current density for a 50 cm active length cell (with thin-wall porous support tube) is shown in figure 12.

SOLID OXIDE FUEL CELL GENERATOR SYSTEMS

To construct an electric generator, individual cells are connected in both parallel and series, as shown schematically in figure 13, to form a semi-rigid bundle that is the basic building block of a generator. Nickel felt, consisting of long nickel metal fibers sinter bonded to each other, is used to provide soft, mechanically compliant, low electrical resistance connections between the cells. This material bonds to the nickel particles in the fuel electrode and the nickel plating on the interconnection for the series connection, and to the two adjacent cell fuel electrodes for the parallel

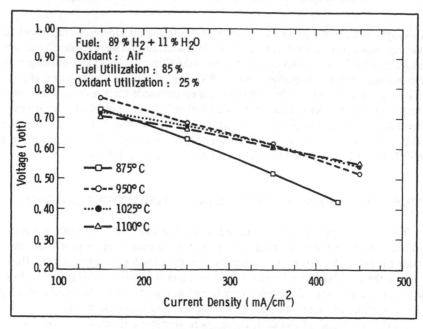

Figure 11. Voltage-current curves at different temperatures
for 50 cm active length thin-wall PST cells.

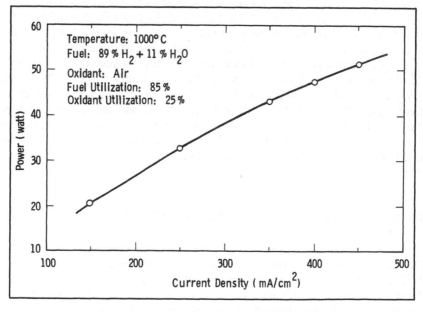

Figure 12. Power output as a function of current density
for a 50 cm active length thin-wall PST cell.

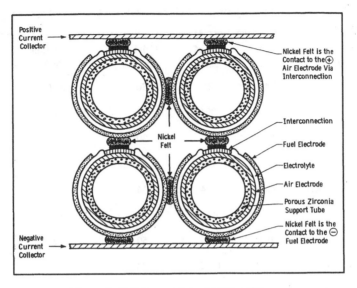

Figure 13. Cell cross-sections showing series
and parallel electrical connections.

connection; such a series-parallel arrangement provides improved generator reliability. The individual cell bundles are arrayed in series to build voltage and form generator modules. These modules are further combined in either series or parallel to form large scale generators.

Figure 14 illustrates the basic design concept of the Westinghouse seal-less solid oxide fuel cell generator [23,24]. Fuel is introduced to a plenum at the bottom and flows upward between the cells. This fuel is electrochemically oxidized producing

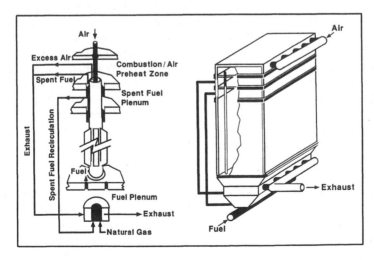

Figure 14. Solid oxide fuel cell seal-less generator concept.

Table I. Summary of Westinghouse SOFC Generator Systems

Customer	Size (kW)	No. of Cells	Cell Length (cm)	Test Time (hrs)	Year
U.S. DOE	0.4	24	30	2,000	1984
U.S. DOE	5.0	324	30	500	1986
TVA	0.4	24	30	1,760	1987
Tokyo Gas (TG)	3.0	144	36	5,000	1987-88
Osaka Gas (OG)	3.0	144	36	3,700	1987-88
GRI	3.0	144	36	5,400	1989-90
U.S. DOE	20.0	576	50	3,355	1990-91
KEPCO/TG/OG	25.0*	1,152	50	--	1992
TG/OG	25.0*	1,152	50	--	1992

* Nominal rating. Capable of producing 40 kW at peak operation.

electricity and heat as it flows past the cells. Spent fuel and oxidation products (e.g., CO_2 and H_2O) enter the spent fuel combustion chamber located above the cells. Air is introduced to a plenum at the top of the generator. From the air plenum, air enters the air injector tubes which are suspended, one per cell, from the air plenum into the cells. After passing inside the cells, the air depleted in oxygen, exits from the open end of the cells into the same chamber as the spent fuel, combusting the spent fuel and producing additional heat which is used to preheat the incoming air. The exhaust products exit the generator from this chamber at temperatures up to $900^{\circ}C$ depending upon operating conditions.

The seal-less generator concept has been tested in a number of different-size generators, as summarized in Table I. In 1984, a 400 W generator consisting of two 12-cell bundles connected in electrical series was tested [23,24] for about 2,000 hours. During operation, this generator was cycled five times to approximately room temperature and back to operating temperature (about $1000^{\circ}C$). The successful test of this generator was followed by the design, construction and test, in late 1986, of a larger 5 kW generator consisting of 324 cells [24]. This generator was tested for about 500 hours, and achieved 5.2 kW at 80 amperes with 6 stoichs air and 85% fuel (66.6% H_2, 22.2% CO and 11.1% H_2O) utilization at $1000^{\circ}C$.

Additional SOFC generators have been designed, built and delivered for field testing by prospective customers of large-scale SOFC power generation systems. The Tennessee Valley Authority (TVA) acquired the first such unit for the purposes of demonstrating the capabilities of the technology and providing independent corroboration of overall performance and reliability. This 400 W generator, consisting of 24 cells arranged in two series connected bundles, was tested [25,26] in early 1987 for about 1,760 hours over a range of currents, fuel utilizations, air flows and temperatures, and demonstrated the capability of the system to satisfy its design requirements for automatic unattended operation and load follow.

Subsequently, Tokyo Gas and Osaka Gas each acquired a 3 kW SOFC generation system in 1987. Each system consisted of an SOFC generator module containing 144 cells, electrical air pre-heaters for generator temperature control, air and fuel handling systems and a computer-based control system [26]. The generators could be run on hydrogen, a mixture of hydrogen and carbon monoxide, or the effluent from a customer supplied fuel gas supply such as a methane reformer. The testing and performance of these systems has been discussed in detail by Veyo [26], Harada and Mori [27], and Yamamoto, et al. [28] In brief, each of these systems produced 3 kW of dc power as designed; the system at Tokyo Gas operated continuously and successfully for over six months. The system at Osaka Gas operated continuously for 3,700 hours before being shut down for external (non-SOFC) causes.

In 1989, another 3 kW generator, again consisting of 144 cells, was fabricated to operate on desulfurized pipeline natural gas; this generator incorporated an integrated, internal reformer and partial recirculation of depleted fuel [29]. The generator was operated successfully for over 5,400 hours, and either met or exceeded the design goals.

In 1990, a 20 kW generator, the first SOFC generator of a commercially meaningful size, was fabricated. This generator consisted of 576 cells of 50 cm active length, and also incorporated an integrated, internal reformer and partial recirculation of depleted fuel. This generator was successfully operated for a total of 3,355 hours on three different fuels - hydrogen, desulfurized pipeline natural gas and naphtha — producing up to 20 kW [30].

In early 1992, a 25 kW dc power generator, consisting of 1,152 cells of 50 cm active length, was fabricated and delivered for testing at Rokko Island, near Osaka, Japan, in a joint program with Kansai Electric Power Company (KEPCO), Osaka Gas Company (OG) and Tokyo Gas Company (TG). The cells in this generator are contained in two independently controlled and operated units. A second 25 kW SOFC system, a cogeneration unit supported by Osaka Gas Company and Tokyo Gas Company, has been fabricated; this system will supply ac power and intermediate pressure steam to another test site in Japan. These two 25 kW units represent a major milestone in the commercialization of zirconia-based fuel cells for power generation.

SUMMARY

High temperature solid oxide fuel cells based on yttria-stabilized zirconia electrolyte offer many advantages over traditional energy systems. These include high conversion efficiency, reliability, modularity, fuel adaptability and virtually unlimited siteability due to very low NO_x emissions [31]. Furthermore, these cells produce high quality exhaust heat which can be used for process heat or a bottoming electric power cycle to further increase the overall plant efficiency. Westinghouse has developed the materials and fabrication processes for these fuel cells, and has successfully employed and tested these cells for power generation in successively larger generators. When fully commercialized, the zirconia-based fuel cell systems are expected to serve a wide range of power and heat applications, such as large-scale power generation by electric and gas utilities and industrial cogeneration.

ACKNOWLEDGEMENTS

The author acknowledges the contributions of his many colleagues whose work is reviewed in this paper. The development of solid oxide fuel cell technology has been supported by the U.S. Department of Energy (DOE), the Gas Research Institute (GRI), the Westinghouse Electric Corporation, and various utility and industry sources.

REFERENCES

1. Singhal, S. C., (ed.), 1989. Solid Oxide Fuel Cells, Pennington, NJ, USA: The Electrochemical Society, Inc., pp. 279-383.

2. Appleby, A. J. and Foulkes, F. R., 1989. Fuel Cell Handbook, New York, NY, USA: Van Nostrand Reinhold, pp. 579-611.

3. Grosz, F., Zegers, P., Singhal, S. C. and Yamamoto, O. (eds.), 1991. Proceedings of the Second International Symposium on Solid Oxide Fuel Cells, Luxembourg: Commission of the European Communities, pp. 1-112.

4. Lessing, P. A., Tai, L. W. and Klemm, K. A., 1989. "Fabrication Technologies for a Planar Solid Oxide Fuel Cell", in Solid Oxide Fuel Cells, S. C. Singhal, ed., Pennington, NJ, USA: The Electrochemical Society, Inc., pp. 337-360.

5. Milliken, C. and Khandkar, A., 1989. "Fabrication of Integral Flow Field/Interconnects for Planar SOFC Stacks", in Solid Oxide Fuel Cells, S. C. Singhal, ed., Pennington, NJ, USA: The Electrochemical Society, Inc., pp. 361-376.

6. Arai, H., 1989. "Solid Oxide Fuel Cells with Stabilized Zirconia Thick Films Fabricated by Various Techniques", in Proceedings of SOFC-Nagoya, Nagoya, Japan: Japan Fine Ceramics Center, pp. 9-14.

7. Minh, N. Q., Horne, C. R., Liu, F., Staszak, P. R., Stillwagon, T. L. and Van Ackeren, J. J., 1989. "Forming and Processing of Monolithic Solid Oxide Fuel

Cells", in Solid Oxide Fuel Cells, S. C. Singhal, ed., Pennington, NJ, USA: The Electrochemical Society, Inc., pp. 307-316.

8. Isenberg, A. O., 1988. "Technology Status of High-Temperature Solid Oxide Fuel Cells and Electrolyzers", in Proceedings International Symposium on Fine Ceramics, Arita, Japan, pp. 107-122.

9. Anderson, H. U., Kuo, J. H. and Sparlin, D. M., 1989. "Review of Defect Chemistry of $LaMnO_3$ and $LaCrO_3$", in Solid Oxide Fuel Cells, S. C. Singhal, ed., Pennington, NJ, USA: The Electrochemical Society, Inc., pp. 111-128.

10. Yamamoto, O., Takeda, Y., Kanno, R. and Noda, M., 1987. "Perovskite Type Oxides as Oxygen Electrodes for High Temperature Oxide Fuel Cells." Solid State Ionics, 22: 241-246.

11. Lau, S. K. and Singhal, S. C., 1985. "Potential Electrode/Electrolyte Interactions in Solid Oxide Fuel Cells." Corrosion 85, Paper No. 345, pp. 1-9.

12. Yamamoto, O., Takeda, Y., Kanno, R. and Kojima, T., 1989. "Stability of Perovskite Oxide Electrode with Stabilized Zirconia", in Solid Oxide Fuel Cells, S. C. Singhal, ed., Pennington, NJ, USA: The Electrochemical Society, Inc., pp. 242-253.

13. Dell, R. M. and Hooper, A., 1978. "Oxygen Ion Conductors", in Solid Electrolytes, P. Hagenmuller and W. van Gool, eds., New York, NY, USA: Academic Press, pp. 291-312.

14. Subbarao, E. C. and Maiti, H. S., 1984. "Solid Electrolytes with Oxygen Ion Conduction." Solid State Ionics, 11: 317-338.

15. Strickler, D. W. and Carlson, W. G., 1964. "Ionic Conductivity of Cubic Solid Solutions in the System $CaO-Y_2O_3-ZrO_2$." J. American Ceramic Soc., 47 (3): 122-127.

16. Isenberg, A. O., 1977. "Growth of Refractory Oxide Layers by Electrochemical Vapor Deposition (EVD) at Elevated Temperatures", in Electrode Materials and Processes for Energy Conversion and Storage, J. D. E. McIntyre, S. Srinivasan and F. G. Will, eds., Princeton, NJ, USA: The Electrochemical Society, Inc., Vol. 77-6, pp. 572-583.

17. Pal, U. B. and Singhal, S. C., 1990. "Electrochemical Vapor Deposition of Yttria-Stabilized Zirconia Films." J. Electrochem. Soc., 137: 2937-2941.

18. Srilomsak, S., Schilling, D. P. and Anderson, H. U., 1989. "Thermal Expansion Studies on Cathode and Interconnect Oxides", in Solid Oxide Fuel Cells, S. C.

Singhal, ed., Pennington, NJ, USA: The Electrochemical Society, Inc., pp. 129-140.

19. Pal, U. B. and Singhal, S. C., 1990. "Growth of Perovskite Films by Electrochemical Vapor Deposition." High Temp. Science, 27: 251-264.

20. Singhal, S. C., Ruka, R. J. and Sinharoy, S., 1985. "Interconnection Materials Development for Solid Oxide Fuel Cells", U.S. Department of Energy Final Report DOE/MC/21184-1985, 54 pages.

21. Ray, E., 1992. "High Temperature Tubular Solid Oxide Fuel Cell Development", in Proceedings Fourth Annual Fuel Cell Contractors Review Meeting, W. J. Huber, ed., Morgantown, WV, USA: U.S. Department of Energy, pp. 144-149.

22. Maskalick, N. J., 1989. "Design and Performance of Tubular Solid Oxide Fuel Cells", in Solid Oxide Fuel Cells, S. C. Singhal, ed., Pennington, NJ, USA: The Electrochemical Society, Inc., pp. 279-287.

23. Zymboly, G. E., Reichner, P. and Makiel, J. M., 1985. "Cell Fabrication, Design, and Operation of a Solid Oxide Fuel Cell Generator", in National Fuel Cell Seminar Abstracts, Washington, DC, USA: Courtesy Associates, Inc., pp. 95-101.

24. Reichner, P. and Makiel, J. M., 1986. "Development Status of Multi-Cell Solid Oxide Fuel Cell Generators", in National Fuel Cell Seminar Abstracts, Washington, DC, USA: Courtesy Associates, Inc., pp. 32-35.

25. Stephenson, Jr., D. R. and Veyo, S. E., 1986. "TVA Test of SOFC Subscale Generator", in National Fuel Cell Seminar Abstracts, Washington, DC, USA: Courtesy Associates, Inc., pp. 36-39.

26. Veyo, S. E., 1988. "SOFC Field Experiments, A Learning Experience", in National Fuel Cell Seminar Abstracts, Washington, DC, USA: Courtesy Associates, Inc., pp. 13-17.

27. Harada, M. and Mori, Y., 1988. "Osaka Gas Test of 3 kW SOFC Generator System", in National Fuel Cell Seminar Abstracts, Washington, DC, USA: Courtesy Associates, Inc., pp. 18-21.

28. Yamamoto, Y., Kaneko, S. and Takahashi, H., 1988. "Tokyo Gas Tests of 3 kW Generation Systems", in National Fuel Cell Seminar Abstracts, Washington, DC, USA: Courtesy Associates, Inc., pp. 25-28.

29. Shockling, L. A. and Makiel, J. M., 1990. "Natural Gas-Fueled 3 kWe SOFC Generator Test Results", in Proceedings of the 25th Intersociety Energy Conversion Engineering Conference, P. A. Nelson, W. W. Schertz and R. H. Till, eds., New

York, NY, USA: American Institute of Chemical Engineering, Vol. 3, pp. 224-229.

30. Dollard, W. J., 1992. "Solid Oxide Fuel Cell Development at Westinghouse." J. Power Sources, 37: 133-139.

31. Singhal, S. C., 1990. "Solid Oxide Fuel Cells for Clean and Efficient Power Generation", in Proceedings of the Fukuoka International Symposium on Global Environment and Energy Issues, T. Seiyama, ed., Fukuoka, Japan: The Electrochemical Society of Japan, pp. 95-104.

Solid Oxide Fuel Cells: Materials, Fabrication Processes and Development Trends

NGUYEN Q. MINH

ABSTRACT

Solid oxide fuel cells (SOFC's), based on stabilized zirconia electrolyte, are currently under development for a variety of power generation applications. In addition to the zirconia electrolyte, the fuel cell employs doped lanthanum manganite as the cathode, nickel/zirconia cermet as the anode, and doped lanthanum chromite as the interconnect. Four common SOFC designs have been proposed: seal-less tubular, segmented-cell-in-series, monolithic, and flat-plate. Fabrication of each design requires different processing techniques for the ceramic structures. At present, the critical issues facing SOFC technology involve the development of suitable materials and fabrication processes. This paper reviews materials and fabrication processes used in the zirconia-based SOFC and discusses development trends.

INTRODUCTION

Stabilized zirconia is a known oxide ion conductor over an extended range of oxygen partial pressures (1 to 10^{-20} atm). High oxide ion conductivity has led to the almost exclusive use of stabilized zirconia as the electrolyte in solid oxide fuel cells (SOFC's). A zirconia-based SOFC (composed of a stabilized zirconia electrolyte sandwiched between an anode and a cathode) produces electricity (and heat) by electrochemically combining hydrogen and carbon monoxide (fuel) with oxygen (oxidant) across the electrolyte. Zirconia cells are commonly stacked in electrical series to produce practical levels of power; an interconnect provides the anode/cathode electrical connection in a stack of cells. To achieve adequate electrolyte conductivity, current zirconia SOFC's operate at about 1000°C. The attractiveness of the zirconia cell relates principally to its solid-state nature, its high efficiency, its potential to reform gaseous fuels within the cell, and its high-quality byproduct heat, which can be used for cogeneration or other purposes[1].

The fabrication and operation conditions of zirconia SOFC's place stringent requirements on cell materials. In addition to having the proper conductivity and stability, cell materials must show chemical compatibility and have thermal expansion coefficients similar to that of the zirconia electrolyte from room temperature to

Nguyen Minh, Allied-Signal Aerospace Company, AiResearch Los Angeles Division, 2525 190th Street, Torrance, California, 90509

1000°C. Thus, development of suitable materials and fabrication processes is key to the technology.

REQUIREMENTS FOR ZIRCONIA-BASED SOFC COMPONENTS

At present, fully stabilized zirconia is preferred as the SOFC electrolyte material because it yields maximum conductivity and avoids phase transformation problems associated with partially stabilized ZrO_2. Although yttria-stabilized zirconia (YSZ) does not have the highest conductivity among the various stabilized zirconia materials, it is most frequently used because of its availability and cost. In an SOFC stack, the YSZ electrolyte and other components must meet certain requirements on conductivity, compatibility, stability, thermal expansion, porosity, and gas permeability. These requirements, based on YSZ as the reference material, are discussed below. It should be noted that these requirements are given here as a guide. They need to be modified depending on the specific cell design.

Electrolyte

Conductivity: The ionic conductivity of YSZ (8 mol% Y_2O_3) is about 0.1 $ohm^{-1}cm^{-1}$ at 1000°C. Due to this relatively low conductivity, the current path in the YSZ electrolyte must be designed to be as short as possible to reduce cell internal resistive losses.

Compatibility: Elemental migration into the electrolyte must be less than 10 percent of the electrolyte thickness during fabrication and operation. Greater penetration may result in unacceptable properties (e.g., a change in thermal expansion or the formation of electronic conductivity). Chemical interactions between YSZ and other cell materials must not form insulating phases at the electrode/electrolyte interface and must not cause a loss in conductivity of more than a factor of 2. Greater conductivity loss may increase cell internal resistance to a level unacceptable for SOFC application.

Stability: YSZ is known to be stable in both the oxidizing environment of the cathode and the reducing environment of the anode. The cubic phase of YSZ is stable from room temperature to its melting temperature.

Thermal expansion: The thermal expansion coefficient of YSZ is about 10.5 x 10^{-6} cm/cm°C (room temperature to 1000°C).

Porosity: The porosity must be less than 6 percent of theoretical density. This limit is set to ensure acceptably low leakage of fuel and oxidant.

Gas permeability: The room temperature gas diffusion coefficient must be less than 10^{-6} cm^2/s. This figure corresponds to fuel/oxidant cross leakage of less than 1 percent.

Anode

Conductivity: The electronic conductivity of the anode must be greater than 50 ohm^{-1} cm^{-1} at 1000°C. This value is set to permit a longer current path in the anode in a stack configuration.

Compatibility: Elemental migration into the anode must be less than 10 percent of the anode thickness during fabrication and operation. Chemical interactions with YSZ and other cell materials must not cause a loss in conductivity of more than a factor of 2.

Stability: The anode must be stable in the fuel environment, including the inlet conditions and the more oxidizing outlet conditions. The anode material must have no disruptive phase transformation (less than 5 percent volume change) and must maintain its microstructure in long-term operation.

Thermal expansion: The thermal expansion of the anode must reasonably match that of YSZ. There must be less than 10 percent change in the thermal expansion coefficient due to changes in oxygen partial pressure or chemical interactions at the temperature of operation and fabrication.

Porosity: The anode must have about 10 to 50 percent porosity. The lower limit is set by mass transport considerations. The upper limit is based on considerations of the strength of the anode.

Cathode

Conductivity: See anode.

Compatibility: See anode.

Stability: The cathode must be stable in the oxidant environment. See anode for other requirements.

Thermal expansion: See anode.

Porosity: See anode.

Interconnect

Conductivity: The electronic conductivity of the interconnect must be greater than 0.1 ohm^{-1}cm^{-1} at 1000°C. The current path in the interconnect must be as short as possible to reduce resistive losses.

Compatibility: See anode.

Stability: The interconnect must be stable in both the reducing atmosphere of the anode and the oxidizing atmosphere of the cathode. See anode for other requirements.

Thermal expansion: See anode.

Porosity: See electrolyte.

Gas permeability: See electrolyte.

MATERIALS FOR ZIRCONIA-BASED SOFC'S

Several anode, cathode, and interconnect materials have been selected and tested for the zirconia SOFC, based on the requirements discussed above. At present, the most commonly used are nickel/stabilized zirconia cermet for the anode, doped lanthanum manganite for the cathode, and doped lanthanum chromite for the interconnect. Table I summarizes the properties of these materials,

TABLE I. PROPERTIES OF ZIRCONIA-BASED SOFC MATERIALS

Property	Material (Component)			
	YSZ (8 mol% Y_2O_3) (Electrolyte)	Ni/ZrO_2 (Anode)	$LaMnO_3$ (Cathode)	$LaCrO_3$ (Interconnect)
Density, g/cm^3	5.90	*	6.84	6.74
Melting point, °C	2700	1453 (m.p. of Ni)	**	2500
Conductivity at 1000°C, $ohm^{-1}cm^{-1}$	0.1	500 (30 vol% Ni, 30% porosity)	75 (undoped) 125 (10 mol% Sr)	1 (undoped) 14 (10 mol% Sr)
Thermal expansion, cm/cm°C (x 10^{-6})	10.5	12.1 (30 vol% Ni)	11.2 (undoped) 12.0 (10 mol% Sr)	9.5 (undoped) 10.7 (10 mol% Sr)
Thermal conductivity, $Wm^{-1}K^{-1}$	3.8	**	**	**
Specific heat at 1000°C ($calmol^{-1}deg^{-1}$)	21.4	**	**	**
Bend strength at room temperature, MPa	300	**	**	**
Fracture toughness R.T., $MN.m^{1.5}$	3	**	**	**
* Dependent on composition				
** Not available				

along with those of YSZ. Examples of the microstructures of YSZ electrolyte, Ni/YSZ cermet anode, Sr-doped $LaMnO_3$ cathode, and Sr-doped $LaCrO_3$ interconnect (prepared by tape calendering and sintered in air at 1400°C) are given in figure 1.

NICKEL/ZIRCONIA CERMET

Nickel has been used as an anode material because the metal is chemically stable in the reducing atmosphere of the fuel gas at 1000°C and compatible with the YSZ electrolyte. Stabilized zirconia is incorporated in the anode to develop a porous substrate on which nickel coating is applied. The zirconia in the anode supports the nickel particles, inhibits coarsening of the metallic particles at the fuel cell operating temperature, and provides an anode thermal expansion coefficient close to that of the YSZ electrolyte. A compromise between conductivity and thermal expansion is required when the nickel loading in the anode is determined. About 30 vol% Ni is needed to maintain the required level of conductivity (> 50 $ohm^{-1}cm^{-1}$), while minimizing the degree of thermal expansion mismatch[2]. The present concerns with the Ni/YSZ cermet anode are sintering of nickel particles in long-term operation, thermal expansion mismatch with the other cell components, and tolerance of sulfur contaminants in the fuel gas.

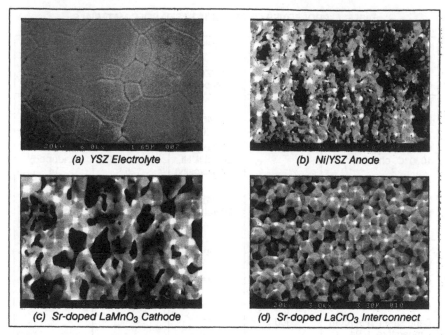

(a) YSZ Electrolyte

(b) Ni/YSZ Anode

(c) Sr-doped LaMnO₃ Cathode

(d) Sr-doped LaCrO₃ Interconnect

Figure 1. Microstructures of SOFC Components

DOPED LANTHANUM MANGANITE

The most common cathode material is presently Sr-doped $LaMnO_3$ (10 mol% Sr). This material has an electrical conductivity of about 130 $ohm^{-1}cm^{-1}$ at 1000°C and is stable in oxidizing atmospheres. Other properties such as thermal expansion coefficient and porosity can be tailored to match those of the YSZ electrolyte. Manganese is known to be a mobile species at high temperatures and can easily diffuse into the electrolyte and other cell components, changing the electrical characteristics or the structure of the cathode or other components. At high temperatures, $LaMnO_3$ can also react with YSZ to produce $La_2Zr_2O_7$[3]. This reaction product is undesirable because the conductivity of $La_2Zr_2O_7$ is about two and a half orders of magnitude lower than that of YSZ. Fabrication temperature is generally limited to below 1400°C to minimize manganese migration and chemical interaction. The present concerns with the strontium-doped lanthanum manganite cathode are long-term stability of cathode morphology and cathode/electrolyte and cathode/interconnect interactions.

DOPED LANTHANUM CHROMITE

Lanthanum chromite is particularly suitable for the zirconia-based SOFC interconnect from the standpoint of stability in the fuel cell environment and compatibility with the other components. Various substitutions (e.g., Mg, Sr, and Ca) in the lanthanum chromite improve the thermal expansion match and electronic conductivity under SOFC operating conditions. $LaCrO_3$, like all chromium compounds, does not densify at temperatures below 1700°C and oxygen activities above 10^{-9} atm. The reason for the low sinterability is that the predominant mass transport during firing is an evaporation/condensation mechanism, which leads to coarsening of the original particles without densification. A common approach is to use liquid phase sintering [4] to enhance the densification of $LaCrO_3$. The main difficulty with liquid phase sintering is that the liquid tends to migrate to other parts of the fuel cell and reacts with other components, causing elemental migration and morphological changes. The present concern with the doped lanthanum chromite interconnect is sinterability of the material in oxidizing atmospheres.

FABRICATION PROCESSES FOR ZIRCONIA-BASED SOFC'S

Fabrication of the SOFC requires incorporation of YSZ and other cell materials into the components of a stack configuration. Each material must be incorporated in a manner that will not cause the properties of the material or any of the cell components to degrade. Successful fabrication of the zirconia SOFC requires that the ceramic fuel cell be produced not only without structural defects, but also with the required physical, chemical, electrical, and electrochemical properties discussed above.

At present, four common stack configurations have been proposed and fabricated for zirconia SOFC's[1]: the seal-less tubular design, the segmented cell in series design, the monolithic design, and the flat-plate design (figure 2). The designs differ in the extent of dissipative losses within the cells, in the manner of sealing between fuel and oxidant channels, and in making cell-to-cell electrical connections in a stack of cells. The fabrication costs and ease of assembly vary among the designs. Table II compares the key characteristics of the four designs. The comparison is relative and qualitative because the differences are hard to quantify at this stage of development.

TABLE II. CHARACTERISTICS OF SOFC DESIGNS

Feature	Design			
	Seal-less Tubular	Segmented-Cell-in-Series	Monolithic	Flat-Plate
Inactive support	Yes	Yes	No	No
Internal electrical resistance	High	High	Low	Medium
Gas sealing	No	Yes	No	Yes
Power density	Low	Low	High	Medium

The fabrication process selected for each design depends on the configuration of the cells within the stack. Current processes can be classified into two groups based on the fabrication approach: the deposition approach and the particulate approach. The deposition approach involves the formation of a thin layer on a porous support by a chemical or physical process. The fabrication of the seal-less tubular SOFC and the segmented-cell-in-series SOFC is based on this approach (electrochemical vapor deposition or EVD and plasma spraying)[1]. The particulate approach involves compaction of ceramic powder into cell components and densification at elevated temperature. The fabrication of the monolithic SOFC and the flat-plate SOFC is based on this approach (tape calendering and tape casting)[1].

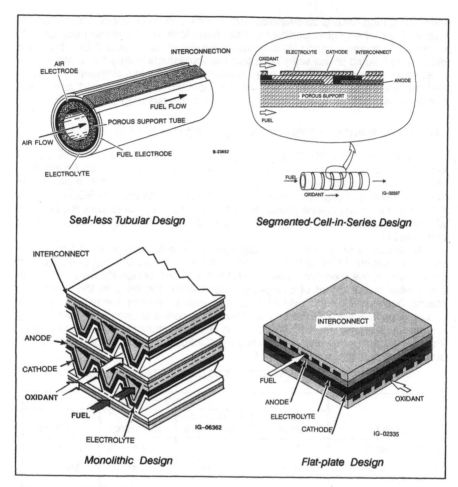

Figure 2. SOFC Designs

ELECTROCHEMICAL VAPOR DEPOSITION

The EVD process is the key fabrication technique in the seal-less tubular SOFC technology. The principle of the process is shown schematically in figure 3. The process involves growing a dense oxide layer on a porous substrate at elevated temperatures and reduced pressures. In the EVD of components for seal-less tubular SOFC's, steam and/or oxygen is fed to the interior of the support tube while metal chloride vapor, hydrogen, and argon are fed to the outside. The growth of the oxide layer takes place in two stages: the formation of the oxide compound by reaction of metal chloride with steam or oxygen, followed by the electrochemical reaction of oxygen ions in the oxide. The EVD process has been used in making the electrolyte and interconnect layers and in fixing the nickel matrix of the anode. The process has been scaled up and has proven to be a reliable method of producing thin, dense electrolyte and interconnect layers for the seal-less tubular design. The main drawback of the process is its high cost.

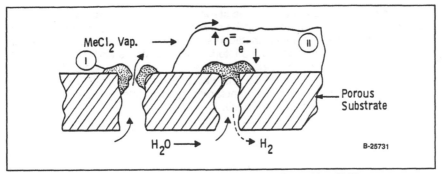

Figure 3. Schematic Diagram of Electrochemical Vapor Deposition (EVD)

PLASMA SPRAYING

Cell components of the segmented-cell-in-series SOFC are made by a process based on plasma spraying. Plasma spraying is a process in which the desired material in powder form is heated above its melting point while being accelerated by a carrier gas stream through an electric arc in a spray gun (figure 4).

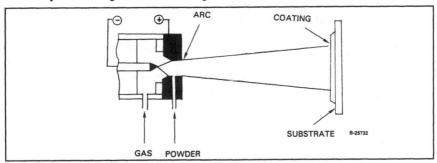

Figure 4. Schematic Diagram of Plasma Spraying

The molten powder is directed at the substrate, and on impact, forms a thin layer on the substrate surface. The spray gun creates a discharge plasma by imposition of a direct current between the two electrodes. Plasma spraying has been used to fabricate the electrolyte and interconnect layers for the segmented-cell-in-series SOFC. During spraying, the support tube rotates around its axis, while the spray gun traverses along the support tube axis. Plasma spraying can be cost effective and has been developed for the fabrication of this type of fuel cell. Layers deposited by plasma spraying must usually be applied at thicknesses greater than 100 μm to eliminate all open porosity.

TAPE CALENDERING

Tape calendering has been developed for the fabrication of the monolithic SOFC. Calendering is the formation of a continuous sheet or tape of controlled size by the squeezing of a softened thermoplastic material between two rolls. In the calendering of ceramic tape (figure 5), ceramic powder, binder, and plasticizer are mixed in a high-shear mixer. The friction resulting from the mixing action heats the batch and softens the binder to form a homogeneous plastic mass. The mass is then calendered in successive steps to a tape of desired thickness. The thickness of the tape is controlled by the spacing of the two rolls. To form multilayer tapes, individual layers are laminated in a second rolling operation as shown in figure 5. Green tapes can be formed into the desired shape and fired at elevated temperature to remove the binder and sinter the ceramic. Calendering is a well-known commercial process for the production of thin plastic sheets and, therefore, can be developed into a manufacturing technique for the ceramic multilayer tapes required in the monolithic SOFC.

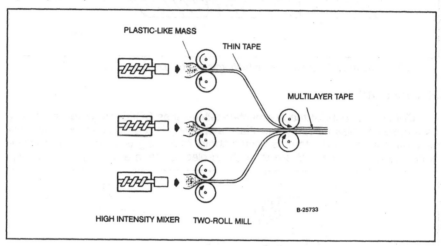

Figure 5. Schematic Diagram of Tape Calendering

TAPE CASTING

Tape casting is a low-cost method for manufacturing thin, flat sheets of ceramics and has been used to make electrolyte and interconnect layers for flat-plate SOFC's. The casting process involves making a layer of slip (ceramic powder suspended in a liquid) and drying this layer on a temporary support. The dry layer can be stripped from the support and fired to form a thin ceramic layer. The most common approach for tape casting is the doctor blade process (figure 6). The process consists of casting a slurry onto a moving carrier surface and spreading the slurry to a controlled thickness with the edge of a smooth blade. Multilayer tapes are fabricated by sequentially casting one layer on top of another. Although tape casting has been commercially used (e.g, in the production of alumina substrate for electronic microcircuits and barium titanate sheets for capacitors), the issue of scale-up for manufacturing flat-plate SOFC components needs to be addressed.

DEVELOPMENT TRENDS

Zirconia SOFC research has received much attention recently, reflecting increasing interest in the potential application of the technology for electric power generation. Recent developments have focused on improving material properties, modifying stack designs, and developing fabrication processes. The objective is to resolve key technical issues to develop a fuel cell system capable of meeting the performance, life, and cost goals for practical applications.

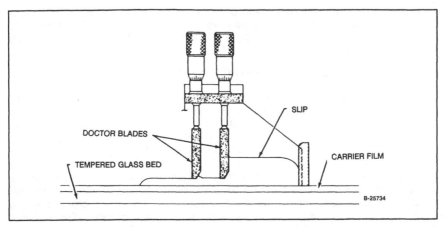

Figure 6. Schematic Diagram of Tape Casting

MATERIALS

Efforts have been carried out to improve the fracture toughness of the YSZ electrolyte[5]. A tougher electrolyte is less sensitive to the presence of flaws and imparts better resistance to the fuel cell during fabrication and operation. The approaches attempted include the use of additives (to create microcracks or act as crack pinning sites) to toughen the electrolyte, and the use of stronger electrolyte materials such as tetragonal zirconia. In other electrolyte-related efforts, reduction of YSZ electrolyte thickness has been proposed as a way to reduce ohmic losses, thus permitting operation of zirconia-based SOFC's at lower temperatures (600° to 800°C)[6]. The benefits of lower-temperature operation include wider choice of cell and ancillary materials, reduced thermal stress, improved reliability, and reduced fuel cell cost. This approach has received much attention recently, and some promising results have been reported in this area.

Work on anode materials has concentrated on developing means to minimize sintering of nickel in long-term operation and to match the thermal expansion between Ni/ZrO$_2$ cermet and other cell materials. For example, zirconia is added to the anode formulation in the form of fiber to provide a better support for nickel metal particles, thus reducing nickel sintering[7]. Minor additives have been used to modify the thermal expansion coefficient of the anode. Work on cathode materials has focused on developing materials with better morphological stability, and investigating the chemical compatibility of cathode materials with the YSZ electrolyte. Chemical potential diagrams have been developed to predict interactions between cathode and YSZ and between cathode and interconnect[8]. Recently, mixed conducting oxides have been investigated as SOFC electrode materials[9]. With mixed conducting electrodes, reactions occur over the entire electrode surface because both electrons and oxygen are mobile in the electrode material. As a result, electrode catalytic activities will be improved significantly, and polarization losses will be reduced. The focus of current interconnect efforts has been to develop methods to enhance the sinterability of LaCrO$_3$ at firing temperatures below 1400°C in oxidizing atmospheres. Several approaches have been pursued[1] such as use of highly reactive starting powders, stoichiometry modification, use of different dopants, and sintering aids. These approaches have been shown to be effective in enhancing LaCrO$_3$ sinterability. However, the material does not densify when fired in contact with the anode or cathode[10]. Interactions of the lanthanum chromite with the electrodes apparently inhibit the interconnect densification. Work is going on to better understand interaction mechanisms and to identify ways of minimizing the interactions.

STACK DESIGNS AND FABRICATION PROCESSES

Efforts on stack designs mainly aim at configuration modifications to improve performance and power density. An example is the current work to replace the zirconia support of the seal-less tubular design with a cathode tube[11]. Other efforts have been concentrated on the flat-plate design. This design is common in other types of fuel cells and has several advantages, including simple fabrication and multiple fabrication options. Due to the nature of their processing and assembly, flat-plate SOFC's can incorporate design and fabrication modifications more readily than other designs. For example, flat-plate SOFC's can incorporate metallic interconnects without major changes in the fabrication processes and stack assembly. The flat-plate design permits a variety of fabrication options. Thus, several alternative fabrication processes have been investigated for making cell components[1]: spray pyrolysis, vacuum evaporation, RF sputtering, metal organic chemical vapor deposition (MOCVD), EVD, CO_2 laser evaporation, and vapor phase electrolytic deposition. To date, only preliminary work has been carried out on these processes. Much development work is needed to verify the feasibility of these techniques for SOFC applications.

CONCLUDING REMARKS

SOFC's of various designs have been fabricated and tested on a laboratory scale, and the results obtained so far have demonstrated the feasibility of the technology. The critical issues facing SOFC technologies are the development of suitable materials and fabrication processes. Much research and development in these areas is required before the SOFC becomes practical. SOFC's allow the clean and efficient use of fuels and provide a number of important operational and economical advantages. Thus, although the technology is still in its early stage of development, it has received much attention recently. SOFC's will be an integral element of future means of generating electricity from a variety of fuels.

REFERENCES

1.   Minh, N.Q., 1993. "Ceramic Fuel Cells." Journal of the American Ceramic Society, in press.

2.   Dees, D.W., T.D.Claar, T.E.Easler, D.C.Fee and F.C.Mrazek, 1987. "Conductivity of Porous $Ni/ZrO_2-Y_2O_3$ Cermets." Journal of the Electrochemical Society, 134:2141-2146.

3.   Westinghouse Electric Corporation, 1984. "High-Temperature Solid Oxide Electrolyte Fuel Cell Power Generation System. Quarterly Technical Progress Summary Report, January 1, 1984 to March 31, 1984," Report DOE/ET/17089-2217, U.S. Department of Energy, Washington, D.C., USA.

4.   German, R.M., 1985. Liquid Phase Sintering. New York, NY, USA: Plenum Press.

5.   Singh, J.P., A.L.Bosak, D.W.Dees and C.C.McPheeters, 1988. "Improved Fracture Toughness of ZrO_2 Electrolyte for Solid Oxide Fuel Cell," in 1988 Fuel Cell Seminar Abstracts. Washington, D.C., USA: Courtesy Associates, pp.145-148.

6.   Negishi, A., K.Nozaki and T.Ozawa, 1981. "Thin-Film Technology for Solid Electrolyte Fuel Cells by the RF Sputtering Technique." Solid State Ionics, 3/4: 443-446.

7.   Murakami, S., Y.Miyake, Y.Akiyama, N.Ishida, T.Saito and N.Furukawa, 1990. "A Study on Composite Anode of Solid Oxide Fuel Cells," in Proceedings of the International Symposium on Solid Oxide Fuel Cells, O.Yamamoto, M.Dokiya and H.Tagawa, eds., Tokyo, Japan: Science House Co., Ltd., pp. 187-190.

8.   Yokokawa, H., N.Sakai, T.Kawada and M.Dokiya, 1990. Chemical Potential Diagrams for La-M-Zr-O (M = V, Cr, Mn, Fe, Co, Ni) Systems: Reactivity of Perovskites with Zirconia as a Function of Oxygen Potential. Denki Kagaku, 58: 489-497.

9.   Liou, S.S. and W.L.Worrell, 1989. "Mixed-Conducting Oxide Electrodes for Solid Oxide Fuel Cells," in Proceedings of the First International Symposium on Solid Oxide Fuel Cells, S.C. Singhal, ed., Pennington, NJ, USA: The Electrochemical Society, pp. 81-89.

10.   Minh, N.Q., T.R.Armstrong, J.R.Esopa, J.V.Guiheen, C.R.Horne, F.S.Liu, T.L.Stillwagon and J.J.Van Ackeren, 1991. "Fabrication Methodologies for Monolithic Solid Oxide Fuel Cells," in Proceedings of the Second International Symposium on Solid Oxide Fuel Cells, F.Gross, P.Zegers, S.C.Singhal and O.Yamamoto, eds., Luxembourg: Commission of The European Communities, pp. 93-98.

11.   Ray, E.R., 1992. "Westinghouse Tubular SOFC Technology," in 1992 Fuel Cell Seminar Abstracts. Washington, D.C., USA: Courtesy Associates, pp. 415-418.

Development of the Solid Oxide Fuel Cell at MHI

Y. YOSHIDA, S. UCHIDA and F. NANJOU

ABSTRACT

Mitsubishi Heavy Industries, Ltd. (MHI) have been developing Solid Oxide Fuel Cell (SOFC), which has high efficiency and the possibility of fuel diversification.

We have been developing three types of SOFC, namely Tubular, MOLB (Mono-block Layer Built) and planar type. The research on a tubular type SOFC was started in 1984 using the plasma spray technology. In 1990, a 1kW module was tested and the maximum power output of 1.3kW was recorded. This module was subsequently moved to a power station for a field test. A continuous operating test of 1,000 hours had been performed until November 1991. MHI has also been developing MOLB and Planar type SOFC with sintering method. A 1kW module test of MOLB type is scheduled in 1992. For Planar type, 10 cells stack of 130mm diameter has been successfully tested and the output of 115W was recorded.

This paper reviews MHI's R & D activities and results for the three types of SOFCs.

INTRODUCTION

TUBULAR TYPE SOFC

MHI started fundamental studies and plasma spraying technology improvement for SOFC production in 1984, and managed to get a cell-stack with sufficient performance and durability for multi-cell-stack test in 1989, and assembled a 1 kW SOFC module including 48 cell-stacks. The 1 kW SOFC module has been operated for more than 2,000 hours in the field.

Y.Yoshida S.Uchida F.Nanjou
  Mitsubishi Heavy Industries Ltd. 2-5-1,Marunouchi,chiyoda-ku,Tokyo,Japan

A Single Cell Stack

Appearance of a typical tubular type cell is shown in Fig 1. On a ceramic support tube which is 21 mm in diameter and 500 mm in length, 15 cells are arranged in series. Fuel is fed to the inside of the tube, and air to the outside. Schematic cell configuration of a cell is shown in Fig. 2. On the porous support tube of calcia stabilized zirconia(CSZ), fuel electrode, electrolyte and air electrode films are deposited and an electro chemical cell are composed. Fuel electrode of a cell and air electrode of the next cell is connected by an interconnector. All films are deposited by plasma or flame spraying process. Typical materials and deposition process of each component are as follows:

Electrolyte : Yttria Stabilized Zirconia (YSZ), 8 mol% Y_2O_3 by Low Pressure Plasma Spraying (LPPS)

Fuel Electrode : Ni by Flame Spraying (FS)

Air Electrode : $LaCoO_3$ by Flame Spraying (FS)

Interconnector : NiAl / Al_2O_3 Cermet by Atmospheric Pressure Plasma Spraying (APPS)

Tubular cell-stack requires interconnector with high electronic conductivity and thermal expansion coefficient which meets that of YSZ. NiAl / Al_2O_3 cermet satisfies this requirements. Microscopic photograph of a cell cross section is shown in Fig. 3. We have been improving performance and durability of cell-stacks. A cell - stack having 15 cells in 400 mm active length has been tested and the power output of 40W was obtained. For a cell - stack having 12 cells in 300 mm active length, fuel utilization of 87% was obtained. The results indicate that the electric conversion efficiency was 38% at high heating value (HHV).

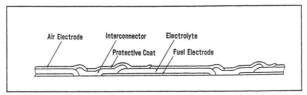

Figure 1. Appearance of Tubular Stacks

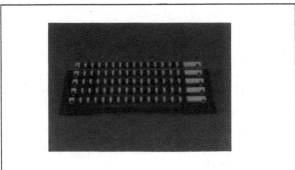

Figure 2. Tubular Cell Configuration

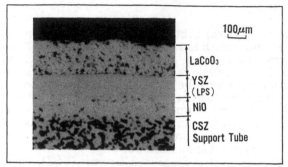

Figure 3. Cross Section of Tubular Cell

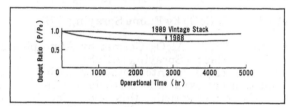

Figure 4. Endurance Test Result of Tubular Stack

Fig. 4 shows the result for durability test of stacks. 5,000 hours continuous operation test, was carried out and the deterioration rate was approximately 3% per 1,000 hours.

The 1 kW SOFC Module

We have assembled a 1 kW SOFC module in 1990. Specifications of the 1 kW SOFC Module is shown in Table I. The appearance and construction are shown in Fig. 5 and Fig. 6. The module consists of fuel inlet / outlet headers, a reaction chamber and air preheater. 48 cell-stacks hang on the upper plenum of the reaction chamber and the thermal expansion of cell - stacks is free. The 1 kW SOFC Module was moved to Wakamatsu Power Station of Electric Power Development Company Ltd. An example of module performance is shown in   Fig. 7. Power output of 1.3 kW was recorded. Fig. 8 shows the system diagram of the 1 kW SOFC module. In this system, remained H_2 fuel and air are exhausted separately from the module without combustion. Remained H_2 fuel gas is recycled to the fuel inlet of the module after separating H_2O drain. This system makes it possible to raise fuel utilization at reacting part, and a gain of around 30 % in fuel utilization was observed. NOx in exhaust air was measured by chemiluminescence method but it was not detected. Since the 1 kW module has no combustor and the operating temperature is no more than 1,000 °C, thermal NOx is not generated. A continuous operating test of 1,000 hours has been performed in Oct. 1991. The result is shown in Fig. 9. During the operation including 3 times thermal cycle, voltage degradation of 3%/1,000 hours was observed. The level of deterioration was same as that of a single cell stack.

Table I. Specification of 1kW Module

Output	1kW (DC)
OCV	180V
Operating Voltage	120V
Operating Current	10A
Number of Stacks	48
Dimension	700 dia.×1625 height (Hot Air Preheater Built In)
Temperature	900°C
Pressure	Atmospheric
Fuel	H2
Oxidizer	Air

Figure 5. Appearance of 1kW Module

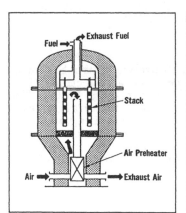

Figure 6. Construction of 1kW Module

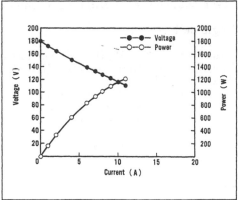

Figure 7. I-V Characteristic of 1kW Module

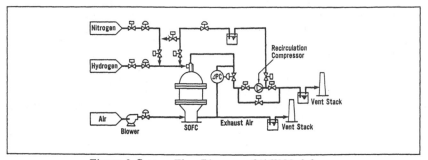

Figure 8. System Flow Diagram of 1kW Module

667

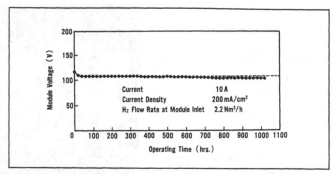

Figure 9. 1000 hours Endurance Test Result of 1kW Module

MOLB TYPE SOFC

MHI have been developing MOLB type SOFC since 1986. Two cells stack of 75mm × 75mm were tested for 2,133hours countinuously. A 1kW module test of MOLB type is scheduled in 1992.

Single Cell

As shown in Fig. 10, the cell consists of electrolyte (Yttria Stabilized Zirconia; YSZ), fuel electrode (Ni/YSZ cermet), air electrode ($La_{1-x}Sr_xMnO_3$; LSM) and interconnector ($LaMg_xCr_{1-x}O_3$; LMC). The corrugated layers on each electrode are the same materials as each electrode, and form the gas paths in a cross flow. For sealing, two types were evaluated. One is fixed type, and the flat YSZ bars adhere to cell and interconnecter. The other is slide type, and a special packing material allows the discrepancy of thermal expantion of cell and interconnecter.

Cell Stack

In fundamental studies on improvements of the electrode behaviour on 23mm diameter single cells, material and polarization characteristics are evaluated. A maximum of 2.8 A·cm^{-2} of short circuit current density (Imax) was obtained. Furthermore, in order to perform the most suitable design of the stack, thermal stress analysis, and sealing test were carried out. Fig. 11 shows the history of stack performance improvement of 60 mm × 60 mm and 75 mm × 75 mm two cell stacks. Other operating conditions were the same as that for single cells. The latest data of Imax, which is not shown in Fig 11, is 1,165 mA·cm^{-2} for the 75 mm × 75 mm cell stack. The two cell stack of 75 mm × 75 mm has been operated for 2,133 hour continuously. Open circuit voltage showed no deterioration, whereas Imax decreased 3% per 100 hours. As fundamental approach to large stack, we have been also improving the performance of large size cells. Imax of 424mA·cm^{-2} is obtained for ten cell stack of 150 mm × 150 mm.

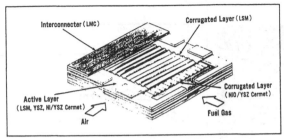

Figure 10. MOLB Type SOFC Structure

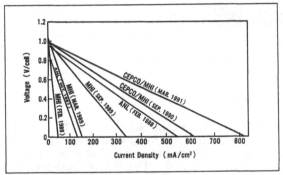

Figure 11. Improvement of Stack Performance of MOLB Type Stack

Figure 12. Test Module of MOLB Type SOFC

Development of large stacks

Fig. 12 shows 1 kW class test module. Using this test apparatus, the performance of 150 mm × 150 mm, 10 cell and 40 cell stacks were verified. Fig. 13 shows appearance of 150 mm × 150 mm, 40 cell stack. The maximum output of 150 mm × 150 mm, 10 cell stack and 40 cell stacks were 101 W and 406 W respectively. Fig. 14 shows the current voltage curve of 150 mm × 150 mm, 40 cell stack. A 1 kW class module consists of three 150 mm × 150 mm, 40 cell stacks. The test will be performed at an early date.

   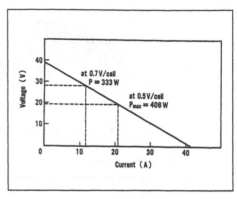

Figure 13. MOLB Type cell Stack   Figure 14. I-V Characteristic of MOLB Stack

PLANAR TYPE SOFC

MHI started R & D on planar type SOFC in 1987. Development was focused on co-sintering technology for 3 thin films; electrolyte, air and fuel electrodes. We have managed to accomplish a co-sintered cell. Fig. 15 shows a cross-section of a cell. The figure indicates good contact between electrolyte and both electrodes.

Co-sintering Technology

To manufacture a co-sintered cell, the following restrictions must be overcome.
- Sintering temperature is limited to below 1,300 °C in order to prevent harmful chemical interaction between electrolyte and air electrode.
- Electrolyte must be dense and gas tight.
- Electrode must be porous.
- Shrinking behavior of electrodes must correspond to that of electrolyte strictly.

To prepare slurry, ceramic powders were dissolved in organic solvent with resin and homogeneously dispersed. Slurry was cast   to green-tapes by doctor blade machine. Green tapes of electrolyte and both electrodes are put together in layers, then dewaxed and sintered. Shrinking behaviour and properties of sintered films were mainly adjusted by slurry compositions, particle size of ceramic powder, quantity of resin, concentration of powder / resin mixture, and so on. Fig. 16 shows a shrinkage profile of a co-sintered cell.

Shrinkage was maximum between 1,100 °C and 1,200 °C and finally reached approximately 30%. Fig. 17 shows large size cell of 170 cm^2, which has been developed.

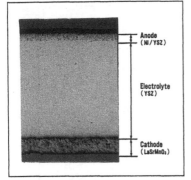

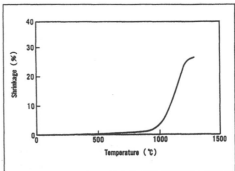

Figure 15. Cross Section of Planar   Figure 16. Shrinkage Profile of Co-sintered Cell
            Cell

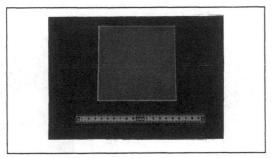

Figure 17. Appearance of Co-sintered Cell [170cm²]

Single cell

By sintering process, dense electrolyte film with high conductivity can be produced. Fig. 18 shows current-voltage-power characteristics of a co-sintered cell. The cell size was 5 cm², and the material is YSZ (Y_2O_3 8 mol%) as electrolyte, Ni 60 wt%/YSZ cermet as fuel-electrode, and $La_{0.9}$ $Sr_{0.1}$ MnO_3 as air electrode. Maximum power density of 0.6 W/cm² was obtained.

Cell stack

In parallel with cell development, we have been improving cell stacking technology; gas sealing, manifolding, current collecting and so on. Fig. 19 shows typical cell-stack configuration. Each cell is set on a ceramic support. Glassy sealant is filled in the clearance between cell and support. Fuel gas and air are fed and exhausted passing through manifolds in the support. A 100 cm²×10 cell stack was assembled and tested, and the power output of 115W was recorded. Appearance of the stack is shown in Fig. 20. Fig. 21 shows current - voltage - power characteristics of the 100 cm²×10 cell stack. This stack has been operated for 1,000 hours at a part load and little degradation was observed. The result is shown in Fig. 22.

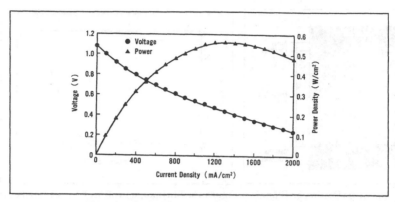

Figure 18. I-V Characteristic of Co-sintered Cell

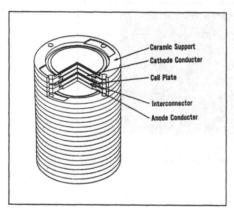

Figure 19. Typical Configuration of Stack

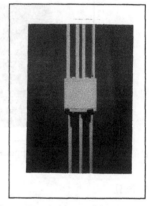

Figure 20. Appearance of 100cm²×10cell Stack

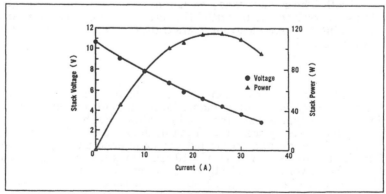

Figure 21. I-V Characteristic of 100cm²×10cell Stack

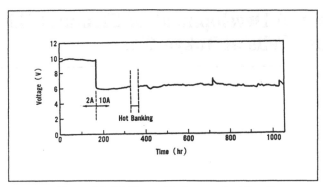

Figure 22. 1000 hours Endurance Test Result
of 100cm^2×10cell. Stack

CONCLUSION

The performance and durability of tubular type SOFC have been evaluated and satisfied in the field test of 1 kW module. MHI will make the next step to improve moduling technology for scale up. Meanwhile, for MOLB type and planar type SOFC, the development of cell - stacking technology is the next step.

ACKNOWLEDGEMENT

The improvement of tubular type cell and planar type cell have been done in cooperation with Tokyo Electric Power Co., Inc.(TEPCO). The moduling technology development of tubular type has been done in cooperation with TEPCO and Electric Power Development Co., Ltd.(EPDC). The improvement of MOLB type cell has been done in cooperation with Chubu Electric Power Co., Inc.(CEPCO). We would like to express thanks to all the people concerned.

Research and Development of Planar Solid Oxide Fuel Cells at Tokyo Gas

T. HIKITA

ABSTRACT

Planar single cells of 5, 10, and 23 cm square have been manufactured and tested. Attention has been focused on controlling the microstructure of the electrodes. A newly-developed electrode preparation process has dramatically reduced the interfacial resistance and has realized excellent results: a power density of 0.65 W/cm^2 in 5 cm square cells and a total power of 97 W in 23 cm square cells. Ten-cell stacks of 5 cm square cells and three-cell stacks of 10 cm square cells incorporating doped lanthanum chromite separators were also successfully manufactured, and their fairly good performance was experimentally demonstrated.

INTRODUCTION

Tokyo Gas has been conducting research and development of planar-type Solid Oxide Fuel Cells (SOFCs), which are some of the most advanced fuel cells and offer potential for the future cogeneration system. An SOFC has many advantages, such as high electric efficiency, high power density, and high-temperature waste heat, which are the features absolutely required for cogeneration equipment. Our efforts have been directed toward achieving a high power density in planar SOFCs. The possibility of attaining this in

Tomoji Hikita, Fundamental Technology Research Laboratory, Tokyo Gas Company, Ltd., 16-25 Shibaura, 1-Chome, Minato-ku, Tokyo, 105 Japan

SOFCs is the remarkable characteristic that makes SOFCs more advantageous than other types of fuel cells.

To attain a high power density, we placed priority on improving the electrodes and we conducted various experiments to clarify how the microstructure and composition of the electrodes affect cell performance. Based on the results of these investigations, we developed a new fabrication process for the electrodes, by which a relevant microstructure could be realized.

CELL STRUCTURE AND MATERIALS

A schematic drawing of a typical structure of the SOFC being developed in our laboratory is shown in figure 1.

Single cells are manufactured by using yttria-stabilized zirconia (YSZ) as the electrolyte plate, Ni/YSZ cermet as the anode or the fuel electrode, and $La(Sr)MnO_3$ as the cathode or the air electrode. The electrodes are applied to both surfaces of the electrolyte plate by means of the "PMSS (Pyrolysis of Metallic Soap Slurry)" method

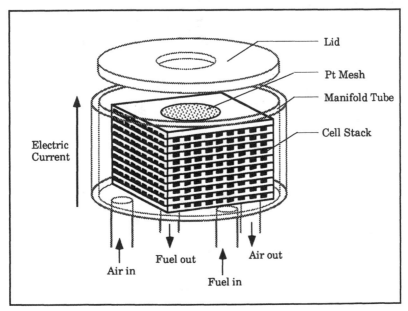

Figure 1.   Schematic drawing of the planar SOFC stack placed in the external manifold tube.

which we developed for this particular purpose. A detailed description of this process for the anode will be given in the following subsection.

The separators are made of La(Ca)CrO$_3$ or La(Sr)CrO$_3$ by the conventional pressing and sintering process. This is followed by mechanical machining which transforms them into a bipolar shape. The density of the sintered material is about 95% of the theoretical value.

A cylindrical external manifold, made of alumina, is used for stack testing. The corners and the side-planes of the stack are sealed with glass-based sealing materials.

IMPROVEMENT OF ANODE

From the correlation between the microstructure and the overpotential of the anode, prevention of agglomeration of fine nickel particles was found to be essential for a high-performance

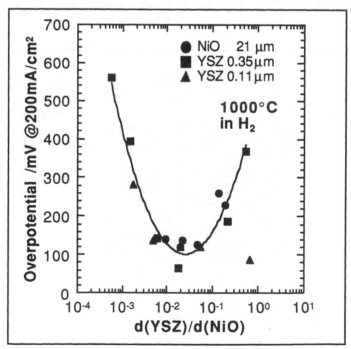

Figure 2.   Relationship between anode overpotential and particle size ratio of the starting oxide powders.

anode.

In preparing the anode by a conventional powder mixing process, the microstructure can be tailored by controlling the particle size of the materials used [1]. As shown in figure 2, the overpotential was strongly dependent on the particle size ratio of YSZ to NiO rather than on the particle size itself, and was minimized at a specific particle size ratio. EDX analyses confirmed that the nickel particle distribution in the YSZ matrix was most uniform at this point.

Further investigation led us to a concept of the anode microstructure in which nickel particles are covered by thin film or fine precipitates of YSZ. Such microstructure will prevent the agglomeration of nickel while also providing sufficient adhesion between the anode and the electrolyte plate. To realize the concept, a solution containing zirconium octylate and yttrium octylate was employed as a starting substance for YSZ in the anode. The solution is mixed with NiO powder, is screen-printed onto the electrolyte plate, and is then fired. The reaction process includes polymerization and thermal decomposition of the metal octylates, followed by an annealing process in which the reaction products are transformed into a cubic solid solution of YSZ. Figure 3 shows the

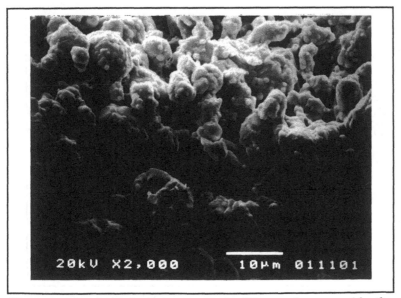

Figure 3. Scanning electron micrograph of the anode prepared by the PMSS method.

microstructure of the anode thus prepared. Fine particles of YSZ are seen precipitating on the surface of the NiO grains.

CELL PERFORMANCE

Figure 4 shows our history of performance improvement in 5 cm square single cells with an effective electrode area of 21cm². All tests were conducted at 1000°C using hydrogen as the fuel and air as the oxidizer. Although the power density was only 0.2 W/cm² in the early development stage, it was increased to 0.65 W/cm² by employing the anode prepared by the PMSS. Single cells as large as 10 cm square and 23 cm square were also successfully manufactured and tested. They produced maximum outputs of 32 W and 97 W, respectively. The 23 cm square cell has an effective area of 400 cm² and is the largest planar-type single cell ever developed as of June, 1992. V-I characteristics of all the three cells are indicated in figure 5.

A ten-cell stack of 5 cm square cells and a three-cell stack of 10 cm square cells were manufactured and tested using bipolar separators made of doped lanthanum chromite. The ten-cell stack,

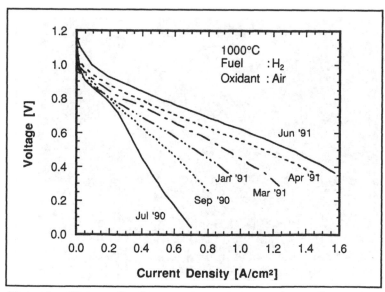

Figure 4.   History of performance improvement in 5 cm × 5 cm single cells.

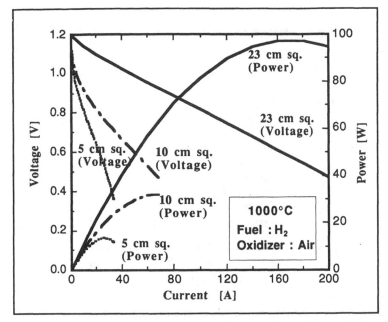

Figure 5.   V-I and P-I characteristics of single cells with three different sizes.

with an active electrode area of 210 (= 10 × 21) cm², produced an
open circuit voltage of 10.2 V and a maximum output of 50 W. The
three-cell stack, with an active electrode area of 264 (= 3 × 88) cm²,
produced an open circuit voltage of 3.1 V and a maximum output of
62 W.Figures 6 and 7 show the performance curves of these two
stacks.

Duration tests were started recently using single cells with an
effective electrode area of 4 cm². The cell is continuously discharged
at a constant current density of 0.3 A/cm² at 1000°C and the
stability of cell voltage is checked. One of the preliminary results is
shown in figure 8. A good voltage stability is obtained in the
continuous run of about 450 hours, though the elapsed time is not
long enough for confirmation of the life. Further verification tests of
the performance stability are currently taking place.

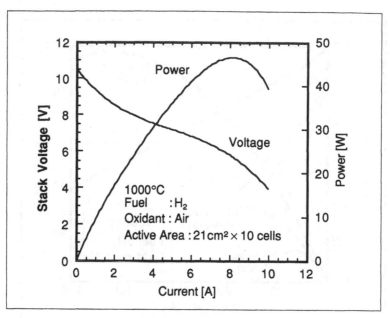

Figure 6.   V-I and P-I characteristics of a ten-cell stack of 5 cm × 5 cm cells.

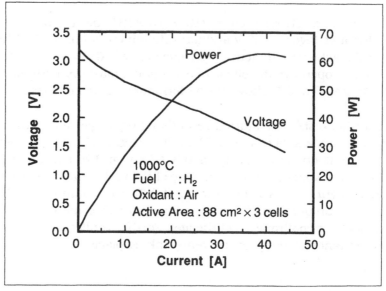

Figure 7.   V-I and P-I characteristics of a three-cell stack of 10 cm × 10 cm cells.

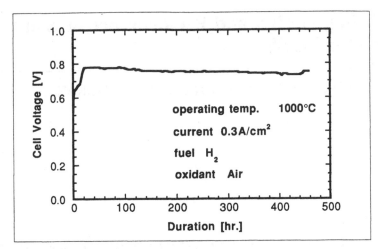

Figure 8.   Time change of the cell voltage of a 2 cm × 2 cm single cell under a constant load.

CONCLUSION

Tokyo Gas has been developing planar SOFCs, placing priority on basic studies that are essential for establishing a basis for a high-performance fuel cell. To date, by carefully controlling the microstructure of the anode through a newly developed "PMSS" process, a high power density of 0.65 W/cm² for a 5 cm square cell was produced, and a total power as high as 97 W was able to be extracted from a 400 cm² cell, which is the largest planar-type SOFC ever developed as of June, 1992. Ten-cell stacks of 5 cm square cells and three-cell stacks of 10 cm square cells were also successfully manufactured and tested. These stacks generated maximum d.c. power of 50 and 62 W, respectively. Tasks to be pursued in the near future include long-duration tests and development of 250 W to 1000 W cell stacks.

REFERENCE

1.   Yasuda, I., T. Kawashima, T. Koyama, Y. Matsuzaki and T. Hikita, 1992. "Research and Development of Planar Solid Oxide Fuel Cells at Tokyo Gas," in Proceedings of the International Fuel Cell Conference, Makuhari, Japan : New Energy and Industrial Technology Development Organization (Japan), pp.357-360.

Modeling, Design and Performance of Solid Oxide Fuel Cells

J. HARTVIGSEN, S. ELANGOVAN and A. KHANDKAR

ABSTRACT

Operating solid oxide fuel cells (SOFC) must reject heat from the stack at a rate similar to that of the power production of the stack. High performance planar SOFCs have both a high volumetric power density and a high volumetric heat generation. Consequently it becomes important to study temperature distributions under various operating and design conditions. The traditional approach to this problem has focused on 2-D thermal conduction in the solid coupled with convection to the reactant gases. Several simplifying assumptions are often made; e.g. uniform solid and gas temperatures, uniform gas phase compositions and adiabatic boundaries. Often, under further simplifying assumptions the problem is rendered one-dimensional.

Thermal radiation, a dominant heat transfer mechanism at the high operating temperatures of SOFCs, plays a major role in determining the thermal condition of the stack and must be taken into account for accurate determination of temperature distributions. Rigorous modeling of SOFCs has been undertaken to include the effects of radiation using a finite element approach. Thermal conduction effects have been coupled with electrical conduction. Local potentials have been rigorously computed. Inclusion of thermal radiation effects provides a significantly different stack temperature distribution compared to results obtained by ignoring radiation effects. The implications of these results on planar SOFC design and performance are discussed.

INTRODUCTION

Modeling and simulation studies are able to provide important input in the development of solid oxide fuel cells (SOFCs). For example, simulation studies can predict the local current densities, voltages and temperatures within a cell. Measurements of these parameters are either very difficult and cumbersome if not impossible to obtain. Such models allow parallel development of SOFC stack components and systems surrounding the SOFC itself. Stack component designs can be optimized on the basis of design studies conducted using simplified cell

J. Hartvigsen, S. Elangovan and A. Khandkar, Ceramatec, Inc.

2425 South 900 West, Salt Lake City, UT 84119, USA.

ohmic loss models. This allows the materials and processing conditions of the components to be selected and developed at an early stage. Design configurations for the thermal system around the stack can also be considered long before the cell and stack materials and processing are refined. Beyond the benefits of early and parallel development of components and systems, a better interpretation and understanding of the experimental results can be achieved.

Typically, single cells and small stacks are tested in electric furnaces where the temperature distributions are driven by conditions external to the cell and not by the cell reaction chemistry. Therefore measurements from small stacks cannot give results for larger systems. These results can be obtained by modeling before testing of a larger system.

TYPES OF MODELS

Several types or levels of models have been used. At the single cell level, using a variety of electrochemical measurement techniques, detailed models for the cell reactions can be obtained. Such models allow one to determine optimal electrode design and to understand fundamentals of cell degradation processes [1,2,3]. Other models involve the prediction of cell potentials, current densities and temperatures as has been done in the present case [4-10]. At the stack level, the interaction of the temperature and flow distributions and their impact on stack performance can be modeled. Finally at the system level, models are used to simulate and predict system characteristics under a variety of load conditions [8].

In the present work, rigorous modeling of SOFCs using a finite element approach has been undertaken to predict the thermal distribution and performance of stacks in two different configurations. In contrast to the previous work in the literature, radiation heat transfer effects have been considered. The effect has been found to play a major role in determining the thermal condition of the stack and must be taken into account for accurate determination of temperature distributions. Thermal conduction effects have been coupled with electrical conduction and local electrical and chemical potentials have been rigorously computed.

FINITE ELEMENT MODELING

The finite element method (FEM) is well suited to analysis of geometrically complex components. The principle alternative for such analysis is the finite difference method. In theory either method could achieve comparable results. In practice it is very difficult to model complex geometries and boundary conditions with a finite difference method. This is due to the data structure of most finite difference codes, which is driven by the equation solution method and tradition. The result is that most FDA codes require a structured mesh and hand coded boundary conditions. SOFC modelers using finite differences avoid these problems by using 2-D, pseudo 3-D or homogenized models [6,7,9]. These approaches have the merits of quick model generation and solution, but eliminate the possibility of including effects such as enclosure radiation which require a detailed geometric representation.

MODELING APPROACH

Comprehensive modeling of SOFC systems is beyond the capability of existing computational methodology due to the wide range of spatial and temporal scales, and physical phenomena which must be considered. Consequently simplified models of certain phenomena and scales must be employed. This work focuses on the detailed spatial resolution of temperature and voltage in a single cell repeat unit using simple models of fluid dynamics, electrode kinetics, and the surroundings.

TOPAZ

TOPAZ is a three-dimensional finite element program for heat transfer analysis from the Lawrence Livermore National Laboratory. Some of the features of TOPAZ include steady state or transient capability, orthotropic temperature dependent material properties, and interfaces to graphical pre and post processors. Boundary conditions include specified temperature, flux, convection, bulk fluid, radiation, and enclosure radiation, with parameters that may vary with time, temperature or position. A significant feature of TOPAZ is its distribution as FORTRAN source which makes it possible to add features needed for fuel cell analysis. Several extensions to the TOPAZ code were required to model SOFCs. Most of the changes are applicable to general heat transfer analysis, but a few are specific to fuel cell modeling.

TOPAZ Extensions for Solid Oxide Fuel Cells

Finite element spatial discretization of the nonlinear system of partial differential equations for coupled thermal and electrical conduction requires repeated solution of systems of linear equations. Solution of these linear systems generally consumes in excess of 90% of the analysis computer time. More importantly, the model resolution is usually limited by the amount of memory required to store the conductance matrix. TOPAZ uses a symmetric skyline solver which stores by column from first non-zero entry in a column to the diagonal. When coupled with a bandwidth minimization scheme, the skyline solver can be quite efficient for high aspect ratio geometries. While when applied to more equiaxed geometries, the skyline matrix has a high degree of sparsity, which equates to excessive memory demand. Several linear equation solution methods from the Sparse Linear Algebra Package (SLAP) were implemented in TOPAZ. The SLAP routines share a common pointer reference scheme in which only the non-zero array entries are stored. The algorithms and data structures needed to generate the reference pointers from finite element connectivity data were developed and implemented in TOPAZ. Application of the sparse storage technique has reduced the memory growth behavior from nodes$^{1.7}$ to nodes^1. A substantial improvement in solution speed also resulted.

A documented but unimplemented feature of TOPAZ is the bulk fluid node. The bulk fluid node models convection from the solid to a fixed mass of fluid, as opposed to the convection boundary condition which represents an infinite amount of fluid. This capability was added to allow thermal coupling with the reactant streams. A two node advection element was also added to enable modeling of streamwise transport between bulk fluid nodes. The advection element generates a non-symmetric matrix due to the directional (upstream to downstream) nature of the transport. The original skyline equation solver, and several of the SLAP solvers are for symmetric matrices only. Modifications to the pointer generation and matrix assembly routines were required to use the non-symmetric SLAP solvers.

TOPAZ solves for a single degree of freedom, the temperature, at each node. In order to model fuel cell performance and the effects of localized ohmic heating an electrical potential degree of freedom was added. An iterative sequential solution method was chosen to couple the two degrees of freedom. The temperature solution is computed using the ohmic heating computed from the voltage distribution of the previous iteration. The electrical solution follows with conductivities evaluated using the temperature distribution of the previous iteration. Boundary conditions for the electrical solution include specified potential, current density, and film resistance.

Analysis of problems involving coupling of radiation in an enclosure with conduction in the solid which encompasses the enclosure can be performed with TOPAZ. This type of analysis requires pre computation and storage of geometric view factors. The view factor F_{ij}, is the fraction of radiant energy emitted from surface i which is incident on surface j. View factors are stored in TOPAZ as a simple array even though most of the array entries are zero, or practically zero. Using a simple array, memory requirements grow with the square of the number of radiating surfaces. A single repeat unit cross flow SOFC model may need more than thirty thousand surfaces to provide appropriate spatial resolution. The memory required for a view factor matrix of this size is one hundred times the memory on a typical workstation. Pointer referenced sparse storage of view factors, similar to that used in SLAP, was implemented in TOPAZ resulting in linear memory growth behavior.

SOFC specific modeling needs were treated with a custom subroutine and input file. This SOFC subroutine reads a file which defines gas flow rates, and associates anode faces and corresponding cathode faces with air and fuel bulk nodes. The routine tracks gas compositions from inlet to exit. Local gas compositions and electrical potentials are used to calculate local current densities and heating rates which are used as boundary conditions in the next iteration.

STACK MODEL

Crossflow and co/counterflow models were developed using the modified TOPAZ program. Both configurations consist of a single cell repeat unit which includes an electrolyte with anode and cathode, and two halves of a bipolar-biflow interconnect. The parallel flow configurations allow reduction in model size by assuming that the solution has channel to channel symmetry. The symmetry assumption implies that the two sides of the stack which are parallel to the flow direction are adiabatic. This may or may not be valid depending on conditions external to the stack.

Stack dimensions, representative material properties and boundary conditions are shown in table I. The temperature dependence of material properties, where known, were used in the model. Single values are reported here due to space constraints. The "effective resistance" of each component (interconnect, anode, cathode, and electrolyte) is computed by the finite element method from actual cell geometry and temperature dependent conductivity. The convective heat transfer coefficients were derived from the laminar flow Nussult number (Nu_T) for a rectangular duct.

Current density is calculated at each electrolyte element from the local electrode potential, overpotential function and local bulk gas chemical potential. Only areas directly exposed to fuel and air were electrochemically active. That is the effects of diffusion under interconnect ribs were ignored. A constant overpotential of 100mV throughout the operating range was used for the cases shown. Any function of local temperature, fuel composition and current density may be input as an overpotential function.

TABLE I - STACK DIMENSIONS AND PROPERTIES

Fuel Channel Depth	0.5 mm
Air Channel Depth	1.0 mm
Channel Width	2.0 mm
Rib Width	1.0 mm
Inter connect Web Thickness	0.3 mm
Electrolyte Thickness	0.18 mm
Electrode Thickness	0.025mm
Electrolyte Ionic Conductivity	4.0e-3 S/mm
Electrolyte Electronic Conductivity	1.0e-8 S/mm
Electrolyte Thermal Conductivity	4.0e-3 W/mm-K
Interconnect Electronic Conductivity Fuel	0.3 S/mm
Interconnect Electronic Conductivity Air	1.6 S/mm
Interconnect Thermal Conductivity	4.6e-3 W/mm-K
Anode Electronic Conductivity	30.0 S/mm
Anode Thermal Conductivity	1.0e-2 W/mm-K
Cathode Electronic Conductivity	10.0 S/mm
Cathode Thermal Conductivity	4.0e-3 W/mm-K
Convection Coefficient Air	3.0e-4 W/mm^2-K
Convection Coefficient Fuel	3.0e-3 W/mm^2-K
Emissivity	0.8

DEVELOPMENT OF THE FINITE ELEMENT MODEL

The finite element meshes of the two models are shown in figure 1. Table II contains information on the models, boundary conditions and computational resources required. It can be seen from the figure and table that the co/counterflow model has benefited from the symmetry reduction by allowing a much finer mesh, while requiring much less computer time and memory. The 360 fold reduction in memory requirements for view factor storage (9432Mb to 26Mb) shows why the sparse storage scheme is essential to radiation modeling in fuel cells. The impact of sparse storage of the global conductance matrices is equally important. A schematic illustration of the equations solved in the solid and at the boundaries is shown in figures 2 and 3.

Radiation boundaries, and enclosures used the specified inlet gas stream temperature as the temperature of the surroundings. That temperature was 1073K for the results presented here. The film resistance coefficient used as the current collection boundary was set to a value which effectively forced the surface to a uniform specified potential. Overall current and energy balances zero satisfactorily. Fuel and air stream inlet conditions are listed in Table III for both crossflow and coflow models.

These models were run on a HP 720 workstation with 64Mb of main memory and 2Gb of disk storage. The 64Mb memory is insufficient for the cross flow model as 113Mb are required when modeling with radiation and 83Mb without radiation. The virtual memory system on the machine allows such models to be run, but at a greatly reduced speed in the case of the radiation model. More memory would be required to develop a multicell model, or a single cell model with higher resolution.

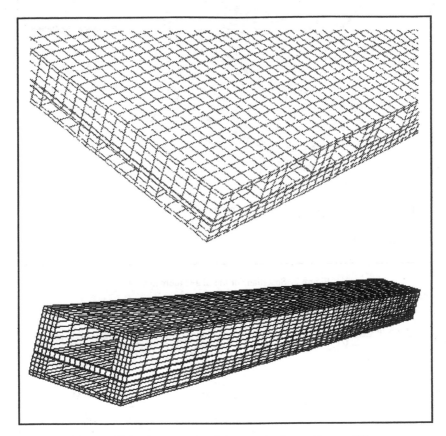

Figure 1. Finite element meshes for cross and co/counterflow conditions

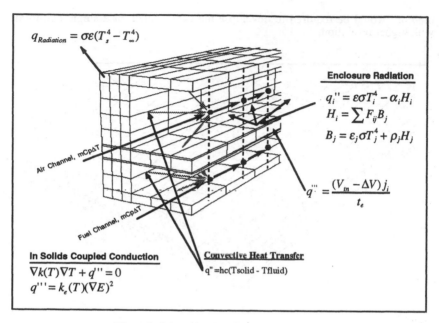

Figure 2. Schematic of the SOFC thermal model

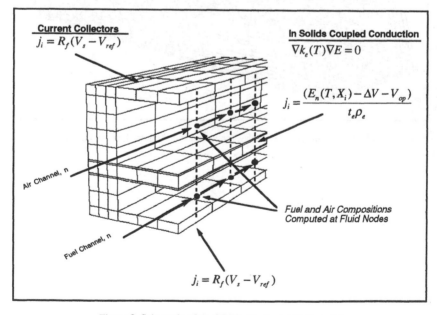

Figure 3. Schematic of the SOFC electrochemical model

TABLE II - FEATURES OF THE FE MODEL

	Crossflow	Co / Counterflow
Nodes	81,508	13,432
Bulk Nodes	2,702	92
Elements	54,000	9,900
Fluid Advection Elements	4,050	135
Bulk Convection Faces	35,100	3,060
Generation Elements	5,400	540
Radiation Faces	840	360
Enclosure Radiation Faces	35,160	3,064
View Factor Storage	26 MB 9,462 MB with conventional storage	2.3 MB 71.6 MB with conventional storage
Current Film Resistance Faces	16,200	1,800
Current Density Faces	10,800	1,080
Iterations No Radiation	8	7
Iterations Radiation	163	20
Solution Time with Radiation	9 hrs (30+ real due to paging)	16 min
Solution Time without Radiation	1.2 hr	12 min
Total Memory with Radiation	113 MB	16.3 MB
Total Memory without Radiation	83 MB	13.6 MB

TABLE III - INPUT OPERATING CONDITIONS FOR THE FE MODEL

Crossflow	
Hydrogen Flow Rate	6.0×10^{-5} mole/sec
Water Flow Rate	1.83×10^{-6} mole/sec
Oxygen Flow Rate	1.56×10^{-4} mole/sec
Nitrogen Flow Rate	6.3×10^{-4} mole/sec
Cell Area	4.5 cm x 4.5 cm
Coflow	
Hydrogen Flow Rate	4×10^{-6} mole/sec
Water Flow Rate	1.22×10^{-7} mole/sec
Oxygen Flow Rate	1.04×10^{-5} mole/sec
Nitrogen Flow Rate	4.2×10^{-5} mole/sec
Cell Area	4.5 cm x 0.3 cm
Fuel / Air Inlet Temperature (Both cases)	800°C

RESULTS AND DISCUSSION

The crossflow model was run for two conditions. The first modeled conduction/convection omitting radiation, the second included radiation. Figure 4 shows the solid temperature distribution of a single SOFC repeat element omitting the effects of radiation. This result is similar to others in the literature, with the maximum temperature occurring at the fuel inlet and air exhaust corner [7,9]. The higher concentration of H_2 at the fuel inlet with negligible change in oxidant concentration along the air channel (due to high air stoich operation) leads to higher current density and hence the higher temperature at the fuel inlet. In addition, air being the primary coolant, the maximum temperature occurs at the air outlet. There is very little contribution to cell cooling from the fuel stream due to their relative flow rates. The current density distribution follows the temperature distribution and the fuel concentration gradient. The average current density for this case was calculated to be 241 mA/cm^2 at 0.6 V.

The temperature distribution when the effects of radiation are included is shown in figure 5. It can be seen that not only is the maximum temperature lower (1136K compared to 1197K for the previous case), the shape of the distribution is significantly different. Strong radiation cooling along the faces and the channels moves the "hot spot" toward the middle of the cell. Isotherms are essentially concentric around the peak temperature. The overall temperature difference is only 46K compared to 98K when the radiation is neglected. The average current density is also slightly lower at 218 mA/cm^2 at 0.6 Volts because of the lower average temperature.

Similar analyses were performed for the coflow configuration for three different conditions, namely, (a) without radiation, (b) radiation from the stack faces alone, and, (c) radiation from the stack faces and the channels. The results are shown in figure 6. As expected, case (c) showed the lowest ΔT as well as the lowest T_{max}. The average current densities in the three cases were 269 mA/cm^2, 253 mA/cm^2 and 244 mA/cm^2 respectively at 0.6 Volts.

CONCLUSIONS

The temperature distribution in an SOFC stack is an important design consideration. Understanding this, one can design to achieve optimum performance without high thermal gradients that can cause mechanical failure and temperature accelerated degradation. Thermally accelerated degradations include coarsening of metal in the cermet anode, cation interdiffusion in the functional ceramics, and deterioration of seals.

A comprehensive SOFC stack thermal model was developed and applied to the crossflow and coflow configurations. Though commonly neglected, thermal radiation was shown to have significant effect on the temperature and current density distributions. The coflow geometry was found to have the lowest temperature variation, and thus is considered to be a desirable geometry for development. This model clearly indicates the need for rigorous modeling to more accurately predict the stack environment.

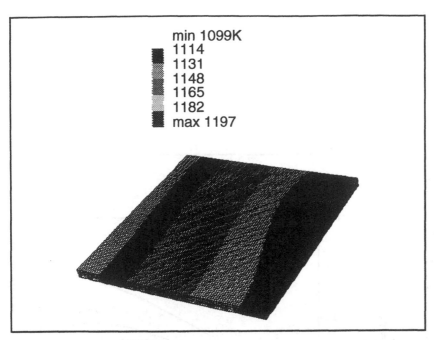

min 1099K
1114
1131
1148
1165
1182
max 1197

Figure 4. Temperature distribution on a planar crossflow cell without radiation

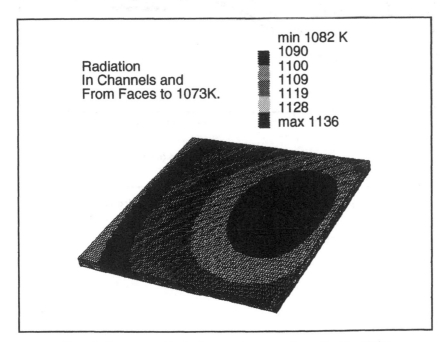

Radiation
In Channels and
From Faces to 1073K.

min 1082 K
1090
1100
1109
1119
1128
max 1136

Figure 5. Temperature distribution on a planar cross flow cell, with radiation

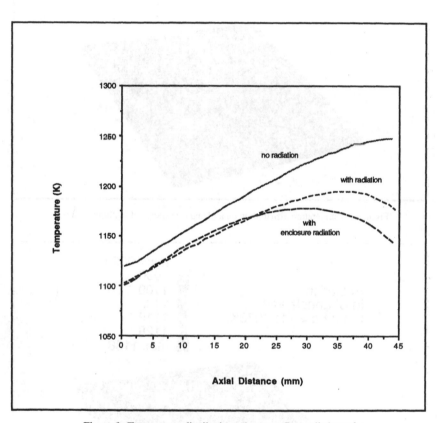

Figure 6. Temperature distributions along a coflow cell channel

REFERENCES

1. Sampath, V., A.F. Sammells and J.R. Selman, 1980. "A Performance and Current Distribution Model for Scaled-Up Molten Carbonate Fuel Cells." J. Electrochem. Soc., 127(1):79-85.

2. Khandkar, A.C., S. Elangovan, M. Liu and M. Timper, 1991. "Thermal Cycle Fatigue Behavior of High Temperature Electrodes," in Proc. Symp. High Temp. Electrode Mater. Charact., D.D. MacDonald and A.C. Khandkar, eds., Washington, DC, USA: The Electrochem. Soc., pp.179-190.

3. Elangovan, S., A. Khandkar, M. Liu and M. Timper, 1991. "Solid State Interface Reactions in SOFCs; an EIS Study," in Proc. Symp. High Temp. Electrode Mater. Charact., D.D. MacDonald and A.C. Khandkar, eds., Washington, DC, USA: The Electrochem. Soc., pp.191-199.

4. Maskalick, N.J. and D.K. McLain, 1988. "Performance Models for Zirconia Electrolyte Cells at Low Current Density." J. Electrochem. Soc., 135(1):6-11.

5. Wolf, T.L. and G. Wilemski, 1983. "Molten-Carbonate Fuel-Cell Performance Model." J. Electrochem. Soc., 130(1):48-55.

6. Yentekakis, I.V., S. Neophytides, S. Seimanides and C.G. Vayenas, 1991. "Mathematical Modelling of Cross-Flow, Counter-Flow and Cocurrent-Flow Solid Oxide Fuel Cells: Theory and Some Preliminary Experiments," in Proc. Int. Symp. Solid Oxide Fuel Cells, 2nd, F. Grosz, P. Zegers, S.C. Singhal and O. Yamamoto, eds., Brussels, Belgium: Commission of the European Communities, pp.281-288.

7. Erdle, E., J. Gross, H.G. Müller, W.J.C. Müller, H.-J. Reusch and R. Sonnenschein, 1991. "Modeling of Planar SOFC Stacks," in Proc. Int. Symp. Solid Oxide Fuel Cells, 2nd, F. Grosz, P. Zegers, S.C. Singhal and O. Yamamoto, eds., Brussels, Belgium: Commission of the European Communities, pp.265-272.

8. Dunbar, W.R. and R.A. Gaggioli, 1988. "Computer Simulation of Solid Electrolyte Fuel Cells," in Proc. Intersoc. Energy Convers. Eng. Conf., 23rd, Vol.2, pp.257-264.

9. Ferguson, J.R., 1991. "Analysis of Temperature and Current Distributions in Planar SOFC Designs," in Proc. Int. Symp. Solid Oxide Fuel Cells. 2nd, F. Grosz, P. Zegers, S.C. Singhal and O. Yamamoto, eds., Brussels, Belgium: Commission of the European Communities, pp.305-312.

10. van Heuwen, J.W., 1991. Proc. IEA SOFC Workshop, Thorstensen, ed., Oslo, Norway, p.33.

ACKNOWLEDGEMENT

This work is supported by Norcell-II managed by Elkem a.s., Oslo, Norway.

Solid Oxide Fuel Cell with a Thin Stabilized Zirconia Film Supported on an Electrode Substrate

KOICHI EGUCHI, TOSHIHIKO SETOGUCHI,
SATOSHI TAMURA and HIROMICHI ARAI

ABSTRACT

Porous electrode support tubes of $La_{0.6}Sr_{0.4}MnO_3$ and Ni-YSZ (yttria-stabilized zirconia) cermet were fabricated by slip casting method. Microstructures of these supports could be controlled by sintering temperature, particle size of powders, and additives. Thin film of YSZ was deposited onto the porous Ni-YSZ cermet supports by slurry coating method. The dense film of YSZ was obtained by repeating the slurry coating and sintering cycles. The H_2-O_2 fuel cell attained 1.05V of open circuit voltage. The fuel cell with $La_{0.7}Ca_{0.35}Cr_{0.95}O_3$ interconnector was fabricated by plasma spraying method. Electrochemical and catalytic activities of fuel electrode have been evaluated as a preliminary study of internal reforming of natural gas. The anodic reaction on Ni-YSZ appears to be limited by the activation step of oxide ion in the electrolyte under H_2-H_2O, CO-CO_2, and CH_4-H_2O atmospheres near the open circuit condition. On the other hand, the kind of fuel strongly affected the polarization conductivity of Pt anode.

INTRODUCTION

The research and development for the solid oxide fuel cells (SOFCs) have been promoted rapidly and extensively in recent years, because of their high efficiency and future potential [1, 2]. The development of components, e.g., component materials, their fabrication techniques, and cell structure, is still a major subject of investigation.

It is important to establish simple and reliable cell design and fabrication technique for the development of SOFCs. Electrochemical vapor deposition (EVD) process is the most successful process for depositing thin and dense YSZ electrolyte film [3]. However, a simpler and cheap fabrication process is required from practical point of view. We have been developing electrode-supported cell structure and fabrication processes for YSZ thin film [4, 5]. Slurry coating and plasma spraying methods are suitable processes for the fabrication of large, thin, and dense film. In this study, tubular SOFC supported on the porous electrode has been developed. Thin films of YSZ were fabricated by slurry coating and plasma spraying processes. Thick film of Ca-doped lanthanum chromite interconnector was deposited on the electrode support

Koichi Eguchi, Toshihiko Setoguchi, Satoshi Tamura, and Hiromichi Arai
Department of Materials Science and Technology, Graduate School of Engineering Sciences,
Kyushu University, 6-1 Kasugakoen, Kasuga-shi, Fukuoka 816, Japan

694

tube by plasma spraying process.

Internal steam reforming of hydrocarbons over an anode material of SOFC is very attractive in improving overall efficiency [6]. In this case, the anode material is employed as a catalyst not only for electrochemical oxidation of fuels but also for steam reforming of hydrocarbons and as a current collector. In addition to these properties, the anode materials are required to possess an appropriate thermal expansion coefficient and to be inactive to stabilized zirconia. Nickel has been most popularly used as an anode material. Electrical conductivity and thermal expansion property of nickel-yttria-stabilized zirconia cermet have been discussed in relation to ratio of nickel by Dees et al. [7]. Kawada et al. [8] have reported the relationship between morphology and polarization property. However, the studies on the activities of anode for electrochemical and steam reforming reactions are still insufficient. In the present study, the anodic polarization conductivity of Ni-YSZ cermet was evaluated at various oxygen partial pressures in different fuel atmospheres. A fuel cell with an internal steam reforming of methane was also examined in relation to power generation characteristics and anodic polarization conductivity.

EXPERIMENTAL

PREPARATION OF POROUS $La_{0.6}Sr_{0.4}MnO_3$ AND Ni-YSZ CERMET SUPPORTS

Tubular electrode supports were prepared by the slip casting method. A perovskite type oxide, $La_{0.6}Sr_{0.4}MnO_3$ (LSM), was used as a cathode in every sample. A powder of LSM was obtained by decomposition of corresponding metal acetates and solid state reaction at $1100°C$ for 1 h. A YSZ powder (average particle size = 45μ m) was mixed with NiO (40wt%). A slip of LSM or NiO-YSZ, which consisted of the above mentioned powder, additives, and distilled water, was cast on a gypsum mold and the dried green compact was sintered at $1400°C$ for 3 h. The size of the experimental piece was generally about 13mm in diameter and 30mm in length. The electrode supports were subjected to the measurements of porosity, N_2 gas permeation coefficient, electrical conductivity, and thermal expansion coefficient. The supports exhibited satisfactory properties as summarized in Table I.

TABLE I   PROPERTIES OF POROUS $La_{0.6}Sr_{0.4}MnO_3$ AND NiO-YSZ CERMET SUPPORT TUBES

	$La_{0.6}Sr_{0.4}MnO_3$	NiO-YSZ(4:6)	YSZ
porosity (%)	29.1	21.5	-
		(33.0 [f])	
N_2 gas permeation	8.25	3.0	-
coefficient ($mm^4g^{-1}sec^{-1}$) [a]		(11.4 [f])	
conductivity ($\times 10^2$ S cm^{-1})	1.65 [b]	3.10 [c]	0.0009 [b]
thermal expansion	11.53	10.0	9.8
coefficient ($\times 10^{-6}$ deg^{-1}) [d]			
mechanical strength (MPa) [e]	26.0	42.0	-
		(27.5 [f])	

a)measured at room temperature, b)measured in O_2 at $1000°C$,
c)measured in H_2 at $1000°C$, d)measured in the temperature range 20-950°C
e)measured by tubular compression fatigue test,
f)measured after reduction in H_2 at $1000°C$

FABRICATION OF YSZ AND LANTHANUM CHROMITE FILMS ON ELEC-TRODE SUPPORTS

A thin film of YSZ was fabricated on the planar Ni-YSZ cermet support by the slurry coating method. We have reported that the slurry coating process is reliable in fabricating dense and thin film of YSZ [5]. The porous Ni-YSZ cermet support tube was dipped into an ethanol suspension of YSZ (6wt%-YSZ) and dried at room temperature. After coating the slurry 10 times, the film was sintered at 1450°C for 1 h. The thin film was sintered at 1450°C for 10 h after 5 times repetition of the coating - sintering cycle. It is difficult to fabricate a YSZ thin film on LSM supports by the slurry coating method because of the chemical reaction between YSZ and LSM during sintering process.

The plasma spraying process (Onoda Cement Co.,Ltd., ASP α -100) was employed for the deposition of YSZ film on the Ni-YSZ cermet support tube. A thick film of $La_{0.7}Ca_{0.35}Cr_{0.95}O_3$ was also fabricated as an interconnector by plasma spraying method. The structure of the fuel cell with the interconnector is shown in figure 1. The interconnector layer is deposited directly onto electrode support tube in the axial direction as a 5 mm wide strip. Then, thin film of YSZ was deposited on the electrode support.

ELECTRICAL MEASUREMENTS

A counter and a reference electrode were attached to the YSZ surface. Electrode materials were mixed with turpentine oil and applied onto the electrolyte.

The SOFC characteristics were measured in a flow system by the previously described method [9]. Gaseous mixtures of H_2-H_2O were supplied to the anode as a fuel, and oxygen or air to the cathode. The oxygen partial pressure, Po_2, at the anode side was controlled between 10^{-9} and 10^{-17} Pa by using H_2-H_2O mixtures. Gaseous hydrogen was humidified by using a water pump (Shimadzu, LC-9A) or by bubbling H_2 through water. Mixtures of CO-CO_2 were also employed as a fuel in the Po_2 range between 10^{-7} and 10^{-11} Pa. A gaseous mixture of CH_4 and H_2O was supplied to the anode to evaluate the internal reforming of CH_4. For the gas tightness between cathode and anode compartments, the cell element was attached to a mullite tube with a molten glass ring. The polarization resistance of cathode and anode were measured by complex impedance analysis.

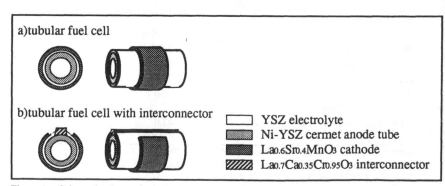

Figure 1   Schematic views of tubular fuel cells with porous electrode supports.

RESULTS AND DISCUSSION

The efficiency of SOFC largely depends upon its internal resistance, which consists of ohmic and non-ohmic components. The ohmic component is attributed to electrical resistances of the electrolyte, electrode, and interconnector, and non-ohmic to the overvoltages of cathode and anode/electrolyte interfaces. To improve the efficiency, it is necessary to reduce the internal resistance of the fuel cell. In the present study, fabrication of thin film of YSZ electrolyte, electrode support, and interconnector was evaluated to reduce the ohmic losses, and anode/electrolyte interface to reduce the overvoltage.

The present cell was fabricated on the electrode supports by eliminating any additional support (figure 1). This structure is helpful in reducing the ohmic loss, because of the thin YSZ layer and thick electrode. LSM is preferably thick as a current collector of tubular cell, because of its low conductivity [5]. On the other hand, thick Ni-YSZ layer promotes internal steam reforming and also permits a high temperature fabrication process.

FABRICATION OF YSZ THIN FILMS BY SLURRY COATING METHOD

The slurry-coated film on Ni-YSZ supports contained a large number of cracks after initial coating - sintering cycle, because of the difference in shrinkage between the YSZ film and the support. The cracks could be eliminated by repeating the cycle. Resulting semitransparent film was dense and free from cracks as can be observed in SEM images (figure 2). Figure 3 shows the current-voltage (I-V) characteristics for the fuel cell consisting of the slurry-coated YSZ film, Ni-YSZ anode support tube, and slurry-coated LSM cathode. The slurry-coated YSZ film was dense enough to attain 1.05V of open circuit voltage, which agreed with the theoretical value estimated from the partial pressures of H_2, O_2, and H_2O. This fuel cell exhibited 0.53 W cm^{-2} of power at 0.7V of terminal voltage.

In the present study, complex impedance analysis was used for the measurement of polarization resistance at the three phase boundary of gas phase/electrode/electrolyte. The complex impedance plots for the fuel cell with slurry-coated YSZ are shown

Figure 2   SEM image of the YSZ film fabricated by slurry coating method.

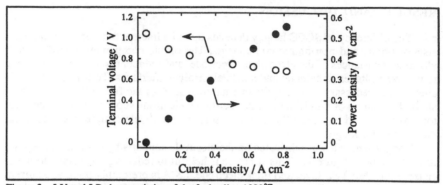

Figure 3   I-V and I-P characteristics of the fuel cell at 1000℃.
H_2+H_2O(Po_2=2.38x10^{-13} Pa), Ni-YSZ cermet support / YSZ / La$_{0.6}$Sr$_{0.4}$MnO$_3$, O$_2$ (Po$_2$'=0.21x10^5 Pa)

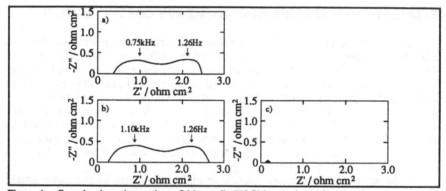

Figure 4   Complex impedance plots of (a)overall, (b)LSM cathode - reference electrode, and (c)Ni-
YSZ support - reference electrode at 1000℃.
H_2+H_2O (Po$_2$=2.38x10^{-13} Pa), Ni-YSZ cermet support / YSZ / La$_{0.6}$Sr$_{0.4}$MnO$_3$, O$_2$ (Po$_2$'=0.21x10^5 Pa)

in figure 4. It was noted that the polarization resistance between cathode and reference electrode (figure 4b)   was always almost equal to overall polarization resistance
(figure 4a). In the present cell with highly conductive Ni-YSZ anode and very thin
YSZ film, the electrical resistance through route 1 is much smaller than that through
route 2 (figure 5). Thus, the polarization resistances of cathode and anode could not
be separated by the reference electrode located on the outer surface of the YSZ film.

FABRICATION OF FUEL CELL WITH INTERCONNECTOR BY PLASMA
SPRAYING METHOD

Plasma spraying method was used to fabricate YSZ and La$_{0.7}$Ca$_{0.35}$Cr$_{0.95}$O$_3$
(LCC) films on Ni-YSZ cermet support. SEM images of YSZ film fabricated by this
process are shown in figure 6. Porous Ni-YSZ support has enough mechanical
strength to endure the thermal shock during plasma spraying process. The plasma
sprayed YSZ layer contained microcracks at the initial stage, but the film quality has

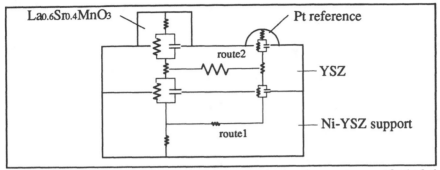

Figure 5 Schematic view of equivalent circuit during complex impedance measurement for the fuel cell.

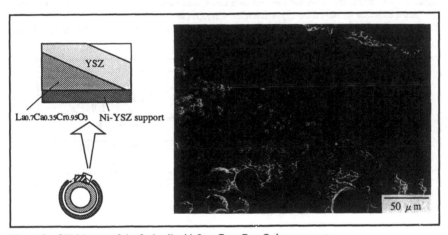

Figure 6 SEM image of the fuel cell with $La_{0.7}Ca_{0.35}Cr_{0.95}O_3$ interconnector.

been significantly improved after modification of spraying process [10].

The I-V characteristics of the fuel cells with plasma sprayed YSZ electrolyte and LCC interconnector are shown in figure 7. Small number of pores were still contained in the YSZ and/or LCC films (figure 6) and the open circuit voltage of the fuel cell was slightly less than theoretical one (1.05V). One current collector was attached to LCC interconnector and the other one directly to Ni-YSZ tube (figure 1). The slope of I-V curve measured through the current collector on LCC was steeper than that on Ni-YSZ. The polarization resistance estimated from the two intercepts of the impedance plots almost agreed with each other as shown in figure 8. However, the ohmic resistance of the cell with LCC current collector was higher than that with Ni-YSZ one. The ohmic loss may include the contact resistance between Ni-YSZ and LCC. The fuel cell with LCC interconnector attained 0.97V of open circuit voltage and 0.22 W cm^{-2} at 0.7V of terminal voltage. It is necessary to reduce the contact resistance between Ni-YSZ and LCC.

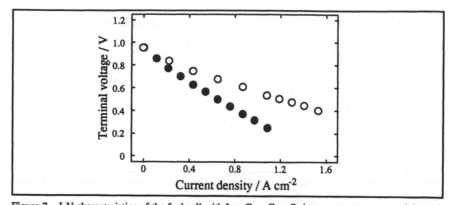

Figure 7   I-V characteristics of the fuel cell with La$_{0.7}$Ca$_{0.35}$Cr$_{0.95}$O$_3$ interconnector measured through the current collector on Ni-YSZ (◯) or La$_{0.7}$Ca$_{0.35}$Cr$_{0.95}$O$_3$ (●) at 1000℃.
H$_2$+H$_2$O (P$_{O2}$=2.38x10^{-13} Pa), Ni-YSZ cermet support / YSZ / La$_{0.6}$Sr$_{0.4}$MnO$_3$, O$_2$ (P$_{O2}$'=0.21x10^5 Pa)

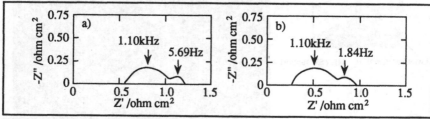

Figure 8   Complex impedance plots of the fuel cell with La$_{0.7}$Ca$_{0.35}$Cr$_{0.95}$O$_3$ interconnector measured through the current collector on a)La$_{0.7}$Ca$_{0.35}$Cr$_{0.95}$O$_3$ or b)Ni-YSZ at 1000℃.
H$_2$+H$_2$O (P$_{O2}$=2.38x10^{-13} Pa), Ni-YSZ cermet support / YSZ / La$_{0.6}$Sr$_{0.4}$MnO$_3$, O$_2$ (P$_{O2}$'=0.21x10^5 Pa)

SOLID OXIDE FUEL CELL WITH INTERNAL REFORMING REACTION OF METHANE

The internal reforming reaction of natural gas is very attractive due to the high overall efficiency resulting from effective use of heat released by ohmic and non-ohmic losses and the TΔS term in the fuel cells. The steam reforming reaction of methane:

$$CH_4 + H_2O \rightleftharpoons 3H_2 + CO \qquad (\Delta H = 206 \text{ kJ mol}^{-1}) \tag{1}$$

accompanies shift reaction:

$$CO + H_2O \rightleftharpoons H_2 + CO_2 \qquad (\Delta H = -41 \text{ kJ mol}^{-1}) \tag{2}$$

and Boudart reaction:

$$2CO \rightleftharpoons C + CO_2 \qquad (\Delta H = -172 \text{ kJ mol}^{-1}). \tag{3}$$

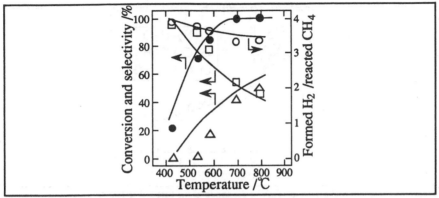

Figure 9   Steam reforming reaction of CH_4 over Ni-YSZ cermet (S/C=5, SV=7000h^{-1}).
Solid lines indicate the equilibrium values.
● CH_4 conversion, △ CO selectivity, □ CO_2 selectivity, ○ formed H_2/reacted CH_4

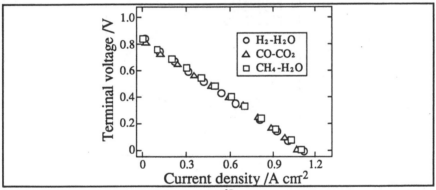

Fig.10   I-V characteristics for the fuel cell at 1000℃.
fuel, Ni-YSZ / YSZ / La$_{0.6}$Sr$_{0.4}$MnO$_3$, O_2 (Po$_2'$=1.01x10^5 Pa)

Catalytic activity of Ni-YSZ cermet for the steam reforming reaction was evaluated in a conventional flow reactor as a preliminary experiment of internal reforming type SOFC. The carbon deposition region in the C-H-O ternary diagram has been reported from equilibrium calculations [11]. From the C-H-O diagram, steam/carbon, S/C, ratios were set within the carbon-free regions, S/C=5. Methane conversion and CO and CO_2 selectivities are defined as follows.

Conversion of CH_4 = reacted CH_4 / supplied CH_4 (4)

CO selectivity = formed CO / (formed CO + CO_2)   (5)

CO_2 selectivity = formed CO_2 / (formed CO + CO_2)   (6)

These parameter for Ni-YSZ cermet agreed with equilibrium values (figure 9) calculated from equations 1 and 2 [12]. Thus, the Ni-YSZ cermet has high activity for the steam reforming reaction of CH_4 above 600℃.

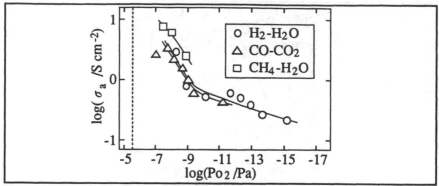

Figure 11   Plots of log(σ_1) vs. log(Po₂) for the anode of the fuel cell at 1000℃.
fuel (Po₂), Ni-YSZ / YSZ / La₀.₆Sr₀.₄MnO₃, O₂ (Po₂'=1.01x10⁵ Pa)

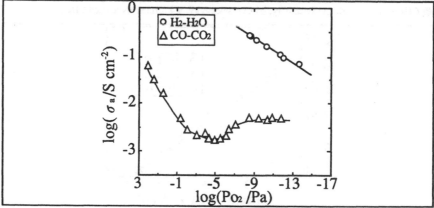

Fig.12   Plots of log(σ_1) vs. log(Po₂) for the anode of the fuel cell at 1000℃.
fuel (Po₂), Pt / YSZ / La₀.₆Sr₀.₄MnO₃, O₂ (Po₂'=1.01x10⁵ Pa)

The fuel cell experiments were carried out under CH_4-H_2O atmospheres at S/C = 3, 6, and 9. The I-V characteristics of the fuel cell at S/C=6 are shown in figure 10. The I-V curve in CH_4-H_2O agreed with those in H_2-H_2O and CO-CO_2 at the same Po₂. Figure 11 summarizes the anodic polarization conductivities of Ni-YSZ under H_2-H_2O, CO-CO_2, and CH_4-H_2O systems. The polarization conductivity is the inverse of polarization resistance, Ra, obtained from the impedance plots: σ_1=1/(Ra A). The values of σ_1 and their Po₂ dependence almost agreed with each other irrespective of the kind of fuel. These results suggests that the overall anodic reaction is limited by the activation of oxide ion. The electrochemical oxidation of H_2 and CO was not affected by the preceding steam reforming step on Ni at the present condition .

The polarization conductivity for Pt anode exhibited contrasting dependence on the kind of fuel (figure 12). The σ_1 values were significantly lower for CO-CO_2 system than for H_2-H_2O system. The activation process of fuel appears to strongly influence the rate of anodic reaction in this case.

CONCLUSIONS

Thin and dense film of YSZ could be fabricated onto the porous electrode support tube by slurry coating method. The fuel cell attained 1.05V of open circuit voltage and 0.53 W cm^{-2} at 0.7V of terminal voltage. The open circuit voltage of the fuel cell with the plasma sprayed $La_{0.7}Ca_{0.35}Cr_{0.95}O_3$ interconnector was 0.95V. The anodic reaction on Ni is not affected by the kind of fuel and expected to be limited by the activation of oxygen ion. On the other hand, anodic reaction on Pt is significantly deteriorated by CO-CO_2 mixture as compared with H_2-H_2O.

REFERENCES

1. Singhal, S. C., Nov. 19-21, 1990. "Solid Oxide Fuel Cells for Clean and Efficient Power Generation." Proc. Fukuoka Int. Sympo. '90 -Global Environment and Energy Issues-, T. Seiyama, ed., Fukuoka, Japan: pp.95-104.

2. Umemura, F., H. Ota, K. Amano, S. Kaneko, T. Gengo, S. Uchida, Nov. 13-14, 1989. "Development of Solid Oxide Fuel Cell." Proc. Int. Sympo. Solid Oxide Fuel Cell, Nagoya, Japan: pp.15-20.

3. Singhal, S. C., July 2-5, 1991. "Solid Oxide Fuel Cell Development at Westinghouse." Proc. 2nd Int. Sympo. SOFCs, F. Grosz, P. Zeager, S.C. Singhal, and O. Yamamoto, eds., Athens, Greece: pp.25-33.

4. Arai, H., Nov. 13-14, 1989. "Solid Oxide Fuel Cells with Stabilized Zirconia Thick Films Fabricated by Various Techniques." Proc. Int. Sympo. Solid Oxide Fuel Cell, Nagoya, Japan: pp.9-14.

5. Arai, H., K.Eguchi, T. Setoguchi, R. Yamaguchi, K. Hashimoto and H. Yoshimura, July 2-5, 1991. "Fabrication of YSZ Thin Film on Electrode Substrate and Its SOFC Characteristics." Proc. 2nd Int. Sympo. SOFCs, F. Grosz, P. Zeager, S.C. Singhal, and O. Yamamoto, eds., Athens, Greece: pp.167-174.

6. Takehara, Z., K. Kanamura and S. Yoshioka, 1989. "Thermal Energy Generated by Entropy Change in Solid Oxide Fuel Cell." J. Electrochem. Soc., 136: 2506-2511.

7. Dees, D. W., T. D. Claar, T. E. Easler, D. C. Fee, F. C. Mrazek, 1987. "Conductivity of Porous Ni/ZrO_2-Y_2O_3 cermet." J. Electrochem. Soc., 134: 2141-2146.

8. Kawada, T., N. Sakai, H. Yokokawa, M. Dokiya, M. Mori and T. Iwata, 1990. "Characteristics of Slurry-Coated Nickel Zirconia Cermet Anodes for Solid Oxide Fuel Cells." J. Electrochem. Soc., 137: 3042-3047.

9. Inoue, T., T.Setoguchi, K. Eguchi and H. Arai, 1989. "Study of A Solid Oxide Fuel Cell with A Ceria-Based Solid Electrolyte." Solid State Ionics, 35: 285-291.

10. Fukami, S., M.Kito, A. Bunya, H. Saito, T. Ito, Y. Kaga, Y. Ohno, K. Eguchi and H. Arai, July 2-5, 1991. "Preparation of electrolyte film by plasma spraying." Proc. 2nd Int. Sympo. SOFCs, F. Grosz, P. Zeager, S.C. Singhal, and O. Yamamoto, eds.,

Athens, Greece: pp.205-214.

11. Lee, A. L., 1987. Final Report of "Internal Reforming Development for Solid Oxide Fuel Cells." DOE/MC/22045-2364.

12. Machida, M., T. Teshima, K. Eguchi and H. Arai, 1991. "High Temperature Steam Reforming of Hydrocarbons over Nickel/Hexaaluminate Catalysts." Chem. Lett., 1991: 231-239.

Development of the Mitsubishi Planar Reversible Cell

FUSAYUKI NANJO, KOUICHI TAKENOBU,
KIYOSHI WATANABE, HITOSHI MIYAMOTO, MASAO SUMI,
SETSUO TOKUNAGA and IKUMASA KOSHIRO

ABSTRACT

We have proposed an energy storage system using SOFC-SOE (Solid Oxide Fuel Cells and Solid Oxide Electrolysis Cells) reversible cells, where the cells work alternately as fuel cells and as steam electrolysis cells. This paper describes the basic test facility and its results. Test cells were 23 mm in diameter and 200μm in thickness. Solid oxide electrolyte, 8 mol% yttria stabilized zirconia (YSZ), was manufactured by a tape casting method. A Ni/YSZ-cermet electrode and a Sr-doped $LaMnO_3$ (LSM) electrode were both screen printed and sintered at elevated temperatures. A single cell was put in an electric furnace and the temperature was maintained at 1000°C. Hydrogen gas and air were provided for fuel cell tests. Conversely, steam and air were provided for the electrolysis cell tests. For both conditions, polarization measurements were carried out using the current interruption method. Effects of electrode compositions on overpotentials and solid electrolyte degradation were also evaluated.

INTRODUCTION

One of Japan's policies calls for an increase in the number of nuclear power plants. As nuclear power generation increases, the surplus power must be stored at night and supplied during the day.

We propose a hydrogen-utilizing electric power storage system [1]. The hydrogen-utilizing electric power storage system consists of SOE/SOFC reversible cells. A unit cell works as an electrolysis cell during periods of low electric power demand and works as a fuel cell during periods of high power demand. The cells work at 1000°C.

Pioneering work on the reversibility of SOE/SOFC was presented by E. Erdle, et al. for a tubular cell [2] and a planar cell [3].

We started with basic tests on the cells [4], because the cells are key component in the proposed system. The purposes of our study were as follows ;

Fusayuki Nanjo, Kouichi Takenobu, and Kiyoshi Watanabe,
Mitsubishi Heavy Industries, Ltd. Kobe Shipyard & Machinery Works,
1-1, Wadasaki-Cho 1-Chome, Hyogo-Ku, Kobe 652 Japan
Hitoshi Miyamoto, Masao Sumi, Setsuo Tokunaga, and Ikumasa Koshiro,
Mitsubishi Heavy Industries, Ltd. Takasago Research and Development Center
2-1-1, Shinhama, Arai-Cho, Takasago, Hyogo Pref. 676 Japan

705

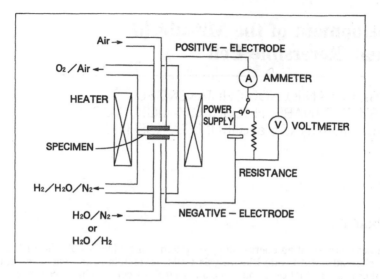

Figure 1. Flow Sheet of Test Facility

(1) Development of materials for the reversible planar cells
(2) To improve cell performance
(3) To identify and to solve the problems of reversible operation.

TEST FACILITY

Figure 1 illustrates the test facility. The test facility consists of an electrical heater which keeps the cell at 1000°C, an air supply system, a hydrogen supply system, a steam supply system, a measurement system for electrochemical cell performance and an exhaust system which can perform gas analysis. The measurement system has a constant voltage-current power supply and a current interruption measurement system. One feature of the test facility is its steam supply system which consists of a mist separator, kept at a desired temperature, and a steam generator. The steam supply system can supply steam at any concentration by controlling the temperature of the mist separator.

STRUCTURE OF THE CELL

We manufactured a planar type single cell for the basic tests. Figure 2 shows the cross-section of a cell with Ni/YSZ-cermet electrode and $La_{0.9}Sr_{0.1}MnO_3$ electrode. Test cells were 23 mm in diameter and 200μm in thickness. Solid oxide electrolyte was 8 mol % yttria stabilized zirconia and was manufactured by a tape casting method. A Ni/YSZ-cermet electrode and a LSM electrode were both screen-printed and sintered at elevated temperatures. Cells with platinum electrodes were also manufactured by screen-printing of the Pt-slurry for comparison.

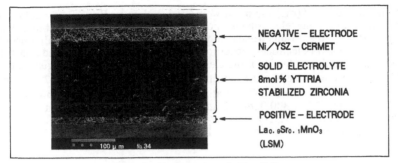

Figure 2.   Cross-section of a Test Cell

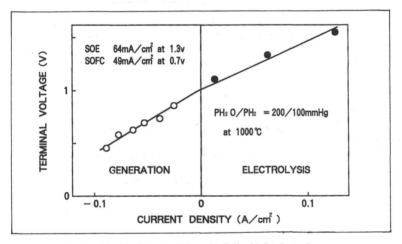

Figure 3.   V/I Characteristics of a Cell with Pt-electrodes

RESULTS

CELL PERFORMANCE

The terms cathode and anode are not used to avoid confusion.   The negative-electrode is the cathode during SOE operation and the anode during SOFC operation.   The positive-electrode is vice versa.   The test cell was put in the electric furnace and the temperature was raised up to 1000°C.   Hydrogen gas and/or steam was directed into the negative-electrode and air was directed into the positive-electrode.   Then the generation test and the electrolysis test were performed.   The characteristics of a cell during generation were measured under constant current load.   While the cell temperature was maintained at 1000°C, the characteristics of the cell during electrolysis were measured under constant current supply.   As can be seen in Figure 3, the cell performance with Pt-electrodes is 49 mA/cm^2 at 0.7 V under fuel cell conditions in the SOFC operating mode at 1000°C, and 64 mA/cm^2 at 1.3 V under P_{H2O}/P_{H2} = 200/100 mmHg and air in the electrolysis mode at 1000°C.   Figure 5 shows the overvoltages of the Pt-electrodes and the Ni, LSM-electrodes cell during electrolysis.   The Pt-electrodes (specifically the negative-electrode) had a larger overvoltage than that of the Ni, LSM-electrodes.   As can be seen in Figure 4, the cell performance with the Ni, LSM-electrodes cell is 680 mA/cm^2 at 0.7 V under fuel cell conditions in the SOFC operating mode at 1000°C and 940 mA/cm^2 at 1.3 V under P_{H2O}/P_{H2} = 300/0 mmHg and air in the electrolysis mode at 1000°C. The overvoltages of the Ni, LSM electrodes, which are shown Figure 5, were far lower than that of the Pt-electrodes.   The current efficiency was 97 % during electrolysis.

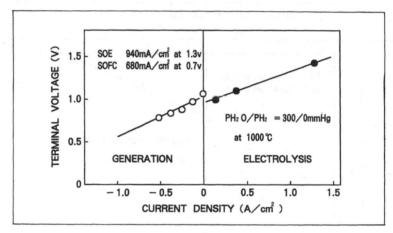

Figure 4. V/I Characteristics of a Cell with Ni/YSZ- and LSM-electrodes

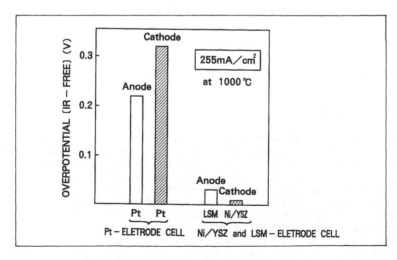

Figure 5. Overpotentials of Electrode in Electrolysis

DEGRADATION OF THE ELECTROLYTE DURING ELECTROLYSIS

Degradation of the electrolyte which destroys the cells during electrolysis under certain conditions was observed.

The signs of the phenomenon include (1) cracking of the solid electrolyte through the grain boundaries, (2) extension of cracks from the negative-electrode and (3) discovery of cracks when the cathode potential (IR free) which is referenced to air is lower than approximately -1.4 V vs. air.

Figure 6 shows a cross-section of the degraded and non-degraded solid electrolytes. Intergranular cracks were observed on the cross-section of the degraded cell with a scanning electron microscope. On the other hand, the non-degraded solid electrolytes crack transgranularly if they are broken artificially. As can be seen in the Figure 7, only cubic phase was detected in the degraded cells by X-ray diffraction measurement. Intergranular cracks were observed on all the current loaded sections of the cell which performed electrolysis at potential

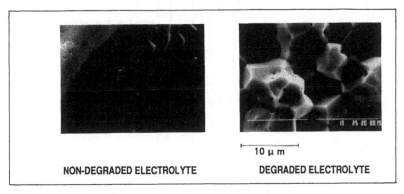

NON-DEGRADED ELECTROLYTE   **DEGRADED ELECTROLYTE**

Figure 6.   Cross-section of a Degraded Cell

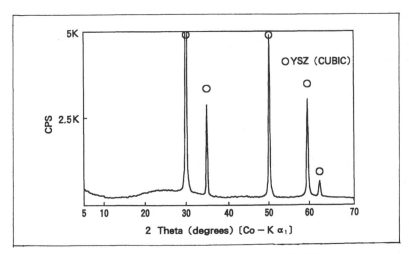

Figure 7.   X-ray diffraction Pattern of a Degraded Cell
(This shows all peaks due to cubic crystal)

lower than-1.4 V vs. air (IR free) for a long time.   Intergranular cracks were observed on the partial cross-section of the cell which performed electrolysis for a short time and the intergranular cracks were initiated from the negative electrode side.

The cell was tested additionally under the same electrolysis conditions except that there was no power supply.   Transgranular cracks were observed in the cell, the same as in a non-degraded cell when it is ruptured by force.   This solid electrolyte degradation occurs during electrolysis.

Figure 8 is a conceptual figure of the test facility which was used to study the relation. between potential at the negative-electrode and degradation of the solid

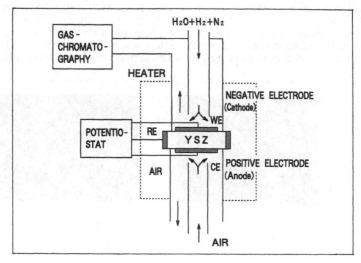

Figure 8.   Test Facility for Degradation Tests

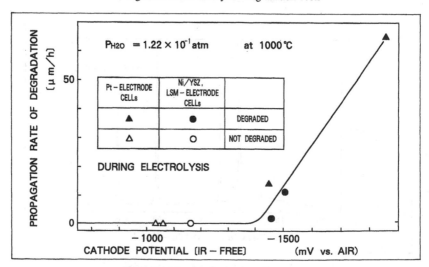

Figure 9.   Degradation vs. Electrode Potential

TABLE 1 TARGETS AND RESULTS

	TARGET	RESULT
POWER REQUIRED FOR ELECTROLYSIS CELLs	$0.59W/cm^2$ ($0.45A/cm^2$ at 1.32V)	$0.5W/cm^2$ ($0.45A/cm^2$ at 1.12V)
POWER GENERATED IN FUEL CELLs	$0.26W/cm^2$ ($0.35A/cm^2$ at 0.74V)	$0.3W/cm^2$ ($0.35A/cm^2$ at 0.85V)

electrolyte.   Figure 9 shows the results.   The rate of degradation was obtained from the depth of the intergranular crack from the negative-electrode side, divided by the electrolysis time.   The degradation occurs when cathode potential (IR free) is lower than -1.4 V vs. air.   This potential can not cause decomposition of solid electrolytes because the decomposition potential of YSZ is lower than that value. So this phenomenon was not caused by the decomposition of the solid electroyte.

DISCUSSION

We manufactured cells which consisted of 8 mol % yttria stabilized zirconia, a negative-electrode of Ni/YSZ cermet and a positive-electrode of Sr-doped $LaMnO_3$. These cells gave far better performance than Pt-electrode cells.

To meet practical system requirements, the power density of the fuel cells during generation should be more than 0.3 W/cm^2 (0.4 A/cm^2 at 0.74 V) and required electrolysis power density should be less than 0.59 W/cm^2 (0.45 A/cm^2 at 1.32 V).   These requirements were obtained through our system analysis[1]. As can be seen in Table 1, the experimental cell performance with Ni, LSM-electrodes is comparable to the target values.

The degradation of the solid electrolyte during electrolysis occurs under conditions when the potential of the negative-electrode (IR free) is lower than -1.4 V vs. air.   A higher potential enables better cell performance and is very important for the development of an energy storage system using SOFC/SOE reversible cells.

REFERENCES

1. Shimaki, R., M.Okamoto, C.Yanagi, Y.Kikuoka, S.Ueda, N.Nakamori, K.Kugimiya, M.Yoshino, M.Tokura and S.Suda, June 1992."Feasibility Study on Hydrogen-utilized Electric Power Storage Systems."Proceedings of the 9th World Hydrogen Energy Conference, Paris, France, pp. 1927-1935.

2.   Erdle, E., W.Dönitz, R.Schamm and A.Koch, July 1990."Reversibility and Polarization Behaviour of High Temperature Solid Oxide Electrochemical Cells." Proceedings of the 8th World Hydrogen Energy Conference, Hawaii, USA, pp. 415-422.

3.   Erdle, E., W.Dönitz, H.G.MüllerandR.Schmidberger, February 1992."Some Aspects of Planar SOFC Development."Proceedings of the International Fuel Cell Conference, Makuhari, Japan, pp. 361-364.

4.   Kusunoki, A., H.Matsubara, Y.Kikuoka, C.Yanagi, K.Kugimiya, M.Yoshino, M.Tokura, K.Watanabe, S.Ueda, M.Sumi, H.Miyamoto and S.Tokunaga, June 1992. "Development of Planar SOE/SOFC Reversible Cell - Fundamental Test on Hydrogen-utilized Electric Power Storage System."Procedings of the 9th World Hydrogen Energy Conference, Paris, France, pp. 1415-1417.

SOFC Design Requirements for Planar Zirconia Electrolyte Components

B. C. H. STEELE

ABSTRACT

The electrical requirements for self-supported and supported zirconia based ceramic electrolytes incorporated into planar SOFC configurations are briefly surveyed. Factors influencing the performance of porous $La(Sr)MnO_3$ cathodes and $Ni-ZrO_2$ anodes are considered to ensure that power densities of $0.2W/cm^2$ ($0.7V$ at $300mA/cm^2$) can be reliably developed at operating temperatures of 950^oC. The mechanical properties of self-supported zirconia components are next discussed and the need for high temperature mechanical property data emphasised so that life-time predictions can be made.

INTRODUCTION

Three main design configurations are being developed for SOFC stacks, namely: tubular, planar and monolithic. The Westinghouse tubular system is the most advanced with a 25 KW unit recently commissioned by Osaka Gas. The tubular design is discussed in detail in a separate paper in these proceedings and except for minor comments will not feature in the present survey. The co-fired monolithic SOFC structure pioneered by the Argonne National Laboratory has encountered problems in scale-up, and as the specification for the zirconia based electrolyte is closely identified with the precise processing conditions for the monolithic structure it is not appropriate to consider the zirconia electrolyte as a separate component. Accordingly emphasis will be given to the requirements for zirconia electrolyte components incorporated into planar configurations.

B.C.H. Steele, Centre for Technical Ceramics, Dept. of Materials, Imperial College, London, SW7 2BP. U.K.

713

DESIGN CONSIDERATIONS

GENERAL

The configuration/fabrication/performance considerations are iterative with many interacting parameters. For example high power density has the benefits of lower total active cell area but requires higher temperatures, and heat dissipation/thermal management becomes more difficult. At present typical power densities of $0.2W/cm^2$ ($0.7V$ at 300 ma/cm^2) are being attained in small planar SOFC stacks and this performance target will provide the basis for subsequent discussions.

POLARISATION LOSSES

Typical polarisation losses for the tubular supported electrolyte configuration, planar self-supported electrolyte and planar supported electrolyte configuration are summarised in Figures 1a, 1b, 1c for current densities of $300mA/cm^2$. It should be emphasised that these data are for single cell arrangements and do not include contact resistances between the PEN (positive-electrolyte-negative) structure and bi-polar plate or interconnect.

The principal polarisation losses for the tubular system are associated with the cathode due to the relatively long current path, and Westinghouse are developing alternative designs to reduce these losses.

For the planar configuration operation at 950°C or higher, examination of Figure 1b indicates that the total polarisation losses are less than $250mV$ for electrolyte thicknesses upto $200\mu m$. For SOFC stacks incorporating La(Ca)CrO$_3$ bi-polar plates operation at temperatures in excess of 950°C is desirable to optimise the power density and so reduce the total area required. However for metallic bi-polar plates it is desirable to reduce the operating temperature to minimise the rate of oxidation. The desired operating temperature will obviously be a compromise involving electrode kinetics, ohmic losses, etc. but a temperature of around 800°C appears a reasonable target. To attain this intermediate temperature it is obviously necessary to reduce the thickness of the zirconia electrolyte and data are provided for $30\mu m$ thick films. These thick films must be supported and the cathode (or anode) becomes the structural component. The data shown in Figure 1c were obtained with a dense impermeable mixed ionic/electronic cathode onto which a $20\mu m$ film of $Zr(Y)O_{2-x}$ had been placed by pulsed laser deposition (PLD).

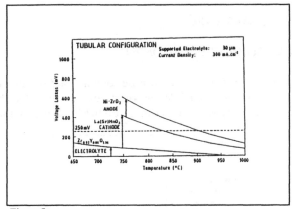

Figure Ia

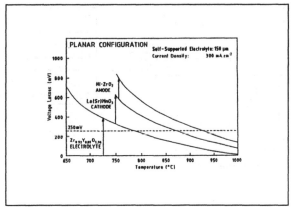

Figure Ib.

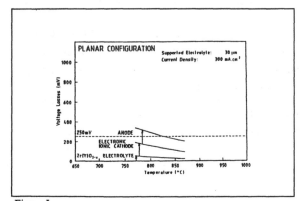

Figure Ic

Figures Ia, Ib, Ic.   Depict polarisation losses as a function of temperature for the tubular, planar
(self-supported), planar (supported) configurations, respectively

715

ELECTRICAL BEHAVIOUR

ZIRCONIA ELECTROLYTE COMPONENT

Assuming that a 50mV ohmic loss can be tolerated across the electrolyte then the relationship between electrolyte thickness and current density for various specific ionic conductivity values is shown in Figure 2. For a minimum thickness

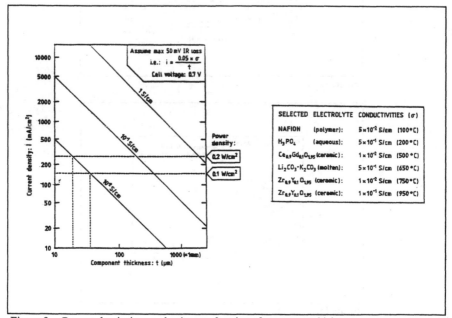

Figure 2   Current density/power density as a function of component thickness

of 167μm and a current density of 300mA/cm^2 then the specific conductivity value must exceed 10^{-1}S/cm. There is general agreement [1,2,3] that at 1000°C specific conductivity values of 1.5×10^{-1}S/cm (10^{-1}S.cm^{-1} at 950°C) can be obtained with a number of commercially available powders. These electrolyte components have now been successfully scaled up to 150x150x0.15mm plates using either tape-casting or calendering process routes.

Although the resistivities of partially stabilized zirconia (PSZ) electrolytes are approximately three times the value for fully stabilised (cubic) zirconia electrolytes certain groups favour the use of PSZ material because of its better mechanical behaviour.

CATHODE RESISTIVITIES

A variety of cathode compositions have been examined but the important parameters influencing the electrode resistances appear to be the powder characteristics and processing variables. For example during the scale up of electrolyte plates to 15x15cm size it was observed that electrode resistivity values were increasing to unacceptable values as shown in Table I. Eventually the degradation in the cathode performance was traced to the fact that alumina-silica fibre boards were being placed on top of the zirconia electrolyte sheets during the sintering stage to ensure that the sheets remained flat. Small amounts of SiO_2 were being transferred to the electrolyte sheets and incorporated into the surface

TABLE I CATHODE RESISTANCE ; INFLUENCE OF PROCESSING

Sample Description	Electrode Resistance at 900ºC ($\Omega.cm^2$) in air.
Pellet Pressed at Imperial College (HSY8)	3*
HSY8 Sheet : 96% Al_2O_3 Boards	35*
HSY8 Sheet : 96% Al_2O_3 (Etched)	10*
HSY8 Sheet : ZrO_2 Boards	10*
HSY8 Sheet : 99% Al_2O_3 Boards	3*
TZ3Y Sheet : 99% Al_2O_3 Boards	3*
HSY8 Sheet : 99% Al_2O_3 Boards	1.5+

Sheets fabricated at ICI (viscous processing/calendering)
*Electrodes : Symmetrical porous Pt, painted and sintered.
+Electrode : Symmetrical porous $La_{0.5}Sr_{0.5}MnO_3$ (tape cast)

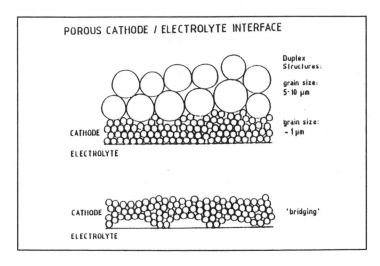

Figure 3   Schematic cathode microstructures

as indicated by SIMS and ESCA examination. Replacement of the Al_2O_3-SiO_2 fibre boards by 'Saffil' Al_2O_3 fibre boards eliminated the problem.

As the Faradaic reaction,

$$\tfrac{1}{2}O_2 + 2e' + V_o^{\cdot\cdot} \rightarrow O_o^x \tag{1}$$

must occur predominately at the restricted three phase interfacial region due to low values of ionic conduction in $La(Sr)MnO_3$ [4] it is also necessary to ensure that this region is optimised. Processing procedures should therefore avoid the formation of 'bridging' microstructures (see Figure 3) which restricts the proportion of three phase region.

ANODE RESISTIVITIES

It is known [5,6,7,8] that the performance of the Ni-ZrO_2 anode is a function of the Ni/ZrO_2 ratio, relative particle sizes of the two components, distribution of the Ni particles, and procedures adopted to reduce the original NiO. In moist hydrogen anode resistances as low as $0.1\Omega\,cm^2$ have been reported but little data are available for CH_4/H_2O fuel gases. The long term stability of the anode structures needs more attention and it is noteworthy that the excellent performance reported for the Westinghouse anode structure is probably due to the fabrication route adopted by this company.

One feature of interest we have observed at Imperial College is that zirconia electrolyte sheets coated with symmetrical anodes can sometimes exhibit relatively high electrode resistances when determined by impedance spectroscopy (see Table II) but the apparent electrode resistivities decrease substantially ($\approx 1\Omega\,cm^2$) when current is passed through the cell. The anomalously resistivities may be due to the presence of adsorbed species which are removed by oxygen ion fluxes.

THICK FILM SUPPORTED ZIRCONIA ELECTROLYTE

The various techniques available to produce thick zirconia based electrolyte films have been summarised by Van Dieten [9]. Undoubtedly the most successful has been the EVD route pioneered by Westinghouse. From published data (e.g. [10]) it appears the EVD route produces material with ionic conductivity values somewhat lower (e.g. $0.8 \times 10^{-1} S.cm^{-1}$ compared to $1.5 \times 10^{-1} S.cm^{-1}$ at 1000^oC) than the sintered sheets. It is not clear at present whether the lower ionic conductivity values are due to the increased fraction of grain boundaries or due to the incorporation of small quantities of chloride ions which could reduce the ionic conductivity according to the following mechanism,

$$Cl_2(g) + V_o^{\cdot\cdot} \rightarrow 2Cl_o^{\cdot} \tag{2}$$

TABLE II. ANODE RESISTANCES

Anode	Resistance ($\Omega.cm^2$) at 900°C
Ni/ZrO$_2$ (Imp. Coll.) HSY8	43
Ni/ZrO$_2$ (ICI) HSY8	25
Pt (Imp. Coll.) TZ8Y	36
Pt (Imp. Coll.) HSY8	40
Pt (Imp. Coll.) HSY8	3 (air)

Anode   : Symmetrical Cells of Ni-ZrO$_2$
Anode Gases : 10% H$_2$ + 3% N$_2$ + 87% N$_2$

MECHANICAL BEHAVIOUR

TUBULAR CONFIGURATION

The absence of seals in the Westinghouse tubular configuration reduces mechanical stresses and the fact that the methane/steam fuel mixtures can be distributed along the length of the tube also ensures that thermo-mechanical stresses due to the endothermic reforming reaction are minimised. The porous zirconia support tube is the principal structural component in the Westinghouse design and published information [11] suggests that even though the mechanical properties of this component are relatively poor, viz.

fracture strength at room temperature   :   45 MPa

fracture strength at 1000°C   :   30 MPa

they are nevertheless sufficient for the small stresses experienced by this component.

PLANAR CONFIGURATION

The presence of seals and localisation of the endothermic steam/methane reforming reaction mean that the planar zirconia electrolyte components have to withstand higher mechanical stresses than is the case for the tubular design. The actual design requirements for fracture strengths and the operating temperature depend upon the configuration but at present most groups assume strengths in excess of 300 MPa and at Imperial College our initial target for cubic zirconia electrolyte sheets was 500 MPa. The room temperature mechanical properties of zirconia electrolyte foils prepared by ICI (viscous processing : calendering route) and by Siemens (tape casting) have been measured at Imperial College using 3-point bend and ball-on-ring methods. A 6mm span was used for the 3-point tests

TABLE III   MECHANICAL PROPERTIES OF ZIRCONIA ELECTROLYTE SHEETS

Material	Mean Strength (MPa)[+]	Weibull Modulus
HSY8	368	6.2 (WP)
TZ8Y (T)	330 (BR)	5.7 (WP)
Tioxide 8	166	2.8
Z–Tech 8	305	3.5
Unitech 8	189	6.0
Unitech 5 (T)	1020 (BR)	13
TZ3Y	803	3.1
TZ3Y	160 (BR)	8.0
TZ3Y (T)	960 (BR)	6.0 (WP)

[+]3-point measurements with 6mm test span unless otherwise stated (15-25 samples)

BR : Ball on Ring

WP : Weibull Plots

Sheets prepared by ICI (viscous processing/calendering) except for samples designated T (tape-cast) which were fabricated and tested at Siemens.

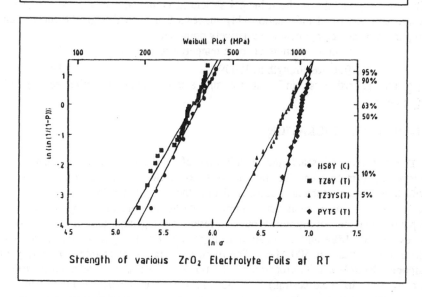

Figure 4   Weibull plots for selected $Zr(Y)O_{2-x}$ electrolyte foils

and 19mm discs for the ball-on-ring. The results are summarised in Table III and Figure 4. For the Siemens electrolyte sheets the best results for the cubic material was obtained with Tosoh powder whereas for the viscous processing route (ICI) the HSY8 powder produced superior mechanical properties, and a mean strength about 15% higher than the tape cast material. The strength improvements noted over the past two years have been principally brought about by changing particle size distribution and agglomerate strength.

It is obviously important to obtain mechanical property data at elevated temperatures but at present little information is available for cubic zirconia electrolyte sheets. Measurements have started at Imperial College and one set of data for the ICI material prepared from HSY8 powder indicates a mean strength of 280 MPa at 900°C compared to 368 MPa at room temperature. This value is close to high temperature strength values ($\approx$ 300 MPa) reported for partially stabilized zirconia.

Other important mechanical properties include fracture toughness values, cyclic fatigue and slow crack growth parameters to enable life-time predictions to be made. For the past two years [12] we have been monitoring slow crack growth and micro-cracking by using A.C. impedance spectroscopy and correlating the changes in impedance with microstructural features. Such techniques have been used on β-Al_2O_3 ceramics [13] and more recently Dessemond et al [14] have shown that cracks produced by indentation hardness measurements can also be correlated with impedance spectroscopic measurements. It is expected that this technique will help to interpret high temperatures failure mechanisms.

ACKNOWLEDGEMENTS

Dr. C.A. Leach, Mr. R.A. Rudkin and Dr. J. Kajda of the Centre for Technical Ceramics have provided major contributions to the work reported in this paper. Financial support for RAR and JK has been provided from CEC contract EN3E-0048 and ETSU contract E/5B/7000/2821, respectively.

REFERENCES

1. Steele, B.C.H., 1990. "Properties and Performance of Materials Incorporated in SOFC System," in Proc. of Intl. Symp. on Solid Oxide Fuel Cells, O. Yamamoto, M. Dokiya and H. Tagawa, eds., Tokyo, Japan: Science House, pp.135-47.

2. Manner, R., E. Ivers-Tiffee and W. Wersing, 1991. "Characterisation of YSZ Electrolyte Materials with Various Yttria Contents," in Proc. 2nd Intl. Symp. on Solid Oxide Fuel Cells, F. Grosz, P. Zegers, S.C. Singhal and O. Yamamoto, eds., Brussels, Belgium: CEC Publ. EUR 13546 EN, pp.715-723.

3.   Poulsen, F.W., P. Buitink and B. Malmgren-Hansen, 1991. "Van der Pauw and Conventional 2-Point Conductivity Measurements on YSZ Plates," in Proc. 2nd Intl. Symp. on Solid Oxide Fuel Cells, F. Grosz, P. Zegers, S.C. Singhal and O. Yamamoto, eds., Brussels, Belgium: CEC Publ. EUR 13546 EN, pp.755-767.

4.   Steele, B.C.H., S. Carter, J. Kajda, I. Kontoulis and J.A. Kilner, 1991. "Optimisation of Fuel Cell Components Using $^{18}O/^{16}O$ Exchange and Dynamic SIMS Techniques," in Proc. 2nd Intl. Symp. on Solid Oxide Fuel Cells, F. Grosz, P. Zegers, S.C. Singhal and O. Yamamoto, eds., Brussels, Belgium: CEC Publ. EUR 13546 EN, pp.517-525.

5.   Dees, D.W., U. Balachandran, S.E. Dorris, J.J. Heiberger, C.C. McPheeters and J.J. Picciolo, 1989. "Interfacial Effects in Monolithic Solid Oxide Fuel Cells," in Proc. 1st Intl. Symp. on Solid Oxide Fuel Cells, S.C. Singhal, ed., Pennington, NJ, USA: Electrochemical Soc. Publ. 89-11, pp.317-321.

6.   Khandkar, A.C., S. Elangovan, M. Liu and M. Timper, 1991. "Thermal Cycle Fatigue Behaviour of High Temperature Electrodes," in Proc. Symp. on High Temperature Electrode Materials and Characterisation, D.D. Macdonald and A.C. Khandkar, eds., Pennington, NJ, USA: Electrochemical Soc. Publ. 91-6, pp.175-190.

7.   Guindet, J., C. Roux and A. Hammou, 1991. "Hydrogen Oxidation at the Ni/Zirconia Electrode," in Proc. 2nd Intl. Symp. on Solid Oxide Fuel Cells, F. Grosz, P. Zegers, S.C. Singhal and O. Yamamoto, eds., Brussels, Belgium: CEC Publ. EUR 13546 EN, pp.553-559.

8.   Murakami S., Y. Akiyama, N. Ishida, T. Yasuo, T. Saito and N. Furukawa, 1991. "Development of a Solid Oxide Fuel Cell with Composite Anodes," in Proc. 2nd Intl. Symp. on Solid Oxide Fuel Cells, F. Grosz, P. Zegers, S.C. Singhal and O. Yamamoto, eds., Brussels, Belgium: CEC Publ., EUR 13546 EN, pp.561-568.

9.   Van Dieten, V.E.J., P.H.M. Walterbos and J. Schoonman, 1991. "Advanced Deposition Techniques for Solid Oxide Fuel Cell Components," in Proc. 2nd Intl. Symp. on Solid Oxide Fuel Cells, F. Grosz, P. Zegers, S.C. Singhal and O. Yamamoto, eds., Brussels, Belgium: CEC Publ., EUR 13546 EN, pp.183-191.

10. de Vries, K.J., R.A. Kuipers and L.G.J. de Haart, 1991. "Planar Solid Oxide Fuel Cells Based on Very Thin YSZ Electrolyte Layers," in Proc. 2nd Intl. Symp. on Solid Oxide Fuel Cells, F. Grosz, P. Zegers, S.C. Singhal and O. Yamamoto, eds., Brussels, Belgium: CEC Publ., EUR 13546 EN, pp.135-143.

11. Rossing, B.R., 1993. "Microstructure-Property Relationships for Porous CaO Stabilized ZrO_2 Support Tubes," in Proc. Conf. on High Temperature Solid Oxide Electrolytes, F.J. Salzano, ed., Brookhaven Nat. Lab., Long Island, USA: Publ. BNL-51728, pp.43-53.

12. Leach, C., 1992. "Microcrack Observations Using A.C. Impedance Spectroscopy." J. Mater. Sci. Lett., 11(5):306-307.

13. Troczynski, T.B. and P.S. Nicholson, 1986. "Sensitivity of the Potential Drop Technique for Crack Length Measurement in a Chevron-Notched Specimen." J. Am. Ceram. Soc., 69(7):C136-C137.

14. Dessemond, L. and M. Kleitz, 1992. "Effects of Mechanical Damage on the Electrical Properties of Zirconia Ceramics." J. Europ. Ceram. Soc., 9:35-39.

Interfacial Evaluation of Solid Oxide Fuel Cell Components Fabricated by the Acetylene Spraying Processes

Y. OHNO, M. HATA, Y. KAGA, K. TSUKAMOTO, F. UCHIYAMA, T. OKUO and A. MONMA

ABSTRACT

The reliability and stability of the solid oxide fuel cell (SOFC) depend mainly on two parameters, namely, the component materials and the mutual relationships between them. The perovskite-type oxides such as $LaCoO_3$, $LaMnO_3$ and $La_{1-x}Sr_xMnO_3$ are the prime candidates for air electrode materials for SOFCs, and their possibilities for application have been examined. The problem contains the following two major themes, adhesion of membranes and mutual ionic movements through the interface. In this study, the adhesion between air electrode/electrolyte and the possible high temperature interactions between perovskite-type oxide membranes, fabricated by acetylene flame spraying was investigated. The disc type SOFCs were classified in two groups and tested in air at 1000 °C for 100 hours in two ways, one with, the other without a conducting current. The SOFC cells were analysed by X-ray diffraction, scanning microscopy and electron-probe, to examine the crystalline changes of the oxides and also the reaction products.

INTRODUCTION

Study of the reliability and stability of the solid oxide fuel cell (SOFC) has become important in recent years. The reliability of SOFC involves many complicated problems such as, interfacial phenomena, due to the migration of ions, and separation of membranes due to the difference in thermal expansion of the electrolyte and electrode materials[1]. One of the most important technological goals to be achieved is to construct tight contacts between electrode/electrolyte membranes fabricated by any kind of process.

Y.Ohno and M.Hata, Kanagawa University,

3-21-1, Rokkakubashi, Kanagawaku, Yokohama, Kanagawa, 221, Japan

Y.Kaga, T.Tsukamoto, F.Uchiyama, T.Okuo and A.Monma, Electrotechnical Laboratory,

1-1-4, Umezono, Tsukuba, Ibaraki, 305, Japan

TABLE I - ACETYLENE SPRAYING PROCESS PARAMETERS

items	numerical numbers
type of spray gun	acetylene gun 6P-II METCO
rotation of sample	300 rpm
traverse speed of gun	3.5 m/min
spray distance	17.5 cm
gas pressures	
O_2	0.22 MPa
N_2	0.1 MPa

One of the major problems is still concerned with the air electrode/electrolyte interface, where perovskite-type oxides such as doped lanthanum manganite and/or cobaltite are still used[2].

Flame spraying processes have been used in the production of SOFC[3] and studied[4]. In this paper, the interfacial adhesion of electrode/electrolyte membranes and the migration of ions between them, are studied.

EXPERIMENTAL TECHNIQUES

PREPARATION OF SAMPLES

13 wt% yttria-stabilized zirconia (YSZ) disc plates, 23 mm in diameter and 2 mm thick were sintered and used as the electrolyte. The air electrode 10 mm in diameter and about $100 \mu m$ thick was fabricated on the YSZ plate by the acetylene flame spray process. Doped lanthanum perovskite-type oxide powders (-200 mesh), $LaCoO_3$, $LaMnO_3$ and $La_{1-x}Sr_xMnO_3$ (x=0.1, 0.2 and 0.3) were used as air electrode. The spray conditions were shown in Table I. Pt paste was coated on the other side of the YSZ and calcined to form the fuel electrode. Both surface of the YSZ plates were blasted (3 kg/cm^2, 20 sec) by

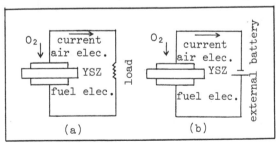

Figure 1. Schematic diagrams of test circuit.
(a) fuel cell operation circuit
(b) equivalent circuit

mixed Al_2O_3 (71 %) /ZrO_2 (25 %) powder to obtain good adhesion of electrode materials on the YSZ plate surfaces.

TEST CIRCUIT AND CURRENT CONDITION

The test samples were tested in atmospheric conditions. Figure 1a shows the fuel cell operation circuit and Figure 1b the equivalent test circuit. The adopted conditions of current densities were 0, 0.5 and 1.0 A/cm^2, respectively, and the duration of current flow was 100 hours for every test sample at 1000 °C. The air electrode was covered with Pt mesh and each electrode was connected with a Pt lead wire.

RESULTS AND DISCUSSIONS

RESISTIVITIES

Figure 2 shows the time resistance variation of test samples for the current density 0.5 A/cm^2. The $LaCoO_3$ electrode shows a high initial resistivity which drops over the first 20 hours. This is due to the poor crystallinity of the as sprayed coating and its recovery of crystallinity during the initial period of heating. Other electrode materials did not show such a drastic crystalline change and resistivity variation.

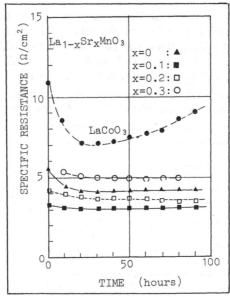

Figure 2.   Resistance change of test cell vs time.
0.5 A/cm^2

In case of $La_{1-x}Sr_xMnO_3$ family, the resistivities are lower than that for $LaCoO_3$ and $La_{0.9}Sr_{0.1}MnO_3$ gives the most preferable characteristics as the air electrode.

ANALYSES OF CROSS SECTIONS

Cross Sectional Observations

From the test samples, the following results were obtained:

(1) $LaCoO_3$ electrode completely separated from the YSZ surface due to the difference of thermal expansion rate.
(2) electrodes of $LaMnO_3$ and $La_{1-x}Sr_xMnO_3$ (x=0.1, 0.2 and 0.3) remained in good contact with the YSZ surface due to the close thermal expansion rates.

Figure 3 shows the cross sectional micrograph obtained from the $La_{1-x}Sr_xMnO_3$ family. $La_{0.9}Sr_{0.1}MnO_3$ shows better adhesion to the YSZ surface than any of the other electrode materials. This is due to the small thermal conductivity of $La_{0.9}Sr_{0.1}MnO_3$ compared with other perovskite-type oxides as shown in Figure 4[5]. As the heat given by the acetylene flame spraying process has been obtained in melted particles still the end of spraying, so they could hold liquid state for a time long enough to contact with substrate surface. A further reason for the good compatibility of $La_{0.9}Sr_{0.1}MnO_3$ and YSZ is their similarity in thermal conductivity which prevents severe thermal temperature gradients between them.

The effect of the surface roughness of YSZ plate on adhesion could not be determined by these experiments. Particles usually have so high velocities of flight that they could reach the surface of substrate materials in fully molten states[6]. So they could be scattered violently to all directions without touching the edges.

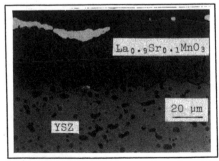

Figure 3.   Photomicrograph of interface between $La_{0.9}Sr_{0.1}MnO_3$/YSZ interface.

0.5 A/cm^2

Cross Sectional Observations by SEM

Figure 5 shows the cross section of the $La_{1-x}Sr_xMnO_3$ electrolyte interface. The surface of the YSZ plate has been roughened to the extent that a clear interface could not be recognized. Many micropores and channels, $La_{0.9}Sr_{0.1}MnO_3$ has shown the good contact with the YSZ plate.

As the particles of electrode material were heated in hot gases (over 2000 °C) during the spraying processes, and were carried to the surface in a hot gas stream, some of them impacted the YSZ plate in a molten state and flowed to the bottom of surface asperities. These asperities may therefore reinforce the adhesion of the electrode to the YSZ plate.

EPMA Analyses

In order to examine ionic transfer at the interface, the cross sections have been examined by EPMA. The diameter of the beam spot is 1 μm. Figure 6 shows the EPMA traces for a test sample of $La_{0.9}Sr_{0.1}MnO_3$ electrode which was polished. The following results were obtained when examine polished and unpolished specimens:

$LaCoO_3$:   could not be examined due to separation of the membranes.

$LaMnO_3$:   La, Mn could be detected in all test samples.

$La_{1-x}Sr_xMnO_3$: La, Mn could be detected in all test samples.

Polished samples, examined to obtain information on the elemental distribution near the surface, provided the following additional results:

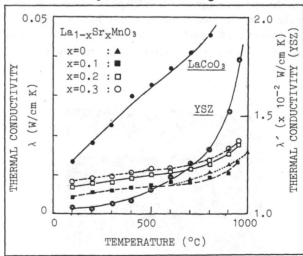

Figure 4.   Thermal conductivities of perovskite-type oxides.

(1) the distributions of Y and Zr in YSZ were much more uniform compared with unpolished samples.

(2) cross over of each element at the interface could not be detected.

X-ray Diffraction

After EPMA examination, X-ray diffraction (diameter beam spot 100 μm, Co target, 8 kW) measurement was carried out to detect any new compounds in the interfacial region.

As already shown in SEM photomicrographs, there were no new white compounds near the interfacial region between the perovskite-type oxides and YSZ. Figure 7 shows XRD results for the $LaMnO_3$ electrode for various experimental conditions.

There is no evidence for $La_2Zr_2O_7$ at the interface. No new compounds were detected in any of the electrode materials for any experimental condition.

DISCUSSION

Resistivity

In spite of having the most preferable intrinsic electrical characteristics, $LaCoO_3$ did no show good adhesion to a roughened YSZ plate. Separation of the electrode only happened in the case of the $LaCoO_3$ electrode. The resistivities of $LaMnO_3$ and the family of $La_{1-x}Sr_xMnO_3$ showed very stable values, because of tight contact with the YSZ plate.

In disc type cells, the edges are free to move, so the thermal expansion dilations occur freely and large shearing stresses may be created at the electrode/electrolyte interface. Nothing could stop the extension, which is in contrast with the tubular cell configuration. The purpose of roughening is to

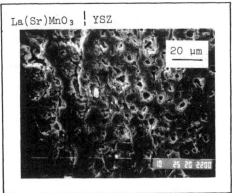

Figure 5. Photomicrograph (SEM) of $La_{0.9}Sr_{0.1}MnO_3$/YSZ interface, dotted line shows the interface.

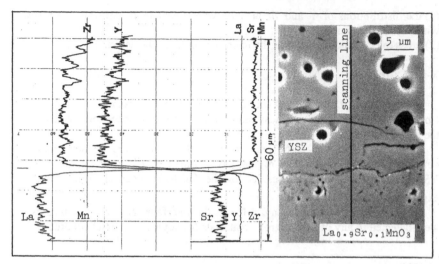

Figure 6.   EPMA trace for $La_{0.9}Sr_{0.1}MnO_3$/YSZ interface.

obtain the so-called anchor effect. As the particles were melted and dispersed on the YSZ plate in small droplets, the asperities on the plate could be covered and droplets combined with each other tightly. In the case of the $LaCoO_3$ electrode, the thermal expansion rate is too large (2.3 times larger than that of YSZ[7]) to ensure the anchor effect. In contrast $LaMnO_3$ and $La_{1-x}Sr_xMnO_3$ family have compatible thermal expansion rates with YSZ, and the anchor effect could be obtained successfully. So in the flame spraying process, the anchor effect would be expected when both materials have compatible thermal expansions.

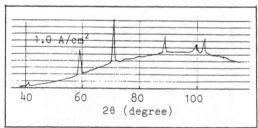

Figure 7.   X-ray diffraction trace of $LaMnO_3$/YSZ interface.

$1.0\ A/cm^2$

Vertical line is an arbitrary.

Interfacial Phenomena

It could be recognized from the detailed observation of $LaMnO_3$ and $La_{1-x}Sr_xMnO_3$ electrode/electrolyte combinations, that contact with the YSZ plate seems to be very tight. EPMA has been carried out for both unpolished and polished cross sections. The concentration of each element changes at the interface, however comparing with many test samples, the slope and the width of the cross over section was almost always the same, which indicates that inter diffusion of the elements does not occur. Usually, in the case of powder-powder reactions such as $LaCoO_3$ and $La_{1-x}Sr_xMnO_3$, lanthanum zirconate could be produced by sintering at high temperature[8]. For $LaCoO_3$, the following reaction is expected:

$$2LaCoO_3 + 2ZrO_2 \rightarrow La_2Zr_2O_7 + 2CoO + 1/2\, O_2 \qquad \cdots\cdots \qquad (1)$$

The CoO phase should be observed if the reaction takes place at the interface. Figure 7 shows no such phase for any of the test samples, in fact no new phases could be detected at all. It appears that the temperatures and times involved in flame spraying were not sufficient for significant interdiffusion to occur. However, the inter diffusion of elements may occur for powder-powder reactions and slurry treatments. It is clear that the flame spraying process does not pose such a problem in the preparation of materials.

Though the measurements on the interface have been restricted only to a narrow section of disc, many samples have been checked by the various techniques and the $La_2Zr_2O_7$ phase could not be recognized in any of them. We believe that adequate temperature regulation during the deposition of the electrode is able to prevent problems due to elemental diffusion across the electrode/electrolyte interface.

CONCLUSIONS

(1) the flame spraying process is suitable for fabricating electrodes on YSZ plate especially electrodes of $La_{1-x}Sr_xMnO_3$ oxides.

(2) with regard to the adhesion to the YSZ plate of the perovskite-type oxides, $La_{0.9}Sr_{0.1}MnO_3$ was the best.

(3) the resistivity of the $La_{1-x}Sr_xMnO_3$ type electrodes is stable for long periods of time at $1000\,°C$.

(4) the formation of new compound oxides, such as lanthanum zirconate, could not be detected.

REFERENCES

1. Lau.S.K and S.C.Singhal, 1985. "Potential electrode/electrolyte interactions in solid oxide fuel cells". Corrosion 85, 345, March, 25-29.

2. H.Tagawa, J.Mizusaki, M.Kitou, K.Hirano, A.Sawata and K.Tuneyoshi, 1991. "On the solid state reaction between stabilized zirconia and some perovskite-type oxides". Proc., 2nd International SOFCs, July, Athenes. p.681-688.

3. Y.Ohno, S.Nagata and H.Sato, 1980. "Development of a high temperature solid oxide electrolyte fuel cell". Proc.,15th Intersociety Energy Conversion Engineering Conference. Aug., Seattle. p.881-885.

4. Y.Ohno, Y.Kaga and S.Nagata, 1986. "A study on the fabrication and the performance of the solid oxide fuel cell". Trans., Journal of Electrical Engineers of Japan. Vol.107, No.1, Aug., p.693-700.

5. Y.Kaga, Y.Ohno, K.Tsukamoto, F.Uchiyama, A.Monma, M.J.Lain, T.Nakajima, F.Mogami and T.Horigome, 1991. "Measurements of thermal conductivities of SOFC air electrode materials". Proc., Joint. Conference of Electrical Engineers of Japan. No.1475.

6. K.Hayashi, K.Kaga, Y.Ohno and H.Yamanouchi, 1987. "Studies on plasma spraying procedures for SOFC materials". Proc., Joint. Conference of Electrical Engineers of Japan. No.1236.

7. Y.Ohno, S.Nagata and H.Sato, 1983. "Properties of oxides for high temperature solid oxide electrolyte fuel cell". Solid State Ionics. 9&10, p.1001-1008.

8. O.Yamamoto, Y.Takeda and T.Kojima, 1989. "Reactivity of yttria stabilized zirconia with $(La_{1-x}A_x)_{1-y}MnO_{3+z}$". Proc., SOFC Symposium Nagoya. 13-14, Nov., p.87-92.

Zirconia Based Oxide Ion Conductors in Solid Oxide Fuel Cells

OSAMU YAMAMOTO, TAKAYUKI KAWAHARA, YASUO TAKEDA,
NOBUYUKI IMANISHI and YOSHINORI SAKAKI

ABSTRACT

The solid solution powders of the $Sc_2O_3-ZrO_2$ system have been prepared by a sol–gel method. The phases and the electrical conductivity of the sintered $Sc_2O_3-ZrO_2$ (ScSZ) and ScSZ–Al_2O_3 have been examined. ZrO_2 with 3 mol% (3ScSZ) and 8 mol% (8ScSZ) Sc_2O_3 showed the single phase of the tetragonal (t) and cubic (c) structure, respectively. The samples with 11 mol% (11ScSZ) and 12 mol% (12ScSZ) Sc_2O_3 showed the rhombohedral structure. The composites of these zirconias had the cubic structure in a content range 0.7–30 wt% Al_2O_3. The electrical conductivity of 8ScSZ and 12 ScSZ was 0.34 S/cm at 1000 °C. The cubic stabilized zirconia with low content of Sc_2O_3 showed a significant aging effect and the conductivity of 8ScSZ decreased to 0.14 S/cm after annealing at 1000 °C for 1000h. The conductivity of the ScSZ and Al_2O_3 composites decreased with increasing Al_2O_3 content. The conductivity of 11ScSZ–20 wt% Al_2O_3 was 0.15 S/cm at 1000 °C and no change with annealing at 1000 °C was observed.

INTRODUCTION

High–temperature solid oxide fuel cells (SOFC) have become of great interest for the electric power generation systems, because of the simplicity of system design, the high energy conversion efficiency, and the availability of high quality by–product heat. Two different types of SOFC are developing, namely tubular and planar type. In the most progressed tubular type configuration by Westinghouse Electric, the active cell components are deposited in the form of thin layers on a porous support tube[1]. The gas tight yttria stabilized zirconia(about 50 μm thick) is deposited by the electrochemical vapor deposition technique. In this configuration, the electrical conductivity for the electrolyte is not such a significant factor to determine the cell performance, because the contribution of the electrolyte resistance to the total is not so high.

Osamu Yamamoto, Takayuki Kawahara, Yasuo Takeda, Nobuyuki Imanishi and Yoshinori Sakaki, Department of Chemistry for Materials, Faculty of Engineering, Mie University, TSU, 514, Japan.

The planar type SOFC's have been developing mainly in Japan and Europe. This structure will be expected to have exceptional potential for high electrical power density and useful preparation techniques can be used to prepare the cell components. However, there are many materials problems to develop a high performance planar cell. Especially, we should provide an electrolyte with high ionic conductivity and good mechanical properties. As the contribution of the electrolyte resistance in the planar SOFC is predominant in the total cell resistance, the high conductivity of the electrolyte improves cell performance.

Previously, yttria stabilized cubic zirconia (YSZ) has been mainly used in SOFC, the conductivity of which is 0.16 S/cm at 1000 °C. The mechanical strength of the c–ZrO_2 was reported to be 230 MPa at room temperature[2]. Because of poor mechanical properties of the YSZ electrolyte, it is difficult to prepare large size thin electrolyte plates. To improve the mechanical properties of the c–ZrO_2, the composite of YSZ and Al_2O_3 was proposed by Ishizaki et al.[3] and Yamamoto et al.[2]. The composite of 20 wt% Al_2O_3 had the conductivity of 0.1 S/cm at 1000 °C and the bending strength of 330 MPa. In the practical development of the planar type SOFC with a size of 169 cm^2 by Sanyo[4], the tetragonal type of ZrO_2 with 3 mol% Y_2O_3(3Y–TZP) was used as the electrolyte, the thickness of which was 0.2 mm. The t–ZrO_2 has the conductivity of 0.050 S/cm at 1000 °C after annealing at 1000 °C for 1000 h and the bending strength of 1200 MPa[5]. The cell voltage of 0.81 V was obtained at 300mA/cm^2, where IR–drop from the electrolyte resistance was more than half of the total cell polarization. To improve the cell performance, the contribution of the electrolyte resistance should be decreased.

We have aimed to develop oxide ion conductors with a high conductivity and good mechanical properties for SOFC. The c–ZrO_2 stabilized with Sc_2O_3 (ScSZ) has been reported to have a higher conductivity than other ZrO_2 based electrolytes[6]. The maximum conductivity in the Sc_2O_3–ZrO_2 system is 0.25 S/cm at 1000 °C for Sc_2O_3 concentrations of 8 to 10 mol%[7]. More recently, Badwal has reported that the conductivity of 7.8mol% Sc_2O_3–ZrO_2 was 0.32 S/cm at 1000°C[8]. The phase diagram for the Sc_2O_3–ZrO_2 system near the ZrO_2 rich region is complex and the observed phases depended on the preparation method[9]. In this study, the Sc_2O_3–ZrO_2 solid solutions in the range 3 to 12 mol% Sc_2O_3 were prepared by a sol–gel method, and the electrical conductivity of the sintered sample was measured as a function of the annealing time at 1000 °C. Additionally also, the composites of ScSZ and Al_2O_3 have been examined. Alumina is added to increase the mechanical strength of the electrolyte matrix. Although investigations into the conductivity of some Sc_2O_3–ZrO_2–Al_2O_3 compositions have been carried out by Badwal[10], the conductivity of the composition with high content of Sc_2O_3 has not yet been fully elucidated.

EXPERIMENTAL

The starting materials used in the preparation of various samples were $ZrO(NO_3)_2 2H_2O$ (Chemical grade, Nakarai Chem. Japan), Sc_2O_3(99.9 %, Furuuchi Chem. Japan), and Al_2O_3 (>99.9 %, Daido Chem. Japan). Most of the ScSZ samples were prepared by a sol–gel method[11]. To prepare the scandium solution, Sc_2O_3 was dissolved in nitric acid. The scandium solution was added into the aqueous solution of $ZrO(NO_3)_2$ in the proper proportion. To the solution with scandium and zirconium were added formic acid and ethylene glycol. This was heated at about 120 °C with stirring, and then fired at 800 °C for 12 h. The obtained powders were pulverized by a planetary–micro–pulverizer (P–7, Fritsch Ltd.) with zirconia vessel and balls for 15 min. with acetone.

The composite powders of ScSZ–Al_2O_3 were prepared by mixing in the required proportion, and calcining at 1100 °C for 1h. The mixture was compacted into bars (for conductivity measurements) and tablets (for XRD measurements) at a low pressure. These were then isostatically pressed at 100 MPa followed by sintering at 1700 °C for 15 h. The relative density of the sintered samples was about 90 % of the theoretical density calculated from XRD data for the samples without Al_2O_3 and about 95 % for these with Al_2O_3.

X–ray diffraction (XRD) patterns of sintered disks were obtained with monochromated $CuK\alpha$ radiation and a scintillation detector. Silicon powder was used as an internal standard. The electrical conductivity measurements were carried out with cylindrical samples of about 0.5 cm in diameter and about 3 cm long with platinum paint electrodes sintered at 1000°C for 1h. The a.c. conductivity was measured using a frequency–response analyzer(Solartron FRA–1250) over a frequency range 10^{-1} to 6.5×10^4 Hz with an applied potential of 0.2 V and a temperature range 250 to 1000 °C . All conductivity measurements were performed in air.

RESULTS AND DISCUSSION

In Table I, the XRD results obtained in this study for the Sc_2O_3–ZrO_2–Al_2O_3 composition are summarized. Badwal reported that XRD patterns of the samples with 8mol % Sc_2O_3 prepared from the oxides heated at 1700 °C showed the presence of monoclinic(m) ZrO_2 and a fluorite–related phase[9]. In our case, also single c–phase of ZrO_2 with 8mol% Sc_2O_3 was observed only by the coprecipitation and sol–gel methods. These results indicate that the homo

TABLE I. OBSERVED PHASE OF THE Sc_2O_3–ZrO_2–Al_2O_3 SYSTEM.

Composition ScSZ		Al$_2$O$_3$	Observed zirconia phase[*]
ZrO$_2$ (mol%)	Sc$_2$2O$_3$(mol%)	(wt%)	
97	3	0	t
95	5	0	t , c
92	8	0	c
92	8	10	c
92	8	20	c
90	10	0	c , r
90	10	20	c
89	11	0	r
89	11	20	c
88	12	0	r
88	12	0.7⁻	c
88	12	20	c

[*]   t = tetragonal   c = cubic   r = rhombohedral

geneous phase could not be obtained at 1700 °C for 15 h from the oxide mixture, because of slow cation diffusion. Bannister and Skilton[12] found that for unknown reason , the presence of Al_2O_3 appeared to hasten reaction between ZrO_2 and Sc_2O_3 . Badwal also observed that the presence of Al_2O_3 had the effect of decreasing the $m-ZrO_2$ in 5.9 mol% $Sc_2O_3-ZrO_2$. It is likely that a difference in the reaction rate was responsible for the effect of Al_2O_3 on the $m-ZrO_2$ content. Mixed oxide samples, 8 mol% $Sc_2O_3-ZrO_2$ and 10 wt% Al_2O_3, sintered at 1700 °C for 15 h showed only $c-ZrO_2$ phase and Al_2O_3. A more remarkable effect of Al_2O_3 addition is observed in 12 mol% $Sc_2O_3-ZrO_2$. Phase diagram studies[13] show that a compound $Sc_2Zr_7O_{17}$, which is known as β phase, a rhombohedral distortion of the fluorite–type structure, has a very narrow range of existence in concentration of 11–13 mol% Sc_2O_3 and in temperature range room temperature to 620 °C. At a higher temperature, it is converted into a cubic solid solution of the fluorite–type. As shown in Table I, the samples with 11 and 12 mol% Sc_2O_3 are indexed as a single phase of β phase. The presence of Al_2O_3 had the effect of stabilizing the high temperature c–phase at room temperature. For example, 12 mol% $Sc_2O_3-ZrO_2$, and 12 mol% $Sc_2O_3-ZrO_2$ + 0.7 wt% Al_2O_3 samples prepared under identical conditions showed different phases, namely rhombohedral and cubic structure. It is surprising that the presence of a mere 0.7 wt% Al_2O_3 should have such a large effect. In the samples with lower content of Sc_2O_3 as 3 mol%, a pure t–phase was observed as reported previously[5]. No m–phase was observed in specimens prepared by the sol–gel method. The mixed oxide samples prepared under the same conditions had 27 % m–phase[10].

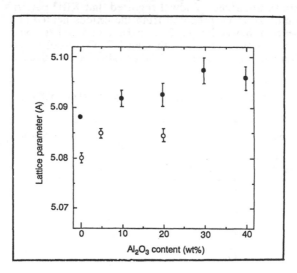

Figure 1 Lattice parameter of cubic phase for the $8ScSZ-Al_2O_3$(●) and $12ScSZ-Al_2O_3$(○) composite.

In figure 1, the lattice constants of the c–phases for the $8ScSZ-Al_2O_3$ and $12ScSZ-Al_2O_3$ composites are shown as a function of Al_2O_3 content. The observed lattice constants of 8ScSZ and 12ScSZ–0.7 wt% Al_2O_3 are 5.088 A and 5.080 Å, respectively, which are in good agreement with reported value for 10ScSZ, 5.085 Å [7]. The lattice constants of 8ScSZ and 12ScSZ increase with increasing Al_2O_3 content in the ScSZ–Al_2O_3 composites. Phase equilibrium studies[12] suggested that Al_2O_3 dissolved sparingly in the Sc_2O_3–ZrO_2 fluorite solid solution. The solubility of Al_2O_3 into 8 mol% yttria stabilized zirconia sintered at 1700 °C was estimated to be 0.5 mol%[14]. The lattice parameter should decrease with addition of Al_2O_3 if Al^{3+} substitute for Sc^{3+} or Zr^{4+}, because of small ionic radius of Al^{3+} compared to those of the others. Additionally, the increase with the Al_2O_3 content could not be explained from the solid solution formation. The increase of the cubic lattice with Al_2O_3 is only explained by the formation of solid solution of Al_2O_3 and Sc_2O_3. If Sc^{3+} in c–ZrO_2 transfers to Al_2O_3, resulting in a decrease of the Sc^{3+} content in ScSZ, the lattice parameter of ScSZ would be increased. However, there is no published information on phase relation in the system Sc_2O_3–ZrO_2–Al_2O_3, and the solubility limit of Sc_2O_3 in Al_2O_3 could not determined by XRD study, because no significant change of XRD patterns was observed.

Figure 2 shows the temperature dependence of the electrical conductivity of some Sc_2O_3–ZrO_2 compositions. The maximum value of the conductivity at 1000 °C was observed for the compositions 8–12 mol% Sc_2O_3. The value of 0.34 S/cm is slightly higher than those reported previously, namely 0.25 S/cm for 10ScSZ[7] and 0.27 S/cm for 8ScSZ[13], and 0.32 S/cm for 7.8ScSZ[8]. The difference of conductivity may be due to the difference of the sample preparation method and the thermal history of the sample after sintering. A conductivity jump near 580 °C is observed for 12ScSZ. According to the Sc_2O_3–ZrO_2 phase diagram, the temperature corresponds to the phase transition of β phase to cubic phase. The sample with 2.9 mol% Sc_2O_3 shows no jump in conductivity corresponding to the t–m transition as reported previously[13]. In figure 3, the electrical conductivity of the composites of ScSZ and Al_2O_3 is shown as a function of temperature. The conductivity of the composites

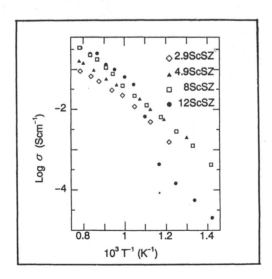

Figure 2 Temperature dependence of the electrical conductivity of some Sc_2O_3–ZrO_2 compositions.

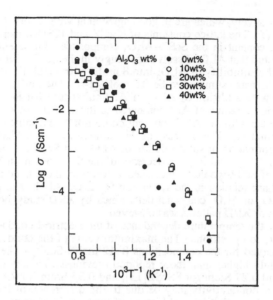

Figure 3 Temperature dependence of the electrical conductivity of the 12ScSZ–Al$_2$O$_3$ composites.

decreases with increasing Al$_2$O$_3$ content. The value of conductivity at 1000 °C for 12ScSZ–20 wt% Al$_2$O$_3$ was 0.15 S/cm, which is lower than that estimated from the volume fraction of the insulating phase of Al$_2$O$_3$.

Among the zirconia based electrolytes, Sc$_2$O$_3$–ZrO$_2$ has the highest ionic conductivity[6]. In spite of its high conductivity, the ScSZ electrolyte is not often used in SOFC, mainly because of its unfavorable aging characteristics. Inozemtsev et al.[15] reported a 50 % decrease in conductivity of 9ScSZ at 800 °C after annealing for 200 h. The decrease was attributed to the formation of an ordered rhombohedral β phase.

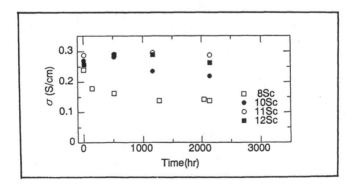

Figure 4 Conductivity change of ScSZ on annealing at 1000°C in air.

In figure 4, conductivity changes of ScSZ at 1000 °C are shown as a function of annealing time at 1000 °C. The conductivity of 8ScSZ decrease with time, and the value after 2000 h was 0.15 S/cm. The β phase was not observed in 8ScSZ annealed at 1000 °C for 2000 h. Similar conductivity decreases with aging time at 1000°C were observed in 7.8mol% Sc_2O_3–ZrO_2[8] and 7mol% Sc_2O_3–1mol%Y_2O_3–ZrO_2[16]. In the case of 7.8mol%Sc_2O_3–ZrO_2, the conductivity of 0.32S/cm before aging decreased to 0.18S/cm after aging for 300h at 1000°C, and X–ray and TEM analysis on the sample did not show the presence of the β phase. The samples with higher Sc_2O_3 content show no significant conductivity decrease with annealing time.

The annealing time dependence of the ScSZ–20 wt% Al_2O_3 composites are shown in figure 5. In this case, a slightly less conductivity decrease with annealing period is observed in 8ScSZ–20 w/o Al_2O_3(8ScSZ20A). The samples of 11ScSZ–20 w/o Al_2O_3(11ScSZ20A) showed no conductivity change after 4000 h annealing. The conductivity at 1000 °C of 11ScSZ20A after annealing 4000 h was 0.13 S/cm and β phase was observed at room temperature. The annealed 8ScSZ20A at 1000 °C for 4000 h showed no β phase . Therefore, the conductivity decrease of 8ScSZ and 8ScSZ20A with annealing at 1000 °C could not be explained by the formation of β phase. In this stage, the conductivity degradation mechanism is not clear. Nevertheless, Badwal suggested from the TEM evidence that the decrease in conductivity with time may be due to slow disproportionation of the tetragonal phase and perhaps some short–range ordering of the matrix[8].

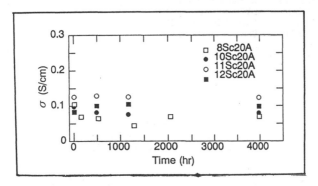

Figure 5 Conductivity change of the ScSZ–Al_2O_3 composites on annealing at 1000°C in air.

CONCLUSION

In Table II, the conductivity data of the zirconia based oxide ion conductors are summarized along with the previously reported data. For an application in SOFC, the long term stability of the conductivity for 50000 h or more operating periods is important. Additionally, the good mechanical properties of the electrolyte are required for the planar type SOFC. A major advantage of the planar SOFC is that it has a capability of high current drain as high as 0.5 A/cm^2 or more. To reduce the contribution of the electrolyte resistance, the conductivity of the electrolyte should be less than 0.1 S/cm at operating temperature. The conductivity data in Table II suggest that 8YSZ without Al_2O_3 and ScSZ with and without Al_2O_3 are available for the electrolyte in SOFC.

TABLE II. ELECTRICAL CONDUCTIVITY OF ZIRCONIA BASED ELECTROLYTES.

Electrolyte*	Conductivity at 1000 °C (S/cm)		Bending strength (MPa)	Ref.
	as sintered	after aging at 1000°C for 1000h		
3Y–TZP	0.056	0.050	1200	[5]
2.9Sc–TZP	0.09	0.063		[5]
8YSZ	0.16	0.12	230	this work
8ScSZ	0.34	0.14		this work
11ScSZ	0.34	0.30		this work
12ScSZ	0.34	0.29		this work
3YSZ20A	0.046		2400	[17]
8YSZ20A	0.09	0.05	310	this work
8ScSZ20A	0.13	0.07		this work
11ScSZ20A	0.15	0.13		this work
12ScSZ20A	0.15	0.10		this work

$$
\begin{aligned}
* \quad 3Y\text{–}TSP &= 3mol\%Y_2O_3 - ZrO_2 \\
2.9Sc\text{–}TZP &= 2.9mol\%Sc_2O_3\text{-}ZrO_2 \\
3YTZ20A &= 3YTZ + 20wt\%\ Al_2O_3 \\
8YSZ20A &= 8YSZ + 20wt\%\ Al_2O_3 \\
8ScSZ20A &= 8ScSZ + 20wt\%Al_2O_3
\end{aligned}
$$

REFERENCES

1. Dollard,W.J., 1990."The Westinghouse solid oxide fuel cell program –A 1989 program report," Proceedings of the International Symposium on Solid Oxide Fuel Cells, eds O.Yamamoto, M.Dokiya and H.Tagawa, Science House Co., p1–11.

2. Yamamoto,O., Y.Takeda, N.Imanishi, T.Kawahara, G.Q.Shen, M.Mori and T.Abe, 1991. "Electrical and mechanical properties of zirconia–alumina composite electrolyte", Proceedings of the 2nd International Symposium on Solid oxide Fuel Cells, eds P.Zegers, et al. EUR p437–444.

3. Ishizaki,F., T.Yoshida and S.Sakurada, 1990."FSZ electrolyte with alumina additions", Proceeding of the International Symposium on Solid Oxide Fuel Cells, eds O.Yamamoto, M.Dokiya and H.Tagawa, Science House Co., p172–176.

4. Murakami,S., Y.Akiyama, N.Ishida, T.Yasuo, T.Saito and N.Furukawa, 1991. "Development of a planar solid oxide fuel cell", Proceedings of the 2nd International Symposium on Solid Oxide Fuel Cells, ed P.Zegers et al. EUR addendum p59–95.

5. Yamamoto,O., Y.Takeda, R.Kanno and K.Kohno, 1990. "Electrical conductivity of tetragonal stabilized zirconia". Journal of Materials Science,25:2805–2808.

6. Etsell,T.H. and S.N.Flengas, 1970. "The electrical properties of solid oxide electrolyte". Chemical Review, 70:339–376.

7. Strickler,D.W. and W.G.Carlson, 1965. "The electrical conductivity Properties in ZrO_2–rich region of several Mn_2O_3–ZrO_2 systems". Journal American Ceramic Society, 48:286–289.

8. Badwal,S.P.S.,1987. "Effect of dopant concentration on electrical conductivity in the Sc_2O_3–ZrO_2 system". Journal of Materials Science, 22:1425–4132

9. Badwal,S.P.S.,1983. "Electrical conductivity of Sc_2O_3–ZrO_2 compositions by 4–probe d.c. and 2–probe complex impedance techniques." Journal of Materials Science, 18: 3117–3129.

10. Badwal,S.P.S.,1983. "Effect of alumina and monoclinic zirconia on the electrical conductivity of Sc_2O_3–ZrO_2 composition". Journal of Materials Science, 18:3230–3243.

11. Pechini,M.P., 1967. "Method of preparing lead and alkaline earth titanates and niobates and coating method using the same to form a capacitor." U.S.Pat.3,330,679.

12. Bannister,M.J. and P.F.Skilton, 1983. "The cubic tetragonal equilibrium in the system zirconia–scandia at 1800°C :effect of alumina." Journal of Materials Science Letters, 2:561–564.

13. Spiribonor,F.M., L.N.Popova and R.Ya.Porilskii, 1970. "On the phase relation an the electrical conductivity in the system ZrO_2–Sc_2O_3."Journal of Solid State Chemistry, 2:430–438.

14. Miyayama,M., H.Yanagida and A.Asada,1985."Effect of Al_2O_3 additions on resistivity and microstructure of yttria–stabilized zirconia" American Ceramic Society Bulletin, 65:650–664.

15. Inozemtser,M.V., M.V.Perfiles and V.P.Gorelov, 1976."Effect of annealing on the electrical and structual Properties of electrolytes based on ZrO_2". Soviet Electrochemistry, 12:1128–1136.

16. Ciacchi,F.T. and S.P.S. Badwal, 1991."The system Y_2O_3–Sc_2O_3–ZrO_2 : phase stability and ionic conductivity studies". Journal of the European Ceramics Society, 7:197–206.

17. Yamamoto,O., T.Kawahara, K.Kohno, Y.Takeda and N.Imanishi, 1991. "Oxide ion conductors for solid oxide fuel cells," in Solid State Materials, eds. S.Radhakrishina and A.Daud, Narosa Publishing House, p366–376.

Selection, Fabrication and Properties of Electrodes Used in High Temperature Solid Oxide Fuel Cells

N. M. SAMMES and M. B. PHILLIPPS

ABSTRACT

A single cell planar solid oxide fuel cell (SOFC) has been fabricated with the aim of using New Zealand natural gas as a fuel. This paper describes the electrode fabrication and properties as a function of their chemical and physical compatibility and conductivity. A series of $La_{1-x}Sr_xMnO_3$ air electrodes were synthesised using a fabrication technique which involved the formation of a polymer pre-cursor. Excellent adherance of electrodes was found, even for relatively high dopant levels. A series of Ni/yttria stabilised zirconia (YSZ) cermets were investigated as anodes as a function of their Ni loading and compatibility with the electrolyte. A single cell was fabricated using the above electrodes and yttria doped fully stabilised zirconia (FSZ) as the electrolyte. Further work is being undertaken to investigate the fuel cell characteristics, initially using methane as the fuel, but later studies will look at natural gas. The long term stability and the fabrication of a stacked planar system are also being investigated.

INTRODUCTION

New Zealand recently faced power shortages due to low levels in the South Island lakes supplying water for hydroelectricity [1]. Both the gas and electric utilities agree that natural gas and coal gas are finite resources and that they need to be better utilised to prolong their availability. Two main gas fields, Kapuni and Maui, produce over 90% of the total gas used in New Zealand. During the commissioning of the Maui gas field, off New Plymouth, a Government White Paper [2] stipulated that electricity generation may not be the most efficient use of Maui gas (Kapuni was already on line), although it does represent the most economic use for the foreseeable future. The Government does agree, however, that a more economic way of utilising these reserves should be pursued and that a research and development programme should be commenced.

N.M.Sammes and M.B.Phillipps
Centre for Technology, The University of Waikato, Private Bag 3105, Hamilton, New Zealand

The main alternatives to gas fired power stations are:

i Gas turbine combined cycle plant (efficiency of approximately 50%)

ii High temperature, high efficiency, simple cycle, gas turbines (approximately 38%)

iii Cogeneration, using gas engines and gas turbines (the precise combination of engines and gas turbines will depend on host plants)

iv Fuel cell technology, particularly for small scale localised cogeneration applications.

Both coal and natural gas could be used in the fuel cell systems, which are significantly more efficient than the current power stations around the country (Huntly is the most efficient at only 35% [3]).

One of the many types of fuel cell systems which may be used is the solid oxide fuel cell. SOFC's are so named because they use a solid oxide material as an electrolyte (usually a system based on FSZ) and convert the chemical energy of the fuel directly to electrical energy. Oxygen reduction occurs at the cathode/electrolyte interface, allowing the fuel to be oxidised at the anode/electrolyte interface, thereby liberating electrons which flow through an external circuit and recombine at the cathode/electrolyte interface.

The SOFC system operates at high temperatures (1000 °C) and is therefore amenable to in-situ reforming [4-5]. However, this high temperature regime requires the ceramic materials to have specific properties such as chemical compatibility, thermal and chemical stability, similar thermal expansivities, and adequate ionic or electronic conductivities.

This paper explores the fabrication of a single cell SOFC system. Using new fabrication routes for the electrode production, a cell was prepared with the required characteristics stipulated above.

EXPERIMENTAL TECHNIQUES, RESULTS AND DISCUSSION

SYNTHESIS OF THE CATHODE MATERIALS

The synthesis of a number of different perovskite materials was undertaken to produce a cathode material which had the properties discussed above. $La_{1-x}Sr_xMO_3$ (where M=Al, Gd, Mn) samples were prepared for values of x from 0 to 0.5. The techniques employed to produce these powders were investigated as a function of their ability to produce a sample which had the required chemical purity, stoichiometry, homogeneity, particle size and surface area at a suitable cost.

Several techniques were employed to produce either a powder, or a precursor for further fabrication into an adherent electrode.

Scanning electron microscopy (SEM), powder x-ray diffraction (XRD) and AC impedance analysis was undertaken on the powder and fired ceramic materials. Impedance spectra of the electrode materials were determined in air and gas mixtures, between 200 °C and 900 °C, for frequencies of 5Hz to 13MHz.

Salts of La {$LaCl_3$ (99.4%), $La(NO_3)_3$ (99%) and La_2O_3 (99%)}, Sr {$SrCl_2$ (99.5%) and $SrCO_3$ (99.5%)}, and Mn {$MnCl_2$ (99.5%) and $MnCO_3$ (99.5%)} or Al {$Al(NO_3)_3$ (99.9%)} or Gd {Gd_2O_3 (99.5%)}, provided by BDH laboratory reagents, were used as the starting materials for all synthesis.

The solid state method of production, (eg [6]), was initially undertaken. However very long firing times were required to obtain a material with the perovskite structure. In all XRD scans executed, secondary peaks were present. It was for this reason that a number of chemical routes were investigated (including co-precipitation and sol-gel techniques), the most favourable being the modified Pechini method.

THE MODIFIED PECHINI METHOD

This method was first performed by Pechini [7] for the preparation of titanates.

Salts of the required cations were dissolved in nitric acid. A mixture of ethylene glycol and ethanol was added to the acid solution forming an ester, which on heating to 50 °C and adding water, polymerised to form a gelatinous liquid. The polymerisation mechanism can be found in references [8-9]. The organics were evaporated leaving a resin. This was charred between 200 and 300 °C, pulverised and fired at 900 °C to form the required perovskite material.

Thermal Gravimetric Analysis and Differential Thermal Analysis (TGA/DTA) was undertaken on the $La_{1-x}Sr_xMnO_3$ (LSM) polymeric precursors using a Shimadzu DT40 Thermal Analyser, interfaced to a Shimadzu CRGA Chromatopac, up to 500°C. This is shown in figure 1 for $La_{0.5}Sr_{0.5}MnO_3$. The first weight loss and endothermic peak occurred between 80 and 120 °C. This represents a loss in water from the system. The major weight loss and exotherm represents the combustion of dehydrated organics, at 240 to 320 °C. All the weight loss was observed at temperatures below 380 °C which was in agreement with the work of Anderson et al. [8], who showed a similar effect for $Pb_3MgNb_2O_9$ perovskite material.

Figure 2 shows x-ray powder diffraction patterns for the $La_{1-x}Sr_xMnO_3$ samples, measured in air at room temperature. The results showed that the samples were all single perovskite phases having orthorhombic or rhombohedral structures [13], although the structural changes are not yet clear. Scan 1 shows improved symmetry with the crystal form closer to a cubic system. The lattice parameters are given in figure 3. These results are in accordance with the published results given by references [19-20]. This work is, however, contrary to the work of Koc et al. [11] who suggested that $LaMnO_3$ has a rhombohedral structure with lattice parameter a = 5.4724Å. The change in the crystal structure of the perovskite $La_{1-x}Sr_xMnO_3$, with increasing x, noted in this work has been observed by Jonker [10] who suggested an orthorhombic structure for $0 \leq x \leq 0.12$, rhombohedral in the range $0.12 \leq x \leq 0.45$ and cubic in the range $0.45 \leq x \leq 1.0$. Further work is currently underway to examine the effect of temperature on the crystal structure of the LSM material, by use of a high temperature powder x-ray diffractometer [21].

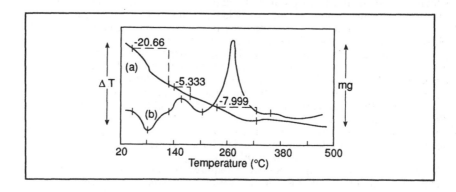

Figure. 1. DTA/TGA plot of $La_{0.5}Sr_{0.5}MnO_3$ up to 500°C. Heating rate of 10°C/min. a = TGA, b = DTA. (numbers on the TGA plot correspond to percentage weight loss).

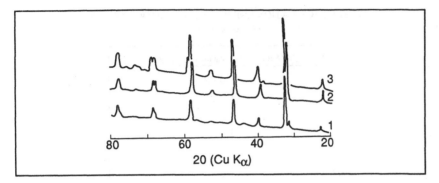

Figure 2. X-ray Powder Diffraction Patterns of the LSM series (1=0.4, 2=0.2, 3=0) in $La_{1-x}Sr_xMnO_3$

The results obtained from the BET isotherms showed that the manganites had a surface area of 12.1 ± 1.0 m^2/g. This value was in the order of that found, for example, by Taguchi et al. [12], who were trying to synthesise a material with very high surface area.

The powders were pressed into pellets and fired at 900 °C. Then their electrical ac and dc conductivities in air were investigated. It has been shown [13] that the manganites and strontium doped manganites have intrinsic p-type conductivity due to the formation of cation vacancies with charge being carried via the small polaron mechanism.

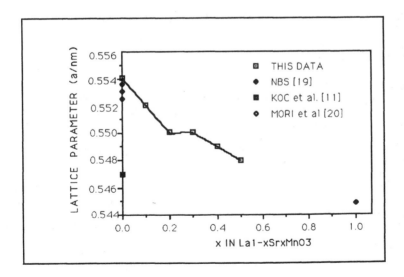

Figure 3. Lattice parameters for the LSM series as a function of x in $La_{1-x}Sr_xMnO_3$

TABLE I ACTIVATION ENERGY AS A FUNCTION OF x IN
$La_{1-x}Sr_xMnO_3$

x in $La_{1-x}Sr_xMnO_3$	Activation Energy (eV)
0	0.09
0.1	0.21
0.2	0.28
0.3	0.34
0.4	0.17
0.5	0.18

The electrical conductivity of the $La_{1-x}Sr_xMnO_3$ series of materials gave a straight line relationship, up to 1000 °C of the type ln (σT) versus 1/T.
This data may be represented by equation 1

$$\sigma = A/T \exp(-E_a/kT) \tag{1}$$

where A is a pre-exponential factor (a function of the charge carrier concentration and a material constant), E_a is the activation energy and k is Boltzmann's constant. The activation energies were calculated and are shown in Table I. These values are higher than the published data [22], although the significance of this is not yet fully understood. A value of 90 ± 5 $(\Omega cm)^{-1}$ was observed for the electronic conductivity of $La_{0.7}Sr_{0.3}MnO_3$ at 900°C.

ADHERENCE OF THE CATHODE MATERIAL TO THE ELECTROLYTE

A number of techniques were undertaken to adhere the perovskite material to the FSZ electrolyte.
1. The polymeric resin was painted onto the substrate, dried and charred at 350°C. This technique caused a frothing of the resin. Unless handled very carefully the frothing was too vigorous and the electrode material poorly adhered.
2. Several routes were undertaken using the fired perovskite powder. The powder was formed into a slurry using:
   i A mixture of trichloroethylene (TCE) and ethanol
   ii A mixture of toluene, methylethylketone, iso-propyl alcohol (IPA) and petroleum spirit
   iii Inorganic acids
   These slurries were either hand painted or air brushed onto the substrate. The samples were dried at 100 °C and sintered at 1200 °C for 1 hour (the minimum firing temperature for producing a well adhered sample). An evenly coated, and well adhered material was obtained using the TCE/ ethanol mix, when 3 to 6 coats were sprayed onto the electrolyte and re-fired after each coat.
3. Using the perovskite material, obtained from the modified Pechini method, a tape could be prepared that showed promise as an electrode material. The oxide was milled for 24 hours with the TCE/ethanol mix, using a soya-oil dispersant. Poly vinyl alcohol (PVA) or Poly vinyl butyral (PVB) binder was then added with the dibutyl pthalate plasticiser. Deairation was performed in an ultrasonic bath and the product was volume reduced for tape/slip casting. The tape

was placed onto a mylar substrate for future firing. Pore formers were used to investigate the effect of increased cathode porosity.

4. Slip casting of the perovskite slurry onto the substrate also showed promise. A mud-flat cracked appearance was seen after firing at 1000°C for 3 hours.

SEM micrographs of the slurry sprayed samples are shown in figures 4a and 4b. The SEM micrograph of a fracture surface of the fired material (figure 4a) shows the FSZ/LSM interface displaying good adhesive properties.

Micro-probe analysis, using the Kevex Microanalyser, confirmed that no Mn was present along the grain boundaries of the system. Mn has previously been shown [14-15] to be the most mobile ion in the system. Zr and Y are not as mobile and have only occasionally been found in the electrode. No interfacial $La_2Zr_2O_7$ phase was found, which has been shown [15] to form in systems annealed at relatively high temperatures. Hence, a cathode material was formed that had excellent adherence to the YSZ electrolyte because of its high surface area and high activity.

SYNTHESIS OF THE ANODE MATERIALS

The preparation of a Ni/YSZ cermet system was investigated using the synthesis techniques described below.

The NiO was prepared by sintering black nickel oxide (99.9%, from Riedel-deHaën) at 1000 °C.

The YSZ material was obtained either as a 3 mol% YSZ sample (commercially available sample from Zircar Products incorporated) or a 8 mol% FSZ sample (prepared in the laboratory).

The preparation of the 8 mol% FSZ material was undertaken using a standard co-precipitation technique [16], using $Zr(NO_3)_2$ and Y_2O_3 (both AnalaR grade >99.5%) as starting materials. The hydroxide powders were calcined at 650°C for 3 hours and fired at 1200°C for 1 hour. XRD on the materials showed a fully cubic phase being present.

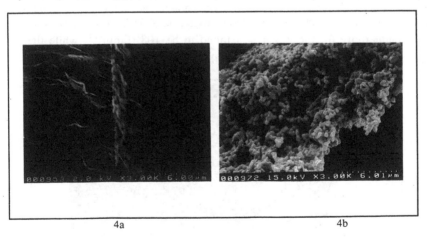

4a	4b

Figure 4. SEM Micrographs of the LSM material adhered to the FSZ ceramic. 4a shows the fracture surface of the fired FSZ/LSM interface (FSZ is on the left). 4b shows the cross sectional image of the fired LSM sample (the FSZ material is in the top left corner). The sample had been fired at 1200°C for 1 hour.

The BET technique gave a surface area for the material of <1 m^2/g. The commercially available 3mol% YSZ powders were found to have a surface area of 6.6 ± 0.3 m^2/g.

NiO and YSZ powders were mixed in various volume proportions (to give between 30 and 45 vol% Ni) and milled, using PSZ media, in ethanol. Samples were then either dried to make a powder, or volume reduced to paint consistency for later spraying or painting onto the substrate.

The material was then air brushed onto the electrolyte substrate and pre-baked at 1200°C for 1 hour, followed by reduction in 5%H_2/95%N_2 for 1 hour at 900°C, forming a Ni/YSZ composite material.

In an attempt to increase the anode pore size, various pore formers (cornflour, graphite and wood charcoal) were added to the composite material.

The loss of the pore formers on ignition was measured and the results showed that the percentage residue left on firing at 900°C was:

Graphite   0.3% residue
Charcoal   35% residue (mineral ash)
Cornflour   0.1% residue

This technique was used to obtain a set of differing cermet materials, containing the 3 pore formers, which were air brushed onto the FSZ electrolyte followed by pre-baking and reduction steps. The effect of the pore-formers on the fired electrode can be seen in the SEM micrographs given in figure 5. The micrographs show the inherent differences that the graphite and cornflour poreformers have on the Ni/YSZ structure, compared to that of the sample with no poreformer. Further work is, however, necessary to characterise these materials more quantitatively.

Pellets of the cermet material were prepared, and fired at 1200°C in a reducing atmosphere. AC and DC conductivity studies were undertaken on the samples to ascertain the effect of nickel content. All samples were measured in a reducing atmosphere. The results are shown in figure 6 and the typical S-plot seen is consistent with the percolation theory [17]. Figure 6 shows the effect of 2 YSZ cermet samples, the high surface area (Zircar) and the lower surface area (FSZ). It is clear that the sample with the lower surface area showed a similar trend to the Zircar material, but had a superior conductivity. At 600°C, for 40 vol% Ni cermet, the conductivity for the Zircar was found to be 108 $(\Omega cm)^{-1}$, while the FSZ material had a conductivity of 1025 $(\Omega cm)^{-1}$.

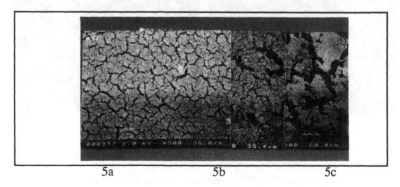

| 5a | 5b | 5c |

Figure 5. SEM micrographs of the three YSZ/Ni cermets, with different poreformers. 5a has no poreformer, 5b is graphite and 5c is cornflour.

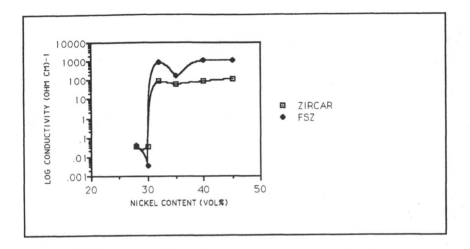

Figure 5. Conductivity of the cermet materials at 600 °C, versus the total volume% Ni present, for two different YSZ samples.

The conductivity was found to be dependent on the percentage of Ni in the cermet and the surface area of the YSZ, which agrees with theory [17-18]. However, conductivity dependence was slightly more complicated, due to the effect of the yttria content in the two YSZ systems. Further work will be necessary to investigate this effect.

FABRICATION OF A PLANAR SINGLE CELL

0.22mm electrolyte sheets (provided by ICI Advanced Materials) were prepared from DK 8 mole% fully stabilised zirconia powder, plastic mixed with a polymer and calender rolled. The electrodes were air-brushed onto the electrolyte material as described above.

Care had to be taken to minimise the stress build up in the electrolyte configuration during firing. It was found that stresses built up on cooling and were at a maximum when the temperature of the sample had equilibrated. If there were any stress concentrations within the sample, then fracture of the electrolyte occurred.

Cracking was experienced in the cathode/electrolyte/anode system upon cooling due to stress concentrations in areas where the electrode coating was thin.

The electrolyte was mounted on a Macor (Machinable Glass Ceramic, Corning Glass) ceramic base after deposition and firing of the electrodes and platinum electrode wires. Channels had been cut into the base and the electrolyte system was adhered to it by use of a high temperature castable ceramic (Aremco Products Inc., Ceramcast, 511) adhesive. This was then fired at 900°C and used as a preliminary single cell SOFC system. The cell was attached to a Pye Unicam Gas Chromatograph, using silica glass as the inlet/outlet tubes and glass/metal seals.

Work is currently underway to investigate the effect of New Zealand natural gas on the long term running characteristics of the cell. A stacked planar cell is now being fabricated with a suitable interconnect material.

ACKNOWLEDGEMENTS

The authors wish to thank Dr Kevin Kendall, ICI, for his assistance and also acknowledge the support of The University of Waikato Research Committee.

REFERENCES

1. Sammes, N.M., 1992. "Fuel Cell Possibilities." New Zealand Science Monthly, 3(9):12-13.

2. New Zealand Government White Paper on the Development of the Maui Gas Field, October 1973.

3. Cornelius, S., July 1992. Private Communication with Electricity Corporation of New Zealand Ltd.

4. Lee, A.L., R.F.Zabransky and W.J.Huber, 1990. "Internal Reforming Development for Solid Oxide Fuel Cells." Ind. Eng. Chem. Res., 29(5):766-773.

5. Omoto, T., N.Iwasa and H.Yamazaki, 1989. "Development of Fuel Cells for Commercial Use under the Moonlight Project." International Gas Research Conference, Tokyo, Japan, Thomas L. Cramer, M.D.Rockville, eds., Pub. Government Institutes, pp.939-947.

6. Yamamoto, O., Y.Takeda, R.Kanno and M.Noda, 1987. "Perovskite-type Oxides as Oxygen Electrodes for High Temperature Oxide Fuel Cells." Solid State Ionics, 22:241-246.

7. Pechini M., July 1967, US Patent No. 3 330 697.

8. Anderson, H.U., M.J.Pennel and J.P.Guha, 1987. "Polymeric Synthesis of Lead Magnesium Niobdate Powders." Advances in Ceramics Volume 21, G.L.Messing, K.S.Mazdiyasni, J.W.McCauley, and R.A.Haber, eds., Ohio, USA:The Am. Ceram. Soc. Inc., pp. 91-98.

9. Anderson, H.U., C.C.Chen, J.C.Wang and M.J.Pennell, 1990. "Synthesis of Conducting Oxide Films and Powders from Polymeric Precursors." Ceramics Powder Science III, Volume 12, G.L.Messing, S.Hirano, and H.Hausner, eds.,Ohio, USA:The Am. Ceram.Soc. Inc., pp. 749-755.

10. Jonker, G.H., 1956. "Magnetic Compounds with Perovskite structure IV Conducting and non-Conducting Compounds." Physica, 22:707-722.

11. Koc, R., A.Harlen and H.Scott, 1989. "Structural Sintering and Electrical Properties of the Perovskite Phase-type $(LaSr)(CrMn)O_3$." Proc. of the Electrochem. Soc., 89(11):220-202.

12. Taguchi, H., D.Matsuda, M.Nagao, K.Tanihata and Y.Miyamoto, 1992. "Synthesis of Perovskite-Type $(La_{1-x}Sr_x)MnO_3$ ($0 \leq x \leq 0.3$) at Low Temperature." J. Am. Ceram. Soc., 75(1):201-202.

13. Otoshi, S., H.Sasaki, H.Ohnishi, M.Hase, K.Ishimaru, M.Ippommatsu, T. Higuchi, M.Miyayama and H.Yanagida, 1991. "Changes in the phases and Electrical Conduction Properties of $(La_{1-x}Sr_x)_{1-y}MnO_{3-\delta}$." J. Electrochem. Soc., 138(5):1519-1523.

14. Taimatsu, H., K.Wada, H.Kaneko and H.Yamamura, 1992. "Mechanism of Reactions between Lanthanum Manganite and Yttria-Stabilized Zirconia." J. Am. Ceram. Soc., 75(2):401-405.

15. Yokokawa, H., N.Sakai, T.Kawada and M.Doikiya, 1991. "Thermodynamic Analysis of Reaction Profiles between $LaMO_3$ (M=Ni,Co,Mn) and ZrO_2." J. Electrochem. Soc., 138(9):2719-2727.

16. Roosen, A. and H.Hausner, 1983. "Sintering Kinetics of ZrO_2 Powders." Advances in Ceramics volume 12, Science and Technology of Zirconia II, N.Claussen, M.Ruhle and H.Heuer, eds., Ohio, USA:Am. Ceram. Soc. Inc., pp. 714-726.

17. Scher, H. and R. Zallen, 1970. "Critical density in Percolation Processes." J. Chem. Phys., 53:3759-3761.

18. Dees, D.W., T.D.Claar, T.E.Eastler, D.C.Fee, and F.C.Mrazek, 1987. "Conductivity of Porous Ni/ZrO_2-Y_2O_3 Cermets." J. Electrochem. Soc., 134(9):2141-2146.

19. International Centre for Diffraction Data (JCPDS), 1987, Inorganic Powder Diffraction File, 1601 Park Lane, Swarthmore Pa., USA.

20. Mori, M., S.Natsuko, T.Kawada, H.Yokokawa and M.Dokiya, 1990. "A New Cathode Material $(La,Sr)_{1-x}(Mn_{1-y}Cr_y)O_3$ ($0 \leq y \leq 0.2$) for SOFC." Denki Kagaku, 58(6):528-532.

21. Sammes, N.M. and G.Gainsford, 1993. Unpublished work.

22. Bossel, U.G., 1992. "Final Report on SOFC Data, Facts and Figures," Programme of R,D and D on Advanced Fuel Cells, International Energy Agency, Swiss Federal Office of Energy, Berne, Switzerland.

Chemical Thermodynamic Stabilities of the Interface

HARUMI YOKOKAWA, NATSUKO SAKAI, TATSUYA KAWADA
and MASAYUKI DOKIYA

ABSTRACT

The stabilities of the interface between various materials have been analyzed in terms of the chemical potentials of all elements involved. The chemical potentials can provide those thermodynamic driving forces for reaction and diffusion which are crucial in understanding what chemical processes take place in the interface. To analyze experimentally observed concentration profiles of interfaces, a simple model using three approximations is proposed: (1) local thermodynamic equilibrium at interface vicinity; (2) infinitely dilute solution for all phases involved; and (3) constituent atoms of a phase and impurities can diffuse along their respective chemical potential gradients. This model makes it possible to draw a plausible reaction path in the chemical potential space which should be compared with an analogous reaction path in compositional phase diagrams. Analyses have been made on interface reactions associated with zirconia.

INTRODUCTION

In recent years there has been a growing interest in the chemical stability of interfaces between many kinds of materials [1]. With regard to zirconia related technology, zirconia is used as a thermal barrier coating and as an electrolyte for high–temperature electrochemical cells. For the former case, zirconia–based ceramics used in gas turbines are exposed to combustion gases from petroleum fuel. A typical contamination in petroleum fuel is vanadium compounds. Therefore, the chemical stability against vanadium oxide is very important [2]. On the other hand, high–temperature electrochemical cells require good chemical compatibilities among cell components [3,4]. Experimental information on such chemical stabilities has been rapidly accumulated due to the recent advance in methods for experimentally analyzing distribution of elements at interfaces. The fundamental understanding, however, on chemical processes involved has been limited to rather simple cases [5,6]. This is because the kinetic properties, such as diffusion coefficients in multicomponent systems, are generally difficult to estimate from the knowledge based on less complicated systems. The present paper aims to bridge

Harumi Yokokawa, Natsuko Sakai, Tatsuya Kawada and Masayuki Dokiya, National Institute of Materials and Chemical Research, Tsukuba Research Center, Tsukuba, Ibaraki 305, JAPAN

such a gap between experimentally accumulated knowledge and theoretical achievement using the thermodynamic approach. This approach is adopted because it is not difficult to extend the thermodynamic considerations to more complicated systems using appropriate estimates. In addition, the chemical potentials derived from the thermodynamic considerations should provide the driving forces for chemical reactions and diffusion at the interface region.

CHEMICAL POTENTIAL AS KEY CONCEPT

Consider two different materials joined together. The chemical stability of interface can be discussed in terms of (1) whether new phases are formed or (2) whether some components in one phase dissolve and diffuse into the other phase. These behaviors are closely related to phase diagrams; thus, the chemical reactivity between two materials has been discussed in terms of reaction paths (diffusion paths) in compositional phase diagrams [1]. This should be also discussed in terms of thermodynamic properties, such as chemical potentials, since the driving forces for chemical reactions and diffusion can be given properly by chemical potentials.

Usually, solid reactions proceed through slow diffusion processes. This implies that the system is no longer in equilibrium *as a whole*. This makes it necessary to introduce a *local* thermodynamic equilibrium approximation. In this approximation, the thermodynamic relations can hold within a small local area. In the adjacent small areas, different equilibria can be established. Thus, the intensive thermodynamic quantities can be assumed to change as a function of location. Among the intensive quantities, the chemical potential is the key property to examine chemical stabilities of interfaces and concentration profiles of the reacted layers.

CHEMICAL REACTIONS VS. PHASE RELATIONS

Consider the A–B–C ternary system which contains AB and AC as binary compounds (see figure 1). A chemical reaction in this system

$$AB + C = AC + B \tag{1}$$

can be examined in terms of the Gibbs energy change for the reaction;

$$\Delta_r G^\circ[\text{eq. 1}] = \Delta_f G^\circ(AC) + \Delta_f G^\circ(B) - \Delta_f G^\circ(AB) - \Delta_f G^\circ(C) \tag{2}$$

The same thermodynamic property can be represented in terms of phase relations. Consider a hypothetical phase combination of AB, C, and AC. When three phases are equilibrated with each other, the following three equations can be set up;

$$\mu(A) + \mu(B) = \Delta_f G^\circ(AB) \tag{3}$$
$$\mu(C) = \Delta_f G^\circ(C) \tag{4}$$
$$\mu(A) + \mu(C) = \Delta_f G^\circ(AC) \tag{5}$$

By solving these equations, the elemental chemical potential, $\mu(B)$, can be obtained as follows;

$$\mu(B) = \Delta_f G^\circ(AB) + \Delta_f G^\circ(C) - \Delta_f G^\circ(AC) \tag{6}$$

A criterion for whether reaction 1 may proceed can be converted to that for whether the phase B can be formed from a mixture of equilibrated AB, AC and C. This can be made by checking the activity of B corresponding to equation 6;

$$RT \ln a(\text{B}) = \mu(\text{B}) - \Delta_f G°(\text{B})$$
$$= \Delta_f G°(\text{AB}) + \Delta_f G°(\text{C}) - \Delta_f G°(\text{AC}) - \Delta_f G°(\text{B})$$
$$= - \Delta_r G°[\text{eq. 1}] \tag{7}$$

When $a(\text{B})>1$, the phase combination of AB, AC and C is thermodynamically unstable against precipitation of phase B; this precipitation should proceed until AB or C is completely consumed through equation 1. As a result, another phase combination of AC, C and B or AB, AC and B can be thermodynamically stable; see tie–lines in figure 1(a). Note that equation 7 is exactly the same criterion as equation 2. Thus, chemical reactions can be represented by phase relations. The most important feature of equation 7 is that the reactions can be described in terms of the chemical potentials of the three elements; this makes it possible to treat the reactions at any place where chemical potentials of all elements involved can be determined.

 Figure 1 shows schematically a three dimensional chemical potential diagram for the A–B–C system. The elements can be represented by those planes which are perpendicular to their respective chemical potential axis and also are perpendicular to each other. The planes for A, B and C form a corner of a cube. The elements A and B react to form the compound AB. This reaction can be geometrically repre-sented by a flat plane AB which forms a facet in the corner of the cube separating the planes A and B. Similarly, the compound AC can be shown by another flat plane separating the planes A and C. A three phase combination of AB, AC and C can be represented by the intersection of the three flat planes AB, AC and C. As described above, this combination is unstable against the phase B precipitation. This is shown by the geometrical feature that plane B cuts off the intersection of the planes AB, AC and C to form the stable intersections of AB, AC and B and AC, B and C.

 The chemical potential diagram is set up by generating all possible phase combinations, checking the thermodynamic stability of each phase combination and deriving valid phase relations [7]. This procedure is equivalent to generating all possible chemical reactions and checking their Gibbs energy change.

REACTION PATH IN CHEMICAL POTENTIAL DIAGRAM

 When a solid–solid reaction occurs, the nucleation of new phases should occur at interfaces along with the associated mass transfer. In the present treatment, kinetic barriers for nucleation are neglected as a first order of approximation; in other words, the most stable phases can be formed without a kinetic barrier. A local thermodynamic equilibria approximation is assumed at interface vicinity; that is, the chemical potentials can change continuously within a phase and have the same value at both sides of phase boundaries. The mass transfer (flux) of species i (J_i) can be described in terms of the chemical potential gradient ($d\mu_i/dx$) and a phenomenological coefficient (L_i) as follows;

$$J_i = - L_i \, (d\mu_i/dx) \tag{8}$$

This equation suggests that the chemical potential of diffusing species i should decrease in its diffusing direction. Since the present system consists of three elements, they can diffuse independently within a particular phase, $A_l B_m C_n$, keeping the following thermodynamic relation

$$\Delta_f G°(A_l B_m C_n) = l \, \mu(\text{A}) + m \, \mu(\text{B}) + n \, \mu(\text{C}) \tag{9}$$

For example, figure 1 shows that in the flat plane of phase B, the chemical potential

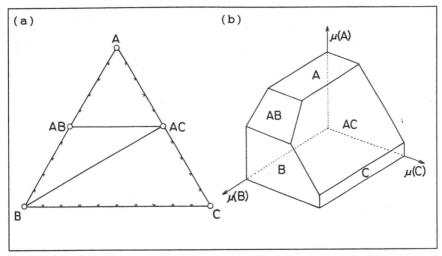

Figure 1. (a) triangle compositional phase diagram and (b) schematic three dimensional chemical potential diagram for the A–B–C system which contains AB and AC as binary compounds.

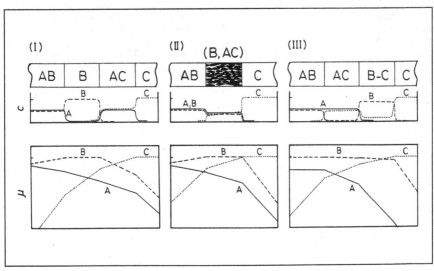

Figure 2. Schematic morphological arrangements, concentration profiles and chemical potential profiles for three plausible reaction paths in reaction AB+C=AC+B:

(I) layered arrangement I: AB/B/AC/C. Atoms A diffuse in reaction product B, while B atoms do not diffuse. After up–hill diffusion in phase B, atoms A react with atoms C to form AC; through phase AC, either atoms C or A diffuse. A typical example is Cu_2O/Ni diffusion couple [6].

(II) aggregated arrangement: AB/(B,AC)/C. Mainly atoms C diffuse to form phases AC and B. Neither atoms A nor B diffuse within phases AC and B. A typical case is SiC/Ni diffusion couple [8].

(III) layered arrangement II: AB/AC/B–C/C. Atoms A cannot diffuse in phases AC and B. Atoms B diffuse in phase AC. In order for atoms B to diffuse in phase AC and to form phase B, phase B should be in alloys with C. Otherwise, atoms B cannot diffuse against the chemical potential hill. Both atoms C and B exhibit up–hill diffusion. A typical example is Cu/Ni_3S_2 diffusion couple [1].

of B is fixed at a(B)=1, while the chemical potentials of A and C can change within that plane; thermodynamically, this means an infinitely dilute solution in which the activity of solvent is equal to unity regardless of the concentration of solutes. Respective points in the plane correspond to dilute solutions at different compositions. Thus, there can arise gradients in concentration and chemical potential of A in phase B which drive the diffusion of A atoms. Similarly, the chemical potential of B can be defined as the dissolved state in phase AC. This allows B atoms to diffuse in phase AC. In principle, all three elements can diffuse in the reaction products AC and B. Even so, the extent of diffusion of elements depends on respective coefficients, L_i. This means that the diffusion path in a particular phase is determined by diffusion properties within the phase and also by chemical potential gradients. Typically, three different morphological arrangements can be developed at the interface due to the difference in diffusion properties: The chemical potential profiles for each morphology were derived from the geometrical feature of the three dimensional chemical potential diagram and are shown in figure 2. All kinds of reaction profiles observed in the ternary systems can be regarded as a combination of the above three types of phase arrangements.

REACTIONS OF ZrO_2-BASED COATING WITH VANADIUM OXIDE

Since materials for practical use often contain many elements as dopants or impurities, the chemical reactivity of the materials should be determined by the behavior of the most reactive component in particular circumstances. For example, zirconia shows relatively good chemical stabilities against sulfur dioxide and vanadium oxides. However, the dopants used to stabilize the cubic phase are unfortunately weak against attack from sulfur dioxide or vanadium oxide. Figure 3 shows phase relations in the ZrO_2-Y_2O_3-V_2O_5 system at 1223 K and the associated chemical potential diagram: since no thermodynamic data is available for ZrV_2O_7, we neglect this phase. In our calculation, the cubic solid solution, $(Zr,Y)O_2$, is in equilibrium with the monoclinic ZrO_2 at this temperature. Figure 3 shows that fully stabilized zirconia (cubic phase) reacts with vanadium oxide to form the monoclinic phase and YVO_4, which is in good agreement with experimental results.

When stabilized zirconia is attacked by S and Na in addition to V, corrosion becomes severe. Jones [2] summarized results of reaction experiments of ceramic oxides with V_2O_5 (mp 954 K), $NaVO_3$ (mp 883 K), and Na_3VO_4 (mp 1539 K). For Y_2O_3 and CeO_2, we construct the chemical potential diagrams at 973 K as shown in figure 4. Y_2O_3 reacts with V_2O_5 and $NaVO_3$ to form YVO_4, whereas Y_2O_3 does not react with Na_3VO_4. This reaction behavior corresponds well to the geometrical features in the chemical potential diagram. On the other hand, CeO_2 does not react with Na_3VO_4 nor $NaVO_3$; it reacts only with V_2O_5 to form $CeVO_4$. This is also shown thermodynamically in figure 4. Although Jones gave an explanation based on the acid–base concept for the reactions above, an alternative explanation can be given in terms of the valence stability and the stabilization energy in double oxides [4,9]. When comparison is made between yttria and ceria in the chemical potential diagrams, the following difference can be derived; that is, the stability field of CeO_2 is wider than that of Y_2O_3, whereas $CeVO_4$ has a narrower field than YVO_4. This is due to the following thermodynamic features about the stabilization. The stabilization energy of $LnVO_4$ (Ln=Rare Earth) from Ln_2O_3 and V_2O_5 increases with increasing ionic size of Ln^{3+}. In this sense, $ScYO_4$ is less stable than YVO_4, and YVO_4 is less stable than $CeVO_4$. The comparison of stability between $ScVO_4$ and YVO_4 is in accordance with the acid–base theory, while however the difference between YVO_4 and $CeVO_4$ contradicts it. Note here that CeO_2 is more stable than Ce_2O_3. This lower valence stability of Ce^{3+} weakens the driving force for the formation of $CeVO_4$ from V_2O_5 and CeO_2. This is the thermodynamic reason why reac-

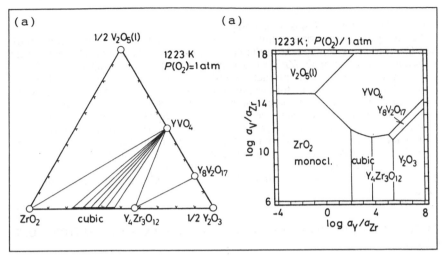

Figure 3. Phase relations in the ZrO_2–$YO_{1.5}$–$VO_{2.5}$ system at 1223 K. (a) Calculated phase diagram; (b) chemical potential diagram.

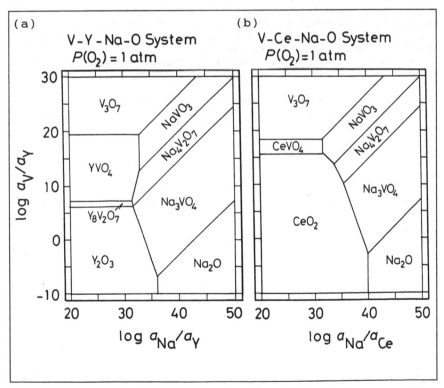

Figure 4. Chemical potential diagrams for (a) Y–V–Na–O and (b) Ce–V–Na–O systems at 973 K and $P(O_2) = 1$ atm.

tivity of CeO_2 with $NaVO_3$ is weak compared with Y_2O_3.

Jones pointed out that although neither ZrO_2 nor CeO_2 alone reacts with $NaVO_3$, the ceria component in the ceria–stabilized zirconia was leached from ZrO_2 and segregated on the ZrO_2 surface as an oxide not as a vanadate. He tried to explain it by the acid–base interaction without success [4]. The thermodynamic explanation can be given as follows: although ceria–stabilized zirconia is metastable at low temperatures, the precipitation of new phases is kinetically hindered. In the presence of the $NaVO_3$ melt, the dissolution–precipitation process can assist the reaction to achieve a true equilibrium. This is why ceria segregated as an oxide. This reaction can be categorized as a different one from that for yttria stabilized zirconia. The present analysis suggests that the reactivity of stabilized ZrO_2 with V_2O_5 can be interpreted in terms of the equilibrium properties described above. This approach can be easily extended to more complicated systems containing sulfur.

ELECTRODE/ELECTROLYTE INTERACTION IN SOLID OXIDE FUEL CELLS

Electrode and electrolyte materials used in electrochemical devices, such as sensors or solid oxide fuel cells, are usually selected based on their electronic and ionic properties. However, the chemical stability between two materials becomes of great importance since high temperature processing is required to achieve good mechanical and electrochemical contact. During this process, there may arise some possibility that chemical reactions take place to form electrically non–conductive substances at the electrolyte/electrode interface.

We made a series of chemical thermodynamic analyses on the reaction between YSZ and $LaMO_3$(M=Transition Metal) [4,10]. The main results can be summarized as follows:

(1) The reactions between the perovskite oxide electrode, $LaMO_3$, and zirconia can be also interpreted thermodynamically in terms of the stabilization energy of $LaMO_3$ and the valence stability between MO and M_2O_3 [4]. For example, $LaCoO_3$ reacts with ZrO_2 as follows;

$$LaCoO_3 + ZrO_2 = 1/2\ La_2Zr_2O_7 + CoO + 1/4\ O_2(g) \qquad (10)$$

Although Co^{3+} ions are stabilized in the perovskite lattice, Co^{3+} ions can be easily reduced to the divalent state in the presence of ZrO_2. On the other hand, Mn^{3+} is stable even in the binary oxide. This difference in the valence stability between Mn^{3+} and Co^{3+} gives rise to the essentially different nature in the $La_2Zr_2O_7$ formation between $LaCoO_3$ and $LaMnO_3$ perovskites [4].

(2) For the case of $LaMnO_3$, two different kinds of reactions should take place with zirconia. One is the reductive reaction which can be written as

$$LaMnO_3 + ZrO_2 = 1/2\ La_2Zr_2O_7 + MnO + 1/4\ O_2(g) \qquad (11)$$

where the formation of $La_2Zr_2O_7$ and the reduction of manganese ions take place simultaneously. Since the valence stability of Mn^{3+} is relatively high, the above reaction can occur only at the reductive atmosphere around 1300 K. The reaction type is the same as equation 10; since this reaction should proceed until one of the reactants is completely consumed, this can be regarded as destructive. The driving force becomes weak with decreasing temperature because of its reductive nature.

Another reaction is related to the A–site deficiency [11]. Since Mn^{4+} ions are relatively stable in the perovskite structure, there arises the lanthanum nonstoichiometry; that is, y in La_yMnO_3 can range over some compositional region. The

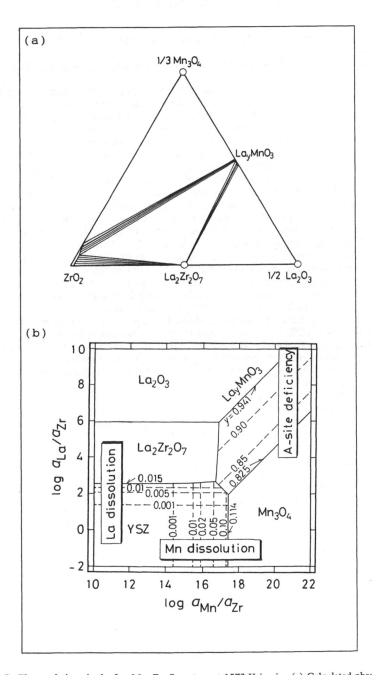

Figure 5. Phase relations in the La–Mn–Zr–O system at 1573 K in air. (a) Calculated phase diagram; (b) chemical potential diagram. The stability region of La$_y$MnO$_3$ can be divided into two regions in terms of the reactivity with ZrO$_2$. Although the A–site deficient manganite does not react with ZrO$_2$, the manganese dissolution may become significant.

second type reaction with ZrO_2 is associated with change in y as follows;

$$La_yMnO_3 + (y-y')ZrO_2 + [3(y-y')/2]O_2(g) = La_{y'}MnO_3 + [(y-y')/2]La_2Zr_2O_7 \quad (12)$$

where change in the oxygen stoichiometry in the perovskite lattice is neglected. This reaction can be characterized as follows; only a small amount of the La_2O_3 component comes out of the perovskite lattice and reacts with ZrO_2 to form $La_2Zr_2O_7$. This is not destructive. On this reaction, the La stoichiometry, y, in the perovskites changes to a smaller value, and the driving force for the $La_2Zr_2O_7$ formation eventually ceases when y reaches the critical value and perovskite, zirconia and $La_2Zr_2O_7$ are equilibrated with each other. These properties are shown as a phase diagram and also as a chemical potential diagram in figure 5. Since this is accompanied by the oxygen absorption, the driving force becomes strong with decreasing temperature in contrast to the first type reaction. The important point of this reaction is that the stability region of the lanthanum nonstoichiometry can be divided into two regions in terms of the reactivity with ZrO_2; that is, the significantly A–site deficient lanthanum manganite should not react with ZrO_2.

(3) The manganese dissolution into YSZ is remarkable [12]. In particular, the manganese activity becomes high in the A–site deficient manganite. When the A–site deficient manganite is in contact with YSZ, manganese dissolution may take place. This gives rise to an increase in the La stoichiometry, y. With manganese dissolution, y approaches that of the three phase equilibrium.

(4) The manganese dissolution and the $La_2Zr_2O_7$ formation may affect electrode activities in different ways. Experimental investigation has revealed that the electrochemical activity at $y=0.9$ is higher than in the vicinity of $y=1.0$, while the manganese dissolution does not cause any serious change in the ionic and electronic conductivities. This suggests that an optimal A–site deficiency level seems to locate around $y=0.9$.

(5) When SrO or CaO are doped to lanthanum manganites, the A–site deficient limit can change as a function of dopant concentration. For the case of SrO doping, alpha–$SrMnO_3$ is precipitated at lower temperatures; this gives the limit of Sr content as a function of temperature. Similarly, zirconates formed in reaction with ZrO_2 change with Sr content from $La_2Zr_2O_7$ to $SrZrO_3$, although the chemical nature related to the A–site deficiency does not change. Figure 6 shows that the perovskites, $(La_{1-x}Sr_x)MnO_3$, can be in equilibrium with either $La_2Zr_2O_7$ or $SrZrO_3$; the perovskite at a particular composition can be in equilibrium with both. This composition should depend on the thermodynamic properties of perovskite solid solution and of $SrZrO_3$ and $La_2Zr_2O_7$. Although phase relations in figure 6 were obtained from the complicated chemical equilibria calculations, a simple analysis may help to understand what kinds of chemical processes are involved. We assume here a simple ideal solution of $(LaMnO_3)_{1-x}(SrMnO_3)_x$ to derive the composition at which the perovskite is in equilibrium with both $SrZrO_3$ and $La_2Zr_2O_7$. The Sr content, x, can be calculated from the following equation:

$$SrMnO_3 + 1/2\ La_2Zr_2O_7 = SrZrO_3 + LaMnO_3 + 1/4\ O_2(g) \quad (13)$$

$$\Delta_f G°(SrMnO_3) + RT \ln x + 1/2\ \Delta_f G°(La_2Zr_2O_7)$$
$$= \Delta_f G°(SrZrO_3) + \Delta_f G°(LaMnO_3) + RT \ln (1-x) + RT \ln P(O_2) \quad (14)$$

$$RT \ln [x/(1-x)] = \Delta_r G°[eq.\ 13]$$
$$= \delta(LaMnO_3) - 1/2\ \delta(La_2Zr_2O_7) - \delta(SrMnO_3) + \delta(SrZrO_3) + \Delta(Mn^{4+},Mn^{3+}) \quad (15)$$

where δ is the stabilization energy of a double oxide from its constituent oxides and Δ is the valence stability between two valence states in binary oxides [4]. Again, the valence stability plays an important role in reactivity of manganites. In figure 7, the left and the right side values in the above equation are plotted separately. Calculated results based on the above equation is compared with those obtained from the

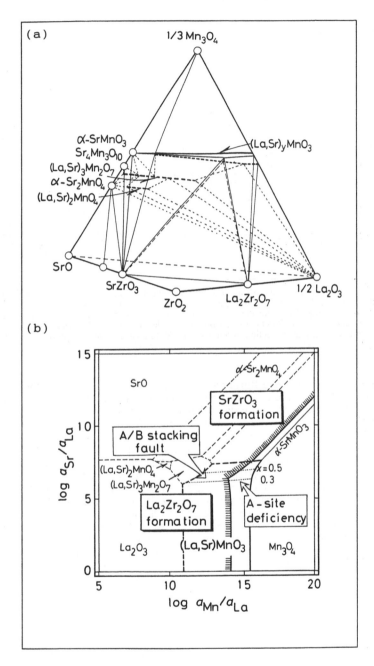

Figure 6. Phase relations in the La–Sr–Mn–Zr–O system at 1573 K in air. (a) Calculated pseudo-ternary phase diagram; (b) chemical potential diagram. An emphasis is placed on the relation between perovskite and $La_2Zr_2O_7$ and/or $SrZrO_3$; phase relations associated with ZrO_2 are omitted for simplicity.

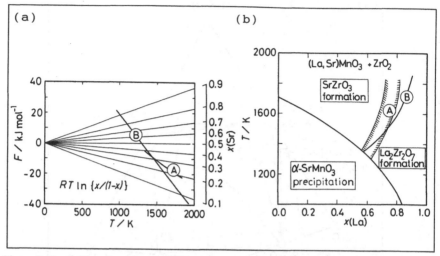

Figure 7. Thermodynamic analyses on the strontium content at which SrZrO$_3$ and La$_2$Zr$_2$O$_7$ are both formed in reactions between ZrO$_2$ and (La$_{1-x}$Sr$_x$)MnO$_3$: (a) relation between the strontium content and the Gibbs energy change for equation 13; Line A designates results of complicated chemical equilibria calculation, Line B being those from the simple analysis based on equation 13. (b) calculated strontium content as a function of temperature. Within zone A, both La$_2$Zr$_2$O$_7$ and SrZrO$_3$ are formed.

complicated chemical equilibria calculations. Note that the magnitude and the fundamental temperature dependence are about the same between the two results; a slight difference in temperature dependence is due to the A–site deficiency and the oxygen nonsotichiometry. Note also that even at low temperatures, the driving force for the zirconate formation does not disappear.

(6) Another aspect appears also in the alkaline earth doping cases; that is, the Ruddlesden–Popper phases other than perovskite phase become more stable. For example, in the La–Mn–O system, La$_2$MnO$_4$ appears only at high temperatures and at reductive atmosphere, whereas in the Ca–Mn–O system, Ca$_{n+1}$Mn$_n$O$_{3n+1}$ phases appear even at low temperatures and in air [4,13]. The Ruddlesden–Popper phases can be characterized as "A–site excess". Although the present calculation was made by neglecting the phases of $n>3$, it is reasonable to assume that the A/B ratio $(n+1)/n$ can continuously change from 1/1 in perovskite to 3/2. This suggests that the alkaline earth doped lanthanum manganites should have lattice defects of A–site vacancies in one side and in the other side of A/B stacking faults. The thermodynamic analysis suggests that the La–A–Mn–O Ruddlesden Popper phases can react with ZrO$_2$ to form the A–site deficient perovskite and La$_2$Zr$_2$O$_7$ and/or SrZrO$_3$.

CONCLUSIONS

The present study has clarified that thermodynamic stabilities of the interface can be readily characterized under the local equilibrium approximation in which the chemical potentials of elements involved play an important role. This approach has been applied to the reactions of zirconia–based ceramics with vanadium oxides and with lanthanum manganites to obtain relations between the reactivity and the thermodynamic properties. The main conclusions are (1) the destabilization of stabilized zirconia can be interpreted in terms of the magnitude of stabilization energy of

$LnVO_4$ phases and of the valence stability of Ln^{3+} ions, (2) the reactions between YSZ and lanthanum manganites can be characterized in terms of the A–site deficiency related to the valence stability of manganese ions and also of the difference in the stabilization energy among $La_2Zr_2O_7$, $LaMnO_3$, $SrZrO_3$, and $SrMnO_3$.

REFERENCES

1. van Loo, F. J. J., 1990. "Multiphase Diffusion in Binary and Ternary Solid–State Systems." Prog. Solid State Chem., 20:47–99.

2. Jones, R. L., 1990. "Oxide Acid–Base Reactions in Ceramic Corrosion." High Temp. Sci., 27:369–380.

3. Yokokawa, H., N. Sakai, T. Kawada and M. Dokiya, 1991. "Chemical Thermodynamic Compatibility of Solid Oxide Fuel Cell Materials," in Proc. Second Internatl. Symp. Solid Oxide Fuel Cells, Luxembourg: The Commission of European Communities, pp. 663–670.

4. Yokokawa, H., N. Sakai, T. Kawada and M. Dokiya, 1992. "Thermodynamic Stabilities of Perovskite Oxides for Electrodes and Other Electrochemical Materials," Solid State Ionics, 52(1):43–56.

5. Wagner, C., 1938. "On the Mechanism for Double Displacement During Solid State Reaction (in German)." Z. anorg. alleg. Chem., 236:320–338.

6. Rapp, R. A., A. Ezis and G. J. Yurek, 1973. "Displacement Reactions in the Solid State." Metal. Trans., 4(4):1283–1292.

7. Yokokawa, H., T. Kawada and M. Dokiya, 1989. "Construction of Chemical Potential Diagrams for Metal–Metal–Nonmetal Systems: Applications to the Decomposition of Double Oxides." J. Am. Ceram. Soc., 72(11):2104–10.

8. Schiepers, R. C. J., F. J. J. van Loo and G. de With, 1989. "The interaction between SiC and Ni or Fe." Defect and Diffusion Forum, 66–69(Pt III):1497–1502.

9. Yokokawa, H., N. Sakai, T. Kawada and M. Dokiya, 1990. "Chemical Potential Diagrams for Rare Earth–Transition Metal–Oxygen Systems:I, Ln–V–O and Ln–Mn–O Systems." J. Am. Ceram. Soc., 73(3):649–658.

10. Yokokawa, H., N. Sakai, T. Kawada, M. Dokiya, 1991. "Thermodynamic Analysis of Reaction Profiles Between $LaMO_3$ (M=Ni,Co,Mn) and ZrO_2." J. Electrochem. Soc., 138(9):2719–27.

11. Yokokawa, H., N. Sakai, T. Kawada and M. Dokiya, 1989. "Thermodynamic Analysis on Relation Between Nonstoichiometry of $LaMnO_3$ Perovskites and Their Reactivity with ZrO_2 (in Japanese)." Denki Kagaku, 57(8):829–836.

12. Yokokawa, H., N. Sakai, T. Kawada and M. Dokiya, "Phase Diagram Calculations for ZrO_2 Based Ceramics: Thermodynamic Regularities in Zirconate Formation and Solubilities of Transition Metal Oxides." in this volume.

13. Yokokawa, H., N. Sakai, T. Kawada and M. Dokiya, 1991. "Thermodynamic Stability of Perovskites and Related Compounds in Some Alkaline Earth–Transition Metal–Oxygen Systems." J. Solid State Chem., 94(1):106–120.

Chemical Stability of Perovskite Interconnects in Solid Oxide Fuel Cell

NATSUKO SAKAI, TATSUYA KAWADA, HARUMI YOKOKAWA
and MASAYUKI DOKIYA

ABSTRACT

The chemical degradations on the surfaces of sintered alkaline earth doped lanthanum chromites were investigated under simulated SOFC operating condition for 200 h. Alkaline earth chromates such as $Ca_m(CrO_4)_n$ or $SrCrO_4$ migrated on the surfaces which were directly exposed to air or fuel gas. The migrated second phases decomposed on the surface which is exposed to fuel gas. The extent of the degradation depends on the amount and the location of alkaline earth chromates. This migration was not observed at the interface between lanthanum chromite and dense platinum electrode.

INTRODUCTION

For recent several years, the main effort of Solid Oxide Fuel Cell (SOFC) development has been on the interconnect materials. To satisfy the requirements such as high electronic conductivity and high chemical stability, alkaline–earth doped lanthanum chromites have been adopted in tubular SOFC. Isenberg reported that $LaCr_{1-x}Mg_xCrO_3$ was used as interconnects in Westinghouse type tubular cell, and Electrochemical Vapor Deposition (EVD) process was used in reducing atmosphere to attach the $LaCr_{1-x}Mg_xCrO_3$ to YSZ porous support tube [1]. Since the area of interconnect is not large in the tubular cell, such a dry process is enough to obtain the gas–tight layer of interconnects. However in planar SOFCs, interconnects are required to have a large area which is comparable to that of electrolyte, and in some cell designs, sufficient mechanical strength is necessary to support other cell components. That is why most of the efforts have been made for the improvement of sinterability of doped lanthanum chromites which are not sintered well in air. Addition of sintering aids is an effective technique because it does not need any special treatment such as hot pressing or sintering in reducing atmosphere. For example, the sintering of calcium doped lanthanum chromites is enhanced by the presence of calcium oxychromate ($Ca_m(CrO_4)_n$) as a second phase [2]. Dense samples were obtained in this case by sintering at 1573 K in air. Strontium doped lanthanum chromites are thought to be more stable materials than calcium doped

Natsuko Sakai, Tatsuya Kawada, Harumi Yokokawa and Masayuki Dokiya, National Institute of Materials and Chemical Research, Tsukuba Research Center, Tsukuba, Ibaraki 305, JAPAN

samples, and they are widely used in planar SOFC interconnects. Since it is generally hard to obtain reproducible good sinterability in $La_{1-x}Sr_xCrO_3$ itself, doping of cobalt components is adopted to enhance the densification [3]. However, there are few reports which discuss the effects of such sintering aids or dopants on chemical stability of lanthanum chromites.

The present paper reports the surface degradation of sintered lanthanum chromites under simulated SOFC operating condition. Effects of sintering aids and dopants are summarized to discuss the compatibility of doped lanthanum chromites as SOFC interconnects.

EXPERIMENTAL PROCEDURE

SAMPLE IDENTIFICATION

The composition and preparation condition of all samples used in the present work are listed in TABLE I. In calcium doped samples, {(La,Ca)CrO_3, samples LCC-1; 2, 3} the amounts of the A–site metals (La+Ca) are set to be larger than that of the B–site metal(Cr) to raise the sinterability in air. Sample LSC-1{(La,Sr)CrO_3} and sample LSCC-1 {(La,Sr)(Cr,Co)O_3} were prepared to be compared with (La,Ca)CrO_3.

CHEMICAL STABILITY TEST

Chemical stabilities of samples were examined under simulated SOFC operating conditions. A schematic view of the experimental arrangement was previously reported, and it is shown in figure 1 [4]. A sintered sample was shaped into a small disk (20 mmφ, 1.5 mm thick), its both surfaces were polished and a platinum electrode (6 mmφ) was attached on each side. The sample was placed between a YSZ tube and an Al_2O_3 tube. Fuel gas mixture {$p(H_2) : p(H_2O) = 98 : 2$} flowed in the YSZ side (40 ml·min^{-1}), and dried air in the other side (100 ml·min^{-1}). An

TABLE I – COMPOSITION, PREPARATION CONDITION AND DENSITIES OF THE SAMPLES.

Sample number	Nominal composition (molar ratio)					Preparing method	Calcining temp.	Sintering temp.	Relative*2 Density(%)
	La	Ca	Sr	Cr	Co				
LCC–1 *1	0.70	0.32		1.00		Oxalic salt [2]	1273 K	1573 K	96
LCC–2 *1	0.70	0.32		1.00		Oxalic salt [2]	1273 K	1873 K	91
LCC–3 *1	0.85	0.18		1.00		Oxalic salt [2]	1273 K	1873 K	84
LSC–1	0.90		0.10	1.00		Pechini [7]	1373 K	1773 K	91
LSCC–1	0.80		0.20	0.90	0.10	Pechini [7]	1373 K	1773 K	91

*1 In LCC–1, 2 and 3, the amount of (La+Ca) was set to about 2 or 3 mol% excess to Cr to obtain the dense sample by sintering in air.
*2 Relative density is d/d_0, where d is the sample density and d_0 is the theoretical density determined from the lattice parameter and the chemical formula of each sample.

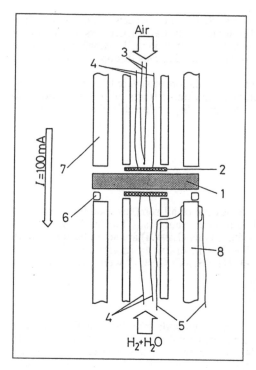

Figure 1. Schematic view of experimental apparatus for chemical stability test [4]: 1– Sample, 2– Pt mesh, 3– Thermocouple, 4 Pt–wire to send an electric current, 5– Pt probes to measure the oxygen partial pressure, 6– Au ring packing, 7– Al_2O_3 tube, 8– YSZ tube. Pt mesh was attached to the sample with Pt paste.

electric current ($j = 350$ mA·cm^{-2}) was sent from the air side to the fuel side. The sample was held at 1273 K for 195 – 205 h. After the test, all samples were examined in SEM/EDX analyses.

RESULTS

The results of chemical stability test are summarized in figures 2 to 6. In any calcium doped samples (figures 2, 3, 4), it is observed that calcium content became high on the surfaces which were exposed to air or fuel gas. In LCC–1, the surface was covered with second phases which were like quenched liquid phase (figures 2–2, 2–3). EDX results of points A and B in figure 2–3 showed that this second phase consisted of mainly calcium and chromium, and sometimes impurities such as aluminum or silicon components were concentrated in it. On the fuel side, the liquid phase appeared to be decomposed to CaO and $CaCr_2O_4$ in some parts, although it is not shown in the figures. The sample exhibited the intra granular fracture (figure 2–4), whereas the as– sintered sample exhibits the typical inter granular fracture (*see* figure 7–1).

In LCC–2, it is observed that both surfaces were covered with second phases, however, their calcium contents were higher than that observed on the surfaces of LCC–1. On the air side, a thick layer of CaO was detected. It should be noticed that second phases which mainly contain calcium and chromium were found at the triple points in the fracture (figure 3–5). As concerns fracture morphology, LCC–2 exhibited the mixture of inter granular and intra granular fractures like the as–sintered LCC–2 sample.

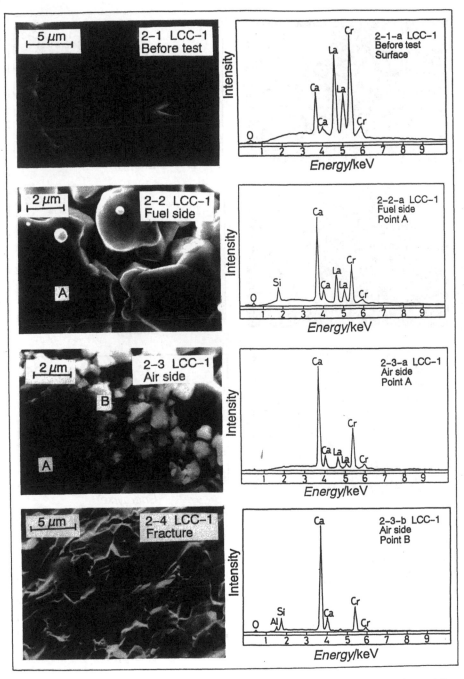

Figures 2. SEM images and composition of the surfaces of LCC–1 before and after chemical stability test.

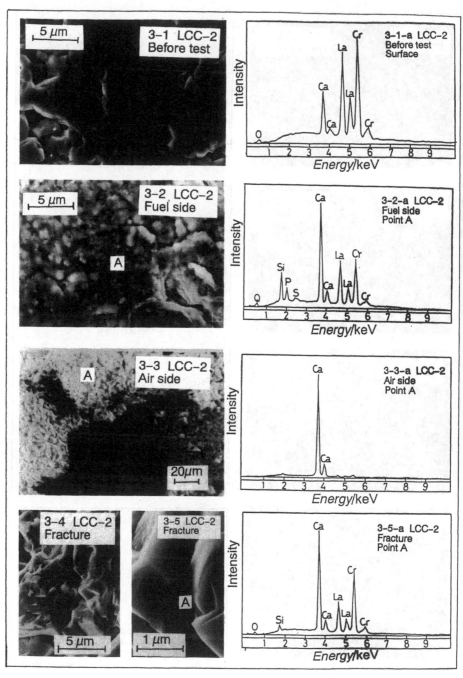

Figures 3. SEM images and composition of the surfaces of LCC–2 before and after chemical stability test.

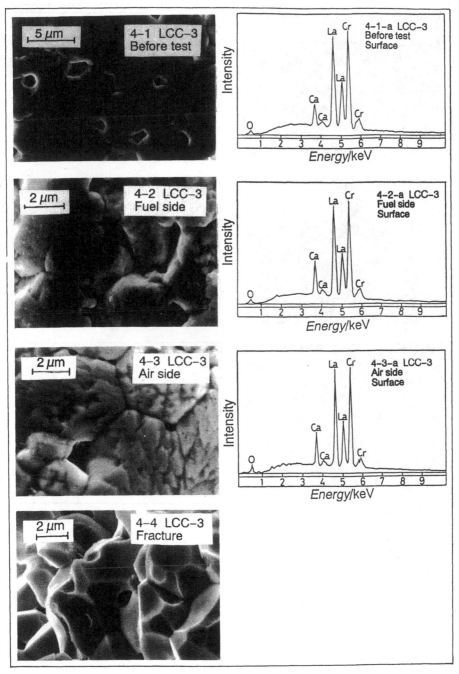

Figures 4. SEM images and composition of the surfaces of LCC–3 before and after chemical stability test.

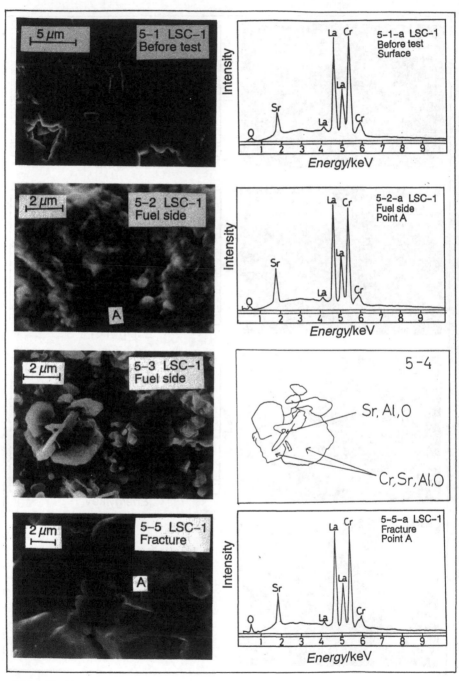

Figures 5. SEM images and composition of the surfaces of LSC–1 before and after chemical stability test.

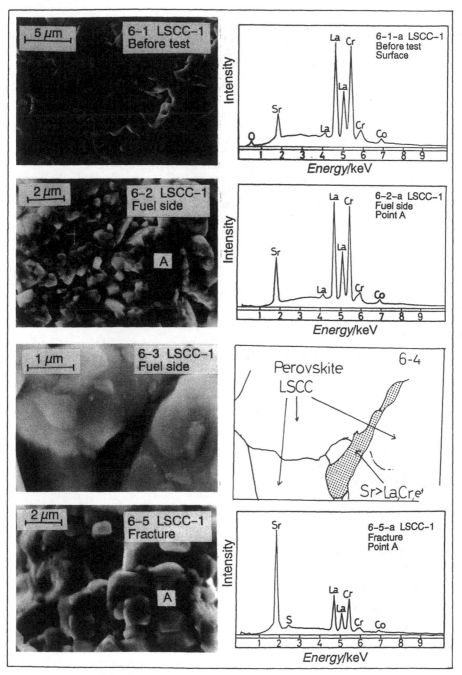

Figures 6. SEM images and composition of the surfaces of LSCC–1 before and after chemical stability test.

In LCC–3, it appeared that almost no second phases were detected on the air side although the calcium content became slightly larger compared to the initial composition (figure 4–3). On the fuel side, CaO and Cr_2O_3 was observed in a small amount, although it is not shown in the figures. There was little second phase in fracture. The morphology of the fracture which has many pores implies that this sample did not sinter very well because of the small amount of second phase.

Concerning LSC–1, strontium contents became larger on both surfaces after the chemical stability test. At several points on the fuel side of LSC–1, the strontium rich compounds were decomposed into strontium oxide and chromium oxide (figure 5–3, 5–4), which implies that it originates from $SrCrO_4$ or its related compounds. In the fracture near the fuel side, second phases that have slightly higher strontium content were found between the LSC perovskite grains (figure 5–5).

In LSCC–1, about the same results as those of LSC–1 were obtained. We could not observe any migration of cobalt component in LSCC–1, whereas the migration of strontium rich compounds via grain boundary was observed in figures 6–3 and 6–4. Strontium rich compounds were also detected between the LSCC perovskite grains in the fracture (figure 6–5).

DISCUSSION

EFFECTS OF SECOND PHASES ON THE CHEMICAL STABILITY

Calcium Doped Lanthanum Chromites

The present experimental results indicate that the degradation of sintered lanthanum chromites are influenced by the existence of alkaline earth chromates. In calcium doped samples (LCC–1, 2, 3), a small amount of calcium oxychromate

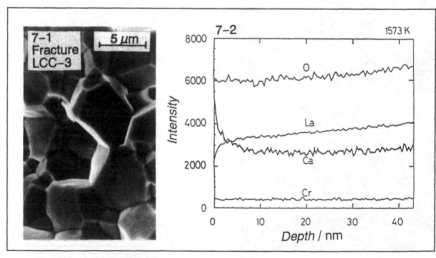

Figures 7. SEM image of general fracture as–sintered (before the test) LCC–1 (7–1) and AES depth profile on a fracture (7–2) [2]. Since LCC–1 exhibits the typical inter granular fracture, the composition of grain boundary may remain on the fracture surface. However, the point on which AES profile was obtained does not exactly correspond to the SEM image. Calcium intensity is abnormally high at the depth = 0 which indicates that some calcium rich compounds remain at the grain boundary, or some compositional distribution is generated as a thin layer.

$(Ca_m(CrO_4)_n, m > n)$ remained. $Ca_m(CrO_4)_n$ forms to compensate the excess of A–site metals, and it may incongruently melt at above 1273 K and may react with $La_{1-x}Ca_xCrO_3$, which enhances the grain growth of $La_{1-x}Ca_xCrO_3$ and results in the good sintering at 1573 K [2]. The Ca/Cr ratio in $Ca_m(CrO_4)_n$ which exists with $La_{1-x}Ca_xCrO_3$ increases with temperature, for example, from $CaCrO_4$ at 1273 K to almost CaO at 1873 K.

Sintering temperature influences the distribution of $Ca_m(CrO_4)_n$ in the sintered sample. In LCC–1 which was sintered at 1573 K, $Ca_m(CrO_4)_n$ exists as a very thin layer at the grain boundaries. The depth profile of Auger electron spectra (AES) on the fracture of LCC–1 shows the existence of calcium rich region near the depth = 0, which indicates that some calcium rich compounds existed at grain boundaries at the sintering temperature (figure 7–2 [2]). In LCC–2 and LCC–3 which were sintered at 1873 K, second phases are thought to be located at triple points as a form of calcium rich $Ca_m(CrO_4)_n$ or CaO. The melting points of $Ca_m(CrO_4)_n$ are rather low, for example, $Ca_{10}(CrO_4)_7$ melts at 1528 K and $Ca_5(CrO_4)_3$ at 1548 K [5]. Hence, $Ca_m(CrO_4)_n$ may migrate with its high cation diffusivity during the chemical stability test. This well explains the present experimental results.

The extent of chemical degradation of sintered LCC samples seems to depend on the amount and location of $Ca_m(CrO_4)_n$ as a second phase in polycrystalline LCC. In LCC–1, second phases exist as a thin layer of calcium rich region at grain boundaries, which results in the smooth migration of $Ca_m(CrO_4)_n$. During the test, the calcium rich region is continuously removed and rearrangement of ions will occur in the vicinity of the grain boundaries. After the test the calcium rich region near the grain boundaries becomes thin, and bonding of perovskite grains becomes tough, which causes the change of fracture morphology.

In LCC–2, second phases are located near the triple points of perovskite grains. Although CaO cannot move by itself because of its high melting point and low diffusivity, figure 3–5 shows that it may move as a form of $Ca_m(CrO_4)_n$ by reaction with surrounding $La_{1-x}Ca_xCrO_3$ perovskite phase. And its composition is rather calcium richer. It takes a long time to migrate from inner triple points to the surface because there is almost no calcium rich region at grain boundaries. Hence, the migration paths will form locally, which is the reason why we did not observe significant changes of fracture morphology after the test in LCC–2.

The results of LCC–3 suggest that chemical degradation is depressed by diminishing the amount of second phases. However, a certain amount of second phase is required to obtain the dense LCC by sintering in air, hence, it is necessary to find the smallest amount of $Ca_m(CrO_4)_n$ which is enough to sinter LCC. This amount will be greatly influenced by the homogeneity and grain sizes of LCC powders.

Strontium Doped Lanthanum Chromites

For calcium doped lanthanum chromites, there are little second phases in calcined powders if they are prepared without any excess or depletion of A–site metals. However, that is not the case for strontium doped lanthanum chromites. Although both LSC–1 and LSCC–1 were prepared without any excess or depletion, $SrCrO_4$ was found in the calcined powders calcined at 1373 K. $SrCrO_4$ is expected to be more stable than $CaCrO_4$, it may remain between the $La_{1-x}Sr_xCrO_3$ grains even after sintering at 1773 K.

The phase diagram of the $SrO–Cr_2O_3$ pseudo binary system by Negas *et al.* [6] helps us to understand the $SrCrO_4$ during chemical stability test. The stability region of liquid $SrCrO_4$ is wide in temperature and composition. This suggests that this liquid may enhance sinterability regardless the A–site excess or depletion. However, with increasing temperature, and lowering oxygen partial pressure, this liquid tends to decompose into Cr_2O_3 and SrO–rich liquid. Hence, it is thought that the strontium rich compound observed on the surfaces and fractures of LSC–1 and LSCC–1 originates from $SrCrO_4$ that remained between the $La_{1-x}Sr_xCrO_3$ grains even after sintering process. Some special preparation techniques or rather high calcination temperature are required to diminish the $SrCrO_4$ formation in air.

EFFECTS OF IMPURITIES

Generally, calcined powders contain a slight amount of impurities such as silicon, phosphorus, sulfur, aluminum, or iron components. They are often contained in starting materials, and sometimes in crucibles, or furnace elements which easily contaminate the powders. In the present experimental results, such impurities were often concentrated in alkaline earth chromates which migrated to the surfaces. The authors have reported elsewhere that silicon enhances the decomposition of $Ca_m(CrO_4)_n$ in reducing atmosphere by the formation of stable Ca_2SiO_4 [4].

COMPATIBILITY AS SOFC INTERCONNECTS

The chemical degradation of alkaline–earth doped lanthanum chromites was schematically clarified in the present experiments. Since the SOFC operating condition is severe, the migration and the decomposition of second phases are not surprising results. The important points are to clarify how this degradation affects the electrical conductivity or mechanical strength of sintered samples which are the main requirements for SOFC interconnects.

Electrical Conductivity

Most of the alkaline earth chromates have low electrical conductivity, and their decomposition products (CaO, SrO, Cr_2O_3, $CaCr_2O_4$, etc.) are insulators. Hence, the formation of these compounds would drastically increase the contact resistance if they form at the interfaces between the interconnect and the other cell components. However, the present study shows interesting results in that such compounds mainly form on the surface which is directly exposed to the air or fuel gas mixture, but not at the interface of the sample and the platinum electrode. As a result, increases of the ohmic resistance was not too high (from 8 to 10 % increase for 200 h treatment).

Reactivity With Other Cell Components

Generally interconnects are in contact with the porous electrodes in planar SOFCs. Interconnects are partially exposed to air and fuel gas, and the migration of alkaline earth chromates will occur in this case. $Ca_m(CrO_4)_n$ may react with the cathode material ($La_{1-x}Sr_xMnO_3$), which results in the dissolution of calcium ions into the $La_{1-x}Sr_xMnO_3$. Furthermore, $CaZrO_3$ may form at the interface between the interconnect and the anode (Ni–YSZ cermet), it may cause the degradation of the electrode activity. However, the authors have not confirmed detailed reactions of interconnects and electrodes at the SOFC operating conditions.

Mechanical Strength

Although the authors did not obtain any data about mechanical strength of the sintered lanthanum chromites, only speculations are shown in this section. It is often observed that the samples exhibited intra granular fracture after the chemical stability test. It is perhaps related to the formation of oxygen vacancies in the perovskite lattice. Alkaline earth doped lanthanum chromites liberate oxygen gas in a reducing atmosphere to form oxygen vacancies, the maximum number of vacan–

cies is related to their dopant concentration. Schafer et al. reported that oxygen vacancy formation in perovskite structure affects their thermal expansion [8]. Hence, some gradient of thermal expansion will occur in a sample under the SOFC operating condition, micro–cracks in the perovskite grains naturally form on heating or cooling the samples. Decreasing the dopant concentration is an appropriate way to overcome this problem. The effects of such micro cracks on the mechanical strength should be clarified by further experiments.

SUMMARY

The present results showed that both $Ca_m(CrO_4)_n$ and $SrCrO_4$ migrate from the inner grain boundaries or triple points to the surface of sintered samples, and they decompose on the surface exposed to reducing gas. The precipitation of such second phases is limited on the surfaces which are exposed to air or reducing gas directly. It is hard to depress this migration in SOFC operating conditions. Its migration behavior depends on the amount and the location of such second phases between the perovskite grains. To diminish the total amount of them is important to avoid degradation, which requires careful control of preparation conditions.

REFERENCES

1. Isenberg, A. O., 1981. "Energy Conversion via Solid Oxide Electrolyte Electrochemical Cells at high Temperatures." Solid State Ionics, 3/4:431–437.

2. Sakai, N., T. Kawada, H. Yokokawa, and M. Dokiya, 1993. "Liquid Phase Assisted Sintering of Calcium Doped Lanthanum Chromites." To be published in Journal of American Ceramic Society.

3. Milliken, C., S. Elangovan and A. Khandkar, 1990. "Interface Reaction and Sintering of Doped $LaCrO_3$ Materials," in Proceedings of the International Symposium on Solid Oxide Fuel Cells, O. Yamamoto, M. Dokiya, and H. Tagawa, eds., Tokyo, Japan: Science House Co. Ltd., pp. 50–57.

4. Sakai, N., T. Kawada, H. Yokokawa and M. Dokiya, 1992. "Air Sinterable $(La,Ca)CrO_3$ for SOFC Interconnectors: Some Problems in Powder Preparation," in Proceedings of International Fuel Cell Conference, Tokyo, Japan: New Energy and Industrial Technology Development Organization (NEDO), pp.365–368.

5. Havlica, J. and V. Ambrúz, 1985. "The Influence of the Partial Pressure of Oxygen on the Phase Stability in the System Ca–Cr–O." Thermochimica Acta, **93**: 357–340.

6. Negas, T. and R. S. Roth, 1969. "The System SrO– Chromium Oxide in Air and Oxygen." Journal of Research of the National Bureau of Standards – A. Physics and Chemistry,73A(4): 431–442.

7. Pechini, M. P., 1967. "Method of Preparing Lead and Alkaline Earth Titanates and Niobates and Coating Method Using the Same to Form a Capacitor." U. S. Patent No. 3,330,697.

8. Schafer, M. and R. Schmidberger, 1987. "Ca and Sr Doped $LaCrO_3$: Preparation, Properties and High Temperature Applications," in High Tech Ceramics, P. Vincenzini, ed., Amsterdam: Elsevier Science Publishers B.V., pp. 1737–1742.

Reaction between Alkaline Earth Metal Doped Lanthanum Chromite and Yttria Stabilized Zirconia

M. MORI, H. ITOH, N. MORI, T. ABE, O. YAMAMOTO, Y. TAKEDA and N. IMANISHI

ABSTRACT

The solid state relations between alkaline earth metal doped lanthanum chromite and 7.5 mole % yttria stabilized zirconia (7.5YSZ) in air have been investigated. A reaction product of the mixture of 7.5YSZ and $(La_{0.7}Sr_{0.3})CrO_3$ heated at 1500 °C for 24 h was found to be $Sr_4Zr_3O_{10}$. $(La_{1-x}Sr_x)CrO_3$ (x = 0.1 and 0.2) showed no reaction product with 7.5YSZ at 1500 °C for 168 h. In the couple of 7.5YSZ and $La(Cr_{0.8}Mg_{0.2})O_3$, $La_2Zr_2O_7$ was observed after heating at 1500 °C for 96 h. No reaction product between 7.5YSZ and $(La_{1-x}Ca_x)CrO_3$ (x = 0.1, 0.2 and 0.3) was found at 1500 °C for 168 h. It was found that all the elements of the perovskites diffused into 7.5YSZ at 1500 °C.

INTRODUCTION

High-temperature solid oxide fuel cells (SOFC) are quite attractive electric power generation systems, because of the simple system design, the high energy conversion efficiency and the various utilization of by-product heat. From the cell-structural point of view, two types of SOFC are being developed. One is a tubular type SOFC; the performance of 25 kW class units of this type cell is demonstrated [1]. Another type is a planar configuration cell; the planar type SOFC has succeeded in generating a power of 1.3 kW in 1991 [2]. However, there are still many material problems to be solved for obtaining a high performance SOFC, especially the interdiffusion of elements in the cell components.

M. Mori, H. Itoh, N. Mori and T. Abe, Advanced Fuel Cell Section, Advanced Energy Department, Central Research Institute of Electric Power Industry, 2-6-1 Nagasaka, Yokosuka, Kanagawa, 240-01, Japan

O. Yamamoto, Y. Takeda and N. Imanishi, Department of Chemistry, Faculty of Engineering, Mie University, Tsu, 514, Japan

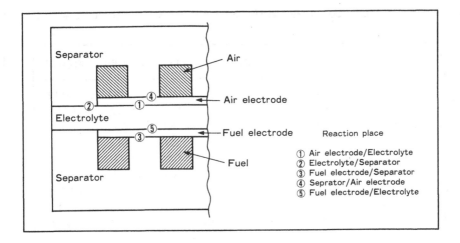

Figure 1. Contacting surfaces of cell components in planar type SOFC.

Many investigations have been reported for reactions among cell component materials [3-5]. Most of them, however, have been studied for the interaction of yttria stabilized zirconia (YSZ) electrolyte and lanthanum manganite air electrode. In SOFC systems, the gastight interconnector is essential for connecting the cells in series. In the case of the planar configuration, the interconnector has twofold functions, that is, separating fuel gas and air, and interconnecting single cells in series. These separators (interconnectors) contact directly with the electrolyte as shown in figure 1. The interface of the impervious separator and the impervious electrolyte plays an important role to seal fuel gas from air. Therefore, the reaction of the separator and the electrolyte should be made clear to obtain a high performance SOFC with long-term stability. Alkaline earth (AE) metal doped lanthanum chromite perovskites are the most promising candidate for the separator in SOFC, since the perovskites have a good electrical conductivity and an excellent chemical stability under oxidizing and reducing atmospheres at high temperature. In this study, we have investigated the reactivity between YSZ, which is widely used as the electrolyte in SOFC, and AE metal doped lanthanum chromite, namely, $La(Cr_{1-x}Mg_x)O_3$, $(La_{1-x}Ca_x)CrO_3$ and $(La_{1-x}Sr_x)CrO_3$ (x = 0.1, 0.2, 0.3). The dependence on the dopants in the lanthanum chromites of the reactivity has been discussed.

EXPERIMENTAL

Lanthanum chromite powders were prepared by a usual ceramic powder preparation method. La_2O_3 (pre-heated at 1500 °C for 1 h), $SrCO_3$, $CaCO_3$, MgO and

TABLE I CHEMICAL COMPOSITION (MOLE %) OF LANTHANUM CHROMITES.

Sample	La	Mg	Ca	Sr	Cr	Al (wt%)	Si (wt%)
$LaCrO_3$	100				100.4	< 0.012	< 0.012
$La(Cr_{0.9}Mg_{0.1})O_3$	100	9.4			93.5	< 0.012	< 0.012
$La(Cr_{0.8}Mg_{0.2})O_3$	100	18.7			18.1	< 0.012	< 0.012
$(La_{0.9}Ca_{0.1})CrO_3$	91.3		9.7		100	< 0.012	< 0.012
$(La_{0.8}Ca_{0.2})CrO_3$	79.5		18.6		100	< 0.012	< 0.012
$(La_{0.7}Ca_{0.3})CrO_3$	70.0		28.5		100	< 0.012	0.02
$(La_{0.9}Sr_{0.1})CrO_3$	91.1			9.8	100	< 0.012	< 0.012
$(La_{0.8}Sr_{0.2})CrO_3$	79.4			19.7	100	< 0.012	< 0.012
$(La_{0.7}Sr_{0.3})CrO_3$	70.4			30.0	100	< 0.012	< 0.012

Cr_2O_3 were purchased from Nakarai Chem. Ltd., Japan and used without further purification. The powders were weighed in selected proportions and mixed by planetary-type partially stabilized zirconia ball mill for 15 min with acetone. After being dried, the mixtures were rapidly heated to 1500 °C in air for one hour and held at this temperature for 24 h. The milling and heating procedures were repeated twice. Table I shows the chemical composition measured by inductively coupled plasma (ICP) analysis. The chemical compositions of the prepared lanthanum chromite are almost comparable to those of the starting materials, and the impurity amounts of Al and Si are negligible. Stabilized zirconia with 7.5 mole % Y_2O_3 (7.5YSZ) was obtained from TOSO, Japan. The zirconia powder was heated at 1500 °C for 1 h before the following reaction tests. The average secondary particle size of the zirconia powder was 29 μm.

The mixtures of 7.5YSZ and lanthanum chromite were ground with a small amount of polyvinyl butyral as a binder, and were pressed into pellets. The pellets were fired in the temperature range 1000 - 1500 °C in air. The reaction products were examined by X-ray diffraction (XRD) technique. The pellets of lanthanum chromites to measure the linear thermal expansion coefficient (TEC) were sintered at 1800 - 1900 °C for 1 h in Ar atmosphere, and their relative densities were above 90 % as calculated. The sintered samples were cut into a rectangular shape of about 3 mm X 3 mm X 13 mm. The linear TEC were measured by TAS-200 (RIGAKU, Japan) with a reference of Al_2O_3 in the temperature region room temperature to 1000 °C at heating rate of 5 °C/min. This measurement was performed in air.

TABLE II CRYSTAL STRUCTURE AND LATTICE PARAMETER
OF LANTHANUM CHROMITES.

x	$La(Cr_{1-x}Mg_x)O_3$	$(La_{1-x}Ca_x)CrO_3$	$(La_{1-x}Sr_x)CrO_3$
0.1	a=5.512 A	a=5.488 A	a=5.504 A
	b=5.479 A	b=5.461 A	b=5.471 A
	c=7.752 A	c=7.728 A	c=7.747 A
	Orthorhombic	Orthorhombic	Orthorhombic
		& Rhombohedral	
0.2	a=5.518 A	a=5.476 A	a=5.501 A
	b=5.455 A	b=5.455 A	c=13.35 A
	c=7.767 A	c=7.716 A	Rhombohedral
	Orthorhombic	Orthorhombic	
0.3		a=5.456 A	a=5.483 A
		b=5.436 A	c=13.30 A
		c=7.690 A	Rhombohedral
		Orthorhombic	

$LaCrO_3$: a=5.518 A, b=7.753 A, c=5.479 A, Orthorhombic

RESULTS AND DISCUSSION

The crystal structure and lattice parameter of the AE metal doped lanthanum chromites prepared in this study are summarized in Table II. Previous papers [6-9] have reported that the crystal structure of Mg and Ca doped lanthanum chromite is based on the $GdFeO_3$ type perovskite, and Sr doped one is based on the $LaCoO_3$ type. In the $La(Cr_{1-x}Mg_x)O_3$ system, we found that $La(Cr_{1-x}Mg_x)O_3$ (x = 0.1, 0.2) showed a single phase of the orthorhombic symmetry, and also Schilling has suggested from thermogravimetry (TG) analysis results that at least 20 mole % magnesium can go into the chromium sites in the perovskite [7]. The XRD results for the $(La_{1-x}Ca_x)CrO_3$ system agree with those by Song et al. [8]. $(La_{0.9}Sr_{0.1})CrO_3$ has two phases with the orthorhombic and the rhombohedral symmetry as reported by Khattak et al. [9].

Figures 2, 3 and 4 show XRD patterns (2θ=28-34°) of the reaction products of the mixture of lanthanum chromite and 7.5YSZ after being heated at 1500 °C for various reaction periods. In the mixture of $LaCrO_3$ and 7.5YSZ, and $La(Cr_{0.9}Mg_{0.1})O_3$ and 7.5YSZ, no reaction product was observed. $La_2Zr_2O_7$ was observed for $La(Cr_{0.8}Mg_{0.2})O_3$ after heating for 96 h. At lower temperature such as 1000 °C and for long heating period, no reaction product between $La(Cr_{1-x}Mg_x)O_3$ (x = 0, 0.1, 0.2) and 7.5YSZ was observed after heating for 1000 h. At 1300 °C,

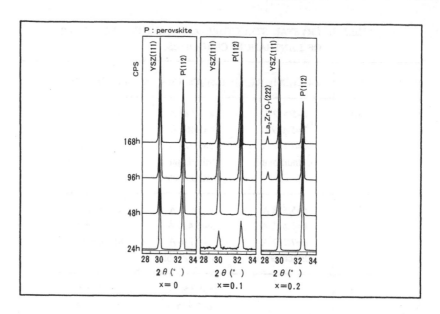

Figure 2. XRD patterns of the mixture of La(Cr$_{1-x}$Mg$_x$)CrO$_3$ (x=0, 0.1, 0.2) and 7.5YSZ after being heated at 1500 °C for various reaction periods (2θ=28-34°).

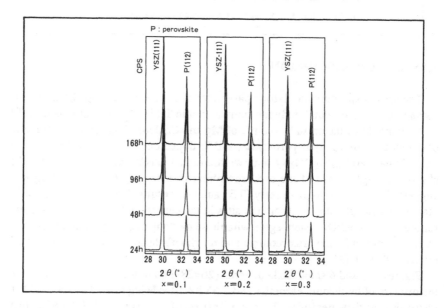

Figure 3. XRD patterns of the mixture of (La$_{1-x}$Ca$_x$)CrO$_3$ (x=0.1, 0.2, 0.3) and 7.5YSZ after being heated at 1500 °C for various reaction periods (2θ=28-34°).

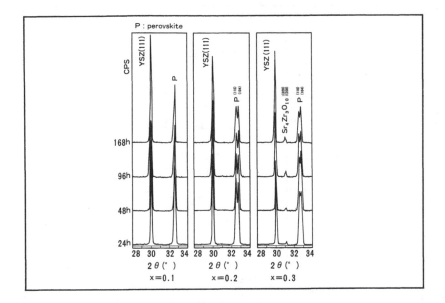

Figure 4. XRD patterns of the mixture of $(La_{1-x}Sr_x)CrO_3$ (x=0.1, 0.2, 0.3) and 7.5YSZ after being heated at 1500 °C for various reaction periods (2θ=28-34°).

$La_2Zr_2O_7$ pyrochlore was found for 500 h for $La(Cr_{0.8}Mg_{0.2})O_3$. For the mixtures of $(La_{1-x}Ca_x)CrO_3$ (x = 0.1, 0.2, 0.3) and 7.5YSZ, no reaction product was found after heating at 1500 °C for 168 h. In the case of $(La_{1-x}Sr_x)CrO_3$, $(La_{0.7}Sr_{0.3})CrO_3$ reacted with 7.5YSZ to produce $Sr_4Zr_3O_{10}$ at 1500 °C for 24 h. However, the low Sr doped $(La_{0.9}Sr_{0.1})CrO_3$ and $(La_{0.8}Sr_{0.2})CrO_3$ were stable and no reaction product was observed at the same temperature even for 168 h. The heavy Sr doped lanthanum chromite was more reactive, and the reaction product of $SrZrO_3$ was observed even at 1000 °C for 500 h in the mixture of $(La_{0.7}Sr_{0.3})CrO_3$ and 7.5YSZ.

To make clear the reaction mechanism between the lanthanum chromite and 7.5YSZ, the change of the lattice parameter of the chromite and 7.5YSZ with reaction time was examined. Figure 5 shows the change of c axis of $LaCrO_3$, $La(Cr_{0.8}Mg_{0.2})O_3$, $(La_{0.8}Ca_{0.2})CrO_3$ and $(La_{0.8}Sr_{0.2})CrO_3$ reacted with 7.5YSZ at 1500 °C as a function of reaction time. The change of c axis is not significant for $LaCrO_3$ and $La(Cr_{0.8}Mg_{0.2})O_3$. Generally, the perovskite oxides show a nonstoichiometry of A-site, B-site or O-site. In the case of $LaCrO_3$, no oxygen deficiency and excess was observed at above 1000 °C in air by chemical and TG analysis, and also La deficient and excess chromites were not observed in the experimental error at 1500 °C by XRD analysis [10]. No change of the lattice parameter in $LaCrO_3$, $La(Cr_{0.8}Mg_{0.2})O_3$ and $(La_{0.8}Sr_{0.2})CrO_3$ by heating with 7.5YSZ suggested that no predominant diffusion of a particular ion from and to 7.5YSZ was occurred. On

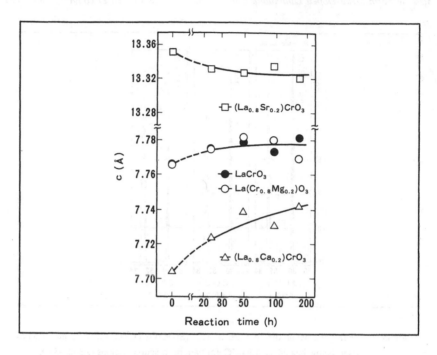

Figure 5. The lattice parameters of lanthanum chromites heated with 7.5YSZ at 1500 °C as a function of reaction time.

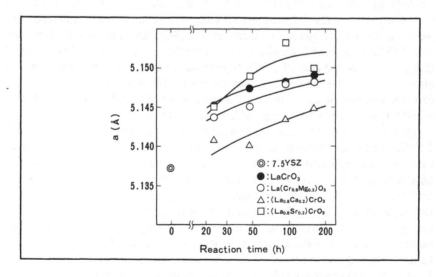

Figure 6. Lattice parameter of 7.5YSZ heated with doped lanthanum chromite at 1500 °C as a function of reaction time.

the other hand, an increase of c axis with reaction time is observed for $(La_{0.8}Ca_{0.2})CrO_3$. For the case of Ca doped lanthanum chromite, the lattice parameter decreased with increasing the dopants as shown in Table I. Therefore, the increase may be explained by the predominant diffusion of Ca^{2+} ions from the lanthanum chromite, if no diffusion of Zr^{4+} and Y^{3+} into lanthanum chromite as observed for $LaMnO_3$ is assumed.

In figure 6, the changes of lattice parameter of 7.5YSZ with reaction time are shown for the mixture of 7.5YSZ and the doped lanthanum chromite, $LaCrO_3$, $La(Cr_{0.8}Mg_{0.2})O_3$, $(La_{0.8}Ca_{0.2})CrO_3$ and $(La_{0.8}Sr_{0.2})CrO_3$ after being heated at 1500 °C. In all the mixtures, a significant increase of the lattice parameter of 7.5YSZ was observed. The lattice parameter of 7.5YSZ increased with introduction of La^{3+} and Ca^{2+} and decreased with that of Mg^{2+}. No significant change of it was observed by introduction of Sr^{2+}. The increase of the lattice parameter of 7.5YSZ by heating with the lanthanum chromites could be explained by the diffusion of La^{3+} and/or Ca^{2+} into YSZ.

It is quite interesting why there is a difference of reactivity in the lanthanum chromite with different dopants. In Table III, the solubility limits of oxides into 7.5YSZ at 1500 °C are shown. The limits were determined from the XRD patterns of mixtures of oxides and 7.5YSZ quenched from 1500 °C. In the case of the systems of 7.5YSZ and the oxides, MgO, CaO, SrO, Cr_2O_3 or La_2O_3, the limits were determined from the change of lattice parameters of 7.5YSZ with the oxide content. However, for the 7.5YSZ and Cr_2O_3 systems, no clear change of the lattice parameter was observed. Therefore, the solubility limit was assumed from the observation of the second phase by scanning electron microscope and energy dispersive X-ray (EDX) analysis. The results of Table III suggest that the solubility limit of $(La_2O_3 + SrO)$ is higher than that of Cr_2O_3. More amount of La^{3+} ions than that of Sr^{2+} diffuses into YSZ, and the excess SrO in the chromite reacts with YSZ to produce strontium zirconate, because the solubility limit of La^{3+} into YSZ was higher than that of Sr^{2+}. In the case of Mg and Ca doped lanthanum chromite, the

TABLE III   SOLUBILITY LIMITS OF OXIDE
IN 7.5YSZ AT 1500 °C.

Oxide	Solubility limit (mole %)
MgO	7
CaO	7
SrO	< 1
La_2O_3	< 2.5
Cr_2O_3	< 1

TABLE IV AVERAGE LINEAR THERMAL EXPANSION
COEFFICIENTS (ROOM TEMPERATURE - 1000 °C).

Sample	TEC ($\times 10^{-6}$ / °C)
$(La_{0.9}Ca_{0.1})CrO_3$	9.7
$(La_{0.8}Ca_{0.2})CrO_3$	10.0
$(La_{0.7}Ca_{0.3})CrO_3$	11.9
$La_2Zr_2O_7$	9.0
$SrZrO_3$	10.8
7.5YSZ	10.0

solubility limits of MgO and CaO are higher than that of La_2O_3. Therefore, excess La^{3+} in the chromite after diffusion of Mg and Ca into YSZ reacts with YSZ to produce $La_2Zr_2O_7$.

For the application to SOFC as the separator, the matching of TEC to the other cell components is quite important. In Table IV, TEC of $(La_{1-x}Ca_x)CrO_3$ and the reaction products, that is, $La_2Zr_2O_7$ and $SrZrO_3$, are shown along with 7.5YSZ. The average linear TEC from room temperature to 1000 °C was listed. Sakai et al. have reported the linear TEC of low-temperature air-sinterable lanthanum calcium chromite with chromium deficit [11]. The chromites in our experiment were stoichiometric and poorly sinterable in air, however their TEC results agree with those of the chromium deficient chromites for the Ca content in the perovskite. TEC increases with increasing Ca content in $(La_{1-x}Ca_x)CrO_3$, and that of $(La_{0.8}Ca_{0.2})CrO_3$ is 10.0 $\times 10^{-6}$ / °C. This value is in good agreement with that of 7.5YSZ. TEC of $La_2Zr_2O_7$ is lower than that of $(La_{1-x}Ca_x)CrO_3$. The formation of $La_2Zr_2O_7$ should cause a gas-leakage at the interface of the lanthanum chromite separator and the YSZ electrolyte because of their TEC mismatch.

CONCLUSION

From the reactivity study of 7.5YSZ and AE doped lanthanum chromite, it is found that Ca doped lanthanum chromite is less reactive with YSZ than Sr and Mg doped materials. The thermal expansion coefficient of Ca doped lanthanum chromite increases with increasing Ca content, and that of $(La_{0.8}Ca_{0.2})CrO_3$ is in good agreement with that of 7.5YSZ.

REFERENCES

1.   Singhal, S. C., 1991. "Solid Oxide Fuel Cell Development at Westinghouse", Proc. Second Internatl. Symp. Solid Oxide Fuel Cells, F. Grosz, P. Zegers, S. C. Singhal and O. Yamamoto, eds., Commission of the European Communities, pp. 25-33.

2.   Sakurada, S. and Yoshida, T., 1991. "Results of Solid Oxide Fuel Cell at Tonen", Proc. Second Internatl. Symp. Solid Oxide Fuel Cells, F. Grosz, P. Zegers, S. C. Singhal and O. Yamamoto, eds., Commission of the European Communities, pp. 45-54.

3.   Yamamoto, O., Takeda, Y., and Kojima, T., 1989. "Reactivity of Yttria Stabilized Zirconia with $(La_{1-x}A_x)_{1-y}MnO_3$ (A=Ca, Sr)", Proc. Internatl. Symp. Solid Oxide Fuel Cells, O. Yamamoto, M. Dokiya and H. Tagawa, eds., Science House, Tokyo, pp. 148-161.

4.   Yokokawa, H., Sakai, N., Kawada, T., and Dokiya, M., 1989. "Chemical Thermodynamic Considerations on Reactivity of Perovskite Oxide Electrodes with Zirconia". Denki Kagaku, 57(8): 821-828.

5.   Labrincha, J. A., Frade, J. R. and Marques, F. M. B., 1991. "Reaction Between Cobaltate Cathodes and YSZ", Proc. Second Internatl. Symp. Solid Oxide Fuel Cells, F. Grosz, P. Zegers, S. C. Singhal and O. Yamamoto, eds., Commission of the European Communities, pp. 689-696.

6.   Berjoan, R., Romand, C. and Coutures, J. P., 1989. "Refractory Oxide for High Temperature MHD Heaters". High Temperature Science, 13: 173-188.

7.   Schilling, D., 1983. "The Solubility of MgO in $LaCrO_3$ as a Function of Stoichiometry", M. Sc. Thesis, University of Missouri-Rolla.

8.   Song, S., Yoshimura, M. and Somiya, S., 1982. "Hydrothermal Synthesis and Properties of $(La_{1-x}Ca_x)CrO_3$." J. Mater. Sci. Soc. of Japan, 19: 49-53.

9.   Khattak, C. P. and Cox, D. E., 1977. "Structural Studies of The $(La,Sr)CrO_3$ System". Mat. Res. Bull., 12: 463-471.

10.   Mizusaki, J., Yamauchi, S., Fueki, K. and Ishikawa, A., 1984. "Nonstoichiometry of The Perovskite-type Oxide $La_{1-x}Sr_xCrO_{3-\delta}$." Solid State Ionics, 12: 119-124.

11.   Sakai, N., Kawada, T., Yokokawa, H., Dokiya, M., and Iwata, T., 1990. "Thermal Expansion of Some Chromium Deficient Lanthanum Chromites". Solid State Ionics, 40/41: 394-397.

The Growth of Electrochemical Vapor Deposited YSZ Films

J. P. DEKKER, V. E. J. VAN DIETEN and J. SCHOONMAN

ABSTRACT

Electrochemical Vapor Deposition (EVD) has become a key technology for depositing gas impervious films of YSZ on porous substrates. The kinetics of film growth of YSZ can be modeled in detail considering the Wagner oxidation process. According to this model the electron diffusion is predicted to be rate limiting. However, the apparent activation energy of the observed growth is not equal to the activation enthalpy for electron conduction in YSZ. This inconsistency can be understood considering the thermodynamic equilibrium at the gas–solid interphases. The calculated thermodynamic equilibrium is used to predict the growth rate. The results show that the EVD growth of YSZ is most likely governed by defect transport in the EVD layer, and a mass transfer limitation at the surface at the metal chloride side.

INTRODUCTION

The synthesis of thin dense films of Yttria Stabilized Zirconia (YSZ) on porous substrates is of great interest for Solid Oxide Fuel Cells and Nernst type gas sensors. Electrochemical Vapor Deposition (EVD) has become a key technology for depositing these kind of layers.

J.P. Dekker, V.E.J. van Dieten, and J. Schoonman, Laboratory
For Applied Inorganic Chemistry, Delft University of
Technology, Julianalaan 136 2628 BL Delft, The Netherlands
J.P. Dekker, Netherlands Energy Research Foundation ECN,
Westerduinweg 3, 1755 ZG Petten, The Netherlands

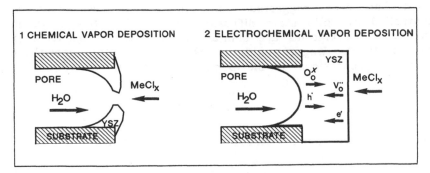

Figure 1. The principle of the EVD process

To our knowledge only Isenberg [1], Pal and Singhal of Westinghouse [2], Carolan and Michaels [3], Schoonman et al. [4-6], and Lin et al. [7-9] have reported on the kinetics of the growth of YSZ by EVD. In order to obtain a better understanding of the growth mechanism, the reported data have been collected and a systematical comparison has been made. A model is presented to explain the observed growth rates.

The steps involved in the EVD process are shown schematically in figure 1. A porous substrate separates the reactant metal chlorides from a mixture of oxygen (O_2) and steam (H_2O). The first step in film formation involves pore closure by a normal CVD type reaction between the reactant metal chloride and steam (or oxygen).

$$MeCl_4(g) + 2\ H_2O(g) \rightarrow MeO_2(s) + 4\ HCl(g) \qquad (1)$$

Once pore closure is complete, the reactants are no longer in direct contact. Film growth then proceeds by oxygen ion diffusion through the film due to the presence of a large oxygen chemical potential gradient across the deposited film. In this step oxygen ion vacancies and electrons formed at the metal chloride side diffuse through the thin metal oxide layer to the oxygen rich side. This results in a net flux of oxygen to the metal chloride side, where the oxygen reacts with the metal chloride vapor to form the metal oxide product.

$$MeCl_4(g) + 2\ O_O^X \rightarrow MeO_2(s) + 4\ e' + 2\ V_O^{\cdot\cdot} + 2\ Cl_2(g) \qquad (2)$$

The overall film growth can be limited by (a) gas diffusion, (b) surface kinetics on either side of the film, or (c) defect transport in the film.

It has been suggested that the growth mechanism is controlled by solid state diffusion in the YSZ film [2–6]. In this case the kinetics are similar to the Wagner oxidation of metals. In the Wagner oxidation the rate of scale growth is inversely proportional to the oxide thickness [10].

$$\frac{dL}{dt} = \frac{K_o}{L} \tag{3}$$

Integrating equation 3 yields the parabolic rate law.

$$L^2 = 2\ K_o t\ +\ C_o \tag{4}$$

Here K_o is the parabolic rate constant, L the film thickness, t the total deposition time of the second stage (EVD growth), and C_o a constant of integration.

This equation can be used to verify as to whether EVD growth of YSZ is controlled by solid state diffusion. The film growth data in the literature can be fitted to equation 4. The fitted parabolic growth rate constants derived from these experimental data can be used for further analysis of the EVD process. The growth rate constant can be calculated considering the Wagner oxidation process [2–9],

$$K_o = \frac{R\ T\ V_m}{C\ F^2}\ \{\ \sigma_p^o\ (\ P_{O_2}^{1/4} - P_{O_2}'^{1/4}\ ) + \sigma_n^o\ (P_{O_2}'^{-1/4} - P_{O_2}^{-1/4}\ )\} \tag{5}$$

where R is the gas constant, T the temperature, V_m the molar volume of YSZ, C the total anion charge per equivalent of YSZ, F the Faraday constant, σ_p^o the electron hole conductivity at an oxygen activity of one, σ_n^o the electron conductivity at an oxygen activity of one, P_{O_2} the oxygen partial pressure on the water side and P_{O_2}' the oxygen partial pressure on the metal chloride side. Substituting values, expected to be typical for EVD experiments (i.e. low oxygen partial pressures) in equation 5 reveals that this equation can be simplified to [2–9]:

$$K_o = \frac{R\ T\ V_m}{C\ F^2}\ \sigma_n^o\ P_{O_2}'^{-1/4} \tag{6}$$

It should be noted that this equation can only be valid as long as the oxygen partial pressure is low enough to maintain an electron controlled semiconductivity in YSZ.

Equation 6 has been used to discuss the observed growth rates [2-9]. However, in order to calculate the parabolic growth rate constant, the oxygen partial pressure on the metal chloride side and the electron conductivity as a function of temperature and oxygen partial pressure have to be known. To our knowledge there are no reports, where the partial pressure of oxygen at the metal chloride side is measured or calculated from thermochemical data.

p'_{O_2} is determined by the thermodynamic equilibrium of the species present at the metal chloride side. The assumption that a gas phase thermodynamic equilibrium is present, is justified as long as the solid state diffusion is rate limiting and all other reaction rates are infinite. Thermodynamic equilibria can be computed with the SOLGASMIX and Chemsage programs [11,12]. The thermochemical data of the species necessary for the calculations are obtained from the JANAF Tables [13].

Electron conductivity data of YSZ can be obtained from the literature [14]. An expression for electron conductivity as a function of its surrounding atmosphere will be derived. From these considerations expressions for the growth rate constant can be obtained. Finally, these calculated growth rate constants are compared with the fitted growth rate constants from the experiments and the results will be discussed.

THERMODYNAMICS

It is common practice to use gas phase equilibria at high temperatures to establish a low partial pressure of a species. For example, the stability of metal oxides can only be studied if the oxygen partial pressure is known (typically between 10^{-4} atm and 10^{-35} atm) [10]. These low partial pressures can be generated by, for example, the CO_2/CO or H_2O/H_2 equilibrium.

It should be noted that, at thermodynamic equilibrium at high temperatures the dissociation of O_2 can not be neglected. Therefore, the equilibrium with atomic oxygen has to be considered too. At low partial pressures of O_2 the partial pressure of O can be high relative to that of O_2. This should always be taken into account, if such a gas mixture is used.

ELECTRON CONDUCTION IN YSZ

Assuming that all defects are fully ionized, the defect reaction for YSZ can be written as in equation 7.

$$Y_2O_3 \rightarrow 2\ Y'_{Zr} + 3\ O^X_O + V^{\cdot\cdot}_O \tag{7}$$

Hence, the oxygen ion vacancy concentration is determined by the yttria dopant concentration. The following non-stoichiometric defect reaction for YSZ at low oxygen partial pressures can be written [14].

$$O^X_O \leftrightarrows 2\ e' + V^{\cdot\cdot}_O + 1/2\ O_2(g) \tag{8}$$

The equilibrium constant K_8 for this reaction is given by,

$$K_8 = [V^{\cdot\cdot}_O]\ [e']^2\ p^{1/2}_{O_2} \tag{9}$$

where $[V^{\cdot\cdot}_O]$ is the concentration of fully ionized oxygen ion vacancies in the oxygen sub-lattice and $[e']$ the electron concentration in the conduction band.

Thus, at a given dopant concentration and temperature, the concentration of electrons is determined by the partial pressure of oxygen.

The electron conductivity is given by,

$$\sigma_n = [e']\ |q|\ \mu_n \tag{10}$$

where q is the electronic charge, and μ_n the electron mobility. If the mobility is independent of the electron concentration the electron conductivity is proportional to the electron concentration. An expression for the electron conductivity can be derived if the equilibrium in reaction 8 is assumed to be present.

$$\sigma_n = \frac{|q|\ \mu_n}{[V^{\cdot\cdot}_O]^{1/2}}\ K^{1/2}_8\ p^{-1/4}_{O_2} \tag{11}$$

Equation 11 is proportional to $p^{-1/4}_{O_2}$. Also other sets of defect reactions, involving electrons and oxygen ion vacancies, instead of reaction 8 can be used to obtain an expression for the electron conductivity. These expressions for electron conductivity will always contain a product of several equilibrium constants and $p_{O_2}^{-1/4}$. Each of these products of equilibrium constants is equal to the square root of equilibrium constant K_8 (in equation 11).

THERMODYNAMIC CALCULATIONS

In order to calculate the growth rate constant from equation 6, the electron conductivity as a function of the metal chloride containing atmosphere has to be known. In principle, this information can be obtained from thermodynamic calculations provided the thermochemical data of the individual defects in YSZ are known. Since these data are not available the calculations can not be executed. Hence, the pre-exponential term of electron conductivity σ_n^o has to be obtained from conductivity data of YSZ from literature.

Only the presence of zirconium containing species will be considered in the following discussion. The presence of yttrium containing species is not taken into account in the derivations and calculations, because this would make the discussion extremely complicated. However, the influence of a possible incorporation of the yttrium containing species will be discussed.

The actual equilibrium between ZrCl4 and ZrO2 (i.e. the EVD film of YSZ) is given by:

$$ZrCl_4(g) + 2\ O_O^X \leftrightarrows ZrO_2(s) + 4\ e' + 2\ V_O^{\cdot\cdot} + 2\ Cl_2(g) \qquad (12)$$

This non-stoichiometric equilibrium in reaction 12 can be expressed as a function of oxygen partial pressure by incorporating the equilibrium of ZrCl4 with oxygen as discussed earlier.

$$ZrCl_4(g) + O_2(g) \leftrightarrows ZrO_2(s) + 2\ Cl_2(g) \qquad (13)$$

The actual partial pressure of oxygen can be governed by several parameters, depending on experimental conditions, which affect the partial pressures in reaction 13.

As long as the conversion of reactant in the EVD process is relatively low, p'_{O_2} can be determined, provided the partial pressure of chlorine is known. Chlorine is not an input species, but it is a product formed by some chemical reaction. So the partial pressure of chlorine will be determined in one of the following ways:

(a) by the conversion of metal chloride reactant at the metal oxide surface. However, this partial pressure of chlorine is determined by the growth rate of the film itself and is dependent on several mass transport processes such as adsorption and desorption and gas phase diffusion.

This complex set of sequential mass transport processes makes it very difficult to obtain realistic values for the chlorine partial pressure.

(b) by the dissociation of $ZrCl_4$,

$$3 \ ZrCl_4(g) \overset{\leftarrow}{\rightarrow} \sum_{n=1}^{3} \left[ZrCl_n(g) + (4-n) \left(\frac{(1-a)}{2} \ Cl_2(g) + a \ Cl(g) \right) \right] \qquad (14)$$

where $a \in [0,1]$.

Because the equilibrium in reaction 13 still holds, p'_{O_2} can be calculated from this equilibrium.

(c) by the presence of oxygen in the carrier gas stream. This will have a major influence on the partial pressure of the species present in the equilibrium 13.

Cases (b) and (c) can be studied by calculation of the thermodynamic equilibrium present in the gas phase at the metal chloride side. It is assumed that the species listed in Table I can be present in the gas phase. The presence of yttrium containing species is not considered because the solid YOCl is a stable species as was shown by thermodynamic calculations, in which yttrium containing species were considered. This makes the interpretation of the results extremely difficult.

TABLE I — SPECIES USED FOR THE THERMODYNAMIC CALCULATIONS.

SPECIES	PHASE	SPECIES	PHASE	SPECIES	PHASE
O_2	gas	Zr	gas	Zr	gas
O_3	gas	Cl	gas	ZrO	gas
O	gas	Cl_2	gas	Ar	gas
ZrCl	gas	ClO	gas	Zr	solid
$ZrCl_3$	gas	Cl_2O	gas	$ZrCl_2$	solid
$ZrCl_3$	gas	ClO_2	gas	$ZrCl_3$	solid
$ZrCl_4$	gas	ZrO_2	gas	$ZrCl_4$	solid

In general, the presence of yttrium containing species will lower the partial pressures of oxygen containing species in the gas phase due to the formation of YOCl. Consequently, the presence of YSZ can not be considered because the mass balance has to be maintained. It should be noted that the presence of YSZ can also have an influence on the partial pressure of oxygen containing species.

The equilibria are computed as a function of temperature and partial pressure of input species. The total pressure was kept constant at 0.01 atm through all calculations. The thermodynamic calculations are split into two sections. The first set of calculations can be used to study case (b), and the latter set to study case (c) for the EVD growth of YSZ.

The first set comprises thermodynamic calculations with only ZrO_2 and $ZrCl_4$ as input species, while a large excess of ZrO_2 with respect to $ZrCl_4$ (i.e. $[ZrO_2] \geq 100 \times [ZrCl_4]$) was chosen in order to prevent depletion of ZrO_2. In figure 2 the equilibrium partial pressures of the main species, with the exception of the input species because of their large excess, are presented as a function of temperature. The equilibrium partial pressures of all species, with the exception of $ZrCl_4$, increase with increasing temperature, indicating that the decomposition of $ZrCl_4$ and $ZrO_2(s)$ becomes more important.

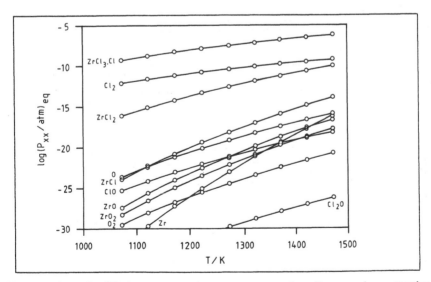

Figure 2. Equilibrium partial pressure of the main species versus temperature, where the input partial pressure of $ZrCl_4$ is 10^{-3} atm, the input amount of ZrO_2 is 1000 mol, and the input patial pressure of oxygen is kept zero.

The thermodynamic calculations as a function of the input partial pressure of ZrCl4 reveal that the equilibrium partial pressures of all species are proportional or inversely proportional to a n-th power of the input partial pressure of ZrCl4 because of its large excess, where n can be 0, 0.5, or 1 respectively. From these results all existing equilibria can be calculated.

Case (c) comprises thermodynamic calculations which are more complicated to comprehend. The variation of the input partial pressure of oxygen may have a large influence on the equilibrium partial pressures of the other species. The thermodynamic calculations as a function of temperature reveal that the equilibrium partial pressure of oxygen is high with respect to the equilibrium partial pressure of oxygen for case (b), where no oxygen as input species is taken into account.

The equilibrium partial pressures of the main species as a function of input partial pressure of oxygen are presented in figure 3. The equilibrium partial pressures of the species are determined by the input amount of ZrCl4 and ZrO2 as long as the input partial pressure of oxygen is kept below 10^{-8} atm, as can be seen in figure 3. However, if the input partial pressure of oxygen is above 10^{-8} atm there is a steep increase in the equilibrium partial pressure of oxygen.

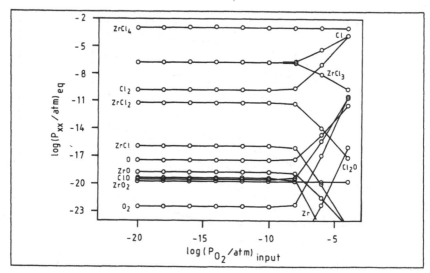

Figure 3. Equilibrium partial pressure of the main species versus the input partial pressure of O2 at a temperature of 1373 K, where the input partial pressure of ZrCl4 is 10^{-3} atm, and the input amount of ZrO2 is 1000 mol.

From these observations it can be concluded that the presence of oxygen in the carrier gas stream has no influence on the equilibrium partial pressures of the gas phase species provided the oxygen impurity in the carrier gas stream is less than 1 ppm (i.e. 10^{-8} atm in the reactor as calculated).

REVIEW OF EVD EXPERIMENTS

We have fitted all the reported data of film thickness as a function of deposition time according to equation 4. This equation was chosen to verify as to whether solid state diffusion is the general rate limiting step in the EVD of YSZ in the reported studies [1–9]. In all EVD YSZ films approximately 8 to 10 m/o yttria was present. All reported EVD experiments could be fitted well by the parabolic growth rate equation 4. A more detailed discussion of this review has been published earlier [15].

The growth rate constants are presented in an Arrhenius plot (figure 4). The strong temperature dependence justifies the assumption that the process is indeed controlled by solid state diffusion. The slope of the fitted line is proportional to the apparent activation energy.

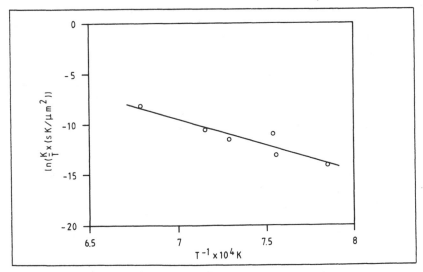

Figure 4. Least squares fit of parabolic growth rate constant as a function of temperature [15].

This apparent activation energy of (4.4 +/- 0.7) eV is a direct function of the activation energy for electron conduction in YSZ, and the change in enthalpy of the equilibrium at the metal chloride side, if solid state diffusion is assumed to be rate limiting and all other reaction rates are assumed to be infinite. In this case the apparent activation energy is not directly proportional to the activation energy of electron conduction as was discussed earlier [15].

MODELING RESULTS OF EVD GROWTH

The growth rate constants are calculated from equation 6. The constants necessary for these calculations are listed in Table II. The equilibrium oxygen partial pressures are obtained from the calculations, where only $ZrCl_4$, ZrO_2 and Ar were used as input species. The results of these calculations for different input concentrations of $ZrCl_4$ are presented in figure 5. The growth rate constant in figure 5 decreases with increasing $ZrCl_4$ input concentration due to the increasing equilibrium partial pressure of oxygen, thereby reducing the oxygen chemical potential across the film.

TABLE II- CONSTANTS NECASSARY TO CALCULATE THE GROWTH RATE CONSTANT, CONSIDERING $ZrCl4$ AND $ZrO2$ AS INPUT SPECIES FOR THERMODYNAMIC EQUILIBRIUM CALCULATIONS.

Symbols:	Data
R (gas constant)	8.312 J/(mol K)
C (anion charge per equivalent YSZ)	3.82
F (Faraday constant)	$9.65 \ 10^4$ C/mol
V_m (molar volume)	$21.5 \ cm^3$/mol
σ_n^o (pre-exponential term of electron conductivity) [14]	$1.31 \ 10^7 atm^{1/4}$S/cm
E_a (activation energy of electron conductivity) [14]	3.88 eV
p'_{O2} (oxygen partial pressure at metal chloride side) obtained from thermodynamic calculations. Case b: no oxygen as input species is considered. Case c: oxygen as input species is considered.	

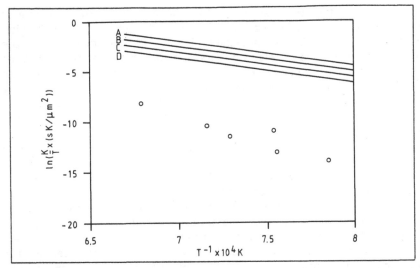

Figure 5. The calculated growth rate constant as a function of temperature, when no O2 is present as input species. A: P_{ZrCl4}(input)= 10^{-6} atm, B: P_{ZrCl4}(input)= 10^{-5} atm, C: P_{ZrCl4}(input)= 10^{-4} atm, and D: P_{ZrCl4}(input)= 10^{-3} atm.

The equilibrium oxygen partial pressures used for the calculation of the growth rate constants are so low that YSZ is no longer in its electrolytic domain. Hence, the oxygen ion vacancy concentration is not constant anymore. The oxygen ion vacancy concentration and the electron concentration then become proportional to the oxygen partial pressure to the power −1/6 [10]. Thus, the electron conductivity data used are not valid anymore. In this region it is possible that the mass transport in YSZ is determined by the oxygen ion conductivity instead of the electron conductivity.

However, these calculations are still indicative and can be used to evaluate the experimental data of EVD growth. The predicted growth rate constant is far too high with respect to the observed growth rate constant. This discrepancy can not be explained with the fact that YSZ is not in its electrolytic domain anymore, because the calculated growth rate constant is still far too high, even in the low oxygen partial pressure limit of the electrolytic domain of YSZ (i.e. $P_{O_2} \simeq 10^{-23}$ atm).

The apparent activation energy of 2.2 eV of these calculated growth rate constants, where YSZ is still in its electrolytic domain, is low compared with the measured apparent activation energy of 4.4 eV. An apparent activation energy of 2.8 eV is obtained if it is assumed that the activation energy for electron conduction remains the same at oxygen partial pressures, where YSZ is not in its electrolytic domain. As mentioned before, yttrium containing species were not considered within these calculations. However, this can not be the explanation for the calculated growth rate constant being too high, because the calculated equilibrium oxygen partial pressure becomes lower if yttrium containing species are taken into account due to the formation of YOCl. This would result in an even higher calculated growth rate constant.

Only the metal chloride and metal oxide were considered as input species for the thermodynamic calculations. However, it is not likely that the carrier gases, argon or helium, used in the experiments are completely free of oxygen. Hence, the influence of small amounts of oxygen in the carrier gas stream on the growth rate has to be investigated.

In order to study the influence of oxygen as an impurity in the carrier gas stream thermodynamic calculations were performed using O_2, $ZrCl_4$, ZrO_2 and Ar as input species. The constants necessary to calculate the growth rate constant from equation 6 are listed in Table II. The results of these calculations are presented in figure 6. If the input partial pressure of oxygen drops below 10^{-12} atm the growth rate constant remains unaffected. The equilibrium oxygen partial pressure in this case is determined by the input amount of $ZrCl_4$ and ZrO_2. Hence, the growth rate constant is the same as in the situation, where oxygen is not present as input species. Thus, line D in figure 5 is equal to lines C,D and E in figure 6.

The temperature dependence of the calculated growth rate constant represented by solid line A in figure 6 seems to describe the experimental growth rate constant reasonably well. This predicted growth rate constant is obtained using an input oxygen partial pressure of 10^{-4} atm. The calculated apparent activation energy of this growth rate constant is 3.45 eV.

Hence, the experimental results can be understood assuming an oxygen concentration in the carrier gas of 10^4 ppm (i.e. in this case 10^{-4} atm in the reactor). However, it is unlikely that such an amount is present in the carrier gas. It is common to have an oxygen impurity on the order of 1 to 10 ppm in the carrier gas.

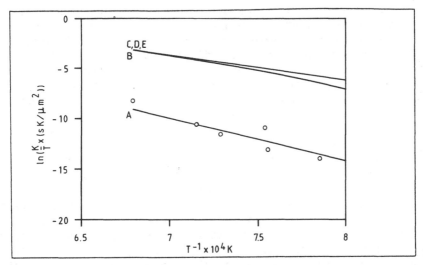

Figure 6. The calculated growth rate constant as a function of temperature at a ZrCl4 input partial pressure of 10^{-3} atm for different input partial pressures of O2. A: p'_{O2}(input)= 10^{-4} atm, B: p'_{O2}(input)= 10^{-8} atm, C: p'_{O2}(input)= 10^{-12} atm, D: p'_{O2}(input)= 10^{-16} atm, and E: p'_{O2}(input)= 10^{-20} atm.

Yet, if parabolic growth is assumed and the oxygen concentration in the carrier gas is on the order of 1 to 10 ppm, the predicted growth rate is too high, as discussed earlier. From these observations it may be concluded that the EVD layer may act as an oxygen selective membrane, and that only a small portion of the total flux of oxygen through the EVD layer contributes to the actual growth, hence leading to large oxygen partial pressures at the metal chloride side.

The ratio between the observed growth rate and the predicted growth rate is proportional to the conversion of transported oxygen to metal oxide. This ratio (i.e. conversion) increases with increasing temperature. The ratio varies from a few tenth of a percent to a few percent typically. This information can only be used in a qualitative way, because the low conversion of oxygen will affect the actual oxygen partial pressure at the metal chloride side, and consequently the oxygen flux through the film. Besides, the assumption that solid state diffusion is rate limiting and that all other transport fluxes are infinite can not be valid anymore. These observations suggest an EVD process which is determined by a combination of solid state diffusion and surface kinetics.

The increase in conversion with increasing temperature might be explained in terms of some limitations in the surface kinetics. However, at this stage this indistinctness of EVD growth can not be solved. To our knowledge there are no reliable data concerning the surface kinetics between oxygen and zirconium tetrachloride present. In principle an EVD growth rate constant can be calculated by a numerical solution of the mass transport through the film and the kinetics at the surface provided that the data necessary are known. Consequently this calculated growth rate constant has to be compared with experimental data of EVD growth to verify as to whether the EVD process is indeed limited by solid state diffusion and surface kinetics.

It should be noted that the electron conductivity data used for these calculations are obtained from sintered YSZ samples [14]. It might be possible that the actual electron conductivity in the EVD layer differs from the electron conductivity data used for the calculations due to the incorporation of chlorine or hydrogen in the film during the EVD process. However, it is not very likely that this possible difference in electron conductivity can explain the discrepancies, as discussed.

CONCLUSIONS

The temperature dependence of the experimental growth rate constant of YSZ was used to verify whether the growth rate is limited by solid state diffusion. The high observed apparent activation energy of 4.4 eV of the EVD growth justifies the assumption that solid state mass transport through the film is the rate limiting step.

However, the calculated growth rate constant is higher than the experimental growth rate constant. Hence, the material constants used for the calculations are not valid for the EVD layer of YSZ or the growth rate mechanism is more complex then originally assumed.

Traces of oxygen have a large influence on the calculated growth rate constant. The experimentally determined growth rate constant can be understood if the presence of 10^4 ppm of oxygen in the carrier gas is assumed. From these observations it might be concluded that the EVD film acts as an oxygen pump and that only a small portion of the oxygen actually reacts with the metal chloride to form the metal oxide.

In principle this complex mass transport phenomenon can be solved numerically. However, at this stage there are not enough data present necessary to solve the equations.

Hence, more experimental work concerning the kinetics of EVD and CVD using $ZrCl_4$ and O_2 as reactants has to be done in order to obtain a full model which can predict the EVD growth rate of YSZ.

REFERENCES

1. Isenberg, A.O., 1977. "Growth of Refractory Oxide Layers by Electrochemical Vapor Deposition," in Proc. Symp. Elec. Mat. and Proc. for Eng. Conv. and Stor., The Electrochem. Soc., 77(6):572-583.

2. Pal, U.B. and S.C. Singhal, 1990. "Electrochemical Vapor Deposition of Yttria Stabilized Films." J. Electrochem. Soc., 137:2937-2941.

3. Carolan, M.F. and J.N. Michaels, 1990. "Growth Rates and Mechanism of Electrochemical Vapor Deposited Yttria Stabilized Zirconia Films." Solid State Ionics, 37:189-195.

4. Dekker, J.P., N.J. Kiwiet and J. Schoonman, 1989. "Electrochemical Vapor Deposition of SOFC Components," in Solid Oxide Fuel Cells, S. Singhal ed., Pennington, New Jersey, USA: The Electrochem. Soc., 89(11):57-66.

5. Kiwiet, N.J. and J. Schoonman, 1990. "Electrochemical Vapor Deposition: Theory and Experiment," in Proc. 25th Int. Soc. Energy Conv. Eng. Conf., P.A. Nelson et al, eds., New York, NY, USA: AIChE, 3:240-245.

6. Schoonman, J., J.P. Dekker, J.W. Broers and N.J. Kiwiet, 1991. "Electrochemical Vapor Deposition of Stabilized Zirconia and Interconnection Materials of Solid Oxide Fuel Cells." Solid State Ionics, 46:299-368.

7. Lin, Y.S., L.G.J. de Haart, K.J. de Vries and A.J. Burggraaf, 1990. "A Kinetic Study of the Electrochemical Vapor Deposition of Solid Oxide Electrolyte Films on Porous Substrates." J. Electrochem. Soc., 137:3960-3965.

8. de Haart, L.G.J., Y.S. Lin, K.J. de Vries and A.J. Burggraaf, 1991. "Modified CVD of Nanoscale Structures in and EVD of Thin Layers on Porous Ceramic Membranes." J. Eur. Ceram. Soc., 8:59-70.

9. Lin, Y.S., 1991. "Chemical and Electrochemical Vapor Deposition of Zirconia-Yttria Solid Solutions in Porous Ceramic Media." PhD. Thesis, University of Twente, The Netherlands, Chapter 6.

10. Kofstad, P., 1988. High Temperature Corrosion, Elsevier Applied Science Publishers LTD.

11. Eriksson, G., 1975. "Thermodynamic Study of High Temperature Equilibria XII. SOLGASMIX, a Computer Program for Calculations of Equilibrium Composition in Multiphase Systems." Chemica Scripta, 8:100-103.

12. Eriksson, G. and K. Hack, 1991. Chemsage (computer program), GTT GmbH, Germany.

13. Stull, D.R. and H. Prophet, eds., 1971. JANAF Thermochemical Tables., 2nd Edition, Washington D.C., USA: U.S. Government Printing Office.

14. Park, J.M. and R.N. Blumental, 1989. " Electron Transport in 8 mole percent $Y_2O_3-ZrO_2$." J. Electrochem. Soc., 136:2867-2876.

15. Dekker, J.P., V.E.J. van Dieten and J. Schoonman, 1992. "The Growth of Electrochemical Vapor Deposited YSZ Films." Solid State Ionics, 51:143-145.

Deposition of YSZ on Porous Ceramics by Modified CVD

G. Z. CAO, H. W. BRINKMAN, J. MEIJERINK, K. J. DE VRIES
and A. J. BURGGRAAF

ABSTRACT

Modified chemical vapour deposition (CVD) has been introduced to deposit yttria stabilized zirconia (YSZ) on porous ceramic media. Experimental results indicate that deposition of YSZ occurred inside pores resulting in a pore narrowing and eventually a pore closure, which was determined by using permporometry. However, deposition on top of porous ceramic substrates will not reduce the average pore size. After pore closure, the following electrochemical vapour deposition (EVD) process results in growing of YSZ layers on top of porous ceramic media.

INTRODUCTION

Among the ionic conducting materials, yttria stabilized zirconia (YSZ) with a fluorite type structure is one of the most attractive candidates for applications, such as electrolytes in solid oxide fuel cells (SOFC) and gas separations. In order to optimize the efficiency of the applications, it is very important to prepare very thin or ultrathin YSZ layers, by decreasing the thickness of the YSZ layer, thus lowering the ohmic polarization loss for SOFC applications and maximizing the oxygen permeation flux, assuming a sufficiently large surface exchange rate. However, it has been very difficult to prepare a very thin and dense YSZ layer on top of a porous substrate by using the conventional methods.

The modified chemical vapour deposition (CVD) and electrochemical vapour deposition (EVD) technique have demonstrated the possibilities for the preparation of very thin, dense, oxygen semipermeable YSZ layers on porous ceramic substrates. In the modified CVD stage the metal reactants (chloride

G.Z. Cao, H.W. Brinkman, J. Meijerink, K.J. de Vries and A.J. Burggraaf, Laboratory of Inorganic Chemistry, Materials Science and Catalysis, Department of Chemical Engineering, University of Twente, POB 217, 7500 AE Enschede, Netherlands

vapours) are delivered to one side of a porous substrate, which separates the CVD/EVD reactor chamber into two compartments, while the oxygen reactant (water vapour and oxygen gas) is supplied to the other side. The reactants counter-diffuse and react to form and deposit solid YSZ inside the pores, causing pore narrowing and eventually pore plugging. After the pore closure the CVD process stops while the EVD process proceeds, i.e. the water vapour and oxygen gas diffuse through the porous substrate to the YSZ plug and are reduced to oxygen ions at the oxygen/YSZ plug interface. The oxygen ions diffuse through the YSZ dense film and then react with the metal chlorides to form YSZ at the metal/plug interface, resulting in growing of the EVD-YSZ layer.

For obtaining an ultrathin YSZ layer the deposition process should stop shortly after the pore closure caused by CVD and the EVD YSZ layer growth process should be suppressed. Thus a good understanding of the CVD process is essential. Previous work in the literature, however, has been mainly focused on kinetic studies of the thin dense YSZ layer growing process with both theoretical analysis and experimental studies [1-7]. The present paper is mainly focused on the CVD stage by studying the deposition profile and pore narrowing and plugging caused by depositing YSZ inside pores for two different pore sizes.

EXPERIMENTAL

CVD/EVD ON POROUS CERAMIC SUBSTRATES

The chemicals used in the CVD/EVD experiments are $ZrCl_4$ (99.9%, 200 mesh, CERAC), YCl_3 (99.9%, 60 mesh, CERAC), H_2O (doubly-distilled), Ar gas (99.999%, UHP 5.0) and air (technical, Hoekloos). Two types of substrates (discs: 12 mm in diameter and 2 mm in thickness) with two different mean pore sizes were used, referred to as type A and B. The type A substrate consists of one layer of α-alumina with a porosity of 50 % and a mean pore diameter of 0.16 μm, determined by using mercury porosimetry. The α-alumina substrate (type A) is also used as a supporting layer for the La-doped γ-alumina ceramic membrane top layer, this two layer ceramic membrane is referred to as the type B substrate. The La-doped γ-alumina layer was made by using the sol-gel technique and dip-coated (2 times) to the α-alumina substrate. The La-doped $\gamma + \alpha$-alumina membrane was then fired at 1100 °C in air for 30 hours to obtain a stabilized pore size and porosity. For the above γ-alumina a porosity and a mean pore diameter are 43 % and 20 nm, respectively (determined using nitrogen adsorption-desorption). The thickness of the La-doped γ-alumina top-layer is approximately 5 μm. The details of preparation of the La-doped $\gamma + \alpha$-alumina membrane have been presented elsewhere [8].

The CVD/EVD experiments (for details, see [2]) were performed in a home-made apparatus, which consists of three sections: a reactor, a vacuum control section and a reactant delivery section. The CVD/EVD experimental conditions used are given in Table I unless otherwise specified. A porous

TABLE I - CVD/EVD EXPERIMENTAL CONDITIONS

Substrate temperature	800 °C
$ZrCl_4$ sublimation bed temperature	160 °C
YCl_3 sublimation bed temperature	640 °C
$YCl_3/ZrCl_4$ ratio in vapour	3/10
Reactor pressure	2 mbar (200 Pa)
Ar gas stream through $ZrCl_4$ bed	3 ml(STP)/min
Ar carrier gas stream through YCl_3 bed	1.5 ml(STP)/min
Total Ar carrier gas stream on chloride side	5.5 ml(STP)/min
Air stream through H_2O sparger	0.7 ml(STP)/min
Water sparger temperature	40 °C
Total pressure in water sparger	150 mbar (15 kPa)

substrate disc (either type A or B) is sealed to a supporting dense alumina tube inside the reactor. From one side of the porous substrate a mixture of the sublimated chloride vapours, diluted and carried by Ar gas, flows and diffuses into the pores of the substrate. At the same time a mixture of water vapour and air is led into the pores of the substrate from the opposite side. Inside the pores the reactants counter-diffuse and react to form a solid, yttria stabilized zirconia (YSZ), deposited on the inner surface of the pores, causing pore narrowing and eventually pore plugging. The following EVD process will result in growing of a dense YSZ layer on the porous substrates.

CHARACTERIZATION

The study on the deposition profile of yttria stabilized zirconia in the porous substrates was conducted mainly by means of Scanning Auger Microscopy (SAM) and Scanning Electron Microscopy (SEM) with Energy Dispersion Spectroscopy (EDS). The substrates with YSZ deposition were put into liquid nitrogen and broken into pieces in air to obtain a smooth cross-section surface with the YSZ deposition remaining undamaged.

The pore narrowing caused by depositing YSZ was determined by using permporometry. This method is based on the controlled blocking of pores by capillary condensation of a vapour phase (cyclohexane) and the simultaneous measurement of the gas diffusional flux (oxygen) through the remaining open pores. Starting with the saturated vapour pressure all pores in the porous substrate are filled and no gas transport through the substrate is possible. When the vapour pressure is reduced, the pores larger than a corresponding size in accordance with the Kelvin equation [9], will open and are available for gas transport. By stepwise decrease of the vapour pressure the gas transport through the substrate is measured, from which the active pore size distribution is calculated. In the present study the pore size distribution of active pores in porous ceramic substrates before and after CVD was measured. The details of the permporometry measurements are presented elsewhere [10].

RESULTS AND DISCUSSION

CVD OF YSZ ON POROUS CERAMIC SUBSTRATES

XRD and XRF analyses show that the deposits on both types of substrates are yttria stabilized zirconia solid solutions with a fluorite-type structure and a composition of $Y_{0.18}Zr_{0.82}O_{1.91}$. However, small amounts of tetragonal and monoclinic zirconia phases were also found in some experiments.

The pore narrowing caused by depositing YSZ during the CVD process on both type A and B substrates is demonstrated in figures 1 and 2, which show the size distribution of active pores determined by using permporometry. Note that both type substrates are thermally stable and chemically inert on the CVD process. Figure 1 shows the size distribution of active pores in the type A substrate after 20 min depositing YSZ. From this pore size distribution one can calculate an average pore radius of the active pores of about 20 nm, which is four times smaller than the original pore size of 80 nm (before CVD, determined by using mercury porosimetry). Figure 2 shows the size distribution of active pores of the type B substrates before and after 20 min depositing YSZ. The average pore radius has been reduced from about 9.5 nm to 6 nm, but the pore size distribution becomes broader. Smaller pores were caused by the deposition of yttria-stabilized zirconia inside the pores and consequently all pores become smaller.

The SEM/EDS and SAM analyses have shown that the deposition of YSZ on both the type A and B substrates can occur inside pores very near to the surface facing the chloride chamber by using the experimental conditions given in Table I. Figure 3 shows the deposition profile of YSZ in the type A

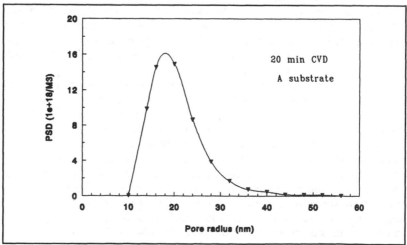

Figure 1.   The pore size distribution of active pores in the type A substrate after 20 min deposition of YSZ.

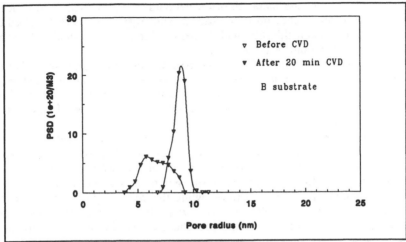

Figure 2. The pore size distribution of active pores in the type B substrate before and after 20 min deposition of YSZ.

substrate. It is seen that the deposition zone is about 4 μm in broadness. A very small amount of YSZ (traces) found deep inside the substrate is probably due to the pre- and post-deposition. In the CVD experiments the metal chloride sublimation beds are inserted into the reactor chamber some half an hour before the deposition starts, and left inside after the CVD experiments end until the reactor cools down. In these periods the evaporation of metal

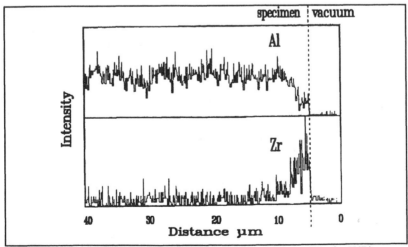

Figure 3. The deposition profile of YSZ in the type A substrate (20 min CVD) determined by EDS.

chlorides occurs and the metal chloride vapours probably then diffuse deep inside the substrates, since no water vapour and oxygen gas were supplied at that time. These metal chlorides will be oxidized in contact with air, after the substrates are taken out of the reactor.

The deposition profile of YSZ in the type B substrate is similar to that in the type A substrate. The broadness of the deposited YSZ is about 4 μm, very similar to that in the type A substrate. It is noted that the deposition of YSZ normally occurs at the γ-alumina top-layer and no deposition was found in the supporting layer.

The deposition location could be changed by varying the concentration ratio of the reactants. Increasing the water vapour concentration and/or decreasing the metal chloride concentration result in a shift of the YSZ deposition towards the surface facing the chloride chamber. When the concentration of metal chloride vapours was reduced 5 times while the water vapour concentration remained unchanged, the YSZ deposition was shifted outside the pores to the top of the substrate facing the chloride chamber. Under such experimental conditions an excess amount of water vapour and oxygen gas diffuse through the porous ceramic substrate and meet the metal chloride vapours outside the pores, and consequently the reaction and deposition occur on top of the substrate facing the chloride chamber. Such a deposition results in a decrease of porosity but not of the average pore size, as demonstrated in figure 4. It is found to be more difficult to get pore closure under these conditions. Note also that the YSZ deposition can be shifted inside the substrate by varying the reactant concentration ratio in the opposite way.

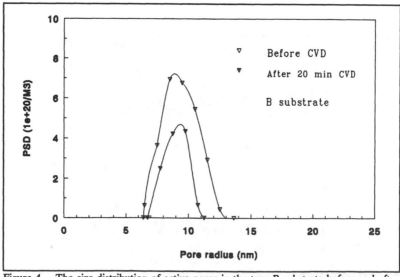

Figure 4.   The size distribution of active pores in the type B substrate before and after 20 min CVD, with an excess amount of water vapour and oxygen supplied.

EVD OF YSZ

After all pores have been plugged by depositing YSZ during the CVD stage, the CVD process stops while the EVD process starts, resulting in growing of the YSZ layer towards the chloride chamber. A gas-tight YSZ layer is formed after about 50 min (on the type A substrates) and 40 min (on the type B ones) deposition, by using the experimental conditions described in Table I. The apparent thickness of YSZ layers is about 1.5 μm and 0.5 μm on the type A and B substrates, respectively, according to the SEM photographs. Such a layer in fact consists of two parts: a microporous layer and a real dense layer. It is very difficult to discriminate between them on SEM photographs.

To avoid forming a thick YSZ layer the CVD/EVD experimental temperature was chosen at 800 °C so that after the pore closure the layer growing of EVD-YSZ proceeds with a very slow rate. This is due to the high activation energy needed for the oxygen ions to diffuse through the YSZ layer. It has been observed that the layer thickness has been increased only little with a 2 hours EVD experiment.

CONCLUSIONS

A deposition of YSZ inside pores of porous ceramic substrates will cause a pore narrowing and eventually a pore closure. However a deposition on top of porous substrates will not lead to a reduction of the average pore size, but can result in pore closure.

The deposition location can be changed by varying the concentration and/or concentration ratio of reactants. The broadness of the YSZ deposition zone is about 4 μm for pores with an initial radius of 10 - 80 nm.

A gas-tight layer with a thickness of 0.5-1.5 μm is formed after 40-50 min CVD/EVD, depending on the substrates (pore size) and experimental conditions used.

REFERENCES

1. Isenberg, A.O., 1977. "Growth of Refractory Oxide Layers by Electrochemical Vapor Deposition (EVD) at Elevated Temperatures," in <u>Proc. Symp. Electrode Materials and Processes for Energy Conversion and Storage</u>, Vol. 77-6, J.D.E. McIntyre et al., eds., New York, NY, USA: The Electrochem. Soc., pp. 572-583.

2. De Haart, L.G.J., Y.S. Lin, K.J. de Vries and A.J. Burggraaf, 1991. "Modified CVD of Nanoscale Structures in and EVD of Thin Layers on Porous Ceramic Membranes." <u>J. Euro. Ceram. Soc.</u>, 8:59-70.

3. Lin, Y.S., L.G.J. de Haart, K.J. de Vries and A.J. Burggraaf, 1990. "A Kinetic Study of the Electrochemical Vapor Deposition of Solid Electrolyte Films on Porous Substrates." J. Electrochem. Soc., 137:3960-3966.

4. Pal, U.P. and S.C. Singhal, 1990. "Electrochemical Vapor Deposition of Yttria-Stabilized Zirconia Films." J. Electrochem. Soc., 137:2937-2941.

5. Carolan, M.F. and J.N. Michaels, 1990. "Growth Rates and Mechanism of Electrochemical Vapor Deposited Yttria-Stabilized Zirconia Films." Solid State Ionics, 37:189-195.

6. Schoonman, J., J.P. Dekker, J.W. Broers and N.J. Kiwiet, 1991. "Electrochemical Vapor Deposition of Stabilized Zirconia and Interconnection Materials for Solid Oxide Fuel Cells." Solid State Ionics, 46:299-308.

7. Dekker, J.P., V.E.J. van Dieten and J. Schoonman, 1992. "The Growth of Electrochemical Vapor Deposited YSZ Films." Solid State Ionics, 51:143-145.

8. Lin, Y.S. and A.J. Burggraaf, 1991. "Preparation and Characterization of High-Temperature Thermally Stable Alumina Composite Membrane." J. Am. Ceram. Soc., 74:219-224.

9. Gregg, S.J. and K.S.W. Sing, 1982. Adsorption, Surface Area and Porosity, 2nd edn., New York, NY, USA: Academic Press.

10. Cao, G.Z., J. Meijerink, H.W. Brinkman, K.J. de Vries and A.J. Burggraaf. "Permporometry Study on the Size Distribution of Active Pores in Porous Ceramic Membranes." submitted to J. Membrane Sci.

Morphology of Very Thin Layers Made by Electrochemical Vapour Deposition at Low Temperatures

H. W. BRINKMAN, G. Z. CAO, J. MEIJERINK, K. J. DE VRIES and A. J. BURGGRAAF

ABSTRACT

Thin zirconia/yttria (ZY) layers on porous α-alumina substrates, prepared by the (Electro)Chemical Vapour Deposition (CVD/EVD) technique at 700-800°C, have been investigated. After about 40 minutes of deposition time the layers are gas tight; the thickness is estimated to be < 1 μm.

Preliminary results indicate that ZY, deposited at 700-800°C, forms polycrystalline layers with a crystallite size of about 20 nm; the crystallite size is slightly increasing with increasing deposition time. The ZY layers grown at 700-800°C for several hours have crystallites equal in size compared to layers grown at 1000°C for several hours; the fraction of monoclinic ZY is larger at low temperatures under current process conditions.

INTRODUCTION

In the past few years, the CVD/EVD technique has become an interesting method to form thin gas tight solid electrolyte (e.g. Yttria Stabilized Zirconia, YSZ) layers on porous substrates for Solid Oxide Fuel Cells (SOFC) [1,2,3]. Since these devices work optimally when the (bulk) ohmic polarization losses are minimal, the electrolyte layer should be as thin as possible. Electrolyte layers with thickness of 2-50 μm, deposited at 1000-1100°C, have been reported [1,2,3].

Another application of CVD/EVD grown layers is that they can be used in oxygen sensors, electrocatalytic reactors and as oxygen separation membranes. In membranes, two parameters are important: the selectivity towards a compound in a gas mixture, and the permeability through the membrane. In the ideal case, in dense oxygen semipermeable membranes (without pinholes or cracks) the selectivity for oxygen should be infinity, since oxygen is the only compound that can permeate through it. In the case that the bulk

H.W. Brinkman, G.Z. Cao, J. Meijerink, K.J. de Vries and A.J. Burggraaf, Laboratories of Inorganic Chemistry, Materials Science and Catalysis, Department of Chemical Technology, University of Twente, P.O.Box 217, 7500 AE, Enschede, The Netherlands

diffusion of oxygen ions is the rate limiting step in the separation process, membrane layers should be as thin as possible to obtain a maximum oxygen permeation rate.

In previous publications [4,5] attention has been focused on the CVD-/EVD process at high temperatures, e.g. 1000°C. Both the (modified) CVD and EVD stages (see below) occur in a well observable way. In this article, some preliminary investigations are presented concerning the deposition of thin zirconia-yttria layers on porous ceramic substrates at 800°C. After some theoretical background of the CVD/EVD process, especially at low temperatures (800°C) results concerning gas tightness, phase- and microstructure are reported. These aspects have great influence on the performance of the above mentioned devices by means of material properties like conductivity and oxygen permeability.

PRINCIPLES OF THE CVD/EVD PROCESS

The CVD/EVD process is an abbreviation for a process in which a (modified) Chemical Vapour Deposition process is followed by an Electrochemical Vapour Deposition process. In the CVD/EVD process reactants are supplied to both sides of a porous ceramic substrate. In the case of depositing YSZ, the reactants are on one hand vapours of $ZrCl_4$ and YCl_3, and on the other hand water which serves as the oxygen source. The only location where the metal chloride vapours can react with the water vapour is in or near the ceramic substrate. The first stage in the process is that the inner pore wall is covered by a thin YSZ layer. When the deposition process proceeds, the second stage occurs in which the pore diameter decreases. The location where this decrease is the largest, is situated at the side of the substrate which is directed to the metal chloride delivery, under current process conditions. When the deposition process is stopped after stage two, it is possible to modify the separation properties of porous ceramic membranes. The second stage results eventually in the third stage in which a plug is formed inside the pore. These three stages are together known as the (modified) CVD stage.

After the (modified) CVD process deposition can only proceed if diffusion of both oxygen ions (or oxygen ion vacancies) and electron holes (or electrons) through the plug is possible. The fourth stage, which is known as the EVD process, is only possible with mixed (i.e. ionic and electronic) conducting materials; under the current process conditions, YSZ is a mixed conductor with low electronic conductivity. The EVD process is influenced by temperature, since it relies on the electronic and ionic conductivities that are strongly temperature dependent. When the limiting step for the growth of the YSZ layer is diffusion through the bulk, layer growth at 800°C is expected to be much slower than at 1000°C. Due to the large value of the activation energy for electronic conduction in YSZ as reported in [6], the diffusion resistance in the YSZ film at 800°C is three orders of magnitude larger than that at 1000°C. This would theoretically imply that when a deposition experiment is performed at 800°C, the YSZ layer will grow very slowly after reaching the stage of pore plugging. This indicates that very thin (< 1 μm) dense YSZ layers can be grown on porous substrates.

EXPERIMENTAL

The CVD/EVD experiments were performed in a home built apparatus which is extensively described in [4]. A porous α-alumina substrate (characterized in Table I) was sealed on the end of a small, dense alumina tube, dividing the reactor into a water chamber (in which a mixture of air and water was introduced) and a chloride chamber. The $ZrCl_4$ and YCl_3 sublimation beds were put inside the chloride chamber just before each experiment. Chloride vapours were generated by heating the beds to the desired temperatures (see Table I); Ar or N_2 carrier gas was passed through the sublimation beds. The temperatures of the sublimation beds and the substrate were controlled by a six-zone furnace. The standard experimental conditions applied during deposition are given in Table I.

The gas tightness of CVD/EVD grown layers was checked in a home made permeation apparatus. A sample was placed in the permeation cell; on one side of the sample (at which the zirconia-yttria layer was deposited) a constant pressure of 4 atm helium is applied; the other side is connected to a bubble flow meter at 1 atm to observe whether there is leakage through the sample. Helium permeation through a coarse α-alumina substrate was also measured.

The phase structure, composition, layer thickness and crystallite size of the CVD/EVD grown zirconia-yttria layers was investigated by XRD (Philips PW 1710, CuKα) and SEM (JEOL, JSM-35CF).

RESULTS AND DISCUSSION

By the CVD/EVD technique deposition experiments were performed at a temperature of 700-800°C. The deposition time at 700°C was 90 minutes. The deposition times at 800°C were 30, 45, 60, 120 and 240 minutes.

TABLE I: TYPICAL CVD/EVD EXPERIMENTAL CONDITIONS

Substrate	α-alumina
Diameter	12 mm
Thickness	2 mm
Mean pore diameter	0.16 μm
Porosity	50 %
Deposition temperature	800°C
Reactor pressure	200 Pa
$ZrCl_4$ sublimation bed temperature	150 °C
YCl_3 sublimation bed temperature	625 °C
Ar carrier gas flow rate through $ZrCl_4$ bed	15 ml(STP)/min
Ar carrier gas flow rate through YCl_3 bed	7.5 ml(STP)/min
Water sparger temperature	40 °C
Air carrier gas flow rate through water sparger	3.5 ml(STP)/min
Total pressure in water sparger	15 kPa
Air/water ratio in vapour	1/1

THICKNESS OF THE LAYERS

Figure 1 shows a SEM photo of an YSZ layer grown on a porous α-alumina substrate. The deposition time was 90 minutes at a temperature of 700°C. The thickness of the dense YSZ layer is between 0.5-1 μm.

By SEM it was also tried to observe any difference in thickness between layers which were deposited at 800°C for 2 hours (sample #92032) and for 4 hours (sample #92047). No difference between these layers could be observed.

As an indication of the amount of ZY that has been deposited on/in the α-alumina (or an indication of the ZY layer thickness), the integrated $K\alpha_1$ intensity ratio of a characteristic ZY peak, which is the (111) peak around 30° 2θ and a characteristic α-alumina peak, which is the (012) around 25.6° 2θ is calculated. Correction for instrumental broadening is applied. The results are shown in Table II. According to this table, a clear distinction (about a factor 2) can be made between the intensity ratio of samples grown for 2 hours and 4 hours. Since the difference in layer thickness can not be seen by SEM, this would mean that the layers are very thin, i.e. < 1 μm, and that the ZY layer of sample #92047 has grown very slowly during the last two hours of deposition. The maximum growth rate of ZY layers at 800°C can be estimated to be several tenths of μm's per hour, while at 1000°C this growth rate is approx. 1.5 μm/h [4].

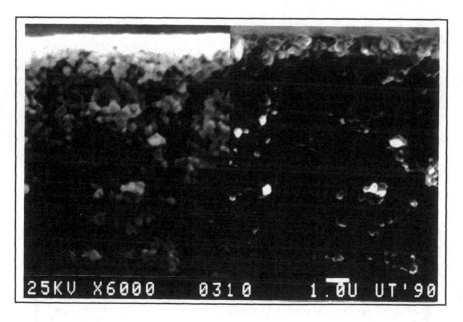

Figure 1:   SEM photograph of YSZ film grown at 700°C on porous α-alumina substrate (left: back-scattered electron image; right: secondary electron image)

GAS TIGHTNESS OF THE LAYERS

In figure 2 the helium permeation through samples grown at 800°C is presented as a function of the deposition time. The pore closure time is estimated around 40 minutes. This value is comparable with the pore closure time for α-alumina substrates, deposited at 1000°C [4].

For some samples it has been observed that the layers are not completely gas tight after longer time deposition; this may have something to do with cracks or pinholes that can be generated during cooling down of the samples (from deposition temperature to room temperature) or by getting the samples off the alumina tube.

PHASE COMPOSITION

In the XRD diagram of porous ceramic substrates on which zirconia-yttria (ZY) is deposited several phases are observed. Beside the α-alumina phase of the substrate, several ZY phases are present: monoclinic, tetragonal and/or cubic. The α-alumina peaks are sharp, while the ZY peaks are rather broad. It is very hard to distinguish between the tetragonal and the cubic phase. In theory, the (400) type reflections in the 2θ region between 73° and 75° are different for the tetragonal and the cubic phase. In the CVD/EVD grown samples this can hardly be seen, since the α-alumina (208) peak is also located in this range, which makes the analysis very difficult. However, both the tetragonal phase and the cubic phase exhibit oxygen ion conductivity, so in this investigation it is not of much importance to find out whether cubic or tetragonal ZY is formed.

The monoclinic ZY phase can be easily distinguished from the tetragonal or cubic phase; the monoclinic (111) and (1̄11) peaks at 28.2 and 31.5° 2θ differ from the tetragonal/cubic (111) peak around 30° 2θ. The nonlinear calibration method [7,8] is used to calculate the volume fraction of

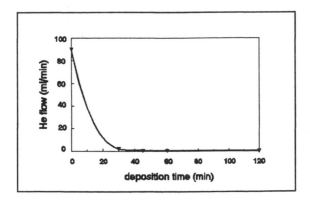

Figure 2: Helium permeation as function of deposition time for CVD/EVD samples manufactured at 800°C

monoclinic ZY. The integrated intensity ratio X_m for monoclinic - tetragonal/cubic ZrO_2 systems is defined as:

$$X_m = \frac{I_m(\bar{1}11) + I_m(111)}{I_m(\bar{1}11) + I_m(111) + I_{t,c}(111)} \tag{1}$$

where the subscript m represents the monoclinic phase; $I_{t,c}(111)$ is the integral intensity of the tetragonal (primitive cell) or cubic (1$\bar{1}$1) reflection. The volume fraction of monoclinic ZY, v_m, is calculated according to [8]:

$$v_m = \frac{P X_m}{1 + (P-1) X_m} \tag{2}$$

For the CVD/EVD grown layers with ~10 mol% Y_2O_3 (see next paragraph), P is taken to be 1.2 in accordance with ref. [8]. In Table II the fraction of monoclinic ZY is shown as a function of the deposition time. As can be seen from the table, the fraction of monoclinic ZY is rather large, between 13 and 28 vol%. It seems that when the deposition process proceeds, the monoclinic phase decreases (13 vol% after 2 and 4 hours deposition). This can imply that during the initial stage of the process (in the first 60 minutes), more monoclinic ZY is deposited than in the layer-growth stage. When the volume fraction of monoclinic ZY, deposited at 800°C, is compared to that deposited at 1000°C under the same conditions (samples #91025 and #91027 in Table II), the volume fraction monoclinic is higher when the temperature is lower under current process conditions.

By least-squares regression of the experimentally measured XRD patterns, the values for the lattice parameters of the cubic phase are roughly estimated. Here it is assumed that - beside the monoclinic ZY phase - only cubic ZY is present. The cubic (111), (200), (220), (311) and (222) peaks, which can be reasonably well distinguished in the XRD pattern, are used for the least-squares regression. Due to the fact that ceramic pills are used here (instead of powders) it is not possible to use an internal standard. Hence, the resulting value for the lattice parameter is not very accurate, but only indicative. The experimentally found relation of lattice parameter vs. Y_2O_3 mol percentage from ref. [9] is used to obtain the mole fraction Y_2O_3 in the assumed-to-be-cubic phase. The Y_2O_3 content in the cubic phase of all investigated samples is between 9-11 mol%.

CRYSTALLITE SIZE

It has been observed that 72-87% of the ZY layers consist of the cubic or tetragonal form. Again, it is assumed that these combined phases only consist of the cubic form. By means of X-Ray Line Broadening (XRLB) the sizes of the cubic crystallites in the layers were obtained. The Scherrer equation [10] is used to calculate the crystallite size D from the YSZ (111) reflection:

$$D = \frac{K\lambda}{\beta_{cs}\cos\theta} \frac{360}{2\pi} \tag{3}$$

TABLE II: EFFECTS OF DEPOSITION TIME ON VARIOUS PARAMETERS FOR CVD/EVD ZY LAYERS

Sample nr.	Deposition time (min)	Vol% monocl. phase	$\frac{I_c(111)^*}{I_\alpha(012)}$	D (nm)
#92030	30	21	0.86	17
#92031	45	19	1.05	18
#92029	60	28	1.16	19
#92032	120	13	1.26	20
#92047	240	13	2.96	25
#91025[**]	120	0	-	-
#91027[**]	120	0	-	-
#9032[**]	210	-	-	23

*   Ratio of XRD integrated $K\alpha_1$ intensity for cubic YSZ (111) reflection to that of the α-alumina (012) reflection

**   Deposition performed at 1000°C, all others at 800°C; data of #9032 is available from [11]

-    Not determined

λ is the wavelength of $CuK\alpha_1$ radiation (0.15405 nm), Θ is the diffraction angle of the (111) reflection, K is a constant (0.94) and β_{cs} is the half width at half maximum of a step-scan profile of the (111) peak after correction for $K\alpha_1$-$K\alpha_2$ separation and instrumental broadening. It is assumed that no line broadening occurred due to local inhomogeneities and strain.

The crystallite sizes of several samples are listed in Table II. These sizes are rather small (about 20 nm), and the size increases slightly with increasing deposition time. This may indicate that in the beginning of the CVD/EVD process, when only zirconia/yttria is deposited inside the pores, smaller crystallites are formed than during the later stages in which a layer is formed on top of the ceramic substrate. In comparison with a layer that was deposited at 1000°C for 3.5 hours, the crystallites of the layer grown at 800°C for 4 hours are equal in size, as can be seen in Table II.

CONCLUSIONS

The CVD/EVD technique is a successful method to modify the internal pore surface of ceramic membranes and to deposit very thin gas tight layers on these ceramic membranes. By performing depositions at low temperatures (700-800°C) it is possible to grow thinner layers than at 1000°C.

SEM, XRD and XRLB analyses show that the layers contain a relative large fraction of monoclinic ZY, compared to layers grown at 1000°C under current process conditions. The cubic YSZ phase consists of about 10 mol% Y_2O_3. The layers are crystalline with a crystallite size of about 20 nm; the crystallite size increases slightly with deposition time.

REFERENCES

1.   Isenberg, A.O., 1977. "Growth of refractory oxides layers by electro-chemical vapor deposition (EVD) at elevated temperatures", in Proc. Symp. Electrode Materials, Processes for Energy Conversion and Storage, eds. J.D.E. McIntyre, S. Srinivasan and F.G. Will, The Electrochem. Soc. Inc., Princeton, NJ, 77(6):572-583.

2.   Dietrich, G., and W. Schäfer, 1984. "Advances in the development of thin-film cells for high temperature electrolysis." Int. J. Hydrogen Energy, 9:747-752.

3.   Carolan, M.F. and J.N. Michaels, 1987. "Chemical vapour deposition of yttria-stabilized zirconia on porous substrates." Solid State Ionics, 25:207-216.

4.   de Haart, L.G.J., Y.S. Lin, K.J. de Vries and A.J. Burggraaf, 1991. "Modified CVD of nanoscale structures in and EVD of thin layers on porous ceramic membranes." J. Eur. Ceram. Soc., 8:59-70.

5.   Lin, Y.S., K.J. de Vries, H.W. Brinkman and A.J. Burggraaf, 1992. "Oxygen semipermeable solid oxide membrane composites prepared by electrochemical vapor deposition." J. Membr. Sci., 66:211-226.

6.   Park, J.H. and R.N. Blumenthal, 1982. "Electronic transport in 8 mole percent Y_2O_3-ZrO_2." J. Electrochem. Soc., 136:2867-2876.

7.   Toraya, H., M. Yoshimura and S. Somiya, 1984, "Calibration curve for quantitative analysis of the monoclinic-tetragonal ZrO_2 system by X-ray diffraction." J. Am. Ceram. Soc., 67:C119-C121.

8.   Toraya, H., M. Yoshimura and S. Somiya, 1984. "Quantitative analysis of monoclinic-stabilized cubic ZrO_2 systems by X-ray diffraction." J. Am. Ceram. Soc., 67:C183-C184.

9.   Ingel, R.P. and D. Lewis III, 1986. "Lattice parameters and density for Y_2O_3-stabilized ZrO_2." J. Am. Ceram. Soc., 69:325-332.

10.   Klug, K.P. and L.E. Alexander, 1974. X-ray diffraction procedures, Wiley & Sons, New York, pp. 687-692.

11.   Lin, Y.S., 1992. "Chemical and electrochemical vapour deposition of zirconia-yttria solid solutions in porous ceramic media", PhD Thesis, University of Twente, p. 78.

The Role of Water Vapour on the Kinetics of H_2 Oxidation on Porous Ni Electrodes at 1000°C

S. P. JIANG

ABSTRACT

NiO/Y_2O_3 - ZrO_2 (Ni/YSZ) cermet is the most common material used as fuel electrodes in solid oxide fuel cells. In this paper, oxidation reactions of H_2 on porous Ni electrodes have been studied in dry H_2 and in 98% H_2 + 2% H_2O at fuel cell operation temperatures. In particular, the role of the presence of H_2O vapour on the mechanism and kinetics of H_2 oxidation reaction has been discussed, based on the reaction models proposed.

INTRODUCTION

In solid oxide fuel cells (SOFC), NiO/Y_2O_3 - ZrO_2 (Ni/YSZ) cermet systems are widely used as fuel electrodes due to their chemical stability under reducing conditions, low volatility and abundant availability. The conductivity and polarization behaviour of the Ni/YSZ cermet anodes have been reported in detail [1-2]. However, compared to oxygen reduction on air electrodes, oxidation reactions of fuel gases such as H_2 and CO on Ni and Ni/YSZ cermet anodes under fuel cell operation conditions have not received much attention. Nguyen *et al.* [3] studied the oxidation reactions of H_2 and some other hydrocarbons on porous Pt and Au electrodes using Sc_2O_3-stabilized ZrO_2 as electrolyte in dry gas, and found that the electrocatalytic activity was the same on porous Pt and Au electrodes and the activity was enhanced considerably by pre-polarization of the zirconia electrolyte. They concluded that the oxidation reaction occurs at active sites on the zirconia electrolyte surface rather than on the metal electrode surface and the reaction is controlled by the charge transfer process. On the other hand, other study [4] shows that in the presence of water vapour, the slopes of the polarization curves for H_2

S.P. Jiang
CSIRO Division of Materials Science & Technology and Ceramic Fuel Cells Ltd.
Locked Bag 33, Clayton, Vic. 3168, Australia

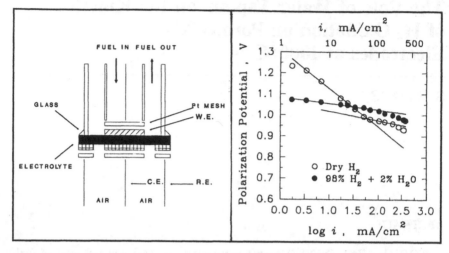

Figure 1. Schematic diagram of the cell configuration.

Figure 2. Tafel plots for H_2 oxidation on a porous Ni electrode, measured at 1000°C in dry H_2 and in 98% H_2 + 2% H_2O.

oxidation approach the resistance of the electrolyte and polarization loss is negligible on porous Pt electrodes. Recently, Setoguchi *et. al.* [5] studied H_2 oxidation reactions on Ni/YSZ cermet anodes and found that the polarization resistance decreased with an increase of the H_2O concentration in the H_2 gas. However, the influence of water vapour on the oxidation reaction of hydrogen in the Ni/YSZ system is still largely unknown. In the present study, H_2 oxidation was investigated in dry H_2 and in 98% H_2 + 2% H_2O at 1000°C on porous Ni electrodes. The role of the presence of water vapour on the kinetics and mechanism of the H_2 oxidation reaction in Ni/YSZ systems has been discussed.

EXPERIMENTAL

Zirconia electrolyte discs were prepared from 3 mol% Y_2O_3 - ZrO_2 (YSZ) powder (TOSOH, TZ3Y). YSZ powder was pressed and sintered at 1500°C for 4 hours into discs of about 19 mm diameter and 0.9 mm thick. NiO paste was applied onto the zirconia electrolyte disc in the shape of a circle (electrode area: ~0.35 cm²) by a slurry painting method, followed by heat treatment at 1500°C for 2hrs in air. Platinum paste (Engelhard) was applied on the other side of the electrolyte disc to make counter and reference electrodes. Platinum mesh was used as the current collectors. The fuel electrode compartment was sealed with a glass packing. Cathodic side of the cell was exposed to air. High purity H_2 (CIG, 99.98%) was used without further treatment and water vapour (~2%) was introduced through a gas bubbler system in a temperature-controlled bath. The flow rate of the fuel gas was 100 ml/min. Reduction of nickel oxide was carried out by passing fuel gas before the measurement. Figure 1 shows the schematic diagram of the cell testing arrangement.

The electrochemical activities of porous Ni electrodes were measured at 1000°C in both dry H_2 and in 98% H_2 + 2% H_2O by using galvanostatic current interruption and impedance spectroscopy techniques. Polarization measurements were carried out in a self-generating mode and maximum current was attained when the ohmic resistance of the cell components caused the cell voltage to approach zero. Impedance measurements were carried out using a Voltech TF2000 frequency response analyzer in conjunction with a potentiostat (Utah Electronics, Model 0152) under open circuit voltage (OCV) at 1000°C. Measurements were made over the frequency range from 1 Hz to 100 kHz with signal amplitude of 20 mV. From the impedance spectra, the overall polarization resistance, R_p for H_2 oxidation was obtained from the difference of the intercepts of the arcs measured at low and high frequencies. The ohmic resistance, R_Ω between the working electrode and Pt reference electrode was measured from the intercepts at high frequencies.

RESULTS

Figure 2 shows the polarization performance for H_2 oxidation on a porous Ni electrode measured at 1000°C in dry H_2 and in 98% H_2 + 2% H_2O. The OCV of the cell was -1.264V in dry H_2 and -1.080V in 98% H_2 + 2% H_2O, respectively. This corresponds to an O_2 partial pressure of 2.00×10^{-16}Pa in dry H_2 and 1.64×10^{-13}Pa in 98% H_2 + 2% H_2O, respectively according to the Nernst equation (equation 2). In the case of dry H_2, the polarization potential drops sharply and falls below that for H_2 oxidation in 98% H_2 + 2% H_2O at current densities higher than about 30 mA/cm². At a current density of 200 mA/cm², the overpotential for H_2 oxidation in dry H_2 is about 0.3V, and in 98% H_2 + 2% H_2O, the overpotential is about 0.07V. The reason for the lower anodic potential in dry H_2 compared to that in 98% H_2 + 2% H_2O at high current densities is not quite clear. However, from the impedance measurements done before and after the tests, the change of the impedance spectra is relatively small, indicating that differences in the electrode morphology are not responsible for such a behaviour.

For 98% H_2 + 2% H_2O, there is only one linear Tafel region with a Tafel slope of ~0.036 V/decade up to a current density of about 200 mA/cm². However, in the case of dry H_2, there are two distinct linear Tafel regions. A very high Tafel slope (~0.182 V/decade) was observed in the low current density range. As current density increases, the Tafel slope dramatically reduced to ~0.044 V/decade. These results clearly suggest that the presence of water vapour promotes H_2 oxidation in the Ni/YSZ system.

The very different polarization behaviour for H_2 oxidation on porous Ni electrodes in dry H_2 and in 98% H_2 + 2% H_2O was also demonstrated by the impedance spectroscopy. Figure 3 gives the corresponding impedance spectra. In the case of 98% H_2 + 2% H_2O, there were two semi-circles or arcs and the overall polarization resistance, R_p, for H_2 oxidation was ~3.13 $\Omega.cm^2$. The high frequency semi-circle is dominant as compared to the one at low frequencies. Again, in the case of dry H_2, the impedance behaviour is remarkably different from that in 98% H_2 + 2% H_2O. The impedance arcs for H_2 oxidation were stretched extensively and also very distorted. By extrapolation of this impedance arc at low frequencies, the R_p for H_2

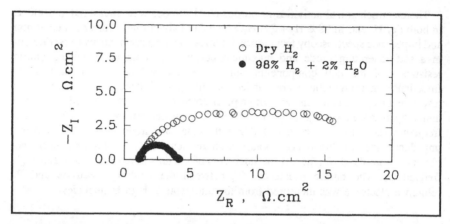

Figure 3. Impedance spectra for H_2 oxidation on a porous Ni electrode, measured at 1000°C and OCV in dry H_2 and in 98% H_2 + 2% H_2O.

oxidation in dry H_2 could be estimated to be at least ~16.5 $\Omega.cm^2$. Such high polarization indicates that the charge transfer reaction for H_2 oxidation in dry H_2 is inhibited compared to that in 98% H_2 + 2% H_2O. The results clearly demonstrate that the presence of even a very small amount of water vapour (2%) in the H_2 fuel gas system can greatly reduce the polarization resistance for H_2 oxidation on Ni/YSZ systems. Another interesting observation from the impedance spectra is the relative increase in the ohmic resistance, R_Ω between the anode and the Pt reference electrode in dry H_2 compared to that in 98% H_2 + 2% H_2O. For the porous Ni electrode studied, R_Ω was 1.08 $\Omega.cm^2$ in 98% H_2 + 2% H_2O and 1.2 $\Omega.cm^2$ in dry H_2, respectively. The impedance observations are in good agreement with the corresponding polarization behaviour for H_2 oxidation on porous Ni electrodes.

DISCUSSION

The remarkable differences in the polarization and impedance behaviour in dry and moist H_2 gas demonstrate that the reaction mechanisms and kinetics for H_2 oxidation on the Ni/YSZ systems are not independent of the presence of water vapour in the system. Under fuel cell operation conditions, oxygen is reduced at the cathode and the electric field present in the electrolyte causes the oxygen ions to diffuse from the cathode to the anode. High resolution transmission electron microscopy studies on directionally solidified eutectic $NiO-ZrO_2(CaO)$ interfaces clearly show that the interface is extremely clean with no evidence of a second phase at the interface [6]. Thus, in the Ni/YSZ system, it could be more appropriate to define the three phase boundary (*tpb*) (where H_2 gas, Ni electrode and YSZ electrolyte meet) as *tpb* lines rather than *tpb* regions. Taking into account that the diffusivity of oxygen in Ni metal is very low [7], the oxidation of H_2 takes place either in the vicinity of *tpb* lines or just on the *tpb* lines. This point will be discussed at a later stage. For H_2 oxidation, the following major reaction steps of the electrode process could be considered: 1) dissociative adsorption of hydrogen molecules at active centres near the *tpb* lines; 2) surface diffusion of the adsorbed species, H_{ads} from the active centres towards the *tpb* lines; 3) charge transfer reaction on or in the vicinity of *tpb* lines; 4) diffusion of H_2O produced on the electrode surface to the bulk gas phase.

In a study of H_2 oxidation on Pt electrodes in H_2-H_2O systems, Etsell and Flengas [8] pointed out the possibility that significant potential drop at low current densities in the case of very low H_2O content is due to concentration polarization because of the buildup of H_2O at the electrode. In a SOFC such as

$$H_2/H_2O, Ni \mid YSZ, O^{2-} \mid Pt, O_2 \qquad (1)$$

the open circuit voltage of the cell can be calculated according to the Nernst equation

$$E_{th} = (\frac{RT}{4F}) \ln (\frac{P_{O_2}{}''}{P_{O_2}{}'}) \qquad (2)$$

where $P_{O2}{}'$ and $P_{O2}{}''$ are O_2 partial pressures at the anode and cathode, respectively. If the cell is slightly discharged, steam is generated and $P_{O2}{}'$ will increase. This will lead to a decrease in cell voltage. Also, H_2O formed on the electrode surface could build up and cause the potential drop, the effect of which depends on the transport rate of H_2O in a H_2-H_2O mixture. An attempt was thus made to estimate the magnitude of the potential drop accompanied by the steam formation and by the possible buildup of H_2O. The $P_{O2}{}''$ at the Pt cathode is ~0.21x10⁵Pa, and the value of $P_{O2}{}'$ can be calculated from the standard Gibbs energy, ΔG^o

$$P_{O_2}{}' = \left[\frac{P_{H_2O}}{P_{H_2}} \exp (\frac{\Delta G^o}{RT})\right]^2 \qquad (3)$$

where ΔG^o is for the following reaction

$$H_2 + \frac{1}{2}O_2 = H_2O \qquad (4)$$

and is given by [9]

$$\Delta G_T{}^o = -58900 + 13.1T \pm 1000 \text{ (cal.)} \qquad (5)$$

In the present study, care was taken to minimize the effect of steam produced on the polarization curve by supplying a sufficiently high flow rate of fuel gas (*i.e.* 100 ml/min) at relatively low current densities. According to the recent results of Nagata and Iwahara [10] on Ni/YSZ cermet anodes, at fuel gas flow rates above 80 ml/min, water vapour pressure in the fuel gas is not affected by the steam produced by the oxidation reaction, *i.e.* the back flow of steam is negligible. First, the potential drop due to the formation of H_2O was calculated according to the Faradaic law and the above relations, based on the assumption that the fuel cell is operating as a concentration cell. When OCV is 1.264V (in dry H_2), it can be calculated that the flux of steam produced at 10 mA/cm^2 is ~0.025 ml/min, which corresponds to 0.025% H_2O in the fuel gas. The corresponding potential drop caused by the steam formation is ~0.005V. The observed potential drop is however ~0.134V at 10 mA/cm^2 in dry H_2. This demonstrates that the contribution of steam formed during discharge to the initial potential drop is negligible.

It is reasonable to assume that only the ordinary diffusion is significant and the effect of porosity on the diffusion process is negligible for H_2 oxidation on the Ni anodes, which could be justified as the present Ni electrode has a very open structure with pore size in the range of 5 to 10µm (see Figure 4). In binary systems such as H_2 - H_2O, the ordinary diffusion coefficient, D_{12}, is independent of the composition and varies approximately with $T^{1.75}$. For H_2-H_2O system, $D_{12} \times 10^{+9}/T^{1.75}$ = 4.107 [11]. Then, the transport rate of H_2O, j_{H2O}, in the H_2-H_2O system can be estimated from Fick's diffusion equation.

$$j_{H_2O} = D_{12} \frac{C^S - C^B}{x} A \qquad (6)$$

where C^S and C^B are the surface and bulk concentration of H_2O, respectively, A the anode surface area and x the thickness of the anode layer. Using conditions that $C^B=0$, $C^S=0.025$ml/100ml at 10 mA/cm^2, $A=0.35$cm^2 and $x=17$µm, the transport rate of H_2O was estimated to be ~34 ml/min, which is far greater than the amount of H_2O produced at the same current density (i.e. 0.025ml/min), indicating the chance of the steam buildup at the electrode surface is also very small.

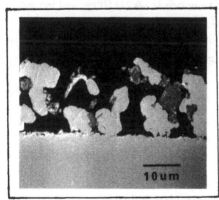

Figure 4. SEM micrograph of the Ni electrode after the fuel cell measurement.

From the above simple calculation, it is clear that the contribution of either the formation of H_2O or the buildup of H_2O to the initial potential drop in dry H_2 is very small. Therefore, it must be the case that H_2O in the H_2 gas plays a catalytic role in the mechanism and kinetics for H_2 oxidation on the Ni/YSZ systems. First, the adsorption behaviour of H_2 molecules in a Ni/YSZ system at fuel cell operation temperature (i.e., 1000°C) will be considered. At low temperatures, dissociative adsorption of H_2 occurs favourably on the metal surface due to the presence of unsaturated metal bonds at the surface [12]. However, at high temperatures, the energy gained in forming two metal atom/hydrogen atom bonds relative to the dissociation energy of a free H_2 molecule could become irrelevant. On the other hand, in the presence of oxide the hydrogen atoms dissociatively adsorbed on the metal surface can diffuse to the adsorption sites on the oxide surface. Such surface diffusion of an active species like hydrogen is called the spillover effect [13]. Kramer [14] has shown that the adsorption of gaseous hydrogen onto alumina is made possible by the presence of platinum, although the adsorption process of H_2 from the gas phase onto Al_2O_3 is kinetically hindered. In the Ni/YSZ system, H_2 could dissociatively adsorb on both the YSZ and Ni surfaces, $H_{ads, YSZ}$ and $H_{ads, Ni}$. In dry H_2, $H_{ads, Ni}$ would diffuse to the adsorption sites close or adjacent to oxygen ions on the YSZ electrolyte surface because of the thermodynamic instability of $H_{ads, Ni}$ atoms. Thus, the charge transfer reaction takes place mainly between $H_{ads, YSZ}$ atoms and oxygen ions. Electrons released have to migrate over the zirconia surface and YSZ/Ni boundary lines to the Ni electrode. Such reaction steps can be simply represented as follows:

Dissociative adsorption of H_2

$$H_2 = 2H_{ads, YSZ} \text{ (on YSZ surface)} \tag{7a}$$

and/or

$$H_2 = 2H_{ads, Ni} \text{ (on Ni surface)} \tag{7b}$$
$$H_{ads, Ni} \dashrightarrow H_{ads, YSZ} \tag{7c}$$

Charge transfer reaction

$$2H_{ads, YSZ} + O_o^x = H_2O_{gas} + V_{\ddot{O}} + 2e \tag{7d}$$

where O_o^x is an oxygen ion on an oxygen lattice site and $V_{\ddot{O}}$ is an oxygen vacancy. The electron transfer path for the above reaction steps could be described by the following scheme.

$$e^-(YSZ) \dashrightarrow e^-(YSZ/Ni) \dashrightarrow e^-(Ni) \tag{7e}$$

The very high reaction resistance for such electron transfer path has been confirmed by the high polarization resistance and high Tafel slope for H_2 oxidation in dry H_2 on porous Ni electrodes. The H_2 oxidation reaction in dry H_2 (*i.e.* low H_2O content) is most likely controlled by this electron transfer process.

In moist H_2, the interaction of H_2O with a metal surface results in the dissociation of the molecule at high temperatures (*i.e.* >350K). Hydrogen gas is released and the surface is covered with a layer of O_{ads} atoms [15]. Also, H_{ads} atoms could be thermodynamically stable at the sites close or adjacent to such O_{ads} atoms on Ni surface. The fact that very low values of the Tafel slope and polarization resistance were observed in the presence of H_2O implies that the resistance for the reaction is very small and the electron transfer process may not involve the YSZ phase as in the case of dry H_2. This could be possible if a redox couple exists between active species on the YSZ and Ni surfaces (most probably O_{ads} atoms). If $O^{2-}_{ads, YSZ}/O_{ads, Ni}$ is a redox couple, possible reaction steps could then be:

Dissociative decomposition of H_2O

$$H_2O_{gas} = O_{ads} + H_{2\,gas} \tag{8a}$$

Dissociative adsorption of H_2

$$H_2 = 2H_{ads,\,YSZ} \text{ (on YSZ surface)} \tag{8b}$$
$$H_2 = 2H_{ads,\,Ni-O} \text{ (on } O_{ads} \text{ covered Ni surface)} \tag{8c}$$

Charge transfer reaction

$$O_o{}^x = O^{2-}{}_{ads,\,YSZ} + V_{\ddot{O}} \tag{8d}$$
$$O^{2-}{}_{ads,\,YSZ} + O_{ads,\,Ni} = O_{ads,\,YSZ} + O^{2-}{}_{ads,\,Ni} \tag{8e}$$
$$O^{2-}{}_{ads,\,Ni} = O_{ads,\,Ni} + 2e \tag{8f}$$
$$2H_{ads,\,YSZ} + O_{ads,\,YSZ} = H_2O_{gas} \tag{8g}$$
$$2H_{ads,\,Ni-O} + O_{ads,\,Ni} = H_2O_{gas} \tag{8h}$$

Obviously, the rate of the redox reaction (8e) is related to the activity of O_{ads} atoms on the YSZ and in particular the Ni metal surfaces. The activity of oxygen species on the metal surface is most likely dependent on the strength of the metal-oxygen bonding, which is related to the heat of oxide formation, $-\Delta H_f^\circ$. From the proposed reaction model, if the reaction is dominated by the step (8e), it could also be expected that the metal with a desirable $-\Delta H_f^\circ$ value is most active for H_2 oxidation as the formation and diffusion processes of MO or $O_{ads,\,M}$ species become most appropriate for the electron exchange reaction (where M represents an active site on metal surface). The proposed reaction model can also explain the volcanic-type dependence of the electrode activities vs $-\Delta H_f^\circ$ for metal anodes as reported by Setoguchi $et\ al$ [5]. For metal with small $-\Delta H_f^\circ$ such as Pt, the electrode activity is low due to the low stability of $O_{ads,\,M}$ on the metal surface. Thus, the overall activity increases with an increase in $-\Delta H_f^\circ$, as the formation of $O_{ads,\,M}$ species becomes easier. On the other hand, if $-\Delta H_f^\circ$ value is too high ($e.g.$ for metals such as Mn), the electrode activity would be reduced again as the diffusion of $O_{ads,\,M}$ on the metal surface becomes more and more difficult, limiting the redox reaction rate.

The catalytic role of H_2O in the H_2 oxidation reaction is also indicated by the shift of Tafel slopes from 0.182V/decade at low current densities to about 0.044V/decade at high current densities in the case of dry H_2. As the discharge current density increases, H_2O is produced and dissociatively decomposed on or near the electrode/electrolyte interface regions. Sufficient surface coverage of O_{ads} atoms on the Ni surface near tpb lines results in the change of the reaction mechanism and kinetics. This is in accordance with the observation that the second Tafel slope of about 0.044 V/decade at high current densities in dry H_2 is very close to the value of 0.036 V/decade observed for the same electrode in 98% H_2 + 2% H_2O.

From both experimental results and the reaction model proposed, it could also be concluded that H_2 oxidation takes place in the vicinities of tpb lines rather just on tpb lines. If the oxidation occurs strictly on the three phase boundary lines, the rate of the charge transfer process would be identical, irrespective of the coverage of O_{ads} atoms on the Ni surface. If the reaction is not controlled by the diffusion of steam as discussed above, the only other factor which could influence the reaction kinetics is the source of H_{ads}, the diffusion rate of which may differ depending on whether it comes from YSZ or oxygen-covered Ni metal surfaces. However, at high temperatures, it would be expected that diffusion rates for $H_{ads, YSZ}$ and $H_{ads, Ni-O}$ are quite similar, as the binding energy for H-O species should be more or less the same in both cases.

Finally, as H_2O is a final product of the reaction, diffusion process of steam produced on the electrode surface to the bulk gas phase will become more and more important as the discharge current densities increase, especially in the case of Ni/YSZ cermet anodes. On the other hand, it is considered that the catalytic role of H_2O is relatively independent of the discharge current as far as there is a sufficient amount of H_2O inside the porous electrodes. The steam in the fuel gas can be added externally or be produced *in situ* by the oxidation reaction. Thus, the role of H_2O on the reaction kinetics and mechanism is also largely dependent on the parameters such as the initial H_2O content and the porous structure of the electrode.

CONCLUSIONS

The mechanism and kinetics for hydrogen oxidation on porous Ni electrodes are strongly dependent on the presence of H_2O in the H_2 fuel gas at 1000°C. It has been shown that the contribution of steam formation or buildup of steam on the electrode surface to the initial potential drop in dry H_2 is very small. Consequently, reaction models have been proposed for H_2 oxidation in both dry and moist H_2. The most important role of H_2O is to provide O_{ads} atoms on the Ni metal surface, leading to the formation of a redox couple ($O^{2-}_{ads, YSZ}/O_{ads, Ni}$) to act as a 'bridge' for the electron transfer from the YSZ phase to the Ni metal phase in the vicinity of *tpb* lines. The kinetics of the H_2 oxidation is most likely related to the oxygen activity on the metal surface. In dry H_2 (*i.e.* very low H_2O content), reaction takes place mainly on the YSZ surface in the vicinity of *tpb* lines and the reaction is most likely controlled by the electron transfer process.

ACKNOWLEDGMENT

NiO pastes and YSZ electrolyte disks were prepared by Ms K.Crane. I would like to sincerely thank Dr. S.P.S.Badwal for his guidance and stimulating discussions during this project, Dr. K.Foger for discussions on the hydrogen spillover effect, and Dr. M.J.Bannister and Dr R.Taylor for invaluable comments on the manuscript.

REFERENCES

1.   Dees, D.W., T.D.Claar, T.E.Easler, D.C.Fee and F.C.Mrazek, 1987. "Conductivity of Porous Ni/ZrO₂-Y₂O₃ Cermets." J. Electrochem. Soc., 134(9): 2141-2146.

2. Kawada, T., N.Sakai, H.Yokokawa and M.Dokiya, 1990. "Characterization of Slurry-Coated Nickel Zirconia Cermet Anodes for Solid Oxide fuel Cells." J. Electrochem. Soc., 137(10): 3042-3047.

3. Nguyen, B.C., T.A.Lin and D.M.Mason, 1986. "Electrocatalytic Reactivity of Hydrocarbons on a Zirconia Electrolyte Surface." J. Electrochem. Soc., 133(9): 1807-1815.

4. Weissbart, J. and R.Ruka, 1962. "A Solid Electrolyte Fuel Cell." J. Electrochem. Soc., 109(8): 723-726.

5. Setoguchi, T., K.Okamoto, K.Eguchi and H.Arai, 1992. "Effects of Anode Material and Fuel on Anodic Reaction of Solid Oxide Fuel Cells." J. Electrochem. Soc., 139(10): 2875-2880.

6. Dravid, V.P., C.E.Lyman, M.R.Notis and A.Revcolevschi, 1989. "High Resolution Transmission Electron Microsropy of Interphase Interfaces in NiO-ZrO_2(CaO)." Ultramicroscopy, 29: 60-70.

7. Ramanaravanan, T.A. and R.A.Rapp, 1972. "The Diffusivity and Solubility of Oxygen in Liquid Tin and Solid Silver and the Diffusivity of Oxygen in Solid Nickel." Metallurgical Transactions, 3(12): 3239-3246.

8. Etsell, T.H. and S.N.Flengas, 1971. "Overpotential Behaviour of Stabilized Zirconia Solid Electrolyte Fuel Cells." J. Electrochem. Soc., 118(12): 1890-1900.

9. Kubaschewski, O. and C.B.Alcock, 1979. Metallurgical Thermochemistry, Oxford, UK: Pergamon Press Ltd., pp.380.

10. Nagata, M. and H.Iwahara, "Measurement of Water Vapour Pressure in the Porous Anode of SOFC during Discharge," this volume.

11. Fuller, E.N., P.D.Schettler and J.C.Giddings, 1966. "A New Method for Prediction of Binary Gas-Phase Diffusion Coefficients," Ind. Eng. Chem., 58(5): 18-27.

12. Tompkins, F.C., 1978. Chemisorption of Gases on Metals, London, UK: Academic Press, p.193.

13. Dowden, D.A., 1980. "The Spillover of Chemisorbed Species," in Catalysis, Vol.8, London, UK: Chem. Soc., pp.136-168.

14. Kramer, R., 1977. "Hydrogen Spillover in the Platinum-on-Alumina System." Naturwissenschaften, 64: 269-269.

15. Benndorf, C., C.Nöbl, M.Rüsenberg and F.Thieme, 1982. "H_2O Interaction with Ni(110): Auto-catalytic Decomposition in the Temperature Range from 400 to 550K." Appl. Surface Sci., 11/12: 803-811.

Measurement of Water Vapor Pressure in the Vicinity of Anode of SOFC during Discharge

M. NAGATA and H. IWAHARA

ABSTRACT

In order to determine the water vapor pressure in the porous anode of a solid oxide fuel cell (SOFC) during discharge, a small steam sensor was attached to the surface of the anode. The sensor is a steam concentration cell and is composed of a stabilized zirconia electrolyte with porous platinum electrodes.

It was observed that the water vapor pressure increased with increasing discharge current of the SOFC. From the steam pressure measured by this method, the hydrogen partial pressure in a porous Ni–YSZ cermet and Pt–plated anode was determined and the anodic overpotential due to concentration polarization was calculated. It was shown that the major part of the anodic overvoltage measured with a dc interruption method was due to concentration polarization of hydrogen.

INTRODUCTION

One of the merits of high temperature solid oxide fuel cells is that, due to high operating temperature, activation polarization is, in general, very small [1]. Therefore, one can obtain large cell currents. In such a case, a large quantity of reactant is needed at the reaction sites in the electrode and substantial product must be removed from the reaction sites. Thus, concentration polarization, caused by slow diffusion of the reactant gas or product gas, is a major source of potential loss during discharge.

In the anode of a SOFC, water vapor forms at the three phase interfaces (gas, electrode and solid electrolyte) as a result of the anode reaction. If diffusion of the water vapor is not fast enough, hydrogen concentration in the vicinity of the reaction sites becomes low compared to that in the bulk of the electrode compartment. Overpotential in the fuel electrode of SOFC is considered to be based on this phenomenon [2].

However, this phenomenon had not yet been experimentally confirmed as the method for measurement of the hydrogen partial pressure in the vicinity of the

M.Nagata, H.Iwahara, School of Engineering, Nagoya University, Furo-cho, Chikusa-ku, Nagoya 464-01, Japan

829

anode had not been developed. In the present study, we measured water vapor pressure in the SOFC anode during discharge using a steam sensor attached to the fuel electrode and studied the effect of water vapor pressure on anodic polarization. Overpotential at the anode of the fuel cell measured by the usual current interruption method was compared with that calculated from the concentration of water vapor and the nature of anodic polarization was discussed. The steam sensor used here was based on the principle of a steam concentration cell using a stabilized zirconia electrolyte [3].

PRINCIPLE OF STEAM SENSOR FOR ANODE OF SOFC

The concept of a steam concentration cell using an oxide–ionic conductor is represented schematically in figure 1 [3]. When hydrogen gas with different humidity is introduced into each electrode compartment, the cell generates an electromotive force (emf) depending on the difference in water vapor pressure. Here, the electrode with higher humidity acts as a cathode. Electrode reactions of this cell are;

$$\text{Anode:} \quad H_2 + O^{2-} \rightarrow H_2O + 2e^- \tag{1}$$
$$\text{Cathode:} \quad H_2O + 2e^- \rightarrow H_2 + O^{2-} \tag{2}$$

Theoretical emf, E, of the cell is given by

$$E = \frac{RT}{2F} \ln \frac{P_{H_2O}(1)\, P_{H_2}(2)}{P_{H_2O}(2)\, P_{H_2}(1)} \tag{3}$$

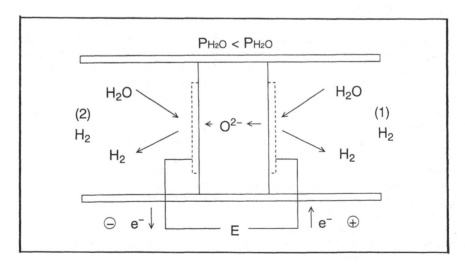

Figure 1. Concept of steam concentration cell using an oxygen ion conductor.

where $P_{H_2}(1)$ and $P_{H_2}(2)$ denote the partial pressure of hydrogen in compartments (1) and (2), respectively, and $P_{H_2O}(1)$ and $P_{H_2O}(2)$ are the partial pressures of water vapor in each electrode compartment. R, T and F have their usual meanings. If the total pressure in each compartment is kept at 1 atm and $P_{O_2} << P_{H_2O}$ and P_{H_2}, then:

$$P_{H_2O}(1) + P_{H_2}(1) = 1 \qquad (4)$$
$$P_{H_2O}(2) + P_{H_2}(2) = 1 \qquad (5)$$

Equation 3 can be rewritten as

$$P_{H_2O}(1) = \frac{P_{H_2O}(2)}{P_{H_2O}(2) + P_{H_2}(2) \exp(-2EF/RT)} \qquad (6)$$

and water vapor pressure $P_{H_2O}(1)$ in compartment 1 can be determined from the emf, $P_{H_2O}(2)$ and $P_{H_2}(2)$ in compartment 2. This type of sensor can be applied to measurement of the water vapor pressure at the anode of a SOFC under operation as described below.

EXPERIMENTAL

Small hydrogen–air fuel cells were constructed using ceramic disks of 8mol% yttria stabilized zirconia YSZ (TZ–8Y) as an electrolyte. On one side of the disk, porous platinum or Ni–YSZ cermet was attached as an anode material by chemical

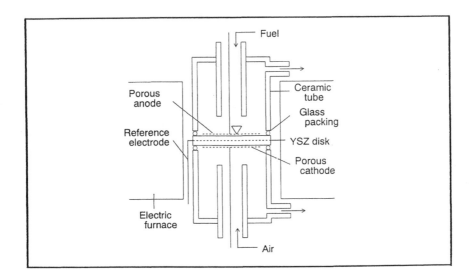

Figure 2. Schematic diagram of a test cell.

plating or screen printing, respectively. The chemical plating of Pt on YSZ was made according to our previous study [4]. The chemically plated platinum was also used as a cathode material. The Ni–YSZ cermet (Ni content; 40 vol%) electrode was prepared by screen printing of a NiO–YSZ powder paste on the YSZ disk [5], baking it at 1400°C for 2 hrs, and reducing it in hydrogen atmosphere at 1000°C. As a screen for printing, a net of 200 mesh with a wire diameter of 47 μm was used. Platinum wire was wound around the circumference of the electrolyte disk as a reference electrode to measure the anode and cathode overvoltages individually.

The electrolyte disk with electrodes thus prepared was sandwiched between alumina tubes using glass gaskets as shown in figure 2. A small square–cone shaped steam sensor element made of YSZ (4 mm height) was attached to the fuel electrode by pressing it with a ceramic tube inside a fuel feeding quartz tube as schematically shown in figure 3. In order to measure the correct value of water vapor pressure and not to impede the diffusion of water vapor from the reaction sites in the anode, an apex of the square cone was used as the sensing electrode by contacting it to the porous anode. On the base of the square cone element, a platinum electrode was deposited by electroless plating to serve as a standard gas electrode.

Overvoltage was measured by the usual current interruption method using the reference electrode wound on to the circumference of the disk electrolyte. Hydrogen gas saturated with water vapor at room temperature (total pressure; 1 atm.) was flowed through the quartz tube.

Water vapor pressure in the anode during discharge of the cell was determined by measuring the emf of the sensor and calculating on the basis of equation 6 where compartment 1 corresponded to the inside of the porous anode to be measured and compartment 2 corresponded to the inside of the fuel feed pipe.

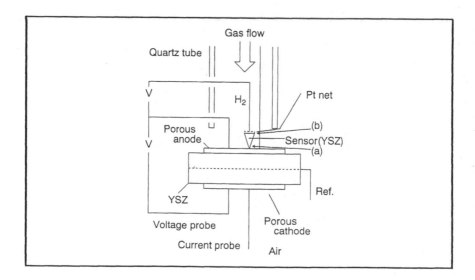

Figure 3. Schematic close–up of the fuel cell arrangement with a steam sensor element.

RESULT AND DISCUSSION

RELIABILITY OF STEAM SENSOR

In order to check the stability and the reliability of the steam sensor, the dependence of its emf on gas flow rate in the fuel cell under operation was first examined. Here, Ni–YSZ cermet was used as an anode material. Figure 4a, shows the result at a current density of 500 mA cm^{-2}. The emf increased with the flow rate to reach a stable and constant value at about 80 ml min^{-1}. This means that, at above this flow rate, feed gas was not contaminated with the steam evolved at the anode ie. the back flow of steam was negligibly small. On the other hand, below this flow rate, the difference in water vapor pressure between the inside of the anode and the bulk of the gas phase near the standard electrode on the base of the square cone decreased with decreasing gas flow rate. Therefore, the following experiments were carried out at a gas flow rate of 100 ml min^{-1}.

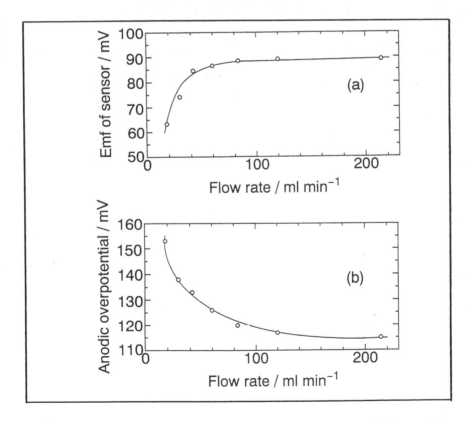

Figure 4. The dependence of emf of the steam sensor (a) and anodic overpotential (b) on gas flow rate, measured at 1000°C in H$_2$ under discharge current of 500 mA cm^{-2}. Anode material: porous Ni–YSZ cermet

Anodic overpotential of the SOFC was also measured in this experiment as a function of gas flow rate, and is shown in figure 4b. The overpotential increases as the flow rate decreases although the emf of the sensor decreases with deceasing flow rate, where the open circuit voltage of the SOFC was independent of the gas flow rate. This indicates that the fuel electrode behaves both as an anode of the fuel cell and as a sensing electrode of the steam sensor, independently. That is, the electric potential of the sensing electrode is not affected by the electrode potential of the anode during discharge.

WATER VAPOR PRESSURE AT ANODE DURING DISCHARGE

The emf of the steam sensor as well as the overpotential of the fuel cell anode during discharge were measured. Using the following equation, similar to equation 6, water vapor pressure $P_{H_2O}(a)$ in the porous anode during discharge was determined from the measured emf and known values of hydrogen partial pressure $P_{H_2}(b)$ and water vapor pressure $P_{H_2O}(b)$ in the bulk of the feed gas.

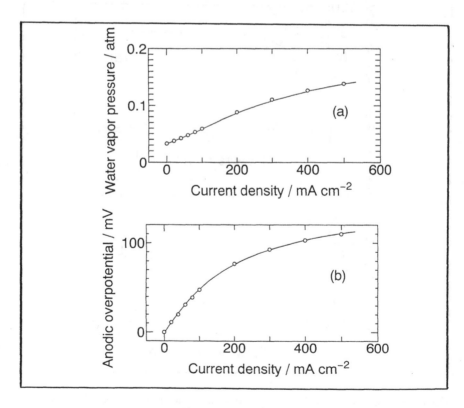

Figure 5. Water vapor pressure (a) and anodic overvoltage (b) during discharge at 1000°C in H_2. Anode material: porous Ni–YSZ cermet

$$P_{H_2O}(a) = \frac{P_{H_2O}(b)}{P_{H_2O}(b) + P_{H_2}(b) \exp(-2EF/RT)} \tag{7}$$

Water vapor pressure in the Ni-YSZ cermet anode thus obtained was plotted against discharge current in figure 5a and was compared with the overpotential of the anode (Figure 5b) measured by the interruption method. It is clear from these figures that the increase in water vapor in the anode causes a marked overvoltage. This interrelation is well confirmed in figure 6, where the anodic overpotential is plotted against water vapor pressure.

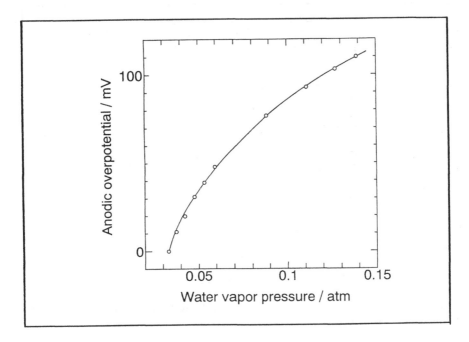

Figure 6. Anodic overpotential vs. water vapor pressure in the porous Ni-YSZ cermet anode at 1000°C in H_2.

DROP IN HYDROGEN PARTIAL PRESSURE IN THE POROUS ANODE DURING DISCHARGE

The partial pressure of hydrogen in the porous anode during discharge was calculated assuming that the total gas pressure in the porous anode is always 1 atm. ie. $P_{H_2}(a) + P_{H_2O}(a) = 1$ and was plotted against discharge current. The result is shown in figure 7 compared with the result in the case of the platinum plated anode reported previously [6]. The partial pressure of hydrogen in the anodes decreased with increasing discharge current.

As shown in this figure, the hydrogen pressure in the Ni–YSZ cermet anode was markedly high compared to that in the Pt–plated anode. In both cases the hydrogen pressure decreased almost linearly with discharge current up to 200–300 mA cm⁻², where the slopes became less steep. This critical value of current density in the cermet anode was lower than that in the Pt–plated anode, and the slope is less steep in the former than in the latter. These results suggest that the Knudsen mutual diffusion of hydrogen and steam in the porous cermet anode is faster than that in the plated Pt–anode. A quantitative discussion of this will require more precise data on diffusion coefficients and a complex mathematical procedure.

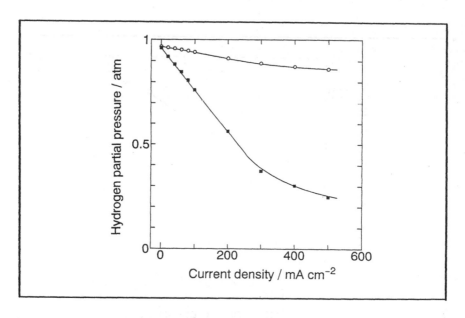

Figure 7. Hydrogen partial pressure in the porous anode during discharge at 1000°C in H_2. ○ : Ni–YSZ cermet   ■ : Plated Pt

DIFFUSION OVERPOTENTIAL

We calculated the diffusion overpotential η_d which was caused by the difference in hydrogen partial pressures between reaction sites in the porous anode, $P_{H_2}(a)$, and the bulk gas phase, $P_{H_2}(b)$, using the values shown in figure 7 and equation 8.

$$\eta_d = \frac{RT}{2F} \ln \frac{P_{H_2}(b)}{P_{H_2}(a)} \tag{8}$$

The values thus obtained were plotted against discharge current and compared with the anodic overpotential of the fuel cell measured by the current interruption method. Figure 8 shows this comparison for both the Ni–YSZ cermet and the Pt–plated anode.

It is clear that the major part of the anodic overpotential can be attributed to diffusion of hydrogen and steam out of the anode in both cases (concentration polarization). However, the concentration overpotential in the cermet anode is markedly less compared to that in the Pt–plated one. Small concentration polarization of the cermet anode compared to that of the Pt–plated one may be ascribed to its higher porosity with fine pores.

The remaining part of the anodic overpotential would be caused by other reaction steps such as charge transfer, chemical reaction of atomic oxygen with hydrogen, dissociation of molecules to atomic hydrogen etc.. Figure 8 also suggests that these kinds of overvoltages are smaller with the Ni–YSZ cermet electrode than with the "Pt–plated" electrode, although the performance of the latter is, in general, much better than that of a "Pt–smeared" electrode [4].

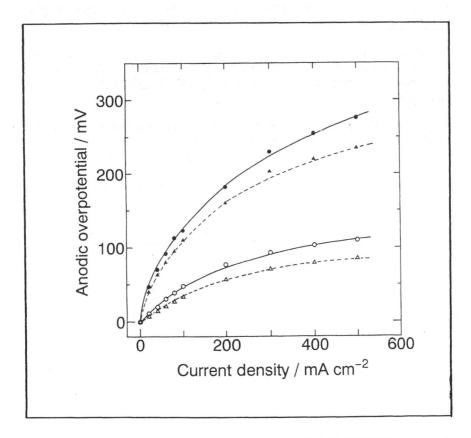

Figure 8. Plots of overpotentials against discharge current at 1000°C in H_2 for Ni–YSZ cermet anode (open symbols) and Pt–plated anode (closed symbols). ○, ● : Total overpotential measured by interruption method. △, ▲ : Concentration overpotential calculated from hydrogen partial pressure using equation 7

CONCLUSIONS

Water vapor pressure in the porous anode of a solid oxide fuel cell during discharge could be measured by a galvanic cell type steam sensor which we devised using a small tip of zirconia as a solid electrolyte. From the water vapor pressure measured by this method, hydrogen partial pressure was determined and the anodic overpotential due to concentration polarization was calculated. Comparing the anodic overpotential measured by the dc interruption method with that calculated from the water vapor pressure, it was confirmed that the major part of the anodic overvoltage was ascribed to hydrogen concentration overvoltage. The anodic overvoltage of Ni–YSZ was markedly smaller than that of a Pt–plated one.

REFERENCES

1. Linkous, C.A., 1985. "Novel Electrode Concepts for Oxide Fuel Cells," in 1985 Fuel Cell Seminar, Tucson, AZ, USA: Abstracts, pp.119-122.

2. Miyamoto, H., M. Ishibashi, H. Nagatani, N. Tomita and H. Iwahara, 1989. "Gas Diffusion in SOFC Elements - Modeling and Experimental Studies on Cell Performance," in Proc. Int. Symp. on Solid Oxide Fuel Cells, Nagoya, Japan, pp.243-252.

3. Iwahara, H., H. Uchida and N. Maeda, 1983. "Relation Between Proton and Hole Conduction in $SrCeO_3$-Based Solid Electrolytes Under Water-Containing Atmospheres at High Temperature." Solid State Ionics, 11(2):117-124.

4. Nagata, M., T. Esaka and H. Iwahara, 1991. "Electroless Plating of Platinum on YSZ and Its Property as a Gas Electrode." Denki Kagaku, 59:707-711.

5. Kawada, T., N. Sakai, H. Yokokawa and M. Dokiya, 1990. "Structure and Polarization Characteristics of Solid Oxide Fuel Cell Electrodes." Solid State Ionics, 40/41(1):402-406.

6. Nagata, M. and H. Iwahara. "The Measurement of Water Vapour Pressure in a SOFC Anode During Discharge," to be submitted to J. Appl. Electrochem.

Anomalous Oxygen Evolution from Zirconia Cells at the Transient State

JUNICHIRO MIZUSAKI, HIDEKAZU NARITA, HIROAKI TAGAWA, MASAAKI KATOU and KATSUHIKO HIRANO

ABSTRACT

When a potential difference is applied to a zirconia cell, an imbalance is some-times observed at the transient state between the amount of oxygen gas incorporated at the cathode and that evolved at the anode. In this paper, the cause of this imbalance was clarified.

Three terminal zirconia cells were prepared using 8 m/o Y_2O_3 doped ZrO_2 with porous $La_{0.6}Ca_{0.4}MnO_3$ as the working electrode and porous Pt as the counter and the reference electrodes. Using the potentiostatic polarization method, measurements were made at 800°C in the stream of Ar–O_2 gas mixtures with controlled oxygen partial pressures of $1x10^5 - 1x10^2$ Pa on the transient current, the electrode impedance at the steady–state, and the oxygen partial pressure of the outlet gas. The amount of oxygen evolved or incorporated by the cell was calculated by the change in oxygen partial pressure of the outlet gas.

Under cathodic polarization, oxygen evolution was found to take place by the change in oxygen nonstoichiometry of $La_{0.6}Ca_{0.4}MnO_3$. Under anodic polarization, the oxygen evolution takes place when the potential between the working and the counter electrodes exceeds the decomposition potential of ZrO_2. The zirconia surface at the counter electrode is blackened due to partial reduction.

INTRODUCTION

The authors and co–investigators have been studying the kinetics of the electrode reaction with oxygen at the porous perovskite–type oxide electrode/zirconia interface as cathode reaction of solid oxide fuel cells (SOFC) [1–5]. For the perovskite–type oxide electronic conductor, we mostly use $La_{0.6}Ca_{0.4}MnO_{3-d}$. For zirconia oxide conductor, we usually employ 8m/o Y_2O_3–doped ZrO_2 [2,3,5].

In the course of studies on potentiostatic polarization, we sometimes observed anomalous oxygen evolution or incorporation by the cell: Typically, we use three terminal method and the electrodes of the cell are exposed to the same gas stream. The oxygen partial pressure, $P(O_2)$, of the gas phase is detected by the zirconia oxygen sensor placed downstream to the cell. Just after a potential is applied to the cell in the Ar–O_2 gas–mixtures of controlled Ar/O_2 ratio, the EMF of the oxygen

Junichiro Mizusaki, Hideki Narita and Hiroaki Tagawa, Institute of Environmental Science and Technology, Yokohama National University, 156 Tokiwadai, Hodogaya-ku, Yokohama 240, Japan
Masaaki Katou and Katsuhiko Hirano, Department of Industrial Chemistry, Shibaura Institute of Technology, Minato-ku, Tokyo 108, Japan

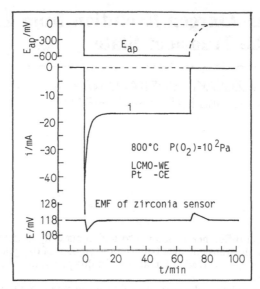

Figure 1. Example of anomalous oxygen evolution and incorporation by zirconia cell.

sensor shows anomalous deviation from the EMF calculated by the Nernst equation. Typical example is shown in figure 1. The deviation indicates that small amount of oxygen is released from the cell. The EMF value returns to the normal value in several minutes. However, when the current is interrupted, the EMF again deviates from the steady–state value, indicating that the cell incorporates small amount of oxygen.

These anomalies would be closely related to the process of the electrode reaction. In this paper, we focus our interest to elucidate the mechanism of this anomalous oxygen evolution and incorporation, and discuss the relation of the phenomena with the cathode reaction of SOFC.

EXPERIMENTAL

Three terminal zirconia cells were prepared using the disks of 8 m/o Y_2O_3 doped ZrO_2 (YSZ) with porous $La_{0.6}Ca_{0.4}MnO_3$ (LCMO) electrode used as working (WE) or counter electrode (CE) and two porous Pt electrodes, one with small electrode area for the reference electrode (RE) and the other as CE or WE. The YSZ pellets (8mm in diameter and 2 – 3mm in thickness) were prepared from the YSZ powder (NZP–D8Y, Nissan Chemical Industries, LTD.) by cold pressing with 4×10^8Pa and sintering at 1400°C for 2h. LCMO powder was prepared by a drip pyrolysis method [5,6], mixed with turpentine oil, painted on one side of the YSZ pellet and fired at 1100°C for 2h. The amount of the painted LCMO was 18 – 25mg/cm² (about 50–70μm in thickness of the porous layer [1]). As the porous Pt electrodes, non–flux type Pt paste was applied and fired at 900°C for 1h. The electrode area of the LCMO was 0.51cm² and that of the larger Pt electrode was 0.28cm², respective- ly. In most of the studies, the LCMO electrode was used as WE, and the larger Pt electrode was assigned as CE (Figure 2a). In some studies, WE and CE were exchanged with each other (Figure 2b). Also, the cell with three Pt paste electrodes was used for reference (Figure 2c).

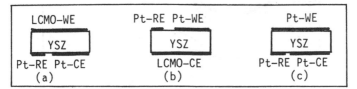

Figure 2. Electrode arrangement for the measurements.

The cell was placed in the stream of Ar–O_2 gas mixtures with controlled oxygen partial pressures of $10^5 - 10^2$ Pa. The flow rate of the gas–mixtures was controlled to be $100 - 105$ cm^3/min at 25°C. Using the potentiostatic polarization method, measurements were made at 800°C on the current, i, the electrode impedance at the steady–state over the frequency range of 100kHz – 1mHz, and the $P(O_2)$ of the outlet gas. The electrode potential, E_{el}, against O_2 gas of 1.013×10^5 Pa was calculated from the applied potential, E_{ap}, between WE and RE by the equation

$$E_{el} = E_{ap} - iR + E_{eq} \tag{1}$$

where R is the effective resistance of YSZ between RE and WE observed by the impedance measurements under the steady state polarization, and E_{eq} is the Nernst EMF between the gas mixtures and O_2 gas of 1.013×10^5 Pa. E_{ap} was maintained so that E_{el} did not exceed −2.0 V, because the electronic conduction of YSZ becomes remarkable for $E_{ap} < -2.0$ V at 800°C. The zirconia oxygen sensor for the $P(O_2)$ measurements was a potentiometric one with air as reference gas, which was held at 750°C. The equipments employed for the electrochemical measurements were the same to those used in the preceding studies [1–5].

RESULTS AND DISCUSSION

STEADY STATE CURRENT AND THE EMF CHANGE DETECTED BY THE ZIRCONIA SENSOR

Figures 3 and 4 show the EMF change of the zirconia sensor for cathodic and anodic polarization under different applied potentials for the cell in figure 2a. The direction of the deviation of EMF is the same irrespective of the direction of the polarization. When the potential is applied, the cell evolves oxygen, and when the current is interrupted, the cell incorporates oxygen.

The steady state current is shown in figure 5 as a function of E_{el}. Since E_{el} is considered to be related to the oxygen activity, a_0^*, at the electrode / YSZ interface [2], a_0^* is indicated on the upper scale.

The deviation of EMF was observed when $P(O_2)$ was 10^2 or 10^3 Pa. However, under the higher $P(O_2)$ atmosphere, we could not detect the deviation of EMF. The change in $P(O_2)$ due to the oxygen evolution and incorporation by the cell may be very small compared to the $P(O_2)$ in the atmosphere.

As shown in figure 5, the anomaly under the cathodic polarization appears only when the $P(O_2)$ is much lower and the overpotential is much larger than the usual conditions for SOFC cathode: Since the overall EMF of H_2–O_2 fuel cells at 800–1000°C is about 1V, the total overpotential must be well below 500mV and the overpotential at the oxygen electrode should be well below 200mV.

Similar experiments were also made on the electrode arrangements shown in figures 2b and 2c, where Pt electrode was used as WE. For these cells, the EMF

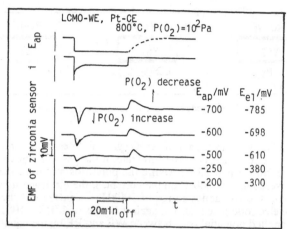

Figure 3. EMF change of the zirconia sensor observed for the cathodic polarization of LCMO electrode. Pt electrode is used as CE.

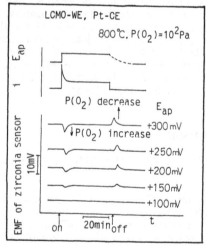

Figure 4. EMF change of the zirconia sensor observed for the anodic polarization of LCMO electrode. Pt electrode is used as CE.

anomaly was observed only for the anodic polarization of the WE. When the anomaly is observed, it always appears that the cell evolves oxygen when the potential is applied and incorporates when the current is interrupted.

In any case when Pt electrode was used as CE and the anomaly was observed, we found that the YSZ at the Pt / YSZ interface of CE was slightly blackened, suggesting that YSZ at the CE interface was partially reduced. The polarization curve for Pt electrode is shown in figure 6, comparing with that of LCMO. The overpotential for cathodic polarization is much larger than that for anodic polarization. With a small anodic polarization at WE, the cathodic overpotential of the CE becomes very large and may exceed the decomposition potential, about 2.3V, of YSZ [7], resulting in the blackening of YSZ.

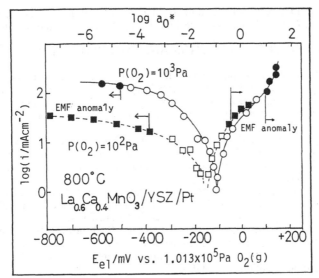

Figure 5. Steady–state polarization current for the LCMO electrode as a function of E_{el}. Closed symbols indicate the condition at which the anomalous EMF change of the zirconia sensor was observed.

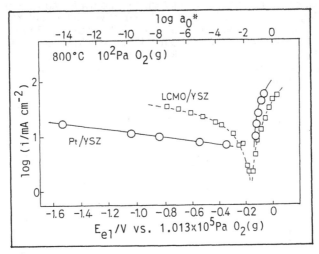

Figure 6. Steady–state polarization current for the Pt electrode. The polarization current for the LCMO electrode is shown for comparison.

When the Pt electrode is used as CE, the EMF anomaly for the anodic polarization of WE, either Pt or LCMO, is considered to be caused by the electrolysis in the initial period and reoxidation in the last period of YSZ.

For the case of the cathodic polarization of LCMO, the anomaly took place at much higher E_{el}, or higher a_O^*, than that corresponding to the decomposition of

YSZ. We consider here that the anomaly of EMF is attributed to the change in the nonstoichiometry of LCMO, when the LCMO is polarized to cathodic direction.

AMOUNT OF OXYGEN EVOLUTION AND INCORPORATION

From the EMF of the zirconia sensor, the $P(O_2)$ in the outlet gas was calculated. Since the gas flow rate is $100 - 105$ cm³/min at 25°C, we can calculate the flux of oxygen passed through the zirconia sensor. From the graphical integration of the calculated oxygen flux as a function of time, we obtain the amount of evolved or incorporated oxygen.

In figure 7, the calculated amount of oxygen, $n(O_2)$, evolved and incorporated by the cell for a unit area of WE is shown as a function of E_{el}. We see that $n(O_2)$ values for the oxygen evolution and incorporation in the same run are essentially the same as each other.

In the figure, $n(O_2)$ values observed in the anodic polarization of the WE for the cells in figure 2b (Pt as WE and LCMO as CE) are indicated by Δ and $\blacktriangle$. The values are much larger than those for the cell in figure 2a (LCMO as WE and Pt as CE). The values (Δ and $\blacktriangle$) are converted to the $n(O_2)$ values for the unit area of LCMO–CE by using the polarization curves in figure 6. The results are indicated by the symbols in ( ). The $n(O_2)$ values thus calculated agree with those observed by the cathodic polarization of LCMO–WE.

YSZ BLACKENING

As to the cases for the Pt cathodic polarization, in order to show the magnitude of YSZ reduction, we transformed $n(O_2)$ into the effective thickness of YSZ which can be reduced to metal. The calculated thickness is less than $2 - 3$ μm. The part of YSZ reduced at the beginning of the cathodic polarization was mostly reoxidized after the polarization. However, there is a tendency that the calculated thickness of the reduced layer is thicker by $0.3 - 0.5$μm than that of the reoxidized one. This difference may result in the blackening of YSZ observed after the measurements.

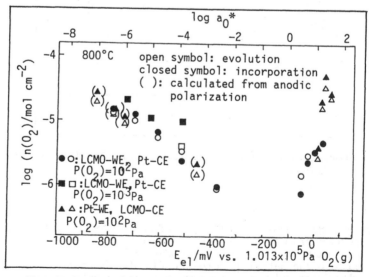

Figure 7. Summary of the calculated $n(O_2)$ shown as a function of E_{el}.

So far, many studies have been reported on the well known zirconia blackening. However, as to the nature of blackened part of YSZ, it seems there still remain some problems to be solved.   If the blackened part contains reduced YSZ, reoxidation to normal white YSZ should be observed during the high temperature electrochemical studies. Moreover, an EMF should be observed between the electrode attached to the blackened part and the one attached to the normal part. Actually neither is observed. When the surface of reduced zirconia is reoxidized, the electronic conductivity, hence the chemical diffusivity, at the surface of zirconia decreases, and the rate of reoxidation would be reduced. Then, the reduced part is left just behind the surface of zirconia while the surface is essentially in equilibrium with the gas phase. The authors think that further investigation is still necessary to give quantitative interpretation as to the nature of the zirconia blackening.

RELATIONSHIP WITH OXYGEN NONSTOICHIOMETRY OF LCMO

From the $n(O_2)$ values for the cathodic polarization of LCMO and the amount of LCMO applied to the electrode, the deviation of oxygen stoichiometry in LCMO during the polarization was calculated and shown in figure 8. In the figure, the nonstoichiometry data for LCMO determined by coulometric titration [8] is also shown. The calculated deviation essentially agrees with the nonstoichiometry data.
This result reveals an interesting feature of the cathode reaction of LCMO. During the cathodic polarization, LCMO is reduced to the state in equilibrium with $a_O{}^*$.

RELATIONSHIP TO THE SOFC CATHODE REACTION

The authors have already shown that the cathode reaction at the porous

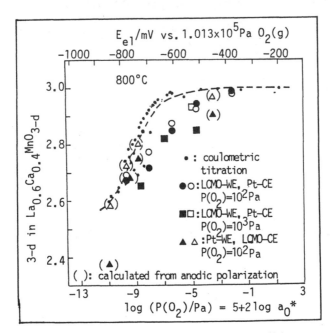

Figure 8. The oxygen deficiency in $La_{0.6}Ca_{0.4}MnO_{3-d}$ calculated from $n(O_2)$, as compared with the nonstoichiometry obtained by coulometric titration.

LCMO/YSZ interface proceeds via the triple phase boundary of O_2 / LCMO / YSZ, at least under low overpotentials and higher $P(O_2)$ conditions. The overpotential at the electrode is attributed to the difference in the oxygen activity between that in YSZ surface at the triple phase boundary and that in the gas phase. The problem was left whether LCMO works as a mixed conductor and what is the oxygen activity in LCMO.

In this paper, although the overpotential is much larger and the $P(O_2)$ is lower than those in the preceding works [1–5], we showed that the oxygen activity in LCMO under the cathodic polarization becomes essentially in equilibrium with YSZ. That is, a large activity gap exists not at the LCMO/YSZ interface but at the interface between the gas phase and LCMO.

Moreover, although the oxygen defect increased due to nonstoichiometry change, the steady state current did not increase remarkably with large cathodic overpotential.

These results indicate that, under large cathodic overpotentials, oxygen transport through LCMO may take place. However, under these conditions, the reaction rate through LCMO is controlled by the surface reaction process at the gas phase / LCMO interface. As observed on the polarization curve, the cathode reaction at the porous LCMO / YSZ is essentially determined by reaction at the triple phase boundary and the effect of oxygen transport through the bulk LCMO is still very small under very large overpotentials.

REFERENCES

1. Mizusaki, J., H.Tagawa, K.Tsuneyoshi, K.Mori and A.Sawata, 1988. "Electrode Thickness, Microstructure, and Properties of Air Electrode for High–Temperature Solid Oxide Fuel Cells." Nippon Kagaku Kaishi, (9):1623–1629.

2. Tsuneyoshi, K., K.Mori, A.Sawata, J.Mizusaki and H.Tagawa, 1989. "Kinetic Studies on the Reaction at the $La_{0.6}Ca_{0.4}MnO_3$/YSZ Interface, as an SOFC Air Electrode." Solid State Ionics, 35: 263–268.

3. Mizusaki, J., H.Tagawa, K.Tsuneyoshi, A.Sawata, M.Katou and K.Hirano, 1990. "The $La_{0.6}Ca_{0.4}MnO_3$–YSZ Composite as an SOFC Air Electrode." Denki Kagaku, 58(6): 520–527.

4. Mizusaki, J. and H.Tagawa, 1990. "SOFC Oxygen Electrodes: Materials & Reaction – A Consideration on the Possibility of the Chemical Diffusion Controlled Reaction," in Proc. International Symp. on Solid Oxide Fuel Cells, O.Yamamoto, M.Dokiya and H.Tagawa eds., Tokyo, Japan: Science House, pp. 107–117. (ISBN4–915572–33–1 C3054)

5. Mizusaki, J., H.Tagawa, K.Tsuneyoshi and A.Sawata, 1991. "Reaction Kinetics and Microstructure of the Solid Oxide Fuel Cells Air Electrode $La_{0.6}Ca_{0.4}MnO_3$/YSZ." J. Electrochem. Soc., 138(7): 1867–1873.

6. Tagawa, H., J.Mizusaki, Y.Arai, Y.Kuwayama, S.Tsuchiya, T.Takeda and S.Sekido, 1990. "Sinterability and Electrical Conductivity of Variously Prepared Perovskite–Type Oxide, $La_{0.5}Sr_{0.5}CoO_3$." Denki Kagaku, 58(6): 512–519.

7. The Soc. Calorimetry and Thermal Anal. Japan, 1985. Thermodynamic Data Base for Personal Computers, Tokyo, Japan.

8. Narita, H., J.Mizusaki and H.Tagawa, unpublished data.

Corrosion of Stabilized Zirconia with Alkali Metal Carbonates

HIROAKI TAGAWA, JUNICHIRO MIZUSAKI, HIDETO HAMANO and MASAHARU SHIMADA

ABSTRACT

During thermodynamic measurements on alkali metal carbonates using a galvanic cell, it was found that calcia- and yttria-stabilized zirconia, CSZ and YSZ, respectively, were corroded by the carbonates even in the solid state. Thus, the reactions of CSZ and YSZ with the carbonates were examined at 650–950°C which includes by SEM, EPMA and BEI, and XRD. The carbonates were impregnated along grain boundaries over the entire temperature range. Below the melting temperature, corrosion mainly occurred at the grain boundaries to form alkali metal zirconate, A_2CO_3 (A = Li, Na and K). The main reaction with the liquid carbonates predominantly proceeded on the surface of the zirconia matrix, and the product was broken away from the surface. Calcium in CSZ was leached into the liquid carbonate phase from the zirconia surface.

INTRODUCTION

Stabilized zirconia doped with calcia and yttria (CSZ and YSZ, respectively), as an oxide–ion conductor for measuring or utilizing the Gibbs free energy change of a reaction, is a useful material for high–temperature electrochemical measurements, electrochemical sensors and solid oxide fuel cells (SOFC).

We needed the thermodynamic properties of simple alkali metal carbonates, Li_2CO_3, Na_2CO_3 and K_2CO_3, and their mixed carbonates at high temperatures to construct phase diagrams in the molten state. For determining the thermodynamic properties, the following cell was prepared: Au, A_2CO_3, C(graphite) / (CSZ or YSZ) / O_2, Pt, where A is Li, Na or K. In the course of measurements, it was found that CSZ and YSZ tubes used were corroded in both the solid and liquid carbonate phases, especially the CSZ tubes which were often broken on raising temperature above 650°C.

Hiroaki Tagawa, Junichiro Mizusaki, and Hideto Hamano
Institute of Environmental Science and Technology, Yokohama National University
156 Tokiwadai, Hodogaya-ku, Yokohama, Japan, 240
Masaharu Shimada, Faculty of Engineering, Yokohama National University
156 Tokiwadai, Hodogaya-ku, Yokohama, Japan, 240

The present work aims at clarifying how stabilized zirconia reacts with alkali metal carbonates, Li_2CO_3, Na_2CO_3 and K_2CO_3, in the solid and liquid states at high temperatures. Scanning electron microscope (SEM), electron probe microanalysis (EPMA) and back electron image (BEI), and X-ray diffraction (XRD) were used as methods. The corrosion of stabilized zirconia is thought to be important when SOFC are operated in the atmosphere including alkali metal carbonates or other alkali metal salts for a long time.

The reactions of stabilized zirconia with solid compounds such as wustite, $Fe_{1-x}O$ [1], $(La,Sr)MnO_3$ [2], and $(La,Ca)CoO_3$ [3], and the reaction of YSZ with molten Na_2SO_4 [4] have been examined. There is no report on the reaction of pure zirconia or stabilized zirconia with alkali metal carbonates.

EXPERIMENTAL

One end–closed stabilized zirconia tubes of ZrO_2–11mol%CaO (CSZ) and ZrO_2–8mol%Y_2O_3 (YSZ) were used: CSZ tubes (Nikkato Co., Osaka, 17mm OD and 4mm thick), and YSZ tubes (Toray Corp., Tokyo, 9mm OD and 1.5mm thick). The specimens were the as–received CSZ and YSZ tubes, rings which were cut from the tubes 4mm in width and which had as–received surfaces, and powders which were pulverized to 0.1mm or less. Alkali metal carbonates of guaranteed reagent grade, were provided by Junsei Chemicals Co., Tokyo.

The reactions of stabilized zirconia with the carbonates were examined by the following procedure. An alkali metal carbonate was put into a CSZ or YSZ tube, or a CSZ or YSZ ring was sunk into the carbonate held in an alumina tube. The specimens were heated for 0.5 to 24h at 650, 800 and 950°C, and other temperatures below and above the melting temperatures of Li_2CO_3, Na_2CO_3 and K_2CO_3, which are 720, 850 and 901°C, respectively. After cooling, the carbonate in the zirconia tubes or on the pieces was mechanically removed or dissolved in warm water or in diluted hydrochloric acid. The surfaces and cross sections of the bulk specimens were examined by SEM, EPMA and BEI.

For identifying reaction products and determining the variation in lattice parameter of the zirconia by XRD, specimens from the reaction of pulverized zirconia with alkali metal carbonates at 650, 800 and 950°C for 24h were also examined.

SEM OBSERVATION OF SURFACE OF CSZ AND YSZ SINTERED BODIES

The bulk surface of CSZ consisted of the sintered grains of 20-30 μm in diameter; grain boundaries and triple junctions were clearly distinguished. A small amount of foreign material was found at some grain boundaries, especially at the triple junctions. Although not analyzed, this might be silica as an impurity or as a sintering aid. If silica is present, these grain boundary films are likely to contain silica as well as CaO or Y_2O_3 stabilizer, and other impurities. The YSZ specimen also consisted of 20-30 μm sintered grains, and grain boundaries and triple junctions were observed, but no foreign material was found in the grain boundaries.

Machined cross sections and fractured sections of the zirconia tubes before and after corrosion were examined by SEM and EPMA. It

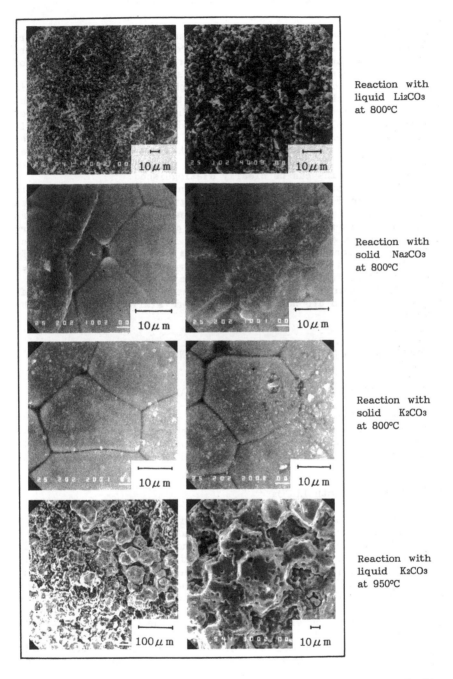

Reaction with liquid Li$_2$CO$_3$ at 800°C

Reaction with solid Na$_2$CO$_3$ at 800°C

Reaction with solid K$_2$CO$_3$ at 800°C

Reaction with liquid K$_2$CO$_3$ at 950°C

Figure 1. SEM images of inner surfaces of CSZ tubes contacted with liquid Li$_2$CO$_3$, and solid Na$_2$CO$_3$ and K$_2$CO$_3$ at 800°C for 24h and with liquid K$_2$CO$_3$ at 950°C for 24h.

was hard to distinguish grain boundaries in the specimens before corrosion, and only isolated closed pores of 2-5 μm in size were found because the grains were well sintered. The cross sections of the CSZ and YSZ specimens appeared very similar.

Figure 1 shows etched surface morphologies of CSZ tubes, in which alkali metal carbonates were heated at 800°C for 24h, and also for K_2CO_3 heated at 950°C for 24h. The CSZ surface after reaction with molten Li_2CO_3 is rough, and looks like a conglomerate of many particles from 0.5 to 5 μm, i.e. smaller than the original grain size. For the tubes contacted with solid Na_2CO_3 and K_2CO_3 at 800°C, as shown in Figure 1, the bulk surface is etched as the original CSZ grains can be clearly distinguished.

The inner surface of a CSZ tube corroded with molten K_2CO_3 is also shown in Figure 1. Here, only grains and isolated closed pores are seen. It is thought that these grains are the same as those of the as-received tubes, and any product formed on the surface has broken away from the CSZ matrix into the molten carbonate. The surfaces corroded with molten Li_2CO_3 and K_2CO_3 are quite different. This is attributed to the difference in the reaction rates because of the different reaction temperatures.

The morphologies of the YSZ surfaces etched with alkali metal carbonates, which are not shown here, were similar to those of the CSZ.

EPMA OF CROSS SECTIONS OF CSZ AND YSZ SINTERED BODIES

REACTION OF CSZ WITH ALKALI METAL CARBONATES

The distribution of each component element in the reactions of stabilized zirconia with alkali metal carbonates was examined by EPMA. Lithium can not be directly demonstrated by EPMA, but its existence is revealed as a difference in the brightness in BEI.

Reaction of CSZ with Li_2CO_3

According to EPMA, the distribution of the elements in cross sections of the CSZ piece in the reaction of CSZ with solid Li_2CO_3 at 715°C is as follows. Zirconium exist homogeneously in the whole area including the grain boundaries, and calcium is also uniformly distributed except near the surface, in which the density of calcium is small. BEI shows that only the surface is blackened in the whole cross section, and the grain boundaries are blackened in a low degree. The blackening is thought to show the existence of lithium. A product layer, which appears on the CSZ surface, may be lithium zirconate, Li_2ZrO_3, from the fact that the solid state reaction of pulverized CSZ with Li_2CO_3 yields Li_2ZrO_3. The results of EPMA and BEI are not shown because they are similar to those for the reaction with Na_2CO_3 as shown in Figure 2.

The reaction of CSZ with liquid Li_2CO_3 at 723°C is essentially the same as that with solid Li_2CO_3 at 715°C. According to EPMA, zirconium is distributed homogeneously in the whole area, and the calcium content is also uniform except at the surface and locally in the interior of the CSZ specimen where it was found in spots.

The calcium will be leached into the liquid Li_2CO_3 phase. Raising the reaction temperature by 8°C makes the blackened layer thicken and increases the blackened gap width between the grains. This means that lithium exists near the surface and at the grain boundaries. These morphology changes in the reaction of CSZ with molten Li_2CO_3 were similar to those in the reaction with molten Na_2CO_3 shown in Figure 3.

Reaction of CSZ with Na_2CO_3

Figure 2 shows the results of EPMA and BEI of a cross section of a CSZ piece in the reaction with solid Na_2CO_3 at 843°C after 24h. Zirconium and calcium are distributed equally in the whole area except for the original closed pores. Sodium mainly exists along the grain boundaries. Thus, it is concluded that the reaction occurs as Na_2CO_3, probably as vapor, impregnates into a CSZ piece along grain boundaries. The original smooth surface of the bulk CSZ is changed into the rugged surface by corrosion. The corrosion product, which is identified as Na_2ZrO_3 by XRD, must be formed on the zirconia surface, but it is not seen clearly on the

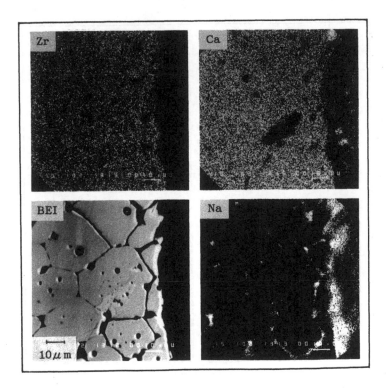

Figure 2. EPMA images of a cross section of a CSZ piece corroded in the reaction with solid Na_2CO_3 at 843°C for 24h. Sodium exists at the surface of the piece and along the grain boundaries.

surface. It is thought that the product was broken away during the reaction or taken off on removing the molten salt.

Figure 3 shows the results of EPMA and BEI of a CSZ piece in the reaction with liquid Na_2CO_3 at 855°C after 24h. Zirconium is not homogeneously distributed, but decreases in concentration at several places. Sodium exists in the interior, while calcium is lost to a depth of several tens of micrometers from the bulk surface. It is thought that calcium is leached through the grain boundaries from the surface layer into the liquid phase. Therefore, cubic CSZ is gradually changed into monoclinic ZrO_2 and a shrinkage in volume occurs because of the loss of calcium.

Reaction of CSZ with K_2CO_3

According to EPMA of a cross section of the CSZ piece after reaction with solid K_2CO_3 at 895°C, the corrosion occurs rather more violently at the CSZ surface layer and through grain boundaries than the reaction with Na_2CO_3 shown in Figure 2. Potassium is found over a wide range, not only in the surface layer, but along the grain boundaries. Zirconium and calcium are seen

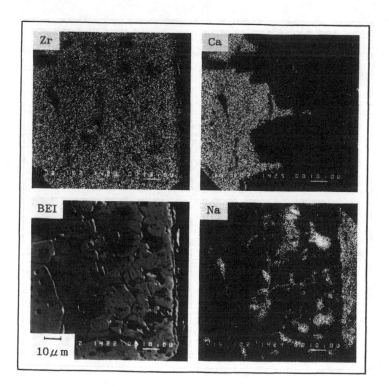

Figure 3. EPMA images of a cross section of a CSZ piece corroded in the reaction with liquid Na_2CO_3 at 855°C for 24h. Sodium exists not only at the surface of the piece, but also along the grain boundaries.

elsewhere.   The reactions of alkali metal carbonates with zirconia can be thermodynamically examined because of high temperature reactions.   The Gibbs free energies of formation for Na_2ZrO_3 and K_2ZrO_3 in these reactions are estimated to be nearly the same negative value at a given temperature, and they have a tendency to decrease with raising temperature.   Thus, the chemical reactivity of K_2CO_3 with zirconia is deduced to be higher than that of Na_2CO_3.

REACTION OF YSZ WITH ALKALI METAL CARBONATES

The reactions of yttria–stabilized zirconia (YSZ), ZrO_2–8mol% Y_2O_3, with alkali metal carbonates were examined by the same procedure as used for CSZ.   From the results of SEM, EPMA and BEI for cross sections of the specimens, the reactions of YSZ with the carbonates were similar to those of CSZ except for the behavior of yttrium.

In the reactions of YSZ with the solid carbonates, alkali metal impregnates along grain boundaries and at/near the surface of the YSZ matrix.

In the reactions with the molten carbonates, it was observed that alkali metal severely impregnated from the bulk surface through grain boundaries into the interior.   Zirconium and yttrium were scarcely found wherever sodium or potassium existed.

REACTIONS OF PULVERIZED CSZ AND YSZ WITH ALKALI METAL CARBONATES

Mixtures of pulverized CSZ and YSZ with the carbonates in 1:1 mole ratio were heated at 650, 800 and 950°C in air for 24h, and the specimens were examined by XRD.   The reactions of CSZ and YSZ with the molten carbonates are always found to form the same compound, A_2ZrO_3, where A = Li, Na and K, and no other compound was found.   When reaction with K_2CO_3 was carried out at 950°C, other compounds including A_4ZrO_4 were produced, but other compounds except A_4ZrO_4 were not able to be identified by XRD.

The lattice parameter of cubic CSZ and YSZ hardly changed with time at temperatures between 650 and 950°C.

REACTION MECHANISM

As shown above, the corrosion of stabilized zirconia with alkali metal carbonate is caused by two processes: One process is the impregnation of the carbonate along grain boundaries and reaction between the materials at the grain boundaries, and the other is the reaction with the carbonate at the surface of the bulk zirconia.   Although the two processes occur in parallel, they do not occur at the same rate, and which process occurs predominantly is decided by the phase of the carbonate.

Corrosion of the zirconia with the solid carbonate is seen to proceed mainly along grain boundaries in a sintered zirconia body by impregnation of the carbonate, although reaction at the surface occurs at the same time.   The solid state reaction rate between the zirconia and the carbonate is smaller than the impregnation rate in the grain boundaries.   The carbonate is vaporized at/near the surface of the zirconia, then impregnates along grain boundaries. Alkali metal zirconate is formed at the grain boundaries and also at/near the surface of the zirconia.

Corrosion of the zirconia with the liquid carbonate mainly occurs at the interface between the carbonate and the zirconia. The alkali metal zirconate which is formed at the zirconia surface is broken away from the surface, consequently the fresh zirconia surface is always exposed to the liquid carbonate. The corrosion proceeds continuously. The surface reaction rate between both the materials is much larger than that in the solid state reaction and the transport of material to the reaction interface will become the rate–determining step. This liquid state reaction is different from the reaction in the solid state. Calcium in the zirconia is leached into the liquid carbonate phase, probably to form a mixed carbonate including calcium, $(A,Ca)CO_3$, from the CSZ surface and through the grain boundaries. The formation of alkali metal zirconate also occurs at the grain boundaries.

Thus, the reaction of CSZ with the carbonates in the liquid phase, A_2CO_3, A = Li, Na and K, at/near the CSZ/carbonate interface is expressed as

$$Ca_x Zr_{1-x} O_2 + A_2 CO_3 \rightarrow A_2 ZrO_3 + Ca_{x'} Zr_{1-x'} O_2 + (A,Ca)CO_3 \qquad (1)$$

or typically expressed as

$$Ca_x Zr_{1-x} O_2 + A_2 CO_3 \rightarrow A_2 ZrO_3 + ZrO_2 + (A,Ca)CO_3 \qquad (2)$$

For YSZ, the reaction is expressed as the following equation, because yttrium is left in the matrix.

$$Y_x Zr_{1-x} O_2 + A_2 CO_3 \rightarrow A_2 ZrO_3 + Y_{x'} Zr_{1-x'} O_2 + Y_2 O_3 + CO_2 \qquad (3)$$

REFERENCES

1. Taimatsu, H., H. Kaneko, S. Mochizuki and F. Nakatani, 1987. "Reactions between Stabilized Zirconia for Oxygen Sensor and Molten Wustite–Based Oxides," in Bulletin of Mining College in Akita University, No.8: 27–33.

2. Kaneko, H., H. Taimatsu, K. Wada and E. Iwamoto, 1991. "Reaction of $La_{1-x}Ca_xMnO_3$ Cathode with YSZ and its Influence on the Electrode Characteristics," in Proceedings of the Second International Symposium on Solid Oxide Fuel Cells, Commission of European Communities, Report EUR 13546EN, pp. 673–680.

3. Tagawa, H., J. Mizusaki, M. Katou, K. Hirano, A. Sawata and K. Tsuneyoshi, 1991. "On the Solid State Reaction between Stabilized Zirconia and Some Perovskite–Type Oxides," in Proceedings of the Second International Symposium on Solid Oxide Fuel Cells, Commission of European Communities, Report EUR 13546EN, pp. 681–688.

4. C. F. Windisch, Jr., J. L. Bates and D. I. Boget, 1987. "In situ Laser Raman Determination of Electrochemical Reactions of Y_2O_3–Stabilized ZrO_2 and Molten Na_2SO_4," Journal of American Ceramic Society, 70(9): C220–221.

Author Index

Subject Index